ADVANCES AND TRENDS IN ENGINEERING SCIENCES
AND TECHNOLOGIES II

PROCEEDINGS OF THE 2ND INTERNATIONAL CONFERENCE ON ENGINEERING SCIENCES AND TECHNOLOGIES, HIGH TATRAS MOUNTAINS, TATRANSKÉ MATLIARE, SLOVAK REPUBLIC, 29 JUNE – 1 JULY 2016

Advances and Trends in Engineering Sciences and Technologies II

Editors

Mohamad Al Ali & Peter Platko

Institute of Structural Engineering, Faculty of Civil Engineering, Technical University of Košice, Košice, Slovakia

CRC Press is an imprint of the
Taylor & Francis Group, an **informa** business

A BALKEMA BOOK

Published by:
CRC Press/Balkema
P.O. Box 447, 2300 AK Leiden, The Netherlands
e-mail: Pub.NL@taylorandfrancis.com
www.crcpress.com – www.taylorandfrancis.com

First issued in paperback 2020

© 2017 by Taylor & Francis Group, LLC
CRC Press/Balkema is an imprint of the Taylor & Francis Group, an informa business

No claim to original U.S. Government works

Typeset by V Publishing Solutions Pvt Ltd., Chennai, India

ISBN 13: 978-0-367-73659-0 (pbk)
ISBN 13: 978-1-138-03224-8 (hbk)

**Visit the Taylor & Francis Web site at
http://www.taylorandfrancis.com**

**and the CRC Press Web site at
http://www.crcpress.com**

Table of contents

Part B: Buildings and structures, Construction technology and management, Environmental engineering, Heating, ventilation and air condition, Materials and technologies, and Water supply and drainage

Part C: *Geodesy, surveying and mapping, and Roads, bridges and geotechnics*

Advances and Trends in Engineering Sciences and Technologies II – Al Ali & Platko (Eds)
© 2017 Taylor & Francis Group, ISBN 978-1-138-03224-8

Preface

The International Conference on Engineering Sciences and Technologies (ESaT 2016) was organized under the auspices of the Civil Engineering Faculty, Technical University in Košice, Slovak Republic and the University of Miskolc, Hungary.

The second annual of the Conference was held from the 29th of June to the 1st of July 2016 in the scenic High Tatras Mountains, Tatranské Matliare, Slovak Republic. ESaT 2016 focused on a wide spectrum of topics and areas of Civil Engineering sciences, facilitating discussions on novel and fundamental progress in the fields of engineering sciences and technologies for scientists, researchers and professionals all around the world. Scientists, scholars, engineers and PhD. students from universities and research institutes worldwide, had the opportunity to present their research activities.

The conference proceedings cover various topics, such as materials and technologies, buildings and structures, reliability and durability of structures, roads and bridges, water supply, heating and air condition, environmental engineering, construction technology and management. The presented analytical methods, theoretical and experimental results are useful for engineers, researchers and professionals, for application of progressive knowledge in the educational process.

Conference Topics

- Buildings and structures
- Computer simulation and modeling
- Construction technology and management
- Environmental engineering
- Geodesy, surveying and mapping
- Heating, ventilation and air condition
- Materials and technologies
- Mechanics and dynamics
- Reliability and durability of structures
- Roads, bridges and geotechnics
- Water supply and drainage

The editors would like to express a special thanks to all reviewers, sponsors and conference participants for their intensive cooperation to make this conference successful. The editors also would like to express their thanks to the members of Organizing and Scientific Committees. A special thanks to the Publisher for the valuable advice and cooperation during the preparation of these Proceedings.

Mohamad Al Ali   Peter Platko

Advances and Trends in Engineering Sciences and Technologies II – Al Ali & Platko (Eds)
© 2017 Taylor & Francis Group, ISBN 978-1-138-03224-8

Committees

SCIENTIFIC COMMITTEE

prof. Stanislav Kmeť
Rector of the Technical University, Technical University of Košice, Slovak Republic

prof. Vincent Kvočák
Dean of the Faculty of Civil Engineering, Technical University of Košice, Slovak Republic

prof. Károly Jármai
Vice Rector of the University of Miskolc, University of Miskolc, Hungary

prof. RNDr. Magdaléna Bálintová
Technical University of Košice, Slovak Republic

prof. Mária Kozlovská
Technical University of Košice, Slovak Republic

prof. Zuzana Vranayová
Technical University of Košice, Slovak Republic

assoc. prof. Ján Mandula
Technical University of Košice, Slovak Republic

prof. Mohamad A.E. Al Ali
University of Aleppo, Syria

prof. Safar Alhilal
Zirve University, Turkey

prof. Ivan Baláž
Slovak University of Technology in Bratislava, Slovak Republic

prof. Abdelhamid Bouchair
Blaise Pascal University, France

prof. Ján Čelko
University of Žilina, Slovak Republic

prof. em. József Farkas
University of Miskolc, Hungary

prof. Jozef Gašparík
Slovak University of Technology in Bratislava, Slovak Republic

prof. Markku Heinisuo
Tampere University of Technology, Finland

prof. Marcela Karmazínová
Brno University of Technology, Brno, Czech Republic

prof. Nicholas Kathijotes
University of Nicosia, Cyprus

Dr. Mohamad Al Ali
Technical University of Košice, Slovak Republic

Dr. Peter Platko
Technical University of Košice, Slovak Republic

Dr. Elżbieta Radziszewska-Zielina
Cracow University of Technology, Poland

ORGANIZING COMMITTEE

Dr. Mohamad Al Ali, Dr. Peter Platko
esat2016@tuke.sk

Dr. Štefan Kušnír, Dr. Eva Singovszká, Dr. Marcela Spišáková,
Dr. Clayton Stone, Eng. arch. Zuzana Poórová, Eng. Katarína Čákyová,
Eng. Stefan Demčák, Eng. Rastislav Gruľ

REVIEWERS COMMITTEE

prof. Vincent Kvočák, *Head of the Committee*
prof. Magdaléna Bálintová
prof. Mária Kozlovská
prof. Zuzana Vranayová
assoc. prof. Ján Mandula

LIST OF REVIEWERS

Al Ali M., Antošová N., Arsić D., Bajzecerová V., Balintova M., Bartoš K., Bašková R., Biela R., Brodniansky J., Bruothová M., Cahcim P., Čarnický Š., Čelko J., Dakowska J., Demjan M., Dubravský M., Farkas J., Fazekašová D., Flimel M., Funtík T., Gajdošík J., Harabinová S., Henková S., Hirš J., Holub M., Hulínová Z., Jandačka D., Jankovichová E., Jármai K., Junak J., Junáková N., Kapalo P., Kaposztasova D., Katunska J., Kmeť S., Koban J., Kormaníková E., Korytárová J., Kota L., Kotrasová K., Kovac M., Kovács G., Kováč M., Kovářová B., Kozlovska M., Krawiec S., Krejčiříková H., Kvočák V., Latka D., Lazarev Y., Lazarová E., Lazić V., Lis A., Lis P., Luptáková A., Mačková D., Makýš P., Mandula J., Markku H., Markovič G., Masarovičová S., Melnikov B., Mesaros P., Miszla M., Molčíková S., Motyčka V., Nikolic R., Novik A., Ordon-Beska B., Orolin P., Pandula B., Panulinová E., Pavlík Z., Petrik M., Plášek J., Platko P., Priganc S., Pukanska K., Radziszewska-Zielina E., Renčko T., Rovňák M., Rovnaník P., Rybakov V., Řezáč M., Salaiová B., Sandanus J., Sedlakova A., Schlosser F., Sičáková A., Singovszká E., Skultetyova I., Soltys R., Somorová V., Spisakova M., Stevulova N., Struková Z., Szamosi Z., Šimčák M., Šlanhof J., Šlezingr M., Špak M., Terpáková E., Tichá A., Tomko M., Tóth S., Ujma A., Úterský M., Vaclavik V., Vatin N., Vavrek P., Vertaľ M., Vičan J., Vilcekova S., Virág. Z., Vodička J., Vojtuš J., Vranayova Z., Vytlačilová V., Wawerka R., Zelenakova M., Zima K.

Part A: Computer simulation and modeling, Mechanics and dynamics, and Reliability and durability of structures

Drift design approach for steel tall buildings according to Eurocode 3

C. Amaddeo
Civil Engineering Department, Faculty of Engineering and Computer Sciences, İzmir University of Economics, İzmir, Turkey

K. Taskin & T. Orhan
Civil Engineering Department, Faculty of Engineering, Anadolu University, Eskişehir, Turkey

ABSTRACT: This paper presents the analysis of a 27-story steel frame tall building under seismic loads according to Eurocode 3. The selected building is located in Izmir, Turkey, that is a high seismic region. To understand this behavior under earthquake loads, a Finite Element Model of the building is created using SAP2000. The structural system is composed of steel frame with concentric bracing system to control the horizontal displacement under earthquake loads. The results from two different bracing configurations are presented, in particular X-braced and K-braced frames, and they are used in order to understand which solution is more effective in reducing the inter-story drift displacement. A non-linear time history analysis is performed to evaluate the inter-story drift considering a spectrum compatible earthquake with a return period of 150 years.

1 INTRODUCTION

1.1 *Introduction*

The design of the building is based on the project of a real concrete tall building realized in Izmir (Turkey). The original design of the building was modified to better fit the new concept in terms of floor heights, as well as the location of staircase and elevator that was moved in the core of the building. The general shape and architectural design of the building wasn't changed. The story height for all stories is 3,8 m and the building has 27 stories (roof included). In total, the height of the building is 102,6 m.

The preliminary design of the structure to be modeled on SAP2000 is shown in Figure 1, Figure 2.a) and 2.b).

1.2 *Selection of structural design*

According to Sarkisian, M. (2012), the design typology with steel frame and shear truss was selected. The shear truss is located in the central core area. According to Sarkisian, a rigid frame structure would be enough for this typology of building, but since the building is located in a high seismicity region, it was seen, during the preliminary analysis, that the horizontal drift values were too high. Hence, the design was changed into rigid frame with shear trusses.

Because the structure is in a high seismicity region, shear trusses were needed to control the horizontal displacements in limit range and resist most of the lateral load. The frame connections are fully fixed which will have full bending moment capacity of the beams. Moment resisting frames are also used to help resisting lateral loads and drift control.

Due to the fact that the building is located in a seismic region, it is important to reduce the total mass of the building, which is directly related to the earthquakes loads acting on

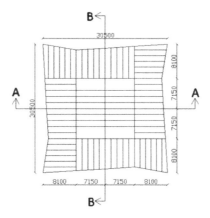

Figure 1.   Plan view of the structure.

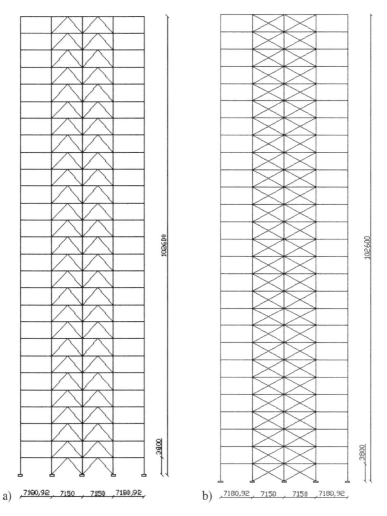

Figure 2.   a) Cut sections A-A and B-B of the K braced structure model, b) Cut sections A-A and B-B of the X braced structure models.

4

the structure. To decrease the total mass of structure, different typology of columns were used at different levels, this will directly effect also the final cost of buildings realization. In particular, different columns were used in the perimeter, defined as Type 1, and in the internal area, defined as Type 2. The type of column section is also changing at different floor levels, in particular the sections are changing at the 4th, 10th and 19th floor levels for both Type 1 and Type 2. All the sections, which are under the maximum load of its type, are designed and verified separately, based on the Eurocode 3 (2005), and according to the corresponding forces coming from structural analysis.

2 STRUCTURAL ANALYSIS

2.1 Load cases

In addition to dead load, according to Eurocode 1 (2002), 1 kN/m² permanent load is defined on each floor for the non-structural wall weight, 500 kg/m² of elevator load on roof was added to corresponding area and 1 kN/m² steel parapet load was defined at the perimeter of the roof.

The project is an office building, therefore according to Eurocode 1 Table 6.1, office areas are defined as Category B. In Table 6.2, the defined imposed loads for Category B is between 2,0 kN/m² and 3,0 kN/m². As a result, 3,0 kN/m² is used for the live load.

According to the acquired wind profile of the region, the calculated wind pressure according to Eurocode 1 2005 is applied on the defined outer shell elements of the structure. W1 and W2 are defined on SAP2000 as the wind loads on x-axis and y-axis respectively.

According to the region calculated snow load according to TS 498 (1997) and Eurocode 1 (2003) is, 0,48 KN/m².

For the seismic analysis of the structure according to Eurocode 8 (2004), first a linear approach is observed. Then a simulated non-linear spectrum is applied to the building for the horizontal drift of the structure whether it's in limit range or not. A real earthquake data for the building location couldn't be obtained, thus a free simulation software by Italian professor Prof. Piero Gelfi, depending on the linear spectrum variables is used (http://dicata. ing.unibs.it/gelfi/software/simqke/simqke_gr.htm). The software transforms a linear spectrum into an estimated non-linear one according to the defined variables, which are the period (T_1 = 3,92 sec.) of the first mode shape obtained from SAP2000 model, the returning period (150 years) for the collapse case (ULS), type of soil (class D), and behavior factor (q = 1,5). Then, it gives a txt file with acceleration and time data as output, which is defined to SAP2000 as a user defined time-history function.

2.2 Verification of elements

The verification of the elements is done according to the worst load combinations which causes the maximum design loads on them. As mentioned above, there are three main load combinations that are SLS (Service Limit State), ULS (Ultimate Limit State) and SD (Seismic Design). SLS load cases are for the element deformation checks according to limits. Hence, the resistance checks are made under ULS and SD load combinations, worst cases of both is checked.

2.3 Columns

The verification of the sections were made according to Eurocode 3 (2005), for columns that has the highest axial load and bending moment values gained from worst of both SD and ULS load cases on the section change levels which are 1st, 4th, 10th and 19th floor for both Type 1 and Type 2 sections. According to FEM analysis, the verified column sections for K and X bracing models are shown in Table 1.

2.3.1 Bracings

In this project, bracings may have both compression and tension. The steel elements are more vulnerable to compression than tension, thus the verification of the bracings will be made according to the design axial load N_{Ed} for the worst compression case which is buckling only, since they have no moment because of their pin-pin connection design.

According to FEM analysis, HEB 300 and TUBO $260 \times 260 \times 14.2$ profiles were selected and verified for **K** and **X** bracing sections respectively.

2.3.2 Beams

As mentioned in clause 1.2, for economic reasons in this project there are three types of section for the beams which are core beams, primary beams and secondary beams.

In this project beams are also designed as fixed-fixed support as mentioned in design section. They will carry shear and bending moment. Core beam elements are under axial compression and bending so they were verified like columns.

According to FEM analysis, the verified beam sections for **K** and **X** bracing models are shown in Table 2.

Table 1. Verified sections for columns of K and X braced models.

Floor number	Column type	Column section
19–27	1	HD 400×237
19–27	2	HEB 260
10–18	1	HD 400×262
19–27	2	HEB 280
4–9	1	HD 400×314
4–9	2	HEB 400
1–3	1	HD 400×551
1–3	2	HEB 400

Table 2. Verified sections for columns of K and X braced models.

Beam type	Verified section
Core	HEB 280
Primary	IPE 300
Secondary	IPE 220

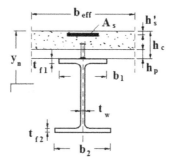

Figure 3. Parameters of the composite slab section.

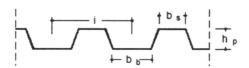

Figure 4. Parameters of the composite slab section.

2.3.3 Composite slabs

For steel structures the slabs have to be composite slabs. The profiled steel decking is connected to the beams with shear studs. In this project for the slab, light weight concrete is used as defined in materials section.

According to FEM analysis, the verified geometric design parameters according to Eurocode 4 (2004) of the slab, which are shown on Figure 3 and Figure 4. Where:

- $h_c = 110$ mm,
- $A_c = 110 \times 1000 = 110\,000$ mm^2,
- $A_s = 0.2\,A_c = 220$ mm^2,
- $b_{eff} = 1520$ m,
- $b_s = 67$ mm,
- $b_b = 105$ mm,
- $i = 225$ mm,
- $h_p = 46$ mm.

3 INTER-STORY DRIFT ANALYSIS

While designing a tall building, the structural load carrying capacity is not the only factor that is necessary to take into account. Due to the high realization cost of the building, a long lifetime and comfort level it is also desired. Thus, for the long life time the rarest earthquake with the highest impact for the location of the building should be taken into consideration. In this project, a real earthquake data for the location couldn't be found so a simulation software was used as mentioned before. For this purpose, when defining the simulated earthquake time, a returning period of 150 years was taken into account, when for normal building a 50 years period is usually considered.

According to Eurocode 8 the limitation for the inter-story drift of the buildings is ranged between 1% and 1,5% of the height for each story. It is considered as 1% for this project as it is the worst case.

According to the FEM analysis, under the worst time history load combinations the total displacement and inter-story drift graphs for K braced and X braced models are shown in Figure 5 and Figure 6.

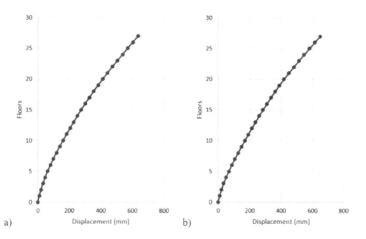

Figure 5.   a) Resultant total displacement graph of K braced model (on the left) and b) Resultant total displacement graph of X braced model (on the right).

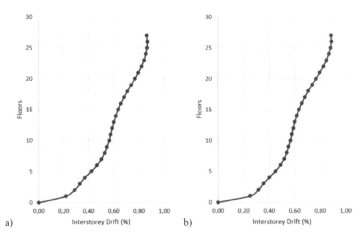

Figure 6.   a) Resultant inter-story drift graph of K braced model, b) Resultant inter-story drift graph of X braced model.

4   CONCLUSION

It is observed from the total displacement graphs in Figure 5.a) and Figure 5.b) that the bracings system is governing the behavior of the structure at the middle level heights of the structure under lateral loads. While the usual moment frame resistance to lateral loads is poor on the middle level heights of the structure but high at the top levels. The braced frame structures, like for shear walls, are strong at the bottom levels but weak at top levels. Thus, it can be concluded that the overall behavior of the structure is both behaviors combined and formed a behavior similar to linear behavior.

In addition, it is observed from the inter-story drift graphs in Figure 6.a) and Figure 6.b) that there are slope changes at levels which frame sections are changed due to the lightweight as expected. However, the inter-story drift values don't exceed 1% limit thus the structural design is adequate.

Also it is observed that X braced model has slightly higher values of displacement, at the same time more material is needed and thus more cost (due to bracing length), this will make also the structure heavier. It can be safely concluded that X braced model is less efficient than K braced model.

REFERENCES

EN 1991-1-1 (2002): Eurocode 1: Actions on structures—Part 1–1: General actions—Densities, self-weight, imposed loads for buildings.
EN 1991-1-4 (2005): Eurocode 1: Actions on structures—Part 1–4: General actions—Wind actions.
EN 1991-1-3 (2003): Eurocode 1: Actions on structures—Part 1–3: General actions—Snow loads.
EN 1998-1 (2004): Eurocode 8: Design of structures for earthquake resistance—Part 1: General rules, seismic actions and rules for buildings. http://dicata.ing.unibs.it/gelfi/software/simqke/simqke_gr.htm.
EN 1990 (2002): Eurocode—Basis of structural design.
EN 1993-1-1 (2005): Eurocode 3: Design of steel structures—Part 1–1: General rules and rules for buildings.
EN 1994-1-1 (2004): Eurocode 4: Design of composite steel and concrete structures—Part 1–1: General rules and rules for buildings.
Sarkisian M. (2012): Designing tall buildings: structure as architecture.
TS 498 (1997): Design loads for buildings.

Advances and Trends in Engineering Sciences and Technologies II – Al Ali & Platko (Eds)
© 2017 Taylor & Francis Group, ISBN 978-1-138-03224-8

Weldability estimation of steels for hot work by the CCT diagrams

D. Arsić & V. Lazić
Faculty of Engineering, University of Kragujevac, Serbia

R.R. Nikolić
Faculty of Engineering, University of Kragujevac, Serbia
Research Center, University of Žilina, Slovakia

B. Hadzima
Research Center, University of Žilina, Slovakia

ABSTRACT: Material's weldability is a very complex property, which shows its ability to form a welded joint, by application of the adequate welding procedure. It can be determined in several ways. Weldability estimate procedure, based on the Continuous Cooling Transformation diagrams (CCT) and the measured temperature cycles, is presented in this paper. The objective was to determine the weldability for the purpose of defining the optimal hard-facing technology of the hot working tool steels. The temperature cycles were measured and then the obtained values for the critical cooling time $t_{8/5}$ were entered into the CCT diagrams to estimate the hardness and microstructure of the welded joint's zones. Those two parameters are indicators of the steel's weldability. Experimental investigation was done on the multi-layered hard-faced samples of steel for the forging dies manufacturing.

1 INTRODUCTION

Weldability, as a very complex technological material property, can be determined by several ways: by calculations, experiments, technological tests and the Continuous Cooling Transformation (CCT) diagram. Considering the fact that the theoretical determination of weldability does not always have to be reliable, the recommendation is to use some other ways to estimate the material's weldability, as well. Theoretical estimate were very reliable for determination of the cooling time $t_{8/5}$, Ito and Bessyo (1972), Lazić et al. (2010), (2014a), since certain output characteristics of the welded joint could be predicted based on that parameter, Arsić et al. (2016). However, even that estimate was not done purely theoretically, since it was verified by either experimental or numerical investigations.

The objective of this paper was to present the CCT diagrams application to estimate the weldability of the hot work tool steels and their reliability in prescribing the technology for hard-facing of the responsible machine parts. Until now, it was shown that various parts could be repaired by the hard-facing, like forging tools parts Arsić et al. (2015a, b), Lazić et al. (2014b), construction mechanization parts, Lazić et al. (2011), or the hydro power plants parts, Arsić et al. (2014), etc. The procedure of using the CCT diagrams to shorten those lengthy and costly experimental procedures is presented in this paper. In order to determine the critical cooling time, as reliably as possible, it is necessary to record the corresponding temperature cycles. Based on those cycles, the distribution of temperature and heat during the welding/hard-facing could be monitored, Arsić et al. (2015c), Murugan et al. (1998), Zimmer (2009), Kumar (2011), Lan et al. (2015) and the certain temperatures could be related to corresponding structures and characteristics of material in individual zones of the joint, Węgrzyn et al. (2007), Galatenu et al. (2015).

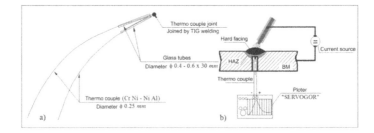

Figure 1.   Scheme of the thermocouple preparation (a) and its position during the hard-facing (b).

Table 1.   Hard-facing parameters.

Thickness s, mm	Electrode diameter d_e, mm	Hard-facing current I, A	Working voltage U, V	Hard-facing speed v_z, mm/s	Heat input, $q_l = \dfrac{U \cdot I}{v_z} \cdot \eta,\ J/mm$
7.4	4	140	25.6	1.62–1.36	1765–2110.1
29	5	210	28.5	2.86–0.98	1673.6–4861

2   THE HOT WORK TOOL STEELS' TEMPERATURE CYCLES

The measurement of the temperature cycles was done using thermocouples in the hard-faced plates of the two thicknesses, 7.4 and 29 mm, made of the 56 NiCrMoV7 (EN 10027-1). The determined temperature cycles and cooling times were entered into the corresponding CCT diagram and the expected micro structure and hardness of steel were obtained. Electrodes of the two different diameters d_e, of 3.25 and 4 mm were used for hard-facing and different preheating temperatures and heat inputs were applied, as well. The steel plates were delivered in the tempered condition, with drilled holes for temperature cycles measurement. The bottom of each hole is at a 4 mm distance from the hard-faced layer, i.e. from the welding arc, to secure that the temperature measurements are as accurate as possible (Figure 1).

The MMA hard-facing parameters are given in Table 1 and present the conditions under which the temperature cycles in the Heat Affected Zone (HAZ) were recorded. In the case of the thick plates a stronger current was applied, since it is considered that the source conducts the heat over the volume and not within the plane as is the case in the thin plates, Lazić et al. (2010).

The given parameters are varying depending on the plates' thicknesses and it is expected that the cooling time between 800°C and 500 °C to be the longest for the largest heat input or the preheating temperature. The electrode UTOP 38 (E Fe 3 according to EN 14700) is aimed for regeneration of parts that operate in the extreme impact pressure loading working conditions. UTOP 38 is the notation by the electrode manufacturer SIJ "Jesenice", Slovenia.

3   RESULTS AND DISCUSSION

For hard-facing of the thin plate ($s = 7.4$ mm) the welding parameters were the following: line energy $q_l = 954.3–1691.1$ J/mm, hard-facing speed $v_w = 2.41–1.36$ mm/s; preheating temperature $T_p = 20–290$ °C; electrode UTOP 38–Ø3.25 mm. For hard-facing of the thick plate ($s = 29$ mm) the parameters were: $q_l = 1650–3273.8$ J/mm; $v_w = 2.58–1.30$ mm/s; $T_p = 20–355$ °C; electrode UTOP 55–Ø5.0 mm. Some of obtained results for the cooling time $t_{8/5}$, determined experimentally and calculated according to three different formulas are presented in Tables 2 and 3, Lazić et al. (2010) Arsić et al. (2016). The corresponding temperature cycles diagrams are shown in Figure 2. Analysis of the obtained temperature cycles has shown that in each diagram one can clearly notice the critical cooling time, which was one of the set experimental tasks. Comparing the diagrams in Figures 2 (a) and (b), the difference in diagrams'

Table 2. Comparative values of the cooling time $t_{8/5}$ (s = 7.4 mm, I = 115 A, U = 25 V, q_{ef} = 2300 W)*.

Hard-facing speed v_z, mm/s	Heat input q_l, J/mm	Preheating temperature T_o/T_p, °C	Cooling time $t_{8/5}$, s				Point/ Layer
			$(t_{8/5})^{I-B}$	$(t_{8/5})^{Slim}$	$(t_{8/5})^{EXP}$	$(t_{8/5})^{R}$	
2.08	1105.8	180	20.10	84.9	27.00	42–50	18/1
1.90	1210.5	178	22.80	100.3	23.00	48–54	11/1
1.86	1219.2	20	13.34	43.63	16.00	24–26	21/1
1.83	1256.8	178	24.05	107.3	19.00	48.5–57	19/1
1.50	1533.3	180	32.90	163.2	24.50	59–68	2/1

*Notation: I-B—calculated according to Ito-Bessyo formula; S_{lim}—calculated according to limiting thickness, Lazić et al. (2010), R—calculated according to Rikalin's formula, EXP—obtained experimentally.
** Shaded fields are the cooling times shown in the CCT diagram in Figure 4.

Table 3. Comparative values of the cooling time $t_{8/5}$ (s = 29 mm, I = 190 A, U = 28 V, q_{ef} = 4256 W).

Hard-facing speed v_z, mm/s	Heat input q_l, J/mm	Preheating temperature T_o/T_p, °C	Cooling time $t_{8/5}$, s				Point/ Layer
			$(t_{8/5})^{I}$	$(t_{8/5})^{Sgr}$	$(t_{8/5})^{EXP}$	$(t_{8/5})^{R}$	
0.258	16500	204	10.43	14.89	12.0	13.5–14.5	6/1
0.185	23000	235	20.20	37.10	20.5	21–23.5	21/1

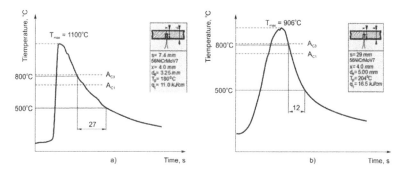

Figure 2. Temperature cycles in the HAZ of the thin plate (a) and the thick plate (b).

appearance, depending on the plate thickness, preheating temperature and quantity of the input heat, can also be noticed. For instance, the maximal temperature T_{max} increases with the input heat for the plate of thickness 7.4 mm, while the maximal temperature for the thick plate (29 mm) is only 906 °C (Figure 2b). The reason for this is the larger mass of the thick plates and faster heat conduction from the heat source towards the periphery.

4 RELATION BETWEEN THE HARD-FACING INPUT PARAMETERS AND OUTPUT CHARACTERISTICS OF THE HARD-FACED LAYERS

For the 56 NiCrMoV7 steel one can find the corresponding TTT and CCT diagrams in manufacturers' catalogues or other reference sources. Those diagrams present the variation of the under-cooled austenite and show the influence of temperature and time on that variation. Although those diagrams were constructed for steel of the certain chemical composition and austenitizing conditions, i.e. in conditions that are different from the hard-facing ones, they can serve in estimates of the output characteristics of the hard-faced layer's zones. This statement was verified several times in various researches involving hard-facing. Considering that this steel belongs into a group of low-alloyed steels with increased carbon content and carbide forming elements

(Cr, V and Mo), such a chemical composition itself causes separation of regions of the perlite and bainite transformation, extension of the austenite stability region and lowered temperature of the martensitic transformation. The CCT diagram for the 56 NiCrMoV7 steel is shown in Figure 3.

From the CCT diagram in Figure 3 one can notice the characteristic temperatures and times of the beginning and ends of individual phase transformations. Separation of the perlite and bainite transformations in the right-hand part of the diagram is a consequence of the steel's chemical composition, i.e. the alloying elements. Some of the most important characteristics of the applied diagrams, which depend on austenitizing temperature, time of heating and time maintained at that temperature, grain size, cooling conditions, and test samples' sizes, are given in Table 4. The newly formed structure and hardness of the HAZ are estimated by entering the cooling time $t_{8/5}$ into the CCT diagram for the given steel. The comparison of structures red-off from the diagrams and structures obtained by the metallography was performed based on the known cooling curves of some characteristic temperature cycles, Figure 3. From the diagrams and Table 4, one can notice that the limiting time $t_{8/5} = t_{100}$ ranges between 500 and 1500 s.

The limiting cooling time $t_{8/5} = t_{100}$ of approximately 500 s could be obtained with preheating to $T_p = 300$ °C and hard-facing parameters $q_l = 6$ kJ/mm for the thin plate $s = 7.4$ mm, namely with $q_l = 15$ kJ/mm for the thick plate $s = 29$ mm. These values of the heat input (driving energy) cannot be achieved by the MMA procedure, which means that, regardless of the hard-facing input parameters (q_l and T_p), one always obtains the martensite-carbide structure of the HAZ, with a hardness between 700 and 743 HV1, Figure 4, Arsić et al. (2015b). Due to that reason one must perform tempering, primarily to reduce the residual stresses and the HAZ hardness and to increase the plasticity of the hard-faced layer's individual zones.

Tempering exhibits a strong influence on the properties of BM hard-faced layers. The hardness decrease is, on average, greater than 200 HV1, while the significant changes of micro

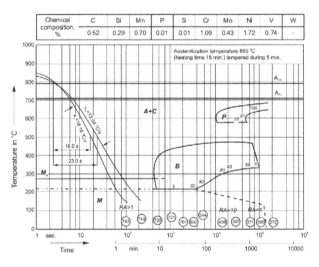

Figure 3.   The CCT diagram for 56 NiCrMoV7 steel.

Table 4.   Austenitizing conditions, critical transformation temperatures and times and maximal hardness, according to different sources.

Literature source	T_{aust}/t_{transf} °C/min	A_{C1}, °C	A_{C3}, °C	M_s °C	t_{100}*s	t_p**s	HV_{max}
Steel plant "Ravne", Slovenia	860/10	713	798	275	1500	6000	708
Atlas zur wärmebehandlung der stähle	880/15	740	780	250	1050	8000	760
Thyssen Marathon Edelstahl	850/5	715	785	275	500	10000	743

*limiting time that corresponds to the end of the pure martensite structure (for each $t_{8/5} \leq t_{100}$).
** time that corresponds to beginning of the perlite transformation.

12

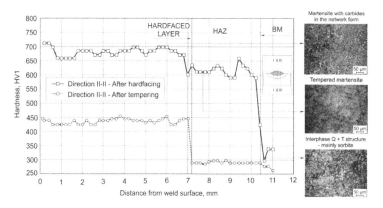

Figure 4.   Hardness distribution and microstructure of the hard-faced layer zones (BM 56 NiCrMoV7—FM UTOP 38 – $s = 7.4$ *mm*), temperature cycle of Figure 2a.

structure were observed, as well (Figure 4). The martensite with carbides, which is present in the weld metal, transforms after the hard-facing into the more stable phases (sorbite and trustite). Tempering also causes significant decrease of residual stresses, Arsić et al. (2015a). This is why one must keep in mind the function of the hard-faced part. Comparison of results obtained by the CCT diagrams and from experiments shows that there is a high correlation of the hard-facing input parameters and the hard-faced layer's output characteristics, while certain deviations arose from inevitable differences in performing the experiments and obtaining the CCT diagrams. Obtained results also show that the high tempering of the hard-faced parts must be performed at a temperature that usually exceeds the working temperature, in order to increase the plasticity, lower the hardness and residual stresses, Rikalin (1951), Mutavdžić et al. (2012).

Based on the experimental results it was concluded that the computational cooling time within the critical temperature interval, could be the most accurately obtained by the Ito-Bessyo formula, Lazić et al. (2010). In that way, the objective of this investigation was reached, since it was shown that the cooling speed can be sufficiently accurately calculated without the expensive experimental procedures. The cooling speed also allows for reading-off the corresponding micro structures and hardness from the CCT diagrams. Knowing the cooling speed and the corresponding transformation diagrams (TTT and CCT), for the considered steel, enables approximate definition of the optimal hard-facing technology.

5   CONCLUSIONS

Based on results presented in this paper, the following conclusions can be drawn:

- Weldability of the hot work tool steels is relatively good, but a series of technological measures must be applied (preheating, tempering) to obtain required properties of the joint;
- Measurement of the temperature cycles can be reliably executed using thermocouples, which is considered the most accurate method;
- Appearance of the martensitic structure cannot be avoided by varying the process parameters, since the time for martensite forming is too long ($\approx$500 s, Figure 3), thus the tempering must be applied;
- Estimation of the hardness and microstructure of hard-faced/welded joint can be done, with high reliability, by the CCT diagrams for the given steel, since, as shown for the 56 NiCrMoV7 steel in this paper, the hardness red-off from the CCT diagram corresponds to hardness measured on the experimental samples and the micro structure is the tempered martensite, what also corresponds to the CCT diagram;
- The weldability estimate can be reliably accomplished with the use of CCT diagrams and thus the expensive and lengthy experiments can be avoided.

ACKNOWLEDGEMENT

This research was supported by the Ministry of Education and Science of Republic of Serbia through Grants TR35024, OI174004 and by European regional development fund and Slovak state budget by the project "Research Centre of the University of Žilina"—ITMS 26220220183.

REFERENCES

Arsić, M., Burzić, M., Karić, R.M., Vistać, B., Savić, Z. 2014. Methodology for repairing defects on internal surfaces of cranks of guide vane apparatus in hydroelectric generating set at hydropower plant Djerdap 1. *Structural Integrity and Life* 14(2): 121–124.

Arsić, D., Lazić, V., Nikolić, R., Aleksandrović, S., Djordjević, M., Hadzima, B., Vičan, J., 2015a. Influence of tempering on the deformation level of the multi-layer hard faced samples, *Procedia Engineering* 111: 49–56.

Arsić, D., Lazić, V., Samardžić, I., Nikolić, R., Aleksandrović, S., Djordjević, M., Hadzima, B. 2015b. Impact of the hard facing technology and the filler metal on tribological characteristics of the hard faced forging dies. *Tehnički Vjesnik-Technical Gazette* 22(5): 1353–1358.

Arsić, D., Lazić, V., Nikolić, R., Aleksandrović, S., Hadzima, B., Djordjević, M. 2015c. The optimal welding technology of high strength steel S690QL, *Materials Engineering-Materialove inzinierstvo* 22(1): 33–47.

Arsić, D., Lazić, V., Sedmak, A., Nikolić, R., Aleksandrović, S., Djordjević, M., Bakić, R. 2016. Selection of the optimal hard facing (HF) technology of damaged forging dies based on cooling time $t_{8/5}$. *Metalurgija-Metallurgy* 55(1): 103–106.

Galatanu, S.V., Faur, N., Pascu, D.R. 2014. Mechanical properties of heat affected zones at macro-microstructural level, using thermal cycle simulation, *Structural Integrity and Life* 14(2): 111–114.

Ito, Y., Bessyo, K. 1972. Weld crackability formula of high strength steels. *Journal of Iron and Steel Institute* 13: 916–930.

Kumar, K.S. 2014. Analytical modeling of temperature distribution, peak temperature, cooling rate and thermal cycles in a solid work piece welded by laser welding process. *Procedia Materials Science* 6: 821–834.

Lan, L., Kong, X., Qiu C. 2015. Characterization of coarse bainite transformation in low carbon steel during simulated welding thermal cycles, *Materials Characterization* 105(7): 95–103.

Lazić, V., Sedmak, A., Živković, M., Aleksandrović, S., Čukić, R., Jovičić, R., Ivanović, I. 2010. Theoretical-experimental determining of cooling time ($t_{8/5}$) in hard facing of steels for forging dies. *Thermal science* 14(1): 235–246.

Lazić, V., Mutavdžić, M., Milosavljević, D., Aleksandrović, S., Nedeljković, B. 2011. Selection of the most appropriate technology of reparatory hard facing of working parts on universal construction machinery. *Tribology in Industry* 33(1): 18–27.

Lazić, V., Aleksandrović, S., Nikolić, R., Prokić-Cvetković, R., Popović, O., Milosavljević, D., Čukić, R. 2012. Estimates of weldability and selection of the optimal procedure and technology for welding of high strength steels, *Procedia Engineering* 40: 310–315.

Lazić, V., Ivanović, I., Sedmak, A., Rudolf, R., Lazić, M., Radaković, Z. 2014a. Numerical analysis of temperature field during hard facing process and comparison with experimental results. *Thermal Science* 18(1): S113-S120.

Lazić, V., Nikolić, R., Aleksandrović, S., Milosavljević, D., Čukić, R., Arsić, D., Djordjević, M. 2014b. Application of hard-facing in reparation of damaged forging dies. Chapter 12. In *Analysis of Technology in Various Industries*, S. Borkowski and R. Ulewicz (eds), Association of Managers of Quality and Production, Częstochowa, Poland, 127–143.

Murugan, S., Kumar, P.V., Raj, B. Bosc, M.S.C. 1998. Temperature distribution during multipass welding of plates. *International Journal of Pressure Vessel and Piping* 75: 891–905.

Mutavdžić, M., Lazić, V., Milosavljević, D., Aleksandrović, S., Nikolić, R., Čukić, R., Bogdanović, G. 2012. Determination of the optimal tempering temperature in hard facing of the forging die, *Materials Engineering-Materialove inzinierstvo* 19(3): 95–103.

Rikalin, N.N., *Computations of the thermal process in welding*, Mashgiz, Moscow, 1951.

Węgrzyn, T., Hadryś, D., Miros, M. 2007. Optimization of operational properties of steel welded structures, *Eksploatacja i Niezawodnosc-Maintenance and Reliability* 9(3): 30–33.

Zimmer, K. 2009. Analytical solution of the laser induced temperature distribution across internal material interfaces. *International Journal of Heat and Mass Transfer* 52: 497–503.

Resistance of flat slabs by different applications of concentrated load

T. Augustín & Ľ. Fillo
Faculty of Civil Engineering, SUT in Bratislava, Bratislava, Slovakia

ABSTRACT: Results of the latest experiments have revealed that the maximum punching resistance defined from the crushing of concrete struts at the perimeter of a column is an insufficient criterion for acquiring a limitation of maximum shear forces. To increase the reliability a further limitation has been introduced in TC 250 SC2 and implemented to national annexes of Eurocodes, STN EN 1992-1-1/NA. The paper will deal with new requirements concerning the maximum punching shear resistance and presents the results of resistance by different applications of concentrated load.

1 INTRODUCTION

The paper deals with the requirements concerning the maximum punching shear resistance which is based on the k_{max} factor and punching shear resistance without shear reinforcement $v_{Rd,c}$. There is also the analysis of minimum flat slab thickness—effective depths, included in the contribution, by taking into account typical configuration of columns in a building.

2 LIMITS OF PUNCHING RESISTANCE

There are two possible ways of structural failure due to punching. The first one is strut diagonal failure (crushing of concrete) at the control perimeter u_0 of the column (Figure 1a). The second one is the shear-tension failure of concrete or transverse reinforcement in area surrounded by the basic control perimeter u_1 (Figure 1b).

Crushing of the struts at the column perimeter is controlled by reduced compressive strength of concrete (1) according to *EN 1992-1-1:2004/AC (2010)*.

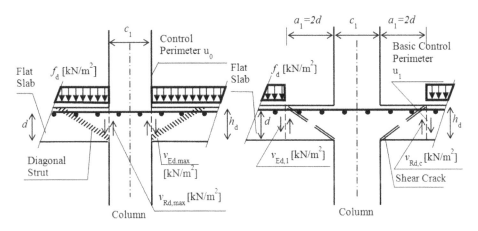

Figure 1. a) Strut diagonal failure model (left), b) model of failure in shear crack (right).

15

$$V_{Ed,max} = \frac{\beta V_{Ed}}{u_0 d} \leq V_{Rd,max} = 0.4 \nu f_{cd}$$

(1)

$$\nu = 0.6\left[1 - \frac{f_{ck}}{250}\right] \quad \text{where } f_{ck} \text{ must be in MPa}$$

(2)

The maximum shear force was limited by the compressive capacity of the struts at the column perimeter. The new limit is based on the punching resistance of a member without shear reinforcement (3) according to *EN 1992-1-1:2004/AC* (2010).

$$V_{Rd,c} = \frac{0.18}{\gamma_c} k \left(100 \rho_l f_{ck}\right)^{1/3} \geq 0.035 k^{3/2} f_{ck}^{1/2}$$

(3)

$$V_{Rd,cs} = 0.75 V_{Rd,c} + \left(\frac{1.5d}{s_r}\right) \frac{A_{sw} f_{ywd,ef}}{u_1 d} \leq k_{max} V_{Rd,c}$$

(4)

However, the latest experiments have also shown that the k_{max} value depends on many factors. The first and most important factor is the type of shear reinforcement and particularly conditions for their anchoring by *Hegger, Siburg* (2010). The second most important factor is rotation ψ of a slab around the supported area according to *Ruiz, Muttoni* (2009).

Double headed studs ensure the best performance of shear reinforcement system ($k_{max} = 1.9$) and also have to meet the requirements introduced in ETA (European Technical Approval) "Punching Prevention System". For other types of shear reinforcement the k_{max} value depends on the effective depth of a slab d. The minimum value is 1.4 if $d \leq 200$ mm and the maximum value 1.7 if $d \geq 700$ mm. For intermediate values of d a linear interpolation can be used.

Punching failure also depends on the position of a column in a plan of a building. There are several types of column—slab connections illustrated on typical flat slab plan (Figure 2b). In the paper we analysed the column connection A, B, C (Figure 2a—C column connection). Zero shear force lines (dotted lines—axis of columns, dashed lines—half of the span, curved lines—$V_{Ed} = 0$) limit the loading areas of the columns (C1, C2, C3 and C4).

The support reaction is eccentric with regard to the control perimeter, the maximum shear stress should be taken from Formula (1), taking into account a coefficient β, given by Formula (5) according to *EN 1992-1-1:2004/AC* (2010).

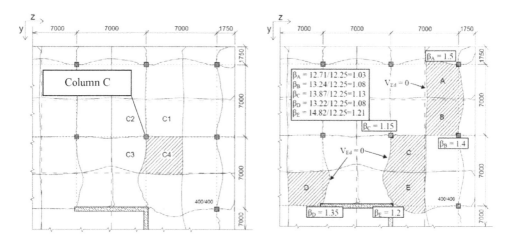

Figure 2.   a) Analysed plan of flat slabs (left), b) Alternative definition of the coefficient β (right).

Table 1. Coefficient k dependence on the ratio of the column dimensions c_1 and c_2.

c_1/c_2	≤0.5	1.0	2.0	≥3.0
k	0.45	0.60	0.70	0.80

Table 2. Comparison of the coefficient β for columns (h_d = 200 mm, slab overhangs—1750 mm).

Column	V_{Ed} [kN]	M_{yEd} [kNm]	M_{zEd} [kNm]	β FEM analysis	β calculation	β recommended
A	413	66.3	66.1	1.64	1.41	1.5
B	539	13.0	72.9	1.31	1.25	1.4
C	691	19.7	14.6	1.07	1.06	1.15

$$\beta = 1 + k \frac{M_{Ed}}{V_{Ed}} \cdot \frac{u_1}{W_1} \qquad (5)$$

Where: u_1 is the length of the basic control perimeter; M_{Ed}—unbalanced bending moment; k is a coefficient dependent on the ratio of the column dimensions c_1 and c_2: this value represents proportion of the unbalanced bending moment transferred by shear into column (according to *EN 1992-1-1:2004/AC (2010)*).

W_1 corresponds to a distribution of shear stresses and is a function of the basic control perimeter u_1. The parameter W_1 is determined by the Formula (6).

$$W_1 = \int_0^{u_i} |e| \, dl \qquad (6)$$

Where: dl—the length increment of the perimeter, e—the distance of dl from the axis about which the moment M_{Ed} acts.

A determination of the coefficient β can be done by the precise definition of the load separation lines ($V_{Ed} = 0$), as described in Figure 2b. The coefficient β can be defined by division of the largest loading area, determined by separation lines ($V_{Ed} = 0$), multiplied by the design uniformly distributed load and the quarter of design shear force (Figure 2b).

3 ANALYSIS OF SLAB EFFECTIVE DEPTH

Coming from above mentioned limits for maximum punching resistance it is possible to determine a slab thickness or effective depths on the basis of a span length, intensity of load and an amount of main reinforcement. The analysis was done for following data: h_d = 200 mm, column span 7000 mm, slab overhang 1750 mm (400 mm alternative), column dimensions 400×400 mm, characteristic permanent load $g_k = 7$ kN/m^2, variable load $q_k = 2.6$ kN/m^2 and reinforcement ratio $\rho = 0.02, 0.015, 0.01$. The bars over column were $\phi20$–25 and concrete cover was 30–35 mm. There are several types of lines in Figure 3–7. Continuous lines represent the failure of diagonal struts. Dashed lines represent shear-tension failure with reinforcement ratio 0.02. Dotted lines represent shear-tension failure with reinforcement ratio 0.015. Dot-dashed lines represent shear-tension failure with reinforcement ratio 0.01.

The nonlinear deflection with cracks and creep of concrete was calculated for quasi permanent combination of load (using the Scia Engineer program). In case of reinforcement ratio $\rho = 0.02$ under the flat slab effective depth 173 mm, is the max total design uniform load limited by deflection. Above 173 mm the crushing of diagonal struts becomes the determining factor (Figure 3).

17

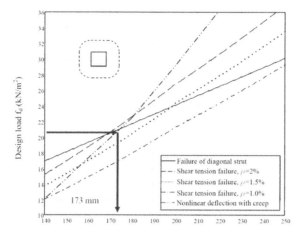

Figure 3. Total designed uniform load in function of effective depth for square column C with dimensions 400×400 mm ($\beta = 1.15$), shear-tension failure lines with kmax = 1.9.

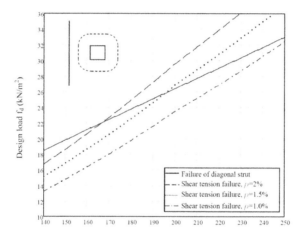

Figure 4. Total designed uniform load in function of effective depth for square column B (1750 mm overhang) with dimensions 400×400 mm ($\beta = 1.4$), shear-tension failure lines with kmax = 1.9.

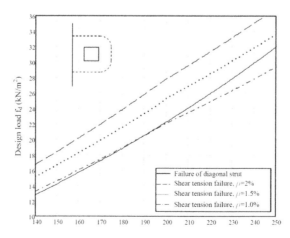

Figure 5. Total designed uniform load in function of effective depth for square column B (400 mm overhang) with dimensions 400×400 mm ($\beta = 1.4$), shear-tension failure lines with kmax = 1.9.

18

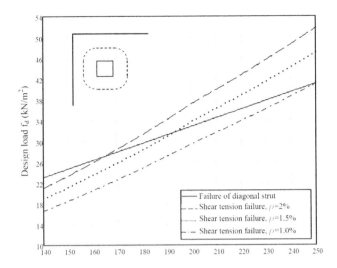

Figure 6. Total designed uniform load in function of effective depth for square column A (1750 mm overhang) with dimensions 400×400 mm ($\beta = 1.5$), shear-tension failure lines with kmax = 1.9.

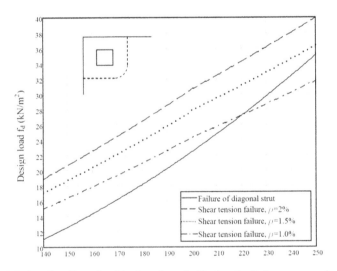

Figure 7. Total designed uniform load in function of effective depth for square column A (400 mm overhang) with dimensions 400×400 mm ($\beta = 1.5$), shear-tension failure lines with kmax = 1.9.

4 CONCLUSIONS

Originating from shear resistance limits based on crushing struts and $k_{max} \cdot v_{Rd,c}$ dependence graphs were created for controlling the minimum required effective depth of flat slab (Figure 3–Figure 7).

It is to say that the limit based on k_{max} factor will govern with higher strength class of concrete, lower amount of main reinforcement and for flat slabs with larger column dimensions. The presented assessments were carried out with k_{max} value of 1.9, where shear reinforcement consists of double headed studs. In case of a different type of shear reinforcement with a different k_{max} the new limit will be decisive for the design of flat slab thickness at the vicinity of the column in almost all cases.

ACKNOWLEDGEMENT

Authors gratefully acknowledge Scientific Grant Agency of the Ministry of Education of Slovak Republic and the Slovak Academy of Sciences VEGA č. 1/0696/14.

REFERENCES

Deutsches Institut fur Bautechnik, European Technical Approval 13/0151-PEIKKO PSB, 2012, Punching reinforcement. Berlin, Germany, May.

EN 1992-1-1:2004/AC(2010), *Eurocode 2: Design of concrete structures—Part 1-1:General rules and rules for buildings.*

Fillo Ľ., Augustín T., Hanzel J., Knapcová V., 2015, *Punching Resistance of Flat Slabs. Engineering Mechanics* 2015. Proceedings of Conference. Svratka 6p.

Hanzel J., Majtánová L., Halvoník J., 2014, *Punching Resistance of Flat Slabs without Shear Reinforcement. In 21. Concrete Days 2014.* Proceedings of Conference. Praha 6 p.

Hegger J., Siburg C., 2010, *Punching—Comparison of Design Rules and Experimental Data*, Proceedings of Workshop "*Design of Con. Structures by EN 1992-1-1*", Prague 113–124.

Ruiz M.F., Muttoni A., 2009, *Application of Critical Shear Crack Theory to Punching of RC Slabs with Transverse Reinforcement, ACI Structural Journal V.106 No.4. 485–494.*

Parameters of various timber-concrete composite connection systems

V. Bajzecerová
Faculty of Civil Engineering, Technical University of Košice, Košice, Slovakia

J. Kanócz
Faculty of Art, Technical University of Košice, Košice, Slovakia

ABSTRACT: Analysis of shear composite connections is a part of extensive research of timber-concrete composite systems. In the first phase of the project, timber-concrete composite beams with a screwed composite connection were investigated. The next phase examined the composite deck system consisting of a vertically laminated nailed timber and concrete using a grooved connection. Finally, composite structural elements with a lightweight concrete and a timber deck system was realized using a special adhesive for wet concrete. In this paper three connection types with different connection stiffnesses are analyzed. To determine the shear parameters of the composite connections, short term shear tests were performed. An experimental estimation of mechanical parameters is presented and the results are finally compared and discussed.

1 INTRODUCTION

For composite connections between the wood and concrete of timber–concrete elements mechanical fasteners are mostly used. The connection provided by these type of connectors is semi-rigid, which has a significant effect on the efficiency of composite cross-sections (Sandanus 2007). Highly rigid connections between the wood and concrete are possible by applying a glued connection. To determine the resistance of timber-concrete composite members some connection parameters are needed, such as the slip modulus or the load-carrying capacity of a given shear plane. For some dowel type fasteners standard EN 1995-1-1 proposes some formulas to calculate these parameters, but not exactly for the connection of timber and concrete. It is particularly preferable to obtain these values experimentally by so called push-out tests.

Analysis of the shear composite connection is a part of the extensive research of timber-concrete composite systems: timber-concrete composite beams with screwed composite connections (Kanócz et al. 2013), deck systems consisting of vertically laminated nailed timber parts and fiber reinforced concrete composite with grooved connections (Kanócz et al. 2014) and finally, structural elements with lightweight concrete and timber deck systems composite with a special adhesive for wet concrete (Kanócz & Bajzecerová 2015). In this paper, the results of the experimental investigation of the three above mentioned connection types are presented and finally compared and discussed.

2 PUSH-OUT TESTS

2.1 *Screwed connection*

Push-out test specimens of screwed connection (denoted as "*S*") consisted of a middle concrete part measuring $200 \times 200 \times 400$ mm; 12.0 mm thick OSB decking sheet and two side timber planks measuring $40 \times 130 \times 400$ mm. The connection between the concrete part and planks was achieved by common steel screws for wood with a 5.0 mm diameter and 120.0 mm length.

Four pairs of screws were driven in at a slope of 45° to the planks top edge. The distance between the pairs of screws along the planks was 150.0 mm.

Fiber reinforced concrete of strength class C20/25 was used for the specimens. A ratio of 15.0 kg per 1 m³ of steel fibers measuring $1.0 \times 0.2 \times 120.0$ mm were added to the concrete mix. Mechanical parameters of concrete were determined according to the relevant standards: mean value of cube compressive strength 31.2 MPa; modulus of elasticity 30.35 GPa. The timber part of specimens consisted of timber with a density of 435 kg/m³, modulus of elasticity 8.44 GPa and mean value of bending strength 25.3 MPa. The following mechanical parameters of OSB3 were certified by manufacturer: density 640.0 kg/m³; modulus of elasticity 4.8 GPa; mean value of bending strength 7.0 MPa and embedment strength 6.8 MPa.

A special arrangement of the shear test setup was proposed to receive a similar strain effect as to that, which would be achieved by beam bending. The specimen was loaded by hydraulic press. On both sides of the timber planks the connection slips were measured by digital gauges (Figure 2). Push-out tests were carried out in standard environmental conditions 64% relative humidity and 23°C. The estimated maximum failure load F_{est} of one specimen was 60.0 kN. The loading schedule was suggested with an unloading process after attaining of 40% of estimated failure load level. After the unloading process the load increased up to the failure of the connection. The failure mode of the screwed connection in all shear test specimens was similar. After attaining of ultimate load level, the shear failure of compressed screws occurred, tensioned screws were withdrawn. Mechanical parameters of one pair of screws determined according to formulas defined in the standard EN 26891 are listed in Table 1.

2.2 Grooved connection

Push-out test specimens of grooved connections (denoted as "D") consisted of a middle concrete part measuring $200 \times 180 \times 500$ mm and of a both side timber parts. The timber part was nailed from 4 planks measuring $40 \times 130 \times 500$ mm together. On each plank a groove

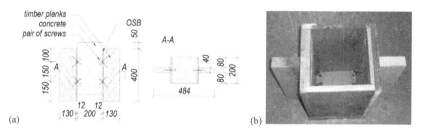

(a)   (b)

Figure 1.   Geometrical parameters of specimens (a), Form prior to casting concrete (b).

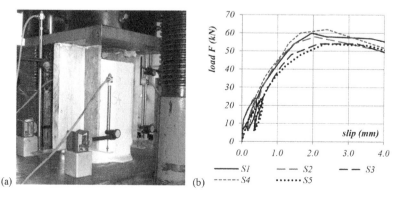

(a)   (b)

Figure 2.   Test set-up (a), Load-slip diagram (b).

22

Table 1. Mechanical parameters of one pair of screws.

Specimen	S1	S2	S3	S4	S5	Average
Instantaneous slip modulus (kN/mm)	13.2	13.6	7.6	14.5	8.7	11.5
Load carrying capacity (kN)	15.0	14.5	13.5	15.5	13.5	14.4

with a depth of 30.0 mm and length of 150.0 mm was created. Grooves were mutually shifted half the length (Figure 3).

Mechanical parameters of the used concrete and timber were identical to the materials of screwed connection specimens. Process of the test was similar. The load action was produced by hydraulic press. Slip of the shear connection was measured by digital gauges on the both sides of the timber part (Figure 4). Tests were carried out in indoor environmental conditions 48% relative humidity and 23°C. The estimated failure load F_{est} of one specimen was 185.0 kN. The loading schedule was suggested with one unloading process. The mode of failure in all shear test specimens of grooved connections was uniform. After attaining of the ultimate load level, the shear failure of concrete occurred. Mechanical parameters of one groove in one plank determined according to EN 26891 are listed in Table 2.

2.3 Adhesive connection

During the third phase of research project bending tests of timber-concrete composite members with adhesive shear connection were realized. Special adhesive for bonding timber and wet concrete SikaDur T35 LVP was used. To determine the shear resistance of adhesive connection shear tests were prepared (Figure 5). Six specimens with an adhesive area 325 × 300 mm were tested.

The timber part of specimens consisted of an 80.0 mm thick glued laminated timber. Mean values of bending strength and modulus of elasticity were estimated experimentally with the value of 54.79 MPa and 13.93 GPa, respectively. The concrete layer was produced using lightweight aggregates—Ceramic Expanded Clay Spheres named "Liapor". Material parameters of used concrete were determined according to the relevant standards: cylinder compressive strength 25.6 MPa, cube compressive strength 29.1 MPa, modulus of elasticity 17.7 GPa and density 1792.0 kg/m³.

The principle of shear tests of glued lines according to EN392 is to expose them to shear stress up to the failure of the connection. Special arrangement of the shear test setup was proposed to receive a uniform plane loading (Figure 6a). Dimensions of the glued area were measured by a caliper with the accuracy of 0.5 mm. Specimens were loaded by hydraulic press parallel to the timber grain. Constant speed of the load acting was carried out, to prevent, that the failure occurred in less than 20 seconds.

Although this standard does not require it, the connection slips were measured by digital gauges. Mutual displacements of the middle points of concrete and timber side surfaces were recorded (Figure 6b). We can expect, that these values were affected by the distance of measured points from glued line and the measured slips were caused by the deformation of the timber and concrete (Negrão et al. 2010). The adhesive connection probably not showed any deformation before failure. After attaining the ultimate load level, brittle shear failure of concrete, partially of the timber occurred. Shear strength of the glued connection obtained from the shear tests according to EN392 are listed in Table 3.

3 DISCUSSION

The above presented investigations allow us to make comparison of the tested types of composite connections from several points of view. The Figure 7a shows comparison of connection flexibility if the resistance of each connection type is equal. The supposed equal value of connection shear resistance 475.0 kN is represented by a bonding area 325 × 300 mm in case

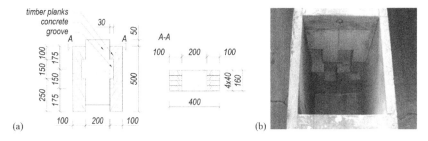

(a)   (b)

Figure 3. Geometrical parameters of specimens (a), Form prior to casting concrete (b).

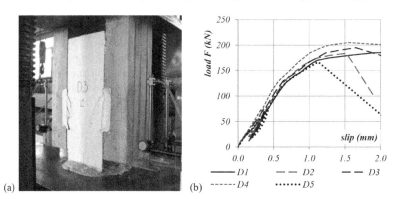

(a)   (b)

Figure 4. Test set-up (a), Load-slip diagram (b).

Table 2. Mechanical parameters of one groove.

Specimen	D1	D2	D3	D4	D5	Average
Instantaneous slip modulus (kN/mm)	20.9	22.2	25.7	30.8	19.1	23.7
Load carrying capacity (kN)	23.1	23.0	24.4	25.6	20.9	23.4

of glued connection, by 20 grooves in case of grooved connection and by 32 pairs of screws in case of screwed connection. This comparison assumes a linear relation between quantity of connectors and the connection stiffness or resistance. It means that the value of the slip modulus is linearly depending on the amount of mechanical fasteners. From the Figure 7a is clear that the load-slip diagram of grooved connection is similar to the diagram of screwed connection. It means that the flexibility of grooves cannot be neglected, even though the grooved connection usually was considered as a rigid connection with high stiffness.

In case of mechanical shear connection the number of usable connectors is determined by the dimensions of the timber-concrete member. Figure 7b shows the comparison of the load-slip diagrams of the investigated connections in which connectors are placed on an area measuring 300 × 320 mm. The surface of a vertically laminated timber slab consisted from 8 pieces of 40.0 mm thick lamellas was mentioned. On this surface is possible to place 8 grooves or 16 pairs of screws in two lines at a distance of 150.0 mm in the lamellas length direction. Comparison shows that the shear resistance and stiffness of 8 grooves is approximately equal to the resistance and stiffness of 16 pairs of screws, but resistance and stiffness of adhesive connection with area 300 × 320 mm is significantly higher.

The presented behaviors of composite connections have significant influence to the global mechanical parameters of timber-concrete composite structural elements (beams, slabs, etc.). Therefore, in design process of these elements it is important to use calculation model which take in to account the real response of the used composite connection. For the timber-concrete composite beams with semi-rigid connection calculation model, so called γ–method is applicable. This model includes the flexibility factor γ according to

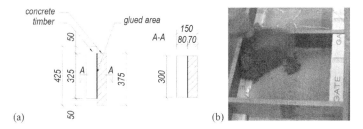

(a)   (b)

Figure 5.   Geometrical parameters of specimens (a), Preparing of specimens (b).

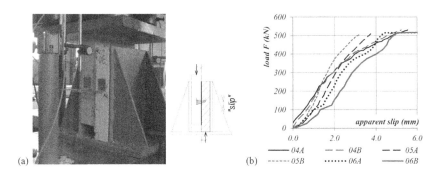

(a)   (b)

Figure 6.   Test set-up (a), Apparent slip caused by deformation of materials (b).

Table 3.   Mechanical parameters of adhesive connection according to EN 392.

Specimen	01	02	03	04	05	06	Average
Shear strength (MPa)	4.84	5.44	4.75	5.75	5.29	5.53	5.27

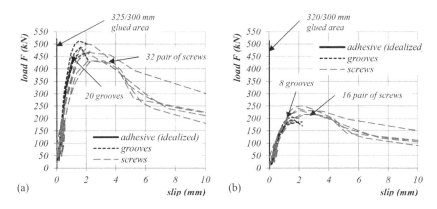

(a)   (b)

Figure 7.   Adapted load-slip diagrams: (a) equal load carrying capacity (b) placeable number of connection.

$$\gamma = \left[1 + \pi^2 E_c A_c s / \left(K L^2 \right) \right]^{-1} \qquad (1)$$

where Ec = Young's modulus of elasticity of concrete in bending; Ac = the cross-section area of the concrete part; s = the spacing between the connectors; K = the slip modulus; L = the span of the beam.

25

For the illustration it was calculated the γ-factor for a reference timber-concrete beam with different composite connections. Because the adhesive connection is the rigid connection; therefore the γ factor is equal to 1.0. In case of the grooved and screwed connections the γ-factor calculated according to (1) is equal to 0.67 and 0.68, respectively. The comparison of these values shows that using screwed or grooved composite connections comparable, but using adhesive connection the higher load-carrying capacity of the timber-concrete composite elements is possible to achieve.

4 CONCLUSION

The behavior of timber concrete composite members is significantly affected by the rigidity of the connection. In this paper, push-out tests of screwed, grooved and adhesive connections was presented and the results of experiments are compared.

Three types of connections were analyzed to show the different stiffnesses between the timber and concrete composite connection. The adhesive connection can be assumed as rigid. Grooves were often considered as a rigid connection, but show some flexibility. Bending tests of real beams confirmed that grooves cannot be considered rigid (Kanócz et al. 2014). Screws show high flexibility, but by placing a sufficient number of pairs of screws a stiffer connection can be achieved. On the other hand the application of a lot of screws can be time consuming.

It can be assumed, that for board materials such as cross laminated timber, the adhesive is a preferable composite connection. But for vertically laminated timber desks the grooves in each lamella are quite easy to realize. In the case of timber beam elements the screwed composite connection is preferable. The screws are the most available solution especially for the reconstruction of beamed ceilings requiring an increase in carrying capacity.

In the field of the timber-concrete composite connections, many push-out tests were carried out all over the world, mostly subjected to short term loading. The long term behavior or the fatigue resistance of most timber-concrete composite connections is unknown. Many different types of connection and various configurations make it difficult to unify some formulas to calculate parameters of composite connection. Therefore more push-out test are required.

ACKNOWLEDGEMENT

Paper is the result of the Project implementation: University Science Park TECHNICOM for Innovation Applications Supported by Knowledge Technology, ITMS: 26220220182, supported by the Research & Development Operational Programme funded by the ERDF. This paper was prepared with supporting of the grant VEGA Project No. 1/0538/16.

REFERENCES

EN 392, 1998: Glued laminated timber. Shear test of glue lines.
EN 26891, 1995: Timber structures. Joints made with mechanical fasteners. General principles for the determination of strength and deformation characteristics.
Kanócz, J., Bajzecerová V. & Šteller, Š. 2013. Timber-concrete composite elements with various composite connections. Part 1: Screwed connection. *Wood research* 58(4): 555–570.
Kanócz, J., Bajzecerová V. & Šteller, Š. 2014. Timber-concrete composite elements with various composite connections. Part 2: Grooved connection. *Wood research* 59(3): 627–638.
Kanócz, J. & Bajzecerová V. 2015. Timber-concrete composite elements with various composite connections. Part 3: Adhesive connection. *Wood research* 60(6): 939–952.
Negrão, J., Leitão de Oliveira, C., Maia de Oliveira, F. & Cachim, P. 2010. Glued Composite Timber-Concrete Beams. I: Interlayer Connection Specimen Tests. *Journal of Structural Engineering*, 136(10).
Sandanus, J. 2007. Parametric study of the factors affecting the resistance of a composite timber-concrete cross-section *Wood Research*, 52(3): 109–114.

Advances and Trends in Engineering Sciences and Technologies II – Al Ali & Platko (Eds)
© 2017 Taylor & Francis Group, ISBN 978-1-138-03224-8

Plastic resistance of IPE-section to interaction of bending internal forces $M_{y,Ed}$, $V_{z,Ed}$ and torsion internal forces B_{Ed}, $T_{\omega,Ed}$ and $T_{t,Ed}$

I.J. Baláž, Mi. Kováč & T.J. Živner
Department of Metal and Timber Structures, Faculty of Civil Engineering, STU in Bratislava, Slovakia

Y.P. Koleková
Department of Structural Mechanics, Faculty of Civil Engineering, STU in Bratislava, Slovakia

ABSTRACT: The resistances of the I- or H-section to interactions of bending and torsion internal forces are missing in the current Eurocodes. New approach is proposed taking into account interaction of bending moment $M_{y,Ed}$, shear force $V_{z,Ed}$, bimoment B_{Ed}, warping torsional moment $T_{\omega,Ed}$ and St. Venant torsional moment $T_{t,Ed}$ acting on I- or H-section. Solutions of various authors, different interaction diagrams and reduction factors are presented and compared. Numerical example shows application of the proposed procedure.

1 INTRODUCTION

1.1 *Rules in Eurocode EN 1993-1-1 for I- or H-section subject to shear due to torsion*

For members subject to torsion for which distortional deformations may be disregarded the design value of the torsional moment at each cross-section should satisfy:

$$\frac{T_{Ed}}{T_{Rd}} \leq 1.0 \qquad (1)$$

where T_{Rd} is the design torsional resistance of the cross-section.

The total torsional moment T_{Ed} at any cross-section should be considered as the sum of two internal effects:

$$T_{Ed} = T_{t,Ed} + T_{\omega,Ed} \qquad (2)$$

where $T_{t,Ed}$ is the design value of the internal St. Venant torsional moment; $T_{\omega Ed}$ is the design value of the internal warping torsional moment.

The values of $T_{t,Ed}$ and $T_{w,Ed}$ at any cross-section may be determined from T_{Ed} by elastic analysis, taking account of the section properties of the member, the conditions of restraint at the supports and the distribution of the actions along the member.

1.2 *Rules in Eurocode EN 1993-1-1 for I- or H-section subject to torsion and bending*

The following stresses due to torsion should be taken into account:

- the shear stresses $\tau_{t,Ed}$ due to St. Venant torsional moment $T_{t,Ed}$;
- the direct stresses $\sigma_{w,Ed}$ due to the bimoment B_{Ed} and shear stresses $\tau_{w,Ed}$ due to warping torsional moment $T_{w,Ed}$.

For the elastic verification the following yield criterion may be applied:

$$\left(\frac{\sigma_{x,Ed}}{f_y/\gamma_{M0}}\right)^2 + \left(\frac{\sigma_{z,Ed}}{f_y/\gamma_{M0}}\right)^2 - \left(\frac{\sigma_{x,Ed}}{f_y/\gamma_{M0}}\right)\left(\frac{\sigma_{z,Ed}}{f_y/\gamma_{M0}}\right) + 3\left(\frac{\tau_{Ed}}{f_y/\gamma_{M0}}\right)^2 \le 1.0 \tag{3}$$

For determining the plastic moment resistance of a cross-section due to bending and torsion only torsion effects B_{Ed} should be derived from elastic analysis.

As a simplification, in the case of a member with open cross-section, such as I or H, it may be assumed that the effects of St. Venant torsion can be neglected.

For combined shear force and torsional moment the plastic shear resistance accounting for torsional effects should be reduced from $V_{pl,Rd}$ to $V_{pl,T,Rd}$ and the design shear force V_{Ed} should satisfy:

$$\frac{V_{z,Ed}}{V_{pl,z,T,Rd}} \le 1.0 \tag{4}$$

in which $V_{pl,T,Rd}$ may be derived for an I- or H-section as follows:

$$V_{pl,z,T,Rd} = \sqrt{1 - \frac{\tau_{t,Ed}}{1,25\left(f_y/\sqrt{3}\right)/\gamma_{M0}}}\, V_{pl,z,Rd} \tag{5}$$

where

$$V_{pl,z,Rd} = \frac{A_{v,z}\left(f_y/\sqrt{3}\right)}{\gamma_{M0}} \tag{6}$$

The shear area $A_{v,z}$ may be taken for rolled I- or H-sections for load parallel to web as follows:

$$A_{v,z} = A - 2bt_f + \left(t_w + 2r\right)t_f \text{ but not less than } \eta h_w t_w \tag{7}$$

The rules in EN 1993-1-1 related to torsion are not sufficient and they are partly incorrect.

2 PROPOSALS

2.1 *Plastic resistance of I- or H-section to bending moment $M_{y,Ed}$ -bimoment B_{Ed} interaction*

The following symbol is introduced for reduction factors $\rho_{IF1,IF2,IF3,IF4,...}$, where the first index *IF1* indicates the internal force which resistance is reduced by one or more other internal forces *IF2, IF3, IF4*. This enables to introduce new reduction factors and to rename reduction factors used in Eurocodes. The advantage is the clarity. Indexes w and f are used for the web and the flange of I- or H-section and index ω is used for torsional internal forces instead of index w used in Eurocodes to avoid the misunderstanding with web. Index ω is omitted at bimoment B. Indexes indicating cross-section axis y and z, which are often in Eurocodes omitted, are used even in the cases when it is clear that bending is about major axis y of I- or H-section.

The relationship between relative bending moment $M_{y,Ed}/M_{pl,y,Rd}$, relative vertical shear force $V_{z,Ed}/V_{pl,z,Rd}$, relative bimoment $B_{Ed}/B_{pl,Rd}$, relative warping torsional moment $T_{\omega,Ed}/T_{pl,\omega,Rd}$ and relative St. Venant torsional moment $T_{t,Ed}/T_{pl,t,Rd}$ may be expressed by the formula derived by the first author. The formula (8) gives the same results as similar formula published in (Streľbickaja, 1954, 1958).

$$C_\sigma^2 + C_\tau = 1.0 \tag{8}$$

where

$$C_\sigma = \frac{\left| \frac{B_{Ed}}{B_{pl,Rd}} - C_2 \frac{M_{y,Ed}}{M_{pl,y,Rd}} \right| + \sqrt{\left(\frac{B_{Ed}}{B_{pl,Rd}} - C_2 \frac{M_{y,Ed}}{M_{pl,y,Rd}} \right)^2 + \frac{C_1}{1.0 - C_1} \left(C_2 \frac{M_{y,Ed}}{M_{pl,y,Rd}} \right)^2}}{2C_1} \tag{9}$$

$$C_1 = 1 - \left(\frac{t_w h_f}{4t_f b} \right)^2, \quad C_2 = 2 \left[\frac{t_w h_f}{4t_f b} + \left(\frac{t_w h_w}{4t_f b} \right)^2 \right]$$

$$C_\tau = \frac{\frac{1}{2} \left(\frac{T_{t,Ed}}{T_{pl,t,Rd}} \right)^2 + \frac{1}{2} \left[C_3 \left(\frac{V_{z,Ed}}{V_{pl,z,Rd}} \right)^2 + C_4 \left(\frac{T_{\omega,Ed}}{T_{pl,\omega,Rd}} \right)^2 \right]}{+ \frac{T_{t,Ed}}{T_{pl,t,Rd}} \sqrt{\left(\frac{1}{2} \frac{V_{z,Ed}}{V_{pl,z,Rd}} \right)^2 + \frac{1}{2} \left[C_3 \left(\frac{V_{z,Ed}}{V_{pl,z,Rd}} \right)^2 + C_4 \left(\frac{T_{\omega,Ed}}{T_{pl,\omega,Rd}} \right)^2 \right]}} \tag{10}$$

where

$$C_3 = \frac{3h_f t_w^2}{2t_f^2 \left(3b - t_f \right) + t_w^2 \left(3h_w + t_w \right)}, \quad C_4 = \frac{6bt_f^2}{2t_f^2 \left(3b - t_f \right) + t_w^2 \left(3h_w + t_w \right)} \tag{11}$$

It was shown in (Baláž et al, 2016) that C_σ^2 may be replaced by the formula which was published already in (Streľbickaja, 1954, 1958):

$$\left(\frac{M_{y,Ed} - M_{pl,y,w,Rd}}{M_{y,f,Rd}} \right)^2 + \frac{B_{Ed}}{B_{pl,Rd}} = 1.0 \tag{12}$$

where

$$M_{pl,y,Rd} = M_{y,f,Rd} + M_{pl,y,w,Rd}, \; M_{pl,y,w,Rd} = W_{pl,y,w} \frac{f_y}{\gamma_{M0}}, \; W_{pl,y,w} = \frac{1}{4} t_w h_w^2 \tag{13}$$

$$B_{pl,Rd} = W_{pl,\omega} \frac{f_y}{\gamma_{M0}}, \; W_{pl,\omega} = \frac{1}{4} t_f h_f b^2 \tag{14}$$

The I- or H-section dimensions with the height h are: the flange breadth b, the flange thickness t_f, the web depth between the flanges h_w, the web thickness t_w and the distance between centers of flanges $h_f = h - t_f$.

Being on the safe side C_σ^2 may be replaced even by the simpler formula:

$$\left(\frac{M_{y,Ed}}{M_{pl,y,Rd}} \right)^2 + \frac{B_{Ed}}{B_{pl,Rd}} = 1.0 \tag{15}$$

2.2 Proposals given in (Mirambell, 2014, 2015)

The original formula proposed in (Mirambell, 2014, 2015) is:

$$M_{c,B,Rd} = \sqrt{1 - \frac{\sigma_{w,Ed.max}}{\alpha f_y}} M_{c,Rd} = \sqrt{1 - \frac{\sigma_{w,Ed,max}}{1.25 f_y}} M_{c,Rd} \tag{16}$$

where according to Mirambell:

$\alpha = 6$ for elastic normal warping stress distribution,
$\alpha = 4$ for plastic normal warping stress distribution,
$\alpha = 5$ for elastic-plastic normal warping stress distribution.

It was proposed in (Mirambell, 2014, 2015) to use the intermediate value $\alpha = 5$. As it is shown below in Figure 1 this choice is questionable.

Formula (16) may be written in more clear and more appropriate form:

$$M_{pl,y,B,Rd} = \sqrt{1 - \frac{6}{\alpha} \frac{B_{Ed}}{B_{pl,Rd}}} M_{pl,y,Rd} = \rho_{My,B} M_{pl,y,Rd} \qquad (17)$$

For $\alpha = 6$ the (16) becomes not only enough correct formula but also simple one and convenient for standard purposes because (18) depends on internal forces and not on stresses:

$$M_{pl,y,B,Rd} = \sqrt{1 - \frac{B_{Ed}}{B_{pl,Rd}}} M_{pl,y,Rd} = \rho_{My,B} M_{pl,y,Rd} \qquad (18)$$

Mirambell's proposal also supposes that bending moment may be reduced by shear force $V_{z,Ed}$ according to clause 6.2.8(4) and (5) EN 1993-1-1 taking into account the shear stresses $\tau_{Vz,Ed}$ due to shear force $V_{z,Ed}$ and shear stress $\tau_{t,Ed}$ due to St. Venant torsional moment $T_{t,Ed}$ in the I- or H-section web. Mirambell's proposal ignores the negligible shear stress $\tau_{\omega,Ed}$ due to warping torsional moment $T_{\omega,Ed}$ in the I- or H-section flanges, but ignores also not negligible shear stress $\tau_{t,Ed}$ due to St. Venant torsional moment $T_{t,Ed}$ in the I- or H-section flanges.

3   NUMERICAL EXAMPLE

Simple supported beam loaded eccentrically by two transverse forces is investigated. The given data are as follows: beam span: $L = 6$ m; rolled section: IPE 300; two vertical transverse forces having the values $F_{Ed} = 49.9$ kN located 1.5 m from supports; eccentricity of the transverse forces: $e = 0.07$ m; steel grade: S235; partial factor for material $\gamma_{M0} = 1.0$; partial factor for action $\gamma_F = 1.5$.

The internal forces of bending and torsion were calculated using solutions of both torsion differential equations: (i) without and (ii) with influence of shear which are given in (Baláž, 2004). It was confirmed the well known fact, that influence of shear for open cross-section is negligible. These analytical results were checked and compared with numerical results obtained by computer program IQ 100. Numerical results of computer program IQ 100 were calculated using analogy between torsion and bending combined with tension force. It was shown that both analytical and numerical results are the same. The internal forces of torsion are in Figures 1, 2, 3 and 4.

Figures 1–4 illustrate that for open section the influence of the shear (dotted lines and the values in the brackets) is negligible. Calculated measure of warping $\mu = 0.997$ and beam parameters:

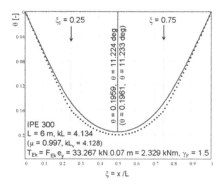

Figure 1.   Twist rotation $\theta(\xi)$ about axis x.

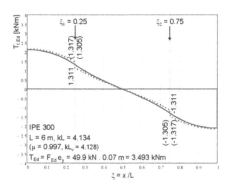

Figure 2.   St.Venant torsional moment $T_{t,Ed}(\xi)$.

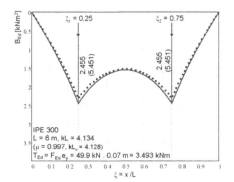

Figure 3. Bimoment $B_{Ed}(\xi)$.

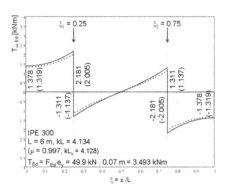

Figure 4. Warping torsional moment $T_{\omega,Ed}(\xi)$.

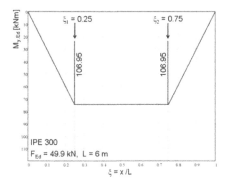

Figure 5. Bending moment $M_{y,Ed}(\xi)$.

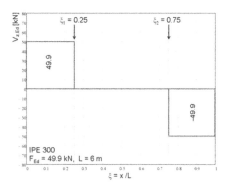

Figure 6. Shear force $V_{z,Ed}(\xi)$.

$$kL = \sqrt{\frac{GI_t}{EI_\omega}} = 4.134 \text{ without and} \, kL_v = \sqrt{\mu \frac{GI_t}{EI_\omega}} = 4.128 \text{ with shear influence} \tag{19}$$

To show very small difference between the solid lines (without influence of shear) and the dotted lines (with influence of shear) the graphical distributions were drawn for $kL_v = 0.9$ instead of correct value $kL_v = 0.997$. Otherwise no difference between solid and dotted lines would be seen in figures. But the related numerical values in Figures 1–4 are correct.

If the criterion defined by Mirmabell's formula (16) is used the action resistance is $F_{Rd} = 49.9$ kN. The action resistance $F_{Rd} = 49.9$ kN becomes $F_{Rd} = 60$ kN when criterion (12) and not criterion (16) is used as a criterion. If the criterion (15) is used the action resistance $F_R = 54.93$ kN (Table 1).

This criteria may be used only if the bending moment cross-section plastic resistance is not reduced due to shear stress $\tau_{Vz,Ed}$ in the web and the bimoment cross-section plastic resistance is not reduced due to shear stresses $\tau_{t,Ed}$ or $\tau_{\omega,Ed}$ in the flanges. In the numerical example $V_{z,Ed}/V_{pl,z,Tt,Rd} = 0.178 < 0.5$ and the bending moment resistance is not necessary to reduce according to EN 1993-1-1. In the section next to the force $F_{Ed} = 49.9$ kN. The shear stress due to shear force $\tau_{Vz,Ed} = 6.48$ MPa and the warping torsional stress in the flanges $\tau_{\omega,Ed} = 7.05$ MPa are negligible. But the St. Venant torsional shear stress in the flanges $\tau_{t,Ed} = 90.53$ MPa is not negligible. Total shear stress in the flanges is $\tau_{Vz,Ed} + \tau_{\omega,Ed} + \tau_{t,Ed} = 104.05$ MPa.

The St. Venant torsional $T_{t,Ed}$ should be taken into account in calculation of action resistance F_{Rd}. If the criterion (8) is used the action resistance $F_R = 54.36$ kN. If in the criterion (8) instead of C_σ^2 the formula (12) is used the calculated action resistance $F_R = 52.44$ kN. If in the criterion (8) instead of C_σ^2 the formula (15) is used the calculated action resistance $F_R = 48.2$ kN.

Table 1. Comparison of action resistances F_{Rd} [kN] calculated from different criteria.

Criterion	(12)	(15)	(8)	(8) using (12) instead of C_σ^2	(16)	(8) using (15) instead of C_σ^2
F_{Rd} [kN]	71.30	68.53	65.60	63.33	62.80	60.9

4 CONCLUSIONS

It was shown (Baláž et al, 2016) that the proposal (Mirambell, 2014, 2015), which was already accepted in SC3 for the next generation of Eurocodes, could be replaced by more exact and more convenient formula (15) based on internal forces and not on stresses.

The formula (15) alone may be used only for the cases when bending moment cross-section plastic resistance $M_{pl,y,Rd}$ and bimoment cross-section plastic resistance $B_{pl,Rd}$ is not necessary to reduced due to vertical force $V_{z,Ed}$ or St. Venant torsional moment $T_{t,Ed}$ or warping torsional moment $T_{\omega,Rd}$. Otherwise the appropriate reduction of cross-section plastic resistance shall be done. The more general loading cases taking into account all 5 internal forces of bending and torsion authors are presented. The action resistance F_{Rd} was calculated from different criteria (Table 1).

The Mirambell's procedure does not take into account in the I- or H-section flanges the shear stresses due to torsion $\tau_{\omega,Ed} + \tau_{t,Ed}$. In some cases the shear stress $\tau_{t,Ed}$ is not negligible.

It is important to note that in some cases when beam is under combined bending and torsion the serviceability limit state may be critical when the size of twist rotation is big and not the ultimate limit state. The second order theory of torsion should be used when the size of twist rotation is big.

The criterion (15) could be chosen for standard purposes if the influence of shear stresses in the I- or H-section on the cross-section resistance is negligible. For such cases the Mirambell's procedure is on the safe side and acceptable. If there are not negligible shear stresses in the flanges the criterion (8) using (15) instead of C_σ^2 could be used. For such cases the Mirambell's procedure could give results on the unsafe side.

ACKNOWLEDGEMENT

Project No. 1/0819/15 was supported by the Slovak Grant Agency VEGA.

REFERENCES

Baláž, I.J. 2004. Thin-walled steel structures. Influence of torsion in large thin-walled bridge structures (in Slovak). ES STU Bratislava, 1st edition, 1984: 1–168. 5th expanded edition. 2004: 1–295.
Baláž, I.J. & Koleková, Y.P. 2015. Resistance of cross-section to bending-shear-normal force interaction. *Eighth International Conference on Advances of Steel Structures, Lisbon, Portugal, July 22–24, 2015.* Paper No. 193: 1–20.
Baláž, I.J., Kováč, M., Živner, T.J. & Koleková, Y.P. 2016. Plastic resistance of I-section to interaction of bending moment and bimoment. *Proceedings of International Conference on Engineering Sciences and Technologies. Tatranská Štrba, High Tatras Mountains—Slovak Republic. CRC Press/Balkema, Taylor & Francis Group, 29 June-1 July, 2016.*
EN 1993-1-1 and Corrigendum AC. Eurocode 3—Design of steel structures, Part 1–1: General rules and rules for buildings. CEN Brussels, March 2005 and February 2006 (AC).
EN 1999-1-1: May 2007 + A1: July 2009 + A2: December 2013 Design of Aluminium Structures. Part 1–1 General Rules and Rules for Buildings. CEN Brussels.
Mirambell, E. & Bordallo, J. 2014. Some considerations on the treatment of torsion and its interaction with other internal forces in EN 1993-1-1. A new approach. report to WG 1.22/10/2014. Berlin.
Mirambell, E. 2015. Amendment to 1993-1-1:2015, WG 1 in EN 1993-1-1, doc. N0086.
Streľbickaja, A.I. 1954. Predeľnyje nagruzki tonkostennych balok pri sovmestnom dejstviji izgiba i kručenija. Sbornik trudov Instituta strojiteľnoj mechaniki Akademii Nauk Ukrainskoj SSR. No.19, (in Russian).
Streľbickaja, A.I. 1958. *Isledovanie pročnosti tonkostennych steržnej za predelom uprugosti.* Izdateľstvo AN Ukrainskoj SSR. Kiev, (in Russian).

Advances and Trends in Engineering Sciences and Technologies II – Al Ali & Platko (Eds)
© *2017 Taylor & Francis Group, ISBN 978-1-138-03224-8*

Plastic resistance of H-section to interaction of bending moment $M_{y,Ed}$ and bimoment B_{Ed}

I.J. Baláž, Mi. Kováč & T.J. Živner
Department of Metal and Timber Structures, Faculty of Civil Engineering, STU in Bratislava, Slovakia

Y.P. Koleková
Department of Structural Mechanics, Faculty of Civil Engineering, STU in Bratislava, Slovakia

ABSTRACT: The resistances of the I- or H-section to interactions of bending and torsion internal forces are missing in the current Eurocodes. New approach is proposed taking into account interaction of bending moment $M_{y,Ed}$ and bimoment B_{Ed}. Solutions of various authors, different interaction diagrams and reduction factors are presented and compared. Numerical example shows application of the proposed procedure.

1 INTRODUCTION

1.1 *Rules in Eurocode EN 1993-1-1 for I- and H-section subject to shear due to torsion*

For members subject to torsion for which distortional deformations may be disregarded the design value of the torsional moment at each cross-section should satisfy:

$$\frac{T_{Ed}}{T_{Rd}} \leq 1.0 \qquad (1)$$

where T_{Rd} is the design torsional resistance of the cross-section.

The total torsional moment T_{Ed} at any cross-section should be considered as the sum of two internal effects:

$$T_{Ed} = T_{t,Ed} + T_{\omega,Ed} \qquad (2)$$

where $T_{t,Ed}$ is the design value of the internal St. Venant torsional moment; $T_{\omega Ed}$ is the design value of the internal warping torsional moment.

The values of $T_{t,Ed}$ and $T_{w,Ed}$ at any cross-section may be determined from T_{Ed} by elastic ana-lysis, taking account of the section properties of the member, the conditions of restraint at the supports and the distribution of the actions along the member.

1.2 *Rules in Eurocode EN 1993-1-1 for I- and H-section subject to torsion and bending*

The following stresses due to torsion should be taken into account:

- the shear stresses $\tau_{t,Ed}$ due to St. Venant torsional moment $T_{t,Ed}$;
- the direct stresses $\sigma_{w,Ed}$ due to the bimoment B_{Ed} and shear stresses $\tau_{w,Ed}$ due to warping torsional moment $T_{w,Ed}$.

For the elastic verification the following yield criterion may be applied:

$$\left(\frac{\sigma_{x,Ed}}{f_y/\gamma_{M0}}\right)^2 + \left(\frac{\sigma_{z,Ed}}{f_y/\gamma_{M0}}\right)^2 - \left(\frac{\sigma_{x,Ed}}{f_y/\gamma_{M0}}\right)\left(\frac{\sigma_{z,Ed}}{f_y/\gamma_{M0}}\right) + 3\left(\frac{\tau_{Ed}}{f_y/\gamma_{M0}}\right)^2 \leq 1.0 \qquad (3)$$

For determining the plastic moment resistance of a cross-section due to bending and torsion only torsion effects B_{Ed} should be derived from elastic analysis.

As a simplification, in the case of a member with open cross-section, such as I or H, it may be assumed that the effects of St. Venant torsion can be neglected.

For combined shear force and torsional moment the plastic shear resistance accounting for torsional effects should be reduced from $V_{pl,Rd}$ to $V_{pl,T,Rd}$ and the design shear force V_{Ed} should satisfy:

$$\frac{V_{z,Ed}}{V_{pl,z,T,Rd}} \leq 1.0 \tag{4}$$

in which $V_{pl,T,Rd}$ may be derived for an I- or H-section as follows:

$$V_{pl,z,T,Rd} = \sqrt{1 - \frac{\tau_{t,Ed}}{1.25\left(f_y/\sqrt{3}\right)/\gamma_{M0}}} \; V_{pl,z,Rd} \tag{5}$$

where $V_{pl,z,Rd}$

$$V_{pl,z,Rd} = \frac{A_{v,z}\left(f_y/\sqrt{3}\right)}{\gamma_{M0}} \tag{6}$$

The shear area $A_{v,z}$ may be taken for rolled I- or H-sections for load parallel to web as follows:

$$A_{v,z} = A - 2bt_f + \left(t_w + 2r\right)t_f \text{ but not less than } \eta h_w t_w \tag{7}$$

The rules in EN 1993-1-1 related to torsion are not sufficient and they are partly incorrect.

2   PROPOSALS

2.1   *Plastic resistance of I- and H-section to bending moment $M_{y,Ed}$—bimoment B_{Ed} interaction*

The following symbol is introduced for reduction factors $\rho_{IF1,IF2,IF3,IF4...}$ where the first index *IF1* indicates the internal force which resistance is reduced by one or more other internal forces *IF2, IF3, IF4*. This enables to introduce new reduction factors and to rename reduction factors used in Eurocodes. The advantage is the clarity. Indexes w and f are used for the web and the flange of I- or H-section and index ω is used for torsional internal forces instead of index w used in Eurocodes to avoid the misunderstanding with web. Index ω is omitted at bimoment B. Indexes indicating cross-section axis y and z, which are often in Eurocodes omitted, are used even in the cases when it is clear that bending is about major axis y of I- or H-section.

The relationship between relative bending moment $M_{y,Ed} / M_{pl,y,Rd}$ and relative bimoment $B_{Ed} / B_{pl,Rd}$ may be expressed by the formula derived by the first author. The formula (8) gives the same results as similar formula published in (Streľbickaja, 1954, 1958).

$$C_\sigma = 1.0 \tag{8}$$

where

$$C_\sigma = \frac{\left|\dfrac{B_{Ed}}{B_{pl,Rd}} - C_2 \dfrac{M_{y,Ed}}{M_{pl,y,Rd}}\right| + \sqrt{\left(\dfrac{B_{Ed}}{B_{pl,Rd}} - C_2 \dfrac{M_{y,Ed}}{M_{pl,y,Rd}}\right)^2 + \dfrac{C_1}{1.0 - C_1}\left(C_2 \dfrac{M_{y,Ed}}{M_{pl,y,Rd}}\right)^2}}{2C_1} \tag{9}$$

$$C_1 = 1 - \left(\frac{t_w h_f}{4t_f b}\right)^2, \; C_2 = 2\left[\frac{t_w h_f}{4t_f b} + \left(\frac{t_w h_w}{4t_f b}\right)^2\right]$$

It is proposed to use in EN 1993 and EN 1999 for Class 1 or 2 I- or H-section the following interaction formula which approximates formula (8) and which was published already in (Streľbickaja, 1954, 1958):

$$\left(\frac{M_{y,Ed} - M_{pl,y,w,Rd}}{M_{y,f,Rd}}\right)^2 + \frac{B_{Ed}}{B_{pl,Rd}} = 1.0 \tag{10}$$

where

$$M_{pl,y,Rd} = M_{y,f,Rd} + M_{pl,y,w,Rd}, \, M_{pl,y,w,Rd} = W_{pl,y,w}\frac{f_y}{\gamma_{M0}}, W_{pl,y,w} = \frac{1}{4}t_w h_w^2 \tag{11}$$

$$B_{pl,Rd} = W_{pl,\omega}\frac{f_y}{\gamma_{M0}}, W_{pl,\omega} = \frac{1}{4}t_f h_f b^2 \tag{12}$$

The I or H-section dimensions with the height h are: the flange breadth b, the flange thickness t_f, the web depth between the flanges h_w, the web thickness t_w and the distance between centers of flanges $h_f = h - t_f$.

Formula (10) enables to derive the following two kinds of reduction factors $\rho_{My,B}$ and $\rho_{B,My}$:

$$\rho_{My,B} = 1 - \left(1 - \sqrt{1 - \frac{B_{Ed}}{B_{pl,Rd}}}\right)\frac{M_{y,f,Rd}}{M_{pl,y,Rd}} \tag{13a}$$

$$\rho_{My,B} = \frac{M_{y,Ed}}{M_{pl,y,Rd}}, \text{ if } \frac{B_{Ed}}{B_{pl,Rd}} = 1 \text{ and } M_{y,Ed} \leq M_{pl,y,w,Rd} \tag{13b}$$

and

$$\rho_{B,My} = 1 - \left(1 - \frac{M_{pl,y,Rd} - M_{y,Ed}}{M_{y,f,Rd}}\right)^2 \leq 1.0 \tag{14}$$

The restrictions in formulae (13b) and (14) are necessary because: (i) for $B_{Ed} = B_{pl,Rd}$ the web is responsible for carrying $M_{y,Ed}$ if $M_{y,Ed} \leq M_{pl,y,w,Rd}$ and (ii) for $M_{y,Ed} < M_{pl,y,w,Rd}$ the factor would be $\rho_{B,My} > 1.0$.

The reduction factors defined by formulae (13) and (14) enable to calculate reduced plastic resistances of Class 1 and 2 cross-sections.

$$M_{pl,y,B,Rd} = \rho_{My,B}M_{pl,y,Rd} \tag{15}$$

$$B_{pl,My,Rd} = \rho_{B,My}B_{pl,Rd} \tag{16}$$

2.2 Proposals given in (Mirambell, 2014, 2015)

The original formula proposed in (Mirambell, 2014, 2015) is

$$M_{c,B,Rd} = \sqrt{1 - \frac{\sigma_{w,Ed.max}}{\frac{\alpha f_y}{4\gamma_{M0}}}}M_{c,Rd} = \sqrt{1 - \frac{\sigma_{w,Ed.max}}{\frac{1.25f_y}{\gamma_{M0}}}}M_{c,Rd} \tag{17}$$

where according to Mirambell
  $\alpha = 6$ for elastic normal warping stress distribution,
  $\alpha = 4$ for plastic normal warping stress distribution,
  $\alpha = 5$ for elastic-plastic normal warping stress distribution.

35

It was proposed in (Mirambell, 2014, 2015) to use the intermediate value $\alpha = 5$. As it is shown below in Figure 1 this choice is questionable.

Formula (17) may be written in more clear and more appropriate form:

$$M_{pl,y,B,Rd} = \sqrt{1 - \frac{6}{\alpha}\frac{B_{Ed}}{B_{pl,Rd}}} M_{pl,y,Rd} = \rho_{My,B} M_{pl,y,Rd} \tag{18}$$

For $\alpha = 6$ the (17) becomes not only enough correct formula (Figure 1) but also simple one and convenient for standard purposes because (19) depends on internal forces and not on stresses:

$$M_{pl,y,B,Rd} = \sqrt{1 - \frac{B_{Ed}}{B_{pl,Rd}}} M_{pl,y,Rd} = \rho_{My,B} M_{pl,y,Rd} \tag{19}$$

2.3 *Comparison and evaluation of different solutions and proposals*

Comparison of various solutions is presented in Figure 1. These results are valid for rolled HE-A 240 profile made of steel S235.

From Figure 1 it is clear that simple formula (10) gives practically the same results as more complicated one (8). It seems that for the Eurocodes EN 1993-1-1 and EN 1999-1-1 could be used the simplest formula (18) with $\alpha = 6$, that means formula (19).

These analytical results are compared in Table 1 with numerical results obtained by DLUBAL program DUENQ (English name SHAPE-THIN) which uses simplex method. Table 1 shows that computer program DUENQ gives the same results as formula (10). Formula (19) gives results which are slightly on safe side. It is interesting to mention that DUENQ gives incorrect results for the values $B_{Ed}/B_{pl,Rd} = 0.9$ and $M_{y,Ed}/M_{pl,y,Rd} = 0.393$.

3 NUMERICAL EXAMPLE

Simple supported beam loaded eccentrically by transverse force is investigated. The given data are as follows: beam span: $L = 6$ m; rolled section: HE-A 240; vertical transverse force in the middle of the span: $F_{Ed} = 71.3$ kN (this is action resistance F_{Rd} if (10) is used as a cri-

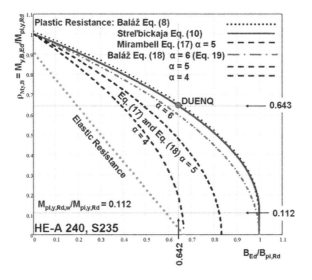

Figure 1. Comparison of various solutions for resistance of rolled H-section under interaction of bending moment M_{Ed} and bimoment B_{Ed}. The results are valid for rolled HE-A 240 profile made of steel S235.

Table 1. Ratio of $M_{y,Ed}/M_{pl,y,Rd}$ as a function of $B_{Ed}/B_{pl,Rd}$. Comparison of analytical and numerical results.

$B_{Ed}/B_{pl,Rd}$	0	0.1	0.2	0.3	0.4	0.5	0.6	0.7	0.8	0.9	0.95	1.0
DUENQ	1.0	0.955	0.907	0.855	0.800	0.740	0.674	0.599	0.509	0.400	0.311	0.112
Eq. (8)	1.013	0.968	0.920	0.869	0.813	0.754	0.687	0.612	0.523	0.407	0.324	0.126
Eq. (10)	1.0	0.954	0.906	0.855	0.800	0.740	0.674	0.599	0.509	0.393	0.311	0.112
Eq. (19)	1.0	0.949	0.894	0.837	0.775	0.707	0.632	0.548	0.447	0.316	0.224	0

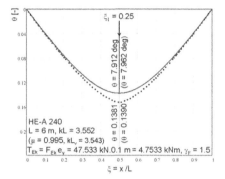

Figure 2. Twist rotation $\theta(\xi)$ about axis x.

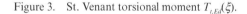

Figure 3. St. Venant torsional moment $T_{t,Ed}(\xi)$.

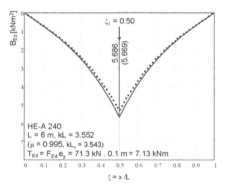

Figure 4. Bimoment $B_{Ed}(\xi)$.

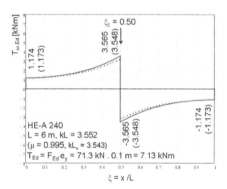

Figure 5. Warping torsional moment $T_{\omega,Ed}(\xi)$.

terion); eccentricity of the transverse force: $e = 0.1$ m; steel grade: S235; partial factor for material $\gamma_{M0} = 1.0$; partial factor for action $\gamma_F = 1.5$.

The internal forces of bending and torsion were calculated using solutions of both torsion differential equations: (i) without and (ii) with influence of shear which are given in (Baláž, 2004). It was confirmed the well known fact, that influence of shear for open cross-section is negligible. These analytical results were checked and compared with numerical results obtained by computer program IQ 100 (created by Prof. Rubin et al at TU Vienna). Numerical results of computer program IQ 100 were calculated using analogy between torsion and bending combined with tension force. It was shown that both analytical and numerical results are the same. The internal forces of torsion are in Figures 2, 3, 4 and 5.

Figures 2–5 illustrate that for open section the influence of the shear (dotted lines and the values in the brackets) is negligible. Calculated measure of warping $\mu = 0.995$ and beam parameters:

$$kL = \sqrt{\frac{GI_t}{EI_\omega}} = 3.552 \text{ without and } kL_v = \sqrt{\mu\frac{GI_t}{EI_\omega}} = 3.543 \text{ with shear influence} \quad (20)$$

37

To show very small difference between the solid lines (without influence of shear) and the dotted lines (with influence of shear) the graphical distributions were drawn for $kL_v = 0.9$ instead of correct value $kL_v = 0.995$. Otherwise no difference between solid and dotted lines would be seen in figures. But the related numerical values in Figures 2–5 are correct.

The values valid for numerical example $B_{Ed} / B_{pl,Rd} = 0.642$ and $M_{y,Ed} / M_{pl,y,Rd} = 0.643$ are shown in Figure 1. For limit state checking of the beam in the numerical example the criterion defined by formula (10) was chosen. This criterion may be used only if the bending moment cross-section plastic resistance is not reduced due to shear stress $\tau_{Vz,Ed}$ in the web and the bimoment cross-section plastic resistance is not reduced due to shear stresses $\tau_{t,Ed}$ or $\tau_{\omega,Ed}$ in the flanges. In the numerical example $V_{z,Ed} / V_{pl,z,Rd} = 0.104 < 0.5$ and the bending resistance according to EN 1993-1-1 is not necessary to reduce. The St. Venant torsional shear stress in the middle of the span $\tau_{t,Ed} = 0$ MPa, the warping torsional stress in the middle of the span $\tau_{\omega,Ed} = 8.5$ MPa is negligible and therefore the bimoment resistance is not necessary to reduce too.

Resistance $F_{Rd} = 71.3$ kN becomes $F_{Rd} = 62.8$ kN when (17) and not (10) is used as a criterion.

4 CONCLUSIONS

It was shown that the proposal (Mirambell, 2014, 2015), which was already accepted in SC3 for the next generation of Eurocodes, could be replaced by more exact and more convenient formula (19) based on internal forces and not on stresses.

The formula (19) alone may be used only for the cases when bending moment cross-section plastic resistance $M_{pl,y,Rd}$ and bimoment cross-section plastic resistance $B_{pl,Rd}$ is not necessary to reduced due to vertical force $V_{z,Ed}$ or St. Venant torsional moment $T_{t,Ed}$ or warping torsional moment $T_{\omega,Rd}$. Otherwise the appropriate reduction of cross-section plastic resistance shall be done. The more general loading cases taking into account all 5 internal forces of bending and torsion authors solved in another paper, which is in print.

Numerical example shows details of application of the proposed rules. The internal forces of torsion were calculated without and with the influence of the shear, which is negligible for open cross-sections.

It is important to note that in some cases when beam is under combined bending and torsion the serviceability limit state may be critical when the size of twist rotation is big and not the ultimate limit state. The second order theory of torsion should be used when the size of twist rotation is big.

ACKNOWLEDGEMENT

Project No. 1/0819/15 was supported by the Slovak Grant Agency VEGA.

REFERENCES

Baláž, I.J. & Koleková, Y.P. 2015. Resistance of cross-section to bending-shear-normal force interaction. *Eighth International Conference on Advances of Steel Structures, Lisbon, Portugal, July 22–24, 2015.* Paper No. 193: 1–20.

Baláž, I.J. 2004. Thin-walled steel structures. Influence of torsion in large thin-walled bridge structures (in Slovak). ES STU Bratislava, 1st edition, 1984: 1–168. 5th expanded edition. 2004: 1–295.

EN 1993-1-1 and Corrigendum AC. Eurocode 3 - Design of steel structures, Part 1–1: General rules and rules for buildings. CEN Brussels, March 2005 and February 2006 (AC).

EN 1999-1-1: May 2007 + A1: July 2009 + A2: December 2013 Design of Aluminium Structures. Part 1–1 General Rules and Rules for Buildings. CEN Brussels.

Mirambell, E. & Bordallo, J. 2014. Some considerations on the treatment of torsion and its interaction with other internal forces in EN 1993-1-1. A new approach. report to WG 1.22/10/2014. Berlin.

Mirambell, E. 2015. Amendment to 1993-1-1:2015, WG 1 in EN 1993-1-1, doc. N0086.

Streľbickaja, A.I. 1954. Predeľnyje nagruzki tonkostennych balok pri sovmestnom dejstviji izgiba i kručenija. Sbornik trudov Instituta strojiteľnoj mechaniki Akademii Nauk Ukrainskoj SSR. No.19, (in Russian).

Streľbickaja, A.I. 1958. *Isledovanie pročnosti tonkostennych steržnej za predelom uprugosti.* Izdateľstvo AN Ukrainskoj SSR. Kiev, (in Russian).

Advances and Trends in Engineering Sciences and Technologies II – Al Ali & Platko (Eds)
© 2017 Taylor & Francis Group, ISBN 978-1-138-03224-8

Experimental investigations of glass plates

Ľ. Balcierák, J. Brodniansky Sr., T. Klas, V. Duchoň, J. Brodniansky & O. Katona
Department of Steel and Timber Structures, Faculty of Civil Engineering, Slovak University of Technology in Bratislava, Bratislava, Slovakia

ABSTRACT: Glass is a fragile elastic material. The impact is one of the most important factors in the design of glass load bearing elements. In glazing terms, the word safety is applied to glass which is capable of reducing the risk of injury from accidental actions of impact, fracture or shattering. Laminated glass consists of a sandwich of glass plates and interlayers that keep the glass fragments in place after the impact, so that the possibility of injury is minimised. This paper focuses on tests of three different glass panels: new laminated glass panels, old laminated glass panels (15 year storage in laboratory) and Float glass panels (only new). All of the testing tools (impact body, steel frame, impact mechanism) used for the presented tests were built by the authors of this paper according to the standard DIN EN 12600.

1 INTRODUCTION

Glass is a fragile elastic material, and therefore the impact is one of the important factors in the design of the supporting glass element. DIN EN 12600 Glass in building—Pendulum test—Impact test method and classification for flat glass (DIN EN 12600) describes the pendulum test of flat glass element and its classification.

2 EXPERIMENTAL INVESTIGATION ACCORDING TO DIN EN 12600

Norm (DIN EN 12600) deals with the testing of a flat glass element, classifying glass behaviour by impact and after breaking the glass element. Classification height of fall of the impact element on the sample is graded according to the energy generated by impact of a moving person at different speeds. The purpose of norm (DIN EN 12600) is to enhance safety, by reducing cuts and stab wounds. Norm specifies the pendulum test method for flat glass panes and provides a classification in three main groups on impact. It does not specify requirements for the design and the durability of the glass element. The test specifies that the sample is either a compact (without any visible cracks) or it breaks into fragments (which can be dangerous for people standing near glass flat glass panels).

2.1 *The test equipment*

The test equipment consists of (see Figure 1):

– Robust main frame (1),
– Chucking frame (2),
– Impact body (3),
– Sample (4),
– Intermediate hanger (5).

Figure 1.   View of the main frame in a laboratory.

Table 1.   The samples.

Sample	Thickness mm	Dimensions mm	Description
Laminated glass	3 + 0.38 + 3	2010 × 840	15 years old, Pendulum test
Laminated glass	3 + 0.38 + 3	2010 × 840	New, Pendulum test
Float glass	3 + 3	2010 × 840	Without PVB, Pendulum test
Float glass	6	2010 × 840	Pendulum test
Float glass	4	2010 × 840	Pendulum test
Float glass	4	2010 × 840	Static test
Float glass	6	1100 × 360	Bending test

2.2   The samples

Norm (DIN EN 12600) recommends the use of a minimum of four samples to fulfil the experiment requirements. Each of the four samples has to have the same thickness. Size of the samples are: width 840 mm, height 2010 mm. At the laboratory of Department of steel an timber structures (SvF STU) were stored exact four 15 years old samples of glass planes. These were samples of laminated float glass consisting of two glass panes of 3 mm thickness and a PVB sheet with a thickness 0.38 mm. Besides that there were ordered four new laminated float glass planes of the same composition and (not laminated) float glass planes of the same dimensions. In total there were tested 20 samples of glass planes. All of the tested samples are in Table 1.

2.3   The course of the experiment

The test begins at the lowest height of the fall, then the fall height increased. The test must be carried out at a temperature of (20 ± 5) °C and with only one impact of a given height. Samples were provided with strain gauges, accelerometers and needles measuring relocation.

All results were evaluated and compared with a calculation model in ANSYS Workbench Explicit Dynamics. Also, the impact was recorded on high-speed cameras.

2.4 *Four-point-bending test by standard EN 1288-3*

For the correct modelling and comparison in the FEM computer program it is necessary to know the Young´s modulus of elasticity (E) of the tested samples. Therefore a four point bending test of glass panes according to EN 1288-3 was executed. Four test samples were prepared with a length of 1100 mm, width of 360 mm and a thickness of 6 mm.

The Young's modulus E of the tested samples is 70.62 GPa. These samples were further tested resistance samples (see Table 2).

3 COMPARISON OF DYNAMIC AND STATIC LOAD OF GLASS PANES OF THICKNESS 4 MM

Effect of the action of glass panels for dynamic and static load is different. For comparison, the test was executed with dynamic (pendulum test by standard) and also with a static load of the glass samples of float glass with a thickness of 4 mm (see Figure 3). The same conditions

Table 2.   The measured maximum stress for the surface and deflection of glass panes.

Allowable maximum stress for the surface of glass panes	The measured maximum stress σ	Deflection w
MPa	MPa	mm
17.51	51.13	25.0

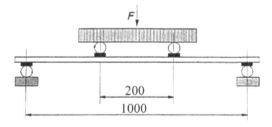

Figure 2.   Scheme of the four point bending test by EN 1288-3.

Figure 3.   Static loading of samples.

as for the impact test were used. The sample is inserted in the chucking frame and using a hydraulic press the imparted power to the tires that have pushed the glass panes. The speed of deformation by means of a hydraulic press was 30 mm/min.

Allowable maximum stress for the surface of glass panes by standard DIN EN 13474-1 is 17.51 MPa. Glass panes were loaded dynamically until they were destroyed. Experiments show (see Table 3) that the glass sample has higher resistance under dynamic load than under static. Horizontal tension under dynamic load was measured to be 41% higher than under the influence of static load and vertically it was up to 100%. Destruction of the sample under static load was pointed on the load of 1.5 kN. At the impact, the glass panel did not resist the fall of 450 mm (representing force 6.6 kN).

4 THE RESULTS FROM AN EXPERIMENT

For samples of laminated glass (15 years old, new glass), destruction took place at the drop of height 450 mm. The difference was in the mode of failure of the sample. In new laminated glass is breached only one board, and by 15 year old samples were breached both panes, with visible hole in shattered glass panel The specimen of pure float glass was destroyed at the drop of height 1200 mm and it was completely destroyed (with dangerous flying fragments, see Figure 5).

A pair of glass panel without **PVB** (without interaction) was tested only for comparison with laminated glass with **PVB** foil.

Table 3.   Maximum values of the experiment 4 mm glass.

Experiment	Vertical stress MPa	Horizontal stress MPa	Deformation mm
Pendulum test	77.38	98.28	34.9
Static test	37.17	69.48	19.2

Figure 4.   Record of high-speed cameras—float glass: a) at the moment of impact. b) after 0.016 s. c) after 0.042 s (fragments speed 18 m/s).

5 CONCLUSION

Preparation and implementation of the experiment the impact test according to DIN EN 12600 is designated to ensure results (stress. deflections. frequency. slow-motion movie) from various samples of float glass (laminated and non-laminated) to help in the creation of fully functional FEM model that could enhance experimental procedures. The results show that time has a significant impact on the life of PVB foil. which no longer has the same ability to ensure such interaction of sheets. From the measured values in the experiment (see Table 4)

Table 4. Maximum values of the experiment—intact samples.

The fall height mm		Sample	Vertical stress MPa	Horizontal stress MPa	Deformation mm
50	A50	Float glass (6 mm)	31.64	46.96	9.24
	B50	Laminated glass (3+0.38+3 mm)	32.10	46.92	8.8
	C50	Laminated glass-15 years old (3+0.38+3 mm)	32.45	47.66	9.1
	D50	Float glass (3+3 mm) without PVB	33.02	53.42	16.0
100	A100	Float glass (6 mm)	40.66	61.53	12.3
	B100	Laminated glass (3+0.38+3 mm)	41.22	61.64	11.9
	C100	Laminated glass-15 years old (3+0.38+3 mm)	43.38	64.61	12.2
	D100	Float glass (3+3 mm) without PVB	47.20	68.72	21.2
150	A150	Float glass (6 mm)	45.99	69.95	15.2
	B150	Laminated glass (3+0.38+3 mm)	47.69	69.92	14.7
	D150	Float glass (3+3 mm) without PVB	52.24	70.58	25.3
190	A190	Float glass (6 mm)	49.55	75.42	16.9
	B190	Laminated glass (3+0.38+3 mm)	50.95	74.49	15.8
	C190	Laminated glass-15 years old (3+0.38+3 mm)	49.01	76.37	17.3
	D190	Float glass (3+3 mm) without PVB		breach	
250	A250	Float glass (6 mm)	52.28	77.45	18.37
	B250	Laminated glass (3+0.38+3 mm)	54.91	77.08	17.8
	C250	Laminated glass-15 years old		breach	
450	A450	Float glass (6 mm)	63.40	88.05	22.49
	B450	Laminated glass (3+0.38+3 mm)	66.61	86.56	21.6

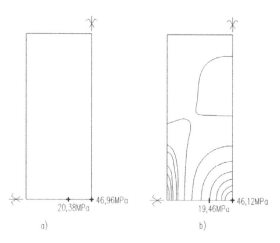

a)   b)

Figure 5.   Normal stress (horizontal): a) The measured values from the experiment. b) Results from ANSYS Workbench Explicit Dynamics.

is concluded following: In the impact test (short load time about 0.2 s) the laminated glass (3 + 3 mm) shows properties almost identically to properties of clear float glass (6 mm). Pair of float glass panes (3 + 3 mm without PVB) were tested to prove interaction between unglued glass panes (this samples were compared with laminated glass samples). All experiments are fully proving fact. that the dynamic load can be considered much higher than the static load. Maximum stress (for example wind load or live loads) allowed (by DIN EN 13474-1) is 17.51 MPa. There is no specification for impact loads which can be much higher (proved in presented experiments). There is also significant difference between measured static load allowable maximum stress according to DIN EN 13474-1 and measured value in Four-Point-Bending Test (this values are strongly influenced by many factor—age. temperature ...). Authors are recommending more experimental test for future revaluation of standards according to impact (dynamic) load tests and introducing values of impact survival values of maximal stressing of glass planes.

ACKNOWLEDGEMENT

The authors thank the Grant agency of the Ministry of Education. Science. Research and Sports of the Slovak Republic for providing a grant from the research program VEGA Nr.1/0747/16.

REFERENCES

DIN EN 12600. Glass in building—Pendulum test – Impact test method and classification for flat glass. April 2003.

DIN EN 1288-3. Glass in building—Determination of the bending strength of glass—Part 3: Test with specimen supported at two points (four point bending).

DIN EN 13474-1. Glas im Bauwesen—Bemessung von Glasscheibgen—Teil 1: Allgemeine Grundlagen für Entwurf. Berechnung und Bemessung.

ANSYS Workbench Explicit Dynamics.

FIGULI. Lucia—MAGURA. Martin—KAVICKÝ. Vladimír—JANGL. Štefan. Application of recyclable materials for an increase in building safety against the explosion of an improvised explosive device. Advanced Materials Research. Vol. 1001 (2014). s. 447–452. ISSN 1022-6680.

SLIVANSKÝ. Miloš. Modeling of reinforced glass beams in MKP (Modelovanie vystužených sklených noníkov v MKP). Modelování v mechanice 2013. Ostrava. ČR. 2013. ISBN 978-80-248-2985-2.

Advances and Trends in Engineering Sciences and Technologies II – Al Ali & Platko (Eds)
© *2017 Taylor & Francis Group, ISBN 978-1-138-03224-8*

The reliability of slender concrete columns subjected to a loss of stability

V. Benko, T. Gúcky & A. Valašík
Faculty of Civil Engineering, Slovak University of Technology, Bratislava, Slovakia

ABSTRACT: The European Standard for the design of concrete structures with the use of non-linear methods shows a deficit in global reliability for cases, when the concrete columns fail by the stability loss before reaching the design resistance in the critical cross-sections. The buckling failure is a brittle failure which occurs without warning and the probability of its formation is markedly influenced by the slenderness of the column. Here the presented calculation results are compared with results from the experiment which was carried out in cooperation with **STRABAG Bratislava LTD** in Central Laboratory of Faculty of Civil Engineering SUT in Bratislava. The following paper aims to compare the global reliability of slender concrete columns with a slenderness of 90 and higher. The columns are designed according to methods offered by STN EN 1992-1-1: namely, a general non-linear method and methods based on nominal stiffness and nominal curvature.

1   INTRODUCTION

The method of reliability of structures according to the Eurocodes is based on the use of partial factors of reliability, when the required probability of failure is ensured by them. In ULS, the effect of loads is increased by partial factor of reliability γ_F and the resistance of materials is reduced by partial factor of materials γ_M.

$$\gamma_F E_k = \frac{R_k}{\gamma_M} \tag{1}$$

Eurocode (EN 1992-1-1, 2004) for the design of concrete structures offers three methods for taking the second order effect into account (Benko, 2001; Moravčík et al., 2012 & Pfeiffer, 2014): The method based on nominal curvature (Chap. 5.8.8), the method based on nominal stiffness (Chap. 5.8.7) and general non-linear method (Chap. 5.8.6).

The buckling failure of compressed slender concrete members can however, overtake the reaching of the material resistance in critical cross-section (Benko et al., 2016). In these cases the definition of partial factor of reliability for buckling failure is appropriate, because the partial factors of materials cannot be applied and do not contribute to the overall reliability of design. So far, only the Austrian NA recommends the partial factor of reliability for stability failure (ÖNORM B 1992-1-1, 2011).

2   EXPERIMENTAL VERIFICATION OF SLENDER CONCRETE COLUMNS

The task of the experiments was to design geometry and reinforcement of columns together with the initial eccentricity of the axial force in such a way that the columns collapse due to stability loss inside the interaction diagram, i.e. before achieving the design resistance in critical cross-section with approximate compressive strain in concrete $\varepsilon_{c1} = 1.5\ ‰$ (Benko et al., 2016).

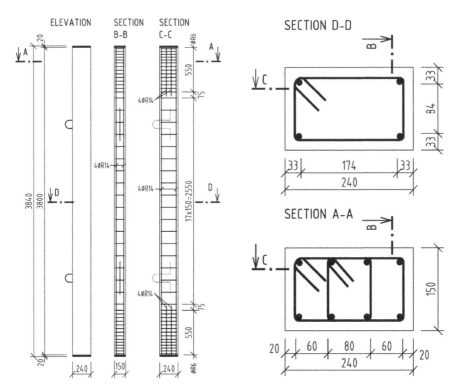

Figure 1.   The shape and reinforcement of columns.

The force and the initial eccentricity e_1 for the buckling failure were determined using non-linear calculations in the Stab2D-NL software. The standard characteristics of the material C45/55 and steel B500B were used in these calculations.

The experimentally verified concrete columns have a rectangular cross-section with the dimensions of 240 × 150 mm. The total length of the columns with steel spread plates is 3840 mm. The columns are reinforced with four bars, Ø 14 mm in diameter. These four bars are supplemented with another four bars with diameter of Ø 14 mm and length of 600 mm on both ends of the columns. The supplementary bars are welded to steel plates with thickness of 20 mm. The transverse reinforcement consists of two leg stirrups with diameter of Ø 6 mm. As the local failure in the ending parts can precede the stability collapse of the columns, the resistance is increased by doubling the transverse reinforcement along the length of the additional bars. In Figure 1 there are presented the geometry and the reinforcement of columns.

3   EXPERIMENTAL RESULTS

After the production of experimental samples and preparation of laboratory conditions, the concrete columns were tested in The Central laboratory of the Civil Engineering faculty SUT Bratislava.

The results of experiments are shown in Figure 2. The diagram presents the M—N relation for the increase of axial force and bending moment in the critical cross-section. Despite the fact that columns were fabricated using the same materials and high attention for accuracy, the differences in results are notable. Difference in the buckling force reaches 15.9%. The measurements were taken on 6 testing samples of slender concrete columns.

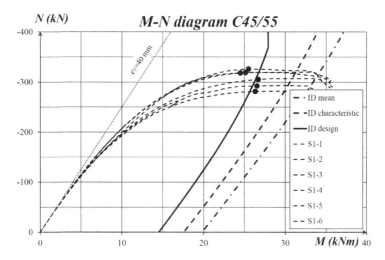

Figure 2. Bending moment—axial force relation.

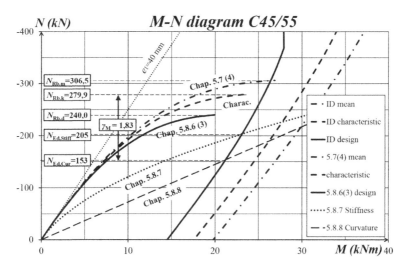

Figure 3. Reliability of the columns by stability loss—$\lambda = 89$ and $e_1 = 40$ mm.

4 COMPARISON OF THE OVERALL RELIABILITY OF THE COLUMNS BY STABILITY LOSS

In Figure 3, there is a comparison of results for the experimentally verified columns. Marked with a dashed thick line with dots is the group of results from the non-linear calculations calibrated to the mean values of material characteristics (EN 1992-1-1, Chap. 5.7(4), 2004) acquired from experiments. The axial force at stability loss is then 306.5 kN. When assuming the characteristic values of material characteristics, the axial force at the stability loss is 279.9 kN. Finally, when assuming the design values of material characteristics (EN 1992-1-1, Chap. 5.8.6(3), 2004) the axial force is 240.0 kN.

According to the method based on nominal stiffness (EN 1992-1-1, Chap. 5.8.7, 2004), the maximal resistance of the column with slenderness of $\lambda = 89$ is 205.0 kN. The maximal resistance is the point where the stiffness curve intersects the design interaction diagram. According to the method based on nominal curvature (EN 1992-1-1, Chap. 5.8.8, 2004), the resistance of the column is 153.0 kN.

Table 1 summarizes the partial factors of reliability for the loads, materials and also the overall factor of reliability.

The overall reliability of the design according to the method based on nominal curvature is 1.57 times higher than the reliability of the non-linear method according to EN 1992-1-1, Chap. 5.8.6(3).

5 PARAMETRIC STUDY OF RELIABILITY

5.1 Columns with slenderness λ = 89, 100, 120, 140, 160 and initial eccentricity e_i=40 mm

The differences in reliability of design methods according to EC2 for slenderness of λ = 160 and the initial eccentricity of e_1 = 40 mm are shown in Figure 4. The resulting values for the column resistance, partial reliability factors for loads and materials together with overall reliability factors of the above stated design methods are in Table 2.

Table 3 shows the comparison of overall reliability factors of slender columns with the slenderness of λ = 89, 100, 120, 140, 160 and the initial eccentricity of e_1 = 40 mm.

5.2 Columns with slenderness λ = 90, 100, 120, 140, 160 and initial eccentricity e_i = 20 mm

In Figure 5 and Table 4, there are resistances of columns according to the non-linear method, the method based on nominal stiffness as well as the method based on nominal curvature for the columns with the slenderness of λ = 90 and the initial eccentricity of e_1 = 20 mm.

Resistances of slender columns (slenderness λ = 90, 100, 120, 140, 160 and initial ec-centricity e_1 = 20 mm) and partial reliability factors of design methods are in Table 5 and Table 6. The overall reliability of the design according to the method based on nominal curvature is 3.23 times higher than the reliability acquired from non-linear method.

Table 1. Comparison of the reliability of columns—λ = 89 and e_1 = 40 mm.

			Axial force [kN]		γ_F	γ_M	γ_o
Overall reliability to the characteristic values of material properties			design	characteristic	load	material	overall
–	–	characteristic	279.9	–	1.40	1.00	–
Section	5.8.6(3)	design	240.0	171.4	1.40	1.17	1.63
Section	5.8.7	stiffness	205.0	146.4	1.40	1.37	1.91
Section	5.8.8	curvature	153.0	109.3	1.40	1.83	2.56

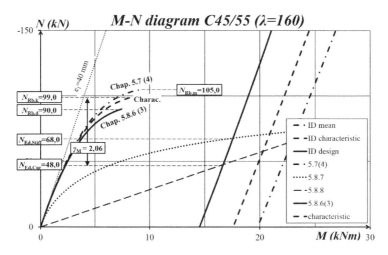

Figure 4. Reliability of the columns by stability loss—λ = 160 and e_1 = 40 mm.

Table 2. Comparison of the reliability of columns—$\lambda = 160$ and $e_1 = 40$ mm.

Overall reliability to the characteristic values of material properties			Axial force [kN]		γ_F load	γ_M material	γ_o overall
			design	characteristic			
–	–	characteristic	99.0	–	1.40	1.00	–
Section	5.8.6(3)	design	90.0	64.3	1.40	1.10	1.54
Section	5.8.7	stiffness	68.0	48.6	1.40	1.46	2.04
Section	5.8.8	curvature	48.0	34.3	1.40	2.06	2.89

Table 3. Overall reliability of columns—$\lambda = 89, 100, 120, 140, 160$ and $e_1 = 40$ mm.

Overall reliability to the characteristic values of material properties			slenderness λ				
			89	100	120	140	160
			overall γ_o				
Section	5.8.6 (3)	design	1.63	1.60	1.54	1.57	1.54
Section	5.8.7	stiffness	1.91	1.89	1.95	2.01	2.04
Section	5.8.8	curvature	2.56	2.64	2.78	2.91	2.89

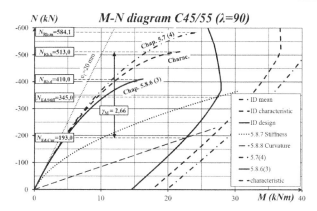

Figure 5. Reliability of the columns by stability loss—$\lambda = 90$ and $e_1 = 20$ mm.

Table 4. Comparison of the reliability of columns—$\lambda = 90$ and $e_1 = 20$ mm.

Overall reliability to the characteristic values of material properties			Axial force [kN]		γ_F load	γ_M material	γ_o overall
			design	characteristic			
–	–	characteristic	513.0	–	1.40	–	–
Section	5.8.6 (3)	design	410.0	292.9	1.40	1.25	1.75
Section	5.8.7	stiffness	345.0	246.4	1.40	1.49	2.08
Section	5.8.8	curvature	193.0	137.9	1.40	2.66	3.72

Table 5. Comparison of the reliability of columns—$\lambda = 160$ and $e_1 = 20$ mm.

Overall reliability to the characteristic values of material properties			Axial force [kN]		γ_F load	γ_M material	γ_o overall
			Design	characteristic			
–	–	characteristic	192.0	–	1.40	–	–
Section	5.8.6 (3)	design	168.0	120.0	1.40	1.14	1.60
Section	5.8.7	stiffness	81.0	57.9	1.40	2.37	3.32
Section	5.8.8	curvature	52.0	37.1	1.40	3.69	5.17

Table 6. Overall reliability of columns—$\lambda = 90, 100, 120, 140, 160$ and $e_1 = 20$ mm.

Overall reliability to the characteristic values of material properties			slenderness λ				
			90	100	120	140	160
			overall γ_o				
Section	5.8.6 (3)	design	1.75	1.71	1.67	1.63	1.60
Section	5.8.7	stiffness	2.08	2.29	2.67	3.02	3.32
Section	5.8.8	curvature	3.72	4.07	4.60	5.02	5.17

6 CONCLUSIONS

The global reliability of slender columns according to the non-linear design method is several times lower than the global reliability of slender columns according to the method based on nominal stiffness or nominal curvature.

The user of non-linear software can cause yet another reduction of the global reliability when entering the input data for non-linear calculations. This applies mainly in cases, when the aim of the design is minimization of dimensions and saving of material. This is often the main criterion of the free market in the European Union. The difference in axial force of predicting results within the planned experiments was 23% and the difference between maximal force within non-linear calculations and experimentally verified columns was 41% (Benko et al., 2016).

In cases, when the buckling failure precedes the failure of critical cross-sections, the non-linear method according to EC2 for slender columns cannot utilize part of the reliability ensured by partial factors of reliability on the side of materials. Therefore the standard EN 1992-1-1 requires revision for this domain with further verification of the reliability index and probability of failure.

ACKNOWLEDGEMENT

Authors gratefully acknowledge Scientific Grant Agency of the Ministry of Education of Slovak Republic and the Slovak Academy of Sciences VEGA No. 1/0696/14.

REFERENCES

2004. EN 1992-1-1:2004 Eurcode 2: Design of concrete structures—Part 1–1: General rules and rules buildings.
2011. ÖNORM B 1992-1-1:2011 Eurocode 2: Bemessung und Konstruktion von Stahlbeton- und Spannbetontragwerken Teil 1–1: Grundlagen und Anwendungsregeln für den Hochbau.
Benko, V. 2001. Nichtlineare Berechnung von Stahlbetondruckglieder. (Nonlinear analysis of reinforced concrete compression members). In.: Innovationen im Betonbau 27. Fortbildungsveranstaltung, OVBB Heft 47, s. 9–12.
Benko, V. et al. 2016. Failure of Slender Columns of Loss of stability. Journal of Composites for Construction ASCE.
Moravčík, M. et al. 2012. Experience with bridges of the older types of precast (Skúsenosti s mostami zo starších typov prefabrikátov). Betonárske dni 2012, zborník prednášok, STU v Bratislave, ISBN 978-80-8076-104-2, s. 439–444.
Pfeiffer, U. 2014. Program: Analysis of reinforced Concrete Structures, TUHH, Version 2.90.

Advances and Trends in Engineering Sciences and Technologies II – Al Ali & Platko (Eds)
© 2017 Taylor & Francis Group, ISBN 978-1-138-03224-8

The cracking of prestressed girders influenced by decreased bond of tendons

V. Borzovič, K. Gajdošová & M. Pecník
Faculty of Civil Engineering, Slovak University of Technology in Bratislava, Bratislava, Slovak Republic

J. Laco
Atkins, Surrey, UK

ABSTRACT: The time between prestressing of tendons in concrete post-tensioned bridges and the injection of ducts usually takes several weeks. Tendons are protected against weathering and atmospheric humidity with various agents, the presence of which influences the bond of tendons with a structure. This bond can also influence the durability of prestressed bridge in view of cracking. Bonded tendons in the tension zone may be assumed to contribute to crack control. This may be taken into account by assuming the ratio of bond strength of prestressing and reinforcing steel. The paper deals with evaluation of crack pattern of experimentally investigated post-tensioned members. These are the two span post-tensioned girders with different bond behavior of prestressing units. Together three different tendon types were investigated. Unbonded tendons, fully bonded tendons and tendons with decreased bond (caused by corrosion protection coatings).

1   INTRODUCTION

The time between prestressing of tendons and the injection of ducts in concrete post-tensioned bridges may take several weeks. During this time, it is necessary to protect tendons against weathering and atmospheric humidity. Tendons should be protected with various agents, which can influence their bond with a structure. Previous research (Lüthi & Breen 2005, Marti & Ullner 2008) has shown significant influence on bond between prestressing unit and concrete or injection grout.

Together 57 pull-out tests were performed at the Slovak University of Technology in Bratislava (Laco 2014, Borzovič 2013). The experimental program was focused on the bond strength of prestressing strands with decreased bond due to corrosion protection coatings. Since the pull-out test results showed a significant reduction of bond capacity of strands coated with corrosion protection, the question arose, whether a girder with such a corrosion protection of prestress will behave more like a girder with a bonded prestress or unbonded prestress. Therefore, the second stage of experimental research (Laco 2014) was focused on the observation of bond influence of post-tensioned girder behavior under acting loads. Experimental girders Ultimate Limit States (ULS) analysis and behavior of secondary effects of the tendons were presented in (Laco et al. 2015). This paper follows on that analysis. However, it is focused on the development of cracks and their width as parameters of Serviceability Limit State (SLS).

2   BOND OF TENDONS

In the posttensioned concrete bridges with bonded prestressing units which are injected with grout inside the corrugated duct, it is usually not necessary to check the bond stress. In case that the tensile stress in concrete is lower than concrete tensile strength (uncracked section), the whole concrete section is participating on shear stress transfer, therefore bond stresses of prestressing tendons are relatively low. When the crack in the cross-section appears (cracked section), bond stress drops at a higher level. In case of reduced bond strength between tendons and surrounding grouting, distance between the cracks extend, thereby crack width is also increasing. The amount of cracks and their size as well as the effect of the outside environment affects the durability of the concrete structure (Koteš & Kozák 2014).

3 EXPERIMENTAL INVESTIGATION

Seven post-tensioned girders were made with a length of 10.5 m. They were acting as two-span continue girders with theoretical span of each field of 5.0 m. The girders had rectangular cross-section with dimensions of 0.40 × 0.25 m. The girders were casted from C40/50 concrete class (characteristic compressive cylinder/cube strength in MPa) and reinforced with passive rebar and also with two post-tensioned seven-wire strands φ15.7/1860. The bonded and partially bonded strands were embedded in HDPE ducts and anchored actively at the prestressing side and passively at the remote forehead of the beam. The unbonded tendons were led without additional HDPE ducts. The strands were prestressed by force of 200 kN and then the ducts, were filled with cement grout. The cross-section of the experimental girders is shown in Figure 1.

The tested girders were supported at three points which provided boundary conditions of a continuous two span girder. Each span was loaded with two increasing concentrated forces *F* supplied by hydraulic jacks. The post-tensioned girders were loaded up to their failure. The girders were placed on the testing setup one after another. They were post-tensioned, grouted and after reaching seven days strength of the cement grout, they were loaded. Force was ascending symmetrically in both spans. Testing arrangement is shown in Figure 2.

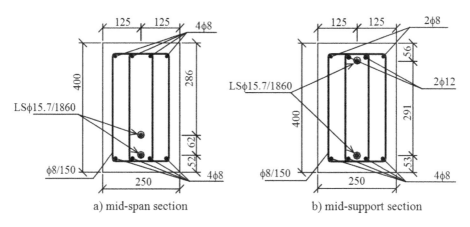

a) mid-span section   b) mid-support section

Figure 1.   Dimensions and reinforcement of investigated girders.

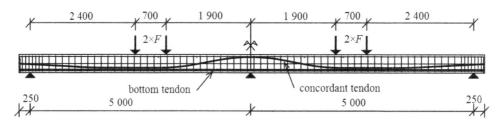

Figure 2.   Longitudinal view of reinforcement arrangement.

The first of the seven girder was used for debugging test assembly. The girders (N1, N2) were prestressed with bonded tendons. Another two beams labeled N3, N4 were prestressed with tendons with decreased bond, due to corrosion protection coatings. The last two girders (N5, N6) were prestressed with unbonded tendons.

4   CRACK PATTERN MEASUREMENTS

The following section presents the differences of cracking between girders with perfect bond prestressing (girder N1), decreased bond prestressing due to corrosion protection coatings (girders N3, N4) and unbonded prestress (girder N6). In Figures 3–4 monitored cracks are presented for two loading steps. The first presented loading step $F = 88.3$ kN (Figure 3) shows the end of crack formation stage. Crack initiation was observed when loading forces F reach approximately 43 kN for all types of girders. The second presented loading stage $F = 113.3$ kN (Figure 4) represents situation, when passive reinforcement reached the yield strength for most sections of the prestressed girders. The spacing between the cracks is evaluated on the dimension lines and summarized in the Table 1.

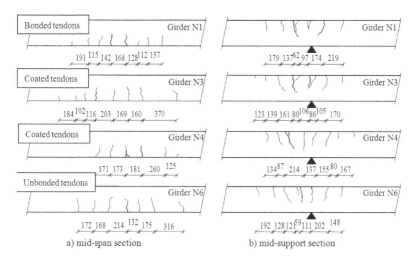

Figure 3.   The cracking of the experimental girders for loading step $F = 83.3$ kN.

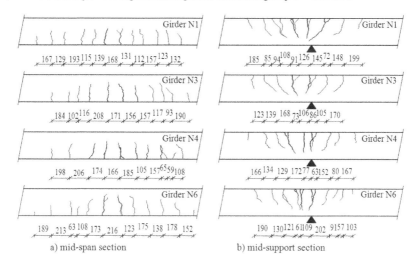

Figure 4.   The cracking of the experimental girders for loading step $F = 113.3$ kN.

The widths of adjacent crack were roughly the same and were substantially increased with increasing load for the girders with bond prestress (girder N1, N2). Together 15 cracks have appeared in the mid-span section and 9 cracks in the inner support section. The behavior of the girders N3 and N4 with tendon coated with corrosion protection emulsion was almost the same as girder N1 with bonded tendons. The differences have been observed only at the end of the loading, prior to bending failure of the section. The width of the main one or two cracks were influenced by bond losing at the highest loading stages. These cracks became wider then adjacent ones. The girders with unbonded tendons (Girder N6) have developed one main crack in the center of tensioned area, the width of which was growing with increasing load. Other cracks near the main one were developed but were significantly narrower to the main crack. Together 12 cracks have appeared in the mid-span section and 8 cracks in the inner support section.

5 CALCULATED CRACK WIDTHS

Bonded prestressing units in tension zone contribute to crack control of the prestressed girders. According to EN 1992-1-1 this can be taken into account by assuming ξ ratio, which is the ratio of bond strength between prestressing and reinforcing steel. For post-tensioned strands and concrete class equal or less than C50/60 ξ ratio is equal 0.5.

From the experimental study of bond behavior of prestressing units coated with corrosion protection agents which were performed at Slovak University of Technology (Laco 2014), the bond strength of specimens with strands coated with emulsifible oil reached 33% of dry one. Therefore ξ ratio for girders prestressed with tendons with decreased bond, due to corrosion protection coatings, is equal $0.5 \times 0.33 = 0.165$. This decreasing of the ξ ratio represents increasing the crack with up to 5% for calculation according to Eurocode 2 and up to 10% according to Model Code 2010. Further increasing of crack width and crack spacing between girders with bonded tendons and coated tendons, which is obvious from following graphs, are due to different redistribution of internal forces in continuous beams.

Table 1. Average values of measured crack spacing s_r [mm].

Section force F	Mid-span		Mid-span	
	88.3 kN	Support	113.3 kN	Support
Girder N1	145	173	142	156
Girder N3	186	161	149	139
Girder N4	182	139	158	127
Girder N6	196	160	157	118

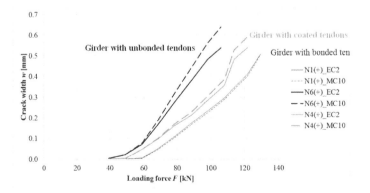

Figure 5.   Calculated width of crack at mid-span cross-section according to Eurocode 2 (continuous line) and *fib* Model Code 2010 (dashed line).

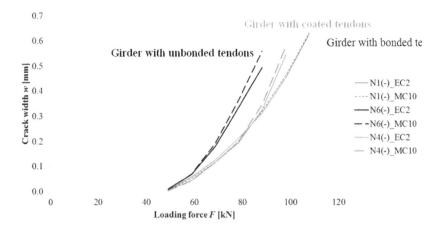

Figure 6. Calculated width of crack at mid-support cross-section according to Eurocode 2 (continuous line) and *fib* Model Code 2010 (dashed line).

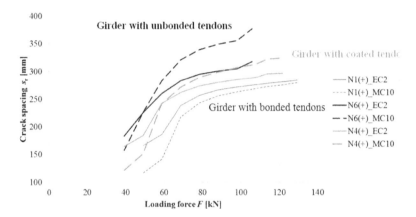

Figure 7. The maximum crack spacing of mid-span section according to Eurocode 2 (continuous line) and *fib* Model Code 2010 (dashed line).

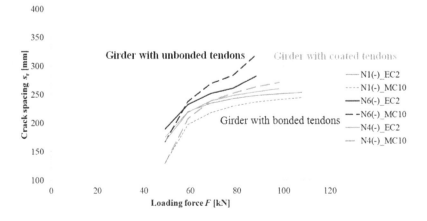

Figure 8. The maximum crack spacing of mid-support section according to Eurocode 2 (continuous line) and *fib* Model Code 2010 (dashed line).

The graphs in Figures 5–6 present the theoretical calculations of the crack width according to Eurocode 2 and *fib* Model Code 2010. The results according to these two predictions are almost coincident. Higher differences are in evaluation of maximum crack spacing (Figures 7–8).

6 CONCLUSIONS

The previous pull-out test experimental study of bond behavior between prestressing units coated with corrosion protection agents have shown significant influence on bond strength of the coated tendons surrounded with cement grouting. Conversely, from the comparison of loading tests of the girders with bonded tendons, unbonded tendons and tendons coated with corrosion protection emulsion, it can be concluded that the behavior of the girders with coated tendons is almost the same as that of girders with bonded tendons. Loss of bond was observed only at the highest loading stages before bending failure occurred.

The theoretical calculation of crack width and crack spacing has shown no major impact of the emulsifiable oil coating on the behavior of the post-tensioned girders. The results for girders with coated tendons are more close to the girder with bonded tendons. From the experimental study of prestressed girders, significant differences of the crack development were observed between the girders with bonded and coated tendons as opposed to unbonded tendons.

The measured crack spacing is halved compared to the calculated value according to Eurocode 2 and Model Code 2010. However, these values represent the maximum crack spacing. Denser development of cracks in fact means a smaller crack width.

ACKNOWLEDGMENTS

This work was supported by the Slovak Research and Development Agency under the contract No APVV-0442-12.

REFERENCES

Borzovič, V. 2013. *Bond stress-slip behaviour of prestressing units coated with corrosion protection*, in: TRANSACTIONS of the VŠB—Technical University of Ostrava: Civil Engineering Series. ISSN 1213–1962, Vol.13, No. 2. pp. 11–18.

Čavojcová, A., Moravčík, M., Bahleda, F. & Jošt, J. 2014. *Experimental verification of reinforced concrete member under cyclic loading*, in: Procedia Engineering, Volume 91, 23rd Russian-Polish-Slovak Seminar on Theoretical Foundation of Civil Engineering, Wroclaw, Poland, ISSN: 18777058, pp. 262–267.

fib Model Code for Concrete Structures 2010 (2013), Ernst & Sohn.

Koteš, P. & Kozák, J. 2014. *Reinforcement corrosion versus crack width*, in: 11th International Conference Binders and Materials 2013, ICBM 2013; Brno; Czech Republic. ISBN: 978-303835026-2, pp. 161–164.

Laco, J., Borzovič, V. & Pažma, P. 2015. *Investigation of bond influence of seven wire strands on behaviour of post-tensioned two span girders*. Concrete—Technology, Construction and Rehabilitation, (Beton TKS), Vol. 15, 2/2015, 55/59.

Laco, J. 2014. *Bond of prestressing units coated with corrosion protection agents* (in Slovak), Dissertation thesis. Bratislava: Slovak University of Technology.

Lüthi, T. & Breen, J. E., et al. 2005. *Factors affecting bond and friction losses in multi-strand post-tensioning tendons including the effect of emulsifiable oils: technical report*. Centre for Transportation Research at The University of Texas at Austin, TxDOT, p. 174.

Marti, P. & Ullner, M., et al. 2008. *Temporary corrosion protection and bond of prestressing steel*. ACI journal, Vol. January–February, pp. 51–59. ISSN 0889-3241.

STN EN 1992-1-1. 2006. Eurocode 2: design of concrete structures—Part 1–1: general rules and rules for buildings.

EUGLI imperfection method for columns with varying cross-sectional parameters – numerical solution

J. Brodniansky, Ľ. Balcierák, V. Duchoň & O. Katona
Faculty of Civil Engineering, Slovak University of Technology in Bratislava, Bratislava, Slovakia

ABSTRACT: EUGLI (Equivalent unique global and local initial) imperfection method is an approach described in the EN 1993-1-1 (+ Slovak National Annex) and the EN 1999-1-1 clause 5.3.2 (11). EUGLI imperfection method allows using the elastic critical buckling mode as the shape of the imperfection with calculated imperfection amplitude. The whole calculation background of this procedure is described in Chladný & Štujberová 2013. This paper focuses on numerical solution of the above mentioned procedure and describes this approach on simple examples of different types of columns, explaining the advantages and disadvantages of EUGLI imperfection method iteration procedures. As a result of research, a computer program developed by the authors is used and verified for calculation.

1   INTRODUCTION

1.1   *Basic information*

According to the standards (EN 1993-1-1 and EN 1999-1-1) and the clause 5.3.2 (11), it is possible to use the elastic critical buckling mode as an imperfection and determine the amplitude+ of this imperfection. Imperfection is called the Equivalent Unique Global and Local Initial Imperfection (EUGLI imperfection). The whole calculation procedure according to the 2nd order theory with EUGLI imperfection is called the EUGLI method. The whole theoretical background of EUGLI imperfection was published in Chladný & Štujberová 2013; Baláž 2008; Dalemulle 2013; Kovač 2012. Because of the complexity of the solved problem, a computer program was developed in MATLAB programming language (MATHWORKS), which is capable of calculating not only the simple beam structures, but a variety of frame structures as well. Some obstacles in the EUGLI method calculation were discovered, which are the main topic of this paper. The most important formulae described in (EN 1993-1-1 and EN 1999-1-1) clause 5.3.2 (11) is in (1). Symbol $\eta_{init,m,max}$ represents the amplitude of the EUGLI imperfection in the shape of the elastic critical buckling mode

$$\eta_{init,m,\max} = e_{0d,m} \frac{N_{cr,m} \left|\eta_{cr}\right|_{\max}}{EI_m \left|\eta_{cr}''\right|_m} \tag{1}$$

Presented explanations cover only a small part of the whole knowledge about the use of the EUGLI imperfection procedure (EN 1993-1-1 and EN 1999-1-1 clause 5.3.2 (11)). Therefore, for a full understanding of EUGLI imperfection method, it is recommended to study the background in Chladný & Štujberová 2013.

1.2   *Iteration procedure*

An iterative approach is used in the EUGLI imperfection method for the determination of the structure's critical cross-section. In the first iteration step, it is necessary to estimate the position of this cross-section. In the calculation of the EUGLI imperfection amplitude (Equation (10a)

$$e_{0d,m} = \alpha(\overline{\lambda}_m - \overline{\lambda}_0)\frac{M_{Rk,m}}{N_{Rk,m}}\frac{1-\dfrac{\overline{\lambda}_m^2\chi}{\gamma_{M1}}}{1-\overline{\lambda}_m^2\chi} = e_{0k}\delta_e$$

	is the design value of the bow imperfection for the equivalent member;
m	index—critical cross-section;
$\overline{\lambda}_m = \sqrt{\dfrac{N_{Rk,m}}{N_{cr,m}}}$	is the relative slenderness;
$M_{Rk,m}$	is the characteristic bending resistance of the critical cross-section;
$N_{Rk,m}$	is the characteristic axial force resistance of the critical cross-section;
χ	is the reduction factor for the relevant buckling curve and relative slenderness;
$N_{cr,m}$	is the elastic critical force, i.e. the axial force in the critical cross-section m under the critical loading of the structure;
η_{cr}	the elastic critical buckling mode;
$|\eta_{cr}''|_m$	second derivation of the elastic critical buckling mode in critical cross-section.

Chladný & Štujberová 2013, Equation (3) Baláž 2008, Equation (5.13) Kovač 2012, this cross-section's parameters are used. The iteration procedure is finished when the position of the cross-section with maximal utilization percentage (by means of the EUGLI method), calculated by the program, is the same in two consecutive iterations. This cross-section is called the critical cross-section. However, in automatic computer calculations, it is possible that this iterative approach does not reach convergence so the program enters a never-ending cycle.

1.3 *Developed computer program*

The computer program is based on the 2nd order theory analysis of imperfect frames with the EUGLI imperfection according the clause 5.3.2 (11) (EN 1993-1-1 and EN 1999-1-1). When necessary, it can also take into account effective cross-section parameters for 4th class cross-sections (this calculation is not used for examples in this paper). It is necessary to mention that the calculation algorithms developed in the dissertation thesis Dalemulle 2013 is implemented in the computer program in a modified version. The computer program is being developed in (MATHWORKS).

1.4 *The matrix calculation procedure*

To avoid the iteration procedure, it is possible to use the so-called 'brute force method'. Every cross-section on the structure is first considered as critical cross-section with parameters used for the EUGLI imperfection amplitude. The best way to align the obtained results is to use a matrix form. Values in the cells along the rows represent the EUGLI method utilization ratios of the structure in cross-sections j, where the row i shows the imperfection amplitude obtained based on the parameters of the cross-section i (i and j are define the position in the matrix). Values on the matrix diagonal ($i = j$) represent the utilization factors in cross-sections used for defining the EUGLI imperfection amplitude. If the maximal utilization ratio in a row lies on the matrix diagonal, $i = j$ is the definition of critical cross-section position. Critical cross-section can only be found by means of the 2nd order axial force and 2nd order bending moment caused by the deformation of the structure into an imperfect shape. The 2nd order theory is used at all stages of calculation and therefore it is possible to add the values of the bending moment caused by the external loading and bending moment caused by the structure deformed into an imperfect shape at the end of the calculation procedure. When

	j										
		1	2	3	4	5	6	7	8	9	10	...j...n
	1	0,89158	1,03622	1,14877	1,23378	1,29544	1,33749	1,36321	1,37542	1,37655	1,36869	
	2	0,79399	0,85899	0,90758	0,94212	0,96478	0,97751	0,98202	0,97978	0,97206	0,95992	
	3	0,76209	0,80104	0,82871	0,84674	0,85665	0,8598	0,85737	0,85041	0,83978	0,82626	
	4	0,74661	0,77293	0,79045	0,80047	0,8042	0,80269	0,7969	0,78764	0,77562	0,76141	
	5	0,73771	0,75678	0,76847	0,7739	0,77407	0,76989	0,76217	0,75159	0,73876	0,72417	
i	6	0,73213	0,74664	0,75468	0,75722	0,75516	0,7493	0,74037	0,72897	0,71563	0,70079	
	7	0,72846	0,73997	0,7456	0,74624	0,74271	0,73575	0,72602	0,71408	0,7004	0,6854	
	8	0,726	0,7355	0,73951	0,73887	0,73436	0,72666	0,71639	0,70408	0,69018	0,67508	
	9	0,72435	0,73251	0,73544	0,73395	0,72878	0,72059	0,70996	0,69741	0,68336	0,66818	
	10	0,72329	0,73058	0,73282	0,73079	0,72519	0,71668	0,70582	0,69311	0,67897	0,66374	
...i...n												

Figure 1.   EUGLI utilization matrix and possible iteration cycles (only diagonal values can serve as possible solutions of the EUGLI method—position of critical cross-section). There is one critical cross-section on the structure. A column example is in section 3.1.

	1	2	6	7	8	9	10	11	12	18	19	20
1	0,90121	1,08072	1,62944	1,72991	1,81782	1,89417	1,95988	2,01583	2,06284	2,19657	2,1997	2,19869
2	0,79824	0,88429	1,1431	1,18931	1,22919	1,26326	1,29198	1,31581	1,33515	1,37723	1,37473	1,3702
6	0,73037	0,75482	0,82255	0,83298	0,84121	0,84741	0,85176	0,8544	0,85552	0,83718	0,83098	0,82412
7	0,72569	0,7459	0,80046	0,80843	0,81448	0,81876	0,82142	0,82262	0,82247	0,79997	0,79351	0,7865
8	0,72225	0,73933	0,78419	0,79034	0,79478	0,79765	0,79908	0,7992	0,79813	0,77255	0,76591	0,75878
9	0,71962	0,73432	0,7718	0,77658	0,77979	0,78158	0,78207	0,78137	0,7796	0,75169	0,7449	0,73768
10	0,71758	0,73042	0,76214	0,76584	0,7681	0,76904	0,7688	0,76746	0,76514	0,73541	0,72851	0,72122
11	0,71595	0,72732	0,75447	0,75731	0,75881	0,75909	0,75826	0,75642	0,75366	0,72248	0,71549	0,70815
12	0,71465	0,72483	0,74829	0,75044	0,75133	0,75108	0,74977	0,74752	0,74441	0,71207	0,70501	0,69762
18	0,71054	0,71699	0,7289	0,72888	0,72786	0,72592	0,72314	0,71961	0,7154	0,67941	0,67212	0,66459
19	0,71022	0,71639	0,72741	0,72723	0,72606	0,72399	0,7211	0,71747	0,71317	0,6769	0,6696	0,66205
20	0,70998	0,71592	0,72623	0,72592	0,72464	0,72246	0,71948	0,71578	0,71141	0,67491	0,6676	0,66005

Figure 2.   EUGLI utilization matrix and possible iteration cycles (only diagonal values can serve as possible solutions of the EUGLI method—position of critical cross-section). There are two possible critical cross-sections on the structure. A column example is in section 3.1.

	1	2	3	4	10	11	12	15
1	0,73661	0,72889	0,70598	0,66856	0,44698	0,76956	0,66177	0,23114
2	0,74445	0,73661	0,71335	0,67537	0,45191	0,78014	0,67073	0,23359
3	0,76916	0,76096	0,73661	0,69685	0,46762	0,81852	0,69897	0,24132
4	0,81493	0,80604	0,77967	0,73661	0,49669	0,87532	0,75126	0,25562
10	0,83609	0,82688	0,79958	0,75499	0,51014	0,90989	0,77543	0,26223
11	0,44541	0,44202	0,43197	0,41556	0,2619	0,37633	0,32904	0,14014
12	0,48679	0,48279	0,47091	0,45151	0,2882	0,43221	0,37633	0,15307
15	1,20119	1,18655	1,14312	1,0722	0,74212	1,39692	1,1926	0,37633

Figure 3.   EUGLI utilization matrix and possible iteration cycles (only diagonal values can serve as possible solutions of the EUGLI method—position of critical cross-section). Critical cross-section does not exist. A column example is in section 3.2.

programming, it is necessary to ensure that the amplitude (based on its parameters) is set up for each cross-section i. For the amplitude, the structural utilization of cross-sections j is evaluated and written into row j. Part of the computer program designed to find the critical cross-section is based on finding the matrix cell with maximal utilization factor on the matrix diagonal (by means of the EUGLI method). All rows must be checked. Parts of such utilization matrices are in Figure 1, Figure 2 and Figure 3 (examples in sections 3.1 and 3.2), together with marked possible iteration steps.

Figure 2 shows the same example of beam as Figure 1, but with two possible critical cross-section positions (both are basically in the same location). It can be assumed that in this case, the critical cross-section exists, but it is not reachable due to mesh settings. Figure 3 shows that the critical cross-section is not found. The iteration cycle enters a never ending loop.

59

	1	2	3	4	5	6	7	8	9	10	11	12	13	14	15
1		0,7289	0,7060	0,6686	0,6178	0,7467	0,6921	0,6227	0,5404	0,4471	0,7694	0,6617	0,5329	0,3877	0,2312
2	0,7444	0,7366	0,7134	0,6754	0,6238	0,7562	0,7008	0,6304	0,5467	0,4521	0,7800	0,6706	0,5399	0,3926	0,2337
3	0,7691	0,7609	0,7366	0,6969	0,6429	0,7862	0,7282	0,6544	0,5669	0,4678	0,8133	0,6988	0,5620	0,4077	0,2414
4	0,8149	0,8060	0,7797	0,7366	0,6781	0,8417	0,7788	0,6990	0,6042	0,4968	0,8751	0,7511	0,6029	0,4358	0,2557
5	0,8907	0,8807	0,8510	0,8025	0,7366	0,9337	0,8629	0,7729	0,6660	0,5450	0,9775	0,8377	0,6707	0,4824	0,2794
6	0,5416	0,5368	0,5225	0,4992	0,4675	0,5101	0,4761	0,4328	0,3813	0,3231	0,5062	0,4389	0,3586	0,2679	0,1703
7	0,5724	0,5671	0,5515	0,5259	0,4912	0,5474	0,5101	0,4627	0,4064	0,3427	0,5477	0,4740	0,3861	0,2868	0,1799
8	0,6210	0,6150	0,5972	0,5682	0,5287	0,6065	0,5641	0,5101	0,4461	0,3736	0,6134	0,5296	0,4296	0,3167	0,195
9	0,6996	0,6924	0,6711	0,6364	0,5892	0,7018	0,6511	0,5866	0,5101	0,4235	0,7194	0,6194	0,4998	0,3650	0,2197
10	0,8358	0,8266	0,7993	0,7548	0,6943	0,8671	0,8020	0,7194	0,6212	0,5101	0,9033	0,7750	0,6216	0,4487	0,2623
11	0,4455	0,4421	0,4320	0,4156	0,3933	0,3935	0,3695	0,3391	0,3029	0,2620		0,3291	0,2726	0,2089	0,1402
12	0,4868	0,4828	0,4710	0,4516	0,4252	0,4437	0,4154	0,3794	0,3367	0,2883	0,4322	0,3763	0,3096	0,2343	0,1531
13	0,5615	0,5564	0,5412	0,5164	0,4828	0,5342	0,4981	0,4521	0,3975	0,3358	0,5330	0,4616	0,3763	0,2801	0,1765
14	0,7180	0,7106	0,6885	0,6525	0,6035	0,7242	0,6715	0,6046	0,5252	0,4353	0,7443	0,6404	0,5163	0,3763	0,2254
15	1,2005	1,1859	1,1426	1,0717	0,9755	1,3096	1,2062	1,0747	0,9187	0,7420	1,3958	1,1917	0,9478	0,6728	0,3763

Figure 4. EUGLI utilization matrix calculated with 'average value' critical cross-section. A cross-section which gives basically the same utilization values in both cross-section 1 and 11 was chosen. (This matrix belongs to the same column example as in section 3.2).

2 PROPOSED SOLUTION

During the numerical research, three different methods for solution of discovered problem were proposed (the critical cross-section was not found). One of them is to choose between cross-sections shown in Figure 20 and 21. The second approach is based on taking into account both cross-sections. The third one is based on finding a section of an average value similar to the critical cross-section. The last one mentioned is considered the best choice here, because it shows reasonable results. However, none of these three methods satisfies the main idea of the EUGLI imperfection method. Figure 4 shows the solution according to the 'average' method (EUGLI utilization matrix).

3 SOLVED EXAMPLES

3.1 Solved example of a column with linearly changing cross-sectional parameters

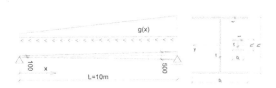

Figure 5. Column scheme Kovač 2012.

$$I_{y,max} = 684,07*10^6 \text{ mm}^4, \quad I_{y,min} = 4,79*10^6 \text{ mm}^4$$

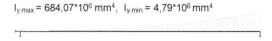

Figure 6. Distribution of the second moment of area.

$\alpha_{cr} = 2,5429$

Figure 7. The buckling mode.

-0.3769

Figure 8. The 2nd derivation of buckling mode.

60

Figure 9. Distribution of the bending moment caused by axial force—buckling mode imperfection has amplitude 1 m (1419.1123 kNm).

Figure 10. The EUGLI imperfection shape and amplitude ($\eta_{init,m,max}$ = 25.77 mm).

Figure 11. The axial force (688.5649 kN).

Figure 12. The EUGLI imperfection bending moment (23.6991 kNm).

Figure 13. Utilization of column and position of the critical cross-section.

3.2 Solved example of the cantilevered column with the step-change of cross-sectional parameters

Figure 14. Column scheme (S355).

Figure 15. Distribution of the second moment of area.

α_{cr} = 1,3130

Figure 16. The buckling mode.

Figure 17. The 2nd derivation of buckling mode.

Figure 18. Distribution of the bending moment caused by axial force—buckling mode imperfection has amplitude 1m (315.1030 kNm).

61

Figure 19.   The axial force (870 kN).

Figure 20.   Utilization of column and position of the critical cross-section no.1 (0.7696); (imperfection amplitude $\eta_{init,m,max} = 230.1$ mm).

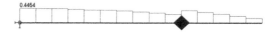

Figure 21.   Utilization of column and position of the critical cross-section no.11 (0.4454); (imperfection amplitude $\eta_{init,m,max} = 100.9$ mm).

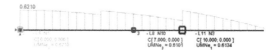

Figure 22.   Solution according to proposed solution in section 4 (0.6210); (imperfection amplitude—$\eta_{init,m,max} = 179.39$ mm).

4   CONCLUSION

This paper presented only two examples of columns which are, in terms of the EUGLI method solution, simple cases. Computer program presented in this paper is primarily used for calculating the frame structures and other types of plane structures. Regardless of the obstacle discovered in the EUGLI imperfection calculation procedure, it can be used very effectively on different types of structures (columns, frames, arches, and special shape plane structures).

ACKNOWLEDGEMENT

The authors thank the Grant agency of the Ministry of Education, Science, Research and Sports of the Slovak Republic for providing a grant from the research program VEGA Nr.1/0747/16.

REFERENCES

Baláž I. "Determination of the Flexural Buckling Resistance of Frames with Members with Non-Uniform Cross-sections and Non-Uniform Axial Compression Forces". In: Proceedings of the 34th Meeting of Experts on Steel Structures, Pezinok, Oct.16–17, 2008, pp.17–22, ISBN 978-80-227-2950-5.

Chladný E, Štujberová M. "Frames with Unique Global and Local Imperfection in the Shape of Elastic Buckling Mode (Part 1, Part2)". Stahlbau Vol.82, No 8, pp.609–617, No 9, pp.684–694. 2013, ISSN: 1437-1049.

Dalemulle M. "Vzperná odolnosť oblúkových konštrukcií v ich rovine" (Buckling Resistance of Arches in their Plane). Dissertation thesis, Slovak University of Technology in Bratislava, 2013 (in Slovak).

EN 1993-1-1: 2005 Eurocode 3: Design of steel structures—Part 1–1: General rules and rules for buildings.

EN 1999-1-1: 2005 Eurocode 9: Design of aluminium structures—Part 1–1: General structural rules.

Kovač M. "Vzperná odolnosť kovových prútov a prútových konštrukcií. Aplikácie nových metód z Eurokódov" (Buckling Resistance of Metal members and Frames. Applications of New Methods from Eurocodes). Bratislava. Slovak University of Technology in Bratislava. 2012. ISBN 978-80-227-3681-7 (in Slovak).

MATHWORKS: Programming language MATLAB v.2009. www.mathworks.com.

Advances and Trends in Engineering Sciences and Technologies II – Al Ali & Platko (Eds)
© 2017 Taylor & Francis Group, ISBN 978-1-138-03224-8

Analysis of passive and active Levy cable dome

P. Cauner & V. Urban
Faculty of Civil Engineering, Institute of Structural Engineering, Technical University of Košice, Košice, Slovak Republic

ABSTRACT: The paper describes analysis of cable dome under static and dynamic load and some types of analyzes with symmetric and asymmetric loads. If a critical load is obtained in Levy cable dome, prolongation of active member can redistribute forces. Cable domes are sensitive to asymmetric load and other changes in prestress. Active cable systems equipped with sensors and actuators provide active control of the shape and can adapt the structure to the various external influences and inducements. The results of this parametric study will be applied as input data for further optimization process using generator of neural network and genetic algorithms.

1 INTRODUCTION

1.1 *Tensegrity and cable dome*

The cable-strut structures are special light weight structural systems suitable for large span space structures. The word "tensegrity" consists of contraction of "tensional" and "integrity". The basic ideas are included in the concept described by the expression "islands of compression in an ocean of tension." Another extended definition is given by Pugh: "A tensegrity system is established when a set of discontinuous compression components interacts with a set of continuous tensile components to define a stable volume in space" (Pugh, A. 1976). Tensile cables and compression struts make up this system with the specific initial pre-stress in a self-equilibrium state. Cable domes belong to the group of hybrid tensegrity systems. The first cable dome was proposed by Geiger in 1986 and first employed in the roofs for the Olympic Gymnastics Hall and the Fencing Hall in Seoul. Since then, cable domes have been intensively studied. Other new forms, such as Kiewitt cable dome, Levy cable dome, various hybrid forms or so called birds´ nests were developed (Motro, R. 2003, Yuan, X. et al. 2007, Kmeť, S. & Mojdis, M. 2015).

2 ANALYSIS OF CABLE DOME

2.1 *Model of cable dome*

Levy cable dome consists of 49 elements (42 cables, 6 compressed elements, 1 actuator). It consists of main ridge cables, hoop cables, diagonal cables and vertical struts.

Levy cable dome is created of circular plan with diameter of 3 m (Figure 1). Main cross characteristics of elements are shown in Table 1.

2.2 *Form—finding*

Into diagonal cables (Set 3) we applied initial forces ($N_0 = 1$ kN to $N_0 = 5$ kN) to find axial forces for chosen geometry. Final axial forces can be seen in Table 2.

Initial axial forces ($N_0 = 5$ kN) were introduced to diagonal cables (Set 3) to obtain self-equilibrium state. Maximum nodal displacement is 3.6027 mm. We need effect approximations to

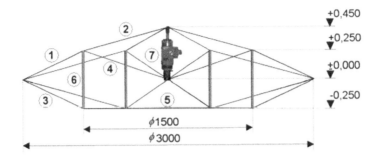

Figure 1.   Geometry of Levy cable dome.

Table 1.   Elements of Levy cable dome.

Elements	Steel	Cross section	Young's modulus
Compression elements	S 235	Φ 30/5 mm	210 GPa
Cables	7x(1+6)	Φ 6 mm	120 GPa

Table 2.   Final axial forces in elements.

	Initial forces in diagonal cables (Set 3) [kN]				
Set of elements	$N_0 = 1$	$N_0 = 2$	$N_0 = 3$	$N_0 = 4$	$N_0 = 5$
1	–	0.3256	0.6791	1.0310	1.3813
2	–	–	0.2251	0.4554	0.6843
3	0.6948	1.0546	1.4176	1.7839	2.1534
4	0.2082	0.3976	0.5861	0.7735	0.9600
5	0.7921	1.2021	1.6160	2.0337	2.4551
6	−0.2362	−0.4211	−0.6064	−0.7921	−0.9782
7	0.2447	−0.1147	−0.4721	−0.8275	−1.1808

Table 3.   Final axial forces.

	Initial forces in diagonal cables (Set 3) [kN]			
Set of elements	$N_0 = 5$ kN	1. app.	2. app.	3. app.
1	1.3813	1.4057	1.4062	1.4065
2	0.6843	0.7017	0.7020	0.7022
3	2.1534	2.1315	2.1312	2.1315
4	0.9600	0.9754	0.9757	0.9758
5	2.4551	2.4311	2.4307	2.4310
6	−0.9782	−0.9816	−0.9817	−0.9819
7	−1.1808	−1.2095	−1.2101	−1.2104

achieve best results. We apply previous data on the original geometry. After 3rd approximation is maximum of nodal displacement 0.00195 mm and according axial forces can be seen
in Table 3.

2.3   *Passive cable dome*

The cable dome was loaded by Vertical Symmetric load (VS-direction Y), Vertical Asymmetric load (VA-direction Y), Horizontal Symmetric load (HS-direction X). Structure is

loaded if forces in the cables are tensioned. Axial forces of each loading scheme are on the Figure 3–Figure 8 (load [kN/m²], axial forces [kN]). Ratio of applied load is shown in Table 4 (Comments: Force in node $F_y = -1$ means 100% of applied load).

A summary of results is given in Table 5. Table involves the group of loads, maximal square loading, maximal nodal displacements and minimal force (cables must be tensioned).

2.4 Active cable dome

Dynamic (transient) analysis consist of 5 Load Steps (LS). Every LS run 10 seconds. Actuator is element number 7 (Figure 1). In the first load step is on the structure applied

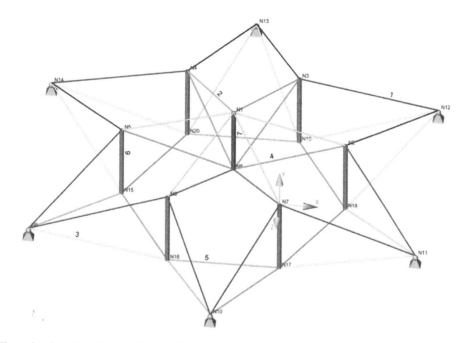

Figure 2.   Loading scheme of Levy cable dome.

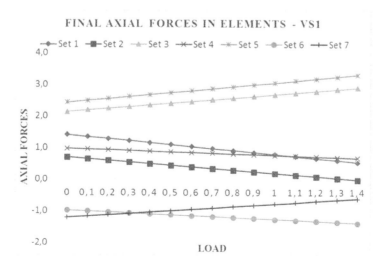

Figure 3.   Vertical symmetric load—VS1 (load [kN/m², axial forces [kN]).

65

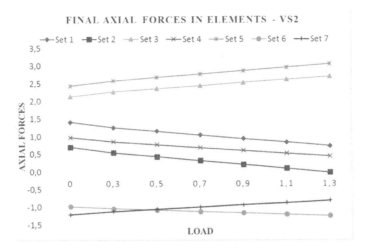

Figure 4. Vertical symmetric load—VS2 (load [kN/m², axial forces [kN]).

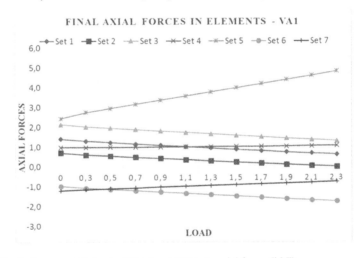

Figure 5. Vertical asymmetric load—VA1 (load [kN/m2, axial forces [kN]).

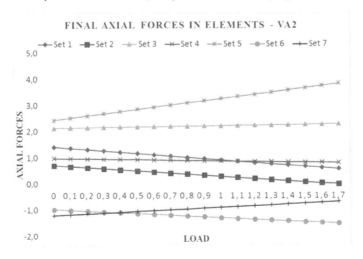

Figure 6. Vertical asymmetric load—VA2 (load [kN/m2, axial forces [kN]).

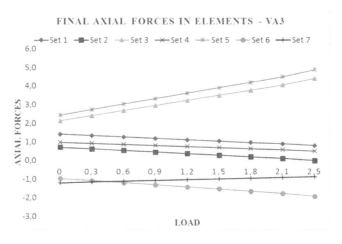

Figure 7.   Vertical asymmetric load—VA3 (load [kN/m2, axial forces [kN]).

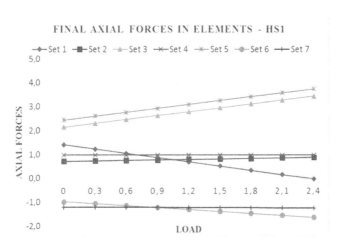

Figure 8.   Horizontal symmetric load—HS1 (load [kN/m2, axial forces [kN]).

Table 4.   Ratio of applied load.

Loading scheme Direction of loading Node	VS1 F_y	VS2 F_y	VA1 F_y	VA2 F_y	VA3 F_y	HS1 F_x
N1	−1	−1/2	−1/2	−3/4	−1/3	1
N2	−1	−1	−1	−1	−1	1
N3	−1	−1	−1	−1	−1	1
N4	−1	–	–	−1/2	–	1
N5	−1	−1	–	−1/2	–	1
N6	−1	−1	–	−1/2	–	1
N7	−1	–	−1	−1	–	1

only self-weight. Load step 2 involves loading to point, when the critical value of tensioned forces are achieved. In the next load step were on the structure applied no other loads. In the LS 4 is beginning of prolongation of actuator (15 mm). In the last ten seconds (LS5) were no other loads applied. Axial forces of each load steps are shown on the Figure 9.

Table 5. A summary of static analysis.

Group of loads	Max. square loading [kN/m²]	Set	Minimal force [N]	Max. nodal displacements [mm]
VS1	1.3	2	2.710	−2.430
VS2	1.2	2	35.850	−1.630
VA1	2.2	2	10.920	−7.779
VA2	1.6	2	3.200	−4.160
VA3	2.4	2	8.630	−6.805
HS1	2.3	1	27.550	5.169

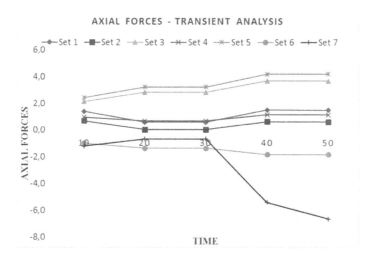

Figure 9. Axial forces of active cable dome (Axial forces [kN], time [s]).

3 CONCLUSION

These results are necessary for further optimization process. Data of this parametric study will be applied as input data in the optimization process including genetic algorithms and generator of neural networks. Necessary quantity of control data are obtained in the analysis of Levy cable dome.

ACKNOWLEDGEMENTS

The paper is carried out within the project VEGA No. 1/0302/16, partially founded by the Science Grant Agency of the Ministry of Education of Slovak Republic and the Slovak Academy of Sciences.

REFERENCES

ANSYS 16.0 2014. ANSYS 16.0 Help. *User's guide in the software ANSYS 16.0 by the ANSYS, Inc.*.
Kmeť, S. & Mojdis, M. 2015. Adaptive Cable Dome. In: *Journal of Structural Engineering.* Vol. 141, Issue 9, Article number: 04014225.
Motro, R. 2003.Tensegrity: Structural system of future. Published in Great Britain and the United States by Kogan Page Science, an imprint of Kogan Page Limited. ISBN 1-903996-37-6.
Pugh, A. 1976. An introduction to tensegrity. University of California Press, Berkeley.
Yuan, X. et al. 2007. Prestress Design Of Cable Domes With New Forms. *International Journal of solids and Structures. Vol. 44, Issue 9, (1 May 2007).* Pages 2773–2782, ISSN 0020-7683.

Advances and Trends in Engineering Sciences and Technologies II – Al Ali & Platko (Eds)
© 2017 Taylor & Francis Group, ISBN 978-1-138-03224-8

Application of system identification on a steel footbridge across the Little Danube River

M. Fábry & R. Ároch
Department of Steel and Timber Structures, Faculty of Civil Engineering, Slovak University of Technology in Bratislava, Bratislava, Slovakia

I. Lipták
Department of Surveying, Faculty of Civil Engineering, Slovak University of Technology in Bratislava, Bratislava, Slovakia

ABSTRACT: The paper is focused on the system identification of bridges by non-destructive dynamic methods. The main idea of this paper is to describe the present state of the issue in Slovakia. The measurements were focused on different bridge types, such as truss bridges, suspension bridges or beam bridges. The main topic of this article is to describe measurements on a suspension footbridge in Bratislava. In the first step the paper describes the development of the modal model. In the next step measurements are made by different types of accelerometers. All measurements are supported by surveying methods. The aim of the research is to develop a system for structural health monitoring. In the end of the article the influence of the parapet on the dynamic stiffness of the footbridge is described.

1 INTRODUCTION

Nowadays, system identification is becoming more important in structural engineering, as there is a constant focus to increase the safety of structures while reducing maintenance costs at the same time. Another reason is that the condition of many bridges requires their reconstruction. There is often a lack of maintenance and regular inspections, which leads to expensive reconstructions. The measurement by non-destructive methods can greatly help to support inspections of a structure. If applied correctly, a continuous health monitoring system itself can trigger a warning that something with the bridge is wrong.

2 NON-DESTRUCTIVE MEASURING METHODS

The methods used to identify the damage offers a possibility of non-destructive evaluation of the structure, in many cases without a previously specified location of the damage. The procedure can be divided into two steps. The load-bearing structure is assessed by means of highly sensitive sensors at first. In the second step, the entire structure is verified by comparing the measured results with either the measured results obtained on the undamaged structure or the numerical results obtained on the modal model.

In Slovakia, the monitoring of bridges is presently predominantly done using traditional diagnostic methods. Since the maintenance of bridges is in many cases neglected the current state of many bridges is alarming. As an outcome of our research, a basis for non-destructive measurement will be created together with a functioning assessment method. Many methods are used for this purpose worldwide (Strauss et al. 2009). For this research the direct stiffness derivation method was used (described in (Maeck & De Roeck 1999)). The eigen-frequencies and mode shapes were verified experimentally and were used to determine the dynamic stiffness. When the distribution of mass is known, there is no need to utilize a numerical model to determine the dynamic stiffness. The method is based on the assumption that the dynamic bending stiffness is equal to the ratio of the bending moment and curvature in each cross-section. Similarly, dynamic torsional stiffness can be determined as being equal to the ratio of the torsional moment and the angular velocity in each cross-section. Dynamic bending stiffness is given by:

$$EI = \frac{M}{d^2\varphi^2/dx^2}$$ (1)

Where EI is the dynamic bending stiffness, M is the bending moment and φ_b is the bending mode shape. The equation of the undamped system can be written as:

$$K_m\varphi_m = \omega_m^2 M_a\varphi_m$$ (2)

Where K_m is the stiffness matrix, M_a is the mass matrix, φ_m is the vector of measured displacements and ω_m is the measured angular velocity. The calculation is based on the assumption that the mass matrix is known. In the mass matrix, concentrated masses are used, being more suitable for the thick measurement grid. φ_m is equal to φ_b for bending shapes, i.e. the vertical or horizontal displacements. If in the equation, φ_m contains only the modal displacements in nodes, then the value of load can be obtained, because the right side of the equation expresses the inertial forces. The stiffness characteristics were estimated at first. The measured and calculated data were used for linear interpolation and a new value of rigidity was calculated by gradual correction.

3 MEASURED DATA FROM THE STEEL FOOTBRIDGE ACROSS THE LITTLE DANUBE RIVER

For measurement we selected a suspension steel footbridge, located in Bratislava (Vrakuňa district). The bridge spans the Little Danube River and connects the free-time zone of Vrakuňa with a residential area. The total span of the load-bearing structure is 54.0 m. The orthotropic deck consists of two IPE 360 beams and two IPE 270 beams in the longitudinal direction, which are connected to IPE 270 cross-beams and to the steel plate bridge deck. The load bearing structure is suspended between two pylons, both 10.0 m high. The deck is supported by nine suspension rods, with 30.0 mm diameter. The main cables have 50.0 mm diameter and are anchored to concrete blocks which are located 15.0 m from the pylons.

The footbridge itself is not used very frequently; the structure is suitable for dynamic measurement because it is easily excitable. The footbridge starts vibrating after being passed by only one person. The numerical model was created in the software Dlubal—RFEM 5 (RFem 2012). The numerical model was created based on (Bock 1992), before the measurement itself took place. The eigen-frequencies and mode shapes were identified in the numerical model by modal analysis. Several numerical models were created to provide a basis for comparison, where the effect of stiffness of the parapet on the dynamic stiffness of the whole structure was investigated. The individual models and the influence of parapets on the frequencies are compared in Table 1.

The models were created with the same boundary conditions. The parapet was the only alternating parameter. The highest accuracy was expected for model A; however, the measurement showed that the frequencies grew when the stiffness of the railing was added to the model. This effect is attributed to the way of modelling the parapet joints in the numerical model. The para-pet is less stiff in the real conditions. The results of the measurement were the same with the model B, except for the first frequency which was modelled without the infill. As for model C, it was created based on the results of measurement, where

Table 1. Comparison the individual models and their frequencies.

Mode shape	Model A—with parapet (Hz)	Model B—without parapet (Hz)	Model C—with parapet without infill (Hz)
1	1.67	1.32	1.43
2	2.41	2.21	2.16
3	2.53	2.30	2.35

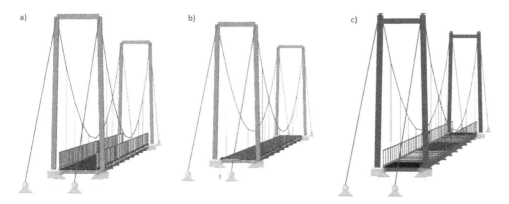

Figure 1. Comparison of models: a) Model A—with parapet, b) Model B—without parapet, c) Model C—with parapet without infill.

only the load-bearing part of the parapet was modelled. Conformity with the results of the measurement which is described in the next section was achieved. Comparison of the first mode shape for the three models is shown in Figure 2.

4 DESCRIPTION OF THE MEASUREMENT

The arrangement of sensors was proposed according to the number of sensors available for the testing. Two types of accelerometers were used for measurement:

– Piezoelectric accelerometers PCB 393B31.
– Optical accelerometers FBG FS6500.

The measurement devices were located at one third of the span, with two accelerometers in each cross-section, so that the torsional mode of vibration could be captured.

The main parameter observed during the measurement was the acceleration of the structure. Maximal dynamic deflection was assumed in the measurement points. The measurement itself was repeated three times. A person was used for excitation, jumping once in one third of the span (for the first mode shape), then in the middle of the span (for the third mode shape). The aim of the measurement was to compare the numerical model results with the ones obtained from the real structure in-situ. The comparison between the Piezoelectric (Piez.) and Optical (Opt.) accelerometers is also included.

Table 2 presents the results from the first measurement, with two persons walking on the footbridge. The second and fourth mode shapes were identified as being torsional. Some of the optical sensors failed to capture the fourth and fifth shape.

Table 3 summarizes vibration of the structure without pedestrian loads. In this case an ambient excitation—the wind—was used. Table 4 shows frequencies of the structure excited by a person jumping in the middle of the span. More result can be found in (Hlavina 2015).

71

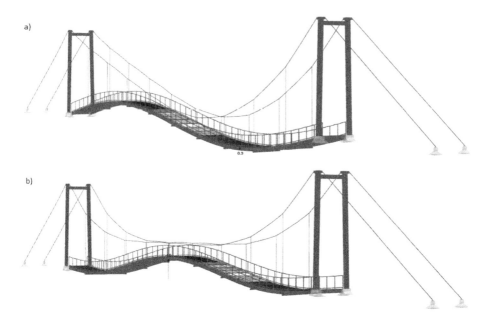

Figure 2.   Comparison of the first mode shape (a), and the third mode shape (b).

Table 2.   Measurement 1—two persons walking on the footbridge.

Mode shape	Frequencies (Hz)	Type	Frequencies in the measurement points (Hz)						Average normalized frequencies (Hz)
			P1	P2	P3	P4	P5	P6	
1	1.43	Opt.	–	–	–	–	–	–	–
		Piez.	–	–	–	–	–	–	–
2	2.16	Opt.	2.173	2.161	2.173	2.173	2.173	2.173	2.173
		Piez.	2.188	2.188	2.188	2.188	2.188	2.188	2.188
3	2.35	Opt.	–	–	–	–	–	–	–
		Piez.	–	–	–	–	–	–	–
4	3.06	Opt.	3.162	3.162	–	3.149	3.162	3.162	3.162
		Piez.	3.150	3.150	3.150	3.150	3.150	3.150	3.150
5	4.03	Opt.	4.077	4.077	–	–	4.077	4.077	4.077
		Piez.	4.075	4.075	4.075	4.075	4.075	4.075	4.075

Table 3.   Measurement 2—ambient load (wind).

Mode shape	Frequencies (Hz)	Type	Frequencies in the measurement points (Hz)						Average normalized frequencies (Hz)
			P1	P2	P3	P4	P5	P6	
1	1.43	Opt.	1.411	1.411	–	–	1.411	1.411	1.411
		Piez.	1.411	1.411	1.411	1.411	1.411	1.411	1.411
2	2.16	Opt.	–	–	–	–	–	–	–
		Piez.	–	–	–	–	–	–	–
3	2.35	Opt.	–	–	–	–	–	–	–
		Piez.	–	–	–	–	–	–	–
4	3.06	Opt.	3.152	–	3.152	3.152	–	3.152	3.152
		Piez.	3.153	3.153	3.153	3.153	3.153	3.153	3.153
5	4.03	Opt.	–	–	–	–	–	–	–
		Piez.	4.091	4.091	4.091	4.091	4.091	4.091	4.091

Table 4. Measurement 3—jump of one person in the middle of span.

Mode shape	Frequencies (Hz)	Type	Frequencies in the measurement points (Hz)						Average normalized frequencies (Hz)
			P1	P2	P3	P4	P5	P6	
1	1.43	Opt.	1.416	1.416	–	–	1.416	1.416	1.416
		Piez.	1.413	1.413	1.413	1.413	1.413	1.413	1.413
2	2.16	Opt.	2.167	2.167	2.167	2.167	2.167	2.167	2.167
		Piez.	2.169	2,169	2,169	2.169	2.169	2.169	2.169
3	2.35	Opt.	–	–	–	–	–	–	–
		Piez.	–	–	–	–	–	–	–
4	3.06	Opt.	3.137	3.137	3.137	3.137	3.137	3.137	3.137
		Piez.	3.138	3.138	3.138	3.138	3.138	3.138	3.138
5	4.03	Opt.	–	–	–	–	–	–	–
		Piez.	–	–	–	–	–	–	–

Figure 3. The steel footbridge in Vrakuňa during measurement.

5 EVELUTION OF THE MEASUREMENT

Based on the obtained results, the following conclusions can be drawn. The comparison of the optical sensor FBG FS6500 and the piezoelectric accelerometers PCB393B31 provided a relatively exact match. In some cases, the frequencies were not captured, owing to lower sensitivity and large nominal range compared to the piezoelectric accelerometers. Based on the results, it can be concluded that also the optical sensors FBG FS6500 are suitable for measurement of the dynamic characteristics of bridges. However, theses sensors are not able to measure the deflections for the lower amplitudes of vibration. The advantage of using optical sensors is the ability to wire the accelerometers in series, which reduces the cost of cabling and simplifies measurement.

6 CONCLUSION

This article deals with measurement of bridges in Slovakia by non-destructive methods. A variety of measurements was initiated on some real bridges and scale models as well. Many types of accelerometers were tested. Numerical models were created of all the measured bridges. The creation of a modal model of the bridge is one of the most important parts of structural health monitoring. The main aim of the model is to portray the real dynamic behaviour of the structure. Many numerical models were created to verify the behaviour of the structure before the measurement itself was performed (Fábry 2015). These models were successively verified by the measurement. It is currently impossible to verify the damage to real structures.

The result of the research is the creation of a base for the system identification of bridges in Slovakia, where non–destructive measurements are preferred. At the present time the measurement system is fully functional.

AKNOWLEDGEMENT

This paper has been supported by the Slovak Research and Development Agency (SRDA)—grant from the program APVV No. 0236-12.

The authors would also like to thank for the financial contribution from the STU Bratislava Grant scheme for Support of Young Researchers.

REFERENCES

Bock, B.: 1992. Project of the Footbridge over the Little Danube. Bratislava: KOVOPROJEKT ES, Bratislava.1992. (in Slovak).

Fábry, M.: Measurement of Damages of Experimental Models and Real Bridges. In: Proceedings of the 17th International Conference of Postgraduate Students "Juniorstav 2015". Brno, Czech Republic, January 29, 2015, Brno University of Technology, 2015, CD-ROM, 7 pages, ISBN 978-80-214-5091-2. (in Slovak).

Hlavina, M.: Determination of Dynamic Properties of Bridge Structures using Optical Sensors. Masters' Thesis, Faculty of Civil Engineering, Slovak University of Technology in Bratislava, 2015 (in Slovak).

Maeck, J., DE Roeck, G.: Dynamic bending and torsion stiffness derivation from modal curvatures and torsion rates, J. Sound & Vibration, V.225(1), pp.153–170, 1999.

RFEM, version 5.01.0013. Ing. Software Dlubal GmbH. 2012.

Strauss, A., Bergmeister, K., Wendner, R., Hoffmann, S., System—und Schadensiden-tifikation von Betontragstrukturen. Chapter VII. In: Beton Kalender. Ernst & Sohn, ISBN 978-3-433-01854-5. 2009.

Advances and Trends in Engineering Sciences and Technologies II – Al Ali & Platko (Eds)
© 2017 Taylor & Francis Group, ISBN 978-1-138-03224-8

Non-linear analysis of concrete slender columns

Ľ. Fillo, V. Knapcová & M. Čuhák
Faculty of Civil Engineering, SUT in Bratislava, Bratislava, Slovakia

ABSTRACT: The paper presents non-linear analysis of slender concrete columns. A step by step analysis with help of the Transfer Matrix Method and modified calculation of rigidities of reinforced concrete cross-sections is carried out. The general non-linear method theoretical result is verified with realized experiments. The general non-linear method is compared with an approximate method of nominal curvature taking into account effects of second order.

1   INTRODUCTION

Design of slender concrete columns requires the analysis of second order effects for taking into account a total bending moment. Approximate methods for calculating this second order moment are not always on the safe side of the design. In the paper we present the possibility of calculating slender columns using the general non-linear method which is based on step by step procedure of Transfer Matrix Method combined with sectional analysis. These calculations were applied on experimental tested concrete columns to verified and calibrate the new method of non-linear analysis. Results of the general method were verified with approximate method of nominal curvature recommended by EN 1992-1-1(2004) and by Model Code (2010).

2   PROCEDURE OF TRANSFER MATRIX METHOD AND SECTIONAL ANALYSIS

The algorithm of transfer matrix method is defined in two principal steps. During the first step the unknown initial parameters defined by the matrix of initial parameters are transferred using the matrix multiplication of transfer and nodal matrices into the end point of the applied simulation. The boundary conditions at the end point, implemented in the matrix of boundary conditions, define the set of algebraic equations for the determination of the unknown initial parameters. In the second step the calculated initial parameters are put into the state vector at the initial point of simulation and its repeated multiplications with transfer and nodal matrices determine the set of resulting state vectors, stress and strain components in nodal points of the used discrete model (Figure 1).

The sectional analysis passes within the frame of state vector calculation by each step of increasing the bending moment after Figure 2. First the equivalence in each cross-section of segments is found in Figure 2—(Equation 1), then the curvature—(Equation 2) and horizontal deformation y is calculated in (Equation 3).

$$N_e; M_e \Rightarrow \sigma_c; \sigma_{s1}; \sigma_{s2}; \Rightarrow \varepsilon_c; \varepsilon_{s1}; \varepsilon_{s2} \tag{1}$$

$$\frac{dx}{r} = \frac{\varepsilon_{s1}dx - \varepsilon_c dx}{r} \Rightarrow \frac{1}{r} = \frac{\varepsilon_{s1} - \varepsilon_c}{r} \tag{2}$$

$$y'' = \frac{1}{r} \Rightarrow \Delta M_2 + M_e \Rightarrow M_{e+1} = \Delta M_2 + M_e + \Delta M; \ N_{e+1} \tag{3}$$

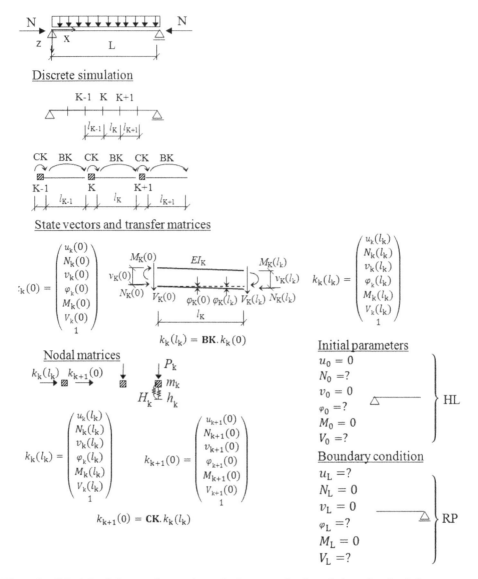

Figure 1. Principle of the transfer matrix method concept for the solution of a simple beam system subjected to bending moment and axial force.

Taking into account the 2nd order effects the new bending moment is calculated after (Equation 3). The next step begins with increasing the bending moment and axial force N_{c+1} (Equation 3).

The second order effects are calculated on the basis of the presented nonlinear analysis accounting for nonlinear behavior of concrete in compression, cracking (Figure 3), creep and shrinkage, reinforcement yielding and other non-linear effects important to the change in behavior over time and loading state.

3 VERIFICATION WITH EXPERIMENTAL RESULTS

Theoretical results of non-linear analysis were compared with experiments realized by Benko et al. 2016—Figure 4. The experimentally verified concrete columns have the rectangular

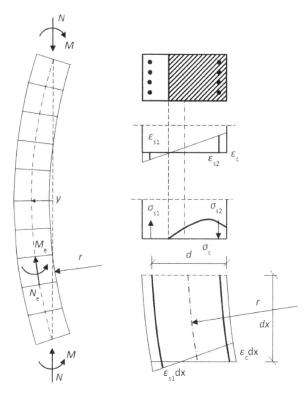

Figure 2. Step by step analysis by transfer matrix method and sectional analysis.

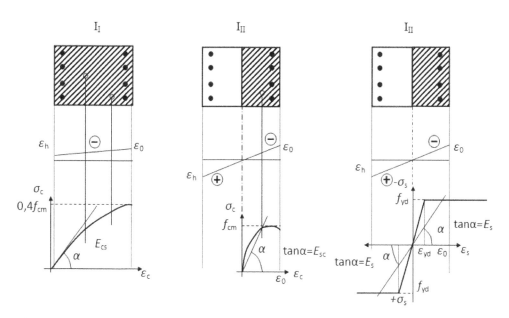

Figure 3. Secant Young modules of concrete and steel by sectional analysis of cross-section (segments).

cross-section with the dimensions of 240 × 150 mm. The total length of the columns was 3840 mm. The columns are reinforced by four bars with diameter of 14 mm. The transferred stirrups φ 6 á 150 mm. Concrete cover 20 mm.

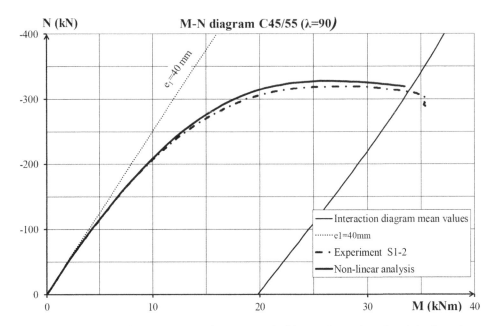

Figure 4. Verification of the general non-linear method with experimental results of slender columns Benko et al. 2016.

4 VERIFICATION OF THE GENERAL AND APPROXIMATE METHOD OF SECOND ORDER EFFECTS CALCULATION

The non-linear analysis is defined in EN 1992-1-1(2004) as a general method, which is based on non-linear analysis principles, including geometric non-linearity i.e. second order effects. The general rules for non-linear analysis are given in the Clause 5.7 EN 1992-1-1(2004). Non-linear methods of analysis may be used for both ULS and SLS, provided that equilibrium and compatibility are satisfied and an adequate non-linear behavior for materials is assumed. The material characteristics which represent the stiffness in a realistic way but take account the uncertainties of failure shall be used. Second order effects cannot be ignored, the design method given in 5.8.6 EN 1992-1-1(2004) may be used. Stress-strain curves for concrete and steel suitable for overall analysis shall be used. The effect of creep shall be taken into account. Normally, conditions of equilibrium and strain compatibility are satisfied in all cross sections.

In a current design an approximate method based on nominal curvature recommended in 5.8.8 EN 1992-1-1(2004) is used. This method is primarily suitable for isolated members with constant normal force and a defined effective length l_0. The method gives a nominal second order moment based on a deflection, which in turn is based on the effective length and an estimated maximum curvature.

The design moment is defined as:

$$M_{Ed} = M_{0Ed} + M_2 \qquad (4)$$

where: M_{0Ed} is the 1st order moment, including the effect of imperfections,
M_2 is the nominal 2nd order moment.

$$M_2 = N_{Ed}e_2 \qquad (5)$$

where: N_{Ed} is the design value of axial force,
e_2 is the deflection $= (1/r) \, l_0^2/c$,

78

1/r is the curvature,

l_0 is the effective length,

c is a factor depending on the curvature distribution.

Differences in use of approximate nominal curvature method by EN 1992-1-1 (2004) and Model Code 2010 (2010) are predominantly in definition of the curvature. EN 1992-1-1 (2004) defines the curvature with respect to the maximum bending moment of resistance $M_{Rd,max}$ and the curvature at the level of yielding state of both reinforcements A_{s1} and A_{s2} respectively (Equation 6).

$$\frac{1}{r_0} = \frac{2f_{yd}}{E_s(d-d_2)}$$ (6)

This basic curvature should be modified by coefficients K_r and K_φ respectively. Model Code 2010 (2010) defines the curvature similarly with simpler definition of K_φ coefficient of creep and shrinkage.

Figure 5 presents results of the non-linear analysis of columns with slenderness $\lambda = 35$ "N-M curves" ($\omega = 0.2$ and $\varphi_{eff} = 0$). For basic eccentricity $e_0 = 40$ and 80 mm with definition of second order moments M_2.

$$\omega = \frac{A_s f_{yd}}{A_c f_{cd}}$$ (7)

$$n = \frac{N_{Rd}}{bh f_{cd}}$$ (8)

$$m = \frac{M_{Rd}}{bh^2 f_{cd}}$$ (9)

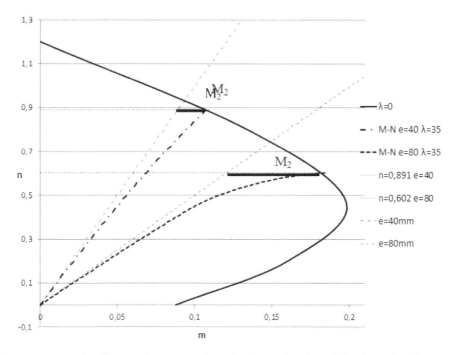

Figure 5.   Interaction diagram for cross-section (slenderness $\lambda = 0$) and the General method results for columns with $\lambda = 35$, $\omega = 0.2$ and $\varphi_{eff} = 0$. Lines represent the N-M relation for $e_0 = 40$ and 80 mm respectively with second order moments M_2.

Results of the general method were verified with approximate method of nominal curvature recommended by EN 1992-1-1(2004). Results did not confirm the fact, that approximate methods gives design on the safe side in all range of interaction diagram—especially for $\lambda = 35n < 0.4$. Inevitably the nominal curvature method requires calibration to be on the safe side in all ranges of the interaction diagram.

5 CONCLUSIONS

The paper presents a non-linear analysis of slender concrete columns. For a discrete model of a concrete member subjected to axial force and bending moment the Transfer Matrix Method and the sectional analysis of segments rigidities calculation were modified. The non-linear analysis accounting for 2nd order effects, nonlinear behaviour of concrete in compression and influence of cracking, creep and shrinkage were compared with experimental results on slender concrete columns realised by Benkoat al. (2016).

This non-linear analysis was used also to compare the general non-linear method with the approximate method of nominal curvature for taking into account effects of second order. The results of this theoretical analysis are compared in Figure 5, and did not confirm the fact, that approximate methods should give results on the safe side in all range of interaction diagram.

ACKNOWLEDGEMENT

Authors gratefully acknowledge Scientific Grant Agency of the Ministry of Education of Slovak Republic and the Slovak Academy of Sciences VEGA č. 1/0696/14.

REFERENCES

Benko, V., Kendický, P., Gúcky, T. (2016) Failure of Slender Columns of Loss of Stability. In: Reliability Aspects in the Design and Execution of Concrete Structures. Key Engineering Materials Vol. 691. pp. 185–194.

EN 1992-1-1 (2004) Design of concrete structures—Part 1–1: General rules and rules for buildings. p. 244.

Fillo, Ľ., Knapcová, V., Augustín, T. (2016) Design of Slender Concrete Columns. In: Reliability Aspects in the Design and Execution of Concrete Structures. Key Engineering Materials Vol. 691. pp. 220–229.

Model Code 2010 (2010) Fib—Volume 2. p. 288.

SIA 262:2003 (2003) Concrete structures. SIA Zurich p. 89.

Advances and Trends in Engineering Sciences and Technologies II – Al Ali & Platko (Eds)
© 2017 Taylor & Francis Group, ISBN 978-1-138-03224-8

Numerical solution for rotational stiffness of RHS tubular joints

M. Garifullin & N. Vatin
Peter the Great Saint-Petersburg Polytechnic University, St. Petersburg, Russia

T. Jokinen & M. Heinisuo
Tampere University of Technology, Tampere, Finland

ABSTRACT: Nowadays optimization of tubular structures plays a significant role in the design of industrial buildings. Particular attention in this regard is paid to welded tubular joints. During optimization a wide range of joint geometries and sections are explored. Determining mechanical properties for each joint requires computationally intensive Finite Element Analyses (FEA) which makes the optimization procedure inapplicable. Application of approximate surrogate models of the joint responses enables to obtain analytical expressions of initial rotational stiffness, leading to a significant reduction of computational efforts. This paper describes a surrogate modeling procedure for initial rotational stiffness of welded RHS Y joints made of HSS. At sample and validation points the joints were analyzed using comprehensive FEA. The surrogate model was determined by the Kriging method.

1 INTRODUCTION

Tubular structures with welded joints are used in a wide range of structural applications (Al Ali 2014). The most typical application is tubular trusses (Al Ali et al. 2014). The structural analysis model is frequently constructed using beam finite elements, and the braces are connected to the chords using hinges. In reality, the welded joint does not behave as a hinge when it is loaded by a moment.

The joint has resistance against the moment, but in the joint area deformations may occur both at the brace and at the chord, so the stiffness against the moment has to be taken into account in the global analysis of the structure (Al Ali & Daneshjo 2012). In (Boel 2010) and (Snijder et al. 2011) it has been shown that the rotational stiffness of the welded tubular joint is the main parameter when considering buckling of members of tubular trusses. The local design model of Y joint based on (Boel 2010) is shown in Figure 1 and the quantity C is the initial rotational stiffness of the joint, denoted as $S_{j,ini}$ in (EN 1993-1-8, 2005).

In (EN 1993-1-8, 2005) the moment resistance is given for the joint where the angle between the brace and the chord is 90 degrees. In (Grotmann & Sedlacek 1998) a method is given to

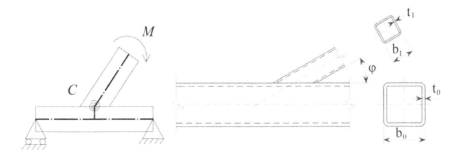

Figure 1. Design model for Y joint and its dimensions.

calculate the initial rotational stiffness for the same case, angle 90 degrees. For other angles there is no design method available.

The rotational stiffness of the joint can be calculated using comprehensive Finite Element Analysis (FEA). In practice, this is impossible, especially when performing optimization of structures when the structural analysis must be done thousands times. In order to avoid these computationally heavy calculations surrogate models (or meta models) have been developed. In (Díaz et al. 2012) the optimum design of steel frames is presented using semi-rigid joints and surrogate models.

The article describes the construction of a surrogate model for initial rotational stiffness of welded tubular Y joints.

2 DESIGN OF EXPERIMENT

2.1 Sampling

The most popular space filling sampling technique is the Latin Hypercube Sampling (LHS) proposed by (McKay et al. 1979). However in this research engineering justification was used for the definition of the sample points (Garifullin et al. 2016).

Every Y joint can be described by the following variables: chord dimensions b_0, t_0; brace dimensions b_1, t_1; angle φ between the brace and the chord; material properties f_{y0}, f_{y1} (Figure 1).

To simplify the surrogate modeling the number of the variables was reduced. The effect of brace thickness t_1 on rotational stiffness was found to be very weak and thus was ignored (numerical values had to be given to t_1 when completing FEA, though). Moreover, in this research we considered joints with only butt welds which have no effect on initial rotational stiffness of joints, so we excluded f_{y0} and f_{y1} from the list of variables. We also replaced b_1 to the relative value $\beta = b_1/b_0$, finally presenting the function of initial rotational stiffness as a function of four variables: b_0, t_0, β, φ.

The sample points were defined so that they met the requirement of Eurocodes and the failure of brace was not critical. There were 3 values (minimum, middle and maximum) for each variable except b_0. Overall, 285 sample points were distributed evenly and covered the whole area of our interest. To calculate the values of rotational stiffness at these points FEA was exploited.

2.2 Requirements of Eurocodes

The (EN 1993-1-8:2005) Eurocodes were used for joints and extension for steel grades up to S700 (EN 1993-1-12:2007). The member sizes were discrete and followed those of Ruukki (Ongelin & Valkonen 2012), meaning cold-formed tubes. Only square sections were considered with chord sizes between 100 × 100 × 4 and 300 × 300 × 12.5. This limits the sizes of braces, because $0.25 \leq b_1/b_0 \leq 0.85$. The ratio b_1/t_1 was limited as $b_1/t_1 \leq 35$ and to the cross-section class 1 or 2. The ratio b_0/t_0 was limited as $10 \leq b_0/t_0 \leq 35$ and, as well, to the cross-section class 1 or 2. We considered steel grades up to S700 and this limited the range of the cross-sections due to cross-section classification. The angle φ between the brace and the chord were due to welding in the range $30 \leq \varphi \leq 90$ degrees.

2.3 Finite element modeling

The Abaqus model was made by using C3D8 brick elements both for the tubes and for the welds. All sections were modeled with round corners, according to (EN 10219-2:2006). Two-layered mesh was created with solid hexahedral elements being refined near the joints, as shown on Figure 2.

The butt welds were modeled as "no weld" by using TIE constraint of Abaqus. The fillet welds were modeled as steel and using TIE constraint where the welds were in contact with chord. The material model was elastic: the modulus of elasticity was 210 GPa and Poisson's ratio 0.3.

Figure 2. FE model for Y joint.

The FEA models were validated with the tests of LUT (Tuominen & Björk 2014) in (Haakana 2014). The verification was done in three steps (Heinisuo et al. 2015): moment load in two opposite directions, use of shell elements instead of brick ones and varying the type of brick elements from 8 to 20 nodes. The proposed FEA model seemed to work well and was used with the fillet welds modeled using the exact geometry.

The joint rotation C was calculated from FEA by extracting the frame behavior from the FEA results, as is given in (Haakana 2014). The list of sample points is provided in Table 1.

3 SURROGATE MODELING

For the validation of the model the following rule has been proposed as a criterion of acceptance: $R^2 \geq 0.85$ for validation points amounting approximately one third of the number of sample points (Díaz et al. 2012). For objective assessment of the model we calculated in Abaqus 48 validation points chosen randomly and different from the sample points.

We started the construction of the surrogate model with a linear regression but the R^2 criterion (Díaz et al. 2012) was rather low. Next we exploited Kriging, as a surrogate model type to approximate deterministic noise-free data. Firstly, we used the DACE toolbox for Matlab (Lophaven et al. 2002) with zero, linear and second order regression and getting our best results with zero order (Heinisuo et al. 2015). The best results (smallest errors) we got for our surrogate modeling using the ooDACE toolbox for Matlab (Ulaganathan et al. 2015) and those results are reported in the article.

We constructed surrogate models of two types: single model (one model for all sample points) and multi-model (with an independent model for every b_0). The idea of implementing the second approach was that the variable b_0 is discrete, getting its values from the Ruukki's catalogue, with no intermediate values among them. Both types gave rather close results to each other and were used for our final model.

The first attempt did not give physically reasonable results using ooDACE, although at the sample points the results were exact; for a detailed discussion, see (Garifullin et al. 2016). In some points the stiffness was "oscillating" with respect to the angle and the symmetry condition at the angle 90 degrees was not fulfilled.

To avoid computationally expensive FEA we implemented "pseudo" sample points. They were defined using polynomial fourth order extrapolations to enforce the stiffness values to behave properly near the boundaries of the variables. These pseudo points were calculated by extending the sample space for the angle φ to 20, 25, 95 and 100 degrees, the variable β to 0, 0.01, 0.9 and 0.95, the variable t_0 to 0, 0.1 mm, and two values over the upper bound. We also calculated a number of interpolated pseudo points in between the existing sample ones: for the angles 45 and 75 degrees, three intermediate β and two intermediate t_0. Overall, we added 1869 pseudo points, resulting with 285 sample points to the total number of 2154 points, see (Garifullin et al. 2016). After adding these pseudo points we managed to construct the physically proper surrogate model (Figure 3).

83

Table 1. Sample points.

b_0 [mm]	β	t_0 [mm]	C [kNm/rad] 30°	60°	90°	t_0 [mm]	C [kNm/rad] 30°	60°	90°	t_0 [mm]	C [kNm/rad] 30°	60°	90°
100	0.400	4	55	27	23	6	174	85	72	10	1082	406	345
100	0.600	4	215	86	68	6	634	262	211	10	4007	1229	1013
100	0.800	4	1135	442	343	6	2847	1107	891				
110	0.364	4	44	23	20	5	83	43	37	6	140	72	62
110	0.545	4	150	63	50	5	272	116	94	6	450	193	158
110	0.818	4	1457	568	439	5	2349	948	751	6	3536	1389	1117
120	0.333	5	70	37	33	7.1	203	106	92	10	638	291	253
120	0.583	5	364	150	121	7.1	1009	422	345	10	3197	1155	953
120	0.833	5	2923	1170	923	7.1	6637	2532	2047				
140	0.286	5	53	30	27	7.1	152	85	76	10	453	231	205
140	0.571	5	353	143	117	7.1	944	399	328	10	2646	1075	891
140	0.786	5	2097	794	618	7.1	4846	1954	1569				
150	0.267	6	81	47	42	8.8	262	146	130	12.5	1004	433	382
150	0.533	6	448	191	158	8.8	1372	593	494	12.5	5046	1897	1586
150	0.800	6	3785	1471	1149	8.8	9697	3742	3034				
160	0.250	6	73	44	39	8.8	236	135	122	12.5	862	397	353
160	0.563	6	559	232	190	8.8	1677	714	590	12.5	5884	2242	1866
160	0.750	6	2493	943	735	8.8	6531	2617	2111				
180	0.278	7.1	148	85	76	8.8	280	158	141	12.5	946	467	415
180	0.556	7.1	896	378	309	8.8	1617	693	571	12.5	5165	2134	1774
180	0.833	7.1	8286	3312	2600	8.8	13374	5401	4322				
200	0.250	7.1	123	74	66	8.8	233	138	124	12.5	750	402	360
200	0.550	7.1	869	367	295	8.8	1566	673	546	12.5	4910	2040	1672
200	0.800	7.1	6690	2589	2020	8.8	10763	4362	3457				
220	0.273	8	203	118	105	10	395	226	203	12.5	840	454	405
220	0.545	8	1157	490	402	10	2131	922	762	12.5	4347	1899	1580
220	0.818	8	11157	4415	3474	10	18372	7443	5944				
250	0.280	8.8	285	164	146	10	416	237	212	12.5	859	472	422
250	0.560	8.8	1718	721	589	10	2429	1034	848	12.5	4815	2094	1731
250	0.800	8.8	12549	4943	3876	10	16880	6768	5384				
260	0.269	8.8	267	155	139	10	389	226	203	12.5	798	448	402
260	0.538	8.8	1496	639	524	10	2118	916	755	12.5	4175	1842	1527
260	0.846	8.8	20765	8271	6441	10	26909	10941	8628				
300	0.267	10	390	227	205	12.5	774	445	401				
300	0.533	10	2136	908	747	12.5	4114	1795	1486				
300	0.833	10	27313	10664	8367	12.5	45620	18624	14858				

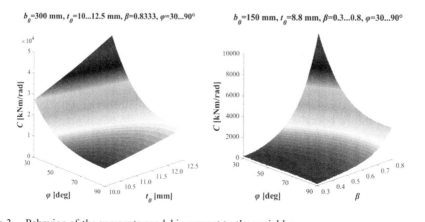

b_0=300 mm, t_0=10...12.5 mm, β=0.8333, φ=30...90°

b_0=150 mm, t_0=8.8 mm, β=0.3...0.8, φ=30...90°

Figure 3. Behavior of the surrogate model in respect to the variables.

We realized soon that R^2 was not proper in our case, all our models had $R^2 \geq 0.95$, but the absolute errors (Eq. 1) were large.

$$Error = \frac{|C_{FEM} - C_{SURR}|}{C_{FEM}} \tag{1}$$

Validation of the resulted showed the following results: average error 8% and maximum error 28%, 16 points with error higher than 10%. This meant that the model required improvements of its accuracy.

Analyzing the validation points we came to conclusion that all inaccurate cases were connected to the points for which β was predicted (only β or together with other variables). Graphical analysis showed that C-β curves had considerable differences from their real values, thus causing serious loss of accuracy, up to 28%. To improve the curves we added extra pseudo points which were determined using polynomial second and fourth order interpolations. The same procedure was conducted also for C-φ curves. Curves improvements helped to increase the accuracy of the model.

To have the best performance, we created a complex model which contained for every chord a surrogate model which suited best. For solving this task every chord was analyzed separately to choose the surrogate model with the best performance. Then all the models were collected in one Matlab file and a complex surrogate model was created.

Validation of the final model demonstrated the average error 4%, maximum 16% and only 4 cases where the error was greater than 10%, see 0. This was the best surrogate model we managed to construct in this study.

The largest errors seemed to appear when the parameter β was predicted, and especially with the large values of β, because the stiffness is increasing rapidly when the value of β is increasing. The use of engineering judgement was, perhaps, not the optimal solution for sampling. The use of LHS (McKay et al. 1979) could have led to better results.

Table 2. Final validation.

b_0 [mm]	t_0 [mm]	β	φ [°]	C [kNm/rad] FEM	C [kNm/rad] SM	Error [%]	b_0 [mm]	t_0 [mm]	β	φ [°]	C [kNm/rad] FEM	C [kNm/rad] SM	Error [%]
100	8	0.800	79	1838	1735	5.6	180	10	0.389	59	405	391	3.4
100	6	0.400	42	115	122	6.2	180	12.5	0.667	33	8504	9029	6.2
100	10	0.400	80	349	348	0.4	180	10	0.500	40	1085	1096	1.1
100	8	0.500	89	300	285	5.0	180	10	0.611	60	1415	1470	3.9
110	5	0.364	34	71	72	1.4	200	8.8	0.250	69	130	129	1.0
110	6	0.364	55	76	75	1.2	200	12.5	0.400	62	855	811	5.1
110	6	0.364	50	82	81	0.9	200	8	0.300	57	138	131	4.8
110	5	0.818	88	746	748	0.3	200	7.1	0.600	66	448	518	15.5
120	5.6	0.583	80	170	171	0.8	220	8.8	0.545	71	569	547	3.9
120	7.1	0.750	83	1098	1111	1.2	220	12.5	0.455	58	1195	1247	4.4
120	8.8	0.833	42	6030	6726	11.6	220	10	0.409	37	703	819	16.4
120	7.1	0.500	39	399	387	3.0	220	8	0.727	62	1757	1974	12.4
140	7.1	0.357	90	106	113	6.5	250	12.5	0.720	60	6397	6163	3.7
140	6	0.786	89	977	1017	4.1	250	10	0.600	90	1073	1086	1.2
140	7.1	0.500	40	393	400	1.7	250	12.5	0.600	87	2199	2170	1.3
140	6	0.500	30	349	366	4.8	250	8.8	0.280	72	151	151	0.2
150	7.1	0.533	47	405	400	1.2	260	10	0.846	79	8820	8802	0.2
150	7.1	0.800	89	1643	1783	8.5	260	12.5	0.577	67	2135	2115	0.9
150	6	0.400	71	84	77	8.0	260	12.5	0.308	73	498	522	4.8
150	7.1	0.267	39	104	105	0.9	260	12.5	0.692	51	6208	6222	0.2
160	8.8	0.750	86	1946	2102	8.0	300	12.5	0.833	33	39163	40996	4.7
160	10	0.313	76	244	243	0.3	300	12.5	0.467	56	1290	1396	8.2
160	8.8	0.500	85	416	438	5.3	300	12.5	0.833	85	14946	14908	0.3
160	8.8	0.438	81	303	321	6.0	300	12.5	0.600	86	2247	2338	4.1

4 CONCLUSIONS

There exists no method to calculate the initial rotational stiffness for welded tubular Y joints for different angles φ, so our sample point data is valuable in the verification of such method. The most potential way to develop such a method is to use the novel component method for tubular joints (Jaspart & Weynand 2015). Before such a method is available the proposed surrogate model might be applied for optimization of tubular frames and trusses.

The best surrogate model was constructed using Kriging via ooDACE toolbox for Matlab. It was essential to use the pseudo points to get the physically reasonable surrogate model. Minimum two pseudo points at the lower and upper limits of the variables were used to extrapolate the sample space and 2–3 additional interpolated ones were used to provide proper curves. The model with the average error of 4% was accepted for further studies.

REFERENCES

Al Ali, M., 2014. The Welding Process as a Local Issue with Global Consequences. *Advanced Materials Research*, 969, pp. 340–344.

Al Ali, M. & Daneshjo, N., 2012. Size and Distribution of Welding Stresses. *Procedia Engineering*, 40, pp. 2–7.

Al Ali, M., Tomko, M. & Demjan, I., 2014. Welding process from the civil engineering point of view. In Civil Engineering and Urban Planning III - Proceedings of the 3rd International Conference on Civil Engineering and Urban Planning, CEUP 2014. World Scientific Publishing Co. Pte Ltd, pp. 541–544.

Boel, H., 2010. Buckling Length Factors of Hollow Section Members in Lattice Girders. Ms Sci thesis, Eindhoven: Eindhoven University of Technology.

Díaz, C., Victoria, M., Querin, O.M. & Martí, P., 2012. Optimum design of semi-rigid connections using metamodels. *Journal of Constructional Steel Research*, 78, pp. 97–106.

European Committee for Standardisation, (CEN), 2006. Cold formed welded structural hollow sections of non-alloy and fine grain steels. Part 2: Tolerances, dimensions and sectional properties (EN 10219-2:2006), Brussels.

European Committee for Standardisation, (CEN), 2007. Eurocode 3. Design of steel structures, Part 1-12: Additional rules for the extension of EN 1993 up to steel grades S 700 (EN 1993-1-12: 2007), Brussels, 2007.

European Committee for Standardisation, (CEN), 2005. Eurocode 3. Design of steel structures, Part 1–8: Design of joints (EN 1993-1-8:2005), Brussels, 2005.

Garifullin, M., Jokinen, T. & Heinisuo, M., 2016. *Supporting document for surrogate model construction of welded HSS tubular Y-joints. Publication 164*, Tampere: Tampere University of Technology, p. 131.

Grotmann, D. & Sedlacek, G., 1998. Rotational stiffness of welded RHS beam-to-column joints. Cidect 5BB-8/98, Aachen: RWTH-Aachen.

Haakana, Ä., 2014. In-Plane Buckling and Semi-Rigid Joints of Tubular High Strength Steel Trusses. Ms Sci thesis, Tampere: Tampere University of Technology.

Heinisuo, M., Mela, K., Tiainen, T., Jokinen, T., Baczkiewicz, J. & Garifullin, M., 2015. Surrogate model for rotational stiffness of welded tubular Y-joints. *Proceedings of the METNET Seminar 2015 in Budapest*, pp. 18–39.

Jaspart, J.P. & Weynand, K., 2015. Design of hollow section joints using the component method. *Proceedings of the 15th International Symposium on Tubular Structures, ISTS 2015*, pp. 405–410.

Lophaven, S.N., Søndergaard, J. & Nielsen, H.B., 2002. *DACE, A MATLAB Kriging Toolbox, Version 2.0, August 1*, Copenhagen: Technical University of Denmark.

McKay, M.D., Bechman, R.J. & Conover, W.J., 1979. A Comparison of Three Methods for Selecting Values of Input Variables in the Analysis of Output From a Computer Code. *Technometrics*, 21(2), pp. 239–245.

Ongelin, P. & Valkonen, I., 2012. *Structural hollow sections. EN 1993 - Handbook 2012*, Rautaruukki Oyj.

Snijder, H.H., Boel, H.D., Hoenderkamp, J.C.D. & Spoorenberg, R.C., 2011. Buckling length factors for welded lattice girders with hollow section braces and chords. *Proceedings of Eurosteel 2011*, pp. 1881–1886.

Tuominen, N. & Björk, T., 2014. Ultimate Capacity of Welded Joints Made of High Strength Steel CFRHS. *Proceedings of Eurosteel 2014*, pp. 83–84.

Ulaganathan, S., Couckuyt, I., Deschrijver, D., Laermans, E. & Dhaene, T., 2015. A Matlab Toolbox for Kriging Metamodelling. *Procedia Computer Science*, 51, pp. 2708–2713.

Advances and Trends in Engineering Sciences and Technologies II – Al Ali & Platko (Eds)
© 2017 Taylor & Francis Group, ISBN 978-1-138-03224-8

Overall imperfection method for beam-columns

G. Hajdú & F. Papp
Széchenyi István University, Győr, Hungary

ABSTRACT: The principles and applications of the overall imperfection method for beam-columns are presented in this paper. The buckling resistance of a beam-column member is determined by the resistance of its critical cross-section taking into account the second-order effect. The maximum amplitude of the initial imperfection, in the shape of elastic buckling mode, is determined from fundamental cases. Fundamental case: simply-supported member with uniform cross-section subjected to uniformly distributed forces. The standardized buckling resistance of this reference member is based on theoretical and empirical background. The proposed overall imperfection method is adequate for computer-aided design procedures which contains an advanced elastic beam-column finite element method. The accuracy of the overall imperfection method is shown by a parametric study.

1 INTRODUCTION

Initial out of straightness and residual stress are appeared in steel elements during the fabrication. These initial imperfections may reduce the elastic buckling strength of steel members. Focusing on EN1993-1-1:2005 to evaluate the reduced strength of beam-columns three approaches are applied: (i) 5.3.2 Imperfections for global analysis of frames, (ii) 6.3.3 Uniform members in bending and axial compression, (iii) 6.3.4 General method for lateral buckling and for lateral torsional buckling of structural components.

In (i) two alternative ways are available to calculate the global buckling resistance:

- Method (A) is a conservative method using simplified initial imperfections with constant amplitude.
- Method (B) computer aided method using equivalent unique global and local geometric imperfection in the shape of the elastic buckling mode.

The principles and applications of Method (B) was presented by Chladný & Stujberová (2013) later Papp (2016) showed the equivalent unique global and local imperfection concept is valid for lateral-torsional buckling and coupled buckling too. The purpose of this article is to present the application of the overall imperfection method and show its accuracy.

2 FUNDAMENTAL CASE OF COUPLED BUCKLING

2.1 Flexural and lateral-torsional buckling

The advantages and applications of the equivalent initial imperfection method for flexural buckling have been presented by Chladný & Štujberová (2013) and an alternative formula was supposed to define the amplitude of v_{init} initial imperfection:

$$v_{init.max} = \frac{\sigma^{II}_{v_{init}.max}}{\sigma^{II}_{v_{cr}.max}} v_{cr.max}. \tag{1}$$

Where:

$\sigma_{v_{cr}.max}^{II}$ is the maximum second-order normal stress from bending as the effect of N_{ed} compressive force with initial imperfection in the shape of elastic buckling mode. In this case the amplitude of the initial imperfection is arbitrary.

$\sigma_{v_{init}.max}^{II}$ is the maximum second-order normal stress from bending as the effect of N_{ed} compressive force with the calibrated value of equivalent amplitude (see Equation (2)).

$v_{cr.max}$ amplitude of the elastic buckling mode.

The calibrated amplitude of the equivalent initial bow imperfection of compressed member was assumed as:

$$e_{0d} = \alpha\left(\overline{\lambda} - 0.2\right)\frac{W}{A}\frac{1 - \dfrac{\overline{\lambda}^2 \cdot \chi}{\gamma_{M1}}}{1 - \overline{\lambda}^2 \cdot \chi} \tag{2}$$

Where α = imperfection factor; λ = relative slenderness; W = section modulus; A = cross-section area; χ = reduction factor for flexural buckling; and γ_{M1} = partial safety factor.

Figure 1 shows the fundamental arrangement of flexural buckling. In this case the amplitude of the initial imperfection is $v_{init.max} = e_{0d}$.

Papp (2016) showed Equation (1) is valid for lateral-torsional buckling too. For lateral-torsional buckling the meanings of the symbols in Equation (1):

$\sigma_{v_{cr}.max}^{II}$ is the maximum second-order normal stress, including bending around z axis and warping, as the effects of $M_{y.Ed}$ bending moment with initial imperfection in the shape of the elastic buckling mode having arbitrary amplitude.

$\sigma_{v_{init}.max}^{II}$ is the maximum second-order normal stress, including bending around z axis and warping, as the effects of $M_{y.Ed}$ bending moment with the reference equivalent amplitude (Equation (3)).

The amplitude of the equivalent initial imperfection for lateral-torsional buckling according to Papp (2016):

$$v_{0d} = \frac{\eta_{LT}(or\,\eta_{LT}^*)}{\dfrac{W_y}{W_\omega} + \dfrac{W_y}{W_z}\dfrac{N_{cr.z}}{M_{cr}} - \dfrac{N_{cr.z}}{M_{cr}^2}GI_t\dfrac{W_y}{W_\omega}} \cdot \frac{1 - \dfrac{\overline{\lambda}_{LT}^2 \cdot \chi_{LT}}{\gamma_{M1}}}{1 - \overline{\lambda}_{LT}^2 \cdot \chi_{LT}} \tag{3}$$

$$\varphi_{0d} = v_{0d}\frac{N_{cr.z}}{M_{cr}}. \tag{4}$$

Where:

W_y, W_z, W_ω = section modulus, regarding the class of cross-section; $N_{cr.z}$ = critical force for flexural buckling; M_{cr} = critical moment for lateral-torsional buckling.

In Equation (3) $\eta_{LT}\left(or\,\eta_{LT}^*\right)$ is the imperfection factor, where the following calibrations are available:

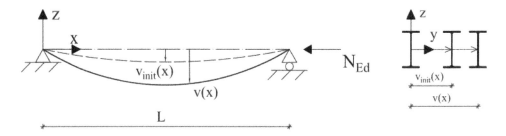

Figure 1.   Deformation of the reference member: fundamental case of flexural buckling.

- In EN1993-1-1:2005 6.3.2.2 as a 'general case': $\eta_{LT} = \alpha_{LT}\left(\overline{\lambda}_{LT} - 0.2\right)$
- Or in the proposal of Taras & Greiner (2010) as a revision of the EN1993-1-1: 2005 specification:

$$\eta_{LT}^{*} = \alpha_{LT.TARAS}\left(\overline{\lambda}_z - 0.2\right)\sqrt{\frac{W_{el.y}}{W_{el.z}} \cdot \frac{\overline{\lambda}_{LT}^2}{\overline{\lambda}_z^2}}.$$ (5)

Figure 2. shows the fundamental arrangement of lateral-torsional buckling. In this case the amplitude of the initial imperfection is $v_{init.max} = v_{0d}$.

2.2 Coupled buckling

Szalai et. al. (2015) have found that the generalized imperfection factor for beam-columns may be expressed by linear interpolation between the above mentioned fundamental cases. Applying this assumption to the equivalent initial imperfection concept Equation (1) could be convenient for coupled buckling. In this case the amplitude of the imperfection is:

$$v_{0d.NM} = \frac{\alpha_{ult}}{\alpha_{ult.N}} e_{0d} + \frac{\alpha_{ult}}{\alpha_{ult.M}} v_{0d}$$ (6)

$$\varphi_{0d.NM} = \frac{v_{0d.NM}}{\dfrac{M_{cr.N}}{N_{cr.z}} \cdot \dfrac{1}{1 - \dfrac{N_{Ed}}{N_{cr.x}}}}$$ (7)

where the minimum load amplifier of the design loads to reach the characteristic resistance of the most critical cross section are:

$$\alpha_{ult.N} = \frac{N_{Rk}}{N_{Ed}}; \, \alpha_{ult.M} = \frac{M_{y.Rk}}{M_{y.Ed}}; \, \alpha_{ult} = \frac{1}{\dfrac{1}{\alpha_{ult.N}} + \dfrac{1}{\alpha_{ult.M}}}.$$ (8)

To apply Equation (6) as an amplitude of the initial imperfection in the shape of elastic buckling mode and after the geometrically nonlinear static analysis the buckling resistance of a beam-column member is appropriate if the utilization of the cross-section will satisfy the following equation:

$$U_{max} = \frac{N_{Ed}}{A \cdot \dfrac{f_y}{\gamma_{M1}}} + \frac{M_{y.Ed}}{W_y \cdot \dfrac{f_y}{\gamma_{M1}}} + \frac{M_{z.Ed}^{II}}{W_z \cdot \dfrac{f_y}{\gamma_{M1}}} + \frac{B_{Ed}^{II}}{W_\omega \cdot \dfrac{f_y}{\gamma_{M1}}} \leq 1.0.$$ (9)

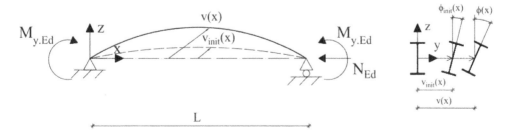

Figure 2.   Deformation of the reference member: fundamental case of lateral-torsional buckling.

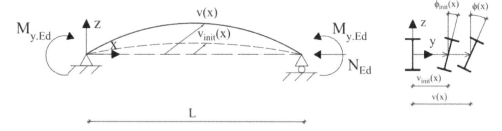

Figure 3. Deformation of the reference member: fundamental case of coupled buckling.

Figure 3. shows the fundamental arrangement of coupled buckling. In this case the amplitude of the initial imperfection is $v_{init.max}=v_{0d.NM}$.

3 GENERALIZED OVERALL IMPERFECTION METHOD

In design practice the above mentioned force distributions are really rare, but Equation (1) could be also used for other types of force arrangement. The solution for non-fundamental case is based on a hypothetical approach, so called equivalent member concept, which has been widely used in structural engineering. For the clarification, the steps of the generalized overall imperfection method according to Papp (2016) are the following:

Step 1: Linear buckling analysis.
    Step 1.1: Critical load amplifier: α_{cr}.
    Step 1.2: Elastic buckling shape with arbitrary amplitude: $v_{cr}(x)$, $v_{cr.max}$.
Step 2: Second-order stress analysis with $v_{cr}(x)$ initial imperfection.
    Step 2.1: Position of the cross section where the second-order normal stress has maximum value (reference point): x_{ref}.
    Step 2.2: Maximum second-order normal stress at the reference point: $\sigma^{II}_{y_{cr}.max}$.
Step 3: Maximum second-order normal stress of the reference member: $\sigma^{II}_{v_{init}.max}$.
    Step 3.1: Parameters of the reference member:
Design forces (taken at x_{ref}): $N_{Ed.ref}(x_{ref})$; $M_{y.Ed.ref}(x_{ref})$
Critical forces:

$$N_{cr.NM.ref} = \alpha_{cr}N_{Ed.ref}; M_{cr.NM.ref} = \alpha_{cr}M_{y.Ed.ref} \tag{10}$$

Step 3.2: Length of the reference member derived from:

$$L_{ref} = \sqrt{\frac{-b \pm \sqrt{b^2 - 4ac}}{2a}} \tag{11}$$

$$a = M^2_{cr.NM.ref} + N_{cr.NM.ref}GI_t - N^2_{cr.NM.ref}r^2_{0.ref} \tag{12}$$

$$b = -\pi^2 EI_{z.ref}GI_t + \pi^2 EI_{z.ref}N_{cr.NM.ref}r^2_{0.ref} + \pi^2 EI_{\omega.ref}N_{cr.NM.ref} \tag{13}$$

$$c = -\pi^4 E^2 I_{z.ref}I_{\omega.ref} \tag{14}$$

Step 3.3: Reduced slenderness for FB and LTB respectively:

$$\bar{\lambda}_{ref} = \sqrt{\frac{N_{Rk}}{N_{cr.z.ref}}}; \bar{\lambda}_{LT.ref} = \sqrt{\frac{M_{y.Rk}}{M_{cr.ref}}} \tag{15}$$

Step 3.4: Equivalent amplitude for FB: applying Equation (2)
Step 3.5: Equivalent amplitude for LTB: applying Equation (3)
Step 3.6: Equivalent amplitudes for coupled buckling: applying Equation (6)
Step 3.7: Maximum second-order stress:

$$\sigma^{II}_{v_{init}.\max} = \frac{M^{II}_{z.\mathbf{max}.ref}}{W_{el.z.ref}} + \frac{B^{II}_{\max.ref}}{W_{el.\omega.ref}} \tag{16}$$

Step 4: Amplitude of equivalent initial imperfection: applying Equation (1)
Step 5: Second-order analysis of the designed member with the equivalent initial imperfection.
Step 6: Check the cross-section resistance of the designed member Equation (9).

4 PARAMETRIC STUDY

The practical usage of this generalized method requires large number of parametric numerical and experimental studies to verify its reliability. This verification is beyond the scope of this article. The primary aim is to show the accuracy of the overall imperfection method. The numerical analyses were carried out by ConSteel Software. For the geometrically nonlinear analysis the initial imperfections were derived from the buckled shape.

Equation (6) is applied to check the coupled buckling resistance of the members which are subjected to uniform compressive force and uniform bending moment (Figure 3). The ultimate forces were taken from GMNIA analysis. The GMNIA analysis was carried out by Luís Simões da Silva and his research group. The cross section of the examinded member is IPE360, its cross-sectional properties were taken from ConSteel Software. The result from advanced numerical simulations (GMNIA) are compared to the Overall Imperfection Method in Figure 4. The normailzed slenderness are decreased from 1.8 to 0.8.

Figure 5 shows scatter-plots where the resistance of the corresponding theoretical r_t and experimental r_e values are compared. These values could be determined from:

$$r = \sqrt{\left(\frac{M}{M_{pl}}\right)^2 + \left(\frac{N}{N_{pl}}\right)^2} \tag{17}$$

The mean value for r_e/r_t is 1.006, the coefficient of variation is approximately 4% which means that the Overall Imperfection Method is stable. The minimum value is 0.965 and maximum value is 1.035.

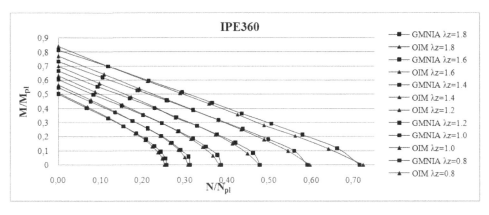

Figure 4.   Theoretical and experimental resistance for IPE360.

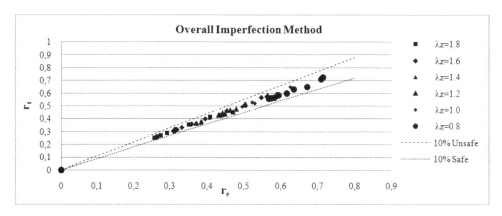

Figure 5. Validation of overall imperfection method.

5 CONCLUSION

In this paper an alternative method was presented to check the stability resistance of steel members. The basic conception of this method was proposed by Chladný & Stujberová (2013) and has been adopted in EN 1993-1-1:2005 5.3.2 (11) as an optional procedure to check the global stability resistance of steel frames. The flexural buckling resistance of the compressed member is determined by its cross-section resistance with take into account the second order effects.

The equivalent amplitude for lateral-torsional buckling assessment of beams was derived by Agüero et al. (2015). For coupled buckling, the equivalent amplitude was derived by Papp (2016). The generalization of the overall imperfection method is derived for the fundamental case and based on the equivalent member concept. This concept is accepted and widely used in structural engineering practice.

The direct conclusion of this work is that the proposed overall imperfection method may be used as an alternative procedure for coupled buckling. This method could be adapted for any design software which applies 14 DOF thin-walled beam column finite element method and possible to run the geometrically non-linear analysis on the buckled shape.

REFERENCES

Agüero, A. Pallerés, L. & Pallarés, FJ. 2015. Equivalent geometric imperfection definition in steel structures sensitive to lateral torsional buckling due to bending moment. *Eng Struct 2015*; 96:41–55

Chladný, E. & Štujberová, M. 2013. Frames with global and local imperfection in the shape of elastic buckling mode (Part 1). *Stahlbau 2013*; 83 (Heft 8):609–17.

ConSteel: Structural design software. ConSteel Solutions Ltd. http://www.consteelsoftware.com/en

EN1993-1-1:2005. Design of steel structures—Part 1–1: General rules and rules for buildings. In: European committee for standardization. Brussels, Belgium.

Papp, F. 2016. Buckling assessment of steel members through overall imperfection method, *Engineering Structures 106* (2016) 124–136.

Szalai, J. & Papp, F. 2010. On the theoretical background of the generalization of Ayrton—Perry type resistance formulas. *J Constr Steel Res 2010*; 66:670–9.

Szalai, J. & Papp, F. & Badari, B. 2015. Overall stability design of members: theoretical basis, Overall Reduction Factor Method (ORFM), and Overall Imperfection Method (OIM). In: Official meeting of ECCS technical committee 8 - structural stability, Budapest, 22 May 2015.

Taras, A. & Greiner, R. 2010. New design curves for lateral-torsional buckling—proposal based on a consistent derivation. *J Constr Steel Res 2010*; 66:648–63.

Punching resistance of column basis

J. Halvonik, J. Hanzel & L. Majtánová
Slovak University of Technology, Bratislava, Slovakia

ABSTRACT: Punching belongs to the most dangerous form of structural failure. Therefore, models for the assessment of punching resistance have to be safe enough in order to prevent such event. Current models are either empirical e.g. EC2 model or mechanical e.g. model based on Critical shear crack theory. Statistical evaluation of the safety plays very important role for calibration of both. The paper deals with evaluation of the safety of several models used for the assessment of punching resistance of column basis using database of 50 experiments on footings without shear reinforcement. Analysis have shown that iterative method for assessment of position of critical control perimeter introduced in current EC2 is correct, however the EC2 model overestimates punching shear resistance of the footings due to high contribution of the struts to the punching capacity. The best results were obtained with new German model (2016) and Model Code 2010.

1 INTRODUCTION

Column basis are structural members which support columns e.g. spread footings or foundation slabs. The major differences from flat slabs are much bigger shear forces and higher effective depths "d" of the foundation members. Shear slenderness of the footings is usually significantly lower than in case of flat slabs. Shear slenderness or shear span-depth ratio a_λ/d is important parameter which have an influence on the punching resistance. The definition of a_λ is clear from the Figure 1. Critical control perimeter is located at distance a_{crit} measured from the face of a column where critical shear crack crosses main reinforcement.

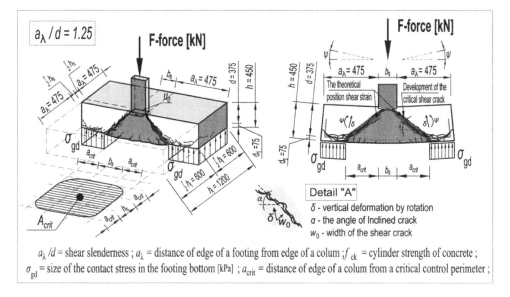

a_λ /d = shear slenderness ; a_λ = distance of edge of a footing from edge of a colum ;f_{ck} = cylinder strength of concrete ;
σ_{gd} = size of the contact stress in the footing bottom [kPa] ; a_{crit} = distance of edge of a colum from a critical control perimeter ;

Figure 1. The punching shear failure of spread footings without shear reinforcement.

With lower shear slenderness the critical shear cracks are steeper and critical control perimeters are closer to the face of a column.

Two ways are currently used for determining of critical control perimeter position. The first way, introduced in EC2, is based on its searching, which means that the position of critical control perimeter is not constant. The second way is definition of single or unique critical control perimeter which is not usually in coincidence with actual one. This control perimeter is mostly assumed at distance $d/2$ from face of a column, e.g. CSCT model (Muttoni & Ruiz 2008, 2012) ACI Model, German update of the EC2 model from 2016 (Hegger et al. 2016) or Cladera & Mari model (Mari et al. 2016). The different actual position of critical control perimeter has to be taken into account by suitable factors incorporated in the model.

The further important parameter is so called size effect. Generally, it is known that shear strength decreases with growing effective depth of a member. However, factors taking into account this effect were usually calibrated for thinner slabs (flat slabs) with depth significantly lower than depth of the column basis.

2 DESCRIPTION OF THE CRITICAL CONTROL PERIMETER

Only current EC2 model requires analysis for determining of the critical control perimeter position a_{crit}. The theoretical values can be found using iterative process where ratio $V_{Rd,c}(a)/V_{Ed,red}(a)$ is minimum. The value $V_{Ed,red}(a)$ is reduced shear force obtained by subtracting of upward force within assumed control perimeter at distance "a" due to the soil pressure and punching shear resistance $V_{Rd,c}(a)$ can be obtained using punching resistance without shear reinforcement $V_{Rd,c}$ multiplied by factor $2d/a$ for $a \leq 2d$, see Figure 1. Position of the critical control perimeter can be determined using values introduced in Table 1. Values in the Table 1 were obtained from the analysis, with assumption of uniformly distributed soil pressure under the footings. The ratio a_{crit}/d depends only on the shear slenderness a_λ/d and effective depth of the footing d. Further parameters that are used for the assessment: strength of concrete f_{ck}, reinforcement ratio ρ_l, an empirical factor $C_{Rd,c}$, shear force V_{Ed} do not influence position of the critical control perimeter.

3 THE ASSESMENT OF PUNCHING SHEAR RESISTANCE

3.1 Empirical models—EC2 and German proposal 2016

Basic principles of current EC2 model has been set by Zsutty (1968). Zsutty proposed empirical approach on the basis of available test data for beams. The model has been later refined and adopted in Model Code 1990. The MC1990 model became the basis for current EC2 model. Growing database of punching tests in last two decades allows for check of model safety and further calibration of EC2 model. Base on this analysis German group proposed several amendments of EC2 model. The first one was reduction of empirical factor $C_{Rk,c}$ from 0.18 to 0.15 for foundation slabs and footings, see German National Annex to DIN EN 1992-1-1. The second one was more comprehensive model, released in January 2016 (Hegger et al. 2016), which is based on new formula for punching resistance (1) with basic control perimeter at distance $0.5d$ from the face of a column for both flat slabs and

Table 1. Relation between parameters a_{crit}/d and shear slenderness for different effective depths "d".

d	a/d	0.80	0.90	1.00	1.20	1.30	1.40	1.50	1.60	1.70	1.80	2.00
200	a_{crit}/d	0.40	0.45	0.49	0.57	0.62	0.66	0.70	0.74	0.79	0.83	0.91
400	a_{crit}/d	0.38	0.41	0.45	0.52	0.56	0.60	0.63	0.67	0.71	0.74	0.82
750	a_{crit}/d	0.36	0.39	0.42	0.48	0.52	0.55	0.58	0.61	0.64	0.68	0.74
1100	a_{crit}/d	0.34	0.37	0.39	0.45	0.48	0.51	0.54	0.56	0.59	0.62	0.68
1450	a_{crit}/d	0.33	0.36	0.38	0.43	0.46	0.49	0.51	0.54	0.57	0.59	0.64

column basis. Still empirical model is more sophisticated because takes into account more factors that have impact on the punching shear capacity. First it is new factor for taking account size effect where previous formula has been replaced by Bazant formula (3). For the second it is factor which takes into account shear slenderness a_λ/d and the length of column periphery u_0. The last change in the model is new value of empirical parameter $C_{Rk,c}$ equal to 1.8 ($C_{Rm,c} = 2.2$, mean value).

$$v_{Rd,c} = \frac{C_{Rk,c}}{\gamma_C} k_d k_\lambda \left[100\rho_l f_{ck}\right]^{0.333}$$ (1)

$$k_d = (1 + d/200)^{-0.5}$$ (2)

$$k_\lambda = \left[\left(\frac{a_\lambda}{d}\right)\left(\frac{u_0}{d}\right)\right]^{-0.2}$$ (3)

3.2 Mechanical models—CSCT, Cladera & Mari

The first one is model based on Critical shear crack theory which has been developed and later refined by Muttoni (1991) and Muttoni & Ruiz (2008). Model has been incorporated in the Model Code 2010 and later adjusted into EC2 format. The CSCT theory is based on the assumption that major part of shear resistance represents aggregate interlocking in critical shear crack and tensile strength of the concrete. Therefore, shear crack width and roughness of the crack surface play the most important role in shear capacity of a member. The shear crack width is directly influenced by rotation of a slab at support. The authors of the model have found out so called failure criterion which is relation between rotation of a slab at support ψ and punching shear capacity $v_{Rd,c}$, see formula (4). They proposed four different ways how to determine load-rotation relationships depending on the level of approximations LoA. The advantage of the model is unique position of critical control perimeter at distance $0.5d$ from column face for both flat slabs and footings. The model takes into account shear slenderness, quality of the concrete, strains in main reinforcement crossing shear crack and aggregate size.

$$v_{Rd,c} = \left(\frac{A}{1 + B k_{dg}\,\psi d}\right)\frac{\sqrt{f_{ck}}}{\gamma_C}$$ (4)

$$\psi = C\frac{r_s}{d}\frac{f_{yd}}{E_s}\left(\frac{m_{Ed}}{m_{Rd}}\right)^{1.5}$$ (5)

Where k_{dg} is factor taking into account aggregate size, for $d_{g,max} = 16$ mm is $k_{dg} = 1.0$, r_s is distance between column axis and sections with zero radial bending moments. Average bending moments per unit length effect/resistance is defined as m_{Ed}/m_{Rd}. For factors A, B and C see in Table 3. Bending moment m_{Ed} has been calculated using simplified formula $m_{Ed} = V_{R,test}/8$ and m_{Rd} is bending capacity of a slab per unit length.

The second mechanical model has been introduced by Mari et al. (2016). The authors attribute the major part of shear resistance to compressive concrete cord. The failure is reached when critical shear crack propagates inside the compressive cord. Therefore, critical control perimeter is located at distance where critical shear crack crosses neutral axis "x". Because inclination of critical shear crack is unknown the basic (critical a_{crit}) control perimeter can be assumed at distance $0.5d$ from the face of a column. Punching shear resistance calculated at basic control perimeter can be assessed by formula (6), where $V_{Ed,red}$ is reduced shear force by upward force due to the soil pressure acting in area surrounded by perimeter at distance $1.0d$ from face of column, A_1 is an area surrounded by considered control perimeter. Parameter a_v is distance from face of column to load center (soil pressure) producing shear. If $a_v < 2.5d$ then adjustment of the neutral axis position x has to be made, instead of x a value of x_1 should be used. Parameter α_e is modulus ratio.

$$v_{Rd,c} = \xi \frac{x_1}{d} \left(0.30 \frac{f_{ck}^{2/3}}{\gamma_C} + 0.85 \frac{V_{Ed,red}}{A_1} \right) \tag{6}$$

$$\xi = \left(\frac{d}{a_v} \right)^{0.2} \frac{2}{(1 + d/200)^{0.5}} \tag{7}$$

$$\frac{x_1}{d} = 0.75 (\alpha_e \rho_1)^{1/3} + \left(1 - \frac{x}{d} \right) \left(1 - 0.4 \frac{a_v}{d} \right) \le 1 \tag{8}$$

4   STATISTICAL EVALUATION OF SAFETY MODELS

4.1   *Description of the statistical analysis of the models P1 to P10*

Statistical evaluation has been performed using results from database which was elaborated by Siburg and Rickert from RTWH Aachen. More than 300 experiments are described in database, but only 50 samples have failed due to the punching. The main statistical variable in analysis was the ratio $P_i = (V_{R,test}/V_{R,c})_i$ where i was number of considered experiment, in our case from 1 to 50. Punching shear resistances $V_{R,c}$ were assessed using 10 models. The P1 model is current EC2 model, P2, P3 and P4 represent amendments of EC2 model and P5 new German update of the EC2 model, see in Table 2. Further four models were based on the CSCT theory, see in Table 3 and finally P10 is Cladera & Mari model. Shear force $V_{Ed,red}$ in (6) have been determined on the basis of $V_{R,test}$. Parameters and material properties used for calculation of $V_{R,c}$ were actual (mean) values tested for the experiment. Partial safety factors were taken equal 1.

4.2   *Discussion about the models safety*

Target value of 5% fractile $P_{k,0.05}$ is 1.0 according to EN 1990. However due to further favorable influences on punching shear resistance, e.g. membrane forces, linear-elastic analysis, value above 0.90 is accepted and model can be assumed reliable. Statistical evaluation has shown that the most reliable is MC2010 model, set P6 for level of approximation *LoA* I., where is used the simplified assumption that bending reinforcement for all experiments have been yielding. Reliability is achieved thanks to high mean value of $P_m = 1.79$ which in turn

Table 2.   Statistical evaluation of the empirical models safety.

Set	Model	a_{crit}	$C_{Rk,c}$	Strut	P_m	COV	$P_{k,0.05}$
P1	EC2	iterative	0.18	$2d/a$	0.781	0.1314	0.604
P2	EC2	iterative	0.18	$1.2d/a$	1.302	0.1314	1.006
P3	EC2	iterative	0.15	$2d/a$	0.937	0.1314	0.735
P4	EC2	iterative	0.15	$1.5d/a$	1.250	0.1314	0.980
P5	New German	$0.5d$	1.80	No	1.231	0.1380	0.952

Table 3.   Statistical evaluation of the mechanical models safety.

Set	Model	A	B	C	m_{Ed}/m_{Rd}	P_m	COV	$P_{k,0.05}$
P6	CSCT/MC2010	0.67	0.60	1.5	1.0	1.790	0.228	1.085
P7	CSCT/MC2010	0.67	0.60	1.5	≤ 1.0	1.333	0.186	0.924
P8	CSCT/MC2010	0.67	0.60	1.2	≤ 1.0	1.027	0.175	0.717
P9	CSCT/EC2	0.70	0.45	1.2	≤ 1.0	1.118	0.174	0.797
P10	Cladera & Mari	–	–	–	–	1.034	0.123	0.813

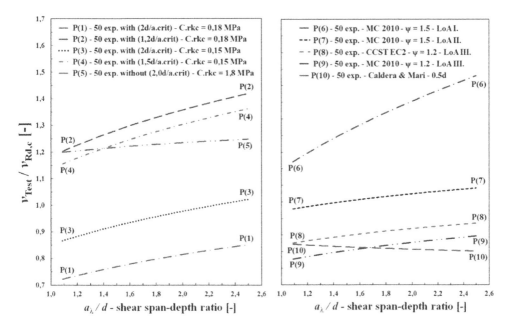

Figure 2. Relation between model safety $v_{test}/v_{R,c}$ and shear span-depth ratio a_λ/d of the footings for model from P(1) to P(10) from Table 2, 3 and 4.

indicates that the model is uneconomical. Therefore, the model is suitable only for preliminary design.

The best calibrated models are new German model P5 and model based on CSCT theory for *LoA* II. P7 with very similar safety level between 0.92–0.95 which is regarded as sufficient. The lowest COV and mean value P_m has shown Cladera & Mari model with values of 0.1234 and 1.034 which resulted in model safety 0.813 which is comparable with the safety of CSCT/EC2 model. The EC2 model is divided to the results, P1 through P4. The analysis have shown, that the current EC2 model P1 is not reliable, with value of 0.60. We suppose that the reason is in overestimation of the compressive diagonal contribution $2d/a$ to the punching resistance $v_{R,c}$. After its reduction to the value of $1.2d/a$, P2 model, the reliability has been increased to the target level 1.0. The same procedure has been applied also for the German amendment of the EC2 model P3, where the reduction of $C_{Rk,c}$ to the value of 0.15 had been used. Adjustment of the strut contribution to $1.5d/a$ ensured adequate safety level.

4.3 *Relations between model safety and selected parameters in the assessments*

Relations between model safety and shear span-depth ratio is shown in Figure 2. Model safety increases with growing shear slenderness for all investigated models except for Cladera & Mari model. The best calibrated model from this point of view is P5 model where the safety is almost the same for each shear span-depth ratio. Very good results provide also models based on CSCT theory for *LoA* II. and *LoA* III and EC2 format and Cladera & Mari model.

5 CONCLUSION

Safety and reliability of empirical model EC2 can be checked only using results of experiments based on their statistical evaluation. Number of the new experiments carried out within last two decades have extended existing database for this purpose not only for flat slabs but also for spread footings and foundation slabs. Safety evaluation of current EC2 model has shown pretty low reliability in case of footings, with mean value deep below 1.0, and characteristic value of the model safety $P_{k,0.05} = 0.603$. It is in spite of low value of COV = 0.13. The reason is in overestimation of the strut contribution in areas where distance of control perimeter is less than $2d$ measured from the face of a column. Amendment of the strut factor by $1.2d/a$ increased EC2 model safety to target value 1.0. Comparison of measured and theoretical position of critical control perimeters a_{crit} also confirmed correctness of EC2 iterative method in case of punching assessment of the footings. Acceptable reliability provides new German model with $P_{k,0.05} = 0.952$ and with favorable value of COV = 0.138.

The highest reliability provides MC 2010 model for LoA I. with the value of $P_{k,0.05} = 1.085$. The best solution, following aspects of safety and economy, provides P7 model with $P_{k,0.05} = 0.924$. The CSCT model written in EC2 format can be well calibrated for flat slabs, but in a case of foundations seems to be unsafe with $P_{k,0.05}$ slightly below 0.80. Cladera & Mari model provides pretty good results. The precision of the model is expressed by the lowest value of COV = 0.1234 in comparison with the other models. However low mean value of $P_m = 1.034$ does not create required safety where $P_{k,0.05}$ is slightly above 0.80. The low safety of the EC2 model for assessment of punching resistance of footings without shear reinforcement is urging necessity to update this model. The easiest way is to change strut factor from value of $2d/a$ to $1.30d/a$ when level of safety can be expressed by acceptable value of 0.93.

ACKNOWLEDGEMENT

This work was supported by the Scientific Grant Agency of the Ministry of Education, science, research and sport of the Slovak Republic and the Slovak Academy of Sciences No 1/0810/16.

REFERENCES

Hegger, J., Siburg, C. & Kueres, D. (2016). Proposal for punching shear design of flat slabs and column bases based on Eurocode 2, RWTH Aachen University, Germany, Presentation from CEN TC250/SC2/WG1/TG4 meeting in Copenhagen, 28.-29. January 2016.

Labudkova, J. & Cajka, R. (2014). Comparison of Measured Deformation of the Plate in Interaction with the Subsoil and the Results of 3D Numerical Model Advanced Materials Research Vol. 1020 (2014) pp. 204–209, Trans Tech Publications, Switzerland, doi: 10.4028/www.scientific.net/AMR.1020.204.

Mari, A., Cladera, A, Bairan, J., Oller, E. & Ribas, C. (2016). Punching shear design proposal based on the CCC mechanical model, Presentation from CEN TC250/SC2/WG1/TG4 meeting in Copenhagen, 28.-29. January 2016.

Muttoni, A. & Schwartz, J. (1991). Behaviour of Beams and Punching in Slabs without Shear Reinforcement, *IABSE Colloquium*, Vol. 62, Zurich, Switzerland, 1991, pp. 703–708.

Muttoni, A. & Fernández Ruiz, M. (2008). Shear strength of members without transverse reinforcement as function of critical shear crack width, *ACI Structural Journal*, V. 105, No. 2, 2008, pp. 163–172.

Muttoni, A. & Fernández Ruiz, M. (2012). The levels of approximation approach in MC 2010: Applications to punching shear provisions, *Structural Concrete*, Ernst&Sohn, Germany, Vol. 13, No. 1, 2012, pp. 32–41.

Muttoni, A. & Ruiz, F. (2015). Proposal for punching shear provisions based on CSCT, Presentation from CEN TC250/SC2/WG1/TG4 meeting in Florence, 12.-13. March 2015.

Siburg, C., Ricker, M. & Hegger, J. (2014). Punching shear design of footings: critical review of different code provisions, *Structural Concrete*, Ernst&Sohn, Germany, Vol. 3, No. 3, 2014, pp. 1–27.

Zsutty, T. (1968). Beam shear strength prediction by analysis of existing data. *ACI Journal* 65(11): pp. 943–951.

Advances and Trends in Engineering Sciences and Technologies II – Al Ali & Platko (Eds)
© 2017 Taylor & Francis Group, ISBN 978-1-138-03224-8

Identifying the causes of failure using Bayesian networks

M. Holický
Klokner Institute, CTU in Prague, Czech Republic

ABSTRACT: During the first steps of sophisticated renovations of a road bridge over rail-way lines the bridge deck suddenly slipped off temporary supports on the railway track. Incidentally at the same time an intercity train approached the location and crashed into the collapsed bridge deck at a high speed. The accident resulted in a number of fatalities, injuries and great economic consequences. Identifying the causes of the collapse was complicated due to a number of uncertainties and additional damage to the bridge caused by the train. The background materials of the investigation were mostly limited to photographic docu-mentation and the presence of a few witnesses. Additional detailed structural analyses did not reveal any convincing evidence. Critical consideration to all possible causes including aerodynamic effects supplemented by Bayesian (causal) networks finally resulted in the iden-tification of the most significant causes including insufficient foundation and overall stiffness of temporary supports.

1   INTRODUCTION

A road bridge was built over an important railway in the 1950s. The triple-span bridge con-sisted of a skew concrete slab lying on steel girders. In 2008, the bridge was subjected to essen-tial repair. The concrete deck located outside the tracks was removed and the bridge partly pulled out of the tracks. The middle part of the deck was then demolished and re-concreted outside the tracks. During the first steps of backward relocation, the bridge suddenly slipped off temporary supports on the tracks. Incidentally at the same time an intercity train passed through the site and crashed into the collapsed bridge at a high speed (see Figure 1). The acci-

Figure 1.   Downfall of the road bridge.

dent resulted in 8 fatalities, 63 injuries and considerable economic consequences. Immediately after the accident two civil engineers were charged with public negligence.

The additional damage of the bridge due to the train collision (see Figure 1) significantly complicated forensic investigation of the collapse causes. Moreover, the remaining structural components of the collapsed bridge and temporary supports were removed shortly after the accident to renew the railway traffic. The background materials of the investigation were mostly limited to photographic documentation and few witnesses.

2 DESCRIPTION OF THE BRIDGE REPAIR

The bridge repair consisted of removal of the old concrete slab, maintenance of the steel girders and casting of the new slab. The old concrete deck located outside the railway track was removed in the original position of the bridge. Then the bridge was pulled out to one side of the railway and remaining middle part of the deck was demolished. After a necessary maintenance of the steel girders, a new concrete deck in the central part of the bridge was cast. During the first steps of backward relocation, the bridge suddenly fell down on the tracks.

Temporary trestles, traction beams and carriages used to transport the bridge were arranged as shown in Figure 2. The bridge was pulled out to the left side and repaired. Temporary supports during the repair and downfall of the bridge are also indicated in Figure 2. The right part of the bridge during the downfall was supported by two supports only. The last right support (not shown in Figure 2) was inactive due to deflection induced by the weight of the concrete deck and only the middle carriages on the traction beams were active.

3 ANALYSIS OF BASIC CAUSES

Detailed analysis of available background documents revealed that the repair procedures suffered from a number of deficiencies. General insufficiencies (lack of systematic surveying and lack of information transfer before backward traction of the bridge) and some less significant insufficiencies (e.g. deformed members of trestles, missing screw in flanges and imperfect tube joints of trestles) are not explicitly included in the following analysis. However, they are taken into account in assessing significance of individual insufficiencies.

Figure 2.   Supports of the bridge during the repair.

Figure 2 clearly indicates that just before the downfall, the most of the bridge weight was transmitted by the middle carriages and nearby trestles. This loading situation was analysed using several models and the capacity of all load bearing members were verified. Results of the analyses showed that all parts of temporary supports had sufficient resistance when ideal conditions (no inclination and correct assembly of temporary structures) were taken into account.

However, the stiffness of the temporary supports appears to be more critical. The complex system of the vertical load-bearing components seems to be relatively flexible. The foundation soil was not consolidated and foundations made of panel packing's were not provided with appropriate gravel sand bedding. The wrestles, tractions beams, crane beams and other components had limited stiffness.

Moreover, the vertical systems were not provided by appropriate bracing as required by assembly instructions for the temporary structures. The trestles were not embedded into stiff foundations and not provided by any bracing. Some bracing appeared to be carried out in transverse direction at the level of tracking beams (see Figure 3).

However, the tracking beams were not firmly connected with the trestles. Insufficient bracing documented in Figure 3 and missing anchorage significantly contributed to the lack of the spatial stiffness.

It appears that the basic technical insufficiencies of the repair can be clustered into the four essential causes considered explicitly in the following:

1. Vertical stiffness of the temporary supports.
2. Cross inclination of the bridge.
3. Horizontal stiffness of the temporary supports.
4. Aerodynamic effects due to high-speed trains.

Note that each of the first three essential causes is dependent on several basic causes (not explicitly treated in this paper). The following basic variables were included in the essential causes:

Figure 3.   Partly repaired bridge before collapse.

Figure 4.   Critical trestles after collapse.

– Vertical stiffness of the temporary supports: stiffness of foundation and vertical elements.
– Cross inclination of the bridge: vertical stiffness, asymmetry of the bridge, weight of the bridge, repair procedures, imperfections and hydraulic system.
– Horizontal stiffness of the temporary supports: anchor of the trestles, bracing of the temporary structures and bedding of the bridge girders.

4 NETWORK ANALYSIS

The Bayesian (causal) network applied in the present forensic investigation (see Figure 5) has been developed using general principles provided by Jensen Finn (1996), Stewart and Melchers (1997) and the software product GeNIe (2007). Basic principles of the Bayesian networks are also described by Holicky (2009). The network in Figure 5 contains three submodels A, B and C yielding probability distribution of the chance nodes 1, 2 and 3. Each random node has two states only (favourable and unfavourable).

Figure 5 shows the names and also resulting probability distributions (degrees of believe) of all the nodes obtained in the direct analysis. In case of the initial node 4 the probability distributions represent assumed initial probabilities (specified by expert judgements based on information provided by detailed analyses).

Results of the inverse analysis given the collapse of the bridge occurred (evidence of the state "Collapsed" of the node 5 is introduced) is indicated in Figure 6. It appears that the Horizontal Stiffness of the temporary supports was the most significant cause of the bridge downfall. The Cross Inclination also seems to significantly contribute to the accident.

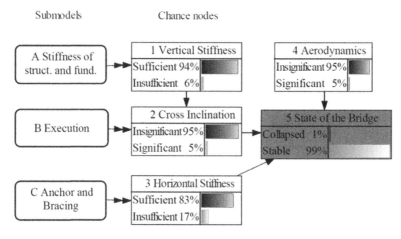

Figure 5.   The Bayesian (causal) network showing results of the forward analysis.

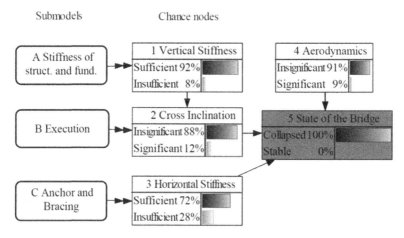

Figure 6.   The Bayesian (causal) network showing results of the inverse analysis given the collapse of the bridge occurred.

Table 1.   Results of the inverse analysis.

Node, essential cause	Expected % of unacceptable state	Relative weight of unacceptable states %	Rank of the basic causes
3 Horizontal Stiffness	28	49	1
2 Cross Inclination	12	21	2
4 Aerodynamics	9	16	3
1 Vertical stiffness	8	14	4
Sum	57	100	

The results of the inverse analysis shown in Figure 6 are reviewed in Table 1. The relative weights of unacceptable states of the essential causes (nodes 1 to 4) clearly indicate significance of the causes contributing to the bridge collapse.

It follows from Table 1 that the essential causes Horizontal Stiffness of the temporary supports and Cross Inclination are the most significant causes for the bridge downfall. Less significant seem to be the Aerodynamic effects and Vertical Stiffness.

5 CONCLUSIONS

During the repair of the bridge a number of insufficiencies occurred. In general, the process of bridge repair suffered of two wide-ranging insufficiencies:

− Lack of systematic surveying.
− Lack of information transfer.

Detail investigation did not reveal any prevailing cause of the collapse. All the potential insufficiencies may be clustered into the four essential causes and their significance may be assessed using the causal network:

− Horizontal stiffness of the temporary structures (49%).
− Cross inclination (21%).
− Aerodynamic (16%).
− Vertical stiffness of the temporary structures (14%).

It appears that the causal networks may provide an effective tool for forensic investigations when analysis of possible causes is partly based on expert's judgment.

ACKNOWLEDGEMENT

This research has been conducted at the Klokner Institute of the Czech Technical University in Prague, Czech Republic, as a part of the research project GAČR 16-11378S, "Risk based decision making in construction" supported by the Czech Grant Agency.

REFERENCES

GeNIe (2007), *Home page: http://genie.sis.pitt.edu, e-mail: genie@mail.sis.pitt.edu*.
Holicky, M. (2009), Reliability analysis for structural design, SUN MeDIA Stellenbosch.
Jensen Finn, V., Introduction to Bayesian networks. Aalborg University, Denmark, 1996.
Stewart, M.G. and Melchers, R.E., "Probabilistic risk assessment of engineering systems." Chapman & Hall, London, 1997.

Advances and Trends in Engineering Sciences and Technologies II – Al Ali & Platko (Eds)
© 2017 Taylor & Francis Group, ISBN 978-1-138-03224-8

Risk-based target reliability

M. Holický
Klokner Institute, CTU in Prague, Czech Republic

ABSTRACT: Codes of practice aim at assuring structures having the risks acceptable to the public and the minimum total costs over a design working life. However, current codified criteria for structural design correspond to a broad range of reliability levels, specified for dissimilar reference periods even though their recalculation for different periods is uncertain due to an unknown dependence of failure events in time. The submitted approach to specifying target reliability levels is based on probabilistic risk optimization considering the objective function as a sum of various costs including the effects of time to failure and discounting. An example of probabilistic optimization of a generic structural member shows the effect of considered input parameters. It appears that the optimum reliability level expressed by reliability index depends primarily on the structural costs, failure costs and costs for improving structural safety. Less significantly it seems to be affected by the time to failure and the discount rate.

1   INTRODUCTION

The target reliability levels in various national and international documents for new structures are inconsistent in terms of the recommended values and the criteria according to which the appropriate values are to be selected. In EN 1990 (2002), the recommended values of reliability indexes are given for two reference periods, 1 year and 50 years (see Table 1), without any explicit link to the design working life that may differ from the reference period. No specific values are given for temporary structures.

It should be emphasized that the reference period is understood as a chosen period of time used as a basis for statistical assessment of the time variant basic random variables, and the corresponding probability of failure. The reference period may therefore be different from the design working life. Confusion is often caused when the difference between these two concepts is not noticed.

It should be recognized that the couple of β values (for 1 year and 50 years) given in Table 1 for each reliability class correspond to the same reliability level. Practical application of the concept of reference period depends on the time period considered in the verification, which may be linked to available probabilistic information concerning time variant basic variables (imposed load, wind, earthquake, etc.). It should be noted that the reference period

Table 1.   Reliability classification in accordance with EN 1990 (2002).

Reliability classes	Consequences of structural failure	Reliability index β for reference period		Examples of buildings and civil engineering works
		1 year	50 years	
RC3—high	High	5.2	4.3	Bridges, public buildings
RC2—normal	Medium	4.7	3.8	Residences and offices
RC1—low	Low	4.2	3.3	Agricultural buildings

Table 2. Examples of life-time target reliability indexes β in accordance with ISO 2394 (1998).

Relative costs of safety measures	Consequences of failure			
	Small	Some	Moderate	Great
High	0	1.5	2.3	3.1
Moderate	1.3	2.3	3.1	3.8
Low	2.3	3.1	3.8	4.3

Table 3. Tentative target reliability indexes β (and associated target failure rates) related to one year reference period and ultimate limit states in accordance with JCSS (2001).

Relative costs of safety measures	Minor consequences of failure	Moderate consequences of failure	Large consequences of failure
Large	$\beta = 3.1 \ (p \approx 10^{-3})$	$\beta = 3.3 \ (p \approx 5 \times 10^{-4})$	$\beta = 3.7 \ (p \approx 10^{-4})$
Normal	$\beta = 3.7 \ (p \approx 10^{-4})$	$\beta = 4.2 \ (p \approx 10^{-5})$	$\beta = 4.4 \ (p \approx 5 \times 10^{-6})$
Small	$\beta = 4.2 \ (p \approx 10^{-5})$	$\beta = 4.4 \ (p \approx 5 \times 10^{-6})$	$\beta = 4.7 \ (p \approx 10^{-6})$

of 50 years is also accepted as the design working life for common structures (see the discussion by Diamantidis (2009)).

For example, considering a structure of reliability class 2 having a design working life of 50 years, the reliability index $\beta = 3.8$ should be used, provided that probabilistic models of basic variables are available for this period. The same reliability level is achieved when a reference period of 1 year, and a target of $\beta = 4.7$ are applied using the theoretical models for this reference period. Thus, when designing a structural member, similar dimensions (reinforcement area) would be obtained considering $\beta = 4.7$ and basic variables related to 1 year or $\beta = 3.8$ and basic variables related to 50-year reference period.

A more detailed recommendation concerning the target reliability is provided by ISO 2394 (1998), where the target reliability indexes are indicated for the whole design working life without any restriction concerning its length, and are related not only to the consequences, but also to the relative costs of safety measures as shown in Table 2.

Similar recommendations are provided in the JCSS (2001) Probabilistic Model Code (Table 3) based on the previous study of Rackwitz (2000). The consequence classes in JCSS (2001) (similar to EN 1990, 2002) are linked to the ratio ρ defined as the ratio $(C_{str} + C_f)/C_{str}$ of the total cost induced by a failure (cost of construction C_{str} plus direct failure costs C_f) to the construction cost C_{str}.

However, it is not clear what is meant in JCSS (2001) by "the direct failure costs". This term indicates that there may be some other "indirect costs" that affect the total expected cost. Here it is assumed that the failure costs C_f cover all additional direct and indirect costs caused by the failure. The structural costs are considered separately and related to the costs needed for an improvement of safety given by the costs C_1 per unit of a decision parameter.

The submitted theoretical study based on probabilistic optimization is supplemented by practical recommendations. This contribution is an extension of the previous study by Holicky and Retief (2011).

2 PROBABILISTIC OPTIMIZATION

Probabilistic optimization is based on a fundamental form of the objective function (not covering monitoring and maintenance) expressed as the present value of the total expected cost $C_{tot}(x,q,n)$

$$C_{tot}(x,q,n) = C_{str} \sum_i P_f(x,i) + C_f \sum_i P_f(x,i)Q(q,i) + C_0 + xC_1. \qquad (1)$$

Here x denotes the decision parameter of the optimization (a parameter of structural resistance), q is the annual discount rate (e.g. 0.03, an average long run value of the real annual discount rate in European countries), n is the number of years to the failure, which may differ from the design working life (specified usually as 50 or 100 years).

Further, $P_f(x,i)$ is the failure probability in year i, $Q(q,i)$ is the discount factor dependent on the annual discount rate q and the year number i, C_0 is the initial cost independent of the decision parameter x and failure (a quantity not affecting the optimization), and C_1 is the cost per unit of the decision parameter x (a structural parameter quantity affecting the structural resistance and optimization).

Note that the design working life may generally differ from the time to failure denoted by the number of years n and considered here as an independent variable affecting the probability of failure. Maintenance and possible repair of the structure not included in the objective function (1), and these aspects are to be considered in further studies. The cost of construction C_{str} is not discounted as it is paid at present.

Assuming independent failure events in subsequent years, the annual probability of failure $P_f(x,i)$ in year i may be approximated by the geometric sequence

$$P_f(x,i) = p(x)(1 - p(x))^{i-1}. \tag{2}$$

The initial annual probability of failure $p(x)$ is dependent on the decision parameter x. Note that annual failure probabilities can be assumed to be independent when failure probabilities are chiefly influenced by time-variant loads (climatic actions, traffic loads, accidental loads). Then the failure probability $P_{fn}(x)$ during n years can be estimated by the sum of the sequence $P_f(x,i)$, that can be expressed as

$$P_{fn}(x,n) = 1 - (1 - p(x))^n \approx n\, p(x). \tag{3}$$

Note that the approximation indicated in equation (3) is fully acceptable for small annual probabilities $p(x) < 10^{-3}$.

The discount factor of the present value of the expected future costs in year i is considered in the usual form as

$$Q(q,i) = 1/(1+q)^i. \tag{4}$$

Thus, the cost of malfunctioning C_f is discounted by the factor $Q(q,i)$ depending on the discount rate q and the point in time (year number defined as i), when the loss of structural utility occurs.

Considering equations (2) and (4), the total costs $C_{tot}(x,q,n)$ described by equation (1) may be written in a simplified form as

$$C_{tot}(x,q,n) = C_{str}np(x) + C_f p(x)\, PQ(x,q,n) + C_0 + x\, C_1. \tag{5}$$

Here the total sum of expected malfunction costs during the period of n years is dependent on the product of the present value of malfunction cost C_f, the annual probability $p(x)$ and a sum of the geometric sequence having the quotient $(1- p(x))/(1 + q)$, denoted as the time factor $PQ(x,q,n)$:

$$PQ(x,q,n) = \frac{1 - \left[\dfrac{1 - p(x)}{1+q}\right]^n}{1 - \left[\dfrac{1 - p(x)}{1+q}\right]}. \tag{6}$$

In general, the total cost $C_{tot}(x,q,n)$ depends on the costs C_0, C_1, C_f, the annual probability of failure $p(x)$, the discount rate q, and the design working life n.

The necessary condition for the minimum of the total cost follows from (1) as

$$\frac{\partial C_{tot}(x,q,n)}{\partial x} = C_{str}\sum_{i=1}^{n}\left[\frac{\partial P_f(x,i)}{\partial x}\right]_{x=x_{opt}} + C_f\sum_{i=1}^{n}Q(q,i)\left[\frac{\partial P_f(x,i)}{\partial x}\right]_{x=x_{opt}} + C_1 = 0. \qquad (7)$$

Equation (7) represents a general form of the necessary condition for the minimum of total cost $C_{tot}(x,q,n)$, the optimum value x_{opt} of the parameter x, and the optimum annual probability of failure $p_{opt} = p(x_{opt})$. The condition (7) can be simplified as

$$\sum_{i=1}^{n}Q(q,i)\left[\frac{\partial P_f(x,i)}{\partial x}\right]_{x=x_{opt}} + (\rho-1)\sum_{i=1}^{n}Q(q,i)\left[\frac{\partial P_f(x,i)}{\partial x}\right]_{x=x_{opt}} = -\frac{C_1}{C_f}. \qquad (8)$$

The optimum probability $P_{fn,opt}$ for the total design working life $T_d = n$ years follows from p_{opt} and equation (3) as

$$P_{fn,opt} = 1 - (1 - p_{opt})^n \approx n\, p_{opt}. \qquad (9)$$

The corresponding optimum reliability index $\beta_{opt} = -\Phi^{-1}(P_{fn,opt})$. Those quantities are in general dependent on the cost ratio C_f/C_1, the discount rate q, and the design working life n.

3 FAILURE PROBABILITY OF A GENERIC STRUCTURAL MEMBER

Consider a generic structural member described by the limit state function $Z(x)$ as

$$Z(x) = xf - (G+Q). \qquad (10)$$

Here x denotes a deterministic structural parameter (e.g. the cross-section area), f the strength of the material, G the load effect due to permanent load and Q the load effect due to variable load. Theoretical models of the random quantities f, G and Q considered in the following example are given in Table 4 (adopted from JCSS (2001) and Holicky (2009)).

Considering the theoretical models given in Table 4, the reliability margin $Z(x)$ may be well approximated by the normal distribution $\Phi_{Z(x)}$ that provides sufficient accuracy. The annual failure probability $p(x)$ is then given as

$$p(x) = \Phi_{Z(x)}(Z(x) = 0). \qquad (11)$$

In equation (11) the normal distribution is evaluated for $Z(x) = 0$; then for $x = 1$ and $n = 50$ the probability $P_{fn}(1,50) \approx 6.7\ 10^{-5}$ and corresponding $\beta \approx 3.8$.

4 THE OPTIMUM RELIABILITY LEVEL

The following example illustrates the general principles, as well as a special case of probabilistic optimization. To simplify the analysis, the total costs $C_{tot}(x,q,n)$ given by equation (5) are transformed to the standardized form $\kappa_{tot}(x,q,n)$ given as

$$\kappa_{tot}(x,q,n) = \frac{C_{tot}(x,q,n) - C_0}{C_1} = p(x)C_{str}/C_1[n+(\rho-1)PQ(x,q,n)] + x. \qquad (12)$$

Here ρ denotes the cost ratio $(C_{str} + C_f)/C_{str}$ of the sum of structural cost and failure costs $(C_{str} + C_f)$ to structural costs C_{str}.

Table 4. Theoretical models of the random variables f, G and Q (annual extremes).

Variables	Distribution	Mean	Standard deviation	Coefficient of variation
f	Lognormal	100	10	0.10
G	Normal	35	3.5	0.10
Q	Gumbel	10	5	0.50

The annual probability of failure $p(x)$ considered here for a general structural member is given by equation (11). However, the following procedure may be applied for any relevant dependence of the failure probability $p(x)$ expressed as a function of a suitable structural parameter x.

In the example illustrated in Figure 1, it is assumed that the discount rate is $q = 0.03$, and the year number when the failure occurs $n = 50$. Under these assumptions, Figure 1 shows the variation of the total standardized costs $\kappa_{tot}(x,q,n)$ (given by equation (12)), and the optimum reliability index β_{opt}, with structural parameter x. The optimal values $x_{opt}(q,n)$ of the structural parameter x, given by equation (8), are shown by the dashed vertical lines. The indicated values of $\beta = 2.6, 3.2, 3.5$ correspond to annual rates 3,7, 4.2 and 4.4 recommended by JCSS (2001) (Table 3) for "normal relative costs of safety measures".

Note that the cost ratio $\rho = (C_{str} + C_f)/C_{str} = 1$ describes the extreme case when the failure costs C_f are negligible, approximately $C_f \approx 0$, and the failure consequences are confined to the structural costs C_{str}. Assuming this cost ratio $\rho = 1$, the discount rate $q = 0.03$, time to failure $n = 50$ and the ratio $C_{str}/C_1 = 100$ it follows from Figure 1 that the optimum design parameter x is about 0.93 and the optimum reliability index $\beta_{opt} \approx 3.3$; if $\rho = 100$, then $x \approx 1.09$ and $\beta_{opt} \approx 4.3$.

Thus, the optimum reliability index β_{opt} corresponds to the minimum cost $\kappa_{tot}(x,q,n)$ given by equation (12). Variation of β_{opt} with the cost ratio ρ for the discount rate $q = 0.03$, the number of years $n = 50$ and the selected ratios $C_{str}/C_1 = 10, 100, 1000$ is shown in Figure 2.

It follows from Figure 2 that the optimum reliability index depends strongly on the cost ratio ρ, particularly for relatively small ratios C_{str}/C_1 (about 10). For $q = 0.03$, $n = 50$ and $\rho = 100$ the optimum reliability index is about $\beta_{opt} = 3.5$.

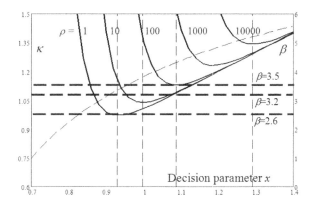

Figure 1.   Variation of the total standardized cost $\kappa_{tot}(x,q,n)$ and the optimum reliability index β_{opt} with the decision parameter x for $q = 0.03$, $n = 50$, $C_{str}/C_1 = 100$, and selected cost ratios ρ.

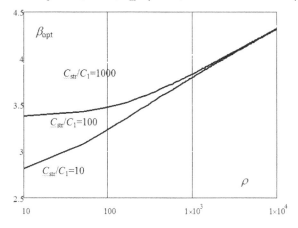

Figure 2.   Variation of the optimum reliability index β_{opt} $(q,n,C_{str}/C_1,\rho)$ with the cost ratio $\rho = (C_{str} + C_f)/C_{str}$ for $q = 0.03$, $n = 50$, and selected ratios $C_{str}/C_1 = 10, 100, 1000$.

The additional analysis of the minimum total standardized cost $\kappa_{tot}(x,q,n)$ given by equation (12) indicates that the optimum reliability index β_{opt} is almost independent of the time to failure n and the discount rate q affecting the failure costs C_f. It appears that the optimum reliability index β_{opt} slightly decreases with increasing n and q. Then, a conservative approximation of β_{opt} can be estimated considering a limited number of years n (say 10) and a small discount rate q (for example 0.01).

5 CONCLUSIONS AND RECOMMENDATIONS

The available documents and codes for structural design do not provide a clear link between the design working life and the target reliability level. As a rule, no recommendations are offered for the specification of target reliability index for temporary or long life structures. Current codified criteria for structural design correspond to a broad range of reliability levels, specified for dissimilar reference periods, even though their recalculation for different periods is uncertain due to unknown mutual dependence of failure events in time.

It appears that the target reliability of structures can be derived using the risk-based probabilistic optimization, considering the objective function as a sum of the following costs: the initial costs C_0, the marginal costs $x \, C_1$ (where x denotes the decision parameter and C_1 the incremental cost of decision parameter x), and the failure consequences consisting of the construction costs C_{str} and failure costs C_f (the loss of structural utility at the time of failure). The failure consequences, the costs C_{str} and C_f, are taken into account by the cost ratio $\rho = (C_{str} + C_f)/C_{str}$. The failure cost C_f is discounted considering an annual discount rate q and the time to failure (number of years) n. In such a way the total cost is also affected by the discount rate q, and the number of years n.

An example of the probabilistic optimization of a generic structural member clearly shows (see Figure 1, and 2) that the optimum reliability level, i.e. the optimum reliability index β, depends primarily on:

– The structural costs C_{str},
– The failure costs (malfunctioning costs) C_f,
– The costs for improving structural safety C_1.

The discount rate q and the time to failure n seem to be less significant. A conservative approximation, the target β, can be determined considering a limited number of years n (for example 10 years) and a small discount rate q (for example 0.01).

ACKNOWLEDGEMENT

This study is an outcome of the research project GACR 16-11378S, "Risk-Based Decision-Making in Construction" supported by the Czech Science Foundation.

REFERENCES

Diamantidis D. (2009). "Reliability differentiation", In.: Holický et al.: Guidebook 1, Load effects on Buildings, CTU in Prague, Klokner Institute, ISBN 978-80-01-04468-1, pp. 48–61, 2009.

Holický M. (2009). "Reliability analysis for structural design", SUN MeDIA Stellenbosch, ZA, ISBN 978-1-920338-11-4, 199 pages.

Holický M. and Schneider J. (2001). "Structural Design and Reliability Benchmark Study", In.: Safety, Risk and Reliability—Trends in Engineering c/o IABSE, ETH Zürich, International Conference in Malta, ISBN 3-85748-102-4, pp. 929–938, 2001.

Holický M. and Retief J. (2011). Theoretical Basis of the Target Reliability. In: 9th International Probabilistic Workshop. Braunschweig: Technische Universität, pp. 91–101. ISBN 978-3-89288-201-5.

ISO 2394. (1998). General principles on reliability for structures, International Organization for Standardization, Geneva, Switzerland, 73 pages.

JCSS. (2001). Joint Committee for Structural Safety. "Probabilistic Model Code", http://www.jcss.ethz.ch/.

Rackwitz R., *Optimization—the basis of code-making and reliability verification*. Structural Safety, 2000. 22(1): pp. 27–60.

Advances and Trends in Engineering Sciences and Technologies II – Al Ali & Platko (Eds)
© 2017 Taylor & Francis Group, ISBN 978-1-138-03224-8

Experimental analysis of a composite steel-concrete slab

A. Hrušovská & Š. Gramblička
Slovak University of Technology, Bratislava, Slovakia

ABSTRACT: The biggest advantage of composite steel-concrete slabs lies not in the usage of steel sheet as a permanent shuttering, but in the constitution of the bar reinforcement designed for a positive bending moment, and the time and money savings connected with it. The calculation of longitudinal shear resistance of composite slabs is the most complicated part of design, due to its determination only by testing on large scale samples. The design of composite slabs is regulated in Eurocode 4 but does not prescribe the design of slabs with plain sheets. The research of smooth sheets, their interaction with concrete, and a combination with end anchorages are necessary due to the worldwide availability of these kinds of steel sheets. In this article we are going to experimentally verify the design of such sheets on six large scale samples by measuring of deflection and slip of concrete against sheet.

1   THE COMPOSITE STEEL-CONCRETE SLABS

The biggest advantage of composite slabs lies in the time saving—due to the absence of additional shuttering—and the more frequent requirement of designers to use steel sheets as a tension reinforcement for positive bending—which can result in monetary savings, too.

The resistance of the composite slab in the direction of longitudinal shear is influenced by many factors and the connection between concrete and steel is determined by three factors: physical-chemical adhesion, friction, and mechanical interlocking. The latter two factors are activated after the appearance of the first micro-cracks. When the slip between steel and concrete is completely prevented, we can speak about a complete interaction, however most shear connectors need to undergo some deformation before they can supply any force. In this case, the interaction is incomplete (partial), which is the most common class of the composite slab stiffness.

A composite action may be ensured by variable types of end anchorage—a local connection between the concrete and the steel sheet (e.g. welded studs; perforated strips; block connectors; angle irons; different types of anchors etc.) or by shaping of the sheet ribs and flanges. According to the connection strength, we can determine whether it is a full or partial shear connection.

According to deformation capacity, flexible and rigid connections can be divided into ductile and brittle (non-ductile) connections. The composite slabs with open-shaped sheets tend to be brittle, while those with re-entrant shaped sheets tend to be ductile. Ductile behaviour can be achieved via mechanical interlock provided by deformations in sheets, embossments, indentations, or by adding connectors.

2   THE USAGE OF PLAIN OPEN-SHAPED SHEET AND END ANCHORAGES IN COMPOSITE STEEL-CONCRETE SLABS

Eurocode 4, chapter 9.7.4 describes the design of the composite slab with end anchorages. The issue of longitudinal shear is defined as follows: Unless a contribution to longitudinal shear resistance by other shear devices is shown by testing, the end anchorage should be designed for the tensile force in the steel sheet at the ultimate limit state. This applies only to slabs connected by adjustments on flanges or ribs of the sheets, or by the re-entrant shape on the ribs. The longitudinal shear resistance can also be calculated by a partial shear connection method—when the shear resistance of the sheet is enlarged by the shear resistance of connectors. In the case of sheets with smooth and open-shaped ribs, adhesion as a part of composite action must be ignored, so that the whole longitudinal shear force must be supplied by shear end anchorage connectors.

2.1 The design of composite slab containing of plain open-shaped sheet and end anchorages

For the design example, a smooth open-shaped steel sheet **RUUKKI** T55-53L-976 with a length of 2,1 m and the **HILTI X HVB 95** end anchorages have been chosen. The thickness of the concrete layer is 110 mm due to standard design recommendations. The deflection and bending moment of the profiled steel sheet in the construction phase were verified. It is necessary to set the number of anchorages before the analysis for the ultimate state and service limit phase is taken. If we would like to design a slab for full shear connectivity, the number of connectors has to be based on full ultimate limit state tensile force:

$$N_p = A_{eff} \times f_{yd} \cong 270 \ kN \tag{1}$$

where A_{eff} = effective area of profiled steel sheeting; f_{yd} = design value of the yield strength of profiled steel sheeting.

The resistance of one anchor is influenced by its direction relative to a rib and by the initial resistance given by the manufacturer. The required number of anchors will be calculated as follows:

$$n_t = Np/P_{Rd} = 270 \ kN/28 \ kN \cong 10 \tag{2}$$

where P_{Rd} = design value of the shear resistance of a single connector.

These 10 anchors need to be placed in each side of a slab with a width of 1 m, which means into four ribs. The minimum distance, which is 40 mm from the anchor edge to the medium rib, is a limiting condition for our set up (Figure 1). It means that only 4 anchorages can be fitted on each side of the slab containing the selected steel sheet. This solution shown is sufficient and the designed slab can supply design load 14 kN/m² at 100% utilization in the longitudinal shear.

2.2 The experimental analysis of composite slab consisting of smooth open-shaped steel sheet and end anchorages

In the experimental program were verified five composite slabs, three in the first phase and two in the second phase. The smooth open-shaped **RUUKKI** T55-53L-976 steel sheet with total length of 2.2 m (2.1 theoretical span), four **HILTI X HVB 95** end anchorages on each side of the slab and the C25/30 class of concrete were used. One sample was concreted without end anchorages. The test was carried out as a four-point bending test.

The loading assembly looked as follows (Figure 2, Figure 3): the steel supports were placed first (1) with the position defined mainly through holes for anchoring rods in the floor (2). Two steel plates in the first phase and three in the second phase were welded per side (3).

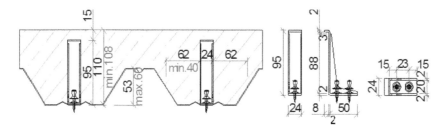

Figure 1.   The installation instruction.

112

Figure 2.   The tested composite slab.

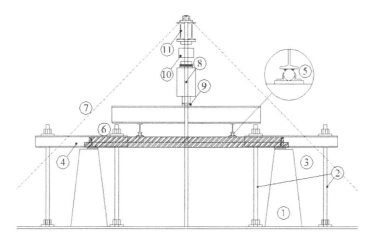

Figure 3.   The loading assembly of the tests.

One steel plate had the dimensions 120 × 1600 × 10 mm. Their position was secured by profiles anchored to the floor (4). The steel sheets were laid on plates and Hilti anchors were nailed with two nails at each anchor. At least 28 days after concrete casting, 2 pieces of plates 80 × 1000 × 10 mm were placed in the gypsum bed with 2 pieces of round bars, Ø30 mm with 1 m length, the location of which was secured by L profiles (5). 2 pieces of IPN100 of 1 m length (6) and 1 piece of IPE200 of length 1.6 m (7) created the loading assembly. The load was transmitted from a hydraulic press with a capacity of 100 kN (8) to a ball-joint (9) and loading beams through two liner loads. The loading level was measured by a load cell (10) placed above the hydraulic press and secured with press anchored bars and stabilizing ropes (11).

The following values were measured on each composite slab: deflection on the bottom edge of the slab in two ribs, deflection on the upper edge in the centre of the slab, the slip of the steel sheet and the concrete on both sides of the slab, and the displacement of supports. The measured values were evaluated graphically with the significant ones being part of this article.

2.3   Measured deflection of composite slabs

Figure 4 describes the comparison between calculated Design Ultimate Load (DUL) and load-deflection curvature, where S1, S2, and S3 label 3 slab samples and the suffixes, B or U, are used for the bottom or upper edge measurements. These 3 slabs were loaded continuously

113

until collapse. All of these 3 samples S1, S2, and S3 were concreted in 20.11.2015. The first phase was tested as follows: 16.12. S1 (27 days), 21.12.S2 (32 days) and 22.12.2015 S3 (33 days).

Figure 5 describe the comparison between calculated Design Ultimate Load (DUL) and load-deflection curvature, where S4 and S5 label two slab samples. These samples were loaded continuously until reaching 32 kN in S4 and 28 kN in the S5 sample. The pressure was then released and, after 5 minutes of rest, the slabs were loaded again continuously until collapse. The behaviour of deflection during loading and deloading is also pictured in Figure 5. All of these three samples, S4, S5, and S6, were concreted in 26.01.2016. The second phase was tested as follows: 29.02. S6 (35 days), 02.03. S4 (37 days) and 03.03.2016 S5 (38 days). The slab sample, S6, was concreted without end anchorages. Until reaching the maximum adhesive force of this slab sample, the deflection rose. After a detach of the concrete layer, it could not be possible to reach higher load and so the slab collapsed.

The deflection capacity of tested slabs with anchorages is markedly greater than the calculated one. The deformation capacity of S1—S5 can be classified as ductile. This is caused by ductile connectors and the behaviour of these slabs resembles re-entrant slabs. The S6 slab sample showed adhesive forces of concrete-steel connection equal to approximately 30 kN.

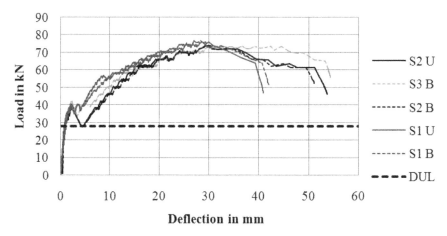

Figure 4. The comparison between calculated design ultimate load ad measured load-deflection curvatures on samples S1, S2, S3.

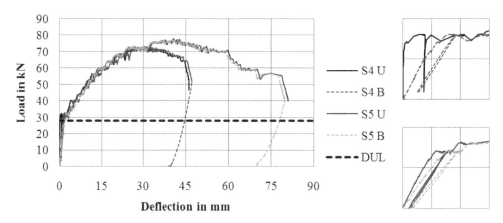

Figure 5. The comparison between calculated design ultimate load and measured load-deflection curvatures on samples S4, S5.

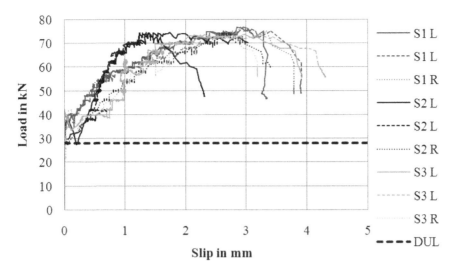

Figure 6. The comparison between calculated design ultimate load and measured load-slip curvatures on samples S1, S2, S3.

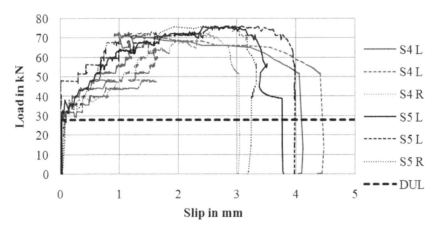

Figure 7. comparison between calculated design ultimate load and measured load-slip curvatures on samples S4, S5.

2.3.1 *Measured slip of concrete against steel sheet*

Figure 6 describes the comparison between calculated Design Ultimate Load (DUL) and load-slip curvature of concrete layer against steel sheet, where the suffixes, L or R, are used for left or right slab side measurements. In fact, zero slip was measured until reaching at least 30 kN loading level in case of S1, S2, S3 samples.

Figure 7 shows the S4 and S5 samples and their slip. In the S4 sample, the load release was realized after initial slip appearance and persisted after release. In the S5 sample, the load release was realized before initial slip appearance and persisted equalled to zero until another loading.

3 CONCLUSION AND SUMMARY

The composite slab cannot be designed without any anchorages, although the adhesion between the steel and the concrete is, surprisingly, high enough to supply the designed load.

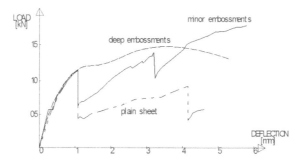

Figure 8. Load-deformation curvature for composite slabs.

Figure 9. First crack appearance and failure of composite slab.

Higher levels of cyclic load or bad surface conditions during the casting may cause adhesion disappearance, and added connectors are necessary. The load-deformation curvatures of composite slabs with profiled steel sheets with minor embossments are similar to those measured experimentally (Stark, 1990) (Figure 8). The process of composite slab collapse started with first crack appearance and separation of both longitudinal edges, which was caused by a thin layer of concrete on both sides (Figure 9).

The experimental analysis showed satisfying result verification with sufficient reserve on the safe side. The partial shear connection method is an important option for the economic use of composite slabs in the building industry, especially when it is practically impossible to place the required numbers of connectors to ensure full shear connection.

ACKNOWLEDGEMENT

This contribution was prepared with the financial support of Slovak Grant Agency VEGA 1/0696/14 and STU Grant scheme for Support of Young Researchers.

REFERENCES

EN 1994-1-1 Design of composite steel and concrete structures, Part 1–1 General rules and rules for buildings, 2004.
Stark, J.W.B. 1990. General Methods of Design of Composite Construction. *IABSE Short Course.* Volume 61: 7–38.

Advances and Trends in Engineering Sciences and Technologies II – Al Ali & Platko (Eds)
© 2017 Taylor & Francis Group, ISBN 978-1-138-03224-8

Reduction of computational operations in the DOProC method

P. Janas, M. Krejsa & V. Krejsa
Department of Structural Mechanics, Faculty of Civil Engineering, VSB-Technical University of Ostrava, Ostrava-Poruba, Czech Republic

ABSTRACT: The application of probabilistic methods for reliability assessment with probabilistic description of random variables in practice is becoming increasingly popular. The new computational methods are still being developed for the calculation of the resulting probability of failure. The paper deals with a new method currently under development: Direct Optimized Probabilistic Calculation (DOProC), which presents a new computational tool for solving probabilistic problems. The method is based on the fundamental principles of probability theory and statistics. Statistically dependent and independent input random quantities (such as load, geometry, material properties, or imperfections) are expressed in the DOProC method by the parametric or empirical distribution prepared in the form of histograms. Calculation of the probability of failure is performed using a specially designed algorithm optimized with numerical integration, which is accurate and efficient for many kinds of probabilistic tasks. This procedure has been programmed in several applications.

1 INTRODUCTION

The load-carrying system of each structure should fulfill several conditions that represent reliability criteria in the assessment procedure. The criteria which govern the design of a structure are either serviceability (criteria which define whether the structure is able to adequately fulfill its function) or strength (criteria which define whether a structure is able to safely support and resist its design loads). A structural engineer designs a structure to have sufficient strength and stiffness to meet these criteria (Cajka et al. 2014, Janulikova & Mynarcik 2014, Vavrusova 2014, Vican et al. 2015).

The structure must be designed so that the resistance of the structure R (e.g. ultimate strength, ultimate strain or resistance to inner force) is higher than the load effect E (stress, strain or inner force in cross-section under assessment). The design process, assessment of the reliability, as well as individual phases of production, assembly or operation of the structure are now affected by many uncertainties that influence reliability of such facilities, and by its random nature, which cannot be neglected (Kotes & Vican 2013). Taking into account all randomness in loads, manufacturing and assembly imperfections and environment properties in which the designed structure performs its operation; resistance R and load effect E are to be considered as random variables. In the case of a probabilistic assessment of the structure, the function of the random variables under analysis is for example defined as:

$$Z = R - E, \tag{1}$$

where Z is the so-called safety margin or a reliability function.

The common notation of the estimated failure probability p_f, evaluated from the criterion of reliability, is defined as:

$$p_f = P(R < E) = \int_{D_f} f(X_1, X_2 \ldots X_n)\, \mathrm{d}X_1, \mathrm{d}X_2 \ldots \mathrm{d}X_n, \tag{2}$$

where D_f is the failure area of the safety margin $Z(X) < 0$, and $f(X_1, X_2 \ldots X_n)$ is the function of joint probability density of random variables $X = X_1, X_2 \ldots X_n$ (Krejsa et al. 2016a). The determination of failure probability p_f, based on the explicit calculation of the integral (2), is generally unmanageable. For solutions of the integral (2) have been developed and is still under development a series of probabilistic methods (simulation—Kmet et al. 2011, or approximation). The proposed method: Direct Optimized Probabilistic Calculation—DOProC, solves the integral (2) in a pure numerical way that is based on basis of probability theory and does not require any simulation or approximation technique. This is a highly effective way of probabilistic calculation in terms of computation time and accuracy of the solution. The novelty of the proposed method lies in an optimized numerical integration. The theoretical background was published in summary e.g. in (Janas et al. 2015a).

2 ESSENTIALS OF THE DOProC METHOD

Similar to many other probabilistic methods, the non-parametric (empirical) distribution of input random quantities in DOProC, such as the load (Jendzelovky & Balaz 2016, Kotrasova et al. 2014, Kralik & Kralik 2014, Lokaj 2015, Salajka et al. 2014), geometry, material properties, or imperfections, can be expressed by means of bounded histograms with empirical probability distribution. It is also possible to use parametric distributions, typically based on observations, often of long-term data (Cajka et al. 2016, Donova & Zdrazilova 2014, Lamich et al. 2016, Yilmaz et al. 2014). In probabilistic tasks input random variables are often statistically dependent—for example cross-section properties, strength and stiffness characteristics of the materials (Major et al. 2014). In calculations, carried out by the DOProC method, statistically dependent input random variables can be expressed by the so-called multidimensional histograms (Janas & Krejsa 2012, Janas et al. 2015b).

The total number of input random variables n and the number of intervals N in histograms of each input random variables are the most decisive factor for the number of necessary arithmetic operations and necessary calculation time. It also significantly affects the accuracy of probabilistic calculation. If there are too many random quantities, the tasks require too much time even if advanced computational facilities are available. Therefore, efforts have been made to optimize calculations in order to reduce the number of operations, while maintaining reliable calculation results:

– The *grouping of input and output variables*. This procedure can be used e.g. in situations where the random variable input or output variables can be expressed using one joint histogram. This leads to a large reduction of computational operations.
– *Parallelization*. The calculation algorithm of the DOProC method is advantageous for use in machines with two or more CPUs or cores. The basic computational algorithm of DOProC can be divided by the number of computational operations up to as many parts as there are available execution units. Combination of partial results serve as an input into the preparation of histogram of resulting variable, e.g. histogram of safety margin Z.
– *Interval optimizing*. The purpose of this computational procedure is to reduce the intervals of each variables involved in the calculation. Since there are input random variables that have lower effect on the outcome of the probabilistic calculation. There is different sensitivity (e.g. Kala 2016). The number of classes can be reduced since there are input variables that less affect the outcoming probability. The custom probabilistic calculation is then carried out with the minimum number of intervals for each input random variable.
– *Zone optimizing*. The *intervals* of each individual histogram are clearly defined during the calculation using one to three types of zones, depending on influence on resulting probability of failure (contribute to failure always, may or may not contribute, contribute never). The calculation then will be limited only to intervals of input random variables, which clearly do not contribute to the resulting value of failure probability.
– *Trend optimizing*. This optimization of probabilistic calculation follows the zonal optimization. It determines the trends of changes in the histograms of input variables when defining the individual zones.

Such procedures can be combined, thereby achieving an even stronger acceleration of the calculation. The algorithm of the DOProC method has been implemented in several software applications, and has been used several times in probabilistic tasks and probabilistic reliability assessments. For the application of the DOProC method, the ProbCalc software package, in which it is relatively easy to implement analytical and numerical transformation probabilistic model of solved tasks, can be used (Krejsa et al. 2014 a, b). The methodology for probabilistic assessment of structures exposed to fatigue, focusing on the determination of acceptable size of fatigue crack and definition of the regular inspection system was published in detail in (Krejsa et al. 2016b). This relatively advanced probabilistic task was solved using the FCProbCalc code, which allows to calculate the probability of fatigue crack progression (Krejsa 2014 a, b) in a user friendly environment. The comprehensive methodology for probabilistic design and reliability assessment of anchor reinforcement in long mining and underground works was utilized, as well (Krejsa et al. 2013).

3 EXAMPLE AND ANALYSIS OF COMPUTATIONAL STEPS

The importance and rate of optimization steps are presented in the following example calculated using the ProbCalc code. The steel load-carrying component made from an HE300B profile is loaded by 2D bending and axial loads (simple compression). The ultimate limit state was assessed considering second-order theory. Attention was also given to the impacts of initial imperfections. The static scheme of the structure under calculation is shown in Figure 1. The example includes ten input random variables (vertical load: dead DL, long-lasting LL and short-lasting SL; the horizontal load: wind WIN and earthquake EQ; variability of cross section variables: cross-section area A, cross-section modulus W_y and moment of inertia I_y, yield stress f_y, and geometric imperfections a), of which seven input random variables are statistically independent and three input random variables (cross-section properties) are correlated. For details, see Table 1.

The calculation was performed for the statistically non-correlated random input variables as well as for the statistically correlated cross-section characteristics. The obtained failure probabilities p_f, machine time needed for calculation, number of computational operations and the final classification into reliability classes, and consequences for calculations with

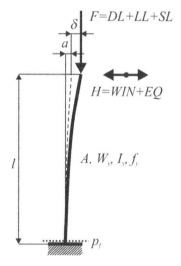

Figure 1.   Static scheme of the column loaded with a system of random variable forces DL, LL, SL, WIN and EQ solved by second-order theory with respect to the deformation δ taking into account the variability of geometric imperfection a, cross-section parameters A, W_y and I_y, and randomness of yield stress f_y. Reliability assessment and calculation of the probability of failure p_f was performed in fixed support.

Table 1. Description of input variables in the example.

Input variable	Minimum	Maximum	N_j	Histogram
Column height l	6 m	–	–	–
Yield stress f_y	200 MPa	435 MPa	217	FY235-01
Dead load DL	260 kN	320 kN	256	DEAD1*
Long-lasting load LL	0 kN	120 kN	256	LONG1*
Short-lasting load SL	0 kN	75 kN	256	SHORT1*
Wind load WIN	−45 kN	45 kN	256	WIND1*
Earthquake EQ	−30 kN	30 kN	256	EARTH*
Geometric imperfections Imp	−30 mm	30 mm	16	IMP016
Variability of cross section properties A, W_y and I_y	–	–	10^3	3DHE300B**
Cross-sectional area A	13076 mm^2	16048 mm^2	10	1DHE300BA
Cross section modulus W_y	$1.44 \cdot 10^6$ mm^3	$1.77 \cdot 10^6$ mm^3	10	1DHE300BW
Moment of inertia I_y	$2.19 \cdot 10^8$ mm^4	$2.70 \cdot 10^8$ mm^4	10	1DHE300BI

*Histograms are taken from (Marek et al. 1995); ** 3D histogram was used for calculation considering the statistical dependence of cross section properties A, W_y, and I_y. Histograms 1DHE300BA, 1DHE300BW and 1DHE300BI are based on this, as well.

Table 2. Analysis of the results for probabilistic reliability assessment of individual types of applied optimization steps considering statistical independence of input random variables.

Optimization used	Calculation time	p_f	RC/CC	Calculation steps
Without optimization	>>24 hours		not performed	$4.13554 \cdot 10^{18}$
Grouping of output quantities	>>24 hours		not performed	$1.75235 \cdot 10^{16}$
Grouping of input quantities	>>24 hours		not performed	$2.27541 \cdot 10^{11}$
Grouping of input variables, zone optimization	>>24 hours		not performed	$1.83501 \cdot 10^{11}$
Grouping of input variables, interval optimization	2:33:22 hours	$5.6736 \cdot 10^{-5}$	RC2/CC2	$4.59571 \cdot 10^9$
Grouping of input variables, interval and zone optimization	2:17:29 hours	$5.5559 \cdot 10^{-5}$	RC2/CC2	$3.38479 \cdot 10^9$
Grouping of input variables, interval, zone and the trend optimization	1:20:43 hours	$5.5559 \cdot 10^{-5}$	RC2/CC2	$2.04303 \cdot 10^9$
Grouping of input and output variables	37:05 min.	$5.1330 \cdot 10^{-5}$	RC2/CC2	$1.04858 \cdot 10^9$
Grouping of input and output variables, zone optimization	28:29 min.	$5.2469 \cdot 10^{-5}$	RC2/CC2	$8.22473 \cdot 10^8$
Grouping of input and output variables, parallelization (2 cores)	9:06 min.	$5.1330 \cdot 10^{-5}$	RC2/CC2	$1.04858 \cdot 10^9/2$
Grouping of input and output variables, interval optimization	4:30 min.	$5.0480 \cdot 10^{-5}$	RC2/CC2	$1.35032 \cdot 10^8$
Grouping of input and output variables zone and interval optimization	3:35 min.	$4.8711 \cdot 10^{-5}$	RC2/CC2	$1.06021 \cdot 10^8$
Grouping of input and output variables parallelization (8 cores)	3:20 min.	$5.1330 \cdot 10^{-5}$	RC2/CC2	$1.04858 \cdot 10^9/8$

statistically non-correlated input random variables are listed in Table 2. A similar analysis was carried out for the calculation with the statistically correlated cross section characteristics, which are expressed using the triple histogram (for results, see Table 3). Both variants of calculation were performed using a DLL library on a PC with the following specifications: an Intel(R) Core(TM) i7-2600 CPU @ 3.40 GHz, MS Windows 7/64-bit/SP1, 8 GB RAM; ProbCalc v.1.5.3.

Table 3. Analysis of the results for probabilistic reliability assessment of individual types of optimization steps used considering statistical dependence of input random variables.

Optimization used	Calculation time	p_f	RC/CC	Calculation steps
Without optimization	>>24 hours		not performed	$9.50648 \cdot 10^{16}$
Grouping of output quantities	>>24 hours		not performed	$5.43227 \cdot 10^{14}$
Grouping of input quantities	3:52:03 hours	$5.2442 \cdot 10^{-5}$	RC2/CC2	$5.68852 \cdot 10^{9}$
Grouping of input and output variables	1:09 min.	$5.2467 \cdot 10^{-5}$	RC2/CC2	$3.25059 \cdot 10^{7}$
Grouping of input and output variables, parallelization (2 cores)	19 sec.	$5.2467 \cdot 10^{-5}$	RC2/CC2	$3.25059 \cdot 10^{7}/2$
Grouping of input and output variables, parallelization (8 cores)	9 sec.	$5.2467 \cdot 10^{-5}$	RC2/CC2	$3.25059 \cdot 10^{7}/8$

It follows from the results that the optimization techniques (such as grouping, interval optimization, zone and trend analysis, and calculation parallelizing) may considerably reduce the computation time at the expense of minor impact on the result without any influence on the class/level of reliability or the class of consequences pursuant to EN 1990.

4 CONCLUSION

The paper discussed the development of probabilistic methods and the use of such methods in the structural reliability assessment. Particular attention was paid to a new method, the DOProC, which is still under development. The DOProC appears to be a very efficient method whose solutions quality is limited by numerical errors and errors resulting from discretization of input and output quantities only.

One shortcoming of the DOProC method is the considerable increase in the required computer time for probabilistic operations for models with many random variables. The maximum number of random variables depends on the complexity of this model and, importantly, whether it is possible to use any of the described optimization steps. The paper focused on the evaluation of the effect of optimization, parallelization and correlation on the computation demand as well as quality of probabilistic reliability assessment in case of selected structure.

ACKNOWLEDGEMENTS

This paper has been completed thanks to the financial support provided to VSB-Technical University of Ostrava by the Czech Ministry of Education, Youth and Sports from the budget for conceptual development of science, research and innovations for the year 2016.

REFERENCES

Cajka, R., Kozielova, M., Burkovic, K. & Mynarzova, L. 2014. Strengthening of masonry structures on the undermined area by prestressing. *Acta Montanistica Slovaca* 19(2): 95–104.

Cajka, R., Mynarcik, P. & Labudkova, J. 2016. Experimetal measurement of soil-prestressed foundation interaction. *International Journal of GEOMATE* 10(4): 2101–2108.

Donova, D. & Zdrazilova, N. 2014. The comparison of the probabilistic calculation of course of temperatures in peripheral construction with actual measured data. *Advanced Materials Research* 1041: 154–157. DOI: 10.4028/www.scientific.net/AMR.1041.154.

Janas, P. & Krejsa, M. 2012. Statistical Dependence of Input Variables in DOProC Method. *Transactions of the VSB-Technical University of Ostrava, Civil Engineering Series* 12(2): 48–58. ISSN: 1804-4824. DOI: 10.2478/v10160-012-0017-3.

Janas, P., Krejsa, M. & Krejsa, V. 2015a. *Direct Optimized Probabilistic Calculation*. VSB-Technical University of Ostrava, p. 191. ISBN 978-80-248-3798-7 (in Czech).

Janas, P., Krejsa, M., Krejsa, V. & Bris, R. 2015b. Structural reliability assessment using Direct Optimized Probabilistic Calculation with respect to the statistical dependence of input variables. In L. Podofillini, B. Sudret, B. Stojadinovic, E. Zio & W. Kröger (eds), *Proc. of Conf. ESREL 2015: Safety and Reliability of Complex Engineered Systems*. CRC Press, 4125–4132. ISBN: 978-1-138-02879-1, eBook ISBN: 978-1-315-64841-5, DOI: 10.1201/b19094-540.

Janulikova, M. & Mynarcik, P. 2014. Modern sliding joints in foundations of concrete and masonry structures. *International Journal of Mechanics* 8(1): 184–189.

Jendzelovsky, N. & Balaz, L. 2016. Analysis of cylindrical tanks under the seismic load. *Key Engineering Materials* 691: 285–296. DOI: 10.4028/www.scientific.net/KEM.691.285.

Kala, Z. 2016. Global sensitivity analysis in stability problems of steel frame structures. *Journal of Civil Engineering and Management* 22(3): 417–424.

Kmet, S., Tomko, M. & Brda, J. 2011. Time-dependent analysis of cable trusses Part II. Simulation-based reliability assessment. Structural Engineering and Mechanics 38(2): 171–193.

Kotes, P. & Vican, J. 2013. Recommended reliability levels for the evaluation of existing bridges according to Eurocodes. *Structural Engineering International: Journal of the International Association for Bridge and Structural Engineering (IABSE)* 23(4): 411–417.

Kotrasová, K., Grajciar, I. & Kormaníková, E. 2014. Dynamic time-history response of cylindrical tank considering fluid—Structure interaction due to earthquake. *Applied Mechanics and Materials* 617: 66–69. DOI: 10.4028/www.scientific.net/AMM.617.66.

Kralik, J. & Kralik Jr., J. 2014. Failure probability of NPP communication bridge under the extreme loads. *Applied Mechanics and Materials* 617: 81–85.

Krejsa, M. 2014a. Probabilistic reliability assessment of steel structures exposed to fatigue. In R.D.J.M. Steenbergen, P.H.A.J.M. van Gelder, S. Miraglia & A.C.W.M. Vrouwenvelder (eds.) *Safety, Reliability and Risk Analysis: Beyond the Horizon*. CRC Press, 2671–2679. DOI: 10.1201/b15938-404.

Krejsa, M. 2014b. Probabilistic Failure Analysis of Steel Structures Exposed to Fatigue. *Key Engineering Materials* 577–578: 101–104. DOI: 10.4028/www.scientific.net/KEM.577-578.101.

Krejsa, M., Janas, P. & Krejsa, V. 2014a. ProbCalc—An efficient tool for probabilistic calculations. *Advanced Materials Research* 969: 302–307. DOI: 10.4028/www.scientific.net/AMR.969.302.

Krejsa, M., Janas, P. & Krejsa, V. 2014b. Software application of the DOProC method. *International Journal of Mathematics and Computers in Simulation* 8(1): 121–126.

Krejsa, M., Janas, P. & Krejsa, V. 2016a. Structural Reliability Analysis Using DOProC Method. *Procedia Engineering* 142: 34–41. DOI: 10.1016/j.proeng.2016.02.010.

Krejsa, M., Janas, P., Yilmaz, I., Marschalko, M. & Bouchal, T. 2013. The Use of the Direct Optimized Probabilistic Calculation Method in Design of Bolt Reinforcement for Underground and Mining Workings. *Scientific World Journal* Article no. 267593. DOI: 10.1155/2013/267593.

Krejsa, M., Kala, Z. & Seitl, S. 2016b. Inspection Based Probabilistic Modeling of Fatigue Crack Progression. *Procedia Engineering* 142: 146–153. DOI: 10.1016/j.proeng.2016.02.025.

Lamich, D., Marschalko, M., Yilmaz, I., Bednarova, P., Niemiec, D., Kubecka, K. & Mikulenka, V. 2016. Subsidence measurements in roads and implementation in land use plan optimisation in areas affected by deep coal mining. *Environmental Earth Sciences* 75(1): 1–11.

Lokaj, A. 2015. Round timber bolted joints with steel plates under static and cyclic loading. *Key Engineering Materials* 627: 29–32. DOI: 10.4028/www.scientific.net/KEM.627.29.

Major, M., Major, I. & Rozycka, J. 2014. Propagation of the surface of a strong discontinuity in the hyperelastic materials. *Advanced Materials Research* 1020: 188–192.

Marek, P., Gustar, M. & Anagnos, T. 1995. *Simulation-Based Reliability Assessment for Structural Engineers*. CRC Press, p. 384. ISBN: 0849382866.

Salajka, V., Kala, J., Cada, Z. & Hradil, P. 2014. Modification of response spectra by probabilistic approach. In *Safety and Reliability: Methodology and Applications—Proc. of the European Safety and Reliability Conf., ESREL 2014*. CRC Press/Balkema, 739–742.

Vavrusova, K. 2014. The bearing capacity of agglomerated wood joints. *Applied Mechanics and Materials* 470: 1077–1080. DOI: 10.4028/www.scientific.net/AMM.470.1077.

Vican, J., Gocal, J., Odrobinak, J., Moravcik, M. & Kotes, P. 2015. Determination of railway bridges loading capacity. *Procedia Engineering* 111: 839–844.

Yilmaz, I., Marschalko, M., Lamich, D., Drusa, M., Machacik, J., Heviankova, S., Kyncl, M., Lackova, E., Bestova, I., Krcmar, D., Stutz, E. & Bednarik, M. 2014. Monitoring of heat transmission from buildings into geological environment and evaluation of soil deformation consequences in foundation engineering. *Environmental Earth Sciences* 72(8): 2947–2955.

Advances and Trends in Engineering Sciences and Technologies II – Al Ali & Platko (Eds)
© 2017 Taylor & Francis Group, ISBN 978-1-138-03224-8

Optimum design of a cellular shell structure of a belt-conveyor bridge

K. Jármai & J. Farkas
University of Miskolc, Miskolc, Hungary

ABSTRACT: The cellular shell consists of two circular cylindrical shells and a longitudinal stiffening welded between them. Halved Circular Hollow Section (CHS) stiffeners are used to ease the welding of outer shell elements. Advantages of cellular shells are as follows: thinner structural parts can be used, which decreases the welding cost, large bending and torsional stiffness can be achieved with small diameter. In the present study a simple supported tubular beam is strengthened in the middle part of the span by cellular shell, which enables to fulfil the strict deflection constraint and the limiting of the shell diameter. The single tube version has too large thickness unsuitable for fabrication and the structural volume is larger than the cellular shell one. The shell thicknesses, dimensions of stiffeners and the length of the stiffening are optimized for minimum cost. The cost function is the cost of material, welding and painting.

1 INTRODUCTION

The economy of welded structures can be achieved by the design of stiffened thin-walled structures instead of unstiffened thick-walled ones. A very efficient structural type is the cellular plate, which consists of two plates and stiffening welded between them (Farkas 1976, 1984) (Farkas and Jármai 1997). Similarly, a cellular shell consists of two circular cylindrical shells and stiffening welded between them. For stiffening longitudinal halved Circular Hollow Section (CHS) stiffeners (EN 10219-2: 2006) are used, which enables the welding of outer shell parts (Figure 1).

The main advantages of the cellular (double) shells over the single shells stiffened by various profiles are as follows: (a) thinner structural parts can be used, which decreases the welding cost, (b) large bending stiffness can be achieved with small diameter, (c) large torsional stiffness enables to consider large torsional moments, (d) the smooth surface enables better corrosion protection.

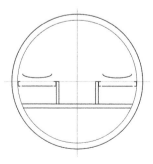

Figure 1. Cross-section of the belt-conveyor bridge with two belt-conveyors and a service walkway in the middle.

One disadvantage of cellular shells is the larger amount of welds. The economy of cellular shells can be achieved in the case, when the cost of welds needed for stiffening is smaller than the cost decrease achieved by the use of thinner shell parts.

Cellular shells are economic in cases of strict constraints on displacements and dimensional constraints. Such a case has been treated in our previous study (Farkas and Jármai 2016). It has been shown that the cellular shell is the only usable structural version for a cantilever tubular column loaded by compression and bending, when the outer shell diameter and the displacement of the column top are limited.

In the present study the cellular shell is applied for a simply supported tubular beam loaded by bending and shear, the maximum deflection and the outer shell diameter are limited.

Given data: beam span length $L = 84$ m. Factored live load is 10 N/m (safety factor 1.5), factored dead load (belts, rollers, service-walkway) is 3 N/m (safety factor 1.35). Yield stress of steel $f_y = 355$ MPa, for thicknesses $16 < t < 40$ mm the allowable stress is $\sigma_{allow} = 314$ MPa, elastic modulus $E = 2.1 \times 10^5$ MPa. In the calculation of displacement the factored loads are divided by safety factors, i.e. the intensity of the uniformly distributed load for deflection is $P = 10/1.5 + 3/1.35 = 8.89$ N/m.

Constraints: limitation of the maximum deflection: $w_{max} = L / \phi, \phi = 1000$ and limitation of the outer shell diameter: $D = 2R = 2800$ mm.

2 THE BEAM CONSTRUCTED FROM A SINGLE CIRCULAR CYLINDRICAL SHELL

From the deflection constraint

$$w_{max} \geq \frac{5pL^4}{384EI_x} \leq w_{allow} = \frac{L}{\phi}, \phi = 1000, I_x = \pi R^3 t, \tag{1}$$

the required shell thickness is

$$t \geq \frac{5pL^3\phi}{384E\pi R^3} = 37.9 \text{ mm}, \tag{2}$$

which is unsuitable for fabrication. Therefore a cellular shell is used for strengthening the beam at the middle span. For a comparison, the structural volume is

$$V \geq 2R\pi t L = 2.800 x 10^{10} \text{ mm}^3. \tag{3}$$

3 THE TUBULAR BEAM PARTIALLY OR TOTALLY STRENGTHENED BY CELLULAR SHELL

3.1 *Geometric characteristics*

The cross-sectional area of a half CHS is (Figure 2)

$$A_s = \pi R_s t_s, R_s = \frac{D_s - t_s}{2}, \tag{4}$$

the distance of its gravity centre is (Figure 3)

$$y_G = 2R_s / \pi, \tag{5}$$

and its moment of inertia

$$I_s = \frac{R_s^3 \pi}{2}\left(1 - \frac{8}{\pi^2}\right). \tag{6}$$

124

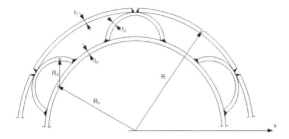

Figure 2. Dimensions of cellular shell.

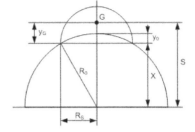

Figure 3. Geometry of cellular shell.

The distance X

$$X = \sqrt{\left(R_0 + \frac{t_0}{2}\right)^2 - R_S^2}.$$

(7)

The radius of the inner shell R_0 can be calculated from the following equation

$$R = X + R_S + \frac{t_1 + t_s}{2},$$

(8)

(t_1 is the thickness of the outer shell)

$$R_0 = \sqrt{\left(R - R_S - \frac{t_1 + t_s}{2}\right)^2 + R_S^2} - \frac{t_0}{2}.$$

(9)

The moment of inertia of n_s stiffeners about the centre of the shell

$$I_{xs} = \left(I_s + A_s s^2\right)\sum_{i=1}^{n_s} \cos^2\left(\frac{2\pi i}{n_s}\right) + \left(\frac{\pi R_s^3 t_s}{2}\right)\sum_{i=1}^{n_s} \sin^2\left(\frac{2\pi i}{n_s}\right).$$

(10)

The moment of inertia of the whole cellular shell (Figure 3)

$$I_x = \pi R_0^3 t_0 + \pi R^3 t_1 + I_{xs}.$$

(11)

Fabrication constraint to enable the welding of the half CHS stiffeners to the inner shell: from

$$\frac{2R_0\pi}{n_s} - 2R_s \geq 2t_s,$$

(12)

the maximum allowable number of half CHS stiffeners

$$n_{s.max} = \frac{\pi R_0}{R_s + t_s}.$$

(13)

3.2 *Constraint on panel shell buckling of the outer shell parts between stiffeners*

The Det Norske Veritas (2002) design rules are used.

In the case of such a very strict displacement constraint the panel buckling constraint is not active. σ_{max} is so small that the effective shell width is equal to the whole width s_0. Calculations show that this constraint is passive.

125

3.3 The deflection constraint

The maximum deflection of a beam strengthened on the middle part loaded by a uniformly distributed normal load of intensity p (the self-mass is neglected) (Figure 4). The maximum bending moment is:

$$M_{max} = \frac{pL^2}{8}.$$

$$(14)$$

Bending moment at distance L_1 from the support

$$M_1 = \frac{pLL_1}{2} - \frac{pL_1^2}{2}.$$

$$(15)$$

The deflection constraint with the maximum deflection at the middle of the beam

$$w_{max} = \frac{1}{E}\left[\frac{QL}{2} - Q_1\left(\frac{L_1}{2} + L_2\right) - Q_2\left(\frac{L_1}{3} + L_2\right) - \frac{2Q_3L_2}{3} - \frac{Q_4L_2}{2} - \frac{Q_5L_2}{3}\right] \le w_{allow},$$

$$(16)$$

$$Q = Q_1 + Q_2 + Q_3 + Q_4 + Q_5,$$

$$(17)$$

$$Q_1 = \frac{pL_1^3}{12I_1}, \quad Q_2 = \frac{M_1 L_1}{2I_1}, \quad Q_3 = \frac{M_1 L_2}{2I_2}, \quad Q_4 = \frac{pL_2^3}{12I_2}, \quad Q_5 = \frac{M_{max}L_2}{2I_2},$$

$$(18)$$

$$w_{allow} = \frac{L}{f}, \phi = 1000.$$

$$(19)$$

I_1, I_2 are the moments of inertia of the beam parts, respectively.

3.4 Stress constraint of the beam cross section at the distance L_1

This constraint is also passive.

3.5 The cost function

The cost is calculated according to the fabrication sequence (Farkas and Jármai 2008, 2013). The beam is divided into 7 units ($84 = 7 \times 12$ m). Two types of units are constructed as follows:

(a) $n_1 = \frac{L_1}{L_0}$ units of simple shell, the length of a unit is $L_0 = 12$ m, $L_1 = 12, 24, 36$ m,

(b) $n_2 = \frac{L - 2L_1}{L_0}$ units of the cellular shell.

Two end plates are welded to each unit and the units are connected with bolted joints using these end plates. The cost of bolted connections is not treated.

3.5.1 Cost of a simple shell unit
Volume:

$$V_1 = 2R_0\pi t_0 L_0,$$

$$(20)$$

The cost of forming plates into curved shell panel 3 m long, curve length $L_c = \frac{2R_0\pi}{3000}$, thickness t_0

$$K_{F10} = k_F\Theta e^{\mu_1}, \mu_1 = 6.8582513 - 4.527217 t_0^{-0.5} + 0.009541996(2R_0)^{0.5},$$

$$(21)$$

The factor of complexity of assembly $\Theta = 2$, fabrication cost factor $k_F = 1.0$ \$/min. Number of curved panels is 12.

Welding of the curved panels into a whole unit with SAW butt welds:

$$K_{F11} = k_F \left[\Theta_1 \sqrt{12 \rho V_1} + 1.3 x 0.152 x 10^{-3} t_0^{1.9358} \left(3L_0 + 3 x 2 \pi R_0 \right) \right], \Theta_1 = 3. \tag{22}$$

Welding of two unit ends to the connecting plates with GMAW-CO$_2$ fillet welds:

$$K_{F12} = k_F \left(\Theta \sqrt{3 \rho V_1} + 1.3 x 0.3394 x 10^{-3} a_w^2 4 x 2 R_0 \pi \right), a_w = 0.4 t_0. \tag{23}$$

Material cost:

$$K_{M1} = k_M \rho_1 V_1, k_M = 1.0\$/kg, \rho_1 = 7.85 x 10^{-5} \text{N/mm}^3. \tag{24}$$

Painting cost:

$$K_{P1} = k_P S_1, k_P = 14.4 x 10^{-6} \$/mm^2, S_1 = 2 x 2 R_0 \pi L_0. \tag{25}$$

Total cost of a unit:

$$K_1 = K_{M1} + 12 K_{F10} + K_{F11} + K_{F12} + K_{P1}. \tag{26}$$

3.5.2 Cost of a cellular shell unit

The costs of forming curved shell panels and that of welding the basic shell are the same as for the simple shell unit. The calculation is similar to that in Section 3.5.1.

Material cost:

$$K_{M2} = k_M \rho_1 V_s, k_M = 1.0\$/kg, \rho_1 = 7.85 x 10^{-5} \text{N/mm}^3. \tag{27}$$

Painting cost:

$$K_{P2} = k_P S_2, k_P = 14.4 x 10^{-6} \$/mm^2, S_2 = 2 \pi L_0 \left(R + R_0 \right). \tag{28}$$

Total cost of a cellular shell unit:

$$K_2 = K_{M2} + 12 K_{F20} + K_{F21} + 6 K_{F22} + K_{F23} + 24 K_{F24} + 6 K_{F25} + K_{F26} + K_{F27} + K_{P2}. \tag{29}$$

3.5.3 Total cost of the whole beam

$$K = 2_{n1} K_1 + n_2 K_2. \tag{30}$$

Total volume

$$V = 2_{n1} V_1 + n_2 V_s. \tag{31}$$

3.6 Optimization and results

In the optimum design procedure the optimal values of variables $\left(D_s, t_s, n_s, t_0, t_1, L_1 \right)$ are determined by a systematic search using a MathCAD algorithm to minimize the structural cost. Table 1 shows the search results.

It can be seen that the optimum solution uses suitable thicknesses. Comparing the volume to that of the single tube structure treated in section 2, it can be concluded that the cellular structure has smaller volume.

Table 1. Optimization results. Dimensions in mm, volume in mm³, cost in $. Optimum is marked by bold letters. It can be seen that in the case of $L_1 = 36$ m the required thicknesses t_0 and t_1 are not suitable for fabrication, this solution is marked by italics. The deflection is limited to 84 mm.

D_s	t_s	t_0	t_1	L_1	w_{max}	$10^{-10}V$	$10^{-6}K$
101.6	6	21	20	0	82.5	4.281	1.170
114.3	**6**	**22**	**21**	**12000**	**82.3**	**4.484**	**1.131**
114.3	10	26	27	24000	83.5	5.551	1.470
114.3	10	39	38	36000	83.1	7.968	2.087

4 CONCLUSIONS

The requirements of a modern load-carrying structure are the safety, manufacturability and economy. This study shows how to design a stiffened shell with thinner elements instead of an unstiffened thick shell, which is unsuitable for fabrication.

A simple supported circular cylindrical shell beam is designed with a strict deflection constraint as well as limitations in diameter and thickness. A single shell is unsuitable for fabrication, since it should have a thickness more than 30 mm. Therefore, a cellular shell is used in the span middle to strengthen the single shell.

The cellular shell consists of the basic shell and an outer shell, the elements welded to the longitudinal stiffeners. Halved CHS stiffeners are used to ease the welding.

The strengthened beam is optimized to minimize the cost. The comparison shows that the volume of the strengthened shell is smaller than that of the single shell.

ACKNOWLEDGEMENTS

The research was supported by the Hungarian Scientific Research Fund OTKA T 109860 project and was partially carried out in the framework of the Center of Excellence of Innovative Vehicle, Mechanical and Energy Engineering Design and Technologies at the University of Miskolc.

REFERENCES

Det Norske Veritas 2002. Buckling strength of shells. Recommended practice RP-C202. Høvik, Norway.
EN 10219-2. 2006. Cold formed circular hollow section profiles.
Farkas, J. 1976. Structural synthesis of welded cell-type plates. ActaTechn. Hung. 83, 1–2, 117–131.
Farkas, J. 1984. Optimum design of metal structures, Akadémiai Kiadó, Budapest, Ellis Horwood, Chichester, UK.
Farkas, J., Jármai, K. 1997. Analysis and optimum design of metal structures. Balkema, Rotterdam-Brookfield.
Farkas, J., Jármai, K. 2008. Design and optimization of metal structures. Horwood, Chichester, UK.
Farkas, J., Jármai, K. 2013. Optimum design of steel structures, Springer, Heidelberg etc.
Farkas, J., Jármai, K. 2016. A new structural version of welded cellular shell for a cantilever column. Welding in the World online first, March, pp. 1–8.
Jármai, K., Farkas, J. 1999. Cost calculation and optimization of welded steel structures. J Constr Steel Res, 50, 115–135.

Advances and Trends in Engineering Sciences and Technologies II – Al Ali & Platko (Eds)
© 2017 Taylor & Francis Group, ISBN 978-1-138-03224-8

Analysis of stress inside the sandwich composite structure

D. Jiroutová
Department of Experimental Method, Klokner Institute, Czech Technical University in Prague, Prague, Czech Republic

ABSTRACT: The aim of the article is to focus on describing the relations of stress inside the sandwich composite structure according to the test conditions. This type of structure is composed of two or more layers having different material and mechanical properties. Abrupt changes of these properties causes interlaminar stress in the structure. A good knowledge about the behavior of sandwich composite structures is important for efficient manufacture techniques, long-term prediction of behavior and also for economics. The experimental test specimens have been made from three layers—outer layers and a core. Test specimens have been loaded by means of a three point bending test. Fiber-optic strain gauges, SOFO SMARTape Compact deformation sensors, have been used for long-time monitoring of total strain in sandwich composite structure. Experimentally obtained data have been used for the creation of a mathematical model of stress inside the sandwich composite structure according to the temperature.

1   INTRODUCTION

The sandwich composite structure belongs amongst perspective materials used in modern engineering structures. This type of structure is composed of two main parts—face sheet and core. The face sheet of a sandwich composite structure is a really thin layer with a very high stiffness. The main function of the face sheet is to carry tension, compression and bending moments having effect on the structure and also protect the sandwich composite structure. The sandwich core is unlike the sandwich face sheet and is thick and lightweight. The function of the sandwich core is to carry transverse shear forces having an effect on the sandwich composite structure. Due to layer composition, the sandwich composite structures have high shear stiffness to weight ratios and high tensile strength to the weight ratios. The sandwich composite structures have high impact strength, high flexural resistance, high flexural stiffness, low thermal conductivity, low acoustic conductivity, and resistance to the corrosion. Due to these properties, the sandwich composite structures are widely used in various industries (for example land transport, marine and aerospace constructions, building structures and so on).

Each layer of the sandwich composite structure has different material and mechanical properties. The sudden change of these properties across the interface of layers causes interlaminar stress in this zone. Interlaminar stress can lead to delamination of the sandwich composite structure. Due to this fact, not only a good knowledge about material and mechanical properties of each layer but also a good knowledge about properties of the whole sandwich composite structure is important. It is required to have a good knowledge of the sandwich composite structure behavior not only in the short-term but also in the long-term. This knowledge is important for designing the sandwich composite structure and its reliability, for efficient manufacture techniques, and for economic aspects.

The monitoring of reliability and durability of "health" of the sandwich composite structure uses different methods. The most important and the most frequently monitored parameters in the structures are strain and deformation. The monitoring of these parameters can be performed in the short-term, middle-term, long-term or during the whole lifetime period of the structure. A whole range of conventional sensors working on various physical principles are used for monitoring of the structure, for example resistance strain gauges, capacity and induction displacement sensors, and video strain gauges and so on (Thomsen & Frostig 1997, Daniel & Abot 1999, Steeves & Fleck 2004). The selection of an appropriate conventional method depends on the application, the measurement range, the desired accuracy and other parameters. A disadvantage of these standard sensors is that they cannot be embedded between layers of the sandwich composite structure.

At present, fiber-optic measuring methods are used to monitor the behavior of the structures. These type of strain sensors uses the capability of optical fibers to transmit optical radiation in the direction of their centerline. The radiation is transmitted by means of the light's reflection at an interface between two environments of a different reflection index (Martinek 2004). The principle of the fiber-optic sensors is derived from various physical phenomena's (sensors on the basis of Fiber Bragg Grating, Fabry-Perot or Michelson interferometer etc.) (Glišic & Inaudi 2007). Compared to traditional methods, among the advantageous of the fiber-optic sensors belongs higher quality of the measurements, higher reliability, insensitivity to electromagnetic field, corrosion resistance, safety in explosive and flammable environments, the possibility of long-term monitoring, easier installation and possibility to install them directly into the structure. The main disadvantage is related to the sensor construction, i.e. fragility and one time application. The relatively high cost of the sensors is compensated by the possibility to use them for long-term monitoring. The fiber-optic technology exhibit unlike the conventional methods a lot of advantageous properties such as higher quality of the measurements, higher reliability, easier installation and maintenance, insensitivity to the electromagnetic field, corrosion resistance, safety in explosive and flammable environments, the possibility of long-term monitoring and lower cost per lifetime period. Fiber-optical sensors can be used for "health" monitoring of the dams, buildings, piping systems, bridges, tunnels and so on (Glišic et al. 2003, Glišic et al. 1999, Inaudi 2003, Mikami & Nishizawa 2015, Kawano et al. 2010).

The aim of the article is to focus on describing the relations of stress inside the sandwich composite structure according to the test conditions. The sandwich composite structure is composed of three layers. The face sheet of the monitored sandwich composite structure has been made from the epoxy-resin-impregnated glass laminates with plain weave and the sandwich core has been made from the light weight foam Divinycell H100. Two SOFO SMARTape Compact deformation sensors were used for long-term monitoring of total strain in each monitored sandwich composite structure. Test specimens were loaded via a three-point bending test for over a year. The loading of sandwich composite test specimens were changed during long-term monitoring. Experimentally obtained data have been used for creation of mathematical model of stress inside the sandwich composite structure according to the temperature.

2 TEST SPECIMEN

Test specimen with foam sandwiched between outer layers were produced. The core of sandwich composite test specimen was made from lightweight foam Divinycell H100 with density 100 kg/m^3 and thickness 20 mm (manufacture by DIAB GROUP). The lightweight foam Divinycell H100 is made from polyurea and polyvinyl chloride. This type of foam core was chosen due to its advantageous such as excellent mechanical properties to low weight, low water absorption, chemical resistance, excellent adhesion and peel strength, and good thermal and acoustic insulation. Lightweight foam core Divinycell H100 was bonded between outer layers. Outer layers were made from epoxy-resin-impregnated glass laminates with plain weave with weight 600 g/m^2 (manufacture by Tomek—fiberglass manufacture). The thickness of outer layer is 0.8 mm. Two component toughened methacrylate adhesive system Araldite® 2021 (manufacture by Huntsman) was used for bonded of outer layers to sandwich lightweight foam core. For long-term monitoring of residual strain inside the sandwich composite structure, seven test specimens with length 700 mm and width 75 mm were manufactured. The thickness of the sandwich composite test specimen is approx. 22 mm.

3 TEST PROCEDURE

For long-term monitoring of strain inside the sandwich composite test specimens, the fiber-optic SOFO® SMARTape Compact deformation sensors (manufacture by SMARTEC S.A.) were used. Due to the small cross section (width 6 mm, thickness 0.3 mm) it is enable to install them between layers of the sandwich composite structure. Amongst advantageous properties of these sensors belong high resolution, insensitive to temperature variations, insensitive to corrosion, vibrations and immune to electromagnetic fields etc. This type of deformation sensor has active part (working temperature range from −55 °C to +110 °C) and passive part (working temperature range from −40 °C to +80 °C).

Two fiber-optic SOFO® SMARTape Compact deformation sensors with active length 600 mm were embedded between upper/bottom face sheets and foam core in all test specimens. Fiber-optic deformation sensors were placed in the middle of the width of the test specimen. For recording of the measurement readings the SOFO measuring system for static quantities with SBD ver. 6.3.53 software from SMARTEC S.A. was used. Photo of the sandwich composite test specimen with embedded fiber-optic SOFO® SMARTape Compact deformation sensors is shown in Figure 1.

A three-point bending test was used for long-term monitoring of the sandwich composite structure. Three-point bending test has been chosen due to its simplicity, possibility to create the complex load in the structure, and suitability for long-term monitoring of the structure. Distance between the supports with a diameter of 50 mm and length 80 mm has been 600 mm. The load has been applied through a steel cylindrical weight with diameter 50 mm and cylindrical weight with diameter 70 mm. All loading weights have a length of 80 mm. Configuration of the long-term monitoring of the sandwich composite test specimen is shown in Figure 2.

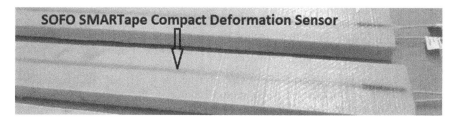

Figure 1.   Sandwich composite test specimen with embedded SOFO® SMARTape Compact deformation sensor.

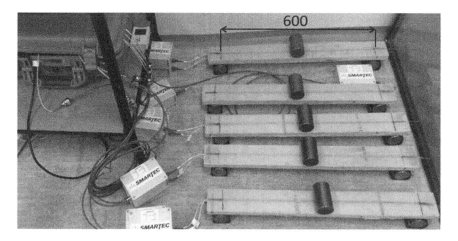

Figure 2.   Long-term monitoring of sandwich composite structure by SOFO® SMARTape Compact deformation sensors loaded in three-point bending test.

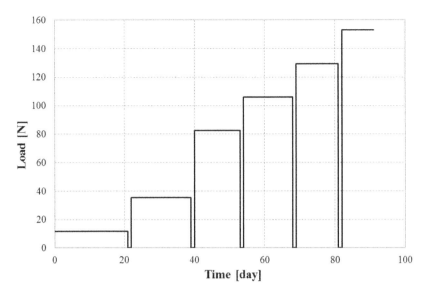

Figure 3.   Loading cycle of the sandwich composite test specimen no. 6 during long-term monitoring.

Long-term monitoring of sandwich composite test specimens was carried out in laboratory conditions. The value of the temperature in laboratory is $(21 \pm 0.7)°C$ and relative humidity is $(41 \pm 5)\%$. The long-term monitoring of sandwich composite test specimens has been taken more than one year.

During the first 265 days of monitoring, five sandwich composite test specimens were loaded by steel cylindrical weight with diameter 50 mm—Figure 2. Then next two sandwich composite test specimens have been added to monitoring, so from this moment seven test specimens have been tested.

From this time, the loading scheme was changed. The value of load has been changed every 14 days. Each sandwich composite test specimen has been unloaded every 14 days for 1 or more days. And then the sandwich composite test specimen has been loaded by higher or same value of load for next 14 days. This cycle has been repeated for more than one year. This cycle has been repeated for more than one year. Illustration of loading cycle of sandwich composite test specimen No. 6 is shown in Figure 3.

4   EXPERIMENTAL RESULTS

Deformation of sandwich composite test specimen has been measured by fiber-optic SOFO SMARTape Compact Deformation sensors with active length LA 600 mm. Strain in the monitored sandwich composite structures has been determined from measuring of these length changes according to equation

$$\varepsilon = \frac{\Delta L}{LA}, \tag{1}$$

where ε = strain [mm/m]; ΔL = length change [mm]; and LA = active length of the SOFO SMARTape Deformation sensor. The strain of the sandwich composite test specimen No. 6 is shown in Figure 4.

The value of strain of the sandwich composite test specimen after it is unloaded is not same as the value of strain at the beginning of the monitoring—see Figure 4. It means, there exists a residual strain inside the monitored sandwich composite structure. Time dependence of this strain over time is shown in Figure 5. The points in the graph corresponds with values

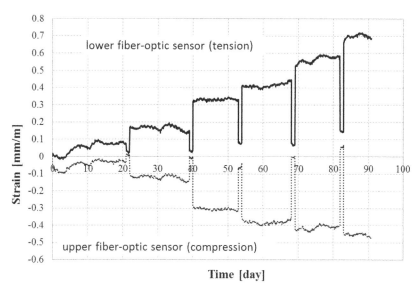

Figure 4. Strain curve of the sandwich composite test specimen no. 6 during long-term monitoring.

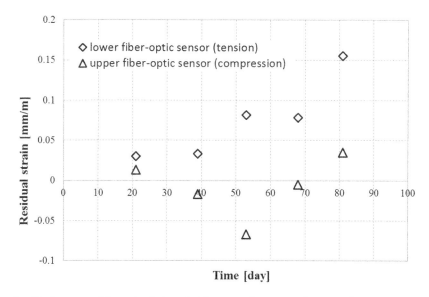

Figure 5. Strain curve of the unloaded sandwich composite test specimen no. 6.

of residual strain which remains in the structure after unloaded. From this graph is evident that the residual strain inside the sandwich composite structure increase with value of load and time. The similar dependences have been obtained for all monitored sandwich composite structures.

5 CONCLUSIONS

The aim of the article is to focus on describing the relations of stress inside the sandwich composite structure according to the test conditions. The sandwich composite structure is composed of two or more layers with different mechanical properties. Sudden changes of these properties across the interface causes interlaminar stress in this zone. Interlaminar stress can lead to delamination and destruction of the sandwich composite structure. It means, a good knowledge about mechanical properties of each layer and whole sandwich composite structure is really important for the designing and reliability of the structure, for efficient manufacture techniques, long-term prediction of behavior and for economics.

Seven experimental test specimens with embedded fiber-optic SOFO SMARTape Compact deformation sensors have been used for-long term monitoring. This type of sensor has been chosen due to its advantageous properties (high resolution, insensitive to temperature and corrosion, vibrations and EM fields, and possibility to install them between sandwich composite layers). Three-point bending test has been used for loading of the monitored sandwich composite test specimens. This test was chosen due to its simplicity and possibility to create complex loading in the structure. During monitoring of the test specimen, the value of the load has been changed. Monitoring of the sandwich composite test specimen lasted over one year.

Results of strain obtained during long-term monitoring of sandwich composite structures show that the residual strain remains in the structure. The course of this residual strain is similar for all test specimens. This result shows that stress inside the sandwich composite structure is a function of not only load, dimensions, material properties, but also of residual strain. Results obtained during long-term monitoring of sandwich composite structure will be used for creation of mathematical model. This model will describe residual stress inside the sandwich composite structure depending on value of load.

ACKNOWLEDGEMENT

This work has been supported by a research project of the Grant Agency of Czech Republic No. 14-35225P.

REFERENCES

Daniel, I.M. & Abot, J.L. 1999. Fabrication, Testing and Analysis of Composite Sandwich Beams. In *Composites Sciences and Technology* 60 (12–13): 2455–2463.

Glišic, B. et al. 1999. Dam Monitoring Using Long SOFO® Sensor. In *Proceedings of the Hydropower Conference*. Gmunden: Aqua Media International. 709–717.

Glišic, B. et al. 2003. Monitoring of Building Columns during Construction. In *Proceedings of the 5th Asia Pacific Structural Engineering & Construction Conference (APSEC)*. Johor Bahru. 593–606.

Glišic, B. & Inaudi, D. 2007. *Fibre Optic Methods for Structural Health Monitoring.* London: John Wiley & Sons Ltd.

Inaudi, D. 2003. State of the Art in Fiber Optic Sensing Technology and EU Structural Health Monitoring Projects. In *Proceedings of the First International Conference on Structural Health Monitoring and Intelligent Infrastructure. Tokyo.*

Kawano, Y. et al. 2010. Health Monitoring of a Railway Bridge by Fiber Optic Sensor (SOFO). In F. Casciati, M. Giordano (ed.) *Proceedings of the Fifth European Workshop "Structural Health Monitoring 2010", Naples, Italy.* DEStech Publications, Inc., Pennsylvania, USA: 1319–1324.

Martinek, R. 2004. *Sensors in industrial practice.* Prague: BEN—Technical literature (in Czech).

Mikami, T. & Nishizawa, T. 2015. Health Monitoring of High-rise Building with Fiber Optic Sensor (SOFO). In *International Journal of High-Rise Buildings* 4 (1): 27–37.

Steeves, C.A. & Fleck, N.A. 2004. Collapse Mechanism of Sandwich Beams with Composite Faces and a Foam Core, Loaded in Three-point Bending. Part II: Experimental Investigation and Numerical Modelling. In *International Journal of Mechanical Sciences* 46(4): 585–608.

Thomsen, O.T. & Frostig, Y. 1997. Localized Bending Effects in Sandwich Panels. Photoelastic Investigation versus High-order Sandwich Theory Results. In *Composite Structures* 37(1): 97–108.

Resistance of timber structures exposed to fire

Z. Kamenická & M. Botló
Department of Metal and Timber Structures, Faculty of Civil Engineering, Slovak University of Technology in Bratislava, Slovakia

ABSTRACT: There are two simplified design methods and advanced design methods for structural fire design of timber structures in the Eurocode 5 (STN EN 195-1-2). The first method is the reduced cross-section method, which is also the recommended procedure. The second method is the reduced properties method. There are differences between the results obtained by these two methods. They should be used for the evaluation of mechanical resistance of timber members, whilst the calculation of the charring depth is to be performed using the simplified rules. Advanced methods deal with the determination of the charring depth, the thermal response model, and the structural response model. They can be used for members, parts of a structure, or entire structures. This paper presents the differences between these simplified and advanced methods and explains the calculation procedures they employ.

1 REGULATIONS IN EUROPE

The maximum height of timber structures according to Slovak national regulations is 2 floors, whereas other countries allow a considerably higher number.

2 THE ZERO-STRENGTH LAYER

The reduced cross-section method includes a zero-strength layer as an additional layer to charring depth. A few years ago, some concerns have been raised about the zero-strength layer being insufficiently conservative in some cases. (Schmid et al., 2014) describe how the zero-strength layer can be determined using advanced numerical simulations, results from fire tests, or both.

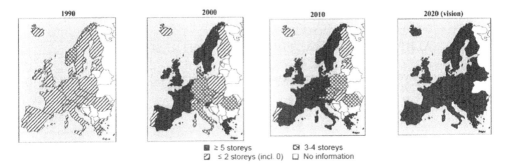

Figure 1. Maximum number of floors in timber load-bearing structures without sprinklers (Östman and Källsner, 2010).

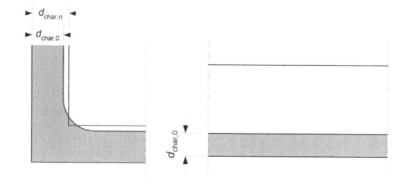

Figure 2a, 2b. Charring depth $d_{char,0}$ for 1-dimensional charring and notional charring depth $d_{char,n}$ (STN EN 1995-1-2, 2011).

3 CHARRING RATE AND CHARRING DEPTH FOR UNPROTECTED TIMBER MEMBERS

Charring is taken into account for unprotected surfaces of timber structural members, or wood-based panels and for protected surfaces where charring occurs during the relevant time of fire exposure. Charring rates are normally different for unprotected surfaces, initially protected surfaces prior to failure of the protection, and for initially protected surfaces after failure of the protection. The position of the char-line is taken as the position at the temperature of 300°C (STN EN 1995-1-2, 2011).

The charring rate for 1-dimensional charring is considered constant in time and is calculated as:

$$d_{char,0} = \beta_0 t \qquad (1)$$

where:

$d_{char,0}$—the design charring depth for 1-dimensional charring
β_0—the 1-dimensional design charring rate under standard fire exposure; it can be found in the table 3.1 in (STN EN 1995-1-2, 2011), or the table 2 in (Kamenická, 2015)
t—the time of fire exposure

The notional charring rate $d_{char,n}$, which includes the effect of corner roundings and fissures, is considered constant in time and is calculated in a same way like 1-dimensional charring, but instead of β_0, the notional design charring rate β_n is used.

4 DESIGN VALUES OF MATERIAL PROPERTIES AND RESISTANCE

Design values of strength properties are calculated (Frangi, König et al., 2010):

$$f_{d,fi} = k_{mod,fi}\, f_{20}/\gamma_{M,fi} \qquad (2)$$

where:

$f_{d,fi}$—the design strength in fire
f_{20}—the 20% fractile of a strength property at normal temperature
$k_{mod,fi}$—the modification factor for fire
$\gamma_{M,fi}$—the partial safety factor for timber in fire = 1.0

5 SIMPLIFIED DESIGN METHODS FOR DETERMINING MECHANICAL RESISTANCE

The recommended procedure is the reduced cross-section method employing the simplified rules (STN EN 1995-1-2, 2011). The second procedure is the reduced properties method.

5.1 *The reduced cross-section method*

The modification factor for fire is $k_{mod,fi} = 1.0$. An effective cross-section is calculated by reducing the initial cross-section using the effective charring depth d_{ef}:

$$d_{ef} = d_{char,n} + k_0 \, d_0 \qquad (3)$$

where:
$d_0 = 7$ mm
$d_{char,n}$—the notional design charring depth which incorporates the effect of corner fillets

The material close to the char-line in the layer of the thickness k_0, d_0 has zero strength and stiffness, whilst the strength and stiffness of the remaining cross-section remain unchanged.

5.2 *The reduced properties method*

This method is only applicable for round or rectangular cross-sections from softwood exposed to fire on three or all sides. For (un)protected members, the modification factor for fire is $k_{mod,fi} = 1.0$ for time $t = 0$. For $t \geq 20$ min, the modification factor for fire $k_{mod,fi}$ is calculated as:

− for bending strength:

$$k_{mod,fi} = 1.0 - 1/200 \; p/A_r \qquad (4)$$

− for compressive strength:

$$k_{mod,fi} = 1.0 - 1/125 \; p/A_r \qquad (5)$$

− for tensile strength and modulus of elasticity:

$$k_{mod,fi} = 1.0 - 1/330 \; p/A_r \qquad (6)$$

where:
p—the perimeter of the fire-exposed residual cross-section [m]
A_r—the area of the residual cross-section [m²]

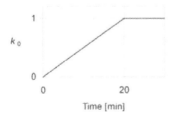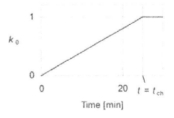

Figure 3a, 3b.   Factor k_0 for unprotected and protected surfaces with $t_{ch} \leq 20$ min; protected surfaces with $t_{ch} > 20$ min (STN EN 1995-1-2, 2011).

6 ADVANCED DESIGN METHODS FOR DETERMINING MECHANICAL RESISTANCE

Advanced methods comprise the determination of the charring depth, the thermal response model, and the structural response model. These methods can be used for individual members, parts of structures, or entire structures.

The ambient temperature is taken as 20°C. Advanced calculation methods are based on the theory of heat transfer and they should consider the temperature-dependent variation of thermal properties of the material.

6.1 *Thermal properties*

The idealisation of thermal conductivity values of the char layer is based on the values measured for charcoal, so that they consider the increased heat transfer caused by occurrence of shrinkage cracks with temperatures exceeding 500°C and the deterioration of the char layer at about 1000°C. Cracks in the charcoal increase heat transfer through radiation and convection.

6.2 *Mechanical properties*

Local values of strength and modulus of elasticity for softwood are multiplied by temperature which is dependent on the reduction factor. The relationship includes effects of transient creep of timber.

7 COMPARISON OF DESIGN METHODS

A case with the following input parameters was examined:

- two cross-sections—120 / 220 mm and 160 / 180 mm,
- three spans—1000 mm, 3000 mm and 6000 mm
- two types of strength class—solid timber C22, and glued laminated timber GL22h,
- time of fire exposure is 45 min,
- axial force 2 kN, load in the z direction 1 kN/m, and load in the y direction 0.5 kN/m
- the reduction factor for load is 0.542,
- the reduction factor due to flexural buckling (for both axes) is 1.0,
- the ratio of the effective length and span is 0.9 according to (STN EN 1995-1-1, 2008) for constant load.

Five static schemes were used for assessment (Fig. 4). Scheme "a" was for bending and shear, scheme "b" for compression and buckling, scheme "c" for compression + bending and buckling + bending, scheme "d" for tension, scheme "e" for tension + bending.

Differences between methods range from 0% to 122% with varying input parameters, i.e. span, cross-section, slenderness, material, etc., and according to assessed load. There are differences in the utilisation of cross-sections (evaluated in percent) or structural members depending on the length, cross-section, and slenderness. Furthermore, with increases in utilisation of the cross-section, the differences between the methods increase as well. In the case of a structural member with high slenderness under load even exceeding its load-bearing capacity, the differences were very significant (more than 50%, sometimes over 100%). On the other hand, in the case of a structural member with lower values of both utilisation (with respect to its dimensions) and the specified load, the differences were minimal (from 0 to 10%). In the case of a member with a moderate utilisation, differences ranged approximately from 5 to 20%. The modification of material (solid or glued laminated timber) was the characteristic with the smallest impact on results. Positive values of results in tables mean that resultant utilisation percentages were higher with the reduced cross-section method and negative values mean higher utilisation with the reduced properties method.

There are differences between the simplified design methods and between the simplified and advanced design methods as well. The charring depth for time of fire exposure 45 min calculated with the reduced cross-section method equals to 43 mm and with the reduced properties method to 36 mm. However, results for advanced method will not be in this publication due to its more limited scope.

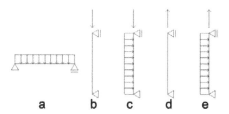

a   b   c   d   e

Figure 4.   Used static schemes.

Table 1.   Formula used for member assessment according to (STN EN 1995-1-1, 2008).

Strain of members	Formula
bending about y	$\sigma_{m,y,fi,d} / f_{m,fi,d}$
bending about z	$\sigma_{m,z,fi,d} / f_{m,fi,d}$
bending about y and z	$\sigma_{m,y,fi,d} / f_{m,fi,d} + k_m \cdot \sigma_{m,z,fi,d} / f_{m,fi,d}$
bending about y and z	$k_m \cdot \sigma_{m,y,fi,d} / f_{m,fi,d} + \sigma_{m,z,fi,d} / f_{m,fi,d}$
tension	$\sigma_{t,0,fi,d} / f_{t,0,fi,d}$
tension + bending	$\sigma_{t,0,fi,d} / f_{t,0,fi,d} + \sigma_{m,y,fi,d} / f_{m,fi,d} + k_m \cdot \sigma_{m,z,fi,d} / f_{m,fi,d}$
tension + bending	$\sigma_{t,0,fi,d} / f_{t,0,fi,d} + k_m \cdot \sigma_{m,y,fi,d} / f_{m,fi,d} + \sigma_{m,z,fi,d} / f_{m,fi,d}$
buckling	$\sigma_{c,0,fi,d} / (k_{c,min,ef} f_{c,0,fi,d})$
compression	$\sigma_{c,0,fi,d} / f_{c,0,fi,d}$
compression + bending	$(\sigma_{c,0,fi,d} / f_{c,0,fi,d})^2 + \sigma_{m,y,fi,d} / f_{m,fi,d} + k_m \cdot \sigma_{m,z,fi,d} / f_{m,fi,d}$
compression + bending	$(\sigma_{c,0,fi,d} / f_{c,0,fi,d})^2 + k_m \cdot \sigma_{m,y,fi,d} / f_{m,fi,d} + \sigma_{m,z,fi,d} / f_{m,fi,d}$
buckling + bending	$(\sigma_{c,0,fi,d} / k_{c,y,ef} f_{c,0,fi,d}) + \sigma_{m,y,fi,d} / f_{m,fi,d} + k_m \cdot \sigma_{m,z,fi,d} / f_{m,fi,d}$
buckling + bending	$(\sigma_{c,0,fi,d} / k_{c,z,ef} f_{c,0,fi,d}) + k_m \cdot \sigma_{m,y,fi,d} / f_{m,fi,d} + \sigma_{m,z,fi,d} / f_{m,fi,d}$
shear in direction y	$\tau_{v,y,fi,d} / f_{v,fi,d}$
shear in direction z	$\tau_{v,z,fi,d} / f_{v,fi,d}$
shear in y and z	$(\tau_{v,y,fi,d}^2 + \tau_{v,z,fi,d}^2)^{1/2} / f_{v,fi,d}$

Table 2.   Differences in results of cross-section or member utilisation between the reduced cross-section method and the reduced properties method, with the 120/220 mm cross-section and varying span.

Dimensions b/h = 120/220; Slenderness λ = 22 ~ 611						
Solid timber			Glued laminated timber			
C22			GL22h			
L = 1000	3000	6000	1000	3000	6000	Strain of members
0.5%	4.3%	17.3%	0.3%	2.9%	11.7%	bending about y
1.8%	16.0%	64.0%	1.0%	8.7%	34.9%	bending about z
1.7%	15.5%	62.1%	1.0%	9.0%	36.1%	bending about y and z
2.1%	19.0%	76.1%	1.2%	10.8%	43.0%	bending about y and z
0.3%	0.3%	0.3%	0.2%	0.2%	0.2%	tension
1.4%	15.2%	61.7%	0.8%	8.9%	35.9%	tension + bending
1.8%	18.7%	75.8%	1.0%	10.6%	42.9%	tension + bending
1.2%	11.6%	46.0%	0.3%	3.8%	15.2%	buckling
−0.1%	−0.1%	−0.1%	−0.1%	−0.1%	−0.1%	compression
1.7%	15.5%	62.1%	1.0%	9.0%	36.1%	compression + bending
2.1%	19.0%	76.1%	1.2%	10.8%	43.0%	compression + bending
1.6%	15.6%	62.4%	0.9%	9.0%	36.4%	buckling + bending
3.4%	30.6%	122.1%	1.5%	14.6%	58.2%	buckling + bending
0.4%	1.3%	2.6%	0.3%	1.0%	1.9%	shear in direction y
0.2%	0.6%	1.3%	0.2%	0.5%	1.0%	shear in direction z
0.5%	1.4%	2.9%	0.4%	1.1%	2.2%	shear in y and z

Table 3. Differences in results of cross-section or member utilisation between the reduced cross-section method and the reduced properties method, with the 160 / 180 mm cross-section and varying span.

	Dimensions b/h = 160/180; Slenderness λ = 29 ~ 280						
	Solid timber			Glued laminated timber			
	C22			GL22h			
L =	1000	3000	6000	1000	3000	6000	Strain of members
	0.4%	4.0%	16.1%	0.3%	3.0%	12.0%	bending about y
	0.3%	2.9%	11.6%	0.2%	2.1%	8.4%	bending about z
	0.7%	6.1%	24.2%	0.5%	4.5%	17.9%	bending about y and z
	0.6%	5.7%	22.9%	0.5%	4.2%	16.8%	bending about y and z
	0.2%	0.2%	0.2%	0.1%	0.1%	0.1%	tension
	0.5%	5.9%	24.0%	0.4%	4.4%	17.8%	tension + bending
	0.5%	5.6%	22.7%	0.4%	4.1%	16.7%	tension + bending
	0.0%	0.9%	3.4%	0.0%	0.4%	1.7%	buckling
	−0.1%	−0.1%	−0.1%	0.0%	0.0%	0.0%	compression
	0.7%	6.1%	24.2%	0.5%	4.5%	17.9%	compression + bending
	0.6%	5.7%	22.9%	0.5%	4.2%	16.8%	compression + bending
	0.6%	6.4%	25.8%	0.5%	4.7%	18.7%	buckling + bending
	0.6%	6.6%	26.3%	0.4%	4.6%	18.5%	buckling + bending
	0.2%	0.6%	1.2%	0.2%	0.5%	1.1%	shear in direction y
	0.1%	0.3%	0.6%	0.1%	0.3%	0.5%	shear in direction z
	0.2%	0.7%	1.3%	0.2%	0.6%	1.2%	shear in y and z

8 CONCLUSIONS

Advanced design methods are difficult to use in common engineering practice—engineers have to know all properties of materials and temperature in time. Although these methods are more time-consuming, they are also more accurate than other design methods. Simplified methods are easy to use, fast, and safe enough in many cases. The reduced cross-section method is more conservative than the reduced properties method, and for this reason it is the recommended one.

REFERENCES

Frangi, A., König, J. et al. 2010. Fire safety in timber buildings: Technical guideline for Europe [online]. 2010 [cit. 2015–09–09]. Retrieved from: http://eurocodes.jrc.ec.europa.eu/doc/Fire_Timber_Ch_5-7.pdf

Kamenická, Z. 2015. Timber structures and their fire resistance. In MMK 2015. 6. Mezinárodní Masarykova konference pro doktorandy a mladé vědecké pracovníky [CD-ROM]: sborník příspěvků z mezinárodní vědecké konference, 14.–18. 12. 2015, Hradec Králové, Česká republika. 1. vyd. Hradec Králové: Magnanimitas, 2015, online, s. 2679–2687. ISBN 978-80-87952-12-2.

Olbřímek, J., Štujberová, M., Osvald, A. et al. 2010. Design of Structures for Fire Resistance According to Eurocodes. Theory. Bratislava: Inžinierske konzultačné stredisko Slovenskej komory stavebných inžinierov, 2010.

Östman, B., Källsner, B. National Building Regulations in Relation to Multi-Storey Wooden Buildings in Europe in 2010?

Schmid, J., Klippel, M., Just, A., Brandon, D. 2015. Comparison of test results and the Reduced Cross-Section Method using a Zero-Strength Layer. In: Proceedings 1st European Workshop Fire Safety of Green Buildings. Berlin: Deutsche Nationalbibliothek, 2015, s. 51–53. ISBN 978-3-8440-3911-5.

STN EN 1991-1-2/AC: Eurocode 1. Actions on structures. Part 1–2: General actions. Actions on structures exposed to fire, 2009.

STN EN 1995-1-1+A1/NA: Eurocode 5: Design of timber structures. Part 1-1: General—Common rules and rules for buildings, 2008.

STN EN 1995-1-2/NA: Eurocode 5: Design of timber structures. Part 1-2: General. Structural fire design, 2011.

Advances and Trends in Engineering Sciences and Technologies II – Al Ali & Platko (Eds)
© 2017 Taylor & Francis Group, ISBN 978-1-138-03224-8

Local buckling of rectangular steel tubes filled with concrete

R. Kanishchev & V. Kvočák
Faculty of Civil Engineering, Institute of Structural Engineering, Technical University of Košice, Košice, Slovakia

ABSTRACT: The paper presents fundamental information about theoretical research oriented towards the local stability of short axially compressed rectangular Concrete-Filled Steel Tubes (CFSTs) with and without welded end-plates. The investigated members were subjected to various forms of compressive loading at its cross section. The influence of different support conditions of loaded edges of the steel section on local buckling deformation of composite structures is also described. The results of this research were obtained by simulating the behaviour of these structures in ABAQUS computational-graphics software.

1 INTRODUCTION

Requirements concerning economy and effectiveness of designed structures force designers to use progressive load-bearing structures. Composite columns made of rectangular concrete-filled hollow sections are definitely regarded as very cost-effective as they enable very fast construction and offer all the advantages of both materials—concrete and steel. These elements have distinct advantages over hollow steel tubes as described in the research works of Kvočák (Kvočák et al. 2012, Kvočák & Kanishchev 2014), Duvanova & Salmanov (2014). However, one of the main structural advantages is its significant resistance to loss of local and global stability, which allows designers to reduce the cross-section of the element.

Authors Yang & Han (2009) presented research aimed at experimentally investigating the behaviour of rectangular CFSTs loaded axially on a partially stressed cross-sectional area. Experimental tests, conducted by Lee (2007) and Storozhenko et al. (2014) on high-strength concrete-infilled steel tube columns subjected to eccentric loads, showed the influence of width-to-thickness ratio, buckling length-sectional width ratio and eccentricity ratio on the behaviour of these structures. Uy (2008), Krishan & Melnichuk (2012), in their scientific works, investigated the ultimate strength, stability and ductility characteristics of rectangular CFSTs subjected to axial compression using the high performance steels and lightweight concrete aggregates. Patel et al. (2012) proposed a multiscale numerical model for simulating the interaction of local and global buckling behaviour of eccentrically loaded high strength rectangular CFSTs with large depth-to-thickness ratios.

Nowadays, standards already exist for the design of the above mentioned structures, which are described by Kanishchev & Kvočák (2015), Kang et al. (2015). The European Union uses Eurocode 4 (2004) to design rectangular CFSTs, but the basic disadvantage of this standard is its limitations regarding the slenderness of the wall of a rectangular cross-section. The design of more efficient or economical composite structures is driving the research on class 4 hollow steel cross-sections, according to EN 1993-1-1 (2005), filled with concrete, which already lie beyond the validity of Eurocode 4.

2 STABILITY OF RECTANGULAR WALLS

The subject matter of local buckling of slender compressed walls was intensively researched by Timoshenko (1971), where he represented a differential equation for a slender wall with a length a, and width b (Figure 1), which is simply supported around its perimeter:

$$C\left(\frac{\partial^4 w}{\partial x^4} + 2\frac{\partial^4 w}{\partial x^2 \partial y^2} + \frac{\partial^4 w}{\partial y^4}\right) + P\frac{\partial^2 w}{\partial x^2} = 0, \tag{1}$$

Where w = deflection of slender walls; P = compression force; and C = cylindrical wall stiffness.

A particular solution of the differential equation is:

$$w = A\sin\frac{m\pi x}{a}\sin\frac{n\pi y}{b}. \tag{2}$$

The given solution satisfies the boundary conditions (Figure 1a, b): for $x = 0$ and $x = a \rightarrow$ $w = 0$ and $G_1 = 0$; for $y = 0$ and $y = b \rightarrow w = 0$ and $G_2 = 0$. The following conditions are fulfilled for stress: $P_1 = -P$; $U_1 = U_2 = P_2 = 0$. Thus, the value of elastic critical stress of the wall:

$$\sigma_{cr} = k_\sigma \sigma_E = \left(m\frac{b}{a} + \frac{1}{m}\frac{a}{b}\right)^2 \frac{\pi^2 E t^2}{12(1-v^2)b^2}, \tag{3}$$

Where k_σ = coefficient of critical stress; E = modulus of elasticity of the steel; t = the wall thickness; v = Poisson's ratio.

The basic principles of designing class 4 cross-sections were stipulated by Bryan (1891), which offers a critical analysis of the elastic stress σ_{cr} for local buckling of long right-angled wall elements. The term of this stress includes various boundary conditions with the aid of the coefficient of critical stress k_σ:

$$\sigma_{cr} = k_\sigma \frac{\pi^2 E}{12(1-v^2)}\left(\frac{t}{b}\right)^2. \tag{4}$$

The minimum values of the coefficient of critical stress k_σ are stipulated in EN 1993-1-5 (2006). This coefficient can be used for hollow rectangular tubes. When the tube is filled with concrete, the standard does not provide a k_σ value.

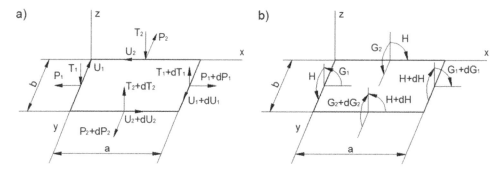

Figure 1. Mathematical model of a slender wall: a) components of force in a unit element of the wall; b) components of the bending moment in a unit element of the wall.

3 SIMULATION OF INVESTIGATED SPECIMENS AND OBTAINED RESULTS

The tube walls of hollow rectangular cross-sections, which are uniformly compressed at the edges, meet the conditions stipulated in section 2 (Figure 2a). Therefore, the simulation of 20 samples of empty rectangular steel tubes (Figure 3a, b) and 40 samples of concrete-filled rectangular steel tubes (Figure 3c–f) was generated in ABAQUS 6.13–4 to determine the effect of filling concrete on local stability of rectangular tubes.

An analysis was carried out on the rectangular cross section of tubes *RHS 20Cx100x3* (Class 4 according to EN 1993-1-1) without imperfections; the length of the column varied between 94 mm and 263 mm. The material characteristics of the steel cross-sections, defined by Eurocode 3: steel class *S235*, elastic modulus *E = 200 GPa*, Poisson's ratio in elastic state *v = 0.3*, were used. The behaviour of the material is modelled as elastoplastic with linear hardening according to EN 1993-1-5.

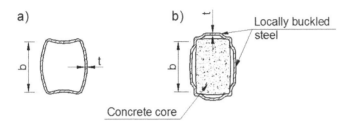

Figure 2. a) local buckling deformation of the cross section of the hollow specimen at the midpoint of the length; b) local buckling deformation of the cross section of the concrete filled specimen at the midpoint of the length.

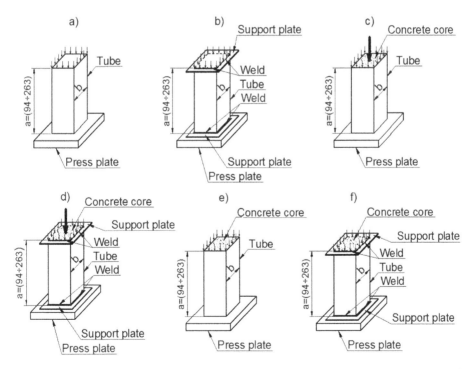

Figure 3. Scheme of the steel-concrete specimens loading: a) on the hollow steel tube; b) on the hollow steel tube through the welded support plate; c) on the steel tube and concrete core; d) on the steel tube and concrete core through the welded support plate; e) on the steel tube; f) on the steel tube through the welded support plate.

143

The core of the hollow section was modelled from concrete C20/25 according to EN 1992-1-1 (2005). A library of ABAQUS elements was used in modelling: the steel section was made from the "shell-type" S4R elements and the concrete core from the "solid-type" C3D8 elements. The simulation of the boundary conditions of the walls of the empty and concrete-filled cross-section has been created in two variants: according to section 2 (hinge supports) and by using the fictive weld of the end (support) plates on both ends of the tube, which means that it was prevented from rotation of its edges.

The interaction of the steel and concrete elements was modelled by means of two components: the "normal", reflected as the compression of concrete on the steel section and "tangential", which represents shear resistance at the steel-concrete interface. The coefficient of friction was regarded as 0.35. Loading of columns was modelled as short-term, rising steadily at a constant rate through the plate of a hydraulic press. Elastic critical stresses of the walls were investigated in the greater wall of the steel cross-section upon attaining first buckling form (Figures 4–6). The results of comparison are shown in Tables 1–2.

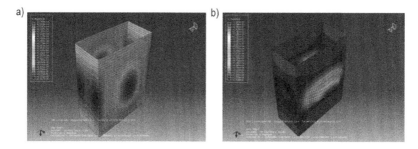

Figure 4.   3D deformation of the hollow tube with length L = 188 mm: a) hinge support of loaded edges; b) clamped support of loaded edges.

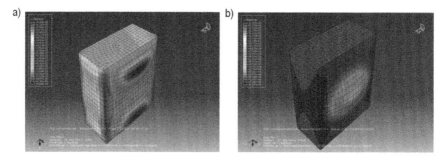

Figure 5.   3D deformation of the CFST with length L = 188 mm: a) hinge support of loaded edges; b) clamped support of loaded edges.

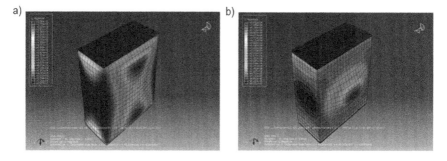

Figure 6.   3D deformation of the CFST when loading of steel part of the composite cross-section with length L = 188 mm: a) hinge support of loaded edges; b) clamped support of loaded edges.

Table 1. Comparison of coefficients of critical stress obtained for hollow tubes.

a [mm]	b [mm]	a/b	Numerical calculation according to section 2 σ_{cr} [MPa]	k_σ	Hollow tubes in ABAQUS σ_{cr} [MPa]	k_σ	Hollow tubes with welded end plates in ABAQUS σ_{cr} [MPa]	k_σ
94	188	0.5	270.16	6.25	221.33	5.12	248.11	–
113	188	0.6	222.09	5.14	194.13	4.49	242.40	–
132	188	0.7	195.85	4.53	179.61	4.16	238.60	–
150	188	0.8	181.66	4.20	176.37	4.08	235.35	–
169	188	0.9	174.83	4.05	180.32	4.17	232.70	5.38
188	188	1	172.90	4.00	185.89	4.30	224.97	5.20
207	188	1.1	174.48	4.04	197.31	4.57	224.01	5.18
226	188	1.2	178.72	4.13	206.02	4.77	226.82	5.25
244	188	1.3	185.08	4.28	223.58	5.17	228.92	5.30
263	188	1.4	193.23	4.47	231.20	5.35	229.64	5.31

Table 2. Comparison of coefficients of critical stress obtained for CFSTs.

a [mm]	b [mm]	a/b	CFST in ABAQUS σ_{cr} [MPa]	k_σ	CFST with welded end plates in ABAQUS σ_{cr} [MPa]	k_σ	CFST with loading of steel part of the composite cross-section in ABAQUS σ_{cr} [MPa]	k_σ	CFST with welded end plates and loading of steel part of the composite cross-section in ABAQUS σ_{cr} [MPa]	k_σ
94	188	0.5	272.87	–	220.94	5.11	273.48	–	258.825	–
113	188	0.6	272.16	–	219.19	5.07	270.81	–	251.203	–
132	188	0.7	271.27	–	218.54	5.06	262.81	–	243.691	–
150	188	0.8	271.59	–	218.79	5.06	258.27	–	235.059	–
169	188	0.9	271.05	–	218.94	5.07	251.67	–	230.264	5.33
188	188	1	270.85	–	216.33	5.00	269.83	–	224.026	5.18
207	188	1.1	276.50	–	226.26	5.23	272.11	–	225.233	5.21
226	188	1.2	277.38	–	226.54	5.24	275.64	–	226.115	5.23
244	188	1.3	277.60	–	227.03	5.25	277.31	–	226.464	5.24
263	188	1.4	281.75	–	227.78	5.27	282.29	–	227.483	5.26

4 CONCLUSION

The minimum theoretical value for the coefficient of critical stress k_σ equals *4* for aspect ratio $a/b = 1$ and the hinge support of a loaded edges (Table 1). This value is given in EN 1993-1-5. The modelling of hollow rectangular tubes in ABAQUS computational-graphics software produced a minimum value of $k_\sigma = 4.08$ for aspect ratio $a/b = 0.8$, which exceeds the theoretical value for 2%. The difference in the aspect ratio shows that the modelling considered a rectangular cross section, where, due to various wall stiffness, the stress-redistribution was attained.

Specimens with welded end plates on the loaded edges have greater stiffness compared to empty rectangular tube without plates. It means, that the local buckling of walls in the elastic area occurs when the aspect ratio $a/b = 0.9$ (Table1) and the minimum value of the coefficient of critical stress in this case is higher for 21%.

The concrete core of the tubes filled with concrete increases the local stability of the steel part of the composite cross-section (Table2) to the extent that the local buckling phenomenon occurs in the field of plastic material behavior. This deformation was most apparent in relatively long tubes closer to the end of the above mentioned elements as is shown in Figure 5a. The change of boundary conditions in the walls of these constructions by welding end plates reduces the critical stress in the elastic field of the material behavior, while the minimum value of the coefficient of critical stress compared to empty tubes without end plates is higher for 18%. In case of the CFSTs with the loading of steel part of the composite cross-section without welded end plates, the critical elastic stresses were not found, because the behavior of the steel material lies in the plastic field. The welding of the end plates decreases the values of critical stresses (see Table 2).

AKNOWLEDGEMENT

The paper presented was supported by the projects: VEGA 1/0188/16 "Static and Fatigue Resistance of Joints and Members of Steel and Composite Structures" of the Scientific Grant Agency of the Ministry of Education, science, research and sport of the Slovak Republic and the Slovak Academy of Sciences.

REFERENCES

Bryan, G.H. 1891. On the stability of a plane plate under trusts in its own plane, with applications to the "bukcling" of the sides of a ship. *Proceedings of the London Mathematical Society* Vol. 22: 54–67.
Duvanova, I.A. & Salmanov, I.D. 2014. Trubobetonnie kolonni v stroitelstve visotnych zdanii i sooruzhenii [Pipe-concrete columns in the construction of tall buildings and structures]. *Construction of Unique Buildings and Structures* 6(21): 89–103.
EN 1992-1-1: 2005. Eurocode 2. Design of concrete structures. Part 1-1: General rules and rules for buildings. CEN, Brussels.
EN 1993-1-1: 2005. Eurocode 3. Design of steel structures. Part 1-1: General rules and rules for buildings. CEN, Brussels.
EN 1993-1-5: 2006. Eurocode 3. Design of steel structures. Part 1-5: General rules—Plated structural elements. CEN, Brussels.
EN 1994-1-1: 2004. Eurocode 4. Design of composite steel and concrete structures. Part 1-1: General rules and rules for buildings. CEN, Brussels.
Kang, W.H., Uy, B., Tao, Z. & Hicks, S. 2015. Design strength of concrete-filled steel columns. *Advanced Steel Construction* 11(2): 165–184.
Kanishchev, R. & Kvočák, V. 2015. Effects of Stability on the Resistance of Composite Concrete-Filled Rectangular Steel Pipes According to World Standards. *Nara: IABSE, Elegance in Structures*: 1–8.
Krishan, A.N. & Melnichuk, A.S. 2012. Prochnost trubobetonnych kolonn kvadratnogo sechenia pri osevom szhatii [The strength of pipe-concrete square columns under axial compression] *Bulletin of the Nosov magnitogorsk state technical university* 3: 51–54.
Kvočák, V., Varga, G. & Vargova, R. 2012. Composite steel concrete filled tubes. *Procedia Engineering* Vol. 40: 469–474.
Kvočák, V. & Kanishchev, R. 2014. Spriahnuté oceľobetónové tlačené prúty a stĺpy vyplnene betónom [Composite steel and concrete compression bars and columns filled with concrete]. *Vedecko-výskumná činnosť UIS 2013*: 107–116.
Lee, S. 2007. Capacity and the moment-curvature relationship of high-strength concrete filled steel tube columns under eccentri loads. *Steel and Composite Structures* 7(2): 135–160.
Patel, V. I., Liang, Q. Q. & Hadi, M. 2012. Inelastic stability analysis of high strength rectangular concrete-filled steel tubular slender beam-columns. *Interaction and Multiscale Mechanics* Vol. 5 (2): 91–104.
Simulia Abaqus 6.13. 2013. *Analysis Users Guide. Volume I.*
Storozhenko, L.I., Ermolenko, D.A. & Demchenko, O.V. 2014. Rabota pod nagruzkoi szhatych trubobetonnych elementov s usilennymi jadrami [The work under load of compressed pipe-concrete elements with strengthened cores]. *Efficiency of resource energy of technology in the construction industry of region* 4: 288–292.
Timoshenko, S. P. 1971. Ustojchivost sterghnej, plastin i obolochek [*Stability of rods, plates and shells*] Moskva: Nauka.
Uy, B. 2008. Stability and ductility of high performance steel sections with concrete infill. *Journal of Constructional Steel research* Vol. 64: 748–754.
Yang, Y. & Han, L. 2009. Experiments on rectangular concrete-filled steel tubes loaded axially on a partially stressed cross-sectional area. *Journal of Constructional Steel Research* Vol. 65: 1617–1630.

Advances and Trends in Engineering Sciences and Technologies II – Al Ali & Platko (Eds)
© 2017 Taylor & Francis Group, ISBN 978-1-138-03224-8

Experimental study of a four-point bending test on CLT deep beams

O. Katona, T. Klas, V. Duchoň, J. Brodniansky, Ľ. Balcierák & J. Sandanus
Department of Steel and Timber Structures, Faculty of Civil Engineering, Slovak University of Technology in Bratislava, Bratislava, Slovakia

ABSTRACT: In the last few years the building product Cross Laminated Timber (CLT) has become very common in timber engineering applications. The number of layers varies normally from three to seven and upwards. Currently, limited information is available for designers engaged in applications that require calculations of the in plane shear strength of CLT beams, deep beams (lintels, headers). This report presents four-point bending tests performed on several CLT deep beams carried at the laboratory of STU in Bratislava, Faculty of Civil Engineering. Two series (3 specimens of each) of CLT deep beams were tested, changing orientation of the outer and inner layers (parallel/perpendicular to the span direction). The experimental campaign's goal is to compare the failure mode, how can the arrangement of lamellas influence the stresses in the cross section.

1   INTRODUCTION

The CLT is a cost-competitive wood-based solution that complements the existing light and heavy-frame options, and is a suitable substitute for some applications which currently use concrete, masonry and steel. The boards within each single layer are placed parallel to each other but orthogonal to the direction of the neighbouring layers. Thus the local mechanical properties of single layers are not the same as the global panels.

The designers, for relatively small and short CLT lintels and beams, can choose the simplified and conservative approach, where only layers running parallel to the span are taken into consideration. However by more exact analysis it is possible to achieve higher resistances, thanks to better stress distribution between lamellas.

2   CLT ELEMENTS UNDER LOADS IN-PLANE

The design methods and the strength properties are given in different technical approvals for the verification of in-plane shear stresses vary significantly, and for most products no information on the shear stiffness in-plane is given at all. The first reason is the complex calculation of shear stresses and deformations in CLT elements compared to traditional timber materials. Nevertheless, in many cases vastly simplified methods are used.

2.1   *Internal in-plane stresses*

An efficient mechanical model for the internal stress pattern in CLT elements has been discussed and presented in different papers by University of Graz (Bogensperger T. 2010) research team.

An elementary Representative Volume Element (RVE) has been introduced, which represent the intersection between orthogonal timber boards in the individual layers. RVE is the smallest unit cell for static verifications and mechanical treatment, whose internal state could describe the global stress pattern of the CLT element. The size of one particular RVE—in thickness is equal to a CLT element and in width and depth to the width of one board plus the half of the width of gaps between adjacent boards. This element can be even further reduced to an Representative Volume Sub-Element (RVSE), which is the smallest possible element for stiffness calculation and bearing verification. RVSE is in width and depth equal to the RVE, the thickness composed of the minimum of the adjacent halved board thicknesses on both sides of the adhesive layer, as plane of symmetry.

A basic assumption in Bogensperger T. 2010, that in principle three types of shear mechanism can be distinguished: Mechanism I "net-shear"—shearing perpendicular to grain of the net cross sections in the controlling plane; Mechanism II "torsion"; Mechanism III "gross-shear"—shearing parallel to grain of the whole CLT element (see Figure 1). The calculation of shear stresses can be divided into two steps: shear stresses in the ideal RVSE—represent an infinite sequence of layers in thickness direction; an extension to a real CLT-element with finite number of layers.

In case of CLT-elements without lateral gluing interfaces at the narrow faces, shear stresses can only appear in end-grain sections, while narrow faces are free of those stresses. Thus the shear forces can only be transferred indirectly across the crossing of two boards in adjacent single layers.

Mechanism I—"neat-shear" (shearing perpendicular to grain) considers the transfer of shear forces via the cross sections of boards within an RVSE. The real shear stress can be calculated with equation 1.

$$\tau_v = 2 \cdot \tau_0 \tag{1}$$

$$\tau_T = \frac{M_T}{W_p} = \frac{\tau_0 \cdot t \cdot a^2}{\frac{a^3}{3}} = 3 \cdot \tau_0 \cdot \frac{t}{a} = \frac{3 \cdot n_{xy}}{n \cdot t} \cdot \frac{t}{a} = \frac{3 \cdot n_{xy}}{n \cdot a} \tag{2}$$

Where M_T = torsional moment; W_p = polar moment of resistance.

As it was mentioned above, the calculation of shear stresses can be divided into two steps. In a second step an extension to a real CLT-element with finite number of layers is carried out. The ideal thicknesses (t_i^*) for each RVSE, can be calculated as shown in Table 1. The overall thickness of all ideal RVSEs Σt_i^* is always smaller than or equal to the geometric overall thickness t_{CLT} of the CLT-element. The symmetry is given in thickness direction, resulting in $t_4 = t_2$, $t_5 = t_1$.

The ideal nominal shear stress can be calculated by dividing the proportionate shear force by the thickness t_i^* of the ith RVSE (see equation 3). The shear stress τ_0^* is constant for all RVSEs.

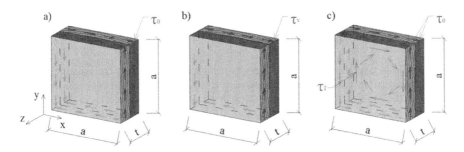

Figure 1. Shear stresses in an RVSE: a) nominal shear stress τ_0 (glued on narrow faces, mechanism III); b) real shear stress τ_v (mechanism I); c) torsional stress τ_T (mechanism II).

Table 1. Geometry.

Ideal thickness t_i^*	$t_1^* = \min(2t_1; t_2)$	$t_2^* = \min(t_2; t_3)$	$t_3^* = \min(t_3; t_4)$	$t_4^* = \min(t_4; 2t_5)$
No. of RVSE	12	2	3	4

148

$$n_{xy,RVSEi}^* = \tau_{0,RVSEi}^* = \frac{n_{xy} \cdot \dfrac{t_i^*}{n-1}}{\displaystyle\sum_{i=1}^{n-1} t_i^*} = \frac{n_{xy}}{\displaystyle\sum_{i=1}^{n-1} t_i^*} = \tau_0^* \tag{3}$$

In the case of in-plane bending action "M" applied to the CLT panel, the bending stress should be calculated taking into account only the boards parallel to the stress direction. The bending stress is defined in relation to the net cross-section modulus W_{net} (according to CUAP document 2005):

$$\sigma = \frac{M}{W_{net}} = \frac{M}{\dfrac{B \cdot H^2}{6}} = \frac{M}{\dfrac{\Sigma b_i \cdot H^2}{6}} \tag{4}$$

3 BENDING TEST CONFIGURATION OF CLT DEEP BEAMS

3.1 *Test materials and geometry*

The experimental test program was carried out at the laboratory of STU in Bratislava, Faculty of Civil Engineering. As it was mentioned above, this study deals with how the arrangement of lamellas and their thickness influence the stresses in the cross section. Two series (3 specimens of each) of CLT deep beams (see Figure 2) were tested under four-point bending tests, changing the orientation of the outer and inner layers (parallel/perpendicular to the span direction). The beams were manufactured from solid wood lamellas of strength class C24, composed of three thick layers (see Table 2). The specimens of length 2,95 m had a cross section of 100×300 mm², composed of three layers 30–40–30 mm glued crosswise to each other. All specimens had non-visible quality (NVI), where dry cracks, knots etc. are permitted.

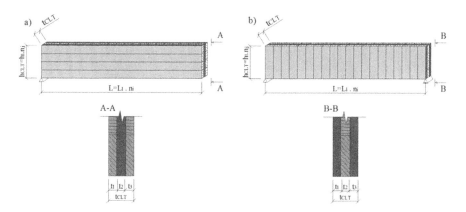

Figure 2. Geometry of CLT beam specimens.

Table 2. Geometry of specimens.

Type	Number of spec.	$\dfrac{L}{m}$	$\dfrac{h_{CLT}}{m}$	$\dfrac{t_{CLT}}{mm}$	$\dfrac{A}{mm^2}$	$\dfrac{I}{mm^4}$
A1	3	2, 95	0, 3	30–40–30 = 100	$3, 0*10^4$	$2, 25*10^8$
B1	3	2, 95	0, 3	30–40–30 = 100	$3, 0*10^4$	$2,25*10^8$

3.2 Test setup

The bending test was carried out under a four-point loading test arrangement according to European Standard EN 408:2005, CUAP 2005. All test beams were simply supported over 2,45 m span and they were subjected to two-point loading at the centre distance corresponding to 1/3 of the beam span, according to the experimental test setup Växjö—Andreolli M., EN 408. The concentrated loads were distributed to the top of the beam via two steel plates (220 mm × 140 mm × 20 mm). The test assemblies with the attached instrumentation during the bending test is shown in Figure 3.

8 LVDT (Linear Voltage Displacement Transducer) were attached in the neutral axis to each test assembly in order to monitor the vertical deflections and to determine the MOE. All measurements were made on both faces of specimens.

The test conducted under displacement control was performed using a hydraulic MTS actuator, able to apply a maximum load of 500 kN. Certain load levels were chosen according to STN EN 380 (Katona O. & Klas T. 2016-Figure 7), where G_2 is the sum of the self weights of the loading construction, of the hydraulic MTS actuator and of the specimen (see Table 3).

After loading the specimens up to $G_2 + n.Q$ and reloading, they were loaded up to failure in order to determine the ultimate strength. All data relating to the vertical deflection and the applied load were registered with a data control unit.

3.3 Experimental results

The values of the maximum loads and maximum deflections at failure for specimens are presented in Table 4. The load-deformation curves for all test specimens are shown in

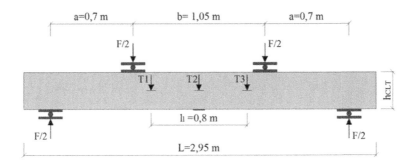

Figure 3. Bending test setup.

Table 3. Loading levels.

Specimen	Number of spec.	G_2 kN	Q kN
A1, B1	6	$2,05 + 0,5 + 0,45 = 3,0$	20

Table 4. Summary of the bending test.

	Spec. 1		Spec. 2		Spec. 3		Mean
	F_{max}	w	F_{max}	w	F_{max}	w	F_{max}
Specimen	kN	mm	kN	mm	kN	mm	kN
A1	118,14	22,08	78,12	13,84	127,44	22,99	107,9
B1	90,6	19,67	94,14	26,42	131,7	22,98	105,48

150

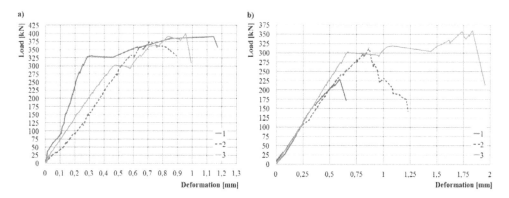

Figure 4. Load-deformation graphs, a) specimens with parallel oriented outer layers to the span direction; b) specimens with perpendicular oriented outer layers to the span direction.

Figure 5. Failure mode for A1, B1 specimen.

Figure 4, where A1(1, 2, 3) specimens represents the beams with parallel oriented outer layers to the span direction and the B1(1, 2, 3) the beams with perpendicular outer layers to the span direction (see Figure 2).

Considering the tested specimens as constituting deep beams, three major types of deformations can be distinguished: bending, shear and local material compression at the loading and supporting areas. The test specimens A1, B1 failed in most of the cases by a tensile bending failure starting from a finger joint or group of knots. Compression failures were almost invisible at the loading, supporting areas and on the upper side of the beams between the points of load application.

3.4 Modulus of elasticity

The Modulus of Elasticity (MoE) has been calculated according to the formula (5) reported in EN 408, taking into account the net moment of inertia of the multilayer section. From the experimental load-deflection curve account is taken of the longest line between $0,1F_{max}$ and $0,4F_{max}$, whose coefficient of correlation is higher than $0,99$ (line included at least between $0,2F_{max}$ to $0,3F_{max}$). The local E modulus can be calculated according the following equation:

$$E_{m,l} = \frac{a \cdot l_1^2 \left(F_2 - F_1\right)}{16 \cdot I \left(w_2 - w_1\right)} \tag{5}$$

where $F_2 - F_1$ = load increase [N]; $w_2 - w_1$ = displacement increase corresponding to $(F_2 - F_1)$ [mm];

151

Table 5. Local E modulus values according to EN 408:2005.

	Specimen A1			Specimen B1		
	Spec. 1	Spec.2	Spec. 3	Spec. 1	Spec. 2	Spec. 3
F_1	23,1 kN	15,75 kN	26,0 kN	20,0 kN	19,0 kN	32,6 kN
F_2	33,0 kN	23,15 kN	37,75 kN	24,05 kN	21,05 kN	39,35 kN
w_1	0,21 mm	0,195 mm	0,3 mm	0,3 mm	0,245 mm	0,305 mm
w_2	0,33 mm	0,324 mm	0,45 mm	0,365 mm	0,277 mm	0,4 mm
$E_{m,l}$	10,87 GPa	7,14 GPa	9,75 GPa	7,75 GPa	7,97 GPa	8,01GPa
Mean $E_{m,l}$	9,253 GPa			7,91 GPa		

Table 6. Experimental stress values for different specimen ($F_{max,mean}$).

	σ [MPa]	$f_{m,CLT,k}$ [MPa]	τ_{zx} [MPa]	τ_{xz} [MPa]	$f_{v,CLT,k}$ [MPa]	τ_T [MPa]	$f_{v,CLT,k}$ [MPa]
A1	31,471	24,0	2,997	4,496	5,0	3,29	2,5
B1	28,324		5,395	8,093		5,921	

a = distance between the load point and the support (Figure 3) [mm]; l_l = reference length for E modulus determination [mm]; and I_{net} = net moment of inertia [mm^4].

The calculated values of MoE and the mean values of the series of tests are shown in Table 5.

From the experimental values of ruptures the following bending and shear stress values have been carried out according to the model described in section 2.1.

4 CONCLUSION

In this work deep beams were mentioned made of CLT, which offers several advantages over solid or glued laminated timber beams due to their typical layup of orthogonally bonded layers. Experimental setup and procedure for four-point bending test according to EN 408 is shown. The types of failure were strongly related to the internal geometry of the specimens. The tested specimens reached a collapse showing failure modes: bending failure starting form finger joint or knot and shear failure mode.

REFERENCES

Andreolli M., Tomasi R. & Polastri A. *Experimental investigation on in-plane behaviour of cross-laminated-timber elements*, Paper and Presentation, CIB-W18/45-12-4, Växjö, Sweden.

Bogensperger T., Moosbrugger T. & Silly G. *Verification of CLT-plates under loads in plane*, Word Conference on Timber Engineering, Riva del Garda (Italy), 2010.

CUAP Common Understanding of Assessment Procedure: *Solid wood slab element to be used as a structural element in buildings*, ETA request No 03.04/06, prepared by Österreichisches Institut für Bautechnik, Schenkenstraße 4, 1010 Wien, Austria, June 2005.

EN 1995. Eurocode 5: *Design of timber structures*.

EN 408:2005, *Timber structures—Structural timber and glued laminated timber, Determination of some physical and mechanical properties*.

Katona O. & Klas. T. *Bending test on CLT deep beams*, Juniorstav 2016, 18. Odborná konference doktorského studia s mezinárodní účastí, Brno, ČR, ISBN 978-80-214-5311-1, 2016.

STN EN 380, *Drevené konštrukcie, Skúšobné metódy, Všeobecné zásady skúšania statickým zaťažením*, SÚTN, 1998.

Advances and Trends in Engineering Sciences and Technologies II – Al Ali & Platko (Eds)
© 2017 Taylor & Francis Group, ISBN 978-1-138-03224-8

Aerodynamic analysis and improvements of a solar car

V. Kiricci, A.O. Celik & C. Insel
Department of Civil Engineering, Anadolu University, Eskisehir, Turkey

E. Guner
Department of Material Science and Engineering, Anadolu University, Eskisehir, Turkey

ABSTRACT: In this study, aerodynamic performance of a solar car (Anadolu Solar Team) is investigated using Computational Fluid Dynamics (CFD). This work represents the current efforts in designing a new vehicle. Aerodynamic concepts were used to modify the previous car, namely the Sunatolia II. A new design is performed by eliminating the aerodynamic efficiency drawbacks of the Sunatolia II. Aerodynamic calculations include two essential force components as drag force and lift force acting on the body (vehicle). The goal is to keep lift force in optimum level and drag at a minimum for safe speeds and with minimum energy consumption. Aerodynamics of five different vehicle geometries is studied (without body accessories/additions, air breather, side bars, back flap, and vehicle with all added). The CFD analyses were performed with proper turbulence closure, mesh independency tests and a transient solution for accuracy. The results represent one of the shortlisted final designs.

1   INTRODUCTION AND AIM

Aerodynamic concerns in automobile design became an important factor and high performance expectations while improving energy efficiency have contributed to the increase in design studies over the past century. The main parameters that dominate the aerodynamic performance of a vehicle can be defined as drag and lift forces. It plays a critical role in keeping these forces in optimum level to improve driving safety, performance and energy efficiency. The drag and lift forces affects the car in different ways. Drag force basically can be expressed as a resistance force to the direction of the vehicle motion and the lift force acts perpendicularly. Lift force in the direction of gravity makes road handling of vehicle and motion on the curve more effective. It increases the frictional force between the road and the tires and also increases the energy consumption. A moderate level positive lift force can be considered as the most favorable condition for a solar car.

Solar cars as alternative energy vehicles provide energy from solar panels covered on a limited area and the panel area is directly related to engine power. Due to these limitations, the panels should be as efficient as possible and the car should be designed in accordance with aerodynamic principles. The specific objective of this study was to show the importance of appropriate aerodynamic design for solar cars and improvement on an existing solar car (Sunatolia II) of Anadolu University Solar Team in Turkey which is the first Turkish solar car to finish the World Solar Challenge in Australia and finished the 2014 Sasol Solar Challenge in second place.

This paper also attempts to represent current efforts in designing a new vehicle. Aerodynamic concepts are used to modify and improve performance of the previous car. The new design is performed by eliminating the aerodynamic efficiency drawbacks of Sunatolia II. The studies to date have tended to focus on major changes in solar car geometries rather than minor details. The results obtained from the preliminary analysis have proven that smallest change on vehicle's geometry affects the vehicle's performance regarding safety, velocity and energy efficiency. For this reason, embedded geometric features (air breather, side bars and back flap) and aerodynamic performance of these additions for the modified new solar car (one of the shortlisted final prototype designs) are investigated using CFD.

2 METHODOLOGY AND STUDIES

The fundamental fluid mechanics equations provide satisfactory results for estimating flow around simple geometries under ideal conditions. The real air flow and geometrical shapes require a much more advanced way of attacking the problem to understand and simulate the dynamic flow behavior. In this study, the aerodynamic performance of a solar car was ana-lyzed using CFD method. CFD is a computer supported numerical method which is based on solution of discredited fluid dynamics equations. ANSYS CFX software is used for the simulations as a powerful CFD solver. Performance of the geometrical modifications on the solar car was modeled separately to control the progress (with and without body accessories/additions) and isolate the individual effects of each modification. The results indicate that simple geometric details might lead to considerable positive and negative impacts on design. Recent advancements in CFD codes and computer technology have allowed to simulate realistic conditions and predict flow patterns around the vehicles accurately. Computational simulation tools can also reduce the time and cost that is needed to test aerodynamic per-formance when compared with the wind tunnel test (Hucho, 1998). Initial methodology was prepared in the view of literature and the experience of Sunatolia II, schematically shown in Figure 1.

High performance race cars are the main area of interest for the aerodynamic improve-ment researches. Focusing on reducing drag is slightly shifted towards reducing lift for the race cars. One of the most popular add-on is the inverted wing. Many researchers tried to reduce lift by using similar details. This kind of addition does not affect only the lift, it also affects the drag negatively (L.J.R. Kloek, 2013).

Dimensions of the solution domain and the location of the car are the important factors that affect the flow behavior. Boundary zones in solution domain must be far enough to eliminate the possibility of affecting flow. It was then inserted in solution domain geometry as shown in Figure 2. The vehicle's Length (L) is 4.5 m, Width (W) is 1.80 m and maximum Height (H) with wheels is 1.1 m.

Generating numerical grid is one of the most important steps for a CFD solution. How-ever, it has generally been assumed that the finer elements provide more sensitive and accu-rate results, the key factor is determining required grid resolution according to the physical conditions of the problem and using the optimum element sizes. Number of unnecessarily small elements will only cause an increase in the computational load without any significant effect on results. Using inflation layers on wall surfaces and local refinements at specific zones

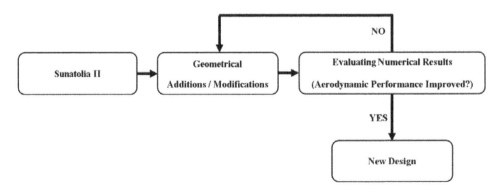

Figure 1.   Design schematic.

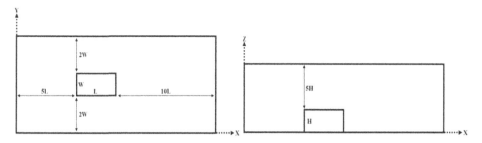

Figure 2.   Geometrical details of the solution domain.

Figure 3.   Numerical grid of the solution domain.

where more detailed and accurate solution is expecting will be a more adequate approach. The solar car's surface is meshed with hexahedral block elements to resolve inner flow layer. Smaller mesh sizes are used for the possible flow separation and negative pressure zones. Maximum mesh element size for the entire solution domain is 0.2 m and size of elements at refined area is 0.04 m. The surface of solar car is meshed with element size of 0.01 m. Total number of elements of vehicle without any aerodynamic additions is approximately 24 million (Figure 3).

In this study, velocity-inlet and pressure-outlet boundary conditions are used. Symmetry condition is applied on side walls of the solution domain to avoid disturbing the wind flowing span-wise direction. Bottom surface of the domain and the solar car surface are defined as no slip wall and top of the domain is defined as free slip wall. Shear Stress Transfer (SST) turbulence model is preferred for the solutions as a powerful turbulence closure for external aerodynamic analysis. It accounts for the transport of the turbulent shear stress and also able to simulate flow separation under adverse pressure gradients (P.B. Meederira, 2015). After solver parameters were adjusted, iterative numerical solution was performed and convergence criteria was determined as 0.0001.According to the results, it is initially observed that the aerodynamic performance of the modified design is better than Sunatolia II in terms of aerodynamic forces acting on car (Table 1).

The solar car model is solved separately with wheels, without wheels and wheels area in mudguard. A significant increase in the drag coefficient is observed in the model with mudguards (Figure 4). This is presumably due to the fact that the trapped air in mudguards creates more drag while decreasing the lift force (Table 1).

Air-breathers are used to turf out entrapped air in mudguards with a cutting angle of 15° (Figure 5). Air breathers decreased the drag coefficient from 0.150 to 0.148. This modification also improved the lift force from 101. 01 N to 105.95 N.

Studies concerning rear screen (a plate behind the vehicle) indicated that the drag could be reduced up to 6.5%. Drag force and the distance between the vehicle and the plate are inversely correlated. Effect of rear plate pushes air flow to separate from the vehicle surface. Vortices are formed away from vehicles rear surface (U.S. Rohatgi, 2012). It also leads

Table 1. Comparing Sunatolia II and the modified new design in 90 km/hr.

Parameters	Sunatolia II	The modified design	Solar car without wheels and wheels area in mudguard	Solar car with wheels and wheels area in mudguard
Drag force (N)	80.57	44.12	35.45	44.12
Drag coefficient	0.231	0.150	0.121	0.150
Lift force (N)	−21.57	101.31	104.182	101.31
Lift Coefficient	−0.00835	0.03378	0.03474	0.03378
Lift (kg)	−2.2	10.3	10.6	10.3

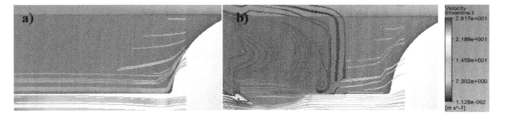

Figure 4. a) Side view of streamlines around the mudguards (without wheels and wheels area in mudguard) b) Side view of streamlines around the mudguards (with wheels and wheels area in mudguard).

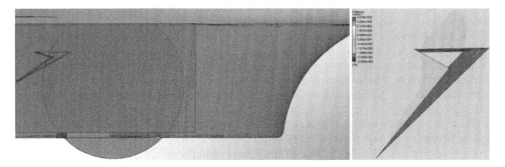

Figure 5. Air-breather on mudguard of solar car and pressure distribution contours around air-breather.

an increase in the length of vehicle. The length of the solar car is limited and cannot be expanded due to certain regulations. For this reason, a back flap with a length of 5 cm was inserted on rear edge of the solar car without violating the regulations. Additionally, the solar cars also require positive lift (opposite direction of gravity) in conjunction with reduced drag. Generally, positive lift is not resorted for conventional automobiles or race cars. Therefore, reverse back flaps or airfoils were used on back side of vehicles to get negative lift force. Particularly, back flap (rear wing) should be inserted with plates on its end. Rear wing with end plates straightens the air flow and generates small vortices. Rear wing without end plates leads large vortices which results an increase in drag and lift forces (S. Chandra et al. 2011). End plates are used on sides of the back flap with 30° between the flap and back of the solar car (Figure 6).

Vehicle with back flap has a drag coefficient of 0.171. Back flap also causes an increase in pressure in the opposite direction of gravity at the back of the car surface. As a result, lift force increased up to about 215,74 N because of the pressure gradient (Figure 7).

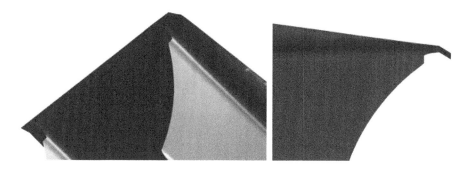

Figure 6.　CAD geometry of vehicle's back sides with back flap.

Figure 7.　Pressure distribution contours at the top and bottom view of back flap respectively.

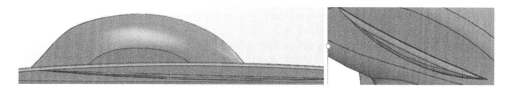

Figure 8.　Side-bar on right side of airfoil.

Table 2.　Change of drag and lift coefficients of all geometries in 90 km/hr.

Vehicles	Drag coefficient	Lift coefficient
Without any additions	0.150	0.03378
With air-breathers	1.7% decrease	4.9% increase
With back flap	13.8% increase	113.4% increase
With side-bars	6.7% decrease	5.3% decrease
With all of additions	36.4% increase	116.8% increase

Steering the air flow could be considered as the main rule of aerodynamic design. Influence of steering is not only at the end of the vehicle; it is also effective on the middle surface of vehicle. This requires an apparatus like a side-bars. In this study, side-bars are also used for increasing the lift force. Length of side-bars used on two sides of car is 2.42 m and the maximum thickness was 0.05 m (Figure 8).

Early separation of the air flow on the side surface towards the upper side of the vehicle reduced the drag coefficient from 0.15 to 0.14. Side-bars caused separation of the air flow from the side surface of mudguards to point of junction of airfoil and mudguards (Figure 9). In addition, lift force is reduced from 101.01 N (lift of vehicle without any additions) to 95.12 N. The last part of the analyses is performed with all of the modifications embedded

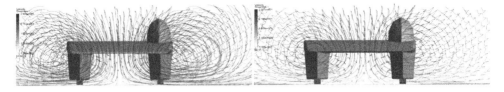

Figure 9. Vortex on back-view of vehicle with and without additional parts.

on the vehicle body. Additions which are in the direction of air flow effect drag negatively. According to the last analysis (with all the modifications included), drag coefficient increased from 0.150 to 0.204, while the lift force also increased from 101.01 N to 219,66 N.

3 CONCLUSION

The purpose of the current study was to improve the aerodynamic performance of an existing solar car. Effects of the applied modifications on to car geometry were investigated using CFD analysis. Drag and lift forces were determined for each isolated geometrical modification and the results were compared. Air-breathers are shown to be useful additions. Trapped air with high pressure came out from mudguards through air-breathers (Figure 5) and caused a decrease in drag by 1.7%. Using back flap increases the lift (kg) by up to 113.4%. The high rate of increase in lift is quite useful for solar cars that compete in world challenges. Side-bars leads to drag reduction as a result of early separation of flow. Air flow on side of the vehicle did not proceed to the region of conjunction between back-side of airfoil and mudguards resulting in lift reduced from 10.3 kg to 9.78 kg.

Results of separate analysis of modifications were promising while the performance of the car has decreased for the case which all accessories were embedded. These results have showed that modifications could affect negatively when used together. Particularly, combination of side-bars and back-flap have caused enlargement and disordering of vortices at the back of vehicle and reduced each others' positive effects (Figure 9). Side-bars, on the other hand, have increased drag coefficient, when used together with back-flap on vehicle. According to all analyses, combinations of additions that create opposite effects was used in the decision making process (a summary is listed in Table 2). Additions that create opposite effect should not be used consecutively (following each other) in the direction of air flow.

REFERENCES

Chandra, S., Lee, A., Gorrell, S. and Jensen, C.G. 2011. CFD Analysis of PACE Formula-1 Car. Computer-Aided Design & Applications, PACE (1), pp. 1–14.
Hucho, W. 1998. Aerodynamics of road vehicles: From fluid mechanics to vehicle engineering (4th Edition). Warrendale, PA: Society of Automotive Engineers.
Kloek, L.J.R. 2013. A Validation Study on The Rooftop Flow Behavior of Road Vehicles; The Aerodynamic Analysis of a Dutch Super Sports Car.
Meederira, P. B. 2015. Aerodynamic Development of a IUPUI Formula SAE Specification car with Computational Fluid Dynamics (CFD) Analysis. Indiana—University Purdue—University, Indianapolis.
Rohatgi, U. S. 2012. Methods of Reducing Vehicle Aerodynamic Drag. U.S. Department of Energy, Brookhaven National Laboratory.

Advances and Trends in Engineering Sciences and Technologies II – Al Ali & Platko (Eds)
© 2017 Taylor & Francis Group, ISBN 978-1-138-03224-8

Evaluation of fibre-reinforced polymer perfobond rib shear connector resistance tests

L. Kolpaský & P. Ryjáček
Faculty of Civil Engineering, CTU in Prague, Czech Republic

ABSTRACT: Fibre-Reinforced Polymer (FRP) is material with strength properties comparable to steel but it is 70–80% lighter than steel. Its weight varies from 1400–2500 kg/m^3. Because of its high-strength and relatively low-weight, FRP is suitable to be used in reconstructions and for the strengthening of structures. Additionally, FRP has high durability and low life-cycle costs and therefore it may compete with steel structural design even in new structures. FRP has, however, one major disadvantage in comparison with steel and it is its low modulus of elasticity. This study describes the interaction between High Performance Concrete (HPC) and handmade FRP plate. Creating a composite action between FRP and concrete produces an additional increase of stiffness of the composite element. The selection of FRP and HPC ensures great performance throughout the lifecycle of the structure placed in aggressive environment. Suitable properties of FRP may be increased by hand lay-up process for connection by perfobond rib shear connectors. Three push tests on fiber-reinforced polymer were performed to investigate behavior of perfobond rib shear connectors.

1 INTRODUCTION

Bridges play a central role as an essential component of the infrastructures network as they do not only cover a large investment volume but also guarantee the smooth transport of goods and passengers. Compared to building, bridges are especially long living structures with a scheduled service life of at least 100 years (Mensinger et al. 2012). Most bridges suffer from inadequate maintenance and therefore lot of them must be repaired more times during their life cycle than it is necessary. Some of them end up being unrepairable and must be removed before their planned service life. Clients start to demand sustainable structures which can survive the whole life cycle without any crucial defects and related high maintenance costs.

Although the traditional material for this purpose is steel, the maintenance expenses are forcing the bridge operators to look for more durable and maintenance free solutions. The main goal is to replace steel elements that require service and coating with more durable FRP systems. The advantages are especially high durability and no additional expenses.

1.1 *Early composite connection research*

Some of the first research conducted on composite FRP-concrete beams was carried out by Descovic. Beam specimens were defined by a composite cross-section with a rectangular FRP tube in the tensile zone and concrete in the compression zone. In their work shear connectors between the upper flange and concrete slab was provided by applying an epoxy resin before pouring the concrete. This solution was not sufficient, because the beam failed suddenly by debonding from the concrete slab. As a conclusion of this experimental study there was a demand for more robust shear connectors (Deskovic et al. 1995).

Next study focused on the feasibility of composite connection by steel bolts (Correia et al. 2009). This connection system showed some level of ductility, but the initial stiffness was very low in comparison with the epoxy adhesive joint.

Cho (Cho et al. 2010) performed tests on pultrude square hollow sections of FRP tubes connected by concrete wedges and coarse sand coating. Connection by coarse sand coating resisted higher maximum load, but the failure was brittle. The advantage of concrete wedge connection was some level of ductility. In fatigue flexural tests two variants of connections were tested. Tests of connection by coarse sand coating and mixed shear connection system (coarse sand coating together with concrete wedge) were performed. Coarse sand coating can resist a fatigue load only at low level loading. The mixed shear system performed a remarkable fatigue resistance for identical fatigue loading.

To obtain the influence of wedge diameter and spacing additional pull-out tests (Cho et al. 2012) of the FRP pultrude plates embedded in concrete block were performed. Pultrude profiles were lubricated to exclude bonding effect. From the extensive research the discrete spring model was derived and proposed for the concrete dowel action including post-failure frictional effects, considering the diameter of the hole and spacing between them as parameters. The resistance induced by concrete dowel action increase practically linearly to the area of the rib hole. The number of rib holes did not increase the resistance linearly due to imperfect reallocation of the load after the occurrence of failure caused by low stiffness of FRP.

Pultrude profiles are reinforced by fibers mostly in the longitudinal direction which leads to very low transverse strengths. In FRP made by hand lay-up process, the amount of fibers and their direction can be adjusted in order to increase suitable properties. This paper focused on the development of a composite FRP-HPC beam, which consists of a pultrude lower flange, hand lay-up FRP webs and a high performance concrete slab. The modulus of elasticity is low, for common profiles between 17–30 GPa which implies that the serviceability limit state is much more important here, than in other usual materials, especially due to deflection requirements and local failures in connections. The disadvantage of low modulus of elasticity is compensated by the use of concrete part in the composite cross-section.

2 PROPOSED FRP-HPC CROSS-SECTION

For this study we propose following composite cross-section (see Figure 1 below). Bottom flange is horizontally placed Glass Fibre-Reinforced Polymer (GFRP) U cross-section, which is glued to two GFRP webs. Webs are 580 mm high and their top part, 90 mm high, is embedded to the HPC slab and this connection provides the composite action.

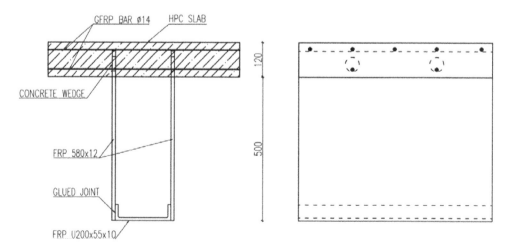

Figure 1. Proposed composite FRP-HPC cross-section.

Concrete is used as the top flange due to its high compressive strength. The use of HPC should limit the initiation and development of cracks. HPC has significant role in aggressive environment for assuring the high durability. Hand lay-up FRP plates are used as webs due to their good properties in transverse direction. The second reason for this choice is that the high pultrude I profiles are still not available in supply of providers. Pultrude profile is used as lower flange due to higher stiffness in longitudinal direction. The CFRP laminate can be added for increasing the stiffness of the beam.

2.1 Setup of the test

Three push-out tests were performed in order to study the behavior of composite system and to compare the influence of diameter of holes and their spacing. The shear performance of rib shear connectors was examined using the 40 a 50 mm diameter with spacing of 250 and 300 mm as it is shown in Table 1. The specimen is illustrated in Figure 2. The load is transmitted by steel fixture to the FRP plates.

GFRP plates with dimensions 580 x 540 mm were connected to steel fixture by 16 bolts of diameter 30 mm. In order to minimize the effect of slip between steel fixture and GFRP plates the holes were drilled with the same diameters as bolts.

Holes with 50 mm or 40 mm diameter were drilled into each side of the plate. After connection of the FRP plate to the fixture, GFRP bars were installed and fixed in the position by steel wire.

On the bottom side of the GFRP plates foam blocks with thickness of 30 mm were inserted to eliminate the effect of supporting the frontal side of the GFRP plate.

Table 1.   Parameters of the push out test specimens.

Designation	Diameter of rib hole mm	Number of rib holes pcs	Spacing of rib holes mm	Thickness of the slab mm
S1	50	2	300	120
S2	50	2	250	120
S3	40	2	250	110

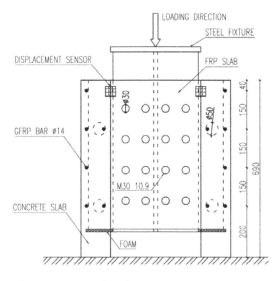 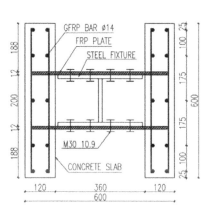

Figure 2.   Shape of push out test specimen S1.

Specimen was placed into the formwork and the concrete was poured to the form and vibrated on the shaking table. The concrete was cured indoors for 28 days.

3 SPECIFICATION OF USED MATERIALS

3.1 *GFRP plates*

The handmade 12 mm thick GFRP plates were supplied by Tatragrate, Slovak republic. Tensile characteristics of the material were obtained by tests carried out in accordance with ASTM D 3039. The strip specimens were cut out from the FRP plate used in push-out specimen. From the strip specimens the maximum failure stress and stiffness of the material were determined. The material properties are the same in the horizontal and vertical direction due to used textile reinforcement. An average tensile strength is 120 MPa and secant tensile modulus of elasticity 8.5 GPa. To obtain the tensile modulus extensometer Epsilon 3542 was used (see figure 3).

3.2 *GFRP reinforcement bars with CFRP core*

GFRP pultruded bars of 14 mm diameter, consist of E-glass fibers with carbon fibers in the middle of the cross section, were supplied by Prefa kompozity, Czech Republic, and were used as shear connectors in this study. These bars with sand coating are commercially available as reinforcement for concrete structures. Tensile tests were provided by the producer according to ACI 440.3R–04. Bars had average tensile strength, modulus of elasticity and ultimate strain of 1050 MPa, 67 GPa and 2.14%, respectively.

3.3 *Concrete slab*

In order to obtain concrete strength class C70/85 for the test specimen, the concrete was mixed and casted in a laboratory. The properties of concrete were determined by six cube tests and three tensile bending strength tests at the time of push out testing. The average compressive strength and the tensile strength were 91.7 MPa and 6.8 MPa, respectively.

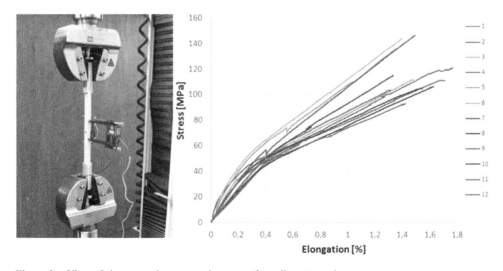

Figure 3.   View of the test and stress-strain curve of tensile test specimens.

4 EVALUATION OF PUSH OUT TESTS

Tests were performed using a hydraulic jack and loading was applied through displacement control at speed of 0.2 mm/min. Four displacement sensors were installed at each composition of FRP plate and concrete slab to measure the relative slip. According to EN 1994-1-1 the specimens were loaded 25x till 40% of the estimated maximum force and then loaded up to the failure.

The maximum force 1000 kN of the jack was reached without failure. The curve S1-1 shows that loading. The following specimens were tested on hydraulic jack with maximum loading force of 2000 kN. Figure 4 shows the comparison of load-relative slip curves of three specimens. In the second loading case of first specimen the stiffness was lower due to debonding between FRP plate and concrete slab during the first loading.

The experiment demonstrated that there were no significant differences between the three specimens. As can be seen in Table 2 the ultimate force was between 1119–1214 kN with a maximum difference of a 8.5%. Figure 4 shows that all three specimens had a high initial

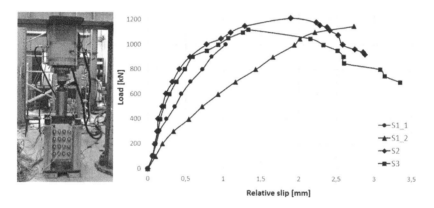

Figure 4. View of the test and load-relative slip curve of push out test specimens.

Table 2. Summary of load and displacement of push out test.

Designation	Diameter of rib hole mm	Number of rib holes pcs	Spacing of rib holes mm	Thickness of the slab mm	Ultimate load kN	Displacement at ultimate load mm
S1	50	2	300	120	1150	2.37
S2	50	2	250	120	1214	2.78
S3	40	2	250	110	1119	3.34

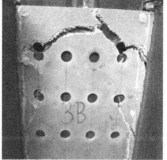

Figure 5. Shear cracks of the push out test specimen.

stiffness of 2.43×10^{-6} kN/m. Near the load 1000 kN the sound of cracking of glass fibers was heard. When the shear resistance was reached, the shear crack developed from the upper rib hole in slope of 45 degrees (see figure 5). When the load gradually dropped by 20%, next shear crack appeared.

5 CONCLUSIONS

A reliable connection between FRP and the concrete slab is important to prevent sudden and brittle failure. The use of perfobond rib shear connectors can reach this robustness. This study demonstrated the feasibility of composite connection by perfobond rib shear connectors between concrete slab and hand lay-up GFRP plates. Concrete wedges together with GFRP bars achieved great properties in performed push out tests.

The effect of the holes diameter on the shear resistance was not observed throughout the tests as all specimens failed by shear crack or debonding. The hand lay-up GFRP plates may be further enhanced by adding suitable reinforcement. Additionally, a design with more wedges of smaller spacing should lead to more uniform stress distribution and increase the overall performance of the tested specimens.

Due to a small modulus of elasticity of GFRP, a composite beam may have problems with serviceability limit states and especially deflection. Therefore the shear connection should be rigid. Connection with rib shear connectors reached ductile behavior during the test. The debonding between surface of GFRP plate and concrete slab occurred before the resistance of shear connectors was reached. In order to increase the stiffness of the connection, coarse sand coating of the GFRP plate surface can be applied.

ACKNOWLEDGEMENTS

Research reported in this paper was supported by Competence Centres program of Technology Agency of the Czech Republic (TA CR), project Centre for Effective and Sustainable Transport Infrastructure (no. TE01020168).

REFERENCES

ACI 440.3R-04, 2004, Guide Test Methods for Fiber Reinforced Polymers (FRP) for Reinforcing or Strengthening Concrete Structures (440.3R-04), American Concrete Institute, Michigan.

ASTM D3039/D3039M-14, Standard Test Method for Tensile Properties of Polymer Matrix Composite Materials, ASTM International, West Conshohocken, PA, 2014, www.astm.org

Cho, K., Park, S.Y., Kim, S.T., Cho, J.R., and Kim, B.S. 2010b. Shear connection system and performance evaluation of FRP concrete composite deck. *KSCE Journal of Civil Engineering*, 14(6): 855–865.

Cho, J., Park, S., Cho, K., Kim, S. and Kim, B. (2012). Pull-out test and discrete spring model of fibre-reinforced polymer perfobond rib shear connector. *Canadian Journal of Civil Engineering*, 39(12), pp. 1311–1320.

Correia, J., Branco, F. and Ferreira, J. (2009). Flexural behaviour of multi-span GFRP-concrete hybrid beams. *Engineering Structures*, 31(7), pp. 1369–1381.

Deskovic, N., Meier, U. and Triantafillou, T. (1995). Innovative Design of FRP Combined with Concrete: Long-Term Behavior. *Journal of Structural Engineering*, 121(7), pp. 1079–1089.

EN 1994-1-1: Design of Composite Steel and Concrete Structures, Part 1–1: General Rules and Rules for Buildings, CEN, Brussels, 2005.

Gai, X., Darby, A., Ibell, T. and Evernden, M. (2013). Experimental investigation into a ductile FRP stay-in-place formwork system for concrete slabs. *Construction and Building Materials*, 49, pp. 1013–1023.

Mensinger, M., Pfaffinger, M. and Schnell, W. (2012). Sustainability Assessment of Road Bridges. *Report (congr. iabse)*, 18(13), pp. 1173–1180.

Vovesný, M. and Rotter, T. (2012). GFRP Bridge Deck Panel. *Procedia Engineering*, 40, pp. 492–497.

Advances and Trends in Engineering Sciences and Technologies II – Al Ali & Platko (Eds)
© 2017 Taylor & Francis Group, ISBN 978-1-138-03224-8

In plane buckling of frames with laced built-up members

Mi. Kováč, Zs. Vaník & M. Magura
Department of Steel and Timber Structures, Faculty of Civil Engineering, Slovak University of Technology in Bratislava, Bratislava, Slovakia

ABSTRACT: For assessment of portal frames with laced built-up members the first order internal forces and the equivalent column method are the mostly used. If the distance between neighboring nodes is used as the buckling length of the chord in the equivalent column method the second order effects with only the bow imperfection between nodes are taken into account. For frames sensitive to buckling in a sway mode the second order effects on structures with initial sway imperfection should be taken into account. In portal frames with the laced compression columns the sway imperfection should be also applied and the second order in frame analysis should be performed. The simplified procedure to check the additive forces, which arise in the chords as the second order effect are proposed.

1 INTRODUCTION

1.1 *The equivalent column method*

The equivalent column method is adopted in section 6.3.1 of norm EN 1993-1-1. The method is commonly used for the individual stability checks of members in frame structures. The bow imperfections of equivalent columns are defined based on the appropriate buckling lengths due to the global buckling mode of the structure. The second order effects on the equivalent column are accounted for by means of a reduction factor. This simplified verification method is used even in some cases of more complex structures, such as frames whose members consist of a laced built-up members.

1.2 *The sway deformation*

For frames sensitive to buckling in a sway mode the second order effects on structures with the initial sway imperfection should be taken into account. In the portal frames with the laced built-up columns, the additional sway deformation due to the second order effect causes additional axial forces in the chords. If the equivalent column method is used for verification of these chords, with the distance between neighbouring nodes as the buckling length, then only the second order effects with the bow imperfection between nodes are taken into account.

The aim of this paper is to find a simple and still sufficiently accurate way of calculating the second order chord axial forces of laced frames with sway imperfections. The norm procedure for stability check of the individual laced compression members, located in the clause 6.4.2 of norm EN 1993-1-1, will be adopted for that. The buckling length of the columns according to the global buckling mode of an equivalent frame structure will be used as the length of an equivalent laced compression member. This equivalent laced column will be checked by the norm procedure. This simplified procedure was firstly used and verified by Kováč & Vaník (2016).

2 THE SIMPLIFIED PROCEDURE

2.1 *The equivalent frame*

The determination of the buckling lengths of laced columns of frames, when the whole laced columns develop a curvature and the frame is under sway deformation, is difficult in frames with a large number of partial members. In such frames a great amount of buckling modes of one particular member under curvature between nodes arises.

The transformation of the laced members of the frame into uniform members with compact cross-sections of equivalent bending rigidity (equivalent frame) is used here. The stability analysis of the equivalent frame will be performed in order to find the global buckling mode (Figure 1b). For stability analysis the loading of the equivalent frame should be equivalent to the loading of the laced frame (Figures 1a, b). From the stability analysis the critical force N_{cr} and appropriate buckling length L_{cr} can be obtained.

2.2 *Equivalent laced column*

The equivalent laced column (Figure 1c) will be defined by the buckling length L_{cr} from the global buckling mode (Figure 1b). The equivalent laced column has the same cross-sections of chords and lacings as the column of a laced frame. It is simply supported on both ends and it is loaded by a centric force which is equivalent with the columns' loading of the given laced frame (Figure 1a). The bow imperfection e_0 of the equivalent laced column will be derived from the global sway imperfection for columns of frames. The second order analysis of this equivalent laced column will be performed.

2.3 *Imperfection and the second order effects*

The norm EN 1993-1-1 defines the global initial sway imperfections of frames by the value of angle Φ, from which the horizontal displacement s in the height h may be determined (Figure 2a). The bow imperfection of the equivalent laced column in the shape of sinus function on the length L_{cr} and with the amplitude e_0 will be used (Figure 2a, b). The amplitude e_0 will be derived in order to obtain the same horizontal displacement s in the height h.

For the second order analysis the norm procedure according EN 1993-1-1clause 6.4 for the design of built-up compression members will be adopted. According to this clause the second order bending moment in the middle of the span of the equivalent laced column can be determined from:

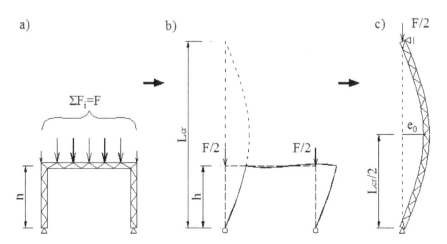

Figure 1.   a) Laced frame, b) Equivalent frame, c) Equivalent laced column.

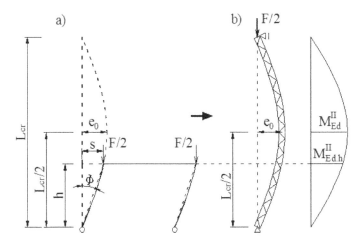

Figure 2. a) Imperfection of the equivalent frame, b) Equivalent laced column.

$$M_{Ed}^{II} = \frac{N_{Ed}e_0 + M_{Ed}^{I}}{1 - \dfrac{N_{Ed}}{N_{cr}} - \dfrac{N_{Ed}}{S_V}},$$ (1)

where M_{Ed}^{I} is the design value of the first order bending moment in the middle of the span; S_V is the shear stiffness of the lacings.

From the second order bending moment M_{Ed}^{II} at the middle of the equivalent laced column the bending moment of the laced frame in the crucial cross-section $M_{Ed.h}^{II}$ can be determined (Figure 2b) and the additive axial forces in the chords of real laced column can be calculated from:

$$\Delta N = \frac{M_{Ed.h}^{II}}{h_0},$$ (2)

where h_0 is the distance between centroids of chords.

3 EXAMPLE

3.1 *Inputs*

The simplified procedure will be presented for the following case of portal two hinged laced frame. The results will be confronted with the second order global in frame analysis of the imperfect laced frame. The shape of the frame from laced built-up members is shown in Figure 3a. All members of the frame have the cross-section of type CHS 139.7/5, steel grade S 235 is used. The frame is simply supported and loaded by forces in the upper nodes of the horizontal member. All member nodes are considered as hinges.

3.2 *Calculation by the simplified procedure*

In the first step the first order member axial forces of the laced frame without imperfection (Figure 3a) has been calculated. The axial forces of chords are shown in Figure 4. Axial force in the most stressed member was $N^I = -431.89$ kN.

The laced frame was transformed into the equivalent frame in Figure 3b from the compact cross-sections with the effective second moment of area $I_{eff} = 1.06 \times 10^{-3}$ m⁴. The stability analysis follows, where the global sway buckling mode and the appropriate buckling length $L_{cr} = 35.045$ m were obtained from the IQ100 program. The bow imperfection of the

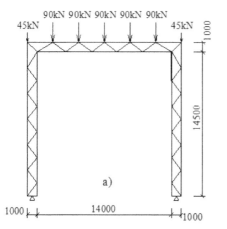
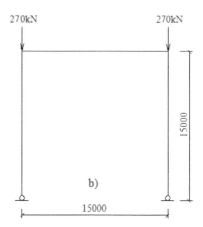

Figure 3. a) Laced frame b) Equivalent frame.

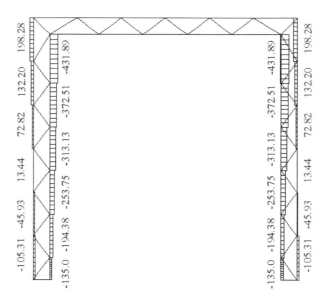

Figure 4. The first order chord axial forces of a laced frame.

equivalent laced column $e_o = 44.433$ mm was determined according to Figure 2a from the sway imperfection $\Phi = 2.887 \times 10^{-3}$ of columns due to 5.3.2 (3) EN 1993-1-1.

In the next step, the additional second order normal force $\Delta N = 13.80$ kN of high-lighted chord in Figure 3a, which is in the most stressed part of the given laced frame, was calculated.

The sum of the second order additive axial force ΔN and the first order axial force N^I is the axial force $N_{SP} = -445.69$ kN (Table 1) by simplified procedure with the influence of the second order and imperfection effects.

3.3 The second order in frame analysis

The imperfect laced frame with the value of horizontal displacement $s = 43.30$ mm at the height h is shown in Figure 5. The appropriate horizontal displacements of other nodes of laced columns are derived from the shape of sinus function, which is defined on L_{cr} depending

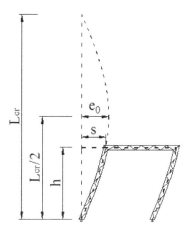

Figure 5.   The imperfect laced frame.

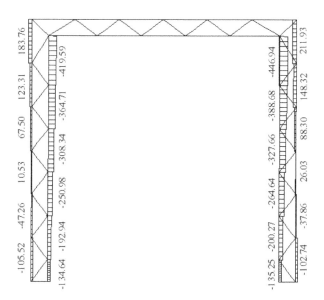

Figure 6.   The second order chord axial forces.

on their heights (Figure 5). The loading, cross-sections and material were used according to section 3.1.

The global second order analysis of imperfect laced frame was performed by the Scia Engineer program. The second order axial forces were calculated by the Newton-Raphson method. The axial forces of chords are shown in Figure 6.

4   CONCLUSIONS AND DISCUSSION

The simplified procedure of chord axial forces calculation with the influence of sway imperfection and the second order theory is presented in this paper. Based on the good agreement (Table 1) between the results of the presented simplified procedure and the second order analysis with imperfection it can be concluded, that the axial forces of the chords of laced columns can be expressed by the simplified procedure with sufficient accuracy.

Table 1. Axial forces and comparison.

Method	Simplified procedure N_{SP}	[kN]	The second order analysis N_{FEA}	[kN]	Error $(N_{FEA}-N_{SM})/N_{FEA}$ Er	[%]
Highlighted member in figures 3, 4, 6	445.69		446.94		0.28	

The aim of presented paper was to find the simple and accurate procedure for calculation of chord axial forces of frames' laced columns. This simplified calculation is effective alternative to the second order analysis of imperfect laced frame. The difference in axial force of the most exposed chord between these two alternatives is only 0.28%.

ACKNOWLEDGEMENTS

The paper was elaborated with support of the research project for excellent teams of young researchers of the SUT, Faculty of Civil Engineering, Department of Steel and Timber Structures.

REFERENCES

EN 1993-1-1. Eurocode 3—Design of steel structures, Part 1–1: General rules and rules for buildings. Brussels: CEN, 2005.

Kováč, M. & Vaník, Zs. 2016. Global Buckling of Frames with Compression Members. *Applied Mechanics and Materials* 837: pp. 103–108.

Advances and Trends in Engineering Sciences and Technologies II – Al Ali & Platko (Eds)
© 2017 Taylor & Francis Group, ISBN 978-1-138-03224-8

Contact model of the fibre-matrix debonding coupling the interface damage and friction

J. Kšiňan & R. Vodička

Civil Engineering Faculty, Technical University of Košice, Košice, Slovakia

ABSTRACT: The plain strain problem of the fibre-matrix debonding coupling the interface damage and the Coulomb friction have been studied. The numerical formulation of the interface contact considers the Cohesive Interface Model (CIM) with the frictional contact. In the study, the influence of friction have been observed during the process of the interface debonding in both phases: the crack onset and the crack growth. The rate-independent process of the interface debonding have been governed by an energy-based formulation that has been implemented. The proposed contact model has been applied and tested in Fibre-Reinforced Composite (FRC) sample in order to demonstrate the impact of the friction effect on the debonding process.

1 INTRODUCTION

1.1 *Cohesive interface model*

In the recent analysis of interface corruption by the methods of computational mechanics, an approach is usually considered which yields a non-linear though continuous response of the interface model, including the so-called *softening period*. Such a model refers to an approach that supposes a non-linear continuous material response of the mechanical stress t and damage parameter ζ known as a *cohesive contact model*. An efficient way to achieve a continuous material response is by means of the energetic formulation of the *stored energy functional* E_S (Roubíček et al. 2014). In the first damage initiation part of the stress strain relation, the stress is a linear function of displacement jump $[u]$ up to achieving its critical value $t_C = (k_1 + k_2 / \sqrt{k_1 + k_2}) \sqrt{2 G_d}$, whereas in the second, damage evolution part of the diagram, the stress evolves non-linearly until it vanishes for the critical relative displacement $u_C = \sqrt{2 G_d / k_1}$. The failure mechanism starts to evolve, when the Energy Release Rate (ERR) reaches activation threshold *fracture energy* G_d. Consequently, at that instant, the mechanical stress t and damage parameter ζ start to decrease: ζ from one non-linearly until it arrives to zero, leading to a continuous response of the model (Roubíček et al. 2014).

1.2 *Interpretation of the friction model*

In this paper, a numerical model that is able to predict interface damage considering the Coulomb friction contact between debonded surfaces is implemented. The effect of the gradual increase of the friction represents the natural outcome of the gradual decrease of the interface damage parameter ζ, from the initial state to the complete decohesion similarly as in (Del Piero & Raous 2010, Raous 2011, Vodička 2013). The process of the crack initiation and the crack propagation at the fibre-matrix composites has been studied by various authors (Correa et al. 2008, Kushch et al. 2008, Mantič 2009, Távara et al. 2011, Mantič & García 2012).

In this study, the contact model is defined by the CIM that governs the damage evolution and for the friction, the simple Coulomb friction law is adopted. The solution of the contact problem is based on the evolution of energies during the loading process: the elastic energy stored in the bulks and the energy dissipated due to damage and friction. The interpretation of the Coulomb friction law in the formulation of the energy dissipation potential is represented by damage dependent friction function (Kšiňan & Vodička 2016). The friction function determines the activation of friction with respect to the decreasing interface stiffness due to damage evolution and characterizes the switch between the interface shear stresses duc to cohesive and friction forces. From the computational point of view, it is necessary to adduce that the friction function should be convex. Summing it up, the friction function $f(\zeta)$ characterizes the process of friction activation and its consequent increasing influence on the interface stiffness degradation; see Kšiňan et al. (2014), Kšiňan & Vodička (2016).

2 MATHEMATICAL MODEL OF THE DEBONDING PROCESS

2.1 *The concept of energetic formulation*

This section discusses the mathematical formulation of debonding process for cohesive model coupling the frictional contact and interface damage. The solution is acquired by variational formulation, which exploits the developed numerical treatment. For defining the energetic conception of the interface damage mechanism, the stored energy functional of the system should be considered (Vodička et al. 2014). Similarly, the potential energy of the external forces and the dissipation potential should be defined reflecting the rate-independence of the debonding process.

The main feature of the proposed mathematical model is that the stored energy functional is separately quadratic both in the $[u]$ and ζ variable. This fact enables to apply very efficient quadratic programming algorithms for solving the minimization problem (Dostál 2009, Roubíček et al. 2014).

2.2 *Computer implementation of the numerical model*

The proposed numerical procedure for solving the minimization problem considers the time and the spatial discretization, separately. The time discretization exploits a suitable semi-implicit formula used also in Kšiňan & Vodička (2016). Similar solutions can be obtained by the Maximally-Dissipative Local-Solution (MDLS) concept (Vodička et al. 2014). Applying the recursive minimisation, the minimisation has been split to minimisations with respect to each interface variable: the displacement jump $[u]$ and the damage parameter ζ. This concept exploits a fractional-step type semi-implicit time discretization (Roubíček et al. 2013), such the problem is split into two quadratic minimizations procedures. The implementation is formulated in terms of the boundary data only, with spatial discretization leading to the Symmetric Galerkin BEM (SGBEM). In the computational procedure, the SGBEM provides a complete boundary-value solution in order to calculate the elastic strain energy stored in the solids. The advantageous properties of the Symmetric Galerkin BEM, based on variational formulation deduced from the potential energetic principles, enable symmetric formulation of the system of integral equations in the numerical solution. This attribute guarantees the positive definite character of the strain energy computed by SGBEM (Vodička et al. 2014).

3 NUMERICAL EXAMPLE

A numerical model of the fibre-reinforced composite with the presence of friction contact is tested in a plane strain problem of the crack onset and growth at the fibre-matrix interface under remote tension. The geometry of the tested composite specimen, has been taken from a glass fibre-matrix composite micrograph, presented in (Távara et al. 2016), see Figure 1.

On the Figure 1 (right), the fibre-matrix debonding problem configuration can be seen. The tested contact model assumes the presence of friction at the interface, in order to capture the influence of the friction in the tested FRC example.

Analysed 2D contact model is represented by a bundle of ten circular inclusions $\eta = 1\ldots10$ with the radius $r = 7.5$ µm located inside a relatively large square matrix with cell of the side length $L = 1,000$ µm. The numerically tested model has been subjected to the tension transverse to the fibres. A uniform displacements representing a tension load prescribed at the top side of the matrix cell as can be clear from the Figure 1 (right). The prescribed displacements $\sigma_y^\infty = w_2$ are applied on the top side of the matrix cell $w_2^k = ut^k$ for $k = 1,2,\ldots120$ with the increment of the external displacement $u = 0.1$ µm and $t^k = kt_0$, $t_0 = 1$s. Totally, the numerical model has been subjected to $k = 120$ load steps. At the fibre-matrix interface it has been considered 90 uniformly distributed boundary elements for each inclusion, therefore the polar angle of each element at the interface is 4°. The numerical analysis of the fibre-matrix composite considers the bimaterial study that consists of the epoxy matrix and glass fibres. The mechanical properties of the fibres and matrix are defined by Young's modulus. $E_m = 2.79$ GPa, $E_f = 70.8$ GPa and Poisson's ratios: $v_m = 0.33$, $v_f = 0.22$. The interface stiffness parameters for the cohesive interface model were defined in the same way as in Kšiňan & Vodička (2016). The parameter that govern the interface damage is the fracture energy $G_d = 2$ Jm^{-2} that corresponds to the critical value of the tension $t_{nc} = 63.8$ MPa. The Coulomb friction coefficient has been assumed for $\mu = 1$.

3.1 Numerical results

The process of the crack nucleation initiates if **ERR** reaches at the part of the interface required amount of the fracture energy per unit length G_d that corresponds to the critical value of the normal stress $t_{nc} = 63.8$ MPa, such damage parameter ζ decreases from 1 to 0 continuously until the total damage of the interface has been occurred. At the zones of the interface debonding (the parts of the interface with decreasing damage parameter ζ) the stresses t_n, t_s gradually decrease with the reducing stiffness of the interface until the state of $\zeta = 0$, when the interface is not able to transver the stresses, see Figure 2, $k = 115$. The presented results confirm a continuous response of the interface model, that corresponds to the characteristic softening stage behind the crack tips located between the angles θ_{c1} and θ_{c2}, see

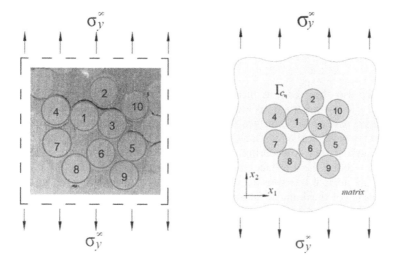

Figure 1.   Tested fibre-matrix sample taken from a glass fibre composite micrograph under applied far field tension load (left), Fibre-matrix debonding problem configuration (right).

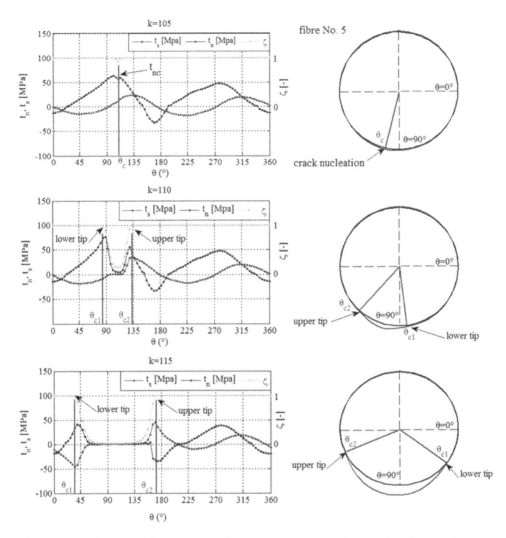

Figure 2. Distributions of the stresses t_s, t_n, damage parameter ζ and the evolution of the crack propagation on the fibre no. 5 for the load steps k: 105, 110, 115.

Figure 2. It can be also observed that during the evolution of interface debonding, the values of the tangential and normal tractions are increasing in the crack edge vicinity.

At the damaged part of the interface θ:(37°–170°), the tangential and normal tractions decrease to zero, however at the locations behind the crack tip the stresses acquire the highest values, see Figure 2. The considering of the friction has influenced the evolution of the debonding process along the interface. It is conspicuous from the Figure 2 that in the area θ: (167°–178°), the normal tractions acquire the negative values $[u]_n^- = -[u]_n$ and in consequence of decreasing of the damage parameter ζ, the friction function $f(\zeta)$ has been activated and the tangential tractions are increasing until the treshold $t_s = \mu|t_n|$ has not been reached. Such the condition for the activation of the energy dissipation due to the friction is satisfied (Kšiňan & Vodička 2016).

The Figure 3 presents the deformed shapes of the fibre-matrix sample for 4 different load steps. The main feature is that the process of fibre-matrix debonding initiates for the fibre No. 5 and in consequence of the stress redistribution the debonding process initiates consequently at the fibres No. 4, No. 6 and No. 1. It is noteworthy that obtained results of

174

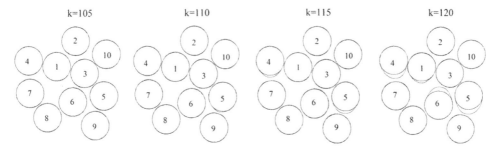

Figure 3. Deformed shapes, at the scale of the fibres multiplied by a factor of five for four different load steps.

deformations reflect the evolution of the crack propagation at the fibre-matrix interface in a similar way as in the studies of Távara et al. (2013, 2016).

4 CONCLUSIONS

A numerical model of the fibre-matrix contact has been analysed in order to clarify the process of the crack initiation and propagation. The sample of the fibre-matrix composite have been subjected to the transversal tension load. The friction effect have been implemented in the model by the damage dependent friction function that determines the activation of friction with respect to the decreasing interfacial stiffness. The achieved results confirm the certain influence of the friction on the process of the fibre-matrix debonding and its evolution. The numerical implementation of spatial discretization via SGBEM enables the whole problem to be defined only by a boundary and interface data. Analysed numerical model proves the models response in accordance with the other studies of the fibre-matrix debonding process. Such numerical studies could further contribute to a better understanding of the FRC behaviour under transverse loads.

ACKNOWLEDGEMENTS

The authors acknowledge the financial support from the Scientific Grant Agency of Slovak Republic. The projects registration numbers are VEGA (Grants No. 1/0078/16 and 1/0188/16).

REFERENCES

Correa, E., Mantič, V., París, F. 2008. Numerical characterisation of the fibre-matrix interface crack growth in composites under transverse compression. *Engineering Fracture Mechanics* 75: 4085–4103.
Del Piero, G., Rauos, M. 2010. A unified model for adhesive interfaces with damage, viscosity, and friction. *European Journal of Mechanics A/Solids* 29: 496–507.
Dostál, Z. 2009. *Optimal Quadratic Programming Algorithms*. Berlin: Springer.
Kšiňan, J., Vodička, R. 2016. An 2-D SGBEM formulation of contact models coupling the interface damage and Coulomb friction in fibre-matrix composites. *Engng Fract Mech*, in press.
Kšiňan, J., Mantič, V., Vodička, R. 2014. A new interface damage model with frictional contact. An SGBEM formulation and implementation. *Advances in Boundary Element and Meshless Techniques XIV, Florence, 15–17 July 2014*. Eastleigh: EC ltd.
Kushch, V.I., Shmegera, S.V., Mishnaevsky L. 2008. Meso cell model of fiber reinforced composite: interface stress statistics and debonding paths. *International Journal of Solids and Structures* 45: 2758–2784.

Mantič, V. 2009. Interface crack onset at a circular cylindrical inclusion under a remote transverse tension. Application of a coupled stress and energy criterion. *Int J Solid Struct*. 46: 1287–1304.

Mantič, V., García, I.G. 2012. Crack onset and growth at the fibre-matrix interface under a remote biaxial transverse load. Application of a coupled stress and energy criterion. *Internal Journal of Solids and Structures* 49: 2273–2290.

Raous, M. 2011. Interface models coupling adhesion and friction. *C. R. Mécanique* 339: 491–501.

Roubíček, T., Kružík, M., Zeman, J. 2014. Delamination and adhesive contact models and their mathematical analysis and numerical treatment. In V. Mantič (ed), *Mathematical Methods and Models in Composites*: 349–400. London: Imperial College Press.

Roubíček, T., Panagiotopoulos, C.G., Mantič, V. 2013. Quasistatic adhesive contact of visco-elastic bodies and its numerical treatment for very small viscosity. *Z. angew. Math. Mech.* 93: 823–840.

Távara, L., Mantič, V., Graciani, E., París, F. 2011. BEM analysis of crack onset and propagation along fiber-matrix interface under transverse tension using a linear elastic-brittle interface model. *Engineering Analysis with Boundary Elements* 35: 207–222.

Távara, L., Mantič, V., Graciani, E., París, F. 2013. BEM modelling of interface cracks in a group of fibres under biaxial transverse loads. *Advances in Boundary Element and Meshless Techniques XIV, Paris, 16–18 July 2013*. Eastleigh: EC ltd.

Távara, L, Mantič, E., Graciani, E., París, F. 2016. Modelling interfacial debonds in unidirectional fibre-reinforced composites under biaxial transverse loads. *Composite Structures* 36: 305–312.

Vodička, R., Mantič, V. 2013. An SGBEM implementation with quadratic programming for solving contact problems with Coulomb friction. *Advances in Boundary Element and Meshless Techniques XIV, Paris, 16–18 July 2013*. Eastleigh: EC ltd.

Vodička, R., Mantič, V., Roubíček, T. 2014. Energetic versus maximally-dissipative local solution of a quasi-static rate-independent mixed-mode delamination model. *Meccanica* 49: 2933–2963.

Advances and Trends in Engineering Sciences and Technologies II – Al Ali & Platko (Eds)
© 2017 Taylor & Francis Group, ISBN 978-1-138-03224-8

Strain of elements reinforced by GFRP at elevated temperatures

Š. Kušnír & S. Priganc
Faculty of Civil Engineering, Institute of Structural Engineering, Technical University of Kosice,
Slovakia

ABSTRACT: This paper deals with the influence of elevated temperatures up to 100°C on strain and stress in symmetrically reinforced concrete samples with different reinforcement ratio. The procedure, results and conclusions of tests conducted on unreinforced and reinforced concrete samples in the laboratory of the Institute of Structural Engineering (ISE), Technical University in Kosice are briefly summarized. It supports a possibility of theoretical prediction of stress caused by elevated temperatures with different kinds of reinforcement (GFRP, steel). The article indicates a possible of application of GFRP reinforcement at elevated temperatures and a contribution for practice.

1   INTRODUCTION

1.1   *Project*

This article deals with the influence of elevated temperatures up to 100°C on strain and stress in symmetrically reinforced samples with a different reinforcement ratio. The research on unreinforced as well as reinforced concrete samples was conducted in the laboratory of ISE. The samples were concreted from three types of concrete and reinforced by GFRP (glass fiber reinforced polymers) and steel reinforcement. For concrete the compressive strength (f_{cm}) were specified as 24,36 MPa—series 1; 51,061 MPa—series 2 and 28,068 MPa—series 3 (STN EN 206,2015). Three different reinforcement ratios (μ) of GFRP reinforcement (Schöck ComBar ϕ8 mm) and steel reinforcement (B500 ϕ8 mm) were used (0,5%; 2,01%; 2,51%) Figure 1. Previously taken measurements were considered in calculations for comparing and calibrating (Šimková 2006), (Priganc 2013). They were marked as series 1* – f_{cm} = 53,5 MPa and series 2* – f_{cm} = 41,71 MPa. The beams in these series were reinforced by steel reinforcement (B500 ϕ8 mm). The twenty-eighth day after concreting, beams were exposed to the temperature in steps of 20°C (20, 40, 60, 80, 100, 120°C).

1.2   *Measurement*

The samples had been exposed to each of the mentioned temperatures for 24 hours and then the length of elements was measured and the specific deformation was determined from it

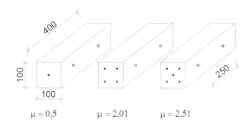

Figure 1.   The way of specimen reinforcing and measuring points distance.

Table 1. Relative length strain of unreinforced and reinforced beams with influence of elevated temperatures series 1.

Temperature °C	μ = 0%	μ = 0,5%		μ = 2,1%		μ = 2,51%	
		G	S	G	S	G	S
20	0,0000	0,0000	0,0000	0,0000	0,0000	0,0000	0,0000
40	0,1642	0,1387	0,1446	0,1369	0,1584	0,1497	0,1675
60	0,3684	0,3374	0,3516	0,3184	0,3713	0,3354	0,3962
80	0,5910	0,5368	0,5743	0,5340	0,6189	0,5423	0,6317
100	0,6817	0,6348	0,6650	0,6415	0,7799	0,6395	0,8045

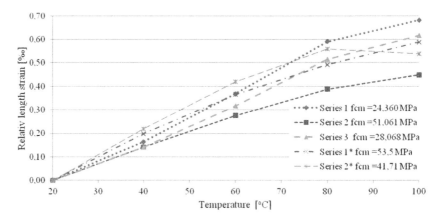

Figure 2. Specific deformation of unreinforced concrete samples with the influence of elevated temperatures (all series).

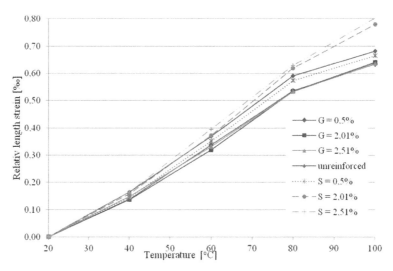

Figure 3. Relative length strain of unreinforced and reinforced samples with the influence of elevated temperatures series 1(G—samples reinforced by GFRP, S—samples reinforced by steel reinforcement).

(Table 1, Figures 2–3). The deformation was measured by deformeter Huggenberger AG-EDU 25 with a 250 mm measuring base. Spacing of targets concreted in the samples was measured. Measurement points were set on two opposite sides of the samples. Minimum three samples in each series were produced. Strain of the samples was measured after being taken out from the heating chamber. The presented results are average from the measured values for each series. The results of all 5 series are in Figure 2. In the following for briefness, only pictures for series 1 are shown.

1.3 Mathematical simulation

The functional dependence between deformation of unreinforced samples and temperature was determined following these measured values (up to 80°C polynomial of 2.degree, above polynomial of 3.degree). Figure 4 shows the comparison of the measured values and results of derived equation. In a similar manner the equations for all 5 series were determined. The approximation arguments of equations for the total relative strain for series 1 are shown in Table 2.

The equations were considered in the form:

$$AT^2 + BT + C \text{ — for polynomial of 2.degree and exactness } R^2 \tag{1}$$

$$AT^2 + BT + C + D \text{ — for polynomial of 3.degree and exactness } R^2 \tag{2}$$

where T = temperature [°C]; A,B,C,D = coefficients given in Table 2.

The total strain coefficient of an element at temperature α_{cc} was derived from equations (1) and (2) in the following way:

$$\alpha_{cc,\tan}(T) = aT + b \tag{3}$$

$$\alpha_{cc,\tan}(T) = aT^2 + bT + c \tag{4}$$

where $\alpha_{cc,\tan}$ = tangential coefficient of thermal expansion; T = temperature [°C]; a, b, c = coefficients given in Table 3.

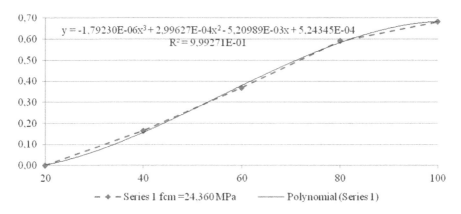

Figure 4. Determination of functional dependence between deformation of unreinforced beams and temperature (series 1—polynomial of 3.degree).

Table 2. The approximation arguments of equations for the total relative strain series 1.

Series 1

Argument		A	B	C	D	R
Polynom	2 st. do 80	3,65300E-05	6,23391E-03	−1,40373E-01		9,99881E-01
		−2,29862E-05	1,17094E-02	−2,40360E-01		9,90158E-01
	3 st. do 100	−1,79230E-06	2,99627E-04	−5,20989E-03	5,24345E-04	9,99271E-01

Final derived approximation arguments of equations for the total relative strain for series 1 are shown in Table 3.

By this coefficient α_{cc}, the modulus of the elasticity, area, reinforcement ratio and other physically required parameters the equations were created (STN EN 1992, 2006), (STN EN, 1992,2007), (Schock, 2013). The equations describe strain and stress in reinforced concrete elements at elevated temperatures.

$$d\sigma_{cs} = dT \left[\alpha_{sT} - \alpha_{cc}(T) \right] \frac{\mu_s E_s}{\mu_s n_s(T)+1} \tag{5}$$

$$d\sigma_{cg} = dT \left[\alpha_{cc}(T) - \alpha_{gT} \right] \frac{E_g}{\mu_g n_g(T)+1} \tag{6}$$

where $n_s(T) = E_s/E_c(T), n_g(T) = E_g/E_c(T)$, σ = stress in concrete caused by reinforcement [MPa]; E = Young's modulus of elasticity [GPa]; T = temperature [°C]; inferior index c = concrete; inferior index g = GFRP reinforcement; inferior index s = steel reinforcement.

This way, theoretically and experimentally defined strain and stress were compared in table 4 and Figures 5–6.

Table 3. Derived approximation arguments of equations for the total relative strain series 1.

Derived series 1					
Argument			a	b	c
Polynom	2 st.	do 80	7,30600E-05	6,23391E-03	
		do 100	−4,59724E-05	1,17094E-02	
	3 st.	do 100	−5,37689E-06	5,99254E-04	−5,20989E-03

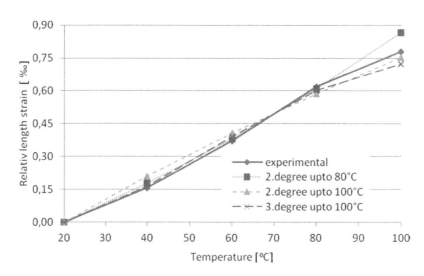

Figure 5. Measured and calculated relative length strain of reinforced samples ($\mu = 2,01$) with the influence of elevated temperatures series 1.

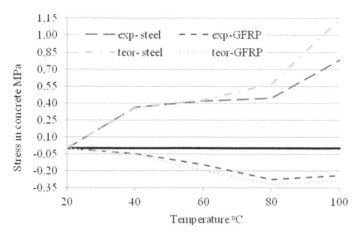

Figure 6. Stress in concrete caused by steel or GFRP reinforcement with the influence of elevated temperatures—samples series 1, $\mu = 2{,}01$.

Table 4. Stress in concrete caused by reinforcement with the influence of elevated temperatures series 1 in [MPa].

Stress in concrete $\mu = 2{,}5\%$

T [°C]	Steel		Glass	
	meas.	calc.	meas.	calc.
20	0,00000	0,00000	0,00000	0,00000
40	0,36466	0,35902	−0,04483	−0,04991
60	0,42102	0,42868	−0,14384	−0,19574
80	0,44379	0,57293	−0,27497	−0,31454
100	0,78170	1,14215	−0,24053	−0,28489

2 CONCLUSION

The influence of elevated temperatures on steel reinforced concrete structures is relevant and not negligible, especially in elements in tension. We can see it in the two series where cracks occurred and divided samples into at least four parts in the longitudinal direction (transversal cracks) already at temperature 80 or 100°C.

GFRP reinforcement, within the experiment, creates more appropriate stress in contrast to the steel reinforcement, which at elevated temperatures still creates tension in concrete (leading to cracks occurring). GFRP reinforcement (according to coefficient of temperature expansion) generally creates compression in concrete and tension in reinforcement.

On condition that the strain of unreinforced elements exposed to elevated temperatures is known, it is possible to determine strain and stress in reinforced elements (reinforced by steel and GFRP reinforcement) with sufficient accuracy, provided that material characteristics are known.

ACKNOWLEDGMENT

This work was supported by the Scientific Grant Agency of the Ministry of Education of Slovak Republic and the Slovak Academy of Sciences under Project VEGA 1/0661/16 The behavior of the structural members from ordinary and from lightweight concrete affected by temperature.

REFERENCES

Priganc, S., Kušnír, Š., Sabol, S. & Kušnírová, D. 2013. Stress and Strain Behaviour of Composite Elements at Elevated Temperatures, In: Procedia Engineering: Concrete and Concrete Structures 2013: 6th International Conference, Slovakia.
http://www.sciencedirect.com/science/article/pii/S1877705813015622
Šimková, S. 2006. Influence of elevated temperatures up to 100°C on length strains of reinforced concrete members with different ratio of reinforcement. (Vplyv teplôt do 100°C na dĺžkové zmeny železobetónových prvkov s rôznym stupňom vystuženia.). Dissertation thesis, Košice.
STN EN 206, SUTN 2015, Concrete. Specification, performance, production and conformity, 2015.
STN EN 1992-1-1, SUTN 2006. Eurocode 2: Design of concrete structures—Part 1–1: General rules and rules for buildings, 2006.
STN EN 1992-1-2, SUTN 2007. Eurocode 2: Design of concrete structures. Part 1–2: General rules. Structural fire design, 2007.
Schöck ComBAR, 2013. Technical Information, Schöck ComBAR, Schoeck Bauteile GmbH, Baden—Baden.

Advances and Trends in Engineering Sciences and Technologies II – Al Ali & Platko (Eds)
© 2017 Taylor & Francis Group, ISBN 978-1-138-03224-8

Design and realization of composite timber-concrete beams

V. Kvočák & M. Al Ali
Faculty of Engineering, Technical University in Košice, Slovak Republic

ABSTRACT: In certain cases, composite beams represent more convenient solution in comparison to classical structures, such as steel or reinforced concrete beams, which led to their widespread use in civil structures in previous decades. Nowadays, timber-concrete composite elements, mainly subjected to bending load, are considered as suitable structural solution for a variety of building and industry fields. This kind of composite structures can be used for new buildings and in the reconstruction of old timber ceilings. Main advantages of these kind of beams are their light-weight and the beneficial combination of concrete and timber material properties. The paper presents a concise information about design and realization of actual composite timber-concrete ceiling.

1   INTRODUCTION

Advantages of shear connection can be quite favorably used in the field of timber-concrete structures mainly in reconstructions of old ceilings, as well as in designing of new structures, Postulka (2003). Different kind of composite connections can be used, Kanocz et al (2013 and 2014). In our case, shear connections were realized by means of nails, which were spaced in specified distances and partially hammered into a timber beams. Their protruding parts were cast-in a reinforced concrete slab, thus creating a composite timber-concrete beam. Shear connectors (nails) have the same effect as shear arbors: they provide transfer of forces between the timber beam and reinforced concrete slab. At the same time, this kind of shear connection provides cooperation between the elements in vertical direction, preventing them from mutual lifting. With regards to material properties of timber and commonly used nails, it is possible to anticipate that this kind of shear connectors will act as elements in tension.

2   THEORETICAL RESULTS OF DESIGNED COMPOSITE BEAM

Parameters and data presented in this paper are a results from previous design of already realized composite timber-concrete ceiling. The considered composite ceiling was composed of timber beams made from wood C20 with length of 3.0 m and cross-section of 120×140 mm

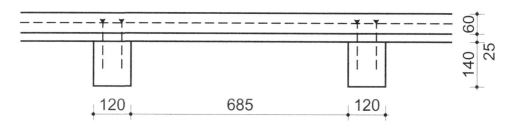

Figure 1.   The shape of considered composite beams.

and reinforced concrete slab made from concrete 20/25 of 60 mm thickness. Timber Beams 0.805 m were axially spaced. Crucial elements in designing of composite structure, composed of timber beams and concrete slab are shear connectors, in our case the nails were used. Proceeding from the above mentioned assumptions, it is possible to determine bearing capacity of the nail as the minimum of the following values:

– Shear bearing capacity of a nail according to metal properties,
– bearing capacity of concrete splicing with a nail,
– bearing capacity of the used nail stressed by shear, in our case 180/6.3 mm.

Minimal bearing capacity of the nail was the shear capacity according to relevant standards, which is 1.5 kN, EN 26891 (1995).

The stiffness of timber formwork made of 25 mm thick boards, which were attached to the timber beams by nails, was disregarded in calculation. From the mentioned ceiling elements and based on the ratio of timber and concrete Young's modules, ideal composite cross-section characteristics were determined. Scheme of the considered composite ceiling is presented in Figure 1.

Following results were obtained by calculation: Maximum deflection of the composite beam 1.83 mm, average shear force 2.6 kN, maximum normal stress in the concrete slab 1.88 MPa and maximum normal stress in timber beam 2.85 MPa.

3 COMPUTER MODEL

For more precisely comprehend behavior of the composite timber-concrete beams, 3D model of the ceiling was realized. The model consisted of hinge-joined timber beams, timber formwork and concrete slab in accordance to the parameters from previous chapter. The used software allowed the creation of contacts between the formwork and the timber beams which have had a mating surfaces. Semi-rigid far-connections were generated between the concrete slab and the timber beams. No connections were generated between the timber formwork and the concrete slab. Some schemes of the computer model are presented in Figures 2 and 3.

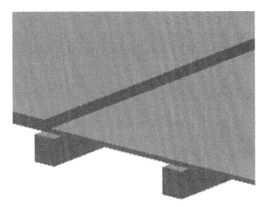

Figure 2.   Configuration of the calculation model.

Figure 3.   Semi-rigid far-connections between concrete slab and timber beams.

The results of presented model proved a favorable effect of the formwork onto the general stiffness of the composite ceiling where the general deflection reached the value 1.033 mm and maximum shear force was 2.73 kN. Maximum value of normal stress was 2.34 MPa in the timber beam and 1.42 MPa in the concrete slab. Moreover, offset of the concrete slab from the timber beams induced additional bending moments that have to be taken into consideration in evaluating the resistance of nails against extraction, see Figure 4.

4 REALIZATION OF THE COMPOSITE CEILING

The composite timber-concrete ceiling was realized to be a first floor of administrative building with floor dimensions of 30×12 m. The building has three load-bearing walls creating two-aisle bearing system of 2×6 m spans. Continuous two-span steel girders, spaced by 3,0 m were set onto the bearing walls, thus forming the main bearing members for the composite ceiling. The timber beams were set onto the lower flange of the steel girders. Timber beams were covered with formwork made from timber boards, see Figure 5.

Holes were then bored into the timber beams by means of prepared steel pattern plate to prevent wood cleavage when hammering nails of dimensions 180/6.3. The holes were 80 mm deep and were bored by a borer of 5.5 mm diameter, so that 1/3 of the nail length could be hammered into non-bored wood and the nails could protrude from the formwork by 40 mm, see Figure 6.

Figure 4. Additional bending moments caused by offset of the concrete slab.

Figure 5. Layout of the steel and timber components.

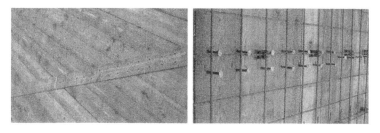

Figure 6. Prepared steel pattern plate and the layout of the nails.

Figure 7. Some steps from the realization of the composite timber-concrete ceiling.

After hammering the nails, welded net reinforcement was placed and the concrete slab was realized. After concrete hardening, individual components were joined and started to act as a composite timber-concrete ceiling structure. The following figures present some steps from the realization of the mentioned structure.

5 CONCLUSION

Regarding above mentioned analyses and calculations, comparison of obtained results from theoretical analyses and computer modeling shows quite good agreement. Application of such composite structures allows the using of favorable properties of used materials, the reduction of concrete slab thickness and amount of the reinforcement. For more exact specification of real behavior of composite timber-concrete beams, it is necessary to carry out experiments to obtain further results which would more precisely demonstrate the effectiveness of timber-concrete shear connection by means of nails or other types of shear connectors.

ACKNOWLEDGEMENT

This paper is prepared within the research project VEGA 1/0188/16, supported by the Scientific Grant Agency of the Ministry of Education of Slovak Republic and by the Slovak Academy of Sciences.

REFERENCES

Postulka, J. 2003. Steel-Timber Construction Finalization of Slimák Department Store in Bratislava. *Proceedings of XXIX Congress of professionals in steel construction field*, pp. 131–136. Lubovnianske kupele 2003.
Kanócz, J., Bajzecerová, V. and Steller, Š. 2013. Timber—Concrete composite elements with various composite connections, Part 1: Screwed connection. *Wood Research*, 58 (4)/2013, pp. 555–570.
Kanócz, J., Bajzecerová, V. and Šteller, Š. 2014. Timber—Concrete composite elements with various composite connections, Part 2: Grooved connection. *Wood Research*, 59 (4)/2014, pp. 627–638.
EN 26891, Timber structures. Joints made with mechanical fasteners (1995). General Principles for the Determination of Strength and Deformation Characteristics.

Advances and Trends in Engineering Sciences and Technologies II – Al Ali & Platko (Eds)
© 2017 Taylor & Francis Group, ISBN 978-1-138-03224-8

An investigation of glued connections with acrylate adhesives

K. Machalická & M. Vokáč
Klokner Institute, Czech Technical University in Prague, Prague, Czech Republic

M. Eliášová
Faculty of Civil Engineering, Czech Technical University in Prague, Prague, Czech Republic

ABSTRACT: Adhesive connections offer a number of benefits in structural applications, especially in the case of brittle adherends such as glass. This paper reports on an experimental analysis aimed at determining the mechanical characteristics of the two types of the same acrylate adhesive product (a new technology adhesive and its predecessor), both produced by the same manufacturer. The adhesive connections were applied in planar connections under shear loading. Finite element analysis shows that it is necessary to use different material models for the two types of adhesive, due to their completely different mechanical properties and behavior under loading. The numerical analysis deals with the application of various material models used for modelling the old and the new types of acrylate adhesives, and provides a comparison of numerical and experimental results.

1 INTRODUCTION

The use of adhesive joints has increased recently in the context of structural glass-metal connections, e.g. in façade systems and in innovative hybrid glass-metal structures. Adhesive joints have several advantages, e.g. a more uniform stress distribution than traditional joining techniques, such as fastening. The stress distribution is influenced by the mechanical properties of the adhesives and adherends, the geometry of the joint (the length of the overlap, the thickness of the bondline and the thicknesses of the adherends) (Petrie, 2007), (da Silva et al., 2006). Significant stress concentrations in adhesive joints can be avoided by applying glue with a low modulus of elasticity and with ductile behavior. Flexible adhesives generally have a low strength, but their ability to distribute the stress more evenly along the overlap may result in higher joint strength than when the rigid adhesives are used. This is because the rigid adhesives are not able to redistribute the stress concentrations at the ends of the overlap (da Silva & Campilho, 2015). The use of adhesives that are relatively strong and, at the same time, flexible, would therefore be ideal for structural applications. For this reason, SikaFast 5211 acrylate adhesive was selected for a study of hybrid steel-glass beams. However, the mechanical characteristics of the currently produced SikaFast 5211 NT acrylate adhesive are different from those of its predecessor. This paper presents a comparison of experimental results for the two adhesives and applies a finite element method for the two adhesively bonded joints. Due to their different behavior under loading, different material models for the glue had to be used.

2 BULK MATERIAL TENSILE TESTS

2.1 *Tensile test procedure*

Tests were performed on bulk material specimens to obtain their adhesive properties without any influence of the adherend. Specimens of the two adhesives were tested under uniaxial tensile loading, according to EN ISO 527 (Plastics—determination of tensile properties). Dumb-bell test specimens, see Figure 1a, were manufactured by casting in a mold. The specimens were subjected to uniaxial loading, see Figure 1b, at a displacement rate of 1 mm/min for the original SikaFast 5211 specimens and at a rate of 1, 2 and 8 mm/min for the new SikaFast 5211 NT acrylate. The tensile strength, the stress-strain diagram and the elongation at break were determined. The results of the tensile tests on the polymeric materials served as an important starting point in the investigation of the adhesives and the numerical modelling.

2.2 Tensile tests results

A comparison of the behavior of the two adhesives under tensile loading is shown in the stress-strain diagram in Figure 2, where SF is the SikaFast 5211 adhesive and SF NT is the new SikaFast 5211 NT. The experimental results showed that the new generation of the adhesive (SikaFast 5211 NT) has the higher initial stiffness and ultimate strength than the original adhesive (SikaFast 5211) for all displacement rates. The initial stiffness does not depend on

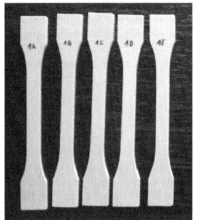

Figure 1.   a) Shape of the test specimens, b) A specimen during a tensile test.

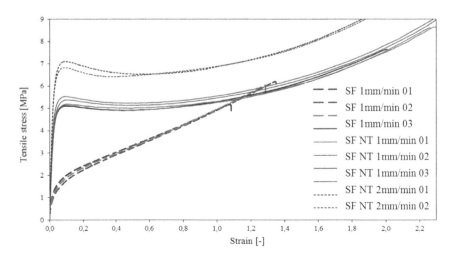

Figure 2.   A comparison of the stress-strain relationships for SikaFast 5211 and SikaFast 5211 NT.

188

Table 1. A comparison of the adhesive characteristics (according to EN ISO 527) at a displacement rate of 1 mm/min.

	SikaFast 5211	SikaFast 5211 NT
Tension strength [MPa]	6.1	7.22
Strain [–]	1.24	2.19
Modulus of elasticity [MPa]	96.94	260.1

the load rate, but there are obvious differences in the yielding stress. SikaFast 5211 NT displayed almost linear behavior up to yielding stress. After that, a significant increase in strain was observed with lower stiffness. The mechanical properties of the two acrylates were tested at a crosshead speed of 1 mm/min, see Table 1.

3 THE SHEAR STEEL-GLASS CONNECTION TESTS

As it was mentioned above, the stress distribution in the joint is influenced by the geometry of the joint and the stiffness of the adhesive and the adherends. Thus, the strength of the bonded joint depends not only on the cohesive strength of the adhesive itself, but also on the geometry and the stiffness of the connection, on the type of loading and the rate of loading, and the degree of adhesion of the glue to the substrate. A properly designed and manufactured joint should not break by the adhesive failure mode. The adhesive failure mode indicates that there was improper surface preparation, or that there were insufficient adhesive forces between the substrate and the glue. This leads to significantly lower ultimate strength and unsafe behavior of the joint during the loading (sudden collapse). By contrast, the cohesive mode of failure is a suitable type of joint breakage, and should be reached in structural adhesive joints. For all these reasons, it was necessary to carry out the shear steel-glass connection tests as a subsequent step in the experimental assessment of the two acrylates.

3.1 The shear test procedure

The shear connection specimens were prepared as a double lap joint consisting of two steel plates 25 mm in thickness and with dimensions of 75 × 50 mm, and two float glass plates 19 mm in thickness with dimensions of 110 × 50 mm. Thus, the bonded area was four times 50 × 50 mm, see the setup scheme in Figure 3. The two acrylates were applied in a layer 3 mm in thickness. The bonded surfaces of the steel plates were treated with abrasive pads. All bonded surfaces (both steel and glass) were cleaned, degreased and activated by Sika ADPrep to provide sufficient adhesion of the glue to the substrates.

The test specimens were loaded by a tensile force with a displacement rate of 1 mm/min, according to the setup scheme in Figure 3. The shear stress in the adhesive layer was reached by a tensile force introduced into the setup by long steel rods. The shear deformation was measured directly by potentiometers for each adhesive layer.

3.2 The shear test results

The overall engineering stress-strain diagram in shear for the two acrylate adhesives is shown in Figure 4. The new type of acrylate in the steel-glass connections proved to have different behavior from that of its predecessor as in the tensile tests. The new acrylate reached the higher shear strength (SikaFast-5211 NT reached up to 7.5 MPa, while SikaFast-5211 reached a maximum of 5 MPa for the same joint and arrangement) and reached the greater shear strain at break. The great initial stiffness in the joints of the new adhesive was not

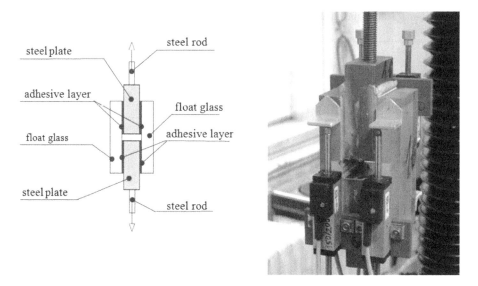

Figure 3.   Setup of the shear steel-glass connection tests.

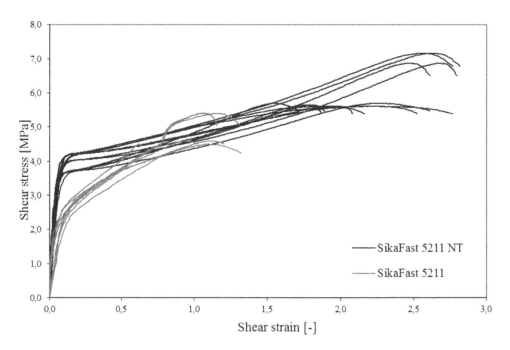

Figure 4.   A comparison of engineering stress-strain diagrams in shear for SikaFast-5211 and Sika-Fast-5211 NT acrylates in a glass-to-steel connection.

changed up to 3.5 MPa, whereas the older type of adhesive had initial stiffness limited to around 2 MPa.

No significant difference was observed for the two glue failure modes. Specimen failure starts predominantly in the manner of adhesion at the glass-to-adhesive interface. However, the rest of the adhesive layer still behaved cohesively and the joint was finally broken by a combination of adhesive and cohesive failure.

190

4 FINITE ELEMENT ANALYSIS OF BONDED CONNECTIONS

The adhesive joints described above can be numerically reproduced using Ansys v. 16.2 software code, which is based on the finite element method. A three-dimensional (3D) model was made, in which glass and steel were defined as linear elastic isotropic materials, as the stress in steel is assumed to be lower than the yielding limit. The steel was determined by Young's modulus of elasticity E = 210 000 MPa and Poisson's ratio μ = 0.3, and the glass was determined by E = 70 000 MPa, μ = 0.23. Different non-linear material models had to be applied for the two acrylates, due to their different behavior, see the description below.

4.1 Modelling the SikaFast 5211 shear connections

The SikaFast 5211 adhesive in a connection showed non-linear elastic behavior, which changes its modulus of elasticity or its shear modulus according to the level of loading. The multilinear elastic material model allows loading to be applied, resulting in higher levels of stress where material non-linearities are incorporated into the calculation. The appropriate material model of SikaFast 5211 was based on the stress-strain curve obtained from material tests. The numerical calculation validated by experimental measurements on glass-steel connections is shown in Figure 5.

4.2 Modelling the SikaFast 5211 NT shear connections

Connections bonded by the older SikaFast 5211showed significantly lower strain (about 100%) than the new SikaFast 5211 NT (up to 250%). SikaFast 5211 NT also displayed a significant yield limit that defines the material state of the transition from elastic to elastic-plastic behavior. Inelastic isotropic hardening plasticity was used to obtain an elastic-plastic character of the new acrylate adhesive in the numerical model. The von Mises yield criterion was applied due to the isotropic material and its recommended applicability for polymers according to the ANSYS Mechanical APDL Material Reference (ANSYS, Inc., 2013). The adhesive layer was modeled both as a bilinear material and as a multilinear material, based on the results of the material test. A comparison of the two numerical simulations with the experimental results is given in Figure 6, where miso means a multilinear model and biso means a bilinear model. Despite its simplification, the bilinear model provides relatively good agreement with the experiments and requires less computation time. By contrast, the multi-linear model needs more computation time, but it provides more accurate results.

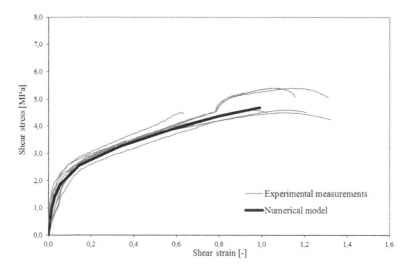

Figure 5.   Experimental and numerical results of the shear stress-strain diagram for SikaFast 5211.

191

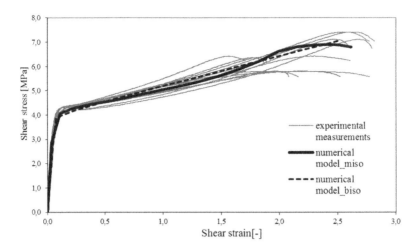

Figure 6. Experimental and numerical results of the shear stress-strain diagram for SikaFast 5211 NT.

5 CONCLUSION

It has been proven that there are major differences in the behavior of two-part acrylate adhesives. The new type of acrylate glue (SikaFast 5211 NT) in steel-glass connections showed completely different behavior under shear loading: there was a significant almost linear part of the stress-strain diagram, with high stiffness up to a stress level of 3.5 MPa. The new type of acrylate also showed increasing shear strain for stress levels higher than the yielding limit. Numerical analyses were performed for the original type of acrylate and for the new type of acrylate in a connection. A multilinear elastic material was chosen for the numerical simulation of SikaFast 5211, while inelastic isotropic hardening plasticity had be used for the new SikaFast 5211 NT adhesive in order to achieve elastic-plastic behavior of the adhesive. The innovative acrylate adhesive represents a new generation of adhesives suitable for load-bearing glued connections, with a high level of strength and initial stiffness up to the yielding point. These properties also provide a safety region above this point and ensure residual capacity (Netušil, M., Eliášová, M., 2014).

ACKNOWLEDGEMENT

This paper has been prepared with support from GAČR grant no. 16 - 17461S.

REFERENCES

ANSYS, Inc. 2003. *ANSYS Mechanical APDL Material Reference.*
da Silva, L. F. M. & Campilho, R. D. S. G. 2015. Design of adhesively-bonded composite joints. *Fatigue and Fracture of Adhesively-bonded Composite Joints—Behaviour, Simulation and Modelling—Vassilopoulos (Ed.)*: 43–71.
da Silva, L. F. M., Rodrigues, T. N. S. S., Figueiredo, M. A. V., de Moura, M. F. D. F. & Chousal, J. A. G. 2006. Effect of Adhesive Type and Thickness on the Lap Shear Strength. *The Journal of Adhesion* 82: 1091–1115.
Netušil, M., Eliášová, M. 2014. Trends and requirements for adhesives with load bearing role. *Proceedings of the Challenging Glass 4 and COST Action TU0905 Final Conference*. Leiden: CRC Press/Balkema, 369–374.
Petrie, E. M. 2007. *Handbook of adhesives and sealants,* McGraw-Hill, New York.

Advances and Trends in Engineering Sciences and Technologies II – Al Ali & Platko (Eds)
© 2017 Taylor & Francis Group, ISBN 978-1-138-03224-8

The maximum punching shear resistance of flat slabs

L. Majtánová, J. Hanzel & J. Halvonik
Slovak University of Technology, Bratislava, Slovakia

ABSTRACT: The paper deals with limits concerning the maximum punching shear capacity of flat slabs with shear reinforcement. The limits are mostly defined as k_{max} multiple of punching shear resistance without transverse reinforcement, with value of k_{max} ranging between 1.4 and 2.0. The second possible limit is determined by the capacity of concrete struts at the column periphery. Results of experimental tests are presented in the paper that were focused on the above mentioned limits, whether failure of the struts can precede any other form of punching failure that is limited by $k_{max} * V_{Rd,c}$. The experiment has shown that limits based only on the k_{max} factors may overestimate actual maximum punching resistances of flat slabs with transverse reinforcement.

1 INTRODUCTION

Punching belongs to the most dangerous forms of structural failure, due to its brittle mode of failure and possible loss of integrity when local failure spreads over the structure, follow by a progressive collapse of the whole structure. In order to improve structural behavior and to increase punching capacity transverse reinforcement is used. Application of transverse reinforcement allows for significant increase of punching shear resistance in comparison with flat slabs without this reinforcement and the failure has more ductile mode. However, many experiments have shown some limits which even very heavy reinforced flat slab is not able to overcome.

Currently these limits are checked by the k_{max} factor, which expresses how many times punching resistance with shear reinforcement can be higher than the resistance without shear reinforcement. The value of k_{max} depends on the type of shear reinforcement (studs, stirrups, cages, bent-down bars), on the effective depth of a slab "d" or on experiences of experts working with the models for assessment of punching resistance. The overview of k_{max} factor values can be found in Table 1.

Based on experiences with beams with high amount of shear reinforcement where shear failure had been reached many times by crushing of the concrete struts, the check of the strut capacity has also been incorporated in current EC2 model for punching, see formula (1).

$$V_{Rd,max} = 0.4 * 0.6 * v * f_{cd} * u_0 * d \qquad (1)$$

This limit is usually decisive for thick slabs or slabs with large amount of bending reinforcement with a small loaded area surrounded by a column periphery u_0, *Einpaul, Bujnak, Ruiz, Muttoni* (2016). The value of 0.4 in (1) has replaced the original value of 0.5 because calculation of β factor is carried out for basic control perimeter u_1, and this value can also be applied for the verification of the strut capacity at control perimeter u_0. Factor β takes into account the effect of unbalanced bending moments. Factor $v = 1 - f_{ck}/250$, f_{ck} and f_{cd} are characteristic and design strength of concrete.

Table 1. Overview of k_{max} factors.

Model	Basic control perimeter	Notes	k_{max}, k_{sys}	Shear reinforcement
Eurocode2	$2d$	Recommended	1.5	all
Eurocode2	$2d$	Germany	1.4	all
Eurocode2	$2d$	UK	2.0	all
Eurocode2	$2d$	Sweden	1.6	all
Eurocode2	$2d$	Austria	1.9	studs
Eurocode2	$2d$	Austria	1.4–1.6	stirrups
Eurocode2	$2d$	Slovakia	1.9	studs
Eurocode2	$2d$	Slovakia	1.4–1.7	stirrups
Eurocode2	$2d$	ETA 13/0151	1.96	studs
GER update of EC2	$0.5d$	Recommended	1.5	stirrups, bents
GER update of EC2	$0.5d$	Recommended	2.0	studs
CSCT/MC 2010	$0.5d$	$s_{max} = 0.75d$	2.0	all
CSCT/MC 2010	$0.5d$	$s_{max} \leq 0.6d$	2.4	stirrups, bents
CSCT/MC 2010	$0.5d$	$s_{max} \leq 0.6d$	2.8	studs
CSCT/EC2	$0.5d$	$s_{max} = 0.75d$	1.4	all
CSCT/EC2	$0.5d$	$s_{max} \leq 0.6d$	1.6	stirrups, bents
CSCT/EC2	$0.5d$	$s_{max} \leq 0.6d$	1.8	studs
ACI-318-08	$0.5d$	–	1.52	stirrups
ACI-318-08	$0.5d$	–	2.0	studs
Mari (Cladera	$0.5d$	strut capacity	–	all

2 EXPERIMENTAL PROGRAM

2.1 Description of samples

Experimental samples consisted of flat slabs with a thickness of 250 mm. The slabs were supported by columns with a small diameter of 180 mm in order to minimize the length of the column periphery, ratio $u_0/d = 2.83$. Slabs were reinforced by bending reinforcement $\phi20/100$ mm made of steel B500B, the average reinforcement ratio was 1.57%. One sample was reinforced by a transverse reinforcement consisting of double headed Peikko PSB studs with diameter of 10 mm made of B500B steel, see Figure 1. Number of studs in one perimeter was 15. The second sample was cast without shear reinforcement.

Slabs were cast at a different time, so the strength of the concrete was a bit different. The proposed strength class of the concrete was C25/30. The actual cylinder compressive strength (mean value) was $f_{cm} = 38.2$ MPa for slab with shear reinforcement and in case of slab without shear reinforcement $f_{cm} = 41$ MPa. The maximum size of aggregates was 16 mm. Slabs were placed on a hydraulic jack and tied to the floor with 8 rods with a diameter of 50 mm. Slabs were loaded under axis-symmetric conditions. The shear-span to depth ratio (shear slenderness) was 5.92. The most important data of the experiment are introduced in Table 2.

2.2 Results

Slab without shear reinforcement failed in a brittle way, with typical cone, where the critical shear crack was inclined approximately at an angle of 25°. The slab failed when shear force reached 903 kN. The slab with shear reinforcement failed in a different way, when the column penetrated into the slab when subjected to the force of approx. 1520 kN and the crushed concrete was observed at column periphery. Damaged part of the slab was concentrated next to the column with a very steep critical shear crack.

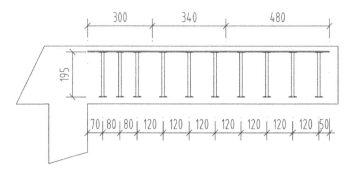

Figure 1. Peikko PSB studs—arrangement of shear reinforcement.

Table 2. Basic parameters and punching shear resistance.

Slab	$\dfrac{d}{mm}$	$\dfrac{\rho}{\%}$	$\dfrac{a_\lambda/d}{-}$	$\dfrac{f_{cm}}{MPa}$	$\dfrac{f_{vm}}{MPa}$	$\dfrac{u_0/d}{-}$	studs	$\dfrac{V_{R,test}}{kN}$
S1-1	200	1.57	5.8	41.0	550	2.827	NO	903
S2-1	200	1.57	5.8	38.2	550	2.827	YES	1520

3 MODELS FOR THE ASSESSMENT OF PUNCHING RESISTANCE

Several models for the assessment of punching resistance have been used and results were compared with tests. The models can be divided into either empirical models e.g. EC2 model or a new German model (2016) and physical models, e.g. models based on critical shear crack theory or Cladera & Mari model.

3.1 Empirical models

Basic principles of current EC2 model were set by Zsutty (1968). Zsutty proposed an empirical approach on the basis of available test data for beams. The model has been later refined and adopted in Model Code 1990. The MC1990 model became the basis for current EC2 model. *Hegger, Siburg, Kueres* (2016), released the new model, which is based on new formula for punching resistance of slab without shear reinforcement (2) with basic control perimeter at distance $0.5d$ from the face of a column. This empirical model introduces several new factors in comparison with EC2 model. First it is factor taking account for size effect with Bazant formula (3). For the second it is factor which takes into account shear slenderness and length of column periphery (4) and the last one is new value of empirical parameter $C_{Rk,c} = 1.8$ ($C_{Rm,c} = 2.2$).

$$v_{Rd,c} = \frac{C_{Rk,c}}{\gamma_C} * k_d * k_\lambda * \left[100 * \rho_l * f_{ck}\right]^{0.333} \tag{2}$$

$$k_d = \left(1 + d_v / 200\right)^{-0.5} \tag{3}$$

$$k_\lambda = \left[\left(\frac{a_\lambda}{d}\right)\left(\frac{u_0}{d}\right)\right]^{-0.2} \tag{4}$$

where d_v = average effective depth; k_d = coefficient of size effect; ρ_l = reinforcement ratio, a_λ = distance between the edge of the loaded area and the line of contra flexure and u_0 = the perimeter of the loaded area (length of column periphery).

3.2 Mechanical models

The first is a model based on the Critical shear crack theory which has been developed and later refined by *Muttoni* (1991, 2008) and *Ruiz* (2008). The model has been used in the Model Code 2010 and later adjusted into EC2 format. The CSCT theory is based on the assumption that the major part of shear resistance is represented by aggregate interlocking in critical shear crack and tensile strength of concrete. The model has position of critical control perimeter at distance $0.5d$ from column face.

$$V_{Rd,c} = \left(\frac{A}{1 + B * k_{dg} * \psi * d} \right) * \frac{\sqrt{f_{ck}}}{\gamma_C} \tag{5}$$

$$\psi = C * \frac{r_s}{d} * \frac{f_{yd}}{E_s} * \left(\frac{m_{Ed}}{m_{Rd}} \right)^{1.5} \tag{6}$$

For the factors A, B and C see Table 4. Average bending moment per unit length was calculated using formula $m_{Ed} = V_{R,test}/8$.

The second mechanical model has been introduced by *Cladera & Mari* (2016). The authors attribute the major part of the shear resistance to the compressive concrete cord. The failure is reached when critical shear crack propagates inside the compressive cord. Therefore, basic control perimeter is located at the distance where critical shear crack crosses neutral axis "x". For simplification a distance $0.5d$ can be used. The punching resistance without shear reinforcement can be assessed by (7) with $D = 0.85$ at the basic control perimeter and $D = 0$ for the other control perimeters.

$$V_{Rd,c} = \xi * \frac{x}{d} * \left(0.30 * \frac{f_{ck}^{2/3}}{\gamma_C} + D * \frac{V_{Rd,c}}{A_1} \right) \tag{7}$$

$$\xi = \left(\frac{d}{a_\lambda} \right)^{0.2} \frac{2}{(1 + d/200)^{0.5}} \tag{8}$$

$$\frac{x}{d} = 0.75 * (\alpha_e * \rho_l)^{1/3} \tag{9}$$

where α_e = modulus ratio; A_1 = area surrounded by basic control perimeter.

4 COMPARISON OF THE THEORETICAL AND EXPERIMENTAL RESULTS

4.1 Slab without transverse reinforcement

The assessments of the punching shear resistance was performed with safety factors equal to 1. Two values of an empirical factor $C_{R,c}$ were used in case of the empirical models. For the first it is mean value $C_{Rm,c}$ and for the second characteristic value $C_{Rk,c}$, see Table 3 and Table 6.

Table 3. Slab without transverse reinforcement $V_{R,test} = 0.903$ MN—empirical models.

Model	$\frac{u_1}{-}$	$\frac{C_{RK,c}(C_{Rm,c})}{-}$	$\frac{V_{R,c}}{MN}$	$\frac{V_{R,c}/V_{R,ctest}}{\%}$
Current EC2	2d	0.18 (0.22)	0.888 (1.085)	98.34 (120.2)
ETA 13/0151	2d	0.159 (0.197)*	0.784 (0.957)	86.84 (106.0)
New GER	2d	1.80 (2.20)	0.692 (0.846)	76.63 (93.69)

*$C_{R,c}(0.1u_0/d + 0.6)$
Note: Values in brackets were calculated with factor $C_{Rm,c}$.

Table 4. Slab without transverse reinforcement $V_{R,test} = 0.903$ MN—mechanical models.

Model	u_1 $-$	A $-$	B $-$	C $-$	$\dfrac{m_{Ed}/m_{Rd}}{-}$	$\dfrac{V_{R,c}}{MN}$	$\dfrac{V_{R,c}/V_{R,test}}{\%}$
MC2010/LoAII	0.5d	0.67	0.60	1.5	0.362	0.625	69.21
MC2010/LoAIII	0.5d	0.67	0.60	1.2	0.362	0.678	75.08
CSCT/EC2	0.5d	0.75	0.47	1.2	0.362	0.744	85.71
Cladera & Mari	0.5d	–	–	–	–	0.832	92.14

Table 5. Slab with transverse reinforcement $V_{R,test} = 1.52$ MN—empirical models.

Model	$\dfrac{C_{Rk,c}\,(C_{Rm,c})}{-}$	$\dfrac{V_{R,c}}{MN}$	$\dfrac{V_{R,s}}{MN}$	$\dfrac{V_{R,cs}}{MN}$	$\dfrac{k_{max}}{-}$	$\dfrac{k_{max}V_{R,c}}{MN}$	$\dfrac{V_{Rm,max}}{MN}$
Current EC2	0.18 (0.22)	0.867 (1.06)	1.06	1.71 (1.86)	1.5	1.31 (1.59)	1.098
ETA 13/0151	0.159 (0.197)	0.765 (0.883)	1.755	1.75	1.96	1.5 (1.73)	–
New GER	1.80 (2.20)	0.676 (0.827)	1.06	1.43 (1.53)	2.0	1.38 (1.69)	–

Table 6. Slab with transverse reinforcement $V_{R,test} = 1.52$ MN—mechanical models.

Model	A $-$	B $-$	C $-$	$\dfrac{m_{Ed}/m_{Rd}}{-}$	$\dfrac{V_{R,c}}{MN}$	$\dfrac{V_{R,s}}{MN}$	$\dfrac{V_{R,s}}{MN}$	$\dfrac{k_{sys}}{-}$	$\dfrac{k_{sys}V_{R,c}}{MN}$	$\dfrac{V_{R,max}}{MN}$
MC2010/LoAII	0.67	0.60	1.5	0.611	0.412	1.373	1.884	2.8	1.153	–
MC2010/LoAIII	0.67	0.60	1.2	0.611	0.466	1.198	1.664	2.8	1.305	–
CSCT/EC2	0.75	0.47	1.2	0.611	0.589	1.198	1.788	1.8	1.060	–
Cladera & Mari	–	–	–	–	1.193	0.594	1.788	–	–	1.098

4.2 Slab with transverse reinforcement

The new German model takes into account only contribution of transverse reinforcement embedded within zone 0.3d to 1.125d measured from edge of a column to the punching resistance in the basic control perimeter. The maximum stresses in reinforcement are limited by a value $f_{ywd,ef}$, see current EC2 model. The punching resistance can be calculated by formula (10) with $\alpha_c = 0.70$ and $\alpha_s = 0.90$.

$$V_{Rd,cs} = \alpha_c V_{Rd,c} + \alpha_s V_{Rd,s} \tag{10}$$

In case of the ETA 13/0151 model only transverse reinforcement from double headed studs located within zone 0.3d and 1.125d are included to the punching capacity at basic control perimeter. The maximum stresses in shear reinforcement are limited by yield strength of the steel. In case of the CSCT models only shear reinforcement within zone 0.3d and 1.0d can be taken into account for the assessment of the punching resistance at basic control perimeter. The shear resistance with transverse reinforcement can be calculated by (10) with $\alpha_c = 1.0$ and $\alpha_s = 1.0$. In case of the Cladera & Mari model only shear reinforcement within zone $(d-x)$ contributes to the punching resistance at basic control perimeter. Overall punching resistance can be assessed using formula (10) with $\alpha_c = 1.0$ and $\alpha_s = 1.0$.

5 DISCUSSION

Current EC2 model with a mean value of empirical factor $C_{Rm,c}$ overestimates punching shear resistance in a case of slab without shear reinforcement. Better results are provided with an application of characteristic values when the assessed punching resistance is below measured value. Generally, model seems to be unsafe. Better results are provided by the ETA 13/0151 model where factors $C_{R,c}$ are adjusted depending on the ratio u_0/d. In a case of mean value, the model still overestimates the punching capacity. The new German model provides very good results when mean value of the assessed punching resistance is slightly below tested value. In a case of mechanical models based on the CSCT theory the level of model safety for LoAIII is comparable with the new German model. The CSCT/EC2 model is more economical but still possesses sufficient safety. The Cladera & Mari model provides very good results that are comparable with results of the new German model. The maximum punching resistance was tested on the slab sample with shear reinforcement. An experimentally determined value of the k_{max} factor was $1.52/0.9 = 1.69$, which is lower value than values used in many standards and models. In a case of the empirical models where the assessment of the maximum punching resistance is based on both limits $k_{max}*V_{Rd,c}$ and $V_{Rd,max}$, see formula (1), the value of the k_{max} factor higher than 1.5 is acceptable. The ETA 13/0151 and New German model with the factors $k_{max} = 1.96$ and 2.0 respectively, may overestimate real maximum punching resistance without verification of the strut capacity. Opposite the German DIN EN1992-1-1/NA with $k_{max} = 1.40$ is safe but is very conservative for the normal values of u_0/d. The model based on the CSCT theory introduced in MC2010 with the highest accuracy in a case of linear structural analysis (LoAIII) provides acceptable safety with $k_{sys} = 2.8$ from the maximum punching resistance point of view. However, in a case of more accurate non-linear structural analysis (LoAIV) the value of 2.8 may overestimate the maximum punching capacity. The model CSCT/EC2 with a value of $k_{max} = 1.8$ provides very safe assessment. The Cladera & Mari model checks the maximum punching resistance only based on the strut capacity therefore for our experiments provides safe solution.

6 CONCLUSIONS

Two ways how to determine the maximum punching resistance of flat slabs with shear reinforcement are currently used. The first way it is the value of $k_{max}*V_{Rd,c}$ where $V_{Rd,c}$ is the punching capacity of a flat slab without shear reinforcement. The second way is based on verification of the strut capacity at the column periphery. Since the check of the strut capacity is used only in current EC2 model, the possible failure due to the crushing of the struts was experimentally tested. One sample was over-reinforced by shear reinforcement in a way that crushing of the struts precedes the shear tensile failure. The second sample was without shear reinforcement.

- The maximum shear capacity obtained from the test was 1.52 MN and punching resistance without shear reinforcement 0.903 MN. The k_{max} factor reached a value of 1.69.
- The k_{max} factors are ranging from 1.5 to 2.0 for slabs with double headed studs in the empirical models. The values higher than 1.5 are justified only if the maximum punching capacity is defined as lower value of $k_{max}*V_{Rd,c}$ and $V_{Rd,max}$. Therefore, we recommend to add second limit which is based on the strut capacity in the ETA 13/0151 model and the new German model respectively.
- In a case of mechanical models based on CSCT theory we recommend to make adjustment of the factor k_{sys} based on the ratio u_0/d. The k_{sys} factor should be lower than 2.8 for lower values of u_0/d. In a case of the CSCT/EC2 model a value of the k_{max} factor 1.8 seems to be very conservative.
- The Cladera & Mari model defines maximum punching resistance only from the capacity of the struts. Based on measured strength of the concrete the theoretical maximum punching capacity calculated using formula (1) for axis-symmetric conditions is only 72.1% of the measured value. This indicates that the strength of a concrete strut under tension in formula (1) calculated with reduction factor $0.6*v$ is very conservative for the assessment of the punching capacity. We recommend to increase the value up to the $0.8*v$.

ACKNOWLEDGMENT

This work was supported by the Scientific Grant Agency of the Ministry of Education, science, research and sport of the Slovak Republic and the Slovak Academy of Sciences No 1/0810/16.

REFERENCES

Muttoni, A. & Schwartz, J. (1991). Behaviour of Beams and Punching in Slabs without Shear Reinforcement, *IABSE Colloquium*, Vol. 62, Zurich, Switzerland, 1991, pp. 703–708.

Muttoni, A. & Fernández, Ruiz, M. (2008). Shear strength of members without transverse reinforcement as function of critical shear crack width, *ACI Structural Journal*, V. 105, No 2, 2008, pp. 163–172.

Einpaul, J., Bujnak, J., Ruiz, M. & Muttoni, A. (2016). Study on Influence of Column Size and Slab Slenderness on Punching Strength, *ACI Structural Journal*, January–February 2016, pp. 135–145.

Hegger, J., Siburg, C. & Kueres, D. (2016). Proposal for punching shear design of flat slabs and column bases based on Eurocode 2, RWTH Aachen University, Germany, Presentation from CEN TC250/SC2/WG1/TG4 meeting in Copenhagen, 28.–29. January 2016.

Muttoni, A. & Ruiz, F. (2015). Proposal for punching shear provisions based on CSCT, Presentation from CEN TC250/SC2/WG1/TG4 meeting in Florence, 12.–13. March 2015.

Mari, A., Cladera, A, Bairan, J., Oller, E. & Ribas, C. (2016). Punching shear design proposal based on the CCC mechanical model, Presentation from CEN TC250/SC2/WG1/TG4 meeting in Copenhagen, 28.–29. January 2016.

Advances and Trends in Engineering Sciences and Technologies II – Al Ali & Platko (Eds)
© 2017 Taylor & Francis Group, ISBN 978-1-138-03224-8

Verification of experiment of vacuum loading on slab

J. Melcher & M. Karmazinova
Faculty of Civil Engineering, Institute of Metal and Timber Structures, Brno University of Technology, Brno, Czech Republic

J. Krsik & J. Krivakova
Faculty of Civil Engineering, Institute of Structural Mechanics, Brno University of Technology, Brno, Czech Republic

ABSTRACT: The aim of this paper is to describe a unique loading method—Vacuum loading, on slab shape construction. The method was originally developed for loading glass or steel plates. This is the first application of method for loading on a pre-stressed concrete slab. Method was proven itself as irreplaceable for the simulation of uniform load over whole slab with linear increase of load. The paper presents the results of the experiment and a comparison with a numerical model from ANSYS program.

1 INTRODUCTION

The aim of this paper is to check the bearing capacity of pair of Spirol type panels under uniform surface load and to summarize the results obtained from the experiment and compare them with numerical model solved by ANSYS program.

Ceiling hollow core Spirol type panels are used for the halls roofing with larger span, for which it is not economical to use an ordinary reinforced concrete ceiling. Approximately 20% of these halls use skeletal bearing systems, in which the panels are supported on girders, which may be deformed under load. This bearing method is modeled in the experiment.

Hollow core panels are manufactured in a continuous extruder on a long production line. The concrete is pressed into the form of the required shape of the panel. The final shape of the panel is made up of concrete ribs, voids and pre-stressing strands. The concentration of normal and shear stresses is in the ribs, to which the effect of transverse normal and shear stresses from the elastic drop of support is added.

Methods of application of uniform surface load by the manual addition of weights to the structure are time consuming; physically demanding and they can't accurately capture the surface effect, because the load is added in steps. Prof. Melcher developed the method of vacuum loading of slab type constructions, as a substitute for these "classical" methods. The method is applicable to various types of structures, especially to the loading of flat glass components (Melcher and Karmazinova 2005), glass fiber plates (Melcher, Karmazinova and Knezek 2002), steel constructions (Melcher 1997) and composite slabs (Karasek and Holomek 2011). This article describes the application of this method for loading concrete slabs assembled from a pair of pre-stressed hollow core Spirol type slabs.

The principle of the method is to create a controlled vacuum under tested construction, which results in a uniform surface load on the upper surface of the construction due to the difference between the atmospheric pressure and the lower pressure under the tested structure. The tested panel must be placed in a timber casing in order to ensure only vertical loading. This casing will take over all the pressures of the atmosphere from other directions than vertical direction. The tested construction, including timber casing is overlapped with a sufficiently large PVC sheet, which is attached to the smooth floor with a tape. It prevents the intrusion of air under the construction.

2 DESCRIPTION OF THE EXPERIMENT

Tested construction is assembled from a pair of Spirol SPG20043 type panels. Thickness of these panels is 200 mm, length is 6 m and width is 1.2 m. Panel is lightened by six approximately circular voids along the length of the panel. The panels are reinforced with four pre-stressing strands with a diameter of 12.5 mm and three pre-stressing strands with a diameter of 9.3 mm. Cross-section scheme of the panel is in the Figure 2. The strands are pre-stressed to 1100 MPa. Panels are made from the concrete class C45/55 and strands are from the steel of Y1860S7_R1 class. The panels are mutually connected by in-situ concrete class C30/37. The same concrete is used on the filling of voids in panels to a depth of 50 mm. The age of panels was 8 months and the age of in-situ concrete was 40 days at the time of the experiment.

The panels are imbedded on both ends on the supporting steel beams IPE 200. These beams are further welded by fillet welds on the shorter beams IPE 200, which stands on the floor. The deflection of the main support beams is allowed under the entire panel foreheads. Short supported beams are interconnected longitudinally and transversely to ensure the stability of the supporting construction. Longitudinal link is performed by square thin-walled profile of 100 × 100 mm with a wall thickness of 3 mm. Transverse link is performed by a pair of L50 × 50 × 5 profiles.

The assembly of experiment is equipped for the measure of the deflections with six digital gauges ID-C1050B Mitutoyo Japan (w1, w2, w3, w7, w8, w9) and with three indication displacement gauges WA200 HMB (w4, w5, w6) with a maximum deflection of 200 mm. Five resistance strain gauges are used to measure the state of stress. Strain gauges T1, T2 are the type 50/120 LY41 HMB with constant $K = 2.06$ and strain gauges T3, T4, T5 are the type 100/120 LY41 HMB with constant $K = 2.10$. The position of displacement gauges and strain gauges are in the Figure 1. The vacuum is read by digital indicator DM9200 MRU.

Data recording from strain gauges T1–T5 and displacement gauges w4–w6 were conducted to a data logger MGCplus HBM (with a 2 seconds recording step). Data recording from displacement gauges w1–w3 a w7–w9 were connected to a data logger DMX-16,

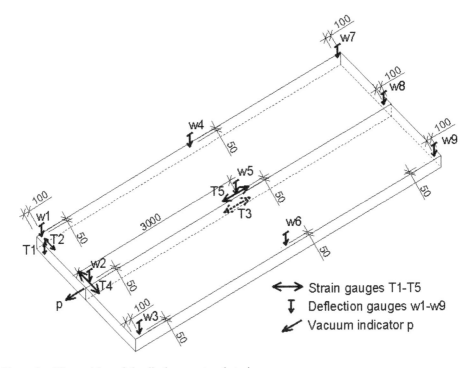

Figure 1. The position of the displacement and strain gauges.

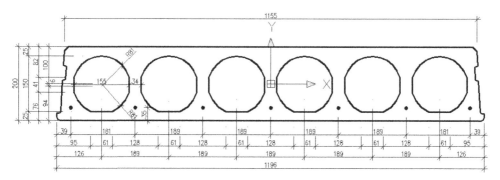

Figure 2. Cross-section drawing of Spirol SPG20043 type panel.

Mitutoyo and to the computer software of **QTREE-DC/DMX**-16 (with 5 seconds recording step).

3 PROGRESS OF THE EXPERIMENT

The automatic data logging of displacements and strains were placed every 2 seconds during the experiment into the table in **MS Excel** software. After reaching the time corresponding to a value of 2 kN/m² load, the pressure increasing was postponed for two minutes in order to stabilize the measured values. When the pressure firstly reaches the value of 6 kN/m², unloading to the value of 2 kN/m² was done. Then the pressure was increased up to the breach of the **PVC** sheet which ensures the vacuum on the load of 20 kN/m².

4 THE EVALUATION OF THE EXPERIMENT

Equations 1 and 2 are used for the recalculation of strains from strain gauges on the stresses. It is necessary to know the values of modulus of elasticity E of concrete, which are obtained from the numerical model in such a way, that is possible to compare the values of stresses from the experiment with the values of stresses from the numerical model. The values of modulus of elasticity used for recalculation of the strains to the stresses are presented in the Table. 1.

$$\varepsilon = \varepsilon_m \cdot \frac{2}{K} \tag{1}$$

where ε_m = measured deflection; K = the constant of strain gauge

$$\sigma = E \cdot \varepsilon \tag{2}$$

where E = modulus of elasticity of concrete; ε = recalculated deflection.

The recalculated values of strains to the stresses, including standard deviations are presented in the Table 2. The diagram of the most significant normal stress on strain gauge T3 is in the Figure 3 (full line).

5 NUMERICAL MODEL

The numerical model was created in **ANSYS** program from the volume finite elements **SOLID** 185 type, which is used for creating of both the concrete model and the support steel beams. Linear finite elements of **BEAM** 188 type are used for the model of pre-stress strands.

203

Table 1. Modulus of elasticity E used for the recalculation of strains to the stresses.

σ	E	ε
MPa	GPa	$1 \cdot 10^{-6}$
2.64	33.0	80
3.84	31.7	121
7.28	28.0	260
8.48	26.5	320
25.05	16.7	1500

Table 2. Recalculated values of stresses on the strain gauges T1–T5.

Load	E	T1	T2	T3	T4	T5
kN/m³	GPa	MPa	MPa	MPa	MPa	MPa
2	33.0	−1.18	0.05	1.39	−0.20	−1.32
		0.24	0.03	0.20	0.05	0.10
4	33.0	−1.99	−0.05	2.43	−0.41	−2.41
		0.04	0.01	0.08	0.02	0.07
6	31.7	−2.59	−0.33	3.66	−0.57	−3.41
		0.04	0.02	0.09	0.02	0.07
8	28.0	−2.86	−0.58	4.69	−0.89	−4.12
		0.03	0.01	0.07	0.01	0.06
10	28.0	−3.60	−0.66	5.85	−1.08	−5.28
		0.10	0.05	0.23	0.03	0.21
12	28.0	−4.12	−1.01	6.94	−1.18	−6.33
		0.02	0.02	0.07	0.01	0.05
14	26.5	−4.33	−1.24	8.09	−1.10	−7.05
		0.01	0.01	0.06	0.02	0.03

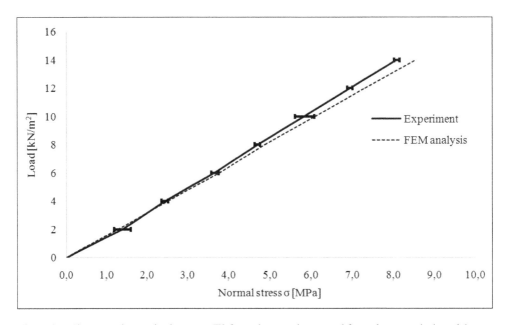

Figure 3. The normal stress in the gauge T3 from the experiment and from the numerical model.

204

Figure 4.   The model in ANSYS program.

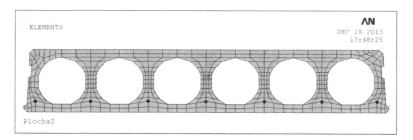

Figure 5.   FEM mesh of the panel's forehead.

Perfect cohesion with concrete is assumed. This assumption is acceptable, as the stresses are evaluated only up to crack appearance and the pre-stress reinforcement was connected to concrete perfectly until that time. The contacts between concrete and steel beams are modeled by pair of finite elements **TARGE 170** and **CONTA 173** type with predefined gap of 1 mm. The surface of panels is covered by surface finite elements of **SURF 154** type, which are used for the distribution of surface load. The density of finite element mesh was chosen as a compromise between the time of calculation (rougher grid) and the precision of required results (smoother grid). The element longitudinal division is 50 mm. Scheme of the model is on the Figure 4. FEM mesh of the panel's forehead is on the Figure 5.

Load steps were set identically to the experiment. Calculated and corrected normal stress in the position of strain gauge T3 was presented in the Figure 3 (dashed line).

6   CONCLUSIONS

A. The paper describes the execution of experiment with uniform surface load on pair of Spirol type panels on the elastic steel rolled profile supports. The large amount of data was obtained from the experiment. Only the most important were presented.
B. The comparison of the normal stress from the experiment and from the model in the position of strain gauge T3 (bottom face of the panel in the middle of the span) was the most important. The values from the numerical model follow the same trend as the measured values with slightly higher normal stress under higher load. It is possible to conclude, that

numerical model correctly describes the real behavior of the experiment and therefore this model can be used for further modeling of different load combinations and supports.

ACKNOWLEDGEMENTS

The project has been realized with financial support of junior specific research project of the Brno University of Technology of FAST-J-16-3687 and with the support of the company GOLDBECK Prefabeton s.r.o.

REFERENCES

ANSYS 13.0 help documentation.
Baloušek, M. (2010), "Předpjaté dutikové dílce v interakci s poddajnou podporou", *Proceedings of the 18th conference "Concrete day 2011"*, Hradec Kralove, Czech Republic.
Bertagnoli, G. and Mancini, G. (2009), "Failure analysis of hollow-core slabs tested in shear", *Structural Concrete*, 10(3), 139–152.
ČSN EN 1992-1-1 (2006), *Eurocode 2: Design of concrete structures—Part 1-1: General rules and rules for buildings*, Prague.
ČSN EN 1993-1-1 (2006), *Eurocode 3: Design of steel structures—Part 1-1: General rules and rules for buildings*, Prague.
ČSN EN 1168/A2:2009, *Precast concrete elements—Hollow core slabs*, Prague.
Hawkins, NM. and Ghosh, SK. (2006), "Shear strength of hollow-core slabs" *PCI Journal*, 51(1), 110.
Karásek, R. and Holomek, J. (2011), "Composite Steel-Concrete Slabs Under Different Load Types", *KONStrukce—Odborný časopis pro stavebnictví a strojírenství* 2011(4), Ostrava, Czech Republic.
Lee, DH, Park, MK, Oh, JY, Kim, KS, Im, JH. and Seo, SY (2014), "Web-shear capacity of prestressed hollow-core slab unit with consideration on the minimum shear reinforcement requirement", *Computers and Concrete"*, 14(3), 211–231.
Melcher, J. and Karmazínová, M. (2005), "The experimental verification of actual behavior of the glass roofing structure under uniform loading", *Proceedings of the 4th European Conference on Steel and Composite structures "Eurosteel 2005"*, Maastrich, Germany, June.
Melcher, J., Karmazínová, M. and Kněžek, J. (2002), "Experimental verification of fiberglassconcrete front waffle members by wind loading", *Proceedings of the 6th conference "New building materials and products"*, Brno, Czech Republic.
Melcher, J. (1997), "Full-Scale Testing of Steel and Timber Structures: Examples and Experience", *Structural Assessment—The Role of Large and Full Scale Testing*, London, United Kingdom, 301–308.
Melcher, J. and Karmazínová, M., "*Glass_ Vacuum loading*", The technology for experimental verification of deformation process and load capacity of structure glass with using loading by vacuum method, (verified technology), Institute of Metal and Timber Structures, BUT Brno, Czech Republic.
Nakamura, E., Avendano, AR. and Bayrak, O. (2013), "Shear Database for Prestressed Concrete Members", *Aci Structural Journal*, 110(6), 909–918.
Navrátil, J. (2014), *Prestressed Concrete Structures*, (2nd Edition), Ostrava: Technical University of Ostrava, Faculty of Civil Engineering, Ostrava, Czech Republic.
Pajari, M. (1998), "Shear Resistance of PHC Slabs Supported on Beams I: Tests", *Journal of Structural Engineering*, 124(9), 1050–1061.
Pajari, M. and Koukkari, H. (1998), "Shear Resistance of PHC Slabs Supported on Beams II: Analysis", *Journal of Structural Engineering*, 124(9), 1062–1073.
Pajari, M. (2009), "Web shear failure in prestressed hollow core slabs", *Journal of Structural Engineering*, 42(4), 207–217.
Pajari, M. (2005), "Resistance of Prestressed Hollow Core Slabs Against Web Shear Failure", *Research Notes 2292*, Helsinki, Finland.
PN SP 01/2009, Company design code, GOLDBECK Prefabeton s.r.o.
Shen, J., Yurtdas, I., Diagana, C. and Li, A. (2015), "Experimental investigation on the shear performance of prestressed self-compacting concrete beams without stirrups", *Materials and Structures*, 48(5), 1291–1302.
Yu, L., Che, Y. and Song, YP. (2013), "Shear Behavior of Large Reinforced Concrete Beams without Web Reinforcement", *Advances in Structural Engineering*, 16(4), 653–665.

Advances and Trends in Engineering Sciences and Technologies II – Al Ali & Platko (Eds)
© 2017 Taylor & Francis Group, ISBN 978-1-138-03224-8

Constitutive equations for selected rheological models in linear viscoelasticity

M. Minárová & J. Sumec
Slovak University of Technology, Bratislava, Slovakia

ABSTRACT: This work deals with mathematical modeling of viscoelastic matters rheological properties. The stress/strain relation in viscoelastic material depends on time and can be performed by using differential or integral-differential operators. However complex, the rheological model consists of basic elements. Two of them are used within the described investigation: Hook elastic matter (H) and Newton viscous liquid (N). (H) and (N) joined serially create Maxwell model {M}; (H) and (N) in parallel connection constitutes a Voigt model {V}. A chain model involves a number of particular members, e.g. Maxwell or Voigt members connected serially or in parallel. A branched chain model is created recursively, parallel and serial connection alternating. Initially (H) and (N) in parallel connection named a first level member is connected e.g. with (H) and entire arising system is connected with (N) in parallel creating a second level member. By repeating this procedure we establish longer and more branched chain.

1   INTRODUCTION

The motivation for the research described in the paper comes from the demand of physical and mathematical models precising of various phenomena focused on the mechanical, thermo mechanical or complex behavior of bodies/materials due to a load. As the experiment comparison with theoretical investigations reveals, some models, need precising by viscoelastic approach.

Viscous material resists shear flow and strain linearly with time when a stress is applied. Elastic material stretches immediately under a load and the strain is removed immediately with the stress being removed. Viscoelastic material couples these properties. The theory of viscoelasticity studies the stress—strain relation in matters, rate of change due to load. This paper deals with the derivation of the constitutive relations for chain and branched chain rheological models. Some applications of introduced models are briefly mentioned Chapter 5.

2   ONE TWO AND THREE ELEMENT RHEOLOGICAL MODELS

The viscoelastic models are represented by structural schemes (Sobotka 1981) with letters (H), (N) representing Hook elastic matter and Newton viscous matter respectively; and signs:| parallel connection—serial connection. By using these shortening tools we can construct the two and three element models as follows:

- (H)|(N) = {V} Voigt model, Figure 1c
- (H)–(N) = {M} Maxwell model, Figure 1d
- (H)–[(H)|(N)] = {PT} Poynting-Thompson model, Figure 1e
- H)|[(H)–(N)] = {Z} Zener model, Figure 1f
- (N)–[(H)|(N)] = {mPT} modified Poynting-Thompson model, Figure 1 g
- (N)|[(H)–(N)] = {mZ} modified Zener model, Figure 1h

   Considering the Constitutive Equations (CE) of basic matters (H): $\sigma = E\varepsilon$ (N): $\sigma = \eta\dot{\varepsilon}$ CE of two and three element models can be derived from these CE heuristically. However heuristics is too cumbersome and in the case of four or more element models' CE

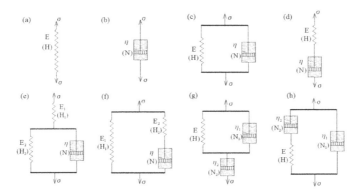

Figure 1. Viscoelastic models: a), b) Hook and Newton elementary matter, c), d) Voigt {V} and Maxwell {M}—two element models, e), f), g), h)—Poynting—Thompson{PT}, Zener{Z}, modified Poynting—Thompson {mPT} and modified Zener {mZ} three element models respectively.

derivation the conditional stiffness concept is used. Due to the presence of time derivative, the conditional stiffness in CE of a complex model will take of the form of a differential operator $E(D)$ with D denoting time derivative of the operand, (Rabotnov 1966, Minárová & Sumec, 2016). Accordingly, $E(D)$ will involve time derivatives of the certain order $D^r = d^r/dt^r$ as well.

$$E(D) = \frac{\hat{A}(D)}{\hat{B}(D)} = \frac{A_0 + A_1 D + ... + A_i D^i}{B_0 + B_1 D + ... + B_k D^k} = \frac{\sigma(t)}{\varepsilon(t)} \tag{1}$$

with $A_l, l = 1,...,i; B_m, m = 1...,k$ being constant coefficients representing physical properties of the matters involved. (1) can be rewritten in equivalent form of ordinary differential equation

$$A_0\varepsilon + A_1\dot{\varepsilon} + ... + A_i\varepsilon^{(i)} = B_0\sigma + B_1\dot{\sigma} + ... + B_k\sigma^{(k)} \tag{2}$$

The order of (2) depends on the number and configuration of the viscous elements within the entire model (Rabotnov 1966, Minárová & Sumec, 2016). Nevertheless, the general rule for the mechanical characteristics combining is valid for q members in parallel connection:

$$\Im = \sum_{i=1}^{q} \Im_i \tag{3}$$

and for q members in the serial connection alike:

$$\Im^{-1} = \sum_{i=1}^{q} \Im_i^{-1} \tag{4}$$

$\circ$ $\{mZ\}: A_0 = 0, \quad A_1 = E_1\eta_2, A_2 = \eta_1\eta_2, \quad B_0 = E_1, B_1 = \eta_1 + \eta_2, A_l = 0; l = 3,...,i,$
$B_m = 0; m = 2,...,k$

3 CHAIN STRUCTURED RHEOLOGICAL MODELS CONSTITUTIVE EQUATIONS

When taking {M} or {V} as an "elementary member" and applying multiple parallel or serial connection, we can build up a number of rheological models. The order of governing differential equation (2) and all its coefficients are determined by the configuration of the entire model and by the physical properties of the basic elements.

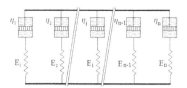

Figure 2.   Maxwell chain rheological model.

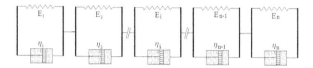

Figure 3.   Voigt chain rheological model.

3.1   *Examples of chain rheological models*

a. Parallel connection of Maxwell models $\{M_p\} = \{[(H_1)–(N_1)] \mid [(H_2)–(N_2)] \mid \dots \}$
Physical and geometrical relations within Maxwell chain model yields

$$\eta_i \dot{\varepsilon} = \sigma_i + n_i \dot{\sigma}, \quad i = 1,\dots,k \tag{5}$$

$$\sigma = \sum_{i=1}^{n} \sigma_i \tag{6}$$

with $n_i = \dfrac{\eta_i}{E_i}$. Then from (5) and (6) follows

$$E(D) = \frac{\sigma(t)}{\varepsilon(t)} = D \sum_{i=1}^{n} \frac{\eta_i}{(1 + n_i D)} \qquad \text{for } \varepsilon(t) \neq 0 \tag{7}$$

b. Serial connection of the Voigt model $\{V_s\} = \{[(H_1)|(N_1)] - [(H_2)|(N_2)] - \dots - [(H_n)|(N_n)] \}$.

Herein, the physical and geometrical relations can be expressed in the form

$$\sigma = E_i \varepsilon_i + \eta_i \dot{\varepsilon}_i, \quad i = 1,\dots,n \tag{8}$$

$$\varepsilon = \sum_{i=1}^{n} \varepsilon_i \tag{9}$$

Then, taking (5) and (6) into account, the conditional stiffness of ith member can be expressed as

$$E_j(D) = \frac{\sigma_j(t)}{\varepsilon_j(t)} \tag{10}$$

and the conditional stiffness of the entire model we have

$$E(D) = \frac{\sigma(t)}{\varepsilon(t)} = \frac{1}{\sum_{i=1}^{n} \dfrac{1}{(E_i + \eta_i D)}} = \frac{1}{\sum_{i=1}^{n} \dfrac{1}{E_i(1 + m_i D)}} \qquad \text{for } \varepsilon(t) \neq 0 \tag{11}$$

4   CONSTITUTIVE EQUATIONS OF BRANCHED CHAIN STRUCTURED RHEOLOGICAL MODELS

A branched chain model is created recursively, with the parallel and serial connection alternating. Initially the first level member, (i.e. (H) and (N) connected in parallel), is connected with basic matter, e.g. (H) and the entire system that arose is connected with (N) in parallel creating a second level member. By repeating this procedure more complex branched chain model is created.

Regarding all considerations above and taking recurrent configuration with one continuous bond we get two types of branched chain rheological structured models, see Figure 4 a), b).

The structural forms of both models are:

a. $\{V_b\} = \{[\ldots[(H_1)|(N_1)-(H_2)]|(N_2)-\ldots-(H_n)]|(N_n)-(H_{n+1})\}$
b. $\{K_b\} = \{[\ldots[(N_1)|(H_1)-(N_2)]|(H_2)-\ldots-(N_n)]|(H_n)-(N_{n+1})\}$

- Conditional stiffness $\bar{E}^V(D)$ of $\{V_b\}$

Parallel connection of a member having equivalent stiffness $E_i(D)$ with the viscous element (N_i) of viscosity η_i yields the resulting particular conditional stiffness

$$E_i'(D) = \bar{E}_i^V(D) + \eta_i D \tag{12}$$

Next, by using (4), we realize the serial connection of the recent member (i-th level of branch) of conditional stiffness $E_i'(D)$ with elastic element (H_{i+1}):

$$\bar{E}_{i+1}^V(D) = \frac{1}{E_i'(D)} + \frac{1}{E_{i+1}} = \frac{1}{\bar{E}_i^V(D) + \eta_i D} + \frac{1}{E_{i+1}} = \frac{\bar{E}_i^V(D) + \eta_i D + E_{i+1}}{[\bar{E}_i^V(D) + \eta_i D]E_{i+1}} \tag{13}$$

and from (13) the conditional stiffness of i-th level of branch can be evaluated

$$\frac{1}{\bar{E}_{i+1}^V(D)} = \frac{[\bar{E}_i^V(D) + \eta_i D]E_{i+1}}{\bar{E}_i^V(D) + \eta_i D + E_{i+1}} \tag{14}$$

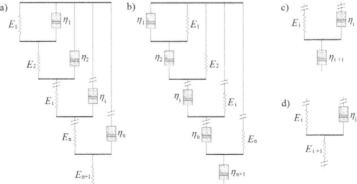

Figure 4.   Configuration of a) Voigt, b) Kelvin branched chain structures; and basic repeating block of c) Kelvin and d) Voigt branched chain structures.

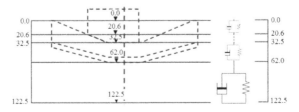

Figure 5.   Concrete hydro power station with the subsoil layers—cross section (left), (Hruštinec & Kuzma 2010) and corresponding chain rheological structure (right).

Putting $\bar{E}_1^V(D) = E_1$ we can claim (16) is the recurrence form for $\{V_b\}$.

- Conditional stiffness of $\{K_b\}$

As branch chain model $\{K_b\}$ uses the same base element likewise $\{V_b\}$, we can use (12) as well for the base of the recurrence of the further form. Realizing the serial connection of the recent member (i-th level of branch) of conditional stiffness $E_i'(D)$ with viscous element (N_{i+1}) yields:

$$\frac{1}{\bar{E}_{i+1}^K(D)} = \frac{1}{E_i'(D)} + \frac{1}{\eta_{i+1}D} = \frac{1}{\bar{E}_i^K(D) + \eta_i D} + \frac{1}{\eta_{i+1}D} = \frac{\bar{E}_i^K(D) + \eta_i D + \eta_{i+1}D}{[\bar{E}_i^K(D) + \eta_i D]\eta_{i+1}D} \quad (15)$$

and from (15) the conditional stiffness of i-th level of branch follows

$$\bar{E}_{i+1}^K(D) = \frac{\bar{E}_i^K(D)\eta_{i+1}D + \eta_i \eta_{i+1}D^2}{\bar{E}_i^K(D) + \eta_i D + \eta_{i+1}D} \quad (16)$$

With $\bar{E}_1^K(D) = E_1$ (16) is the recurrence formula for Voigt branch chain rheological model.

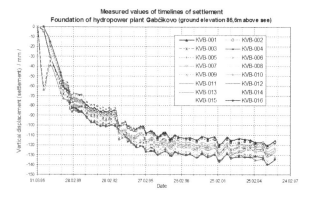

Figure 6. Time lapse of subsoil settlement—measured values in altimetric points KVB001-016 (Hruštinec & Kuzma 2010).

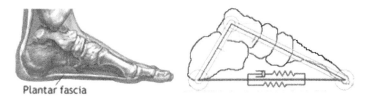

Plantar fascia

Figure 7. Plantar fascia (left); biomechanical model of foot with viscoelastic model of plantar fascia (right), (Konvičková & Valenta 2000).

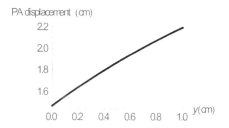

Figure 8. Displacement of plantar fascia (plantar aponeurosis tendon) in horizontal direction due to vertical deflection (settlement) of the ankle arc.

211

5 EXAMPLES OF MODEL UTILIZATION IN CIVIL ENGINEERING AND BIOMECHANICS

5.1 *Hydro power station subsoil modeling*

Horizontally layered subsoil behavior beneath a construction can be represented by chain rheological models, see an example of hydro power station and the scheme of corresponding rheological model in Figure 4. Obliquely embedded layers are then modeled by branched chain models.

5.2 *Biomechanical example—Plantar fascia (aponeurosis)*

All parts of the human body exhibit a more or less viscoelastic behavior. A simplified model of a flat tendon situated beneath the plantar arch which plays an essential role while walking is aponeurosis. In Figure 7 the one member of the chain structure of rheological model is depicted. The viscoelastic part of the flat tendon is in parallel connection with the stiff part of the foot skeleton. Hereafter, Figure 8 performs the displacement in horizontal direction of the plantar aponeurosis with respect to the rotation origin of ankle joint.

6 CONCLUSION

Viscoelastic chain and branch chain models are introduced. The need of such an investigation comes from practice and finds the application in practice as well. A lot of structural changes are determined by temperature differing, irreversibility of the deformations due to aging, influence of solidification and crystallization, influence of molecular and macromolecular structure. Regardless of the mechanical and physical properties of these materials, their time dependence is apparent. From the practice it is evident that the classical three element rheological models are not sufficient for representing the rheological matters and so their generalization is inevitable.

In the paper the chain and branched chain rheological models including Maxwell, Voigt and Kelvin groups are explored. These types of models can be utilized to model layered structures of constructive systems as ceiling constructions, roadways, runways, subsoil beneath the constructions and biomaterials.

ACKNOWLEDGEMENTS

This work was supported by grants VEGA 1/0544/15 and VEGA 1/0696/14.

REFERENCES

Hruštinec, Ľ. & Kuzma, J. 2010. The Analysis of Deformations in the Subsolil of the Hydropower Plant in Gabčíkovo. In *XIVth Danube-European Conference on Geotechnical Engineering.*

Konvičková, S. & Valenta, J. 2000. *Biomechanics of human joint and prothesis.* Praha: Štrofek (in Czech).

Minárová, M. 2014. Mathematical Modeling of Phenomenological Material Properties—Differential Operator Forms of Constitutive Equations. *Slovak Journal of Civil Engineering* 22(4): 19–24.

Minárová, M. & Sumec, J. 2016. Application of More Complex Rheological models in Continuum Mechanics. *Transaction of the VSB—Technical University of Ostrava* 15(2). 69–78.

Rabotnov, R. J. 1966. *Creep of Structural Elements.* Moscow: Nauka (in Russian).

Sobotka, Z. 1981. *Rheology of Materials and Structures.* Prague: Academia (in Czech).

Sumec, J. & Minárová, M. 2013. Mathematical Modeling of Viscoelastic Continua—Constitutive Equations. *Transaction of the VSB—Technical University of Ostrava* 13(2). 179–188.

Sumec, J. & Hruštinec, Ľ. Time Dependent Response of Mass Continuous Solid Phase Media by Integral Form of Constitutive Equations—Mathematical Modeling. *New Trends in Statics and Dynamics of Buildings.* Bratislava: STU.

Advances and Trends in Engineering Sciences and Technologies II – Al Ali & Platko (Eds)
© 2017 Taylor & Francis Group, ISBN 978-1-138-03224-8

Simulation modeling in pipeline safety assessment

L.V. Muravyeva
St. Petersburg State Polytechnical University, St. Petersburg, Russian Federation, Russia

ABSTRACT: Assessment of the risk degree and determination of the safety level are not the ultimate goals in risk analysis. The quantitative risk factors should be used to ensure optimal operation of the facility. For this purpose, optimal smanagement decisions are developed in order to bring the risk down to an acceptable level; then engineering measures are taken to ensure the required safety. In the safety context, operation of main pipelines includes the following stages recurring in cycles: Determination of all possible risk scenarios; Probabilistic assessment for each kind of a pipeline failure when the responsibility is not economic; Quantity analysis for the effects of a no economic failure; Risk assessment and overall risk summary; Comparison of the estimated risks with the acceptable or standard risks and determination of the safety degree for the system.

1 INTRODUCTION

1.1 *Term reliability*

In practice, the term reliability is usually considered as a combination of two elements. Reliability is defined as the probability that the system will perform the prescribed functions under certain conditions; it is expressed through certain quantitative values and assessed with the help of statistical methods (Borodavkin 1986, Lalin 2010 and Iavarov 2011). In a narrow sense, it is understood as freedom from failure (reference guide under editor, Berdichevsky 1970). The other element in defining reliability is "certain conditions" (autors term—Muravieva 2004, 2012) under which a system should be operated. This is an aggregate of the properties of an engineering system (main pipeline) which ensure that performance characteristics are maintained within certain limits (under the prescribed operation conditions during the time period being considered).

2 STOCHASTIC SYSTEM

2.1 *Linear section of a gas main pipeline*

The linear section of a gas main pipeline is considered as a stochastic system whose operation is defined by the following input random characteristics: $\tilde{p}, \Delta \tilde{t}$ —random values, $\tilde{q}(x)$ — stationary random function of position (Muravieva 2004). Bending deformation of the pipeline under the influence of a process load with a random variation of the $\tilde{p}$ parameter (internal pressure), the random $\Delta \tilde{t}$ parameter of temperature conditions, and the random $\tilde{q}(x)$ function of pressure distribution of the backfill soil, is a stochastic nonlinear problem (Borodavkin 1986). The aim of this research is to determine the area of failure-free operation for the gas pipeline portion on the basis of probability calculations and reliability assessment. In this case, a failure will be associated with excessive values (out of the acceptable area Ω_0) of output random characteristics of the system which describe the behavior of the pipeline structure (Muravieva 2004, 2012).

3 THE FUNCTION OF RELIABILITY

3.1 *Output random characteristics of the given system*

The $\tilde{f}(x)$ movement and longitudinal forces $\tilde{N}_y(x,\Delta\tilde{t},\tilde{p})$ in the pipeline cross-sections, stresses and forces: $\tilde{\sigma}_{eq},\tilde{\sigma}_1,\tilde{\sigma}_2,\tilde{M}_x,\tilde{M}_y,\tilde{M}_{xy},\tilde{N}_x$.

3.2 *Reliability level of the system*

The function of reliability for the given system (1) at the designated reliability level of the system (Muravieva, 2012)

$$P_t = P\left[N_y(\tilde{p},\Delta\tilde{t},\tilde{q}(x))\in\Omega_0\right]=1-\frac{1}{2}\int\limits_\beta^\infty\exp\left(-\frac{\beta^2}{2}\right)d\beta \qquad (1)$$

where output parameter = longitudinal force $\tilde{N}_y(x,\Delta\tilde{t},\tilde{p})$; the values of operation parameters = $(\Delta p, \Delta t, \tilde{q}(x))$.

In order to determine the area of failure-free operation for the pipeline structure, one can describe the system analytically, for the purpose of defining the correlation between the function variable (output parameter and the values of operation parameters with the help of simple linear relations. This problem, as a system in the form of a discrete spatial model, is solved at the level of mathematical expectations using the finite element method. Probability calculation for the system is done using the statistical modeling method. Random parameters of the system $(\Delta p, \Delta t, \tilde{q}(x))$ are taken as being distributed under the normal probability law.

It should be noted that solving such problems with account of statistical variation in input parameters (the load) demands plenty of machine time and requires powerful computers. Since the method of statistical modeling is a simulation of some real experiment, with complex systems it is quite efficient to use experiment design algorithms and data processing algorithms (reference guide under editor, Berdichevsky 1970 and Sovetov with Yakovlev 1989).

3.3 *Mathematical methods of experimental design*

Mathematical methods of experimental design are based on a discrete presentation of the experimentation process; the most appropriate model for this process is an abstract scheme like $y = f(x_1,x_2,...x_n)$. This method is based on the assumptions which are usually fulfilled. Even if they are fulfilled incompletely, the results obtained by this method are closer to reality than those obtained by the method consisting in summation of deviations for a complex multifactor model.

The aim of conducting a computer experiment with the model is to assess the characteristics of the system under study during operation and to detect the impact of the factors which are in a functional connection with the sought characteristic.

For this purpose, the following factors are selected: internal pressure of the product being transported $(x_1 = \tilde{p})$, thermal action of the product being transported $(x_2 = \Delta\tilde{t})$, weight force of the backfill soil $(x_3 = \tilde{q}(x))$. Polynomial models are the most suitable for the problems of statistical modeling. Each fixed set of levels of factors corresponds to a certain point in the multidimensional space which is called a *factor space*. Fulfillment of these problems is not formalized in experimental design. In order to determine the pipeline area of failure-free operation, the longitudinal force is represented in an analytical manner, as a function of random values: $\tilde{N}_y(x,\Delta\tilde{t},\tilde{p})$.

The probability calculation for the system is done by the statistical modeling method. Random parameters of the system $(\tilde{p},\Delta\tilde{t},\tilde{q}(x))$ are taken as being distributed under the normal probability law (Zeldovich and Myshkis 1977, Sovetov and Yakovlev 1989, Palmer and Ellinas 1990). The analytical expression for the function variable is as follows: $y = f(x_1, x_2, ... x_n)$, where f is a *reaction function*, $X = (x_1, x_2, ... x_n)$ is a set of vectors of input independent variables (*factors*). The modeling is conducted on the computer in order to determine the correlation between the levels of factors and the reaction of the system.

Factors of the system:

- Internal pressure of the product being transported $x_1 = \tilde{p}$;
- Thermal action of the product being transported $x_2 = \Delta \tilde{t}$;
- Weight force of the backfill soil $x_3 = \tilde{q}(x)$.

When modeling is performed at the i-point of the factor space, the variation of the x_3 factor throughout the length of the linear section of the gas main pipeline is taken into consideration.

In order to determine the f correlation, a mathematical (analytical) design model is built, as a first order polynomial (2):

$$y = b_0 + b_1\tilde{x}_1 + b_2\tilde{x}_2 + b_3\tilde{x}_3 + b_{12}\tilde{x}_1\tilde{x}_2 + b_{23}\tilde{x}_2\tilde{x}_3 + b_{123}\tilde{x}_1\tilde{x}_2\tilde{x}_3 \qquad (2)$$

The selected model includes linear polynomial components and their products. Design of the fractional factorial experiment, type 2^{3-1}, is used to estimate coefficients of the model equation. Table 1 shows the design matrix of the fractional factorial experiment. The local subarea of the experimental design is determined by selecting the basic (zero) level x_{i0} and variability intervals for each selected factor.

The $x_N = \|x_{1N}, x_{2N}, \ldots, x_{nN}\|$ factors are set at the levels corresponding to coordinates of the x_N point; after this, the reaction of the system (y_{Ni}) is measured. With the use of the least squares method, coefficients of the regression model are determined: $b_j = \sum_{i=1}^{N} x_{ji} y_i / N, j = \overline{0, k}$, where N—the number of trials; j—the number of factors.

3.4 Example

The calculation has been done for the section of a gas main pipeline with a length of 900 m. Pipe material: steel 10G2FB (density—7,850 kg/m³, Young's modulus E_0—206,000 MPa, linear expansion coefficient α -0,000012 degree^{-1}, Poisson ratio at the elastic stage of metal work— 0.3, yield point $\sigma_{0.2}$—4500 kgf/cm², $x_1(0) = 7,6$ MPa, $x_2(0) = 50°C$, $x_3(0) = 0,0147$ MPa.

To ensure operational reliability of a gas pipeline, all force and deformation factors affecting it should be assessed. In this regard, among the issues being considered are the serviceability and operational suitability of the pipeline structure, its service time to failure. Deformations of the buried pipeline structure were studied and analysis of the three-dimensional pipeline structure for deterministic load was presented. To solve the differential equation of longitudinal and transverse strains for the pipeline of any geometry, the finite element method was used allowing for the representation in matrix form, which is convenient for software. The solution was obtained by comparing the alternate analytical models, i.e. the straight-line section of the pipeline structure and the curved one in order to better reflect the stress state of the section under consideration.

A comprehensive technology has been developed for assessing the condition of sections of main gas pipelines based on numerical methods used for solving 3D nonlinear problems in continuum mechanics designed by Muravieva (Lalin, Iavarov 2010, Muravieva, 2004). All calculations of SSS (Stress-Strain State) are carried out using the finite element method. Analysis based on beam model is a way to construct an overall picture of SSS of the pipeline structure, to identify the most loaded sections and to determine forces and moments at the boundaries of these sections. The calculations are carried out on the assumption of linear-elastic behaviour of the pipe material. Plastic models are considered in the accident analysis.

Table 1. The design matrix of the fractional factorial experiment.

Number of the trial N_i	$x_3 = x_1 x_2$, $x_0 = x_1 x_2 x_3$					
	x_0	x_1	x_2	x_3	y	N_y
1	+1	−	−	+	y_1	N_{y1}
2	+1	+	+	+	y_2	N_{y2}
3	+1	+	−	−	y_3	N_{y3}
4	+1	−	+	−	y_4	N_{y4}

Refinement calculations are performed for the most loaded pipeline sections with the use of shell and solid finite-element models Figure 1.

To ensure operational reliability of a gas pipeline, all force and deformation factors affecting it should be assessed. In this regard, among the issues being considered are the serviceability and operational suitability of the pipeline structure, its service time to failure.

To ensure operational reliability of a gas pipeline, all force and deformation factors affecting it should be assessed. In this regard, among the issues being considered are the serviceability and operational suitability of the pipeline structure, its service time to failure.

With the analysis of the resulting expressions, the conclusion is made that the reaction of the system (longitudinal force Ny) is influenced mostly by the $X_0 = x_1 x_2 x_3$ factor; the impact of the first and the second factor is the same. Signs before the X_1 factor show that, when this factor increases, the reaction of the system is increased, too, while with the increased X_2, X_3 factors the reaction is decreased. With regard to this parameter, values of the X_3 coefficient have an impact of the next lower order on the system under study. This is because the linear section of the gas main pipeline under study lies in the area where the soils have low lifting properties.

Variation of coefficient values throughout the length of the pipeline section under consideration is shown in Figure 2.

On the basis of the results which model the system behavior, the diagrams are plotted by the failure points to show the areas of failure-free operation. This plotting defines independent acceptable limits for the variation of each factor.

In order to determine the correlation φ, coefficient values are inserted in the selected experimental design model:

$$\tilde{y} = -6{,}681.84 + 1{,}084.40 \cdot \tilde{x}_1 - 1{,}216.91 \cdot \tilde{x}_2 + 102.3 \cdot \tilde{x}_3 \tag{3}$$

where $x_1(0) = 7.6$ MPa, $\sigma_{X1} = 30$ t/m²; $x_2(0) = 50°C$, $\sigma_{X2} = 1.6°C$; $x_2 = 45 \div 50 \div 55$; $x_3(0) = 0.0147$ MPa, $\sigma_{X3} = 0.108$t/m²; $x_3 = 1.142 \div 1.471 \div 1.799$.

Values of the factors are assigned with the cumulative probability being equal to three standards:

$$x_1(0) \pm 3\sigma_{X1}; x_2(0) \pm 3\sigma_{X2}; x_3(0) \pm 3\sigma_{X3}. \tag{4}$$

Statistical methods can be used for an analytic definition of deviations from the assumed values of factors (reference guide under editor, Berdichevsky 1970 and Sovetov with Yakovlev 1989); an adequate accuracy can be achieved when applying these methods to solve most problems on condition the following procedure is followed.

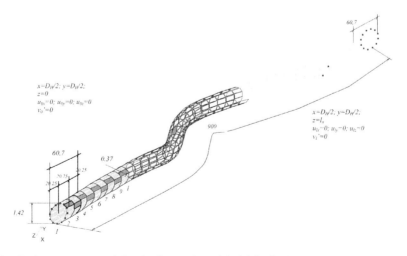

Figure 1.   Basic arrangement of the pipeline section with rigidly fixed ends.

$$\beta = \frac{\bar{R} - \bar{F}}{\sqrt{(\sigma_R^2 + \sigma_F^2)}} \qquad (5)$$

where $\bar{R}$—mean value of the longitudinal critical force obtained as the result of statistical modeling [5]; $\bar{F}$—mean value of the estimated longitudinal force N_y in (MPa) based on the results of the model calculation for the considered alternatives of variation in factors; σ_R and σ_F—standard deviation of the $\bar{R}$ and $\bar{F}$ values, $\beta = 0,11$.

The data for plotting the diagrams of areas with failure-free operation are developed in the following manner. The x_1 factor is assigned a value; the x_2 factor is given discrete increments.

The reliability level of the structure is assigned as the probability of failure-free operation. $R_{\lim} = \bar{R} - 3\sigma_R = 24,55$ MPa; $R_{\lim} = \bar{R} - 2\sigma_R = 36,82$ MPa; $R_{\lim} = \bar{R} - \sigma_R; 49,09$ MPa;

The areas of failure-free operation are determined in accordance with the assigned reliability level of the pipeline structure.

The diagrams are plotted for each effective cross-section. The complete area of failure-free operation of the system is shown in Figure 2 in accordance with the data contained in

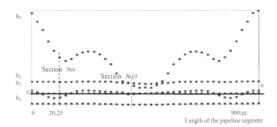

Figure 2.   Values of coefficients of the regression model.

Table 2.   Acceptable limits for the variation of each factor.

Factors	Variation limits,%	
	High	Low
$\tilde{x}_1$	$x_1(0) + 3\sigma_{X1}$	$x_1(0) - 3\sigma_{X1}$
$\tilde{x}_2$	$x_2(0) + 3\sigma_{X1}$	$x_2(0) - 3\sigma_{X1}$
$\tilde{x}_3$	$x_2(0) + 3\sigma_{X1}$	$x_3(0) - 3\sigma_{X1}$

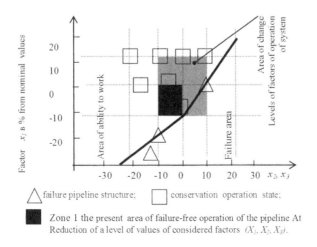

Figure 3.   The complete area of failure-free operation of the main gas line.

217

Table 3. The resulting data.

Factors	Variation limits,%	
	High	Low
$\tilde{x}_1$	+13.5	−13.5
$\tilde{x}_2$	+10	−10
$\tilde{x}_3$	+22.5	−22.5

Table 3. Each line represents a system failure caused by the variation in the set of factors $(\tilde{p}, \Delta\tilde{t}, \tilde{q}(x))$ and any specific factor. The study of the area of failure-free operation is done when only two factors change at one time is shown in Figure 2.

It follows from Figure 2 that:

The X_2 factor is correlated with the X_3 factor throughout the length of the gas main pipeline section under study.

The X_3 factor is independent of X_1 and will cause a failure of the system when its nominal value is increased by 23%. The resulting data are shown in Table 3.

The operation area of the section in Zone 1 is optimal when the variations of the factors are within the specified limits.

4 CONCLUSIONS

The presented technique enables the user to determine easily enough the risk areas for the parameters of the operated gas pipeline section.

REFERENCES

Borodavkin, P.P. 1986. Soil Mechanics in Pipeline Construction. Moscow: Nedra.
Iavarov, A.V., Lalin, V.V. 2011. K voprosu postroeniia konechno-elementnoi obolochechnoi modeli podzemnoi prokladki magistralnogo truboprovoda, *Problemy prochnosti materialov i sooruzhenii na transporte*, *PGUPS, St. Petersburg (2011)*. 106.
Lalin, V.V., Iavarov A.V. 2010. Raschetnoe obosnovanie konstruktcii nadzemnogo uchastka gazoprovoda v usloviiakh Krainego Severa, *Izvestiia VNIIG, 257 (2010)*. 112–115.
Lalin, V.V., Iavarov, A.V. Modern calculation methods of trunk pipelines, Magazine *of Civil Engineering, 3 (2010)*. 43–47.
Muravieva, L.V., Ovtchinnikov, Igor, G., Pshenichkina, V.A. 2004. Estimation of reliability a pipeline construction with operational damage. Saratov.
Muravieva, L.V. 2012. Safety and durability of pipeline designs at dynamic influences. Lambert Academic Publishing.
Reliability Reference Guide. 1970. Vol. 1, 2 under the editorship of Berdichevsky, B.E. Moscow: Nauka.
Sovetov, B. Ya., Yakovlev, S.A. 1989. Modeling of Systems. Moscow: Vysshaya Shkola.
Palmer, A.C., Ellinas, C.P., Richards, D.M., Guijt, J. Design of Submarine Pipelines Against Upheaval Buckling, Proceedings of the 22nd Offshore Technology Conference, Houston, 1990.
Zel'dovich, YA.B., Myshkis, A.D. EHlementy prikladnoj matematiki, Nauka, Moscow, 1977.

Advances and Trends in Engineering Sciences and Technologies II – Al Ali & Platko (Eds)
© 2017 Taylor & Francis Group, ISBN 978-1-138-03224-8

Analysis of secondary effects due to prestressing at continuous concrete beams at ultimate limit state

P. Pazma, J. Halvonik & V. Borzovič
*Department of Concrete Structures and Bridges, Faculty of Civil Engineering, Slovak University
of Technology in Bratislava, Bratislava, Slovak Republic*

ABSTRACT: The effects of prestressing on a structure are closely bound with the redundancy of a structural system. Imposed deformations due to the prestressing are restrained by hyper static restraints, which lead to the development of additional reactions. Those reactions, may significantly influence the internal forces in a structure. They are called secondary effects of prestressing. Regarding the above mentioned principle, there is a question of what will happen with those effects after changing the structural system e.g. due to development of plastic hinge or after development of a kinematic mechanism. This paper deals with investigation of behavior of post-tensioned beams with significant secondary effects of prestressing subjected to the ultimate load. Six two span beams prestressed by two single strand tendons were subjected to external load that has changed the structural system up to the kinematic mechanism. Results have shown that secondary effects did not disappear and they represent permanent part of the beam action.

1 INTRODUCTION

The first methods for designing concrete structures were based on the theory of allowable (permissible) stresses. Controlling of stresses in structural materials were used for both safety verification and for control of serviceability. Stresses were checked with load combinations based on unfactored loads with the assumption of elastic behaviour of structural materials. With the introduction of the theory of limit states in building codes the design of structures became more complex. The serviceability and durability are still controlled by effects of load combinations composed of service loads, however, the structural safety is verified with design combinations of actions. These combinations include factored loads and the effects of actions are verified with resistances that are assessed with the assumption of plastic behaviour of structural materials. This assumption many times leads to development of plastic hinges in a concrete structure and to the change of structural scheme at the ultimate limit state (Mattock 1983).

Tendon layout in prestressed beams normally complies with distribution of internal forces, e.g. tendons are located in the bottom part of simply supported beams or in the case of continuous beams they usually have polygonal arrangement, which allows for having suitable position of the tendon in a structure. They are at the bottom in areas with sagging moments and in the top in areas with hogging moments. It is because the bending moments due to prestressing are proportional to the distance "e" between the prestressing unit and the center of the gravity of beam. The product $P \times e$ is then the primary bending moment due to prestressing. Prestressing usually generates additional, the so called secondary effects (bending moments, shear forces and reactions) in statically indeterminate structures. These internal forces may significantly influence stresses in a structure and many times in the wrong way, therefore they are also called parasitic effects (Navratil 2014). The secondary effects depend on the structural system and on the geometry of a tendon. They can be equaled to zero in the case of suitable tendon geometry, this tendon is known as the concordant. However the secondary effects accompany primary effects of prestressing in the most cases. Because the secondary effects depend on structural system they should also be influenced by a change of the structural system e.g. due to development of plastic hinges. This raises question whether the secondary effects would remain unchanged in a structure after change of its structural system, or should they be omitted in an assessment of bending capacity. In order to give an answer about behavior of the hyperstatic structures prestressed by tendons producing the secondary effects, subjected to ultimate load that is changing the structural system up to the terminal stage of a kinematic mechanism following experimental program has been carried out.

2 DESCRIPTION OF THE EXPERIMENTAL PROGRAM

Six normal and one trial beams were cast for experiment with same cross-section dimensions 0.25×0.4 m and length of 10.5 m. Concrete strength class of C40/50 has been used. Beams were produced in specialized precast factory, then they were transported to the laboratory, installed on three supports equipped by load cells (dynamometers) then prestressed and grouted.

Beams were prestressed by the two one-strand tendons ϕLs15.7 mm/1860 MPa (sectional area of 7 wire strand was 1.5 cm^2).Tendons had different geometry, see Figure 1 and Figure 5. The first tendon had polygonal lay-out producing minimum secondary effects. The second one was straight and located at the bottom. Geometry has been proposed with intention to produce maximum secondary effects. Beams were prestressed by tendons with different bond. There were three groups of samples. The first group consisted of two beams prestressed by bonded tendons, the second group, two beams prestressed by tendons coated with oil emulsion for protection against corrosion with lower bond, and the last group, beams prestressed by unbonded tendons. Each tendon was prestressed with force $P_0 = 200$ kN. Elasto-magnetic sensors embedded on each tendon were used for recording of prestressing forces. Measuring devices used for each beam are displayed in Figure 1. Loading device consisted of two hydraulic cylinders, one for each span. Force from jacks has been divided into two forces. Reactions were monitored with load cells. Strains were measured at mid-span sections and at intermediate supports.

3 RESULTS

3.1 *Material properties*

The following mechanical properties of structural materials were tested: cube strength of concrete, modulus of elasticity of concrete, yield strength and tensile strength of reinforcing steel, modulus of elasticity of reinforcing steel, strength, 0.1% proof-stress and modulus of

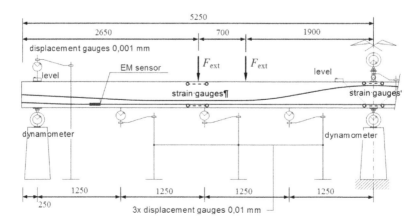

Figure 1.   Tendon layout and layout of measuring devices arrangement.

elasticity of prestressing steel. Material properties of concrete and grouting mortar were tested at the time of the experiment. Results of the tests are introduced in Tables 1 and 2.

Each beam was prestressed in the same order of tendons, first the deviated one and then the straight one. Beams were prestressed in 4 steps up to the force 200 kN. Immediate pre-stress losses due to the slip at anchorages and losses caused by elastic shortening of concrete were evaluated from elastomagnetic sensors record, for results see Table 3. Measured results from EM sensors show good agreement between experimental values and values declared by the producer, see Table 3. Producer of the prestressing system, VSL Company, declared the average anchor slip 6 mm. This value is comparable with 5 mm, what is the average experi-mental value. Measured friction coefficient was 0.135. For this type of duct (VSL PT-PLUS) it is declared to be 0.12–0.14. Correctness of prestressing system installation has been con-firmed by these basic results and obtained results are usable for further analysis.

3.2 Internal forces and reactions

The internal forces in the experimental beams were easily reconstructed using reactions meas-ured in each support, see Table 4. Based on these values the secondary bending moments due

Table 1. Measured material properties of concrete—mean values.

Beam	N1	N2	N3	N4	N5	N6
$f_{c,cube}$ [MPa]	53.93	53.32	60.0	66.1	60.78	61.48
E_c [MPa]	33.16	34.13	35.58	37.78	37.41	36.58

Table 2. Material properties of reinforcing and prestressing steel—mean values.

Reinforcing steel					Prestressing steel			
ϕ [mm]	f_y [MPa]	f_t [MPa]	E_s [GPa]	ε_u [%]	f_p [MPa]	$f_{p0,1}$ [MPa]	ε_u [%]	E_p [MPa]
8	567	649	200	8.0	1862	1516.3	6.0	195.7
12	508	601	212	11.0	–	–	–	–

Table 3. Results from elasto-magnetic sensors—beams prestressed by bonded tendons.

Average force on active side of polygonal tendon after the prestress transfer	177.7 kN
Average force on passive side of polygonal tendon after the prestress transfer	187.5 kN
The average anchor slip	5 mm
Measured friction coefficient for bonded strands	0.135

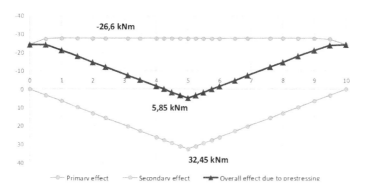

Figure 2. Bending moments caused by straight tendon on tested beams N1.

221

Table 4. The reactions and corresponding bending moments.

Beams with	Reactions		Bending moment at	
	R1*	R2*	Mid-span section	Intermediate support
Bonded tendons	kN	kN	kN.m	kN.m
Self-weight g_0	6.09	12.58	6.34	−3.58
Loading devices	0.41	1.19	0.98	−0.21
Secondary effects	6.45	−12.90	15.41	32.10
External force 2×259.05 kN	69.97	370.06	167.93	−196.3
Partially bonded				
Self-weight g_0	7.27	10.57	9.05	3.39
Loading devices	0.18	1.65	0.42	−1.37
Secondary effects	6.57	−13.14	15.77	32.85
External force 2×259.05 kN	69.91	361.38	167.78	−192.1
Unbonded tendons				
Self-weight g_0	8.40	8.24	11.78	9.13
Loading devices	0.36	1.28	0.86	−0.45
Secondary effects	6.57	−13.14	15.77	35.45
External force 2×243.3 kN	58.69	306.08	140.86	−182.8

*Note: R1—end supports; R2—intermediate support

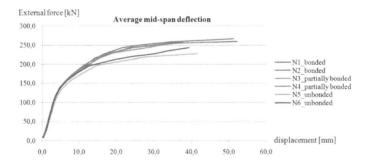

Figure 3. Diagram mid-span displacement vs. external load.

to prestressing were evaluated. They represented almost 122% of the primary moments at intermediate support, see Figure 2.

3.3 *Measured displacements and deformations*

All beams failed due to the crushing of concrete in the compression cord at the sections within plastic hinges without respect to the bond strength of the tendons. Prior to the concrete crushing significant increment of mid span displacement without growth of experimental load was observed, see Figure 3. This indicated development of kinematic mechanism in the beams.

4 PLASTIC ANALYSIS

Experimental samples represented continuous two-span beams. All beams were working in an elastic state after transfer of prestressing force. With step-by-step increase of external force, the bending capacity of section at intermediate support has been reached which resulted in development of the first plastic hinge. The statically indeterminate structure (Figure 2a) has been transformed into sequence of two simply supported beams (Figure 2b). Further

222

a) elastic state - continuous beam

b) plastic hinge at intermediate support

c) development of plastic hinges
at mid-span sections

Figure 4.   Change of structural system from a statically indeterminate beam to a kinematic mechanism.

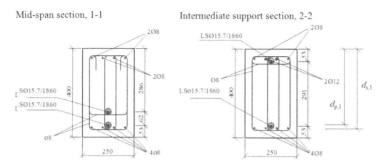

Figure 5.   Layouts of the critical cross-sections.

increment of the external load led to the formation of second plastic hinge (Figure 2c) at mid-span sections and development of a kinematic mechanism.

Bending capacity of the critical cross-sections, where the plastic hinges had been developed were needed for further analysis. The bending capacity was determined with assumption of ultimate compressive strain of concrete 0.0035 and with actual material properties that were obtained from the tests, see Tables 1 and2. Stresses in bonded tendons and reinforcement at ULS have been determined using the bilinear stress-strain diagrams that we reconstructed from an actual one. The non-linear stress-strain diagram from EN1992-1-1 for structural analysis has been used for stresses in concrete under compression. Position of neutral axis "x" has been found from horizontal forces equilibrium condition (1), where N_p is axial force due to prestressing, F_c internal force as a result of stresses in the concrete under compression, z_c is a distance of F_c from edge of section under compression, $F_{s,i}$ are forces in reinforcement and $\Delta F_{p,i}$ is an increment of prestressing force due to deformation of sections in an individual tendon.

The bending capacity has been assessed from moment condition to the center of concrete cross-section (2). Theoretical bending resistance does not include secondary moments, because they are considered as an action. Moment M_p in (2) represents primary moment due to prestressing. Experiments have shown that application of oil emulsion for protection of strands from corrosion did not significantly influence behavior of the beams. Therefore the model has been applied for beams prestressed by both bonded and bonded tendons with anti-corrosion emulsion. In the case of beams prestressed by monostrands, forces $\Delta F_{p,i}$ and M_p were evaluated from directly measured stresses in tendons by EM sensors. The bending capacity of mid-span sections of beams with bonded tendons was 192.3 kNm and for beams with unbonded tendons167.9 kNm. In case of sections at intermediate support it was 150.8 kNm for bonded and 136.3 kNm for unbonded tendons.

$$F_c - \sum F_{s,i} - \sum \Delta F_{p,i} - N_p = 0 \tag{1}$$

$$M_R = F_c\left(h/2 - z_c\right) + \sum F_{s,i}\left(d_{s,i} - h/2\right) + \sum \Delta F_{p,i}\left(d_{p,i} - h/2\right) + M_p \tag{2}$$

5   DISCUSSION

The bending moments in the beams were constructed with measured reactions introduced in the Table 4. Two values of the moments were evaluated in the critical cross-sections. First, moments

Table 5. Comparison between theoretical and experimental results.

Type of beam	Cross-section	Theoretical bending capacity	Experimentally obtained with secondary effects	Bending moments without secondary effects
N1; N2	1–1	192.3 kNm	190.7 kNm	174.9 kNm
	2–2	–150.8 kNm	–168.8 kNm	–200.9 kNm
	1–1	100%	99.2%	91.0%
	2–2	100%	111.4%	133.2%
N3; N4	1–1	192.3 kNm	193.0 kNm	177.3 kNm
	2–2	–150.8 kNm	–157.2 kNm	–190.1 kNm
	1–1	100%	100.4%	92.2%
	2–2	100%	104.3%	126.0%
N5; N6	1–1	167.9 kNm	170.5 kNm	153.5 kNm
	2–2	–136.3 kNm	–138.3 kNm	–173.2 kNm
	1–1	100%	101.6%	91.4%
	2–2	100%	101.5%	127.1%

that include secondary effects of prestressing, see the forth column in the Table 5 and second, moments without considering of the secondary effects, see the fifth column in the Table 5. The moments were compared with the theoretical bending capacities that were assessed based on the assumptions introduced in chapter 4, see the third column in the Table 5. The experimental moments with secondary effects are much closer to the bending capacities than moments without this effect. This comparison proved, that the secondary effects have not disappeared, neither after development of plastic hinge at intermediate support (two isostatic beams), nor after development of the second plastic hinge when kinematic mechanism was achieved.

6 CONCLUSIONS

Comparison of bending moments with and without secondary moments with assessed bending capacities in critical cross-sections has shown that the secondary effects have not disappeared after transition of structural system from hyperstatic to isostatic or to the kinematic mechanism. For instance in the case of beams with bonded tendons the difference between theoretical bending capacity and obtained moment without the secondary effects was –9.0% in mid-span cross-section, while in the case of assuming the secondary effects the difference was only –0.8%. For section at intermediate support, the differences were even more prominent. In the case of moments without the secondary effects the difference was 33.2%, while with secondary effect, the difference was 11.4%. Similar results were obtained for the other beams. Based on these results it can be concluded that the secondary effects of prestressing are permanent action on a structure. Therefore, they have to be assumed in all combinations used for verification of structures.

ACKNOWLEDGEMENTS

This work was supported by the Slovak Research and Development Agency under the contract NO APVV-0442–12. Authors gratefully acknowledge provider of prestressing anchors VSL/CZ.

REFERENCES

Mattock, A. H. (1983). *Secondary moments and moment redistribution in ACI318-77 Code*. In M. Z. Cohn (Ed.), Int. Symposium Nonlinearity and Continuity in Prestressed Concrete, Volume 3, pp. 27–48. University of Waterloo, Ontario, Canada.
Navrátil, J. (2014). *Prestressed concrete structures*. 2nd ed. Ostrava: Technical University of Ostrava, Faculty of Civil Engineering, 2014, 220 p. ISBN 978-80-248-3625-6.

Advances and Trends in Engineering Sciences and Technologies II – Al Ali & Platko (Eds)
© *2017 Taylor & Francis Group, ISBN 978-1-138-03224-8*

Adaptive tensegrity module: Description of an adaptation process

P. Platko & S. Kmeť
Faculty of Civil Engineering, Institute of Structural Engineering, Technical University of Košice, Košice, Slovakia

ABSTRACT: This paper describes an approach for determining the size of the movement of an action member of an adaptive tensegrity module in the form of a double symmetrical pyramid. The module is formed by thirteen members—four top and four bottom pre-stressed cables, four edge compressed struts and one centrally positioned compressed strut which is designed as the action member. Elongation or contraction of the action member allows a level of pre-stressing or tensile forces in the cables to be controlled in order to meet selected reliability criteria (such as maximum and minimum tensile forces in the cables).

1 INTRODUCTION

Tensegrity systems are spatial structures, based on the effective combination of only tensioned members—cables and compressed members—struts (e.g. Fuller 1975, Motro 2003). Some members of tensegrity systems can simultaneously work as load-carrying members, sensors and actuators (Skelton, Sultan 1997). Therefore, their implementation is very promising in projects that require adaptive structural systems.

Tensegrity systems geometry and initial pre-stresses applied to members of the tensegrity system at its self-equilibrium state significantly influence the behaviour of a loaded structure and greatly contribute to its stiffness and stability. Adaptive tensegrity structures equipped with sensors and actuators have the ability to alter its geometrical form and stress properties in order to adapt its behaviour in response to current loading conditions (Figure 1).

In the case when an excessive external load is applied on the adaptive tensegrity structure the existence of a pre-stress in its members allows for the modification or adaptation of

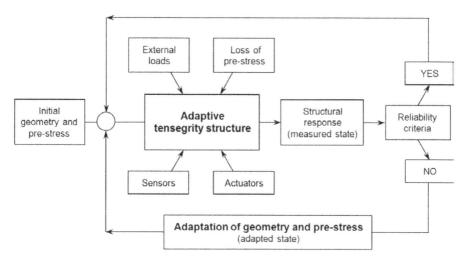

Figure 1. Process of the monitoring and adaptation of the adaptive tensegrity structures.

geometry and stiffness of the structure in order to meet selected reliability criteria. Reliability criteria may be selected from the ultimate and serviceability limit states such as

– maximum normal force in the members (they cannot be overloaded),
– minimum normal tension force in the cables (to prevent cables slacken),
– maximum deflection of the structure or movement of a selected node,
– or another reliability criteria.

This paper provides a brief description of processes during the adaptation or modification of an adaptive tensegrity structure. The presented tensegrity structure, or better said module or cell has the ability to alter its geometrical form and stress properties in order to adapt its behaviour in response to current loading conditions (see also Kmeť, Platko 2014a, 2014b).

2 CHARACTERISTICS OF THE MODULE

A full-scale prototype of the adaptive tensegrity module was developed at the Institute of Structural Engineering of the Faculty of Civil Engineering at the Technical University of Košice with the cooperation of INOVA Praha Ltd.

The elementary shape of the chosen tensegrity module can by described as double symmetrical pyramid with a square base and it is also known as a tensegric unit cell of type I (Saitoh, M. 2001). Tensegrity modules with this shape can be used in the creation of line structures or plate structures with a straight or curved central line.

The theoretical dimensions of the square base of the unit are 2.000×2.000 mm and its theoretical height is $L_{AM,0} = 750$ mm. This basic structural bearing system consists of thirteen members (Figure 2, Figure 3 and Table 1)

– four circumferential compressed members,
– four bottom cables,
– four top cables,
– one central compressed strut—an actuator.

The module is also equipped with six strain gauges and four force transducers. All members are mutually connected in nodes by hinge joints.

The initial self equilibrium state (state $i = 0$) of the module was determined by using the dynamic relaxation method and ΔDRM software developed by Dr. Marek Mojdis (2011). In this state compression force in the action member measure $P_{AM,0} = 1000$ N, tension force in the top cables is $N_{t,0} = 816$ N and in the bottom cables is $N_{b,0} = 1154$ N.

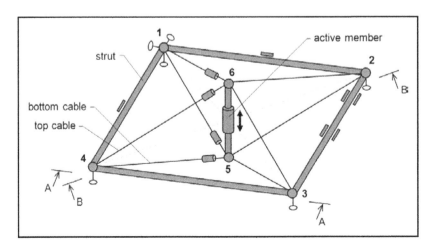

Figure 2. Isometric diagram of the tensegrity module.

226

Table 1. Members of the adaptive tensegrity module and their properties.

Member	Cross-section	A [mm²]	Material	E [GPa]
Bottom cables	ø 6 mm	$A_c = A_b = 15.14$	Steel cable 7 × 7*	$E_c = E_b = 120$
Top cables	ø 6 mm	$A_c = A_t = 15.14$	Steel cable 7 × 7*	$E_c = E_t = 120$
Circumferential members	ø 51/3.2 mm	$A_s = 475.9$	Steel S 235	$E_s = 210$
Central active member	–	–	–	$E_s = 210$

*7 strands with 7 wires per strand, stainless austenitic steel 1.4401.

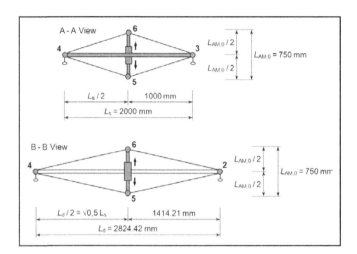

Figure 3. Geometry of the tensegrity module in A-A view and B-B view.

3 ADAPTATION OF THE MODULE

An approach for determining the size of the movement of the action member (its elongation or contraction) so that the required level of pre-stressing or tensile forces in the cables can be generated is presented in this section.

3.1 Pre-stressed state

In order to obtain stability and stiffness of the tensegrity module initial pre-stressing forces are introduced into each cable by a corresponding elongation of the action member. The level of the pre-stressing forces in the bottom cables N_b are slightly higher than the level of the pre-stressing forces in the top cables N_t. These small differences are caused by the influence of the self-weight of the active member.

The following approach assumes that the *i)* asymmetry caused by the own weight is neglected, *ii)* material and cross-sectional properties of the bottom and top cables are same, iii) nodes 2 and 4 are unmoved and *iv)* Hooke's law is valid.

Pre-stressing forces $N_{c,i} > N_{c,0}$ in the cables with the modulus of elasticity E_c and the cross-sectional area A_c, induce the corresponding strains $\varepsilon_{c,i} > \varepsilon_{c,i}$ of the cables. This can be written as

$$\varepsilon_{c,i} = \frac{N_{c,i}}{E_c \cdot A_c} = \frac{L_c(\Delta_{AM,i}) - L_{c,0}}{L_{c,0}} \tag{1}$$

where $L_{c,0}$ is the initial theoretical length of the cables and $L_c(\Delta_{AM,i})$ is the length of the cables after deformation when the active member is moved. Using notation in Figure 4 and after an application of the cosine law the new cable length may be calculated from equation

$$L_c\left(\Delta_{AM,i}\right)=\sqrt{L_{c,0}^{2}+\frac{1}{4}\Delta_{AM,i}^{2}-L_{c,0}\cdot\Delta_{AM,i}\cdot\cos\left(\alpha_{c,0}+90^{\circ}\right)} \tag{2}$$

where $\Delta_{AM,i}$ is the movement of the active member in a pre-stressed state i. The value of angle $\alpha_{c,0}=\alpha_{b,0}=\alpha_{t,0}$ can be calculated from the geometry of the module in the state $i=0$ (Figure 3).

Substituting equation (1) into equation (2) and performing the necessary arrangement, the following quadratic equation for $\Delta_{AM,i}$ is found as

$$0.25\cdot\Delta_{AM,i}^{2}-L_{c,0}\cdot\Delta_{AM,i}\cdot\cos\left(\alpha_{c,0}+90^{\circ}\right)-\varepsilon_{c,i}^{2}\cdot L_{c,0}^{2}-2\cdot\varepsilon_{c,i}\cdot L_{c,0}=0. \tag{3}$$

Solving the equation (3) the necessary movement of the active member to obtain the required pre-stressing forces $N_{c,i}>N_{c,0}$ in the cables in a pre-stressed state i is obtained in the form

$$\Delta_{AM,i}=2\cdot L_{c,0}\left[\cos\left(\alpha_{c,0}+90^{\circ}\right)+\sqrt{\cos^{2}\left(\alpha_{c,0}+90^{\circ}\right)+\varepsilon_{c,i}^{2}+2\cdot\varepsilon_{c,i}}\right] \tag{4}$$

Pre-stressing forces in the cables are controlled by measuring the tensegrity system.

3.2 A load state

The basic reliability condition for the tensegrity system is the existence of tensile axial forces in cable members during the required service life of the structure. The reliability condition can be expressed in the form

$$N_{b,i,j}(t)\wedge N_{t,i,j}(t)>0 \tag{5}$$

where $N_{b,i,j}(t)$ are the tensile forces in the bottom cables and $N_{t,i,j}(t)$ are the tensile forces in the top cables of the pre-stressed (state i) and loaded (state j) tensegrity system. Generally, slackening of cables (mainly of top cables) of the tensegrity system may be caused by influences such as external loads, temperature effects, creep of cables and likewise.

If the condition (5) is not fulfilled or the decrease of the tensile forces in the top cables of the tensegrity module due to the applied external load $P_{LC,j}$ is significant, the active member must be elongated in order to increase the level of tensile forces in the top cables.

The required elongation of the active member $\Delta_{AM,i+1}$ necessary to obtain the target increased forces $N_{t,j,tar}$ is calculated from the following relationship (Figure 5)

$$\Delta_{AM,i+1}=2\cdot L_{t,j}\left[\cos\left(\alpha_{t,j}+90^{\circ}\right)+\sqrt{\cos^{2}\left(\alpha_{t,j}+90^{\circ}\right)+\left(\frac{N_{t,j,tar}}{E_{t}\cdot A_{t}}\right)^{2}+2\cdot\frac{N_{t,j,tar}}{E_{t}\cdot A_{t}}}\right] \tag{6}$$

where $L_{t,j}$ is the length of the top cables after deformation under load $P_{LC,j}$ given by the formula

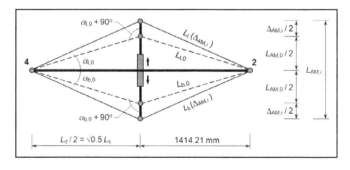

Figure 4.   Geometry of the module before (state 0) and after (state i) the elongation of the active member.

$$L_{t,j} = \sqrt{0.5 \cdot L_s^2 + \left(0.5 \cdot L_{AM,i} - w_j\right)^2} \tag{7}$$

where $L_{AM,i}$ is the length of the active member after the initial pre-stressing forces were introduced into the system (Figure 4) and w_j is the vertical deflection at the midpoint of the loaded tensegrity module. The angle $\alpha_{t,j}$ is given by the formula

$$\alpha_{t,j} = \sin^{-1}\left[\left(0.5 \cdot L_{AM,i} - w_j\right)/L_{t,j}\right]. \tag{8}$$

For the Ultimate Limit State (ULS) the design value of the tension force in the bottom cables $N_{b,j}$ from the applied external load $P_{LC,j}$ shall satisfy the reliability condition

$$N_{b,j} \le F_{Rd} \tag{9}$$

where F_{Rd} is the design value of the tension resistance of the bottom cable. The value of tension force $N_{b,j}$ can be obtained from a structural closed-form or FEM analysis and also by measuring the cable forces by force transducers.

If the reliability condition (9) is not fulfilled, the action member must be shortened in order to achieve a decrease in the forces of the bottom cables. The required reduction of the active member $\Delta'_{AM,i+1}$ necessary to obtain the target reduced forces $N_{b,j,tar}$ is calculated from the following relationship (Figure 6)

$$\Delta'_{AM,i+1} = 2 \cdot L_{b,j}\left[\cos\left(90° - \alpha_{b,j}\right) + \sqrt{\cos^2\left(90° - \alpha_{b,j}\right) + \left(\frac{N_{b,j,tar}}{E_b \cdot A_b}\right)^2 + 2 \cdot \frac{N_{b,j,tar}}{E_b \cdot A_b}}\right] \tag{10}$$

where $L_{b,j}$ is the length of the bottom cables after deformation given by the formula

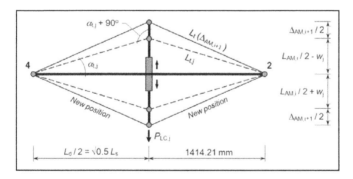

Figure 5. Geometry of the loaded module before and after the elongation of the active member.

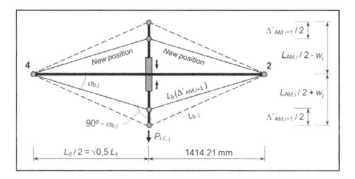

Figure 6. Geometry of the loaded module before and after the shortening of the active member.

229

$$L_{b,j} = \sqrt{0.5 \cdot L_s^2 + \left(0.5 \cdot L_{AM,i} + w_j\right)^2} \tag{11}$$

where $L_{AM,i}$ is the length of the active member after the initial pre-stressing forces were introduced into the system (Figure 4) and w_j is the vertical deflection at the midpoint of the loaded tensegrity module. The angle $\alpha_{t,j}$ is given by the formula

$$\alpha_{b,j} = \sin^{-1}\left[\left(0.5 \cdot L_{AM,i} + w_j\right) / L_{b,j}\right]. \tag{12}$$

After the necessary system adaptation has been performed and relevant values are calculated and measured, condition (8) is checked again.

In the case of Serviceability Limit States (SLS) the limiting force is specified to keep deflections of the system under the defined vertical deflection limit w_{lim}.

4   CONCLUSION

An approach for determining the size of the movement of the action member of the adaptive tensegrity module in the form of a double symmetrical pyramid was presented in the paper. Elongation or contraction of the action member allows a level of pre-stressing or tensile forces to be controlled in the cables in order to meet selected reliability criteria.

To prevent slackening of cables (mainly of top cables) of the tensegrity system due to the applied external load the active member must be elongated in order to increase the level of tensile forces in the top cables. In case that the cables (mainly of bottom cables) of the tensegrity system may be overloaded, the action member must be shortened in order to achieve a decrease in the forces of the bottom cables.

ACKNOWLEDGEMENT

This work is part of Research Project No. 1/0302/16, partially founded by the Scientific Grant Agency of the Ministry of Education of the Slovak Republic and the Slovak Academy of Sciences.

REFERENCES

Fuller, R.B. (1975). *Synergetics explorations in the geometry of thinking.* Collier Macmillan Publishers, London.

Kmet, S. & Platko, P. (2014a). "Adaptive tensegrity module. Part I: Closed-form and finite element analyses." *Journal of Structural Engineering,* 10.1061/(ASCE)ST. 1943–541X.0000957, 04014055.

Kmet, S. & Platko, P. (2014b). "Adaptive tensegrity module. Part II: Tests and comparison of results." *Journal of Structural Engineering,* 10.1061/(ASCE)ST.1943–541X.0000958, 04014056.

Mojdis, M. (2011): "Analysis of adaptive cable domes." PhD. Thesis, Technical University of Kosice, Slovakia (in Slovak).

Motro, R. (2003). *Tensegrity: Structural systems for the future.* Hermes Science Publishing, Penton Science, U.K.

Saitoh, M. (2001): "Beyond the tensegrity, a new challenge toward the tensegric world," *Theory, design and realization of shell and spatial structures*—H. Kunieda (ed.), International Symposium of IASS, Nagoya, Japan, IASS, TP 141.

Skeleton, R.E. & Sultan, C. (1997). "Controllable tensegrity, a new class of smart structures." *Proc., Conf. on Mathematics and Control in Smart Structures*—*Smart Structures and Materials 1997,* Vol. 3039, SPIE, Bellingham, 166–177.

Advances and Trends in Engineering Sciences and Technologies II – Al Ali & Platko (Eds)
© 2017 Taylor & Francis Group, ISBN 978-1-138-03224-8

Dynamic identification of composite beam pedestrian bridges

G. Poprawa, M. Salamak & P. Klikowicz
Silesian University of Technology, Gliwice, Poland

ABSTRACT: In this paper simple beam structures of double span footbridges are presented. Those are the objects that are not distinguished by sophisticated form or a design scheme. Despite that, field test done by authors shown that these structures are also sensitive to the dynamic effects. The aim of this paper is to draw attention of designers to the problems associated with dynamics of simple beam footbridges in which it is strongly recommended to use a modal analysis at the design stage. Its results can be used in planning activities aimed at improving the durability and comfort of the structure. As an example four similar double span composite structures had been chosen. All of them were tested under static and dynamic loads by authors. Results are presented and discussed.

1 INTRODUCTION

Footbridges give designers greater opportunities to create a distinctive structure with sophisticated architectural form, than road or railroad bridges (Kuras 2012). Many of these footbridges became specific icons, characteristic for places where they were built. However, not always the most important criterion for selecting a solution for a pedestrian bridge is its shape, which is eye-catching sight of observers. In many cases costs of construction and speed of execution are crucial factors. In such case contractors choose usually the simplest systems of beam structures (Figure 1).

The turn of the century became the pretext for the construction of very unusual and unique footbridges. Designers are competing in finding the most sophisticated forms, applying bizarre static schemes, and using of unconventional or prototype materials (Salamak 2011). Also in Poland, the rapid development of the road network is taking place in recent years, this results into demand of building many new footbridges. This of course gave the designers wide opportunities to manifest their expression in designing landmark structures. On this occasion wide wave of discussion on these structures have swept through the world of engineering. Most often in many papers structures of distinctive shape and complex structural system were shown. Attention was given to many problems associated with the design of those objects and their dynamic properties which are often characterized by increased susceptibility to the interaction with a pedestrian traffic (Salamak 2008). Among many presented examples very few were simple footbridges, which in terms of static scheme can be classified as a group of typical beam bridge solution. But we should not forget that these

Figure 1.   Double-span beam footbridge during static test load.

structures, thanks to their simplicity, appropriate slenderness and the shape of grade line, can also cause a pleasant aesthetic experience. Mainly because of its lightness against adjacent much heavier road or rail overpasses. Less often literature takes the issues of dynamic properties of these structures, engineers in an unwritten way assume, that with so little exposed objects and so simple schemes, they will not be dynamically prone structures. However, it turns out that they also can have adverse dynamic properties, what authors have found out realizing load tests of several such structures. It is worth to note that we have the opportunity to shape the dynamic parameters of these structures already at the design stage what is often forgotten. This is particularly important because diagnosed problem of the vulnerability in already existing object is then difficult to correct.

2 BEAM FOOTBRIDGES UNDER STUDY

The authors examined several beam footbridges, which usually were located over dual carriageway roads. From among them only four were selected for comparison. Structures were limited to double span solutions with steel girders with composite reinforced concrete slab (Table 1). For the purpose of the paper, footbridges have been marked from the K-1 to K-4. Their location and information about the designer and contractor were skipped. The description is focused only at the geometrical properties, materials, and structural solutions.

Structures selected for the analysis are characterized by a similar static scheme of double span continuous beam. The superstructure is made as steel-concrete composite twin plate girders with reinforced concrete deck slab. K-1 (Figure 2) and K-2—steel girders have a

Table 1. Summary of the discussed structures.

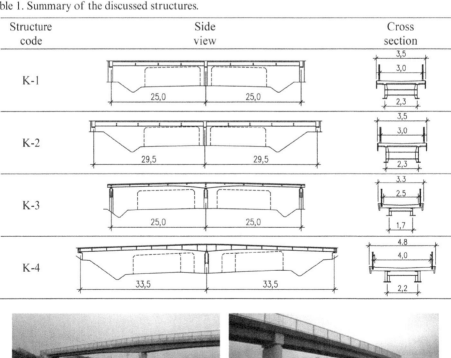

Figure 2.   Footbridge K1 and K4 during tests.

constant height along the length—1.0 m and 1.1 m respectively. The main beams are braced transversely with steel I-beams in the span, and with the reinforced concrete crossbeams at the supports. Total thickness of the concrete deck is between 0.21–0.25 m. K-3 and K-4 (Figure 2)—steel girders have a variable height—K-3 from 0.55 to 1.20 m and K-4 from 0.80 to 1.60 m. The main beams are braced transversely with a steel I-beams. Total thickness of the deck slab is between 0.19–0.30 m. All of superstructures are based at the supports by means of elastomeric bearings.

3 COMPUTATIONAL MODELS AND RESULTS OF MODAL ANALYSIS

Described structures have been modeled in the same way. Their basic computational model was a flat grillage. Individual beams had cross section parameters assigned according to the commonly agreed methodology of modeling composite structures, however reduction of the width of cooperating reinforced concrete slab in sections near supports was omitted. Weight of superstructure and additional equipment have been taken into account by relevant parameters of beam elements. The first three mode shapes and corresponding frequencies are presented in Table 2.

The model created at the design stage is a certain idealization of the structure. This is reflected in discrepancies between the calculated and measured deflections. The key parameters affecting the uncertainty of the model are: the actual modulus of elasticity of concrete slab and—in the case of footbridges K-1 and K-2—the degree of cooperation of precast reinforced concrete planks and the overlying monolithic slab. Other important factor could be a real range of cracked concrete zone over intermediate support.

4 METHODOLOGY AND SCOPE OF THE STUDY

Load test have been done by the laboratory of Silesian University of Technology in 2014. During all of the tests weather conditions were favorable, there was no excessive sun exposure and measured air temperatures were in the range of 3 to 6°C, thus influence of temperature changes was skipped. The tests were carried out in two stages. In the first stage static load scheme of the structure was completed, the aim of which was to verify the stiffness of computational model. This was done by measuring the vertical displacement of selected points under a known load and compare these results with those obtained from the calculations. Results from static tests are not presented. In the second stage the dynamic tests were performed. The dynamic forces were generated by a single person or group of pedestrians. Measured time histories of displacements and accelerations were used to identify the numerical model with the determination of natural frequencies and mode shapes and corresponding damping.

Table 2. Results of modal analysis from a design stage, frequencies in Hz.

Mode	Mode shape	Frequency			
		K-1	K-2	K-3	K-4
1		3.48	2.74	2.03	1.64
2		4.92	3.88	3.55	2.82
3		7.68	6.23	5.70	3.72

4.1 Tests under dynamic load

Dynamic trials were carried out in accordance to the tests project, which included customary accepted ways of vibration excitation starting from free walk of single person and ending with synchronized jumping of ten people group. The main purpose of testing is to determine the actual modal parameters and the level of comfort for footbridge users.

To record the time histories of displacements LVDT (Linear Variable Differential Transformer) transducers were used. Vibrations in two directions (vertical and horizontal) were also measured using capacitive accelerometers. The location of measuring points was chosen based on maximal expected amplitudes of displacements and accelerations (Figure 3).

During the measurements, the following naming of extortion schemes were adopted: CS-1(CS-G)—free walk of single person (group of people), BS-1(BS-G)—free run of single person (group of people), BZ-1(BZ-G)—run of single person (group of people) synchronized to the first bending mode, S-1(S-G)—jumps of single person (groups of people) synchronized to the first bending mode. Implementation of schemes referred as synchronized was possible due to the use of loudspeakers and properly tuned sound system. The first load scheme was usually synchronized jumps of single person. This approach allowed the initial identification of the first resonant frequency used consistently later in subsequent trials as excitation frequency.

Free walk and run of single person generates very similar, relatively low, accelerations in three of analyzed footbridges (Table 3), the exception is the structure K-3. In contrast, a person jumping, but in a way synchronized with the first natural frequency, causes vibrations many times higher and with amplitudes greatly diversified between the various structures. In the K-1, because of the high values of the first natural frequency, synchronization was not possible. Clearly, the footbridge K-3 is the most sensitive to dynamic actions.

During tests only the number of people in loading party was controlled, mass and even more important engagement of people involved were largely varying. This causes variations in input force thus causing uncertainty in results comparison. Issue of people engagement strongly influences trials where synchronization in group is required.

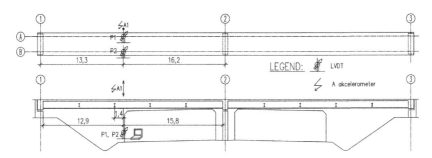

Figure 3.   Example of distribution of measuring points during dynamic tests K-2.

Table 3.   Selection of dynamic tests with maximum values of vertical accelerations and displacements.

Scheme	K-1		K-2		K-3		K-4	
	Disp. [mm]	Acc. [m/s²]	Disp. [mm]	Acc. [m/s²]	Disp. [mm]	Acc. [m/s²]	Disp. [mm]	Acc. [m/s²]
BS-1	0.176	0.130	0.170	0.076	0.420	0.816	0.311	0.273
BS-G	0.840	0.301	2.115	0.498	–	–	1.643	1.063
BZ-1	0.169	0.120	1.015	0.394	–	–	–	–
BZ-G	1.648	0.846	14.198	4.578	13.281	3.796	1.221	1.060
CS-1	0.053	0.059	0.042	0.025	0.145	0.733	0.411	0.181
CS-G	0.463	0.114	0.670	0.131	–	–	1.650	0.603
S-1	0.195	0.260	1.285	0.462	2.406	1.793	–	–
S-G	1.234	0.799	5.645	1.975	12.180	5.198	8.456	3.282

Table 4. Summary of calculated and identified (measured) natural frequencies and damping.

Mode	K-1 f[Hz] calc.	m.	ζ [%]	K-2 f[Hz] calc.	m.	ζ [%]	K-3 f[Hz] calc.	m.	ζ [%]	K-4 f[Hz] calc.	m.	ζ [%]
1 bending	3.48	3.76	2.42	2.74	2.97	0.68	2.03	2.32	0.69	1.64	1.76	0.60
2 bending	4.92	5.66	1.12	3.88	4.48	0.71	3.55	4.19	1.55	2.82	3.14	0.69
3 twisting	7.68	8.56	0.56	6.23	5.92	0.60	5.70	–	–	3.72	–	–

The recorded time signals were transformed to frequency domain and spectral analysis was used to estimate frequencies of vibration associated with the mode shapes (Table 4). In addition to the frequencies and mode shapes the damping was also estimated which is the most difficult to clearly identify parameter characterizing the structure (Salamak 2007). Please note the differences between the calculated and measured frequencies. The difference magnitude is similar and within 15% of the measured value regardless of structure.

As mentioned earlier, each identified natural frequency has assigned fraction of critical damping ζ determined by a logarithmic decrement of damping (Table 4). Knowledge of the level of damping is of key importance because it allows to estimate the risk of the occurrence of excessive vibration (Salamak 2007, Markocki 2014). In most cases we are dealing with a very small damping of less than 1%, which is typical for light footbridges.

5 EVALUATION OF PEDESTRIANS COMFORT

Test load under dynamic loading can be used to verify the level of pedestrians comfort when using the footbridge. From the structure users point of view, the vibration in the vertical and horizontal (lateral) direction should be small enough. These vibrations can generally be caused by the movement of pedestrians—their interaction with the structure—or, for example, the influence of wind (vortex shedding). The specificity of described here footbridges makes that the horizontal vibration modes are characterized by high frequencies—far more than 2.5 Hz—thus they seem immune to crowd induced vertical vibrations.

In the literature a number of comfort criteria proposed by different researchers can be found (Tilly 1984, Bachmann 1986, Pańtak 2013). For this analysis, it was decided to use only two of these proposals. According to Eurocode EN 1990:A2, vertical acceleration of the deck should be checked at the design stage, when the first natural frequency of vertical vibrations is lower than 5 Hz. During the normal use value of the acceleration cannot exceed a = 0.7 m/s². This value should be treated as an elementary pointer during the design of the structure and its testing after construction. The criterion of comfort must be related to the activity undertaken by users of the footbridge. Another tolerance to the vibrations demonstrate pedestrians standing still on structure, different walking along it and running yet another. Accordingly, the level of user comfort experience is a result of several components. Which is a compromise between the values tolerated by various groups. According to the classification of Hawryszków (2011) studied bridges may be included to the group with comfort level in the range a = 0.7–2.3 m/s², where the limits are respectively defined as maximal and minimal level of comfort acceptable.

Examined structures should be classified as providing users with a high or acceptable level of comfort. This conclusion arises after the analysis of results presented in Table 3. The walk and run of unsynchronized people generates acceleration, which in principle does not exceed a = 1.0 m/s². From the number of recorded time histories meaningful were those where there was no synchronization. Trials in which the maximum acceleration exceeds this threshold are generally the result of synchronized actions of the group. On the basis of presented results, we can assume that described footbridges are sensitive to synchronized or malicious excitation in the frequency range induced by pedestrians. The design of K-1, which is by far the most rigid, can be regarded as the least susceptible to this type of extortion.

6 CONCLUSION

The paper presents issues of the double-span beam footbridges. These objects are not distinguished by sophisticated form or structural system yet still needs attention in the area of their dynamic behavior. They tend to be more rarely subject of papers as less attractive and research as apparently more resistant to dynamic influences. Meanwhile, the authors' dynamic tests have shown that these structures also tend to be dynamically prone. Modal analysis is a must in the design stage. Its results can be used in the actions planning to improve the durability and comfort of the structure related to its dynamic properties.

The paper is based on the results obtained during the load test of four typical double-span composite objects. Impact of basic physical and geometrical properties on the dynamic response of footbridges is illustrated. This is evident in the structures K-1 and K-2 that are characterized by much greater rigidity of the superstructure relative to the rest of the analyzed objects. Their resonant frequencies are outside the range of impact caused by the movement of pedestrians.

Based on the experience gained in the examination of several beam footbridges, we can make the following conclusions: obtaining a first vertical vibration frequency greater than 3 Hz enable us to achieve the structure almost not susceptible to vertical dynamic extortion introduced by pedestrians; theoretical modal analysis led mostly to an underestimation of the frequency of vibration. In the examples cited here, differences reached up to 15%; once expecting particularly large stream of pedestrians, it can be considered to tune the structure by changing the stiffness or span the length or to provide additional means of vibration reduction.

The authors plan to continue studies of further footbridges with a similar beam superstructure. It is planned to use methods of operational modal analysis, as well as experimental method with artificial excitation by electrodynamic vibration exciters.

At the end it should be noted that the biggest responsibility for dynamic behavior of footbridge lays upon designer. It is his obligation to know and understand phenomenon of crowd induced vibration in this kind of structures.

REFERENCES

Bachmann, H. & Ammann, W. 1986. Vibrations in Structures Induced by Man and Machines. Structural Engineering Documents, Vol. 3e, International Association of Bridge and Structural Engineering (IABSE). Zurich 1986.

Hawryszków, P. 2011. Analysis of dynamical sensitivity and comfort of footbridges. Materiałykonferencji 4th International Conference Footbridge 2011, 6–8 July 2011, Wrocław, Poland.

Kuras, P., Owerko, T., Ortyl, Ł., Kocierz, R., Sukta, O. & Pradelok, S. 2012. Advantages of radar interferometry for assessment of dynamic deformation of bridge. Bridge maintenance, safety, management, resilience and sustainability. Proceedings of the Sixth International IABMAS Conference. Stresa, Italy, 8–12 July, pp. 885–891.

Markocki, B. & Salamak, M. 2014. Durability of stress ribbon bridge checked during load test, Journal of Civil Engineering and Architecture, Apr. 2014, Volume 8, No. 4 (Serial No. 77), pp. 470–476.

Pańtak, M. 2013. Dynamic characteristic of the medium span concrete footbridges, 9th Central European Congress on Concrete Engineering CCC2013 Concrete Structures in Urban Areas, Wrocław, 4–6.09.2013, 423–426.

Salamak, M. 2007. Vibration damping identification maximizing adjustment to viscous model in civil structures, Archives of Civil Engineering, LIII, 3, 2007., s. 497–518.

Salamak, M. & Łaziński, P. 2008. Experimental identification of the dynamic properties of three different footbridge structures, Third International Conference footbridge 2008, Footbridges for Urban Renewal 2–4 July, Porto s. 319–320.

Salamak, M. 2011. Inspirations in footbridges designing, 4th International Conference footbridge 2011, Attractive Structures at reasonable costs, 6–8 July, Wroclaw, Poland, s. 132–133.

Tilly, G.P., Callington, D.W. & Eyre R. 1984. Dynamic Behavior of Footbridges. IABSE Surveys S-26/84 May 1984, 13–24.

Advances and Trends in Engineering Sciences and Technologies II – Al Ali & Platko (Eds)
© 2017 Taylor & Francis Group, ISBN 978-1-138-03224-8

Evaluation's methods of the seismic stability of buildings and structures in the Gaza Strip

D. Rashid, E.S. Kolosov, N.B. Kolosova & T.N. Soldatenko
Peter the Great Saint-Petersburg Polytechnic University, St. Petersburg, Russia

ABSTRACT: The developed methods of evaluation seismic stability of buildings and structures in the different countries (Turkey, Israel, Slovenia and etc.) are described and analyzed in this paper for the further usage in the building construction in the Gaza Strip. In this paper the authors determine main factors that increase or decrease the stability of buildings in the earthquake conditions. They describe the methodology for risk ranking and offer a universal approach to the evaluation of seismic resistance buildings and structures in Gaza considering totality of considered factors.

1 EXISTING METHODS OF ASSESSMENT

1.1 *The preliminary Turkish method*

This method has been described by Yakut Ahmet in his works (Yakut 2004, 2008). In his monograph Yakut A. indicates the capacity index (CPI) as the main factor for valuation.

The given method classifies buildings and constructions either as dangerous, or as safe.

1.2 *The detailed Turkish method*

Further the authors of the previous method improved their earthquake resistance valuation method for constructions, completing it with such factors as: the number of stores (n); minimal normal lateral stiffness index (mnlstfi); minimal normal lateral force index (mnlfi); normal redundancy standard (nrs); cellar condition index (cci) and the overhang ratio (or).

Taking the given factors into consideration, the authors classify the constructions into three categories: dangerous, moderately dangerous and safe.

1.3 *Method of long-beach valuation*

American researchers Wiggins J.H. and Noran D.F. developed and described the classification system for earthquake resistance valuation for residential buildings in Long-Beach, California. The system consists of the simple paper check, carried out by the professional engineers following the constructed model of an earthquake. It has the scale of 0 to 180 points and determines three levels of the construction risk: low, medium or high risk level (Wiggins, 2008).

1.4 *The Japanese method of valuation*

The Japanese researchers evaluated the earthquake resistance of buildings and constructions analyzing various factors, but first of all the ground foundations, elastic properties of materials and architectural and engineering properties of buildings. In 2008 Professor Jalal Namer Al-Dabbek determined the additional criteria (factors) of valuation, sufficient for earthquake resistance determination for buildings and constructions (height to width ratio, presence of bay windows, long-span walls, asymmetric weight distribution within the construction, absence of aseismic joints, etc.) (Al-Dabbek 2008).

1.5 The Israel method of valuation

This method was developed by Scalat A.S. It allows quick, but approximate earthquake resistance evaluation of a building (Scalat 2007). The main value is determined by the type of a building and its production technology. It is also related to the seismic zone, determined by the country seismic regionalization. Having determined the structural value, one can classify the building either as sufficiently earthquake-resistant, or insufficiently earthquake-resistant.

1.6 Yugoslav valuation method

The method introduces the concept of relative seismic risk V_u (Sheppard, Lutman 2007). The main parameters of seismic vulnerability are determined by means of weighting parameters, which in their turn depend on the engineering features of walls and columns, the state of a building and architectural features. The method takes a number of additional factors into account: the number of floors, seismic activity in the area, ground properties and other. Having calculated the value V_u, one determines the level of the relative seismic risk: superlative, high, medium or low.

1.7 Risk level valuation for a standard building in the Gaza Strip

To define a new method for assessing the earthquake resistance of buildings in the Gaza Strip we have evaluated the risk level of a standard four-story building with the basement in the state of disrepair, by means of the abovementioned methods.

The valuation results are shown in Figure 1.

As seen on the diagram the most valid estimation of the given model is the Israel method. Though this method is appropriate only for a quick and approximate valuation.

Taking this matter into account, the authors consider that the more valid valuation for a building in the Gaza Strip can be obtained by means of a more objective method of earthquake resistance valuation for buildings, given that the main factor system is completed with the additional factors, directly relevant to the Gaza Strip territory. Such an approach for

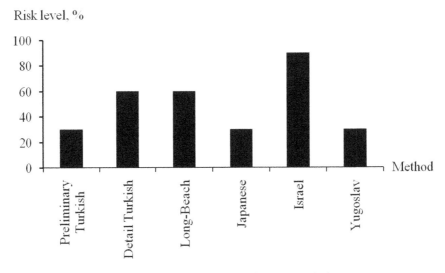

Figure 1. The risk level valuation of a standard building in the Gaza Strip.

the earthquake resistance valuation, in the authors' opinion, is the method of risk factors identification.

2 METHOD OF RISK FACTORS IDENTIFICATION

2.1 *General provision*

All methods, used for risk valuation of the earthquake resistance of a building may be conventionally divided into qualitative and quantitative (Soldatenko 2011).

In the particular case, due to the insufficiency of statistical data, only the qualitative risk valuation methods should be taken into consideration. The expert evaluation method will be applied for the risk valuation of the earthquake resistance for the Gaza Strip buildings. The results will be demonstrated by means of diagram approach in the "risk rose".

Risk valuation of the earthquake resistance for buildings and constructions should be carried out step by step. As a prototype the model of the same building will be used. Its danger level had estimated in the first part of this work.

2.2 *The first stage*

At the first stage the experts will be ranged and chosen. This procedure is crucial for obtaining the authentic results of the earthquake resistance valuation for buildings.

The choice of experts is carried out in the light of competency coefficient K_c, comprising of two components: coefficient of familiarity with similar situations K_1 and qualification coefficient K.

The competency coefficient is determined according to the formula:

$$K_c = \frac{K_1 + K_2}{2} \qquad (1)$$

To solve the given matter ten experts were ranged in competency coefficient. The results of this ranking are demonstrated in the Table 1.

The obtained data shows the first five experts, whose competency coefficient nearly equals one, as the most competent.

2.3 *The second stage*

At the next stage the competent experts determined the most significant risk factors for earthquake resistance valuation of the building in question from the Gaza Strip.

Table 1.　The results of experts' ranking in competency level.

Expert's number in ranking	Conventional expert's number	Competency coefficient value
1	9	1.060
2	1	1.055
3	4	1.045
4	6	1.040
5	2	0.965
6	8	0.950
7	3	0.795
8	7	0.725
9	5	0.685
10	10	0.510

Here it should be pointed out that for the following estimation the selected risk factors should be reduced to their relative values. It is impossible to work with absolute values of the factors, as they describe various features of a building (Povzun, Kolosov 2013).

Knowing the actual and terminal values of a factor it is fairly easy to reduce it to the relative value, dividing the former by the latter. The calculations results are presented in the Table 2.

2.4 The third stage

Now the risks of earthquake resistance of the building will be evaluated using the obtained calculations of the relative values of the factors. The risk valuation will be carried out in two approaches, and each one will have a "risk rose" drawn based on the results.

2.4.1 Approach 1

Let us assume that all factors have equal relevance. This value is determined by the formula:

$$v_k = \frac{1}{K}$$

(2)

where K = the total number of factors (in our case $K = 13$).

Then we shall determine the value of each factor, using the formula:

(3)

$$R_k = v_k \cdot w_k$$

where w_k = the relative value of a factor.

The obtained calculations are shown in the Table 3.

On the ground of the obtained calculations the corresponding "risk rose" of earthquake resistance of the building are constructed (Figure 2).

2.4.2 Approach 2

The relevance of risk factors varies. In this case the earthquake resistance valuation will be determined by means of Fishbern's method.

The factors will be ranged according to their relevance for the earthquake resistance valuation of a building with the aid of experts.

The level of relevance will be determined for each factor by means of the ratio:

$$v_k = \frac{2 \cdot (K - k + 1)}{K \cdot (K + 1)}$$

(4)

Table 2. Relative values of the risk factors.

No.	Risk factor title	Relative value of the fact or w_k
1	Number of floors	75
2	Structural system	70
3	Seismic activity in the area	30
4	Modification coefficient	35
5	Minimal lateral stiffness index	60
6	Minimal lateral force index	60
7	Redundancy standard	90
8	Cellar condition index, foundation condition	50
9	Overhang ratio	90
10	Geological aspect	70
11	Housing density	20
12	Presence of columns	50
13	Horizontal and vertical level symmetry	50

Table 3. The risk factor with equal relevance.

No.	Risk factor title	Factor relevance	Risk factor value
1	Number of floors	0.077	5.769
2	Structural system	0.077	5.385
3	Seismic activity in the area	0.077	2.308
4	Modification coefficient	0.077	2.692
5	Minimal lateral stiffness index	0.077	4.615
6	Minimal lateral force index	0.077	4.615
7	Redundancy standard	0.077	6,923
8	Cellar condition index, foundation condition	0.077	3.846
9	Overhang ratio	0.077	6.923
10	Geological aspect	0.077	5.385
11	Housing density	0.077	1.538
12	Presence of columns	0.077	3.846
13	Horizontal and vertical level symmetry	0.077	3.846

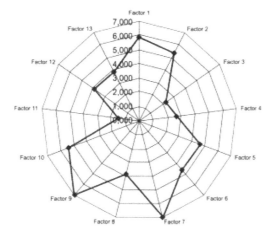

Figure 2. Risk rose with equal factor relevance.

Table 4. The risk factor with varied relevance.

Rating scale	Risk factor	Rank sum values	Factor relevance	Risk factor value
3	Seismic activity in the area	8	0.143	4.286
10	Geological aspect	13	0.132	9.231
5	Minimal lateral stiffness index	22	0.121	7.253
8	Cellar condition index, foundation condition	23	0.110	5.495
2	Structural system	26	0.099	6.923
4	Modification coefficient	33	0.088	3.077
6	Minimal lateral force index	35	0.077	4.615
1	Number of floors	36	0.066	4.945
7	Redundancy standard	40	0.055	4.945
12	Presence of columns	51	0.044	2.198
11	Housing density	53	0.033	0.659
13	Horizontal and vertical level symmetry	56	0.022	1.099
9	Overhang ratio	59	0.011	0.989

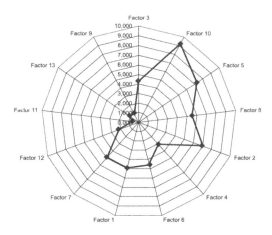

Figure 3.   Risk rose with equal varied factor relevance.

where $k = 1(1)K$. Having determined the level of relevance, we shall calculate the value of each factor (Table 4).

On the ground of the obtained calculations the corresponding "risk rose" of earthquake resistance of the building are constructed (Figure 3).

3   CONCLUSIONS

Analyzing the given diagrams, the most and the least dangerous factors can be determined.

Thus, in the first case, given the equal factor relevance, the most dangerous factors are No. 7 (Redundancy standard) and No. 9 (Overhang ratio). In the second case, given the varied factor relevance, the most dangerous factor is No.10 (Geological aspect).

In authors' opinion, the application of the risk factor identification method for earthquake resistance valuation of the buildings in the Gaza Strip produces more valid results. Moreover, this makes possible not only the general valuation of the danger level of a building, but also the determination of the factors, relevant for the building safety.

REFERENCES

Al-Dabbek, J.N. 2008. Dead Sea Earthquakes. In Earth Sciences and Seismic Engineering Center. Al Najah National University Palestine Nablus Research center. *www.eerj.org/Palstinian February.*

Povzun, A.O. & Kolosov, E.S. 2013. Criteria for selecting foundations depending on conditions of construction and type of object. *Construction of Unique Buildings and Structures* 10(15): 2–14.

Scalat, A.S. 2007. Evaluation of existing building in Israel for seismic hazard. In Wiley Inter Science. *Earthquake Engineering Dynamics*, Volume 6, Issue 3: 317–325.

Sheppard, P. & Lutman, M. 2007. Estimation of expected seismic vulnerability. A Simple Methodology. In Zag Ljubljana Lento Porocilo. *www.Zag.Si/dl/lento, access date, October 2007.*

Soldatenko, T.N. 2011. Expert-statistical method for estimating the parameters of control actions on the engineering networks of buildings in conditions of uncertainty. *Magazine of Civil Engineering* 5(23): 60–67.

Wiggins, J.H. & Noran, D.F. 2008. Earthquake safety in the city of Long Beach. California. *Amazon. com. www.amazon.com/Earthquake, access date May 2008.*

Yakut, A. 2004. Preliminary seismic performance assessment procedure for existing RC buildings. In Elsevier. *Engineering structures* 26 (2004): 1447–1461.

Yakut, A., Ozcebe, G. & Yucement, S. 2008. Seismic vulnerability assessment using regional empirical data. In Wiley InterScience. *www.interscience.wiley.com April 2008.*

Advances and Trends in Engineering Sciences and Technologies II – Al Ali & Platko (Eds)
© 2017 Taylor & Francis Group, ISBN 978-1-138-03224-8

Verification of reinforced concrete slab model using MSC Patran/Marc

V.A. Smirnov
MGSU, Moscow, Russian Federation

ABSTRACT: Nonlinear finite element analysis of reinforced concrete slabs using modern software packages poses many numerical difficulties. Besides, numerical modelling of slabs critical behavior are increasingly important in solving non-trivial engineering tasks, mainly because the possibilities of analytical approaches, implemented in existing building codes are not sufficient. That's why exceptional attention has to be paid to the verification of numerical models with experimental ones taking into account theoretical and practical knowledge of cracking behavior of concrete itself. This article presents results of comparative studies of Reinforced Concrete (RC) slab deflection analysis. This slab is considered as 3D solid model for concrete and 1D beam elements for reinforcement. The main emphasis is put on application of different material models, which are currently available in MSC Patran/Marc software package, like Drucker-Prager and Buyukozturk concrete. The results of numerical model nonlinear analysis is presented and compared with the results of an experimental study.

1 INTRODUCTION

The application of commercial Finite Element (FE) packages in design process of reinforced structures is a must for detailed and in-depth analysis of stress-strain conditions for different structural members. However, the engineers have to be very confident, that existing FE software produce reliable and accurate results. As most of the commercial FE packages, like ANSYS, ABAQUS, MSC Patran/Nastran, MSC Marc, etc. are aimed at solution of wide variety of multidisciplinary problems, they present a whole bunch of different solution approaches, as well as material models and modelling tools. That is why an appropriate treatment of solution and mesh parameters and selection of solvers takes the center stage. Moreover, each material model has to be verified before the analysis of a real structure in order to find proper problem-sensitive parameters, like mesh size, boundary conditions applications, material model or type of solver.

This paper presents the use of MSC Marc FE package and advantages of using this software for non-linear analysis of RC slabs.

The majority of the available general purpose commercial FE packages are not suitable for non-linear post-yielding analysis of concrete members because concrete shows strain-softening behavior once it is yielded (Sreekanta, et al. 1996). Figure 1 shows a typical stress-strain curve for concrete. This figure shows that the stress-strain relationship follows a downward path after its yielding. This fact possess many difficulties for standard Newton-solution methods as there exist a point (ultimate stress—point A or B) where the tangent line to the stress-strain curve is parallel to the strain axis. When this point is reached in one of the elements, the solution usually stops with lack of convergence. To overcome this problem, the arc-length method is applied with Ricks, Crisfield or Ramm criteria (Ragon, et al. 2000). All of these methods are applied in MSC Marc and provide a wide range of adjustment features to get a stable converged solution of RC slab post-critical behavior.

The choice of both an adequate material model and the solution procedure for numerical modelling of RC member is the most important aspects in FE analysis. The implemented

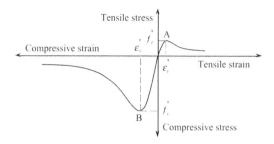

Figure 1. Stress-strain curve for concrete both in tension and in compression.

material models in MSC Marc, like Drucker-Prager or Buyukozturk can handle concrete and other low-tension materials. The latter model proposed by (Moussa et al., 1990) adopts a scalar representation of the damage related to strain and stress states of the material. The bounding surface in the stress space shrinks uniformly as the damage due to strain softening and/or tension cracks accumulates. The material parameters depend on damage level, the hydrostatic pressure and the distance between the current stress point and the bounding surface.

In order to perform an accurate analysis of RC slab, a combination of three material models were used. This combination consists of an elastoplastic material model (Drucker-Prager or Buyukozturk) that accounts for yielding properties of concrete, cracking material model, that accounted for crack developing and propagation through the structure and damage material model, that was capable of excluding failed elements under certain stress/strain conditions or other failure criteria. The last model was used in analysis together with the adaptive remeshing criteria because the initial mesh was too coarse to model a crack.

2 RESEARCH OBJECT

There exists a great amount of experimental data available investigating RC slabs deflection under point loading with different boundary conditions. The slab, studied by McNiece (McNiece, 1967) is used in this paper. Recently, many scientists used this model to verify different FE packages, that is why it is so interesting to perform an analysis of this slab in MSC Marc in order to compare its solution to other FE solutions.

Figure 2 plots the schematics of McNiece slab. It is a square $0,9 \times 0,9$ m slab with constant 44,4 mm thickness supported at four corners and loaded with a point load 18,4 kN at its center. The reinforcement ratio is 8.5×10^{-3} in each direction. The material properties of concrete are taken from (Gilbert and Warner, 1978). Some of the data are assumed values, because they are not available for the concrete used in the experiment. The work made by (Thiagarajan and Sujata, 2005) also investigated four different values for the strain beyond failure at which all strength is lost: $5\varepsilon_{cr}$, $10\varepsilon_{cr}$, $20\varepsilon_{cr}$ and $50\varepsilon_{cr}$, where $\varepsilon_{cr} = 1 \times 10^{-4}$. It was made to illustrate the effect of tension stiffening parameter on the response of the slab. Material parameters are presented in Table 1.

Another important aspect is the problem of correct reinforcement simulation. MSC Marc provides three major options for reinforcement modelling. First one, which is mostly used in practice (e.g. Dearth, 2013; Saifullah, et al., 2011) is replacing reinforcement with 1-D elements, e.g. truss (element 9), which conceives only tension/compression and solid section beams (element 52), which carry not only axial, but shear load also as well as moment. The major difficulty concerning this approach is linking of concrete 3-D elements to reinforcement 1-D element.

The second approach that is extremely complex is reinforcement simulation from 3-D solid elements. The first difficulty concerning this approach is an accurate meshing and linking between reinforcement and concrete meshes. And the second problem is correct choosing of contact interaction between reinforcement and concrete. MSC Marc provides two major

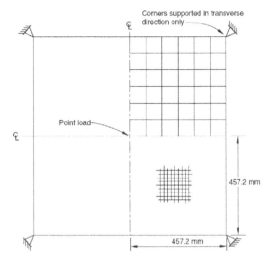

Figure 2.   McNiece slab geometry.

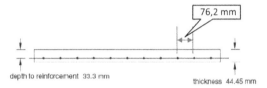

Figure 3.   McNiece slab steel reinforcement location.

Table 1.   Material parameters for McNiece slab.

Material name	Property parameter	Value
Steel	E_s	$1,9 \times 10^{11}$ N/m^2
	v	0,3
	Yield stress, σ_y	$4,14 \times 10^8$ N/m^2
Concrete	E_b	$2,86 \times 10^{12}$ N/m^2
	v	0,15
	Critical cracking stress, f_r	$3,17 \times 10^6$ N/m^2
	Tension softening strain at failure, ε_{cr}	$5\varepsilon_{cr}$
		$10\varepsilon_{cr}$
		$20\varepsilon_{cr}$
		$50\varepsilon_{cr}$
	Strain at failure, ε_{cr}	1×10^{-4}
	Compressive failure stress, f_c	$3,83 \times 10^7$ N/m^2
	Crushing strain	0,003
	Fracture energy, G_f	127,8 N/m

types of contact interactions as well as plain mesh linking. The latter method lacks for physical interpretation, but the other two—glued and friction contacts possess a whole new bunch of problems, because the task of concrete to steel interaction simulation is underexplored.

And the third approach (e.g. Barbosa, 1998) is utilization of special rebar elements (e.g., element 23, 46, 147), in which user can place single strain members such as reinforcing ribs or cords. The element is then used with conjunction with 2-D or 3-D continuum element. This technique allows the rebar and the filler to be represented accurately with respect to their stress distribution, so that separate constitutive theories can be used in each (e.g., cracking concrete and yield rebar). The key advantage of the method is the lack of necessity to mesh and link 1-D and 3-D meshes.

In this paper, we employ the first and third approaches and investigate the proper geometric property for steel reinforcement—truss or solid section beam in MSC Marc. As MSC Marc employs the smeared crack model, which is mesh-sensitive, we run the analysis on three meshes and try two different stepping procedures—the "Adaptive Multi-Criteria" and the "Arc Length".

3 NUMERICAL MODEL

Due to symmetry of the model and reinforcement, only a quarter of the slab was analyzed. Three FE models of the slab are shown on Figure 4 for different mesh seeds—$12 \times 12 \times 3$, $18 \times 18 \times 3$ and $36 \times 36 \times 6$.

Arrows represent different types of boundary conditions—applied force (green arrow), symmetry conditions (purple and orange arrows) and fixed edge (purple arrow).

Figure 5. plots the load versus deflection (N vs m) curve for each of the slabs with reinforcement modelled as 2-node solid section beam (element 98), 2 node truss (element 9), rebar plain strain membrane (element 147) and concrete as 8-node hex brick solid (element 7).

Figure 6 plots the load-deflection curve for three slabs with different mesh sizes and different reinforcement modelling techniques.

Figure 7 plots the load-deflection curve versus three mesh sizes and four different strain at failure values (ε_{cr}).

All results are also compared to the experimental ones.

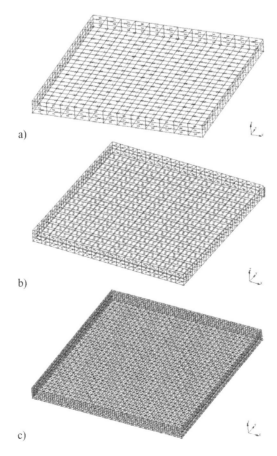

a)

b)

c)

Figure 4.   RC slabs with different mesh size: a) 12×12; b) 18×18; c) 36×36.

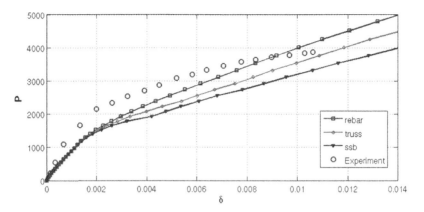

Figure 5. Load-deflection curve for three types of slabs.

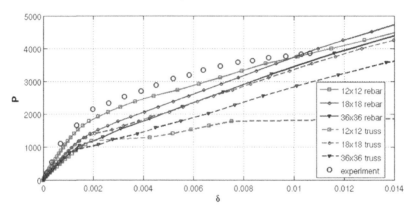

Figure 6. Load-deflection curve for different mesh size.

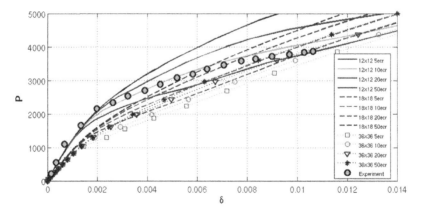

Figure 7. Load-deflection curve versus ε_{cr}.

4 DISCUSSION AND CONCLUSION

In virtue of obtained results we conclude, that the load (stepping scheme) the "Adaptive Multi-Criteria" and the "Arc Length" have minor influence on the load-deflection curve. For slightly nonlinear problems, the "Adaptive Multi-Criteria" scheme converges with lesser steps, than the "Arc Length", but the difference between the curves is negligible. However, "Constant step" scheme failed to converge in some of the cases.

The results of comparative study revealed that the solid section beam for reinforcement produce the most reliable results as the beam conceives shear and moment forces opposite to truss or rebar elements. However, this model is more sensitive to both the stepping procedure and the adequate material parameters selection. It can be seen from Figure 5 that the load-deflection curve for solid section beam has a clear flat section for 1800 N load that is similar to the experimental flat section for 2200 N load. The mesh size dependency, plotted in Figure 6, shows that the most accurate results are obtained on $12 \times 12 \times 3$ mesh with rebar elements. This is because coarse mesh is less "rigid" that the fine one and that inaccuracy introduced by rebar element is scaled, so it takes less external load to deflect the slab.

The analysis of Figure 7 shows that the material parameters have a great influence on the results. The most accurate solution, agreeable with the experimental one, was obtained with $12 \times 12 \times 3$ mesh and $10\varepsilon_{cr}$ tension strain value. Moreover, all results, obtained on this mesh have similar qualitative shape as the experimental results. One of the possible reasons that explains less accurate results, obtained on finer mesh, is the border effect—load and boundary conditions are applied on smaller regions of the mesh, so the stresses in such smaller elements becomes higher. Moreover, the linear shape function (for 8-node brick) on a big element cuts such distortion between constrained edge and the others, so the strains and, consequently, stresses becomes smaller which eventually results in lower deflections of the slab.

An accurate analysis of RC slab possesses many difficulties concerning material plasticity, cracking model and adequate mesh size. MSC Marc proposes a reliable concrete and steel material model with adjustable parameters, which has to be fixed in a number of test runs. The results of comparative study shows that the most accurate results were obtained on the coarsest $12 \times 12 \times 3$ mesh with rebar elements. The models with finer mesh are more material and mesh-sensitive than the first, so more studies have to be performed prior to the engineering analysis of that slab. The most affluent parameter is the tension strain value ε_{cr}, which unfortunately is quite complex to obtain from experimental results, so its impact on the model has to be studied with three or four test runs.

As for the future work, the study of higher-order elements on the load-deflection curve for MacNeice's slab is of great interest. Moreover, in order to reduce the border effect the application of contact boundary conditions with local or global mesh refinement is suggested.

REFERENCES

Barbosa, A.F. & Ribeiro, G.O. 1998. *Analysis of reinforced concrete structures using Ansys nonlinear concrete model.* Computational Mechanics. New Trends and Applications. CIMNE, Barcelona, Spain.

Dearth, D.R. 2013. *Analysis of Reinforced Concrete Beams Using Nonlinear Finite Element Techniques.* Proc. of 2013 MSC Regional User's Conference.

Gilbert, R.I. & Warner, R.F. 1978. *Tension Stiffening in Reinforced Concrete Slab.* Journal of the Structural Division, American Society of Civil Engineers, Vol. 104, ST12, pp. 1885–1900.

McNeice, G.M. 1967. *Elastic-Plastic Bending of Plates and Slabs by Finite Element Method.* Thesis Submitted to University of London for Degree Doctor of Philosophy, Department of Civil and Municipal Engineering University College of London.

Mousa, R.A. and Buykozturk, O. 1990. *Bounding surface model for concrete.* Nuclear Endineering and Design. Vol 121, pp. 113–125.

Ragon, S., Gurdal, Z. & Watson, L. 2000. *A comparison of three algorithms for tracing nonlinear equilibrium paths of structural systems.* 41st Structures, Structural Dynamics, and Materials Conference and Exhibit, Structures, Structural Dynamics, and Materials and Co-located Conferences.

Sreekanta Das, Muhammad, N.S. Hadi, 1996. *Non-linear finite element analysis of reinforced concrete memebers using MSC Nastran.* Proc. of 1996 MSC World User's Conference.

Saifullah, I., Nasir-uz-zaman, M., Uddin, S.M.K., Hossain, M.A. & Rashid, M.H. 2011. *Experimental and Analytical Investigation of Flexural Behavior of Reinforced Concrete Beam.* International Journal of Engineering & Technology; Vol. 11 Issue 1, pp. 146–151.

Thiagarajan, G. & Sujata, R. 2005. *Finite Element Modelling of Reinforced Concrete Bridge Decks with ABAQUS.* Center for Infrastructure Engineering Studies/UTC program. Washington, USA.

Advances and Trends in Engineering Sciences and Technologies II – Al Ali & Platko (Eds)
© 2017 Taylor & Francis Group, ISBN 978-1-138-03224-8

Searching for optimally corrected shape of a chosen tensegrity system

M. Spisak & S. Kmeť
Faculty of Civil Engineering, Institute of Structural Engineering, Technical University of Kosice, Kosice, Slovak Republic

ABSTRACT: Behaviour of tensegrities has been investigated since the end of the first half of the 20th century with an effort to better understand their unique structural and mechanical features as well as explore their potential in almost all spheres of science. In spite of existing structures mirroring a tensegrity principle, practical implementation of such notion in civil engineering and architecture is also nowadays innovative in comparison to commonly used rigid structures. In addition, tensegrities are perfect candidates to become adaptive with regard to surrounding environment because they can be relatively easily handled, in terms of geometry and flexibility, by means of actuators' equipped compressed or tensioned members (active members). The contribution deals with exploration of shape correction of an active tensegrity system regarding to defined criteria. The aim of the article is to examine a shape correction's possibility of loaded system. Results showed a satisfactory level of correctness.

1 INTRODUCTION

A principle of tensegrity, coming from a connection of two words tension and integrity (Fuller 1975), may be defined by a harmony between discontinuous internal compression surrounded by continuous external tension with invariable location of tension and compression (Motro 2003).

Progressive idea of a tensegrity principle, since its inception, has been of the interest already for more than 60 years. Even nowadays has found its place in many spheres of research and development in order to deeply investigate properties and capabilities of a practical tensegrity systems' application.

Elementary tensegrity systems are usually cells which have modularity features. Therefore, they may be assembled into various interesting geometrical shapes. One type of such cell tensegrities are tensegrity pyramids (T-pyramids—more accurately Sn T-pyramids with n number of struts) pertaining to pure tensegrity systems (Burkhardt 2008, Motro 2003, Platko 2015). Tensegrities are interesting candidates for creation of active or even intelligent (adaptive) systems because of their structural and mechanical properties, such as flexibility, kinematics, light weight, pin-jointed members (Adam & Smith 2008, Domer & Smith 2005, Platko 2012).

The matter of the contribution is investigation of a girder's shape correction created from five four-strut tensegrity pyramids (S4 T-pyramids) assembled together with all compressed members chosen to be active (i.e. 20 active members—linear actuators). With the aim to find optimal length modifications (strokes) for active members a non-linear static analysis coupled with design exploration procedure with regard to defined goals has been utilised.

2 INSIGHT INTO THE EMPLOYED DESIGN EXPLORATION PROCEDURE

In order to achieve a product with a high-class quality on a low economic cost or to find an adequate control decision one is faced with solving an optimisation problem (Singiresu 2009).

Regarding to a size of a design space and difficulty of an optimisation problem, running a numerical computer simulation for each possible solution as well as its consequent investigation could be awfully time-consuming.

A more sophisticated and time-efficient approach is to utilise a Design Optimisation (DO) for beforehand formulated problem with expected convergence until a global optima is reached or a Design Exploration (DE) for design development based on a belief that system evolves during searching but without any proof about convergence (Jenkins 2014).

Before starting a DE, one has firstly to define a numerical model undergoing computer simulations with its input (design) and output (state) variables of the interest.

Searching for a solution to a problem which represents global optima as exactly as possible is conducted by a convenient optimisation algorithm with desired objectives and constraints controlling an evolution of the design and objective space. In the optimisation, samples (design points) are gained by means of approximative metamodel created effectively from as low number of samples as possible or by realising a direct solution through numerical simulations. Both approaches have some pros and cons. The choice of a reasonably applicable procedure depends on a particular optimisation problem.

In the contribution, values of output variables were obtained by means of a computer simulation, i.e. static structural analysis using Finite Element Method (FEM). Geometric non-linearities (large displacements and strains), stress-stiffening as well as strokes were included in the FEM calculations. Equilibrium in a geometrically non-linear static analysis is expressed by a formula in global coordinates (Bathe 1996).

$$K_T \cdot u = F, \tag{1}$$

where K_T = tangential stiffness matrix (sum of elastic and geometric stiffness matrices); u = vector of nodal displacements; and F = vector of nodal loading forces.

In the contribution, shape correction of a tensegrity system was investigated, hence, strokes of active members were chosen for DE's inputs and the outputs of the interest were represented by minimal/maximal nodal displacements as well as normal forces in members.

As in the real life one has often to take into account with more than one factors in order to reach the best solution to a certain problem, a necessity for a Multi-Objective (MO) optimisation approach arises.

A Multi-Objective (MO) optimisation problem is defined by a following definition: Given an i-dimensional design variable vector $x = \{x_1,..., x_I\}$ in a design space X, find a vector x^* that minimises a given set of N objective functions $f(x^*) = \{f_1(x^*),...,f_N(x^*)\}$. The design space may be restricted by a series of constraints, such as inequalities $g_j(x^*) \leq b_j$ and/or equalities $h_k(x^*) = b_k$ for $j = 1,..., J$, $k = 1,..., K$ and bounds $x^{(L)} \leq x^* \leq x^{(U)}$ on design variables (Konak et al. 2006).

An optimal solution to a MO problem may be formulated as follows: If all objective functions are for minimisation, a feasible solution x_1 is said to dominate another feasible solution x_2 $(x_1 \succ x_2)$, if and only if, $f_n(x_1) \leq f_n(x_2)$ for all $n = 1,..., N$ and $f_n(x_1) < f_n(x_2)$ for at least one objective function $n = 1,..., N$. A solution is said to be optimal (Pareto optimal) if it is not dominated by any other solution from the design space (Konak et al. 2006).

Improving of a Pareto optimal solution according to any objective leads to worsening of at least one another objective in the objective space. Hence, mathematically in MO optimisation there are more equally optimal solutions (Pareto optimal set). For such solutions objective functions form a Pareto front. Therefore, a major goal of an MO optimisation is to uncover the best-known Pareto set which represents the Pareto optimal set as much as possible (Konak et al. 2006). Choosing the best solution among all Pareto optimal solutions is then a question of compromise. In the article, for optimisation purposes, due to its effectiveness a Multi-Objective Systematic Hybrid Exploration that is Robust, Progressive, and Adaptive (MO-SHERPA), included in Hierarchical Evolutionary Engineering Design System (HEEDS) for solving multi-objective optimisation problems (HEEDS 2015).

MO-SHERPA introduces a hybrid and adaptive search strategy by means of which during a single parametric optimisation study, the features of multiple search methods are used simultaneously in a unique mixed manner. Such conception attempts to take advantage of the best attributes of each method. A combination of global and local search methods is used which internal tuning parameters are modified automatically during the search with respect to the nature of the evolving design space. In addition, gained knowledge about the design space determines when and to what extent each approach contributes to the search. Therefore, MO-SHERPA efficiently learns about the design space and adapts itself so as to effectively search all kinds of design spaces. MO-SHERPA is a direct optimisation algorithm in which all function evaluations are performed using the numerical model as opposed to an approximate metamodel. Multiple objectives in MO-SHERPA are treated independently of each other to provide a set of solutions, each of which is optimal in some sense for one of the objectives. MO-SHERPA employs a non-dominated sorting scheme to rank designs similarly to a Non-dominated Sorting Genetic Algorithm II. (NSGA-II), see (HEEDS 2015, Chase et al. 2009).

3 REACHING A CORRECTED SHAPE OF ACTIVE TENSEGRITY SYSTEM

DE procedure was utilised for revealing an optimal shape of a girder assembled from 5 tensegrity units. Each unit is represented by one counter-clockwise rotated S4 T-pyramid. A bottom base of each unit is of a theoretical size 2×2 m, a top base of 1.414×1.414 m and height of 1.5 m. Theoretical span of the whole girder is 10 m. Tensegrity girder consists of 28 nodes and 76 members (56 cables and 20 active struts). Spatial model of analysed girder is shown in Figure 1.

Geometrical and Finite Element (FE) model of the chosen tensegrity system were created in the software ANSYS 16.0 Mechanical APDL. For all of the cables, finite elements

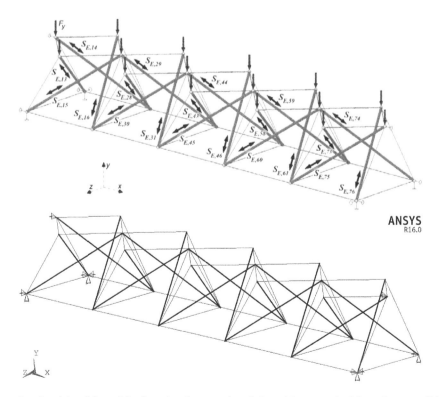

Figure 1. Spatial solid model of analysed tensegrity girder with earmarked boundary conditions, nodal forces and active members with double-sided arrows (above), finite element model with highlighted active members (below).

251

LINK180s (compression-tension elements) with both tension and compression property were put to use with keeping compressive forces as infeasible during the entire optimisation procedure. For active struts, finite elements LINK11s (linear actuators) allowing a change in an axial length were employed. Both kinds of elements incorporate only axial forces once they are loaded (i.e. no bending stiffness is supported), for more information see ANSYS 16.0 (2014) or Platko (2013). Considered movement restrictions in the nodes are shown in Figure 1.

Cables were chosen regarding to CarlStahl Architektur (2016), with metallic cross-sectional area of 288.9×10^{-6} m^2 (construction 1×19, nominal diameter of 22×10^{-3} m) and Young's modulus of 130×10^9 Pa (material stainless austenitic steel 1.4401). Active members were considered with mass of 64.1 kg, stiffness of 2.4×10^8 Nm^{-1} and material steel S235, approximately equal to properties of a strut with theoretical length of 2.693 m (compressed member of S4 T-pyramid which is a part of analysed girder), cross-sectional area of 3040×10^{-6} m^2 (cross-section ($\varnothing$ 83/14) $\times 10^{-3}$ m) and Young's modulus of 210×10^9 Pa.

Tensegrity girder in its initial geometrical configuration was pretensioned with 100,000 N of normal force in all the cables in the same manner and loaded by its self-weight. Subsequently, acquired configuration was loaded by 10 equal increments $F_{y,i} = -200$ N of downwardly oriented loading forces in the nodes of the top layer with total value of $F_y = -2000$ N, see Figure 1. Consequently, strokes of active members were gradated by 1/10 of maximal stroke value regarding to each active member independently until a total value of stroke was reached. Employed structural analysis, as a numerical simulation essential for optimisation purposes, was a geometrically non-linear and physically linear static analysis realised in ANSYS 16.0 Mechanical APDL running from a script with defined input as well as output variables.

The whole DE procedure was accomplished using HEEDS 2015.04.2 software automatically calling ANSYS for executing a numerical simulation and acquiring results. DE was guided by a following multi-objective formulation:

$$\min U_{x,y,z,\max,abs}\left(\mathbf{S}_E\right) = \text{abs}\left(UL_{x,y,z,\max}\left(\mathbf{S}_E\right)\right) \text{ [m]}, \tag{2}$$

$$\min U_{y,TL,diff}\left(\mathbf{S}_E\right) = \sum_d^D \frac{(D-1)\cdot\text{abs}\left(U_{y,d}\left(\mathbf{S}_E\right)\right)+1}{\text{abs}\left(\sum_d^D U_{y,d}\left(\mathbf{S}_E\right)-U_{y,d}\left(\mathbf{S}_E\right)\right)+1} - D, \tag{3}$$

$$\min S_{E,\max\,mabs}\left(\mathbf{S}_E\right) = \text{abs}\left(S_{E,\max}\left(\mathbf{S}_E\right)\right) \text{ [m]}, \tag{4}$$

subject to

$$U_{x,y,z,\max,abs}\left(\mathbf{S}_E\right) \le L/250 = 0.04 \text{ [m]}, \tag{5}$$

$$0 \le N_{c,\min(\max)}\left(\mathbf{S}_E\right), N_{am,\min(\max)}\left(\mathbf{S}_E\right) \le 0 \text{[N]}, \tag{6}$$

$$N_{c,\max,abs}\left(\mathbf{S}_E\right) = \text{abs}\left(N_{c,\max}\left(\mathbf{S}_E\right)\right) \le 280,000 \text{ [N]}, \tag{7}$$

$$N_{am,\max,abs}\left(\mathbf{S}_E\right) = \text{abs}\left(N_{am,\max}\left(\mathbf{S}_E\right)\right) \le 400,000 \text{ [N]}, \tag{8}$$

$$-0.05 \le S_{E,i} \le 0.05; \quad i = 13, \dots, 76 \text{ [m]} \tag{9}$$

Where $\mathbf{S}_E$ = design variable vector of strokes $\{S_{E,13},\dots, S_{E,76}\}$ for the 20 active members; $U_{x,y,z}$ = nodal displacements in x, y and z directions; d = unconstrained nodes of the top layer in an y direction; D = maximal number of unconstrained nodes of the top layer in an y direction; L = theoretical span of the girder; $N_{c,\min}/N_{c,\max}$ = minimal/maximal force in the cables; and $N_{am,\min}/N_{am,\max}$ = minimal/maximal force in the active members.

Objectives are for decreasing of maximal nodal displacements in x, y and z directions separately (2), balancing of nodal displacements of the top layer in an y direction (3) and minimising of maximal strokes for active members (4). Constraints (5–8) are for keeping of all nodal displacements under 0.04 m (1/250 of the theoretical girder span which is a deflection limit for trusses), regulating of minimal/maximal normal forces to achieve tension in cables and

252

compression in active members and controlling of minimal/maximal normal forces below their resistance respectively (breaking force in the cables is 280,000 N and buckling resistance for active members is −400,000 N). Bounds on design variables are highlighted in (9).

After finishing of searching for optimal strokes of active members was successfully observed that shape of analysed tensegrity girder was corrected on a satisfactory level. Nodal displacements were significantly reduced with respect to chosen constraints. Obtained results are sorted in Table 1 and Table 2. Vertical deflection of uncorrected and corrected shape can be seen in Figure 2.

Table 1. Nodal displacements and normal forces in members obtained by means of optimisation.

Shape correction	$U_{x,min}$ (x 10^{-3}) m	$U_{x,max}$	$U_{y,min}$	$U_{y,max}$	$U_{z,min}$	$U_{z,max}$	$U_{y,TL,diff}$
NO	−0.6(0.6)	−44.6(44.6)	−2.6(9.2)	−94.6(9.2)	−0.7(0.7)	−47.9(47.9)	0.466
YES	−1.0(2.2)	−38.8(26.9)	−1.0(1.0)	−35.1(33.0)	−1.2(0.6)	−27.9(31.3)	2.35

	$N_{c,min}$ N	$N_{c,max}$	$N_{am,min}$	$N_{am,max}$
NO	58,373	123,496	−130,241	−168,868
YES	12,466	267,338	−78,578	−371,972

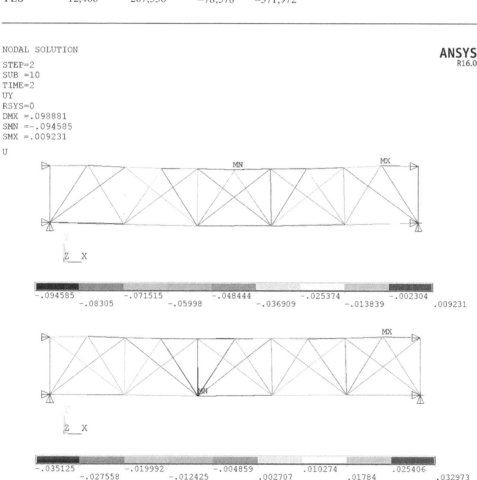

Figure 2. Vertical deflection of uncorrected (above) and corrected (below) shape of a tensegrity girder.

Table 2. Optimal calculated strokes for active members.

$S_{E,13}$	$S_{E,14}$	$S_{E,15}$	$S_{E,16}$	$S_{E,28}$	$S_{E,29}$	$S_{E,30}$	$S_{E,31}$	$S_{E,43}$	$S_{E,44}$	$S_{E,max,abs}$
(x 10^{-3}) m										
6.5	−9.0	8.0	22.0	1.0	6.0	−12.5	16.0	7.5	−3.0	22.0
$S_{E,45}$	$S_{E,46}$	$S_{E,58}$	$S_{E,59}$	$S_{E,60}$	$S_{E,61}$	$S_{E,73}$	$S_{E,74}$	$S_{E,75}$	$S_{E,76}$	
4.5	20.5	−7.0	18.5	−17.5	4.0	15.0	21.5	−6.0	−0.5	

4 CONCLUSIONS

Tensegrity systems consist of tensioned and compressed members which synergy defines a self-equilibrium. Under loading they behave relatively flexibly, therefore a question of meeting a serviceability criteria is mandatory. In the paper is presented a shape correction of a tensegrity girder with respect to defined goals. Resulting optimised strokes were successfully calculated with significant reduction of nodal displacements. Results of such optimisation procedures comes into attention in the case of adaptive tensegrities' training which should be capable of quickly finding a convenient control command.

ACKNOWLEDGEMENTS

The paper is carried out within the project VEGA No. 1/0302/16, partially founded by the Science Grant Agency of the Ministry of Education of Slovak Republic and the Slovak Academy of Sciences.

REFERENCES

Adam, B. & Smith, I.F.C. 2008. Active tensegrity. A control framework for an adaptive civil-engineering structure. *Computer & Structures* 86(23–24): 2215–2223.
ANSYS 16.0 2014. ANSYS 16.0 Help. User's guide in the software ANSYS 16.0 by the ANSYS, Inc..
Bathe, K.J. 1996. *Finite Element Procedures*. New Jersey: Prentice-Hall.
Burkhardt, R.W. 2008. *A Practical Guide to Tensegrity Design*. Cambridge: Cambridge University Press.
CarlStahl Architektur 2016. *I-SYS, Stainless steel wire rope system*, CARL STAHL ARC GMBH.
Chase, N., Redemacher, M., Goodman, E., Averill, R., Sidhu, R. 2009. *A benchmark study of optimisation search algorithms*, BMK-3021, Michigan State University, Red Cedar Technology.
Domer, B. & Smith, I.F.C. 2005. An active structure that learns, *Journal of Computing in Civil Engineering* 19(1): 16–24.
Fuller, R.B. 1975. *Synergetics Explorations in the Geometry of Thinking*. London: Collier Macmillan Publishers.
HEEDS 2015.04.2 2015. User's manual by the Red cedar technologies, Inc..
Jenkins, B. 2014, *Design Exploration. vs. Design Optimization*. Raleigh: Ora Research LLC.
Konak, A., Coit, D.W., Smith, A. E. 2006. Multi-objective optimization using genetic algorithms: A tutorial, *Reliability Engineering and System Safety* 91(9): 992–1007.
Motro, R. 2003. *Tensegrity: Structural Systems for the Future*, London: Kogan Page Science.
Platko, P. 2012. Výpočet odozvy aktívnej tensegrity jednotky pomocou MKP modelu. *9. International Scientific Conference FCE TUKE: The 35th Anniversary of the Faculty of Civil Engineering and The 60th Anniversary of the Technical University of Košice*: 1–6 (CD-ROM).
Platko, P. 2013. Elementy LINK10 a LINK11 a ich použitie pri modelovaní regulovateľných tensegrity konštrukcii. *Oceľové, kompozitné a drevené nosné konštrukcie a mosty: 38. aktív pracovníkov odboru oceľových konštrukcií*: 169–174.
Platko, P. 2015. Možnosti spájania tensegrity priziem a tensegrity pyramíd. *Vedecko-výskumná činnosť ÚIS 2014*: 89–92.
Singiresu, S.R. 2009. *Engineering Optimization. Theory and Practice, Fourth Edition*. New Jersey: John Wiley & Sons.

Advances and Trends in Engineering Sciences and Technologies II – Al Ali & Platko (Eds)
© 2017 Taylor & Francis Group, ISBN 978-1-138-03224-8

Composite timber-concrete floor of a family house

L. Surovec & M. Slivanský
Department of Steel and Timber Structures, Faculty of Civil engineering, Slovak University of Technology, Bratislava, Slovakia

ABSTRACT: This paper contains results of experimental research executed on a composite timber-concrete floor of a reconstructed family house. The composite floor was built on the first storey on the original brick walls after removal of the original roof structure. The new floor consists of timber beams with dimensions of 100/220 mm, with regular spacing at 740 mm intervals and span of about 5.3 m. A concrete slab with a thickness of 70 mm is coupled with the timber beams by inclined self-tapping screws with full thread (Würth ASSY plus VG 8 × 220 mm). Measurement of deformations and stresses was performed immediately after removing temporary supports. Obtained experimental results were compared to the theoretical calculation models. Basic model represents design calculations based on Eurocode 5, Annex B. The other model is a modification of the first one considering only the effective compressed height of the concrete to global stiffness of composite T-section. Also two "Finite Element Method" (FEM) models were used.

1 INTRODUCTION

Reconstruction of the family house located in Stupava involves a complete reconstruction of a load-bearing structure and also secondary parts of the building including a technology equipment and a partial extension of an attic. These changes also include complete replacement of the floor structure and the framework. The original timber beam floor system was replaced with the composite timber-concrete floor. Coupling between timber and concrete was ensured by inclined Würth ASSY plus VG self-tapping screws.

Connection of timber and concrete produce many benefits comparing with the timber floors or the reinforced concrete floors. Regarding the mechanical properties, timber is placed in the tensioned area of the section. Because of the higher stiffness and the sufficient strength in compression concrete is placed in the compressed area of the cross section. This allows to form a high capacity and adequately rigid cross section with minimizing self-weight. Also uncomfortable vibrations that are often associated with the timber floors are minimized. Besides that, more benefits are produced in the field of building physics. For example acoustic properties and the thermal accumulation are much better than in the timber beam floors.

A shear connection of these materials is characterized by a semi-rigid behaviour. It is caused by a compliance of the used connections (Agel & Lokaj 2014, Čajka & Burkovič 2013). This phenomenon has to be considered in calculation.

2 LOAD BEARING STRUCTURE OF THE FLOOR

As mentioned above, the floor structure of the reconstructed family house located in Stupava was realized as the composite timber-concrete floor. It consists of the timber beams with the dimensions of 100/220 mm and regular spacing at 700–800 mm intervals. These beams are made of KVH (from German "Konstruktionsvollholz") and their span ranges from 4.95 m to 5.3 m. The timber beams are placed at the top surface of reinforced concrete bond beams

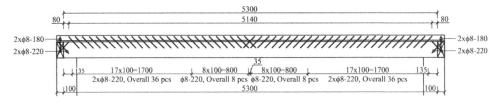

Figure 1. Longitudinal spacing of the screws.

of the original brick walls and they are connected to timber facing beams. As a decking large-format cement bonded particleboards CETRIS Basic with a thickness of 18 mm and dimensions of 3350/1250 mm were used. Constructional connection of these plates to the timber beams was realized by nails. Joints of the plates with each other were sealed with the technical tape Rothoblaas Flexiband. When the decking was ready, reinforcement of the bottom surface of the concrete was laid. Seventy millimetres thick concrete slab is reinforced with welded wire meshes made of steel bars with diameter 6 mm and regular spacing of 150 mm intervals in two perpendicular directions. As the shear connection between timber and concrete Würth ASSY plus VG self-tapping screws with diameter of thread 8 mm and length 220 mm were used. These screws were inclined under 45° and were threaded 120 mm to the timber (Figure 1).

3 SHORT-TERM, NON-DESTRUCTIVE TESTS OF THE FLOOR

3.1 *Preparation of the tests*

Based on the geometry of the floor (Figure 2), two beams ("Beam 1" and "Beam 2") with the span of about 5.3 m were chosen to be measured. Experimental measurement took 8 days in March, 2015. The strains of the timber beams at the both (bottom and top) surfaces as well as the strains of the decking at the bottom surface and the strains of the concrete at the top surface were measured using strain gauges. Also the deformations of the loaded beams and the deformations of two beams next to the loaded beams ("Beam 3" and "Beam 4") were measured using displacement sensors.

Collecting and writing the data (16 gauging nodes overall) was performed using digital gauging equipment Spider8 connected to the laptop.

The loading and the measuring of the composite floor was performed as a short-term, non-destructive tests and it was realized approximately two months after the slab was concreted. The main goal of the measurements was to obtain instant increment of the strains and deformations with the loading/unloading the structure. Because of that, resistance strain gauges with no thermal compensation were used.

Additional loading of the floor was caused by gravel packs (Figure 3). Level of the loading was set to 100% of the loading used in the calculation for "Serviceability Limit State" (SLS) and it represents 0.5 kN/m² of the layers of the floor, 2.8 kN/m² of the utility load including light partitions. Considering this load, approximately 2800 kg of the gravel was concentrated above the measured beams ("Beam 1" and "Beam 2"). The measurement was divided to three stages—removing the temporary supports (measurement of the increments by self-weight), loading of the beams by the gravel packs and unloading the structure.

3.2 *Results of the tests*

During the measurements no damages, such as cracks or excessive deformations were observed. Detected deflections of the structure were approximately equal to the deflections determined by a preliminary structural analysis. Maximal immediate measured deflection of the beam by the full load in the middle of the span (after the subtraction of the compression in the support areas) achieved the value 5.3 mm. After unloading the structures, behaviour

256

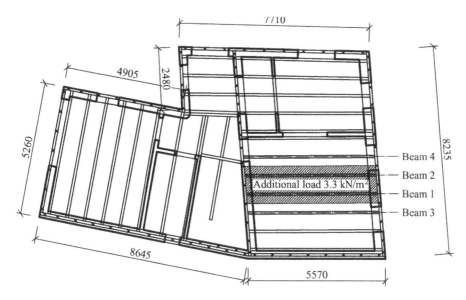

Figure 2. Geometry of the floor.

Figure 3. Additional loading of the floor by the gravel packs.

of the structure was almost ideal elastic and no additional massive creep during the next four days was observed. From the flowchart displayed at the Figure 4 there is relatively easy observable influence of the temperature to a variation of the deflection (moderate fluctuating course).

As it was expected before the tests, positive transverse distribution of the load to the unloaded beams was observed. Work of the loaded beams on the overall transmission of the loading was approximately 45–50%.

Global evaluation of the measured strains is relatively complicated task. Influence of the temperature changes to the strain changes was testified immediately and it was very significant. The most significant influence was observed on the strain gauges glued to the concrete slab. It is expected that this phenomenon relates mainly with two factors—firstly, the space under the roof, where the concrete slab is situated, was not heated during the tests and the temperature of the internal environment was the same as the external temperature. Secondly, wiring scheme of the strain gauges was without the compensation of the thermal changes. The temperature changes influenced mainly the strain gauges (that is measurement error) and not the floor structure itself.

257

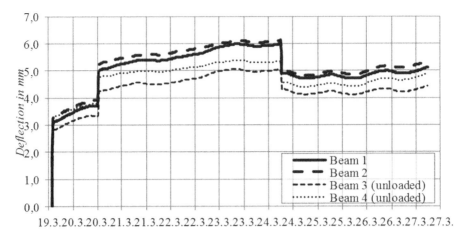

Figure 4. Deflections of the beams during the measurement.

4 STRUCTURAL ANALYSIS OF THE FLOOR

To compare the measured values of the deformations and the strains with the theoretical results, three dimensional FEM model in software Scia Engineer was created. The model was loaded by two load cases—self weight and the strip load with the value of 3.3 kN/m² over the two monitored beams. Using this model, it is able to compare the results with the measured values for each beam. Besides that, FEM model of the single span beam loaded by equivalent loading was created.

When the composite timber-concrete floor is designed, only effective "T" section (Figure 5) is considered. This calculation approach is elected in technical standard EN 1995-1-1/Annex B. This method, also known as γ-method, express a rate of the shear connection using coefficient γ. If timber and concrete have no shear connection, coefficient γ has value 0. If they are totally rigidly connected, coefficient γ has value 1. It means, that for all real cases the value of the coefficient γ lies between 0 and 1. This calculation method has some limits. It is possible to use it for a single span beams. Formulas are also valid for continuous beams and cantilevers but real span width L must be replaced by the value of $4/5L$ for continuous beams and $2L$ for cantilevers. This method allows tensional stress in the concrete part of the cross section. In fact, when some level of tensional stress in concrete is reached a crack occurs and it will spreads up to the neutral axis. It leads to a redistribution of a bending stiffness of the beam. To consider this phenomenon it is possible to modify this calculation model and take into account only effective compressed height of the concrete (Figure 6) (Surovec 2015).

Based on the methods mentioned above four models of the composite timber-concrete floor (beams) were made:

- 3D FEM model of the whole floor (results of this analysis are marked as "FEM Beam 1" for "Beam 1" and "FEM Beam 2" for "Beam 2")
- FEM model of single span beam (results of this analysis are marked as "FEM Single beam")
- Simplified calculation according to EN 1995-1-1/Annex B (results of this analysis are marked as "EC5 beam")
- Modified calculation according to EN 1995-1-1/Annex B considering only effective compressed height of the concrete (results of this analysis are marked as "EC5 Modified beam")

All models had equivalent parameters as the realized composite floor of the family house located in Stupava and were loaded according to this chapter. Course of the deflection through the time is shown in the Figure 7. Comparison of the increment of the stress in timber and concrete for each load case is stated in Tables 1 and 2. Stresses of the measured beams were evaluated based on the Hooke's law—measured values of the strains were multiplied by

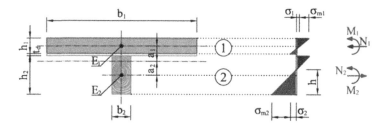

Figure 5.   Calculation scheme according to EN 1995-1-1/Annex B.

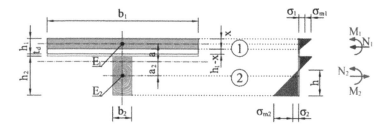

Figure 6.   Modified calculation scheme considering effective compressed height of the concrete.

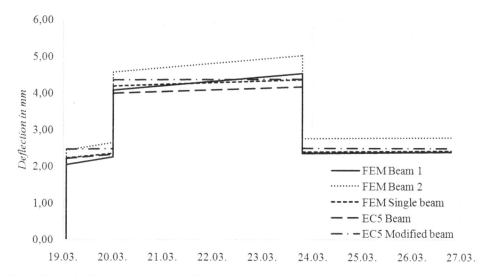

Figure 7.   Calculated deflection of the floor through the time for different methods.

Table 1.   Stress of the floor by self-weight.

Measured beam/ Calculation method	Stress of the bottom surface of the timber MPa	Stress of the top surface of the concrete MPa
Beam 1	1.52	−1.33
Beam 2	1.83	−2.05
FEM Beam 1	1.47	−1.16
FEM Beam 2	1.77	−1.37
FEM Single beam	1.81	−1.28
EC5 Beam	1.94	−1.29
EC5 Modified beam	2.05	−1.27

Table 2. Stress of the floor by the additional strip-load with the value of 3.3 kN/m^2.

Measured beam/ Calculation method	Stress of the bottom surface of the timber MPa	Stress of the top surface of the concrete MPa
Bcam 1	0.68	0.68
Beam 2	0.79	−0.67
FEM Beam 1	1.30	−1.10
FEM Beam 2	1.37	−1.15
FEM Single beam	1.49	−1.05
EC5 Beam	1.47	−0.98
EC5 Modified beam	1.57	−0.96

Young's modules of materials. Material tests were not realized, Young's modules of timber and concrete were determined only based on the technical standard properties.

5 CONCLUSION

From the comparison of the flowcharts displayed in the Figures 4 and 7 it is obvious that all calculation methods used have a relatively good accuracy compared with the measured values of the deflections. The difference is the most visible during unloading and it is very challenging to represent it accurately. Comparing the increment of the stresses caused by the self-weight it is possible to also observe a very good accuracy with the measured values. The only marked difference is observable comparing the calculated value of the stress on the top surface of the concrete with measured value of the "Beam 2". Measured values of the increment of the stress by the additional strip load are much smaller than a values determined by the calculation. This difference may be caused by a measurement error because of the thermal changes compensation that was not applied. It would be also appropriate to modify material properties in computational models according to real properties determined by material tests. Results stated in this paper are evaluated based on the material properties stated in technical standards.

REFERENCES

Agel, P. & Lokaj, A. 2014. Semi-rigid joint of timber-concrete composite beams with steel plates and convex nails. *Wood research* 59(3): 491–498.
Čajka, R. & Burkovič, K. 2013. Technologie spřažení dřevo betonových stropních konstrukcí pomocí lepených smykových lišt. *Sborník vědeckých prací Vysoké školy báňské—Technické univerzity Ostrava* 1. 8p.
Dias, A.M.P.G. & Van De Kulien, J.W. & Lopes, S. & Cruz, H. 2007. A non-linear 3D FEM model to simulate timber—concrete joints. *Advances in Engineering Software* 38: 522–530.
European technical approval ETA-13/0029 2013. Self-Tapping Screws for use in Wood-Concrete Slab Kits. Würth Self-Tapping Screw. ETA-Danmark. 17p.
Slivanský, M. & Surovec, L. 2015. Experimentálne overovanie spriahnutého drevo-betónového stropu na RD v Stupave. In *Zborník prednášok zo 40. aktívu pracovníkov odboru oceľových konštrukcií: Oščadnica, SR, 22.–23. 10. 2015. 1. vyd.* Kysucké Nové Mesto: Miroslav Gibala—Reklamný servis, pp. 247–252. ISBN 978-80-89619-01-6.
Surovec, L. 2015. Interaction of timber and concrete in composite beams = Pôsobenie dreva a betónu v spriahnutých nosníkoch. In *Juniorstav 2015: 17. odborná konferencia doktorského studia. Brno, ČR, 29. 1. 2015 [elektronický zdroj]. 1. vyd.* Brno: Vysoké učení technické v Brně, 2015, CD-ROM, 7p. ISBN 978-80-214-5091-2.
Surovec, L. - Slivanský, M. 2015. Composite timber-concrete structures. In *ISeC 2015—Interdisciplinary Scientific eConference* [elektronický zdroj]: *proceedings of the International virtual conference.* Bratislava, SR, 20.–24. 7. 2015. Bratislava: NEXSYS, online, [7] s. ISBN 978-80-972051-0-2.
STN EN 1995-1-1: Eurokód 5. Navrhovanie drevených konštrukcií. Časť 1-1: Všeobecne. Všeobecné pravidlá a pravidlá pre budovy. 2008. 114 s.

Experimental investigation of the effects of tensile force on the characteristics of turbulence downstream of a novel active grid

N. Szaszák, P. Bencs & Sz. Szabó
Department of Fluid and Heat Engineering, University of Miskolc, Miskolc, Hungary

ABSTRACT: In this study the investigation of the effects of the tensile force on turbulent properties of flow downstream of a novel type active grid is described. Active grids are applied to achieve flows with high turbulence intensities in wind tunnels, nevertheless they require significant expense. Our novel active grid offers a more affordable option. It contains metal tapes—set perpendicularly to the main flow—which flutter caused by the flow, itself. The moving boundary layers create rapid changes in the flow parameters thus producing a high level decaying turbulence downstream of the grid. Previous investigations verify the applicability of the grid with respect to increased turbulent properties. However the magnitude of tension of the tapes has detectable effects on the flutter of the tapes. Based on this the characteristics of turbulence was investigated in case of different tension forces. Due to its beneficial properties a Constant Temperature Anemometry was applied.

1 INTRODUCTION

1.1 *Turbulence grids*

Grids for turbulence generation in wind tunnels are of wide interest both in scientific applications as well as in practical engineering. Primarily, these grids are applied to generate turbulence downstream of their cross-section. However, when the inlet flow is of high turbulence intensity and the velocity distribution is rather non-homogeneous over the cross-section, grids are also employed in order to improve flow quality. Passive grids simply represent an obstacle to the flow, thus in this sense they are passive indeed. However, they can even be regarded as active grids in a general dynamic sense (Gad-el-Hak & Corrsin1974): e.g. the boundary-layer separation lines move unsteadily. Almost all experiments on isotropic turbulence, beginning with the very first (probably Simmons & Salter 1934) until several decades ago, have been carried out using the passive form of turbulence grids. The role of contraction in the isotropy of the turbulence induced by passive grids was investigated by Comte-Bellot & Corrsin (1966). Probably the most important publication presenting grid experiments is that of Comte-Bellot (1971). Flow measurements in grid-generated turbulence are the most important experiments for turbulence theories and models, delivering suitable boundary conditions and validation data. The grid (or rather the Taylor) Reynolds number in the case of passive grids is dependent on the wire dimensions and should be high enough to generate turbulence (Kurian & Fransson 2009). Therefore, these kinds of grids are limited, since either the test-section dimension or the flow velocity should be quite high.

1.2 *Active grids*

For this reason, experiments applying active grids with moving boundaries (e.g. Ling & Wan 1972 and Sato & Saito 1974 used vibrating grids) or capable of adding mean momentum to the fluid (Ozono et al. 2007) are rather typical in the last several decades. Moreover, grids which eject secondary fluid jets into the mean flow, called 'jet grids', have promising potential in active grids. These kinds of jet grids generate turbulence with higher turbulence intensities, but still preserve a reasonable level of homogeneity, requiring a reasonable size of wind tunnel and mean flow velocities. Jet grids also result in higher Reynolds numbers and thus enable the extension of previous existing experiments (Kang et al. 2003). However, it should be mentioned that jet grids have some disadvantages. By utilizing secondary air to produce jets, the amount of the air in the closed-circuit wind tunnels increases over time, which, without proper treatment, leads to an undesirable pressure rise in the tunnel. This problem specifically appears in case of measurement techniques which apply flow motion indicative particles (PIV, LDA, LDV, etc.). In addition, the jets need pressurized air, thus such grids require expensive compressors.

Recently the most common active grids are those with rotating vanes, primarily introduced by Makita & Miyamoto (1983). Afterwards a more detailed study described the performance of the active grid and showed some characteristics of the induced homogeneous, quasi-isotropic turbulence (Makita 1991). The properties of decaying turbulence behind an active grid (following that of Makita 1991) were compared with the results of large-eddy simulation by Kang et al. (2003), moreover their results updated the results of Comte-Bellot & Corrsin (1971). Larssen & Devenport (2011) have built such a grid, probably the largest ever developed, and not only provided a comprehensive overview on the different stages of the development of turbulence grids, but also reported extensive experiments for several operating conditions.

Our aim was to construct a cost effective active grid which does not require external energy input to maintain itself. For this reason a grid containing non-rigid, adjusted tensioned steel tapes was built into a wind tunnel upstream of the test section. This article presents this novel type of active grid. Among others, the grid construction, the experimental setup and the investigation of effect of the tensioning force of the tapes are presented.

2 EXPERIMENTAL SETUP AND GRID DIMENSIONS

2.1 *The wind tunnel and the novel grid*

The investigated turbulent air flow was generated in a wind tunnel of the University of Miskolc. The wind tunnel is an open circuit blower-type wind tunnel with test section dimensions of $400 \times 400 \times 800$ mm. The mean speed of the air flow (0–6 m/s) is controlled by a radial fan equipped with an AC/AC inverter.

Our aim was to construct a cost-effective, active turbulence grid which is able to utilize the energy of the mean flow to produce motion, along with forced boundary-layer separation. For this purpose non-rigid steel tapes were used with an active length of 400 mm, width of 6 mm and thickness of 0.05 mm. The plane of the tapes was parallel to the main flow itself, while the plane of the grids was built perpendicular to the flow at the starting point of the test section. The mesh spacing (M) was set to $M = 25$ mm, thus the grid is comprised of 15 tapes. The tapes pass through the prepared gaps on the top and bottom walls of the wind tunnel and are indirectly connected to a rigid aluminum frame. Means to measure and adjust the tensioning force of each tape are built into the frame of the grid. In order to adjust the tensioning force, a threaded tensioning device with coil spring was applied and attached to the lower horizontal frame section. Additionally the individual measurement of the tensioning force is necessary at the same time as force is adjusted in case of each tape. Therefore digital dynamometers were connected between the top ends of the tapes and the upper horizontal frame section.

The operation of this kind of grid is derived from the main flow itself. While the flow passes among the tapes, it causes a fluttering motion of the tapes. Thus a moving boundary layer is created, which causes increased turbulent properties of the fluid downstream of the grid. The layout of the built-in grid and the wind tunnel test section can be observed in Figure 1.

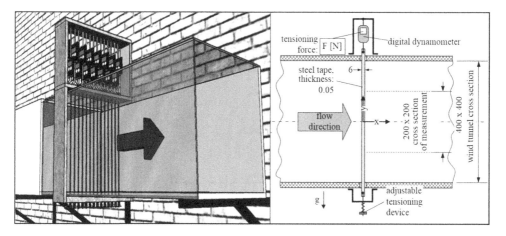

Figure 1.   The apparatus of the novel grid built into the test section of the wind tunnel.

2.2   Applied measurement techniques, measurement positions

The investigated grid was placed 800 mm upstream of the exit of the wind tunnel, so that the flow properties can be measured downstream of the grid. The grid turbulence is considered to be decaying and quasi-isotropic turbulence. Based on Kolmogorov's theory of turbulence (Pope, 2000), in case of fully developed turbulence the Turbulent Kinetic Energy (TKE) decays downstream of the grid, as following the tapes there is no more eddy production but energy transfer develops between the different turbulence scales, and then during viscous dissipation in small scale turbulent eddies. Kolmogorov envisioned that turbulent mixing occurs over a range of turbulent eddies. Dimensional analysis shows (Pope, 2000) that in the inertial range the slope of the energy density per unit wave number follows the exponent of –5/3. A theoretical turbulent energy spectrum with the three sections (production, inertial subrange, dissipation) can be seen in Figure 2.

In the present investigation our aim was to observe the effects of the tensioning force F of the tapes on the turbulent properties downstream of the grid at 3 different mean velocities of 4, 5, 6 m/s and for 3 different tensions of the tapes of 5, 10, 15 N. Based on these settings, 9 different cases were investigated. Note that in all 9 cases the tape flutter was caused by the main flow.

Due to its advantages constant temperature hot wire anemometry (CTA) was applied. This measurement technique is highly recommended for turbulence investigations since the probes have high spatial resolution and the sampling rate is relatively high (some kHz). For these reasons, it is possible to detect small-scale eddies and fast changes in the velocity of the flow. Thus beside the statistical flow properties, spectral analysis could be carried out, as well.

A Dantec 2-channel CTA device with two-dimensional 55R51 X fiber film probe, attached to a computer-controlled lightweight traverse positioning device with the Streamware 4.10 software were applied following the velocity and the directional calibration of the probe, so that the velocity component (u) parallel to the main flow and the traverse component (v) can be measured properly. The data acquisition was performed by an NI MIO 16 bit A/D converter. Two kind of measurements were done.

For spectral analysis, once the built-in square wave test had been performed, the sampling frequency was set to 1.5 kHz. The measurement points were placed in the centerline of the wind tunnel, downstream of the grid at the dimensionless distances of $10M$, $15M$, $20M$, $25M$, and $28M$ (M is the grid spacing). In each point the sampling interval was set to 60 s.

For statistical properties (turbulence intensity, turbulent kinetic energy, mean velocity, etc.), three vertical planes perpendicular to the mean flow were defined in the central 200 mm × 200 mm part of the cross section with grid spacing of 20 mm × 20 mm, in distances from the grid of $10M$, $20M$ and $28M$. In this case, the frequency of the acquisition was set to

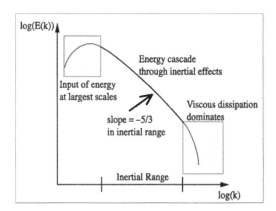

Figure 2. The theoretical turbulent kinetic energy spectrum as a function of wave number, with its 3 different sections: vortex production, energy cascade, energy dissipation.

80 Hz (based on the integral length scale from correlation) with the number of samples being 400 at each point. Following the measurements, the data with instantaneous velocity values was processed in the MATLAB software.

3 RESULTS

3.1 Energy spectra

For energy spectrum calculations, a large amount of instantaneous velocity data with high temporal resolution was provided by the CTA technique. Following data export, the files were imported into MATLAB. The built-in functions *spectrum.welch* and *psd* (power spectral density) were applied to calculate the spectral energy of each cases. The Welch method is a non-parametric explicit estimation for calculating the spectrum of the dataset without using a model. For weight function, a Hanning window with 50% overlapping was applied. Non-dimensional units were used for independence of the calculated energy spectrum from the system. Instead of the wave number κ [1/m], $\kappa\eta$ [–] was used as the horizontal axis, and instead of the energy density per unit wave number E [m³/s²], $E/(\eta u_\eta^2)$ [–] was used as the vertical axis, where η is the Kolmogorov length scale at which the eddies dissipated and u_η is the velocity scale of the dissipative range (Pope, 2000). The energy spectra derived from our measurements show the effect of the tensioning force of the tapes at 3 different fixed mean velocities at the distance from the grid of $15M$ (Figure 3).

By observing Figure 3, several remarks can be made. Firstly, since the scales of the corresponding axes are the same in case of the three mean velocities, it is possible to compare the magnitude of the *TKE*, which is proportional to the area under the curve. If we look at the curves corresponding to the tension of 5 N (dotted line) on all three diagrams, it can be stated that the *TKE* increases with the mean velocity. It is worth noting that at lower mean velocities only the setting with the lowest tensioning force follows the theoretical curve, a −5/3 exponent slope. At 4 m/s mean velocity, increase in the tension causes an increase in deviation of the curve from the theoretical tendency. Furthermore, as can be seen on the diagram for 4 m/s, in case of $F = 15$ N a peak in the curve is formed. A possible reason for this is the phenomena of the resonance of the linear system of the tape and the spring: possibly at this tensioning force the tapes have a resonant frequency close that of frequency caused by boundary layer separation. Proof of this requires further investigations.

Another turbulent quality, namely the turbulence Reynolds number (or Taylor Reynolds number) R_λ was calculated for each case in all positions by:

$$R_\lambda = u' \, \lambda / v, \qquad (1)$$

where u' is the r.m.s. of the velocity component parallel to the mean flow, λ is the Taylor scale, ν is the kinematic viscosity of the air. Thus the turbulence Reynolds number is calculated based on the Taylor microscale λ, which is a measure of the size of eddies in the inertial sub-range and can be calculated by Equation 2. Consequently, λ is the length scale limit: at lower scales than λ, the fluid viscosity affects the dynamics of the turbulent eddies; however, scales larger than λ are not strongly affected by viscosity.

$$\lambda = (10\nu TKE / \varepsilon)^{1/2}, \tag{2}$$

where ε is the dissipation rate of the TKE, which can be calculated from the energy spectrum (Pope, 2000). The effect of the tension on the Taylor Reynolds numbers at constant mean velocities can be seen in Figure 4.

As can be seen, in all cases the Taylor Reynolds number decreases with distance. Mainly this is caused by the decreasing velocity fluctuations (u') with distance. It was possible to reach $R_\lambda \approx 30$ at the distance of $10M$. Similar results were obtained by investigation of a vibrated grid performed by Ling & Wan (1972). The higher the turbulence Reynolds number value, the more developed turbulence evolves, and in grid experiments a high R_λ value is a desirable trait.

In addition to the one-dimensional measurement in the center line, a further three different planes were investigated for time averaged analysis. Since the time-averaged data processing requires non-correlated samples, based on preliminary calculations the time between samples was set to more than two times larger than the integral time scale of the velocity fluctuations. Thus 80 Hz was chosen as the sampling rate. In Table 1 the time-averaged flow quantities belonging to the plane at a distance of $20M$ from the grid are presented.

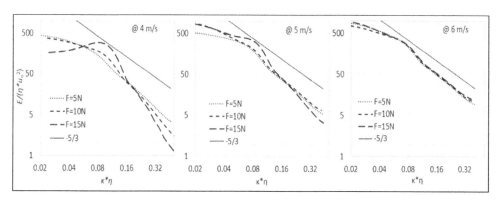

Figure 3. The effect of the tensioning force on the energy spectrum at the distance from the grid of $15M$, for mean velocities 4, 5, and 6 m/s.

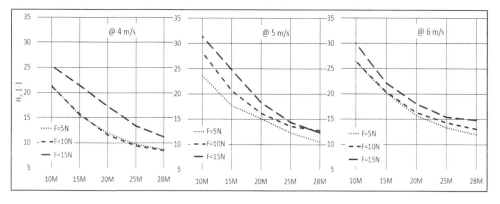

Figure 4. Taylor Reynolds numbers in the centerline downstream of the active grid.

Table 1. The mean values of fluid properties at the $20M$ plane.

Tens. force [N]	mean velocity [m/s]	Turbulence intensity [%]	turbulent kinetic energy [m²/s²]	standard dev. of u [m/s]	standard dev. of v [m/s]
5	4	8.55	0.199	0.398	0.345
10	4	9.10	0.198	0.381	0.350
15	4	7.79	0.168	0.293	0.327
5	5	8.97	0.347	0.513	0.463
10	5	9.33	0.328	0.530	0.431
15	5	9.50	0.355	0.517	0.469
5	6	8.53	0.427	0.559	0.519
10	6	9.41	0.490	0.636	0.536
15	6	9.43	0.485	0.644	0.527

4 CONCLUSIONS

A relatively large-scale turbulence field was created in a small wind tunnel by employing the newly designed metal tape active grid. In most measurement cases it was possible to create fully developed turbulence with an appreciable internal subrange. Based on the energy spectra data, the spectrum is affected by the tensioning force only at lower mean velocities. In case of a mean velocity of 4 m/s and tension of 15 N, a peak can be observed on the energy spectrum which could be caused by resonance. This behavior of the grid demands further investigations. Taylor Reynolds number R_λ reached about 30 (at the distance of 10 M). In most cases the inertial subrange covers about one order of magnitude in wave number in the spectrum.

REFERENCES

Comte-Bellot, G. & Corrsin, S. 1966. The use of a contraction to improve the isotropy of grid-generated turbulence. *Journal of Fluid Mechanics* 25 (04): 657–682.
Comte-Bellot, G. 1971. Simple Eulerian time correlation of full- and narrow-band velocity signals in grid-generated, isotropic turbulence. *Journal of Fluid Mechanics* 48 (part 2): 273–337.
Gad-el-Hak, M. & Corrsin, S. 1974. Measurements of the nearly isotropic turbulence behind a uniform jet grid. *Journal of Fluid Mechanics* 62 (01): 115–143.
Kang, H.S., Chester, S. & Meneveau, C. 2003. Decaying turbulence in an active-grid-generated flow and comparisons with large-eddy simulation. *Journal of Fluid Mechanics* 480: 129–160.
Kurian, T. & Fransson, J.H.M. 2009. Grid-generated turbulence revisited. *Fluid Dynamics Research* 41, 021403.
Larssen, J.V. & Devenport, W.J. 2011. On the generation of large-scale homogeneous turbulence. *Experiments in fluids*, 50(5), 1207–1223.
Ling, S.C. & Wan, C.A. 1972. Decay of isotropic turbulence generated by a mechanically agitated grid. *Physics of Fluids* 15, 1363.
Makita, H. & Miyamoto, S. 1983. Generation of high intensity turbulence and control of its structure in a low speed wind tunnel. *Proc. 2nd. Asian Congress on Fluid Mechanics*: 101–106.
Makita, H. 1991. Realization of a large-scale turbulence field in a small wind tunnel. *Fluid Dynamics Research* 8: 53–64.
Ozono, S., Miyagi, H. & Wada, K. 2007. Turbulence generated in active grid mode using a multi-fan wind tunnel. *Journal of Fluid Science and Technology* 2 (3): 643–654.
Pope, S.B. 2000. *Turbulent Flows.* Cambridge University Press, Cambridge.
Sato, H. & Saito, H. 1974. *Proc. 6th Turbulence Symposium*, 103.
Simmons, L.F.G. & Salter, C. 1934. Experimental investigation and analysis of the velocity variations in turbulent flow. *Proc. of the Royal Society of London. Series A* 145 (854): 212–234.

Advances and Trends in Engineering Sciences and Technologies II – Al Ali & Platko (Eds)
© 2017 Taylor & Francis Group, ISBN 978-1-138-03224-8

Crack initiation in concrete specimens based on alkali-activated binders

H. Šimonová, L. Topolář, I. Havlíková, I. Rozsypalová, B. Kucharczyková &
Z. Keršner
Faculty of Civil Engineering, Brno University of Technology, Brno, Czech Republic

V. Bílek Jr.
Faculty of Chemistry, Brno University of Technology, Brno, Czech Republic

ABSTRACT: The aim of this paper is to quantify crack initiation in two types of concrete containing alkali-activated binder. Beam specimens with a stress concentrator were tested via three-point bending after 28 and 90 days of maturing. The fracture tests were recorded in the form of load versus mouth crack opening displacement (*P–CMOD*) diagrams, which were evaluated using the *Double-K* fracture model. This model allows the quantification of two different levels of crack propagation: initiation, which corresponds to the beginning of stable crack growth, and the level of unstable crack propagation. The initiation of cracks during the fracture tests was also monitored via the acoustic emission method.

1 INTRODUCTION

Binders based on Ordinary Portland Cement (OPC) are the most common binders used in concrete production. However, cement manufacturing processes significantly contribute to global emissions of CO_2. On average, approximately one ton of CO_2 is liberated per ton of cement produced (Aïtcin & Mindess 2011). There has therefore been an increasing effort over recent decades to develop alternative binders, including alkali-activated binders.

Alkali-Activated (AA) binders are composed of aluminosilicate-based material, such as Blast Furnace Slag (BFS) or Fly Ash (FA), combined with an alkaline activator and water. The activating solution promotes the dissolution of aluminosilicate glass, which results in the nucleation and growth of hydration products and a subsequent increase in the strength of the binder (Provis & van Deventer 2014, Shi et al. 2006). BFS and FA have specific disadvantages with regard to alkaline activation, but these may be partially overcome by blending them (Marjanović et al. 2015, Chi & Huang 2013, Aydin, S. 2013). This paper is therefore focused on the determination of the fracture parameters of AABFS/FA concrete, as there is a lack of studies dealing with this topic. A concrete made with Cement Kiln bypass Dust (CKD) was also tested. CKD is a by-product from Portland clinker manufacturing.

The values of the mechanical fracture parameters of quasi-brittle materials are usually determined via the evaluation of the results of experiments performed on specimens with stress concentrators. For the purposes of this paper, three-point bending fracture tests were conducted on specimens made from the above-mentioned materials. The results were evaluated using the *Double-K* fracture model (Kumar & Barai 2011). Specimen response during fracture tests was also monitored by means of acoustic emission.

2 EXPERIMENTAL PART

2.1 *Materials*

Two different concrete mixtures with alkali-activated binder were prepared. They both contained BFS, which was partially replaced by low calcium FA (50%) in the first case and CKD (25%) in the second case. Ground granulated BFS with a majority of glassy phase and a Blaine surface of 400 m^2/kg was used. Besides the glassy phase, the BFS contained small amounts of melilite and mervinite, while the major crystalline phases of the FA were mullite and quartz. The CKD was composed of potassium salts (KCl, K_2SO_4), free CaO and larnite. The binder was activated by liquid sodium silicate with a silicate modulus of 2.3. Its dose was adapted to the 8% by mass Na_2O content of the binder.

Natural sand and gravel were used as fine and coarse aggregate, respectively. Maximum aggregate size was 16 mm. The binder: aggregate: water ratio was 1: 3: 0.52 in the case of the FA concrete and 1: 4: 0.47 for the CKD concrete. Two types of chemical admixture were also applied to improve the performance of the prepared concrete: Lignosulphonate-based plasticizing agent was used at 0.2% by mass of the binder in both types of concrete, while a shrinkage-reducing admixture consisting of 5-ethyl-1,3-dioxane-5-methanol, 2 ethylpropane-1,3-diol and 2 ethyl-2-(hydroxymethyl)propane-1,3-diol was used at 1.0% for the FA concrete and 0.5% for the CKD concrete.

2.2 *Specimens*

Six specimens with nominal dimensions of 75 × 75 × 295 mm were prepared from each concrete mixture. After 24 hours, these specimens were demolded and cured under dry laboratory conditions, i.e. at approximately 23°C and 45% relative humidity. After that, the three-point bending test was performed on these specimens at the age of 28 days in order to obtain the mechanical fracture parameters.

Three slab specimens (75 × 295 × 600 mm) were also prepared from each mixture. These specimens were stored in air for 42 days and then tested via the vacuum test method. After the vacuum tests had been performed, two specimens of the nominal dimensions 75 × 75 × 295 mm were cut from the undamaged end parts of each slab. One half of these specimens were also tested in three-point bending at the age of 90 days. The second half have been stored exposed to the air under dry laboratory conditions and will be tested in the future at the age of 1 year.

2.3 *Fracture tests*

The aforementioned three-point bending tests were performed on beams with a central edge notch (Fig. 1(left)); span length was 245 mm. The initial notch was made by a diamond blade saw. Note that the depth of the notches was about 25 mm. Fracture tests were carried out using a Heckert FPZ 100 testing machine within the range of 0–10 kN at a laboratory at the Institute of Building Testing, Faculty of Civil Engineering, Brno University of Technology.

If displacement increment loading is performed, it is possible to record load versus displacement (*P–d*) and load versus crack mouth opening displacement (*P–CMOD*) diagrams during the tests. Also, acoustic emission signals were recorded during the fracture tests.

The data from the *P–CMOD* diagrams were used as input data for the *Double-K* fracture model (Kumar & Barai 2011), which was used to determine the mechanical fracture parameters of the tested materials. The advantage of this model is that it describes different levels of crack propagation: an initiation part which corresponds to the beginning of stable crack growth (at the level where the stress intensity factor, K_{Ic}^{ini}, is reached), and a part featuring unstable crack propagation (after the unstable fracture toughness, K_{Ic}^{un}, has been reached).

3 METHODS

3.1 *Double-K fracture model*

The mechanical fracture parameters of the tested materials were determined using the *Double-K* fracture model (Kumar & Barai 2011). Input data for this model was obtained from the above-mentioned *P–CMOD* diagrams namely maximum load P_{max} and its corresponding critical crack mouth opening displacement $CMOD_c$, while load P_i was obtained from the linear part of diagram and its corresponding crack mouth opening displacement $CMOD_i$; see Figure 1 (right).

The *Double-K* model combines the concept of cohesive forces acting on the faces of the effective crack increment with a criterion based on the stress intensity factor. The unstable fracture toughness K_{Ic}^{un} was determined first, followed by the cohesive fracture toughness K_{Ic}^{c}. When both of these values were known, the following formula was used to calculate the initiation fracture toughness K_{Ic}^{ini}:

$$K_{Ic}^{ini} = K_{Ic}^{un} - K_{Ic}^{c}. \tag{1}$$

Details regarding the calculation of both unstable and cohesive fracture toughness can be found in numerous publications, e.g. in Kumar & Barai 2011.

Finally, the value of the load P_{ini} was determined according to the following definition. This value can be defined as the load level at the beginning of stable crack propagation from the initial crack/notch:

$$P_{ini} = \frac{4WK_{Ic}^{ini}}{SF_1(\alpha_0)\sqrt{a_0}}, \text{ where } F_1(\alpha_0) = \frac{1.99 - \alpha_0(1-\alpha_0)(2.15 - 3.93\alpha_0 + 2.7\alpha_0^2)}{(1+2\alpha_0)(1-\alpha_0)^{3/2}}, \tag{2}$$

W is the section modulus determined as $W = 1/6BD^2$, B is the specimen width, D is the specimen depth, S is the load span, $F_1(\alpha_0)$ is the geometry function, α_0 is the ratio a_0/D, and a_0 is the initial notch length.

3.2 *Acoustic emission method*

The Acoustic Emission (AE) method is a non-destructive testing technique (Grosse & Ohtsu 2008) that can be used to monitor the fracture process during the three-point bending test. Unlike other techniques, AE is a passive monitoring method that can examine the entire volume of a structure, structural member, or—as in this case—of a specimen. The advantage of AE over other defectoscopy methods is that it enables the continuous monitoring of the studied object, and saves time in comparison with sequential testing by other methods. However, the AE method only detects active failures, doing so while they are developing during

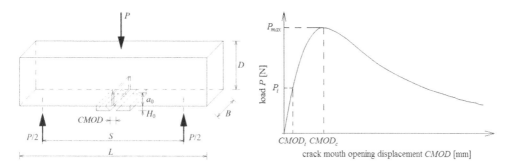

Figure 1.   Three-point bending fracture test geometry (left), acquisition of input data for the *Double-K* model from the *P–CMOD* diagram.

the test. The AE method monitors acoustic emissions occurring during the release of energy because of stimulation by internal or external stress.

4 RESULTS

The following mean values (and coefficients of variation) of the selected mechanical fracture parameters obtained using the *Double-K* fracture model at the age of 28 and 90 days are summarized in Table 1: elasticity modulus E; fracture toughness K_{Ic}^{un}, the K_{Ic}^{ini}/K_{Ic}^{un} ratio (i.e. the ratio expressing the resistance to stable crack propagation), and maximum crack tip opening displacement w_c.

As can be seen from Table 1, higher values of the monitored mechanical fracture parameters of alkali-activated concrete with BFS were achieved with the replacement of BFS by low calcium FA in comparison with replacement by CKD. The modulus of elasticity values are about 23% lower for CKD concrete in comparison with FA concrete. Similar results were also obtained for unstable fracture toughness. In the case of the ratio expressing the resistance to stable crack propagation, the value is lower (it is only about 65% in the case of CKD concrete).

The modulus of elasticity and unstable fracture toughness values increase by about 25–40% with specimen age for both types of concrete. On the other hand, the value of the ratio expressing the resistance to stable crack propagation decreases by about 25–30%.

The results obtained for selected specimens from three-point bending tests (*P–CMOD* diagrams), coupled with AE results (AE counts) and outputs from the *Double-K* fracture model (load level P_{ini}), are shown in the following Figures 2 to 5.

Table 1. Mean values of selected parameters (coefficients of variation in %).

Parameter	FA		CKD	
	28 days	90 days	28 days	90 days
E [GPa]	5.4 (33.1)	7.2 (3.8)	4.2 (22.2)	5.3 (8.1)
K_{Ic}^{un} [MPa·m$^{1/2}$]	0.529 (13.1)	0.680 (15.4)	0.382 (14.2)	0.521 (18.6)
K_{Ic}^{ini}/K_{Ic}^{un} [–]	0.411 (26.8)	0.311 (30.0)	0.139 (29.5)	0.097 (27.2)
w_c [mm]	0.223 (21.5)	0.233 (17.8)	0.177 (17.8)	0.233 (16.5)

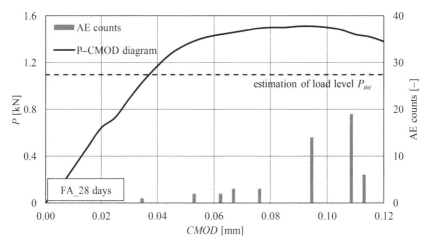

Figure 2. *P–CMOD* diagram with AE counts and estimated load level P_{ini} for selected FA concrete specimen after 28 days of curing.

270

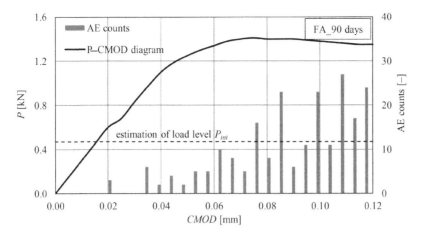

Figure 3. *P–CMOD* diagram with AE counts and estimated load level P_{ini} for selected FA concrete specimens after 90 days of curing.

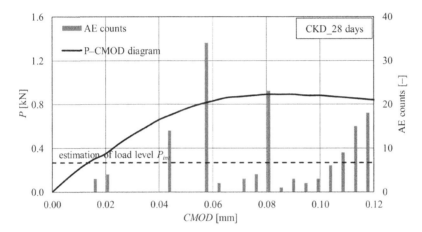

Figure 4. *P–CMOD* diagram with AE counts and estimated load level P_{ini} for selected CKD concrete specimens after 28 days of curing.

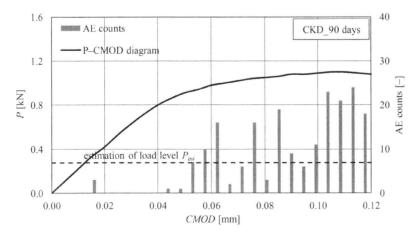

Figure 5. *P—CMOD* diagram with AE counts and estimated load level P_{ini} for selected CKD concrete specimens after 90 days of curing.

5 CONCLUSIONS

The aim of this study was to quantify the influence of Fly Ash (FA) and Cement Kiln bypass Dust (CKD) on the mechanical fracture parameters of alkali-activated BFS concrete, and their comparison for different maturing ages of specimens.

The mechanical fracture parameters of concrete based on alkali-activated binders were higher in the case of the substitution of FA for BFS in comparison with CKD. With the exception of resistance to stable crack propagation, the monitored parameters increased with specimen age. Stable crack propagation starts at a lower load level for specimens at the age of 90 days for both types of AA binder-based concrete. The remaining specimens will be tested at the age of one year in order to monitor the development of selected mechanical fracture parameters over time.

From the Acoustic Emission (AE) results it is evident that during the first higher number of AE counts, stable crack propagation begins from the initial crack/notch. This means that this moment should correspond to the value of load P_{ini}—we can see this in Figures 2 to 5, which show the results for selected specimens for both types of alkali-activated binder-based concrete. It can be seen that it almost corresponds to the value predicted by the *Double-K* fracture model.

ACKNOWLEDGEMENT

This outcome has been achieved with the financial support of the standard specific research program at Brno University of Technology, project No. FAST-S-16-3265.

REFERENCES

Aïtcin, P.C. & Mindess, S. 2011. *Sustainability of concrete*. New York: Spon Press, xxv, p. 301. *Modern Concrete Technology*, 17.

Aydin, S. 2013. A ternary optimisation of mineral additives of alkali activated cement. *Construction and Building Materials* 43: 131–138.

Chi, M. & Huang, R. 2013. Binding mechanism and properties of alkali-activated fly ash/slag mortars. *Construction and Building Materials* 40: 291–298.

Grosse, Ch.U. & Ohtsu, M. 2008. *Acoustic Emission Testing*. Berlin: Springer-Verlag, p. 416.

Kumar, S. & Barai, S.V. 2011. *Concrete Fracture Models and Applications*. Berlin: Springer, p. 406.

Marjanović, N., Komljenović, M., Baščarević, Z., Nikolić V. & Petrović, R. 2015. Physical–mechanical and microstructural properties of alkali-activated fly ash–blast furnace slag blends. *Ceramics International* 41: 1421–1435.

Provis J.L. & van Deventer J.S. (editors) 2014. *Alkali activated materials: state-of-the-art report*, RILEM TC 224-AAM. Dordrecht: Springer.

Shi, C., Krivenko, P.V. & Roy, D. 2006. *Alkali-activated cements and concretes*. London: Taylor & Francis. ISBN 04-157-0004-3.

Advances and Trends in Engineering Sciences and Technologies II – Al Ali & Platko (Eds)
© 2017 Taylor & Francis Group, ISBN 978-1-138-03224-8

A finite element algorithm of cables on pulleys

H. Štekbauer & I. Němec
Faculty of Civil Engineering, Institute of Structural Mechanics, Brno University of Technology, Brno, Czech Republic

ABSTRACT: This paper presents a new algorithm for the calculation of cable-pulley systems. This brand new algorithm considers the magnitude of the radius of each pulley. The procedure is suitable for the application of general FEM codes and offers higher accuracy of results at reasonable additional cost, compared to radii neglecting approaches. It facilitates the modelling of cable passing through multiple pulleys and includes internal dry friction in arbitrary pulley axis. It has been implemented in the RFEM program because of its accuracy and high efficiency compared to the previous algorithm. The paper contains the comparison of the commonly used algorithm, which ignores pulley radius, and the new algorithm.

1   INTRODUCTION

Pulleys are used for their mechanical advantage. That is why pulleys are a part of many building and machine structures, they can be seen e.g. at crane structures, cableways and even in car engines. The set of pulleys is often a complex system too complicated for hand calculation and therefore there is an effort to enable calculation of such structures with the aid of computer programs. Manual calculation is completely impossible in case of pulleys which are a part of load bearing structures and dynamic effects or when large displacements and rotations are taken into account.

For that reason a number of algorithms were developed in order to offer sufficient solutions for this nonlinear problem (Ju & Choo 2005 a, b; Aufaure 1993). This paper treats the given issue in a way neglected by commonly used algorithms.

2   ALGORITHMS FOR THE CALCULATION OF CABLES ON PULLEYS

Besides simulators like MapleSim, which do not allow detailed static and dynamic calculation of pulleys and structures connected to them, there are two possible approaches for solving cables on pulleys.

The first approach is a connection of pulleys using 1D elements only in the centre of each pulley (Figure 1a) or at the intersection of real cable sections (Figure 1b) and defining behaviour of such system to find an equivalent solution. One of many solutions is e.g. the superelement presented by F. Ju and Y. Choo in Ju & Choo (2005 a,b). Such a solution lacks any consideration of the radii of pulleys, which inevitably leads to inaccurate results. In the first case (Figure 1a) spatial geometry of cables in the model is changed and in the second case (Figure 1b) reaction occurs in the node located out of the structure. This approach is used in programs such as LS-DYNA or RFEM.

The second approach is modelling of each pulley and cable using 3D elements and between these elements adding other contact elements for the correct behaviour of the multibody system. This solution is indeed accurate for the correct model, but it is extremely laborious to create a mathematical model and the calculation is also extremely time-consuming. Because of the difficulty mentioned above, use of this approach is impossible for large structures. Such approach can be used with programs like ANSYS, Nastran and Abaqus.

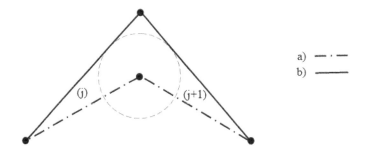

a) — · —

b) ——

Figure 1. Connection possibilities of two cables in node for the first approach.

3 THE CURRENT ALGORITHM

The current algorithm, used in the RFEM program, utilizes the first approach (Figure 1a). It is based on two basic assumptions using just normal forces N, change of normal force dN, lengths of each elements between pulleys L and its change dL. The first assumptions considers equality of normal forces in elements with one shared pulley

$$N^{(j)} + dN^{(j)} = N^{(j+1)} + dN^{(j+1)}.$$ (1)

The second assumption is preservation of whole system length

$$\sum dL^{(j)} = 0.$$ (2)

Equation 2 can be written as

$$\sum \left(dN^{(j)} \bullet L^{(j)} \right) = 0.$$ (3)

Equation 1 can be edited to

$$N^{(j)} - N^{(j+1)} = dN^{(j+1)} - dN^{(j)}.$$ (4)

Equation 3 and Equation 4 form the system of equations

$$\begin{pmatrix} N^{(j)} - N^{(j+1)} \\ 0 \end{pmatrix}_{((n-1)\times 1)} = \begin{pmatrix} -1 & 1 \\ L^{(j)} & L^{(j+1)} \end{pmatrix}_{((n-1)\times(n-1))} \begin{pmatrix} dN^{(j)} \\ dN^{(j+1)} \end{pmatrix}_{((n-1)\times 1)}$$ (5)

The change in the inner forccalculated in Equation5 is transformed to a global coordinate system and added to the actual inner force at each iteration step using the Newton-Raphson method.

This solution has a drawback, for small input forces and large stiffness of cable elements too many iterations are required to find a solution, which leads to long computing times. That is why the new algorithm was developed.

4 THE NEW ALGORITHM

The new algorithm, implemented in the RFEM program, combines advantages of the previous two approaches mentioned above. These advantages are simplicity and speed of pre-processing and processing in the first approach and more accurate geometry in the second approach. This enables fast and accurate solution of pulleys using 1D elements (Figure 2).

Each pulley is a set of two parts. These parts are the pin of the pulley and its rotating part (wheel). A pulley is defined by its centre, radius, contact points of cables and transformation

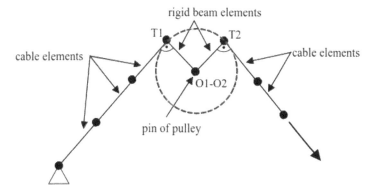

Figure 2. New algorithm geometry.

Figure 3. Shift of nodes.

matrix. The transformation matrix is corresponding to the axis of pulley (node O1). The pulley axis is also the axis of rotation. Node O2 is located on node O1 and it is connected with the rotating pulley wheel. Between the nodes O1 and O2 friction may be defined. Node O1 is supported or it is a part of another structure. The pulley wheel is modelled by rigid beams T1-O2 and O2-T2 where the nodes T1 and T2 are the contact points of the cables. Rigid beams of pulley should always be perpendicular to the cable elements. Cable elements are not wrapped around the wheel, but for simplification these cable elements are considered only in sections between the pulleys and optionally between the pulley and other parts of the structure. Coordinates of T1 and T2 are recalculated at the end of each iteration to fulfil the orthogonality condition of the cable elements to rigid beams.

This new algorithm enables the calculation of static analysis using Newton-Raphson method and dynamic analysis using explicit method. Both of them enable consideration of all nonlinearities of the structure and have been integrated to the RFEM program.

There is also possibility to define friction in the pin of pulley. For that reason a nonlinear hinge is located between O1 and O2. Behaviour of the nonlinear hinge can be described by a graph. Which one?

4.1   Calculation of correct node positions

Due to the rotation of a pulley nodes T1 and T2 can be shifted (Figure 3). That is why correct nodes positions have to be calculated after every iteration.

The position of node T1 is calculated with the basic use of analytical geometry. First plane *p1* perpendicular to the vector N1-T1 and passing through node O2 is found (Figure 4a). Node T1 is shifted to the intersection of plane *p1* with vector N1-T1 (Figure 4b), then is T1 shifted again along vector O2-T1 so that

$$\|O2T1\| = radius. \tag{6}$$

The last shift of node T1 in each iteration is very small (Figure 4c) so any changes in the angle can be neglected.

275

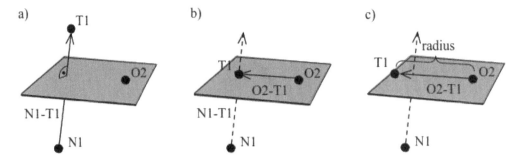

Figure 4. Correction of node T1.

Position of node T2 is calculated same way as for T1 only instead of the node N1, the node N2 used.

After geometry correction, transformation matrices have to be transformed for rigid beam and cable elements connected to T1 and T2. The following procedure is used for all these elements.

To determine the required rotation of transformation matrix T, the calculation of the angle between original and new elements is needed. For this reason, the applied formula to calculate the angle is subtended by two vectors

$$\theta = arccos \left(\frac{\vec{u} \cdot \vec{v}}{\|\vec{u}\| \cdot \|\vec{v}\|} \right) \tag{7}$$

Now angle calculated in Equation 7 is used for definition of the rotation tensor using Rodrigues' rotation formula (Murray 1994):

$$R = \begin{pmatrix} \cos\theta + \omega_x^2 - C & \omega_x \omega_y - C - \omega_z \sin\theta & \omega_y \sin\theta + \omega_x \omega_z - C \\ \omega_z \sin\theta + \omega_x \omega_y - C & \cos\theta + \omega_y^2 - C & -\omega_x \sin\theta + \omega_y \omega_z - C \\ -\omega_y \sin\theta + \omega_x \omega_z - C & \omega_x \sin\theta + \omega_y \omega_z - C & \cos\theta + \omega_z^2 - C \end{pmatrix} \tag{8}$$

where

$$C = (1 - \cos\theta) \tag{9}$$

and ω is unit vector of rotation in R^3

$$\omega = (\omega_x, \omega_y, \omega_z) \tag{10}$$

New transformation matrix is calculated as

$$T_{new} = R^T T. \tag{11}$$

5 A COMPARISON OF THE COMMONLY USED AND THE NEW ALGORITHM

The example demonstrates the difference obtained by negligence of the pulley radius. The structure consists of two beams connected by cables on pulleys (Figure 5). The flexible support also prevents beams from rotating in the plane.

The beams are of I-section 120 and cables of section R 8. All components are from steel S 235.

5.1 Calculation by the new algorithm

The input for the new algorithm better matches the scheme of the model. Calculated deformations u_z and reactions are presented in Figure 7. The supports on sides are loaded almost symmetrically, asymmetry is given by absence of a pulley in node 10. The internal force in the cable elements is exactly 10 kN. Displacement in node 5 is 1519.3 mm.

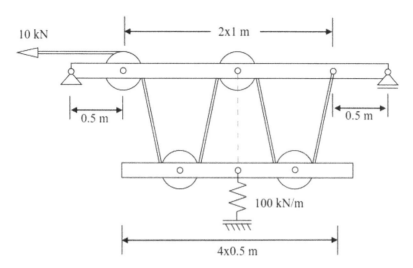

Figure 5.   Scheme of the model.

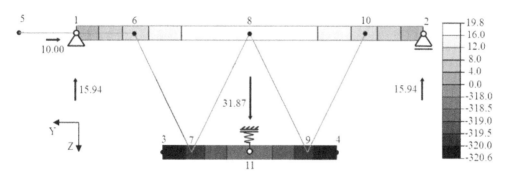

Figure 6.   Results of the current algorithm [kN, mm].

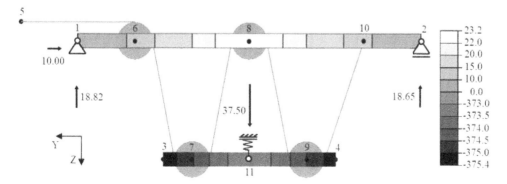

Figure 7.   Results of the new algorithm [kN, mm].

5.2 Calculation by the new algorithm

The input for the new algorithm better matches the scheme of the model. Calculated deformations u_z and reactions are presented in Figure 7. The supports on sides are loaded almost symmetrically, asymmetry is given by absence of a pulley in node 10. The internal force in the cable elements is exactly 10 kN. Displacement in node 5 is 1519.3 mm.

5.3 Comparison

Comparing the results of both solutions shows a striking difference in the use of the current and the new algorithm. This difference is determined by taking into account the radii of the pulleys. Due to pulleys radii in the new algorithm there is another input geometry. In this example, the main change is in inclinations of cable elements, which leads to steeper inclination of cable elements. This results in larger action force in the direction of the Z axis.

6 CONCLUSION

A new approach to numerical analysis of cables on pulleys is proposed using the finite element method. Pulleys are considered to be a composition of two rigid beams connecting the pulley centre with the tangent points of the cable and doubled centre node with possibility of considering the friction law in the pin of the pulley. The proposed approach is used in an example and is found to be able to provide a more accurate and realistic result. Numerical results show the importance of considering pulleys radii in structures with significant ratio between length of the cable segment and the pulley radius.

REFERENCES

Aufaure, M. 1993. A finite element of cable passing through a pulley. *Computers & Structures*. Vol. 46, no.5, 807–812.
Ju, F. & Choo, Y. 2005. Dynamic Analysis of Tower Cranes. *Journal of Engineering Mechanics*, Vol. 131, no.1, 88–96.
Ju, F. & Choo, Y. 2005. Super element approach to cable passing through multiple pulleys. *Mechanics International Journal of Solids and Structures*, Vol. 42, no.11–12, 3533–3547.
Murray, R.M., Li, Y. & Sastry, S.S. 1994. *A Mathematical Introduction to Robotic Manipulation*. Boca Raton, FL: CRC Press.

Advances and Trends in Engineering Sciences and Technologies II – Al Ali & Platko (Eds)
© 2017 Taylor & Francis Group, ISBN 978-1-138-03224-8

Modeling of the airflow in the small-scale model of ventilated air channel

L. Tazky
Institute of Architectural Engineering, Kosice, Slovakia

A. Sedlakova
Department of Building Physics, Kosice, Slovakia

ABSTRACT: Evaluation of validity is realized by comparing results of computer model to data measured on physical full-scale model. Two characteristic parameters are evaluated, ratio of stagnation pressure in the outer stream of the outlet and in the inlet of outer stream. Used CFD system is Ansys CFX. System Ansys CFX is applied with turbulence model of k-epsilon SST with structured numerical grid with detailed resolution in the volume of boundary layer. Computational results confirm capability of current CFD systems to accurately predict parameters of air channel, without accounting for design details and as well to accurately predict the effect of losses on power parameters, provided the numerical grid which is used properly for solving details of the boundary layer. This work is very useful for practical communication in industry, because it specifies conditions for reliable application of CFD methods in computation of air channel around the historical building.

1 INTRODUCTION

Moisture is a major source of damage in historic solid masonry. Rising damp is a well-known phenomenon around the world and it occurs when groundwater flows into the base of a construction and is allowed to rise through the pore structure. From practical experience it is known that many factors may play a role regarding permeability problems in masonry (Burgova, Matejcek, 2008.). The amount of possible causes of moisture problems in historic masonry underlines the complexity of this phenomenon. Evaporation is an important factor in rising damp. The surface of an affected wall contains moisture that has risen from the ground and this moisture is then subject to evaporation. The factors controlling evaporation include: temperature, humidity, air movement and surface (Vlcek, Benes, 2006).

2 HISTORY OF THE CHURCH

2.1 *Visual survey*

During the visual survey it was found that the moisture content of the internal masonry pillars reaches up to the height at around 1700–1900 mm. The situation was the worst at the apse of the east side of the church. On the inner surface of the plaster there was visible efflorescence and mildew occurrences especially in the higher parts of the plinth. On the outside of the walls there are still visible lichens, mosses and algae (Havel, 2012). During the reconstruction of the 80es a gutter walkway was built around the church which only worsened the situation. The biggest problem is the concrete sidewalk on the west side of the church, which compresses the water towards the walls and concentrate at the foundations (Horanska, Dvorakova, 2011).

Figure 1. Left—the church Gemerský Jablonec in nowadays, right—church in year 1932.

Samples were taken from the perimeter walls for moisture laboratory evaluation. Each sample is taken from the plaster or mortar from the peripheral wall. These samples were collected from the bottom of the wall. The highest value of moisture at around 9% have been measured on the east side of the outer walls of the apse (Young, 2008).

In the summer of 2013, work began on the dehumidification of masonry which started by removing of the original plaster to the height at around 1500 mm and replaced by the appropriate remediation plaster (Balik, Solar, 2011) (Figure 1).

3 GEOMETRY OF THE VENTILATED AIR CHANNEL

3.1 Detail of the channel

The church is a protected cultural monument and a historic value should be particularly sensitive to any remediation done. To improve the technical condition of the church minor structural modification is required (Rehak, 1993). The drain pipes will be replaced with outdoor air channels. The channel must be masoned of ceramic burned bricks with lime-cement mortar to ensure the natural evaporation of moisture from the soil through the brick masonry. The bottom of the air channel will be filled with gravel; the drain pipe will be placed in this layer to ensure the drainage of water. The next step will be the coverage of the channel with precast concrete panels. These panels are perforated to ensure the natural evaporation from the channel (Figure 2).

3.2 Simulated variants

We designed the cross-sections dimensions of ventilated air channel in 10 variants. The size and values of air pressures and air velocities are shown in the Table 1. The air pressure and air velocity values are from the second simulation. In the second simulation we used the values of air pressure what we obtained from the first simulation. We designed the church and terrain in the first simulation to get the properties of the wind on the surface church (Alexander, Armando, 2007).

3.3 Numerical model

The first model was realized with an air flow of 3 m/s, during the summer with air temperatures approximating 25°C. In the second model we used obtained values of pressure in monitored points, what represented the location of the inlets on the wall in the first model (Figure 4, 5).

3.4 Boundary conditions of the environment

In this case we used a SST (Shear Streas Transport) numerical model. Temperature of the overall model is 25°C (summer temperature). Apparent density of the air is 1.1845 kg/m3 and

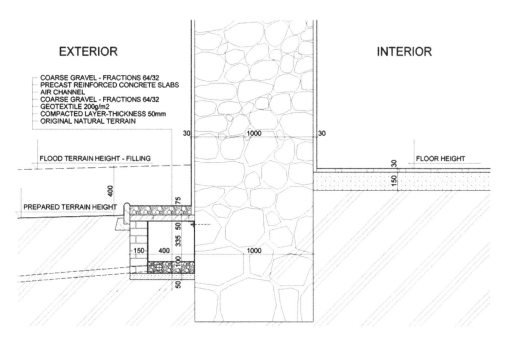

Figure 2.   Detail of the foundation with air channel.

Table 1.   Air velocity and air pressure in the air channel for each variants.

Versions	Dimensions (*mm*)	Velocity (*m/s*)	Pressure (*Pa*)
Version 1	450 × 400	0.163681	0.0133150
Version 2	450 × 450	0.153650	0.0134039
Version 3	450 × 500	0.142533	0.0121934
Version 4	450 × 550	0.132798	0.0111474
Version 5	450 × 600	0.125289	0.0104080
Version 6	**500 × 400**	**0.165185**	**0.0155124**
Version 7	500 × 450	0.149221	0.0121549
Version 8	500 × 500	0.137618	0.0110277
Version 9	500 × 550	0.128301	0.0101979
Version 10	500 x 600	0.121408	0.0094466

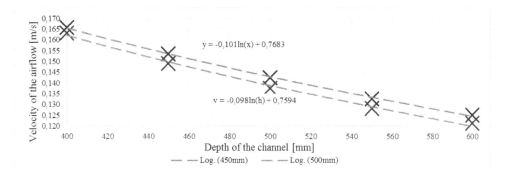

Figure 3.   Dependence of the velocity and depth of air channel.

281

the dynamic viscosity of the air is 1.86159.10-5 kg/m.s. All section models of the channels are simulated by a wind flow of 3 m/s (Brestovic, Jasminska, Kubik, 2013).

4   RESULTS OF THE SIMULATION

In the Figure 3 is displayed dependence of the air velocity on the depth of the air channel. If the depth of the air channel is higher air velocities are lower. Therefore in the final numerical

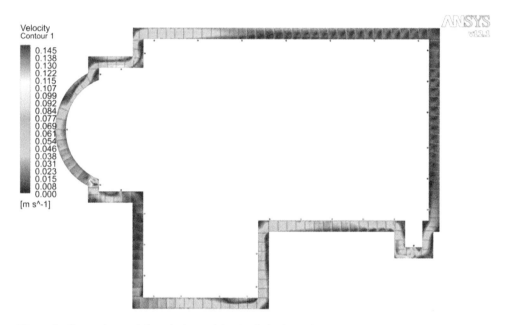

Figure 4.   Dependence of the velocity and depth of air channel.

Figure 5.   Simulation model of the church and air pressure on surface.

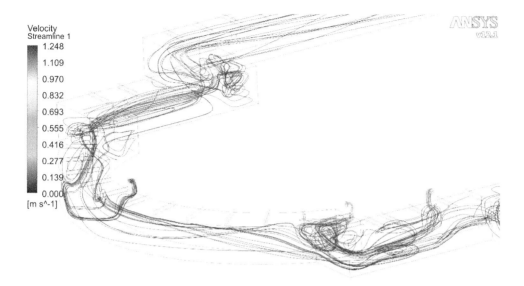

Figure 6. Streamlines coloured by air velocities in the variant 6 around the apse of the church.

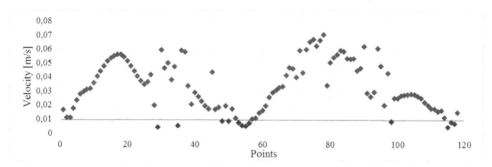

Figure 7. Air velocities in the air channel with the dimensions 500×400 mm (Variant 6).

simulation we used lower depth of the air channel with the enlarged width. Final cross-section dimensions of the air channel are 500 x 400 mm (Variant 6).

The Figure 6 displayed streamlines of air in the air channel around the apse of church. The streamlines air is coloured according to air velocity. In this part of the church is the worst situation, because water content in the masonry is around 9%. (Rencko, Sedlakova, 2013) The tested samples have been collected from the bottom part of the masonry wall. Results obtained from the numerical simulation in the software ANSYS SFX are satisfactory in the complicated parts of the church as well as at the apse (Balik, et.al. 2008).

The graph in Figure 7 displayed air velocity in the monitored points 1–118 in the second model. Location of the monitored points is in the center of the channels in the distance 300 mm.

5 CONCLUSION

The air channel around the church in Gemerský Jablonec is simulated in many different variants, see Table 1. Results obtained from the numerical simulation in the software ANSYS CFX showed cross-section 500×400 mm (Variant 6) as the best for this church. In this case it is air velocity in the air channel that is the highest as well as maximal drying effect of the masonry. Final results showed average air velocity in the air channel on the value 0.032 m/s, (Tazky, Sedlakova, 2015).

ACKNOWLEDGEMENTS

This article was written as a project solution entitled "Numerical analysis and modeling of interactive problems multilayered composite structural members" Project code: VEGA 1/0477/15.

REFERENCES

Alexander, L.B., Armando, M.A. 2007. *Aerodynamic analysis of buildings using numerical tools from computational wind engineering.* Córdoba: AAdeMC.

Balik, M. et.al. 2008. *Dehumidification of buildings* Vol. 2. Praha: Grada.

Balik, M., Solar, J. 2011 100. *Traditional details of buildings—rising damp.* Prague: Grada Publishing.

Brestovic, T., Jasminska, N., Kubik, M. 2013. *Development of software support for ANSYS CFX.* Zilina, Slovakia.

Burgova, E., Matejcek, R. 2008. *Proposal for remediation of damp masonry church. Giles in Mliesk* [online]. Praha: ČVUT, SvF. [2014-5-15]. <http://www.asb.sk/stavebnictvo/rekonstrukcia/navrh-sanacie-vlhkeho-muriva-kostola-sv.-jilji-vmilevsku>.

Havel, M. 2012. *Procedure for remediation of damp masonry locks on the upper Vimperk,* [online]. Bratislava. [2014-5-15]. Dostupné na internete: < http://www.asb.sk/stavebnictvo/rekonstrukcia/postup-sanacie-vlhkeho-muriva-horneho-zamku-vo-vimperku>.

Horanska, E., Dvorakova, V. 2011 *National cultural heritage, importance and reconstruction a financial support.* Bratislava: Jeka Studio.

Rehak, J. 1993. *Survey of drainage tunnels in the monastery Milevsko,* [online]. Praha, [2014-5-15]. <http://old.speleo.cz/soubory /speleo/sp16/ milevsko.htm>.

Rencko, T., Sedlakova, A., 2013 A. *Assessment of underfloor ventilation of historic buildings using Ansys CFX.* Krakow.

Tazky, L., Sedlakova, A., 2015. *Method of modeling the different solution of ventilated air channels for historical buildings,* online, International journal of Engineering and Innovative Technology Vol. 5. ISSN 2277-3754.

Vlcek, M., Benes, P. 2006. *Disturbances and reconstruction of buildings—Modul1.* Brno, Czech Republik.

Young, D. 2008. *Salt attack and rising damp—A guide to salt damp in historic and older buildings.* Melbourne: Red Rover.

Advances and Trends in Engineering Sciences and Technologies II – Al Ali & Platko (Eds)
© 2017 Taylor & Francis Group, ISBN 978-1-138-03224-8

Reduction of ceiling vibrations by stiffness modification of a load-bearing structure

M. Tomko, R. Soltys & I. Demjan
Faculty of Civil Engineering, Technical University of Kosice, Kosice, Slovakia

ABSTRACT: The paper is a structural static and dynamic review of an existing steel ceiling which has been conducted after two-years of service and the completion of subsequent construction work. The construction work involved the reconstruction of a steel structure and structural modifications to the ceiling, which is part of the production hall. The construction work was realised in order to reduce the vibration response of the ceiling. The interior of the steel structure was supplemented by steel structural elements to share the load transfer of automatic washing machines. The structural behaviour of the ceiling was monitored before and after the modification. Accelerations of the ceiling were time-domain-monitored at specific locations. Measured data was transformed to the spectral-domain which is represented by power spectral densities.

1   INTRODUCTION

A static-dynamic assessment of a load bearing steel ceiling structure has been conducted after two years of operation and the completion of subsequent static/construction modifications of a steel production hall functioning as a factory tool room. The structure is located at a height of +10.5 m and is complemented by structural steel components for the load transfer of tested automatic washing machines.

2   FACILITY CHARACTERISTICS

From a structural point of view, the steel production hall consists of three-aisles with a floor size consisting of 7 fields of 6 m/(10.225 m + 9.945 m + 9.730 m).

Frame girders of rolled HEA 340 sections lie across central steel columns consisting of HEA 300 profiles. The girders are anchored to rolled HEA 200 profiles at an axial distance of 2 m which supports Rannila RAN-85B profiled cladding having a thickness of 0.88 mm, which is covered by a layer of concrete with dispersed reinforcement.

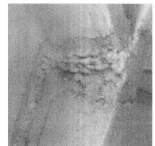

Figure 1.   Discontinuities of some welds observed after the renovation work and completion of structural adjustments of the steel hall structure.

Foundations of the original reinforced concrete skeleton hall consist of foundation footings which are supported by concrete piles, Figure 5. The interior of the indoor spaces are supported by steel columns anchored using steel M36 anchors onto plain concrete footings with an area of 2.0 m × 2.0 m and a depth of 1.0 m. The footings are placed on a compacted gravel bed with a thickness of 300 mm.

3 INSPECTION OF HALL CEILING STEEL STRUCTURES

A visual inspection of the concrete floor-ceiling (trapezoidal concrete slabs) revealed visible cracks at different locations some; having a width of up to 5 mm.

After renovation works and structural modifications of the steel structure located at a height of +10.5 m were completed in (2013), the hall was used for two years as a laboratory space designed for testing automatic washing machines.

4 GEODETIC SURVEY OF STEEL CEILING ELEMENTS

In 2015 a geodetic survey revealed deformations of steel ceiling elements (Kasprišin, M., 2015). The geodetic measurements observed a maximum vertical deformation of 22 mm. Permissible vertical deformation is defined in Standard (EN 1993-1-1, 2005) as 1/250 for a 9000 mm span. In this case, the standard permits a maximum vertical deformation of 36 mm. Structural steel ceiling in Serviceability Limit State (SLS) terms.

The load bearing steel ceiling in terms of SLS assessment satisfies of the permissible maximum vertical deformation criteria is satisfied and the steel beam stress lies in the elastic range.

5 CONTROL OF SELECTED WELDS AND SCREWS

In order to determine the current state of the ceiling steel structure diagnostics of welded joints and bolted joints were realized in 2015. The magnetic particle method detected discontinuities, weld defects and other undesirable defects according to EN 17638 while evaluation standards EN ISO 5817 intended for finished structures, Figure 1, did not reveal any fatigue cracks, (Horban, Š., 2015).

6 EXPERIMENTAL MEASUREMENT OF VIBRATIONS

To comprehend the dynamic behaviour of the steel ceiling structure before the execution of structural modifications (initial state, 2013) and post static/renovation (2015), local experimental vibration measurements were carried out "in situ".

Figure 2.   Photo documentation of experimental "in situ" measurements.

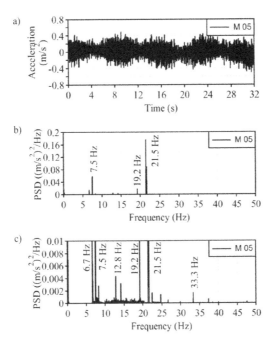

Figure 3. Measurement 05 (M 05), year 2015, a) time domain log of vibration acceleration, b) power spectral density of acceleration, c) power spectral density of acceleration—detail.

Diagnostic vibration measurements were carried using PULSE instruments and software from Brüel & Kjær, Figure 2. The vertical direction of vibration acceleration of the ceiling construction was recorded at a sampling frequency of 128 Hz.

6.1 The results of dynamic response experimental measurements of the ceiling

The results of "in situ" experimental vibration measurements of steel ceiling elements are presented in the form of a time-domain log of vibration acceleration and corresponding Power Spectral Density (PSD), Figure 3 and Figure 4.

From the attained PSD vibration accelerations it can be assumed, that for power spectral density responses of load bearing steel ceiling constructions the dominant frequency was in the range of 6–13 Hz and 19–25 Hz.

7 ANALYSIS OF THE DYNAMIC RESPONSE OF LOAD BEARING STEEL CEILING CONSTRUCTIONS ON THE BASIS OF THEORETICAL AND EXPERIMENTAL APPROACHES

For static and dynamic analysis of the structure, mathematical models of the wall-panel and line elements were created. The computational model contained 6694 nodes, 5600-wall plate elements and 3121 line elements, Figure 5.

The eigenfrequency and associated mode shapes were obtained using a computational model. Forced vibration was computed using the modal decomposition method and direct numerical integration method, Figure 6.

Forced vibration analysis was performed by the modal decomposition method. Theoretical computations were monitored for 10 seconds of vibration velocity with a time step of 0.01 seconds. Logarithmic decrement of attenuation was assumed to be 0.1. Harmonic excitation was simulated with a variable force $F(t)$ defined by $F(t) = F.sin(\omega t + \varphi)$. The amplitude of the harmonic variable force $F(t)$ was assumed to be 0.2 kN. The exciting angular vibration frequency ω was considered for frequencies of 13 Hz, 14 Hz, 15 Hz, 16 Hz, 17 Hz, 18 Hz, 19 Hz, 20 Hz, 21 Hz, 22 Hz, 23 Hz. Phase shift φ was assumed to be 0° and 90°.

After performing static structural modifications to the building construction (installation of HEA 300 and 200 elements) the dynamic response of the building structure's vibrations decreased by approximately 36%.

7.1   *Analysis of the dynamic response of the steel ceiling structure*

Vertical vibrations of ceiling structures with large spans (≥ 5 m) degrade a person's sense of comfort. Movements of 2 Hz to 35 Hz are caused by a variety of mechanical devices, or by the movement of people.

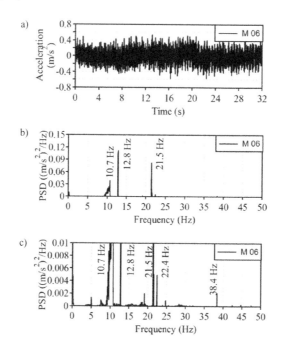

Figure 4.   Measurement 06 (M 06), year 2015, a) time domain log of vibration acceleration, b) power spectral density of acceleration, c) power spectral density of acceleration—detail.

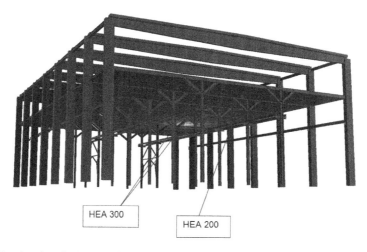

Figure 5.   Rendered static-dynamic diagram of the steel hall structure.

To assess the dynamic flexibility of the ceiling structure we use the ratio of the speed of movement of the harmonic excitation force.

This ratio is also a function of damping vibration of the structure, which depends not only on the system and construction materials, but also on the degree of occupancy of rooms.

According to the measured results of the reinforced concrete ceiling construction with spans of 5 m to 10 m, the maximum deflection peaks are in the range of 10^{-5} to 10^{-6} (ms^{-1}.N^{-1}).

Critical peak values of deflection according to the standards stipulate a range of 0.00001–0.000001 ms^{-1}.N^{-1}.

The deflection was calculated for the ceilings original condition (year 2013), where Root Mean Square (RMS) vibration velocities measured 3.0 mm.s^{-1} and the amplitude of the harmonic excitation variable force is 200 N. The calculated ceiling deflection is 0.000015 ms^{-1}.N^{-1} and lies at the threshold permitted by relevant standards.

Figure 7 shows experimentally observed frequency spectra of the vibration velocity of the building structure at selected measuring points obtained in 2013, prior to structural modifications of the ceiling structure.

13,100 Hz   17,06 Hz   22,097 Hz

Figure 6.   Selected eigenfrequencies and corresponding shapes of the building structure's computational model.

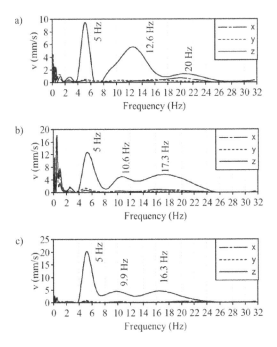

Figure 7.   Experimental power spectral density vibration rate of the building structure at selected measurement points, prior to structural modifications of the ceiling (original condition), year 2013, a) beam HEA 340, at the mid-span of the beam, point 2, b) beam HEA 200, at the mid-span of the beam, point 3, c) sheet, point 4.

Deflections were calculated for the structurally modified ceiling where RMS vibration velocity is 1.9 mm.s^{-1} and the amplitude of the harmonic excitation variable force is 200 N.

The calculated deflection of the ceiling is 0.0000095, which is the value of deflection located outside the limit range permitted by standards.

The post dynamic response of the structure exhibited a 36% reduction in deflection.

8 CONCLUSIONS

Based on a geodetic measurement of the deformation of steel elements supporting the ceiling, it was found that the maximum observed vertical deflection (22 mm) does not exceed that stipulated by standard STN EN 1993-1-1 Eurocode 3 permissible maximum vertical deflection (36 mm), i.e. structural steel ceiling in SLS terms.

SLS assessment of the permissible maximum vertical deformation is suitable and the steel beam ceiling lies in the elastic stress range.

The magnetic particle test of the welds and bolts in selected locations revealed that the welds and bolts do not exhibit cracks and remain intact. Some welds were not properly executed, are intermittent or do not meet the standards stipulated by STN EN 17638 and evaluation EN ISO 5817. However, the current state of the welds does not significantly affect the carrying capacity of the building structure.

Experimental vibration measurements at selected points of the steel ceiling support structure in the dynamic response frequency spectrum showed a dominant frequency range between 6 Hz to 13 Hz and 19 Hz to 25 Hz.

By comparing the experimentally measured frequency spectra dynamic response of the steel ceiling structure in 2015 with theoretically calculated values for the calculation model and experimentally measured frequency of the spectra dynamic response of the steel ceiling structure in 2013 it can be stated that dominant frequencies of vibration are diminished and shifted to a lower frequency range.

Based on the results and knowledge obtained it can be stated that the structural steel ceiling of the factory building at a height of +10.5 m, is safe and stable and can serve its intended purpose (laboratory space designed for testing automatic washing machines) over the next 2 years.

ACKNOWLEDGEMENTS

The paper is carried out within the project No. 1/0302/16, partially founded by the Science Grant Agency of the Ministry of Education of Slovak Republic and the Slovak Academy of Sciences.

REFERENCES

EN 1993-1-1: Eurocode 3, 2005: *Design of steel structures.*
Horban, Š. 2015. *Report on Non-destructive testing of welds*, Nekom—NDT.
Kasprišin, M. 2015. Geodetic *Surveying of steel ceiling elements*, ZEKA, 2015.
Li, J. & Chen, J. 2009. *Stochastic Dynamics of structures*. John Wiley and Sons.
Mandula, J. & Salaiová, B. 2003. The utilizing of dynamic penetration test results in diagnostics and design of sleeper bearing. In: *Construction* (3–4): 237–241.
Simiu, E. 2010. *Chaotic Transitions in Deterministic and Stochastic Dynamical System*. Princeton an Oxford: Princeton University Press.

Advances and Trends in Engineering Sciences and Technologies II – Al Ali & Platko (Eds)
© 2017 Taylor & Francis Group, ISBN 978-1-138-03224-8

The acoustic non-destructive methods and their application to stress situation construction materials

L. Topolář, L. Pazdera, K. Timčaková & D. Štefková
Faculty of Civil Engineering, Brno University of Technology, Brno, Czech Republic

ABSTRACT: The Acoustic Non-Destructive Testing methods are a forceful tool for the determination of the lifetime of technical construction and engineering structures. The study of its properties for civil engineering is as important as the study of steel for mechanical engineering. An understanding of microstructure–performance relationships is the key to a true understanding of material behaviors. Nevertheless, application of Non-Destructive Testing methods in civil engineering area is not easy because many building materials are heterogeneous. Verification of application possibilities of Acoustic Emission and Impact-echo Methods during stress situation will be the aim of this article. Both methods can describe material changes during its lifetime. Acoustic Emission Method monitors composite structure continuously and Impact-echo Method monitors one in discrete time. Changes in the whole specimen are recorded.

1   INTRODUCTION

Currently, the preservation of already built structures is of equal importance as the control of new constructions. Assessment of the health condition of existing structures is becoming more and more important. In civil engineering, materials exposed to stress or strain states a quantitative evaluation of damage which is of great importance due to the critical character of these phenomena, which at certain point abruptly turns into a devastating failure (Aitcin 2005). Efficient non-destructive quality control plays an important role in the optimization of resources for manufacturing, maintenance and safety.

1.1   *Acoustic Emission method*

The Acoustic Emission (AE) method offers the opportunity of continuous and in situ monitoring capabilities and the possibility to examine the whole volume of a structure simultaneously. The classic sources of AE's are defect-related deformation processes such as crack growth and plastic deformation. Without stress, there is no emission (Iturrioz et al 2014, Pollock 1989). Therefore, an AE inspection is usually carried out during a controlled loading of the structure. The released energy induces elastic waves that propagate inside the structure (Pazdera et al 2015, Ohtsu 1996). The emitted waves can be recorded at the surface of the structure by piezoelectric sensors and converted into acoustic signals. The AE signals (Figure 1) from one or more sensors are amplified and measured to provide data for display and interpretation.

Three main types of data analysis have been researched so far (Surgeon and Wevers 1999):

– An AE activity analysis focuses on measuring the amount of AE signals produced by a specimen or a structure. It primarily provides information about the initiation and the evolution of damage throughout a test or during the service life of a component.

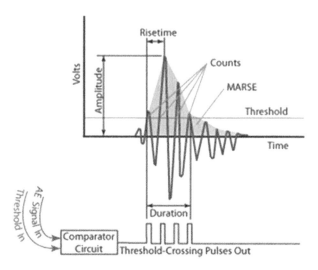

Figure 1. Typically AE signal (Iwanami et al 1997).

A good example is the well-known Felicity ratio analysis which is nowadays being used for structural health monitoring of fiberglass tanks, pressure vessels and pipes.

– An AE frequency analysis uses Fast Fourier Transform (FFT) techniques to calculate the frequency spectrum of AE waves. This approach is used for discrimination purposes, based on the assumption that different damage phenomena will produce signals with different frequency contents.

– An AE parameter analysis makes use of signal parameters like ring down counts, amplitude, energy and duration. This type of analysis has been mainly used for evaluating the severity of damage or for discrimination between different damage types and has been used in this paper.

AE activity is attributed to the rapid release of energy in a material, the energy content of the AE signal can be related to this energy release. The true energy is directly proportional to the area under the AE waveform (Sagar 2009):

$$AE\ energy_i = \int_{t_0}^{t_1} V_i(t)^2\, dt \tag{1}$$

where i = the recorded voltage transient V(t) of a channel; t_0 = the starting time of the voltage transient record; and t_1 = the ending time of the voltage transient record.

Various types of cracks generate different types of AE signals with varying parameters (e.g. frequencies, amplitudes etc.). These differences can be related to the degree of damage to the structure or material composition. Micro-cracks generate a large number of small amplitude events while macro-cracks generate fewer events but with higher amplitude. When the cracks are opening up, as most of the energy has already been released, many events are created, but with a small amplitude. Furthermore, tensile cracks spawn large amplitude events while shear cracks create smaller amplitude signals (Iwanami et al 1997, Li and Xi 1995). It is obvious that the crack propagation starts from the pre-defined notch tip.

1.2 *Method of Impact-Echo*

The Impact-Echo (IE) method is a useful a non-destructive technique for flaw detection in concrete. It is based on monitoring the surface motion resulting from a short-time mechanical impulse. This method overcomes many of the barriers associated with flaw detection

292

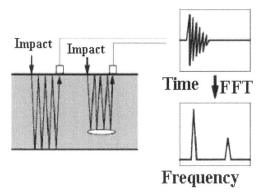

Figure 2. Principle of Impact-Echo method.

in concrete which occur at ultrasonic methods. One of the key features of this method is the transformation of the recorded time domain waveform of the surface motion into the frequency domain. The impact gives rise to modes of vibration and the frequency of these modes is related to the geometry of the tested object and the presence of flaws.

A short-time mechanical impulse, generated by tapping a hammer against the surface of a concrete structure (Figure 2), produces low-frequency stress waves which propagate into the structure (Kucharczyková et al 2010). Thus generated waves propagate through the specimen structure and reflect from the defects located in the volume of the specimen or on the surface. Surface displacements caused by the reflected waves are recorded by a transducer located adjacent to the impact (Carino 2001). The signal is digitized by an analog/digital data system and transmitted to a computer memory. This signal describes transient local vibrations, which are caused by the mechanical wave multiple reflections inside the structure. The dominant frequencies of these vibrations give an account of the condition of the structure, which the waves pass through.

The signal analysis from the impact-echo method is the most frequently performed by frequency spectra obtained from the FFT. Fourier analysis converts time to frequency and vice versa. An FFT is an algorithm to compute the Discrete Fourier Transform (DFT) and it is inverse. Results of FFT are widely used for many applications in engineering, science, and mathematics. The Eq. 2 is the expression of Fourier transform for a continuous function:

$$x(t) = \frac{a_0}{2} + \sum_{n=1}^{\infty} \left[a_n \cdot \cos\left(\frac{2\pi nt}{T}\right) + b_n \cdot \sin\left(\frac{2\pi nt}{T}\right) \right] \tag{2}$$

where a_n and b_n may be calculated from function $x(t)$ using following the relations

$$a_n = \frac{2}{T} \cdot \int_0^T x(t)\cos\left(\frac{2\pi nt}{T}\right) dt \quad b_n = \frac{2}{T} \cdot \int_0^T x(t)\sin\left(\frac{2\pi nt}{T}\right) dt \quad \omega = \frac{2\pi}{T} \tag{3}$$

and T is the time period.

2 EXPERIMENTAL PART

2.1 Acoustic emission method

During tests, an acoustic emission activity was recorded. Four acoustic emission sensors were attached to the surface by beeswax. Acoustic emission signals were taken by measuring equipment DAKEL XEDO with four acoustic emission sensors IDK-09 with 35 dB pre-amplifier.

The nominal dimensions of the beam specimens were $100 \times 100 \times 400$ mm; span length was 300 mm. Specimens had an initial central edge notch. The initial notch was made by a diamond blade saw before testing. Its depth was about 33 mm. Fracture tests were carried out using a Heckert FPZ 100/1 testing machine within the range of 0–10 kN at a laboratory of the Institute of Building Testing, Faculty of Civil Engineering, Brno University of Technology.

2.2 Method of impact-echo

For the IE method, a short-time mechanical impulse (a hammer blow) was applied to the surface of the specimen during the test which was detected by means of a piezoelectric sensor. The impulse reflects from the surface but also from micro-cracks and defects present in the specimen that is under investigation. The frequency analysis can be carried out from the response signal by means of FFT and thus dominant resonant frequencies are found. A MIDI piezoelectric sensor was used to pick up the response and the respective impulses were directed into the input of an oscilloscope TiePie engineering Handyscope HS3 two-channel with resolution 16 bits.

3 RESULTS

3.1 Acoustic emission method

During the three-point bending test from the viewpoint of acoustic emission method, a greater amount of water in the mixture is present so that the decreasing number of events and concurrently increasing the value of the amplitude of AE signals. Micro-cracks generate a large number of small amplitude events while macro-cracks generate fewer events but with higher amplitude.

3.2 Method of impact-echo

The graph shows that the addition of carbon nanotubes has a positive effect on the development of micro-cracks during setting alkali-activated mixture. The cracking is represented, decreased dominant frequency, it is considered that the higher the dominant frequency the more compact the structure of this type of sample.

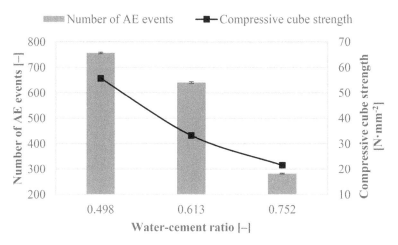

Figure 3.   The dependence of number of AE events and compressive cube strength on water-cement ratio.

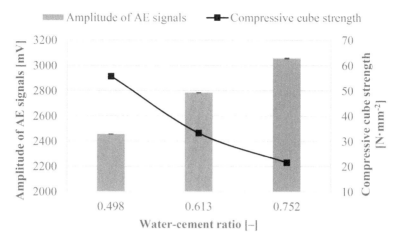

Figure 4. The dependence of amplitude of AE signals and compressive cube strength on water-cement ratio.

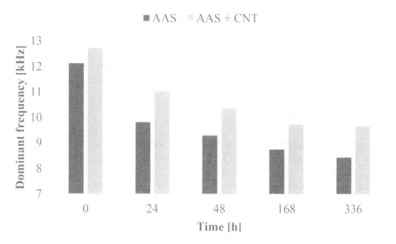

Figure 5. The evolution of the value of the dominant frequency during setting.

4 CONCLUSIONS

In this work, several experimental tests carried out on concrete specimens loaded up to failure are analyzed. Specimens with central notches were subjected to the three-point bending test. During the experiments, the AE technique was used to monitor the damage process taking place in the specimens. The value of water-cement ratio has a significant influence on monitored parameter signals of AE. In summary, the results of the experiments can be used to infer properties of materials as well as the specific characteristics of microcracks. The properties of the microcracks can ultimately be linked to the overall fracture behavior of the materials.

Concurrently was studied the application of the impact-echo method for detection of defects in composite materials during the different way of curing. It is known that the impact response signal of a specimen is composed of frequencies corresponding to the modes of vibration of the specimen. A shift of the dominant frequency to a lower value is a key indication of the presence of the flaw. From the results obtained in the framework of our research group and the results demonstrated in this paper it can be summarized that the frequency inspection carried out by means of the Impact-echo method makes a convenient tool to assess the quality and life of these composite materials when exposed to a stress situation.

ACKNOWLEDGEMENT

This paper has been worked out under the project GAČR No.16-02261S supported by Czech Science Foundation and the project No. LO1408 "AdMaS UP—Advanced Materials, Structures and Technologies", supported by Ministry of Education, Youth and Sports under the "National Sustainability Programme I" and under the project No. S-16-2967supported by Faculty; of Civil Engineering of BUT.

REFERENCES

Aitcin, P.C. 2005. *High Performance Concrete*, Informační centrum CKAIT, Prague, ISBN 80-86769-39-9.

Carino, N.J. 2001. Structures Congress and Exposition 2001, *Proceedings American Society of Civil Engineers*, May 21–23, Washington, DC, pp. 1–18.

Iturrioz, I. et al. 2014. Acoustic emission detection in concrete specimens: Experimental analysis and lattice model simulations, *International Journal of Damage Mechanics*, Vol. 23 (3), pp. 327–358, ISSN: 1056–7895, DOI: 10.1177/1056789513494232.

Iwanami, M. et al. 1997. Application of Acoustic Emission Technique for Crack Monitoring in RC beams, JCA Proc, *Cement and Concrete* (1997) 51, pp. 192–197.

Kucharczyková, B. et al. 2010. Determination and evaluation of the air permeability coefficient using Torrent Permeability Tester, *Russian Journal of Nondestructive Testing*, 46(3), pp. 226–233, DOI:10.1134/S1061830910030113.

Li, Z. and Xi, Y. 1995. Application of acoustic emission technique to the detection of concrete cracking and rebar corrosion, Berlin, *NDT-CE: Int. Symposium Non-Destructive Testing in Civil Engineering*, pp. 613–620.

Ohtsu, M. 1996. The history and development of acoustic emission in concrete engineering, *Magazine of Concrete Research*, 48 (1996), pp. 321–330.

Pazdera, L. et al. 2015. Nondestructive Testing of Advanced Concrete Structure during Lifetime, *Advances in Materials Science and Engineering*, vol. 2015, Article ID 286469, p. 5, DOI: 10.1155/2015/286469.

Pollock, A.A. 1989. Acoustic Emission Inspection, *Metals Handbook*, Ninth Edition, Vol. 17, ASM International, pp. 278–294.

Sagar, R.V. 2009. An experimental Study on Acoustic Emission Energy and Fracture Energy of Concrete, India, *National Seminar & Exhibition on Non-Destructive Evaluation*, pp. 225–228.

Surgeon, M. and Wevers, M. 1999. Modal analysis of acoustic emission signals from CFRP laminates, *NDT & E International*, Vol. 32 (6), pp. 311–322, ISSN: 0963–8695, DOI: 10.1016/S0963-8695(98)00077-2

Advances and Trends in Engineering Sciences and Technologies II – Al Ali & Platko (Eds)
© 2017 Taylor & Francis Group, ISBN 978-1-138-03224-8

University BIM distance learning course for secondary school students

K. Usanova & N. Vatin
Peter the Great Saint-Petersburg Polytechnic University, St. Petersburg, Russian Federation

ABSTRACT: The Building Information Model Distance Learning Course was created for University applicants. The course is a part of pre-university education and a tool of University online marketing. We use Peter the Great Saint-Petersburg Polytechnic University Virtual learning environments that based on Moodle. This allows secondary school students to prepare themselves to the University's learning process and get them an outlook on the right choice of major. They use Autodesk Revit software for building and construction design. The results of the course are a three-dimensional model of a small two-storey residential building, presentation graphics, and a set of drawings in accordance with Russian construction standards. The training mission is actually a simplified version of the educational task of the first year students in Computer graphics discipline. Course graduators subsequently become successful Civil Engineering students.

1 INTRODUCTION

The construction industry needs graduate engineers with knowledge and skills in Building Information Modeling (BIM) (Cheng et al., 2016). However, not all of universities have included BIM topics into their curricula for construction engineering students (Pikas et al., 2013). A BIM course is significant for civil engineering majors (Shenton et al., 2014). Teaching BIM is important in students research projects, PhD theses and MSc dissertations (Sampaio, 2014, Kim, 2014). Recent surveys showed that university level BIM education of the construction engineering increased in the past several years (Usanova et al., 2014, Liu and Hatipkarasulu, 2014, Shenton et al., 2014).

The article is devoted to BIM Distance Learning Course not for university students but for university applicants (secondary school students). Previously we report about "Academy of Construction for University Applicants" as a tool of university online marketing (Gamayunova O., 2012). It is a distance course in Peter the Great St. Petersburg Polytechnic University (Russian Federation). The course posted on university Virtual learning environment (http://dl.spbstu.ru) based on Moodle (Tuchkevich et al., 2015).

2 DISTANCE LEARNING COURSE DESCRIPTION

The university organized "Academy of Construction" as a free distance-learning course. The course based on Autodesk Revit, which is actively used in the construction industry (Razumova and Slabnova, 2015). Students study the 3D BIM-technology parametric modeling of building on the example of a two-story residential building in this course. The creator of the course and the director of Academy of Construction is Kseniia Usanova.

Students start their work with a motivational letter that they send to the teacher. Teachers verify that the applicant is really a high school student and that he is motivated to study in the course. Thereafter, the student receives detailed instructions to install Revit program on his home computer. If necessary, the teacher gives advice on installation. Before the beginning of individual work student has the opportunity to download examples of the 3D models and samples of project presentations.

A students sequentially performs the assignments one after the other:

– creation of a new project;
– project levels;
– grid axes plan;
– multilayer external walls and partition walls of the first floor;
– foundation and slab;
– windows and doors;
– stairs; second floor (similar to the first);
– balcony rails;
– roof (including over porches and terraces, canopies);
– roof windows and ventilation ducts;
– decorative and additional elements (roofing system, rust, architraves, cornices);
– landscaping;
– photorealistic image and the final project presentation.

The student has the opportunity to perform a new assignment only after the teacher approves the previous assignment. The student receives teacher's comments on mistakes in the assignment. The student could also ask questions in the forum, which is a part of the Distance Learning Course.

The course concludes with a final test and a feedback form.

Course graduators receive a certificate of attendance. Those who have the opportunity to come to St. Petersburg could receive a certificate at the university in a festive atmosphere. Authors of the best projects receive awards. The Construction Committee of St. Petersburg Government supports Academy of Construction. Author of the best project will receive a certificate from the hands of the vice-governor of St. Petersburg. Information about the authors and the projects is placed on the social networks.

Two samples of building project photorealistic images are on Figure 1.

We view Academy of Construction not only as a marketing tool to attract students for university, but as tool to select and attract the most motivated persons.

Our team looked at the group of 120 Academy of Construction enrolled students. The analysis includes a look at the number of students who passed certain assignments. They analyzed the results of students' final test of the Academy of Construction course.

After that, we looked at the educational results of the first year students on the Civil Engineering BSc program in Peter the Great Saint-Petersburg Polytechnic University. Some of those students are Academy of Construction graduators.

The analysis based on the so-called "total score" of the Russian Unified National Exams (UNE). These exams are required for secondary school graduating classes' students. The admission to the Civil Engineering BSc program is done by the UNE results in mathematics, physics, and Russian language. Each exam is estimated on a 100-point scale. Thus, the maximum score was 300, the minimum amount of enrolled Civil Engineering BSc students is 125 points. Those enrolled with a score of 236 or higher study the Civil Engineering in university at the expense of the Russian Federation federal budget. Those with a lower score—at their own expense or at the expense of the federal subsidies for multiple-child families.

The Ministry of Education and Science of the Russian Federation uses the UNE average score as the estimation of admission marketing appeal of the Bachelor program and the university (Gamayunova, 2015, Tatuev et al., 2015).

We analyzed 12 academic groups of first-year students of the academic year 2015/2016. All of these students studied in the same department. The name of the department is Construction of Unique Buildings and Structures. The sum-total number of students is 234 people. Students of all groups have equal opportunities for education (list of discipline, assignments, teachers, number of hours, etc.). We analyzed how successful was their passing of the first semester depending on whether they graduated from the Academy of Construction.

Figure 1. Two samples of a building project photorealistic images.

3 RESULTS OF EDUCATION

Distance Learning Course is an effective filter, allowing to pass through only to the most motivated students. Only a small part of the students originally enrolled in the course were able to finish it. As students progressed through the course, their number decreased. It can be seen from the following data of students in the control group. From the total number of 120 initially enrolled in the course; 40 students—passed the 50% mark; 30 students—passed the 75% mark; only 20 students—graduated (Figure 2).

Distance Learning Course graduates achieve good results of the final test of the Distance Learning Course (Figure 3). Data on the distribution of students on the UNE scores presented on the Figure 4. Evidence approximated by the normal law of distribution. The x-axis shows the amount of UNE score.

Curve 1 shows the initial distribution of students on a score with highs in the value of 220. After the first semester 32 students (14% of the total number) were expulsed from the university by the result of training. The reason was the failure in university exams. Curve 2 shows distribution of expelled students with a distribution maximum at 177. Curve 3 shows

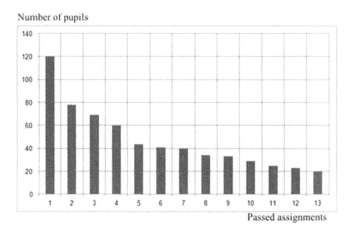

Figure 2. Passed assignments of the distance learning course.

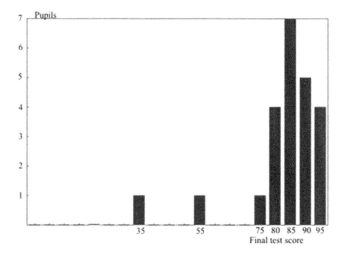

Figure 3. Results of the final test of the distance learning course.

the final distribution after expulsing with a maximum allocation at 226. It is evident that students with a low number of UNE scores are more prone to dismissal. Curve 4 represents the distribution of UNE scores of students who graduated from the Academy of Construction. None of these students was expulsed from the university by the result of training in the first semester. The average UNE score of students is 246. All students who have graduated from the Academy of Construction completed training in Engineering Graphics with good results.

As seen in Academy of Construction assignments listed above, achieving mastery of the BIM Distance Learning Course for University Applicants is not a means of preparing for exams in mathematics, physics and Russian language and does not directly affect the results of these exams. Therefore, the construction of the Academy has proved effective as a means of selecting and collecting applicants that have good training in physics and mathematics and high motivation to learn.

In 2015 (as in previous years), the Polytechnic University won the first place in Russia on the enrolled students' average score of the Russian Unified National Exams (UNE) in a branch of Civil Engineering. Of course, it is not only the result of the "Academy of Construction".

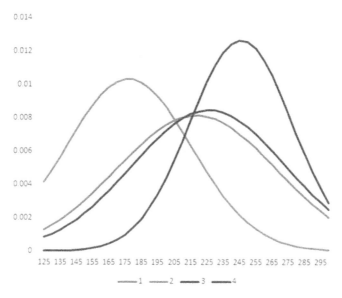

Figure 4. Evidence approximated by the normal law of distribution. The x-axis shows the amount of UNE score. 1—All enrolled students, 2—Expulsed students, 3—Continuing education in 2nd semester; 4—Academy of Construction graduates.

4 CONCLUSIONS

We have come to the following conclusions:

- BIM Distance Learning Course for University applicants can be successfully used for secondary school students.
- The course is an outstanding marketing tool to attract applicants to the Institute of Civil Engineering.
- Graduates of the course are among the most successful students of the university first year education.

ACKNOWLEDGEMENT

The work reported in this paper has been supported by the project: MARUEEB ERASMUS + PROJECT 561890-EPP-1-2015-1-IT-EPPKA2-CBHE-JP.

REFERENCES

Gamayunova, O. 2015. The role of Civil Engineering Institute in increasing the international competitiveness of the St. Petersburg State Polytechnical University C3—Procedia Engineering. 117, 1070–1077.

Gamayunova, O., R. A., Usanova K. 2012. Fundamental and polytechnical experience of construction education with using Moodle. *Construction of Unique Buildings and Structures*, 2, 6–17.

Kim, J. L. 2014. Effectiveness of green-BIM teaching method in construction education curriculum C3—ASEE Annual Conference and Exposition, Conference Proceedings.

Liu, R. & Hatipkarasulu, Y. 2014. Introducing building information modeling course into a newly developed construction program with various student backgrounds C3—ASEE Annual Conference and Exposition, Conference Proceedings.

Pikas, E., Sacks, R. & Hazzan, O. 2013. Building information modeling education for construction engineering and management. II: Procedures and implementation case study. *Journal of Construction Engineering and Management*, 139.

Razumova, T. V. & Slabnova, N. N. 2015. Putting into practice 3D designing on the base of information modeling in "Gipronickel Institute" OJC. *Tsvetnye Metally*, 2015, 95–98.

Sampaio, A. Z. 2014. The disclosure of the BIM concept in civil engineering training C3—High Value Manufacturing: Advanced Research in Virtual and Rapid Prototyping—Proceedings of the 6th International Conference on Advanced Research and Rapid Prototyping, VR@P 2013. 549–554.

Shenton, H. W., Iii, Conte, P. R. & Bonzella, J. 2014. A first course in BIM for civil engineering majors C3—Structures Congress 2014—Proceedings of the 2014 Structures Congress. 1097–1105.

Tatuev, A. A., Edelev, D. A. & Zhankaziev, A. H. 2015. The state unified exam as a requirement in Russia's new economic relations. *Asian Social Science*, 11, 176–184.

Tuchkevich, E., Rechinsky, A., Vysotskiy, A., Zolotova, J. & Tuchkevich, V. 2015. ADN and AP programs for civil engineering students C3—Procedia Engineering. 117, 1142–1147.

Usanova, K., Rechinsky, A. & Vatin, N. 2014. Academy of construction for university applicants as a tool of university online marketing. *Applied Mechanics and Materials*, 635–637, 2090–2094.

Advances and Trends in Engineering Sciences and Technologies II – Al Ali & Platko (Eds)
© 2017 Taylor & Francis Group, ISBN 978-1-138-03224-8

The numerical analysis of deteriorated steel elements reinforced with CFRP

J. Vůjtěch, P. Ryjáček & M. Vovesný
Faculty of Civil Engineering, Czech Technical University in Prague, Prague, Czech Republic

ABSTRACT: As the current transport infrastructure grows older, there is a rising need to find an easy and effective way to rehabilitate existing steel bridges and to extent their fatigue life. As corrosion is almost always present on old bridges, the stress increases close to the corrosion pits and can lead to the creation of fatigue cracks. One of the most promising new method of repair is the use of Carbon Fiber Reinforced Polymers (CFRP). The paper describes the results of the numerical analysis of the deteriorated steel elements reinforced with CFRP. Several numerical models in Abaqus software were created with different level of corrosion. The difference in the stress level and distribution in the specimens with and without the CFRP will be analyzed. Even a small decrease can significantly extend the remaining fatigue life. The results will serve as a basis for further experimental research works.

1 INTRODUCTION

There is a rising need to find affordable solutions for repairing, retrofitting and strengthening old steel structures. This is especially the case with old steel bridges as many of them have already passed their originally expected service life. The considerable age brings a variety of issues for the infrastructure owner. In many cases, existing corrosion increases the stresses in the steel structure. This often results in the creation and propagation of the fatigue crack, which can be potentially extremely dangerous. If conventional repair (welding, replacement) is not possible for various reasons, the member can be reinforced in order to increase the remaining life time.

The fast repair of the crack is usually difficult, especially when the structure has to be repaired under load. This is when externally bonded FRP can be fast and easy solution, on how to extend the service life of the structure, especially when welding of loaded steel element or replacement is not possible, as shown in Wu et al., 2012; Pipinato et al., 2012; Colombi et al., 2003; Colombi et al., 2006; Zhao et al., 2007 and Tavakkolizadeh et al., 2003. Although several experimental studies on strengthening steel members were evaluated, only little is known about that strengthening behaviour over long term period, to which especially bridges are exposed.

The FRP material consists of two main components, whose parameters together form a new material of very high strength and durability. The matrix is usually formed by polyester, vinyl-ester or epoxy glue. The matrix is reinforced by fibres (glass, carbon, kevlar etc.). For the reinforcement of steel members, only carbon fibres are suitable, because their modulus of elasticity is similar to that of steel. However, steel carbon has to be separated by epoxy resin or other means in order to prevent galvanic corrosion.

2 NUMERICAL STUDY OVERVIEW

This numerical study forms as a basis and a preparation for series of experiments. The goal of those experiments is to study the influence of carbon FRP reinforcement on fatigue life of deteriorated steel. The state of the steel can play a significant role in establishing the fatigue life. As is shown in experiments, e.g. Macho et al., 2016, corrosion pits lead to localised plasticization in the defect even for relatively small loads.

This topic is of course not entirely new, there were already several studies into the effect of CFRP reinforcement on remaining fatigue life and even several national design guidelines such as CIRIA Design Guide (CIRIA Design Guide), US Design Guide (Schnerch et al., 2007) and document CNR-DT 202/2005 (CNR, 2005). However not everything was covered in those and much additional research is needed to fully understand the behaviour of reinforced elements in all possible scenarios and under severe environment. It is also worth considering, that mainly modern steel was used in the existing research.

There are two types of elements in this study. Each one with different type of fault. They correspond to two common types of damage that can be found on old steel structures. The first is heavy corrosion in the form of corrosion pits. In this case the corrosion is idealised so the corrosion pits are symmetrical. The second is heavy corrosion in the form of corrosion hole. That forms either with corrosion going through the whole element or around a rivet.

The elements are plain steel plates reinforced on both sides. The reinforcement is covering the faults and it is not prestressed. Although as shown in Taljsten et al., 2009 and in Pipinato et al., 2012 the prestressing of the FRP can have huge positive impact on the fatigue life, it is not included in the following experiments. The reason being that we are looking for method of repair that is as easy and uncomplicated as possible. The stress peak in corrosion pits and on the side of the corrosion hole is measured and then the stresses with and without the reinforcement are compared. Reduction in the stress is than compared with the appropriate Wöhler curve and the conclusion is made about how effective is the reinforcement. This is done for different volumes of reinforcement and different types of carbon fibre.

3 PREPARED ELEMENTS

The real dimensions and materials of the test specimens were used for the numerical study. The overall geometry is same for both types of specimens. The specimen is a steel plate 500 mm long,

Table 1. Material properties of used steel material.

Material	Elastic modulus (GPa)	Shear strength (MPa)	Tensile strength (MPa)
Pre 1900 Mild steel	200	77	320

Table 2. Material properties of used carbon fibre textiles.

Textile	Area weight (g/m²)	Thickness (mm)	Binding	Tensile strength (MPa)	Elastic modulus (GPa)
CC160P	160	0.3	plain	3400	230
CC160T	160	0.34	twill	3400	230
CC200P	200	0.32	plain	3400	230
CCA200T	200	0.38	twill	3800	240
CC600T	600	0.9	twill	3400	230
Style 461	80	0.12	plain	3400	230
Style 493	68	0.1	plain	3400	230

Table 3. Material properties of used resin.

Material	Elastic modulus (GPa)	Tensile strength (MPa)
Epoxy	90–110	3.5

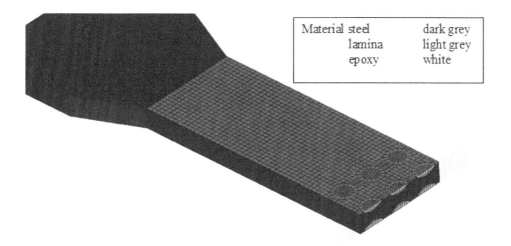

Material steel	dark grey
lamina	light grey
epoxy	white

Figure 1. The numerical model for fault type 1, element cut in half.

12 mm thick and with width of 60 mm in the tested part and 95 mm at the ends. Both faults are in the middle of the specimen. Fault 1 consists of 9 simulated circular corrosion pits 3 mm deep with diameter of 15 mm on both sides of the specimen (see Figure 1). Fault 2 is a single circular hole with surface diameter of 41 mm and inner diameter of 22 mm.

The FRP reinforcement is applied on both sides of the specimen. The wet layup process is used. The FRP patches are 300 mm long and 60 mm wide.

4 NUMERICAL MODEL DESCRIPTION

The numerical model is created using the **ABAQUS** program. There are two sets of models created, each set for different fault type. In each model there are 3 different materials see Figure 1 (ilustration of type 1). The main material is steel, it is modeled as an isotropic elastic material and as a 3D solid 6-node linear triangular prism elements (C3D6). The epoxy that fills the faults (pits or hole) is also modeled as a 3D solid 6-node linear triangular prism elements (C3D6) and material is also isotropic and elastic. The lamina is created as a 2D linear 4-sided shell element (S4R). Its material type is LAMINA with different number of layers depending on volume of reinforcement. The laminate is connected to the steel rigidly. Although this does not represent the real behavior accurately as it does not take into account slip resistance and failure by debonding, it is a close enough estimation of the fatigue behavior. It occurs with small strain, that does not activate the slip or debonding resistance.

One side of the element is fixed and the other side is used for the force application. The force differs according to the fault type. It is 100 kN for type 1 and 32 kN for type 2 respectively.

5 MAIN RESULTS OF THE ANALYSIS

5.1 *Behaviour of the reinforced elements*

The stress distribution on unreinforced elements for the fault type 1 can be seen in Figure 2. There are peaks in stress at the bottom of corrosion pits almost reaching yield strength of the steel. Similarly, the stress peaks are present also for fault type 1, the corrosion hole. When the reinforcement is applied, a substantial reduction in stress peaks is observed, as shown in Figure 3. Here the element is reinforced with 5 layers of CFRP roughly 2.5 mm thick on both sides. The stress is reduced to ca 75% of its original value for the fault 1. In total numbers the reduction is 55 MPa. As the results for fault type 2 are nearly identical (fault type 2 seeing

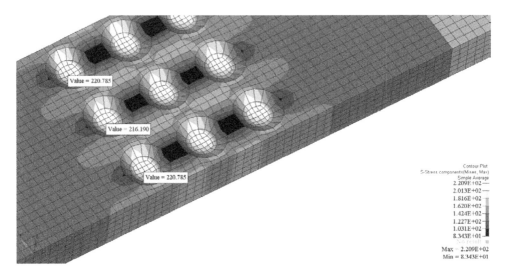

Figure 2.   Stress distribution for fault type 1 element, no reinforcement.

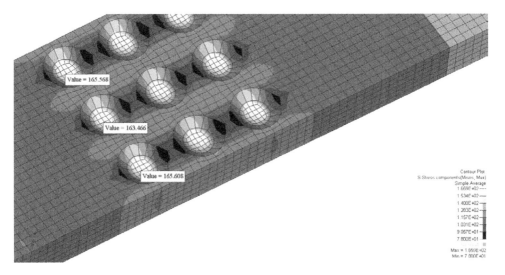

Figure 3.   Stress distribution for fault type 1 element, reinforcement 5 layers of CC160P.

slightly bigger reduction in stress), the conclusions about behaviour of the elements are valid for both types of faults.

5.2   *Reduction in stress*

The total reduction in stresses based on the level of reinforcement for the fault type 1 can be seen in Figure 4. On the *x* axis is ratio (SR) between the volume of reinforcement normalized to the elastic modulus and the cross-sectional area of the element.

$$SR = \frac{A_{CFRP} * E_{CFRP}}{A_S * E_S} \tag{1}$$

Where A_{CFRP} is the area of the CFRP strengthening, E_{CFRP} is modulus of elasticity of strengthening, A_S and E_S are the area and modulus of elasticity of steel respectively.

306

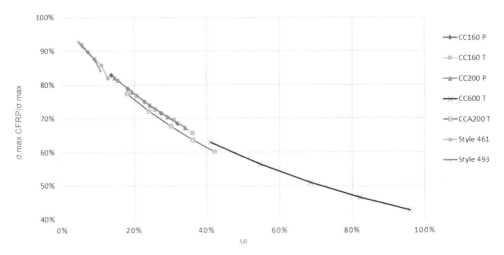

Figure 4. Fault type 1—reduction in maximum stress based on volume of reinforcement.

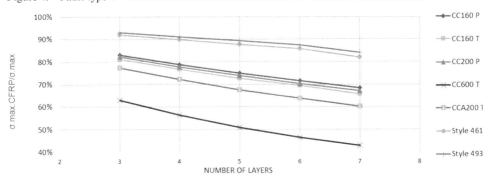

Figure 5. Fault type 1—number of layers.

On the y axis is the ratio of peak stresses of reinforced and unreinforced element. Each line represents different type of CFRP textile. Those differ in the amount of carbon used, in the elastic modulus of fibres or in the orientation of the fibres. Most of the textiles are two directional with plain binding.

5.3 *Estimated extension of fatigue live*

For determining the estimated life extension, the UIC 778-2 R leaflet is used. Its topic is the determination of remaining fatigue life for existing steel bridges. The adequate fatigue prone detail for this experimental setup and type of steel is detail WII with permissible stress range of 85 MPa. As an example to show the expected life extension a specimen with fault type 1 and reinforced with 3 layers of CC160 textile on both sides (corresponds to ca 14% of steel area). This specimen is chosen as the textile is one of the most commonly used and thus available.

The original stress range is 100 MPa and that corresponds to 1.2 mil. load cycles. After the reinforcement the maximum stress lowers to ca 83% of the original, 83 MPa in this case. The stress range of 83 MPa corresponds roughly to 2.3 mil. load cycles. This represents increase of 1.1 mil. load cycles and should effectively double the life expectancy.

6 CONCLUSION

In this paper the stress distribution on deteriorated steel plate with or without CFRP reinforcing was analysed. The commercial software **ABAQUS** was used to perform the analysis.

There are two sets of three dimensional models based on the configuration of two types of future test specimens. This numerical analysis will serve as a basis for future experiments and some simplifications have been made to cut down on the complexity, deemed unnecessary at this stage of the project. The reduction of peak stresses in or around the faults depending on the volume and material of the carbon fibre composite was analysed and the following conclusions can be made.

Out of the two fault types the more significant is type 1—corrosion pits. The type one is usually more common in real life. Corrosion is scarcely allowed to develop to full corrosion hole. Also, as mentioned in section 5.1, the results of both types are nearly identical so conclusions are made for both of the fault types.

The stress reduction is not linearly dependant on the volume of the reinforcement. The relation is more exponential. As seen in Figure 4 and Figure 5, with the growth in of reinforcement the value of max stress ratio gradually flattens. So using more reinforcement makes sense only up to a certain value of the volume ratio. This value should be around 40%, because higher values mean using either much more or a much higher quality of fibres and subsequent rise in costs.

The cheaper and most common carbon fibre textiles offer good results. They offer stress reduction to values between 80% to 65% of the original (that is for type 1). As was shown in section 5.3, reduction of peak stress can lead to doubling of the fatigue life expectancy.

ACKNOWLEDGEMENTS

Research reported in this paper was supported by COST CZ (LD) project of the Ministry of Education, youth and sports LD (No. LD15131) "Fatigue behaviour of FRP reinforced steel members under severe environment".

REFERENCES

CNR Italian Advisory Committee on Technical Recommendations for Construction. Guide for the designand construction of externally bonded FRP systems for strengthening existing structures. FRP systems for strengthening existing structures. Preliminary study. Metallic structures. 2005.

Colombi, P., Bassetti, A., Nussbaumer, A.: Analysis of cracked steel members reinforced by pre-stress composite patch. Fatigue & Fracture of Engineering Materials & Structures. Vol. 26. pp. 59–66. 2003.

Colombi, P., Poggi, C.: Strengthening of tensile steel members and bolted joints using adhesively bonded CFRP plates. Construction and Building Materials 20, pp. 22–33. 2006.

Macho, M., Ryjáček, P.: The impact of the severe corrosion on the structural behavior of steel bridge members (2016) Advances and Trends in Engineering Sciences and Technologies—Proceedings of the International Conference on Engineering Sciences and Technologies, ESaT 2015, pp. 123–128. 2016.

Pipinato, A., Pellegrino, C., Modena, C.: Fatigue Behaviour of Steel Bridge Joints Strengthened with FRP Laminates. Modern Applied Science; Vol. 6, No. 10; 2012.

Ryjáček, P., Vovesný, M.: Application of FRP composites for decks of temporary bridges Proceedings of the 12th International Conference on Steel, Space and Composite Structures, pp. 319–325. 2014.

Taljsten B, Hansen C.S., Schmidt J.W.: 'Strengthening of old metallic structures in fatigue with pre-stressed and non-prestressed CFRP laminates', Construction and Building Materials, 23(4), pp. 1665–1677. 2009.

Tavakkolizadeh, M., Saadatmanesh, H.: Fatigue Strength of Steel Girders Strengthened with Carbon Fiber Reinforced Polymer Patch. Journal of structural engineering. pp. 186–196. February 2003.

Tavakkolizadeh, M., Saadatmanesh, H.: Galvanic corrosion of carbon and steel in aggressive environments. J. Compos. Constr., vol. 5, pp. 200–210. 2001.

Xiao-Ling Zhao, X., Zhang, L.: State-of-the-art review on FRP strengthened steel structures. Engineering Structures 29, pp.1808–1823. 2007.

Wu, Chao, Zhao, Xiaoling, Duan, Wenhui, Emdad, Mohammad R., Al-Mahaidi, Riadh.: Fatigue of center cracked steel plates with UHM CFRP plate strengthening. International Institute for FRP in Construction; 2012.

Basic features of aggregate-matrix-interface fracture of concrete: Pilot modelling

M. Vyhlídal, H. Šimonová, V. Veselý & Z. Keršner
Faculty of Civil Engineering, Brno University of Technology, Brno, Czech Republic

J. Klusák & L. Malíková
Institute of Physics of Materials, Academy of Sciences of the Czech Republic, Brno, Czech Republic

ABSTRACT: In this paper, the attention is paid to investigation of the importance of the Interfacial Transition Zone (ITZ) in concrete for the global fracture behaviour. A simplified cracked geometry (consisting matrix, ITZ and aggregate) is modelled by means of the finite element method with a crack terminating at the matrix-ITZ interface. Numerical studies assuming two various ITZ thicknesses and several various ITZ elastic moduli are performed. Based on the values of the opening stress ahead of the crack tip (its average value and stress range) a few conclusions are discussed. The pilot analyses dealing with the effect of ITZ on the stress distribution should contribute to better description of toughening mechanisms in silicate-based composites.

1   INTRODUCTION

Among the most frequently used building materials one can primarily name silicate-based, particularly cement-based composites (Nevile, 2011). Thanks to their character and to the production technology, these traditional materials are very adaptable and utilizable for a wide range of applications. As it is well known, matrix of these composites—hardened silicate/cement paste—shows nearly brittle behaviour, while concrete like composites show significant non-linear response. This is caused by the presence of aggregates in the matrix resulting in activation of various toughening mechanisms. This paper is focused on the investigation of the influence of the Interfacial Transition Zone (ITZ) between Matrix (MTX) and Aggregate (AGG) on the stress stateto better understanding of fracture processes of these quasi-brittle materials. As Scrivener et al. (2004) mentioned, the non-linear response of concrete is due to the development of multiple microcraking predominantly in the ITZ.

2   INTERFACIAL TRANSITION ZONE

The importance of the interfacial transition zone between cement paste and aggregates and its influence on the behaviour of concrete has already been studied for decades from different points of view both experimentally and numerically. Geometric and materials properties of the ITZ influence on the mechanical and transport properties of concrete are well documented in the literature; see e.g. (Rilem, 1998; Scrivener et al., 2004; Ollivier et al., 1995; Prokopski & Halbiniak, 2000).

   Number of publications concerning the experimental and numerical investigations of ITZ is connected with homogenization of overall mechanical properties of concrete on mesoscale level. Lesser of them are devoted to toughening mechanisms; see e.g. (Wriggers & Moftah, 2006).

Investigation of concrete fracture is connected with the recognition that its structural behaviour is controlled not only by compressive strength but also by other independent material parameter—fracture toughness; see e.g. Li & Maalej (1996), Shah (1990), Merchant et al. (2001). Fracture mechanics based analytical-numerical approaches, mainly connected with finite element codes are widely used to simulate structural response by explicitly accounting for cracks development in the structure (Li & Maalej, 1996). The investigation of toughening mechanisms of cement-based composites is motivated by the need to control the toughness property of the material under consideration, especially in connection with the brittleness associated with high strength concrete (Shah, 1990).

Summary of toughening mechanisms in cement-based composite is given e.g. in paper by Li & Maalej (1996), where three basic types of fracture manners, namely frontal, crack tip and wake processes are considered and their contribution to concrete toughness is assessed. In the case of concrete, the frontal processes may include microcrack-shielding, the wake processes include mainly crack-face pinning and aggregate-ligament bridging. Typical crack tip toughening processes correspond to crack deflection and crack front trapping, see e.g. Cox & Marshall (1994), Bower & Ortiy (1991), Zhang & Xu (2007). Usually, the wake processes provide the greatest amount of toughening (Li & Maalej, 1996).

One of the most important features of ITZ is its higher porosity. The increase in porosity leads to the decrease in strength which is presented in numerical model by Young's modulus. It means that the value of Young's modulus of the ITZ should be lower than the modulus of the bulk paste. It leads to the common view of the ITZ as the weak link in concrete and it is useful to deal with this area.

The influence of ITZ on crack behaviour can also be studied by means of generalized fracture toughness approaches. (Leguillon, 2002; Kotousov, 2005) The problems of the interaction of crack and interfaces can be evaluated by means of controlling quantities as described in Knésl et al. (2007–2008), Klusák et al. (2013).

3 NUMERICAL MODELLING

To determine clearly the impact of the ITZ, a simplified model of the cracked specimen was considered. Particularly, the effects of the vertical position of inclusion, size of the aggregate

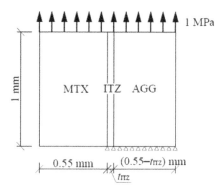

Figure 1. Schema of the cracked specimen investigated; the crack terminates at the interface between MTX and ITZ.

Table 1. Elastic constants of materials.

Label	Material	E [GPa]	v [-]
AGG	Aggregate (basalt)	60	0.2
MTX	Matrix (cement paste)	30	0.2
ITZ	Interfacial transition zone	10–60	0.2

and its circular shape are ignored; see the schema in Figure 1. Note that the configuration assumed is based on the three-point bending fracture test of a beam with central edge notch.

Numerical model was created in ANSYS software. Materials were modelled as linear, elastic and isotropic, which are represented by their elastic constants, i.e. Poisson's ratio and Young's modulus (Table 1).

The crack was modelled by the boundary conditions with its tip at the interface between MTX and ITZ (Figure 1). Moreover, two submodels were considered with various values of the ITZ thickness (50 and 32 μm).

4 RESULTS

For quantitative description of the influence of ITZ on the stress state in the crack tip vicinity the opening stress σ_{yy} is observed and evaluated. The mean stress $\overline{\sigma}_{yy}$ and the stress range $\Delta\sigma_{yy}$ are calculated

$$\overline{\sigma}_{yy} = \frac{1}{d} \cdot \int_0^d \sigma_{yy}(x, y = 0)dx \text{ and } \Delta\sigma_{yy} = \sigma_{yy,\max}^{AGG} - \sigma_{yy,\min}^{ITZ},$$

where d is a size of region, where the stress is averaged.

4.1 The ITZ thickness 50 μm

The dimension 50 μm represents an average ITZ's thickness as observed in Scrivener et al. (2004). In Figure 2, there isthe opening stress σ_{yy} shown in dependence on the distance from

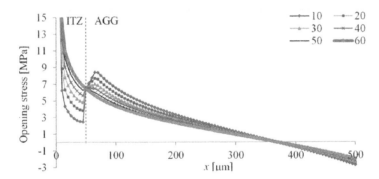

Figure 2. Distribution of the opening stress σ_{yy} in ITZ and AGG for the specimen with the crack tip at the MTX-ITZ interface for various E_{ITZ} values (between 10 and 60 GPa); ITZ thickness of 50 μm.

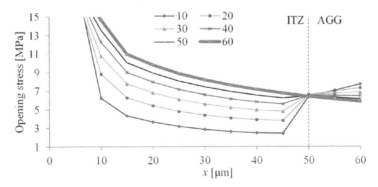

Figure 3. Detail of the opening stress σ_{yy} near the ITZ-AGG interface for various E_{ITZ} values (between 10 and 60 GPa); ITZ thickness of 50 μm.

the crack tip. The value $x = 0$ refers to the crack tip. The step change of the stress σ_{yy} is apparent at the interface between ITZ and AGG. Detailed view of the distribution near the interface is shown in Figure 3.

As the stress at the crack tip goes to infinity, the average value of σ_{yy} can give good description of the crack severity. Here the average value σ_{yy} is evaluated over the whole ITZ thickness ($d = 50$ µm) and is stated in Table 2. There is also the stress range which describes the influence of the interface between ITZ and AGG to which the crack approaches. Relative values (dimensionless) of average stress are also included in Table 2—reference value 1.00 represents the case of $E_{ITZ}/E_{MTX} = 1$.

4.2 Comparison of results and their discussion

In the following, the results obtained for both configurations with various ITZ thicknesses are compared and discussed. The graphical expression of the dependences can be seen in

Table 2. Stress values for various elastic mismatches considering the ITZ thickness of 50 µm.

E_{ITZ}/E_{MTX}	Average stress [MPa]	Stress range [MPa]	Relative value of average stress [–]
0.3	4.9	5.9	0.57
0.7	7.3	3.9	0.85
1.0	8.6	2.2	1.00
1.3	9.4	0.9	1.10
1.7	10.2	0.2	1.18
2.0	10.8	0.0	1.25

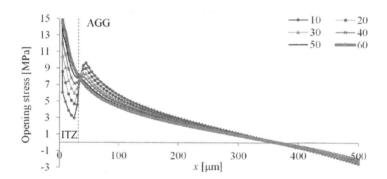

Figure 4. Distribution of the opening stress σ_{yy} in ITZ and AGG for the specimen with the crack tip at the MTX-ITZ interface for various E_{ITZ} values (between 10 and 60 GPa); ITZ thickness of 32 µm.

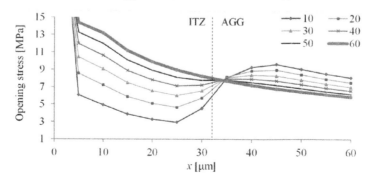

Figure 5. Detail of the opening stress σ_{yy} near the ITZ-AGG interface; ITZ thickness of 32 µm.

312

Table 3. Stress values for various elastic mismatches considering the ITZ thickness of 32 μm.

E_{ITZ}/E_{MTX}	Average stress [MPa]	Stress range [MPa]	Relative value of average stress [–]
0.3	5.5	6.7	0.58
0.7	8.1	4.3	0.84
1.0	9.6	2.4	1.00
1.3	10.5	0.8	1.09
1.7	11.7	0.0	1.22
2.0	12.3	0.0	1.28

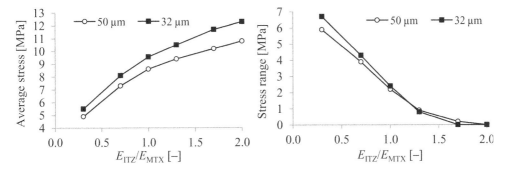

Figure 6. Values of the mean stress and stress range for various elastic module ratios and ITZ thicknesses.

Figure 6 where the values of the average stress and stress range, respectively are plotted in dependence on the elastic mismatch between ITZ and MTX.

The plots in Figure 6 clearly show that there exist differences between the stress values when various ITZ thicknesses are considered. It holds, the thinner ITZ the higher average stress level, which can be explained by the shorter distance ahead of the crack tip where the opening stress can decrease. Similarly, it can be observed that the stiffer ITZ leads to the higher opening stress. Therefore, the more compliant ITZ seems to be more advantageous; it decreases the stress level but in a general cracked configuration it can also attract the crack in order to arrest it or at least blunt it as a kind of the toughening mechanisms. Thus, the elementary conclusions of these pilot numerical simulations shall be used for future more advanced assessment of fracture behaviour of silicate-based composites with a crack near the aggregate-matrix interface.

5 CONCLUSIONS

The authors focused their attention on the evaluation of the influence of the Interfacial Transition Zone on the stress distribution ahead of a crack tip approaching the aggregate. The crack tip was located at the interface between matrix and ITZ. As the average stress is used for stability criterion suggestion in cases of general singular stress concentrators (Knésl, 1991), it can be used for quantification of the severity of the crack with its tip at a bi-material interface. Therefore, the opening stress was observed and the average stress ahead of the crack tip calculated. Not only the influence of the various ITZ thicknesses, but also of various elastic module of ITZ on the near-crack-tip stress field was studied and discussed. Knowledge of the effect of the ITZ on the stress distribution will contribute to better understanding of the toughening mechanisms of the ITZ and aggregate in silicate-based composites.

ACKNOWLEDGEMENTS

This outcome has been achieved with the financial support of the Czech Science Foundation, project No. 16-18702S (AMIRI).

REFERENCES

Bower, A.F. & Ortiy, M. 1991. A three-dimensional analysis of crack trapping and bridging by tough particles. *J Mech Phys Solids* 39(6): 815–858.

Cox, B.N. & Marshall, D.B. 1994. Concept for bridged cracks in fracture and fatigue. *Acta Metal Mater* 42(2): 341–363.

Klusák, J., Profant, T., Knésl, Z. & Kotoul, M. 2013. The influence of discontinuity and orthotropy of fracture toughness on conditions of fracture initiation in singular stress concentrators. *Eng Fract Mech* 110: 438–447.

Knésl, Z., Klusák, J. & Náhlík L. 2007, 2008. Crack initiation criteria for singular stress concentrations, Part I–IV. *EngMech* 14: 399–408, 409–422, 15: 99–114, 263–270.

Knésl, Z.1991. A criterion of V-notch stability. *Int J Fracture* 48(4): R79–R83.

Kotousov, A. 2005. On stress singularities at angular corners of plates of arbitrary thickness under tension. *Int J Fract* 132(3): 29–36.

Leguillon, D. 2002. Strength or toughness? A criterion for crack onset at a notch. *Eur J Mech—A/Solids* 21(1):61–72.

Li, C.V. & Maalej, M. 1996. Toughening in Cement Based Composites. Part I: Cement, Mortar and Concrete. *Cem Concr Comp* 18(4): 223–237.

Merchant, I.J., Macphee, D.E., Chandler, H.W. & Henderson, R.J. 2001. Toughening cement based materials through the control of interfacial bonding. *Cem Concr Res* 31(12): 1873–1880.

Nevile, A.M. 2011. *Properties of Concrete*. Harlow: Pearson Education Limited.

Ollivier, J.P., Maso, J.C. & Bourdette, B. 1995. Interfacial Transition Zone in Concrete. *Advn Cem Bas Mat* 2(1): 30–38.

Prokopski, G. & Halbiniak, J. 2000. Interfacial transition zone in cementitious materials. *Cem Concr Res* 30(4): 579–583.

Rilem 1998. A. Katz, A. Bentur, M. Alexander, G. Arliguie, (eds.) *Second International Conference on the Interfacial Transition Zone in Cementitious Composites*. New York: E & FN Spon.

Rossignolo, J.A. 2007. Effect of silica fume and SBR latex on the paste-aggregate interfacial transition zone. *Materials Research* 10(1): 83–86.

Scrivener, K.L., Crumbie, A.K. & Laugesen, P. 2004. The Interfacial Transition Zone (ITZ) between Cement Paste and Aggregate in Concrete. *Interface Sci* 12(4): 411–421.

Shah, S.P. 1990. Fracture toughness of high strength concrete. *ACI Mater J* 87(3): 260–265.

Wriggers, P. & Moftah, S.O. 2006. Mesoscale models for concrete: homogenisation and damage behaviour. *Finite Elem Anal Design* 42(7): 623–636.

Zhang, X. & Xu, S. 2007. Fracture resistance on aggregate bridging crack in concrete. *Front Archit Civ Eng China* 1(1): 63–70.

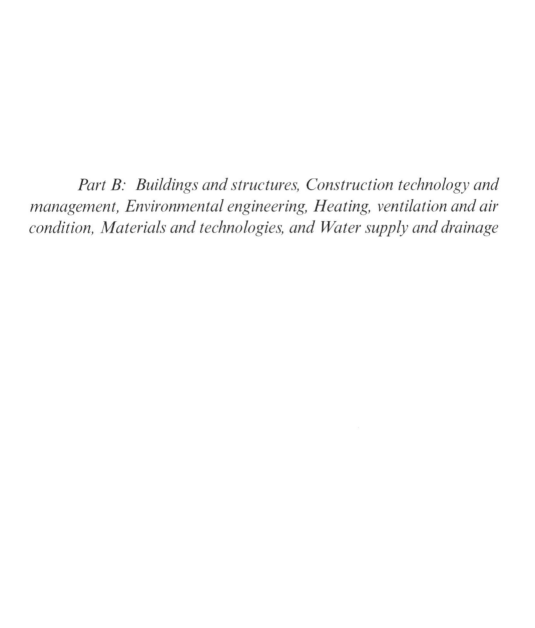

Part B: Buildings and structures, Construction technology and management, Environmental engineering, Heating, ventilation and air condition, Materials and technologies, and Water supply and drainage

Advances and Trends in Engineering Sciences and Technologies II – Al Ali & Platko (Eds)
© 2017 Taylor & Francis Group, ISBN 978-1-138-03224-8

Diagnostics to determine the cause of occurrence of cracks on ETICS

N. Antošová, M. Ďubek & M. Petro
Slovak University of Technology in Bratislava, Bratislava, Slovakia

ABSTRACT: The occurrence of failures in the construction of buildings, either older or newer, can have several causes. It can be due to poor-quality design, realisation, material used, and also neglected maintenance of existing buildings. An actual review of housing stock must provide the basic requirements for static, hygiene, fire, acoustic and thermal protection, health, and reduce energy intensity. The article concerns the diagnosis of the causes of failures of a multifunctional object.

1 INTRODUCTION

Application of External Thermal Insulation Composite Systems is an effective way improving the thermal insulation of the external cladding of a building. However, in a relatively short time of usage, we encounter failures in already realised insulation systems. The most common symptoms of failure are cracks, crevices, microorganisms, mechanical damage, humidification, leakage into the system and the pillow effect. These failures can be caused by different factors. Those factors may include incorrect application, substitution of materials, failure to comply with PD, and improper maintenance. If these errors are not detected in time then can lead to a significant breach or total collapse of the insulation system. Therefore correct and early diagnosis is very important because it would clarify the cause and subsequently the appropriate remediation technologies can be proposed.

The problem arises, however, if the analysis of the ETICS failures is not exactly systematised. Certain diagnostic methods are specified only for some of the tests listed in the STN 73 2901 and STN 73 2902. Further tests are carried out by professionals, owners of insulation systems or construction companies, but not without a general (single) technology process. In a study of this issue we have not met a single methodology, which would determine the diagnostic procedure for ETICS faults. Therefore, it is very difficult for the owner of the damaged insulation to determine a suitable methodology for his damaged system. Thus, for the owner, the price of diagnostic intervention is essential as it is the most important criterion by which he chooses a method of diagnosis.

Using an appropriate diagnosis can be accurately determined by the type of failure and its causes. Based on this information, the designer can propose the correct technological procedure of remediation failure.

With the correct diagnosis we can eliminate faults in time and avoid further disorder.

2 WORK METHODOLOGY

In this work we try to describe all available forms of diagnostic methods of thermal insulation faults. Based on all the available methods (test) which are summarised for application on the ETICS construction, they are divided into:

– Non-destructive method
– Destructive method

Each of the described methods is written in detail so even a home owner could study what the basic principles of realisation and execution for each method are (Antošová, 2014). Based on these facts and the necessary practice, the capable designer and owner of insulation can determine the extent of the failure and its consequences for ETICS construction.

The aim of this work is to analyse the diagnosis of faults of thermal insulation composite systems with brief instructions applicable in practice. According to received information, we will try to demonstrate methods of diagnosis in practical examples.

3 DIAGNOSTIC METHODS APPLICABLE TO THE ETICS CONSTRUCTION

After analysing the causes of ETICS failures it can be said: "Whatever the amount of identified faults emerging or demonstrating on the ETICS, they are in principle appearing in similar ways." (Baliak et al, 2007; Hobst et al, 2005) These visual signs can be classified or identified as:

– Cracks
– Microorganisms
– Mechanical damage
– Pillow effect
– Humidification and leakage into system
– Total collapse of system

Based on these findings it was necessary to find the most suitable method by which the causes of failures of ETICS constructions could be found. Individual tests were divided in terms of intervention in the construction using non-destructive and destructive methods.

Non-destructive techniques belong to the non-invasive tests which do not cause a breach of the insulation system tested. Their main advantage is that they can be carried out repeatedly as necessary without subsequent reparation of performed tests. Another advantage is that we have in a relatively short time a general picture about the condition of the insulation. The disadvantage might be that these methods do not always give us all the information necessary, about the condition of construction which could have a considerable significance when choosing the correct diagnosis.

Destructive methods are those in which the results are obtained at the cost of permanent damage to the examined object. In doing so, the damage may not be so severe that the studied object would not enable another test immediately afterwards. It means that interfering with construction does not need any special repair if we want to put it back to its original state. We consider this destructive as such a method results in acceleration of degradation, but it does not affect the functional capacity of the object. Despite the fact that this method can slightly damage the ETICS, it can still precisely diagnose the problem, and after the evaluation of the test it can determine how it is to be repaired.

The classification of individual tests is as follows:

3.1 Non-destructive methods

– Visual inspection
– Thermography
– Hygrometer
– Planennes of surface plaster (facades)
– Identification of microorganisms

3.2 Destructive methods

– Pull-out (tension) test of the spacing anchors
– The thickness of the thermal insulation layer

– Bond strength of adhesive mortar to substrate
– Bond strength of adhesive mortar to thermal insulation
– Adhesion of surface plaster
– Strength of plaster
– Capacity of plaster of ETICS base
– Bulk density TI
– Surface density of reinforcing grid
– Determination of water vapour permeability of coatings
– Adhesion pullout test of the paint
– The thickness of reinforcing layer

This number of tests shows that it may be very difficult to choose the right test that will help us to determine the correct diagnosis of ETICS failures.

4 DECISION-MAKING PROCESS WHEN ANALYSING FAULT DIAGNOSTIC

The decision-making process for diagnosing ETICS faults in the construction is processed in the scheme, see Figure 3 on the next page.

The basic scheme consists of a collection of available data on the subject in the form of the project documentation, site diary, information about performed maintenance, repairs or anomalies in a given area, etc. Based on the clarification of inputted data, it is possible to approach the local inspection and investigation. Based on the well-known pathological

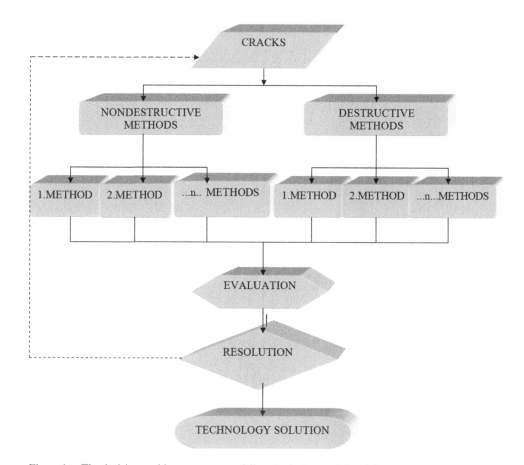

Figure 1. The decision-making process on a failure in the form of "crack".

manifestations in the form of cracks, the pillow effect, etc, we can precisely determine the cause of the ETICS breach with the help of destructive and non-destructive diagnostic methods. After determining the cause of the problem we can decide on the appropriate technology solution.

With the help of all the diagnostics we can identify the symptoms that were assigned to each basic existing demonstration (cracks, microorganisms, mechanical damage, the pillow effect, humidification and leakage into the system, the total collapse of the system) with subsequent analysis of different diagnostic methods suitable for an application to the damaged contact insulation system.

A decision-making process has been drawn up for each failure, see Figure 2. Based on these processes we concluded that for each symptom there are several tests which can determine the cause of the ETICS breach (see Figure 1). For this reason, it may be difficult for inexperienced people to decide which of the diagnostics to utilise and in what order.

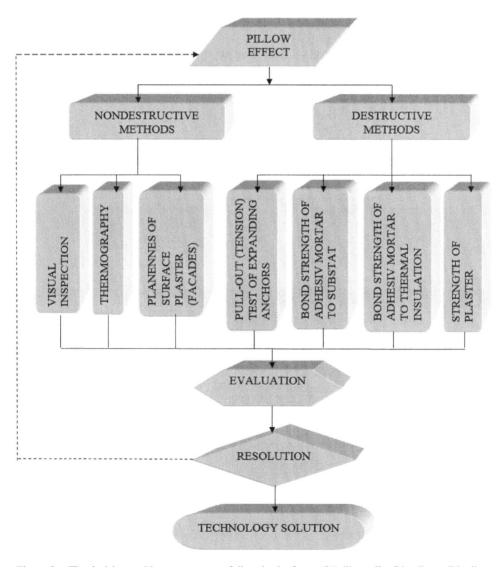

Figure 2.   The decision-making process on a failure in the form of "pillow effect" in all possible diagnostic methods.

320

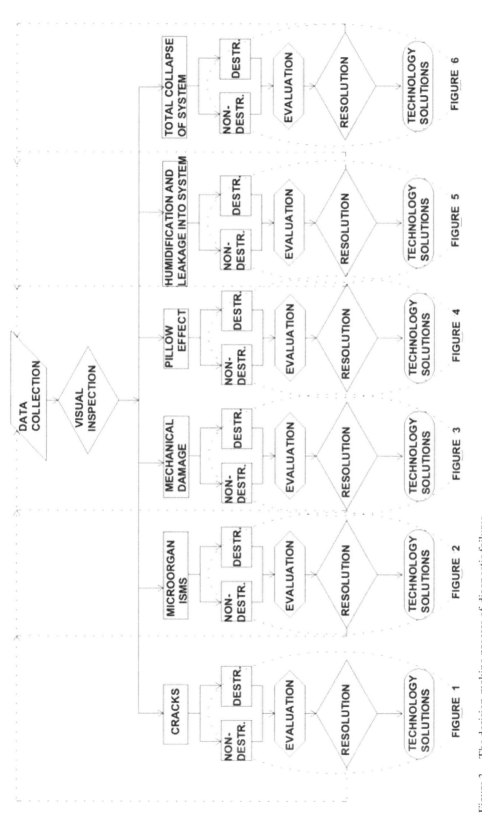

Figure 3. The decision-making process of diagnostic failures.

321

5 CONCLUSION

It may not be necessary to carry out all the tests of all these methods, because it also depends on the professional experience of the investigator of the issue. For this reason, we see the lack of a technological process (methodology) when diagnosing a damaged thermal insulation composite system, under which the expert, administrator or insulation owner would know how to follow.

In this article, we managed to gather all available diagnostics that can be applicable to damaged insulation systems. With the help of all the diagnostics we can identify the symptoms that were assigned to each basic existing demonstration (cracks, microorganisms, mechanical damage, the pillow effect, humidification and leakage into the system, the total collapse of the system) with analysis of diagnosis of failures for the damaged contact insulation system. Besides the familiar diagnosis is needed a certain professional experience with this topic, which will ensure that the person conducting the tests does not necessarily carry out all the tests mentioned for the symptom. It would be helpful for a layman because then he can easily diagnose the cause of the ETICS breach.

For this reason we see room for a deeper analysis of the diagnostic methods, which would then attach a certain importance. Based on the significance of different diagnostic methods it can be easier to determine the sequence of the relevant diagnostic methods for the symptom.

We will be able to see that methodology is applied in practice in the form of manuals for ETICS owners or even apply this methodology to a certain standard-setting treatment.

REFERENCES

Antošová, N.: Biokorózia kontaktných zatepľovacích systémov. Bratislava, Vydavateľstvo STU, 2007, ISBN 9788022727860.

Antošová, N.: Analýza poznania príčin a technológií riešení biokorózie ETICS a model zabezpečenia ich rezistencie. Bratislava, Vydavateľstvo STU, 2014, ISBN 978-80-227-4302-0.

Antošová, N.: Patológia stavieb I. Vybrané časti prednášok. Brno, Tribun EU, 2013, ISBN 978-80-263-0345-9.

Baliak, F. a kol.: ZNALECTVO bulletin článkov z odboru STAVEBNÍCTVO 1996–2006. Žilina, Žilinská univerzita, 2007, ISBN 978-80-8070-650-0.

External thermal insulation composite systems: Critical parameters for surface hygrothermal behavior (2014). Advances in Materials Science and Engineering, 2014, art. no. 650752.

Hobst, L. a kol.: Diagnostika stavebních konstrukcií. 2005, Vysoké učení technické v Brně, Fakulta stavební.

Inspection and diagnosis system of ETICS on walls (2013). Construction and Building Materials, 47, pp. 1257-1267.

Petro, M.: Metodika výberu technológie riešenia porúch pre kontaktné zatepľovacie systémy. Dizertačná práca, Bratislava 2013, SvF-13419-18598.

STN 73 2901: 2015: Zhotovenie vonkajších tepelnoizolačných kontaktných systémov (ETICS).

STN 73 2902: 2012: Vonkajšie tepelnoizolačné kontaktné systémy (ETICS). Navrhovanie a zhotovovanie mechanického pripevnenia na spojenie s podkladom.

STN EN 1542: 2001: Výrobky a systémy na ochranu a opravu betónových konštrukcií. Skúšobné metódy. Meranie prídržnosti pri odtrhových skúškach.

STN EN ISO 2409: 2013: Náterové látky. Skúška mriežkovým rezom (ISO 2409: 2013).

STN EN ISO 4624: 2002: Náterové látky. Odtrhová skúška priľnavosti (ISO 4624: 2002).

Stains in facades' rendering—Diagnosis and maintenance techniques' classification (2008). Construction and Building Materials, 22 (3), pp. 211–221.

Advances and Trends in Engineering Sciences and Technologies II – Al Ali & Platko (Eds)
© 2017 Taylor & Francis Group, ISBN 978-1-138-03224-8

Verification of the technology choice repairs ETICS

N. Antošová, M. Ďubek & M. Petro
Slovak University of Technology in Bratislava, Bratislava, Slovakia

ABSTRACT: In the past several works have been published by various authors at various scientific levels, which deal with the relationship between the process of realization of thermal insulation systems, the occurrence of errors and failures and feedback—prevention—in the design and realization of insulation. Knowledge is systematic at this stage; however, there is no opinion or choice of systematic solutions to the errors and faults in the process of maintaining the building structure—insulation. The article concerns the summary of basic theoretical knowledge of technology repairs ETICS. Based on the knowledge of the problems of the current situation and the requirements of building practice a methodology solution repair ETICS is developed.

1 INTRODUCTION

In the past the articles were elaborated by many different authors, which deal with the relationship between the design and process of realization of ETICS. They also deal with errors in the design and realization and the following incorporation of learned mistakes into preventive measures in the design and realization of ETICS. The given issue is systematized into the indications. We noted certain reserves in the process of maintenance and repair, i.e. the removal of already existing faults on implemented systems. However, there is a lack of a systematic methodology that could not only record, analyze and evaluate the failures but also solve the consequences of such technology of maintenance, repair, reconstruction or redevelopment.

2 THE PRESENT STATE OF THE ISSUE IN TECHNOLOGY REPAIRS ETICS

Basically, except for Germany, the faults of thermal insulation composite system are considered as a necessary concomitant of these systems.

Germany is very intensively engaged in home renovation, especially the former GDR. Expanded polystyrene with a large thickness, typically from 120 mm or more is typically used as a thermal insulator. Compare to the insulator with smaller thickness, this results in a more favourable distribution of stress at the thin-film plaster surface and thus in a smaller failure rate for the final layer of ETICS. Statistically, in these systems occur such a significant number of plaster failures (Pašek, 2007), which creates the need for technology solutions.

German author of the publication (Blaich, 2001) states: "For thermal insulation of composite systems there are two major problems given by system:

The first problem results from the fact that the outside plaster is applied to a very soft surface.

The second problem is associated with the fact that the layer of plaster can be quickly and very strongly heated and cooled. On the one hand it is due to the small volume and on the other hand it is due low thermal conductivity of the plaster carrier."

Measurements of the thermal insulation composite systems have shown that influence of sunlight is warming the plaster above the air temperature and on the contrary on clear nights the temperature of the plaster drops below the air temperature.

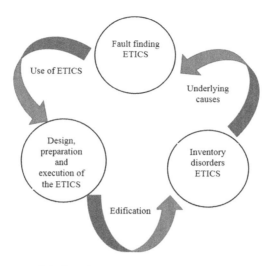

Figure 1.   The present state of the issue in technology repairs ETICS.

The author also identified basic failures that were observed after the realization of thermal insulation in composite systems, and divided them into five groups:

– irregular cracks in the shape of a net,
– cracks in the place of contact of thermal insulation discs,
– pillow effect,
– changing colours,
– failures in surface plaster.

The Czech professional public also deals with the occurrence and analysis of causes of failures thermal insulation composite systems. It is "Association for Thermal Insulation", which tries to address the issue in a comprehensive and systematic manner, along with the development of new materials and technologies. The authors (Šála et al, 2002) identified almost identical thermal insulation composite systems failures as authors in other countries. Greater attention is paid mainly to avoid excessive stress on the substrate as an outcome of crack formation in the plaster.

In principle, the authors collect and identify the same types of failures, analyze their causes in term of realization as well as in term of shortcomings of design. However, none of the authors pays attention to the elimination of failures, they do not deal with the consequences and they do not address functional requirements security during the expected durability of the ETICS.

The current view on solving failures of ETICS is narrowed to the field of identification failures, finding the causes formation of errors and following learning from mistakes, form of adjustment of the normative principles (adjustment of the principles of design, technological regulation adaptation, adjustment the principles of realization, maintenance ...).

3   THE ANALYSIS OF INFORMATION, PERFORMANCE AND CAUSE OF THE ETICS FAILURES IN TERMS OF TECHNOLOGICAL STAGES OF THE REALIZATION

Based on meetings and analyzing the existing evidence, categorization of failures and short-comings of ETICS were processed. Processed information from failures is available by other authors and based on common features they are grouped into seven main groups.

Classification of the technological stage is borrowed from the source (Sloboda, 2007). The classification into the technological stages corresponds with the basic definition of techno-logical stage which is given in the terminology of this work. The essence of technological stage is not determined by any technological norm or regulation (Fickuliak et al, 2004). They are based on a detailed analysis of the construction technology and partial process suggested by its technologist. The result is a comprehensive part of the construction process, which technically allows independent use (Fickuliak et al, 2004). Based on the analysis and documents analysis (Sloboda, 2007 and Greguš, 2010) we agree with the categorization and we use this classification further in this article.

Failure classification according to technological stages:

- TE 1—cleaning and layout of the substrate,
- TE 2—bonding insulator,
- TE 3—anchorage of insulator,
- TE 4—preparation of the substrate under the reinforcing layer,
- TE 5—making the reinforcing layer,
- TE 6—preparation of the substrate under the surface layout,
- TE 7—making the final surface layout.

Failure analysis is further processed in the four stages of development.

The first step is the actual performance, followed by a logical consequence of this performance and possible cause.

If there is well-known repair solution, then a simple way of technology solutions—removal in the last step of process is presented.

The symptoms, causes and consequences are processed in the following chapters according to the source (Sloboda, 2007). By means of targeted exploration and research work, the technological solutions were identified. These technologies were subsequently attributed to various causes, expressions and consequences. Terminology and classification into compact technological stages is taken from the works of authors (Sloboda, 2007 and Greguš, 2010) and the actual assignment of already existing technology solutions is referring to a published classification of ETICS failures.

We considered the assigning of targeted possible solutions to classified failures as the first partial output from the targeted research.

There are two basic alternatives from the data summary about potential technology solutions of ETICS failures:

Alternative I—conservative repair technology, which includes removal of failures of ETICS construction, relieving symptoms—failures.

Alternative II.—radical repair technology, which includes removal of causes of unwanted condition—ETICS construction faults, which usually includes replacing the entire ETICS or parts of ETICS group of strata, for example, the exchange of surface, the final layer, or supplementing the parts of the group of strata.

Based on these symptoms, catalogue list was processed in article as basic documents for the optimal choice of technology. In each catalogue list, failure demonstration is described with its visual demonstration, followed by brief technological process solutions with general principles and measures. Finally, each catalogue list is completed by assumed durability of intervention and economic indicator like the price of 1 m² action or work expenditure. When compiling catalogue lists, we focused on the most important data from the perspective of the ETICS owner, which capture by a simple but yet concise form a brief description of the technology solutions and also indicate the cost and labour intensity per 1 m² of thermal insulation. The extent and form of processing is intentionally simple, objective and general due to the smooth orientation for lay people.

We have eight technology repairs based on the identified failures:

1. Sealing of cracks + new facade coating
2. Application of biocidal products + coating
3. Exchange of reinforcing layer + plaster
4. Incision of reinforcing layer (raster 100×100 mm) + new reinforcing layer + plaster
5. Thermal insulation
6. Anchorage

7. Local exchange of group of strata
8. Total remediation (replacement of the entire ETICS)

All technological repair procedure from catalogue lists form database of available technologies. Catalogue lists repairs are primarily designed for companies, professional public, administrators or owner of the insulation system. Based on these catalogue lists, we supposed simple orientation of the lay and professional public in identifying signs of faults that occur in the insulation and the subsequent selection and realization of repairs under this technological process.

To determine the appropriate decision-making model in choosing appropriate technology solutions of ETICS failures, we studied all the faults that may arise during durability of thermal insulation. From our obtained expertise and foreign literature, as well as based on our practical experience, we have processed own scheme of failures, which summarizes all the errors that arise during the realization of thermal insulation.

After analyzing the causes ETICS failures and synthesis of learned lessons, identified by professional public either in our country or in the world, it can be said: "Whatever amount of the indentified faults emerging or demonstrating on the ETICS, they are in principle reflected in a similar way." These forms of expression can be classified or identified as: cracks, microorganisms, mechanical damage, the pillow effect, humidification and leakage in the system, the total collapse of the system. Under the term demonstration/expression we understand the first visual fault demonstration, perceived with the naked eye, without a deeper diagnosis.

4 METHODOLOGICAL PROPOSAL FOR DECISION-MAKING PROCESS OF SELECTION FOR TECHNOLOGY SOLUTIONS OF ETICS REPAIRS BASED ON DEFINITION OF CRITERIA

From the gathered data for the creation of a database repairs and applications for the demonstration we conclude that for one type of ETICS fault is in many cases possible to use multiple variations of technology solutions, whether in the alternative I (conservative) or II (radical).

However, while deciding from the individual technology solution, it is not clear when it is appropriate, possible or necessary to use alternative no. I or II. and their variant solutions. The owner of ETICS asks many times whether a violation of the existing insulation system can be suitable and also more economically viable to repair, redevelop, or it is necessary to consider its overall exchange. From the perspective of the building owner, the contractor, manufacturer of insulation may be a number of different criteria for the decision, however, the greatest importance attaches to the necessary costs incurred to repair. The decision-making process includes several factors such as: the intervention price, labour intensive repairs, remediation, fulfill the KZS functional requirements with the disorder, the range of disorders, age of KZS, the lifetime of the technology repair of failures, economic efficiency etc.

Despite the processed database of solutions, the decision-making processes become confusing and complicated. The basic scheme of decision making is branched, while some technological procedures and solutions are useful in many cases of demonstration.

It was necessary to process the basic scheme of decision making procedure of technology solutions of ETICS repairs according to identified five basic failures. We found out that on one symptom can be applied several variants of remediation solution.

Selection of the appropriate remediation technologies for specific problem of ETICS failure and its subsequent use depends on several factors. Criteria of decision that each owner, promoter, manager have to take into account can be divided into three main (basic) groups: physical, technological and economic.

Each of these criteria requires deeper analysis, clear expression of content with option to correspond on the fulfilled criteria and express the importance of criteria—the meaning of criteria, to determine the degree of preference and achieve an objective or an acceptable compliance.

The entry conditions for proposal on the issue of the use of possible remediation technologies are defined in the submitted catalogue lists and related decision-making processes. They depend on the extent of damage, the actual properties of technical thermal failure, ability to achieve the quality of realization, compliance with the climatic conditions during the repair realization, security, or prolong the durability, and achievement of the desired or improved technical thermal properties. However, the wider range conditions will meet the proposal technology repair, the greater functionality of technology solutions will be. From the above it is clear that is not always a simple matter to choose the appropriate remediation according to predefined requirements of user. Therefore it is necessary to note the large number of factors that influence it. The mono criterial (one-sided) evaluation cannot be applied in the selecting, i.e. based on only one of the possible criteria. Therefore is necessary to pay close attention to the right choice. To help to decide, it is possible to use any of the available methods of evaluation, which are very often used in the fields of construction, and also in logistics. In case when a person depends on several criteria, the methods of multi-criteria evaluation are the most commonly use.

During the experiment Fuller's method was used. The application of this method to any decision problem (available at least two convenient options) allows you to specify a broader perspective and to choose effective option on multiple sites.

The knowledge of solutions and selection of criteria corresponding to these solutions still remains questionable. In case of solutions of ETICS failures, the aim is to ensure the expected durability and function for reasonable costs so that returns on investment will not extent in realization and maintenance of thermal insulation. Since the decision-making system is open and there is a possibility to enter it even at a later time, the specified criteria according to the degree of knowledge can be reevaluate (for example, focus on labour content, or use the impact criteria of remediation on the environment, or pick up the rate of realization conditions when it comes to tough conditions on the construction site chosen technology). By

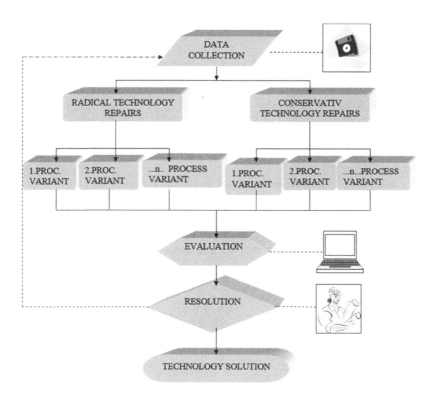

Figure 2.   The decision-making process in the technology solutions of ETICS repairs.

reevaluation of established variants according to new goals and assessment of the previous decision is possible to create a new objectified decision.

It can be concluded that a wider application of this method in general, reduce the process of decision. Self-handling by this method is not difficult and does not require the continuation of a deep analysis, as it operates according to best variant, the introduction of which is possible in the system to look for opportunities for improvement while minimizing risk.

5 CONCLUSION

The construction area devoted to repairs, remediation of existing thermal insulation is now high on the agenda. In professional circles this issue is discussed the most, particularly in view of the requirements of security standards and requirements of legislation on reducing energy intensity. So far, the scientific conclusions are partial and disorganized and they point out the necessity to pay attention, especially to systematic approach for ETICS solution repairs. The fact that to increase living comfort and reduce heating costs of owners of flats and houses were invested considerable funds it is certainly necessary to pay due attention to the repair and remedial actions. The reasons are clear, i.e. providing the expected durability, ensure the return on the investment, and also the extent of the durability and improving existing properties of ETICS.

REFERENCES

Antošová, N.: Biokorózia kontaktných zatepľovacich systémov—analýza príčin a technológia dekontaminácie, Bratislava 2005.

Antošová, N.: Životnosť a údržba zateplenia obvodového plášťa. Vedeckoodborná konferencia. ISBN 978-80-232-0301-1, Realizácia a ekonomika stavieb. Október 2009 Trenčianske Teplice.

Blaich, J.: Poruchy stavieb. Bratislava, Vydavateľstvo JAGA, 2001. ISBN 80-88905-49-4.

Machatka, M. Šála, J. Svoboda, P.: Kontaktní zateplovací systémy—Příručka pro navrhování a provádění, Praha 1998.

Pašek, J.: Spolehlivost a trvanlivost kontaktních zateplovacích systémů obvodových plášťů panelových budov, Ostrava 2007.

Raschle, P. Büchli, R.: Algen und Pilze an Fassaden Ursachen und Vermeidung 2., vydanie, 2006, 109 S., Fraunhofer IRB Verlag ISBN 978-3-8167-7051-0.

Sloboda, P.: Nejčastejší nedostatky při provádení vnějších tepelně izolačních kontaktních systému (ETICS) a jejich následné poruchy. S-Therma Olomouc, 01/2007.

Somorová, V.: Optimalizácia nákladov spravovania stavebných objektov metódou facility managmentu. (Optimalisation of expenses used for maintenance of constructions through facility management) ISBN 978-80-227-2782-2.

Sternová, Z.: Zatepľovanie budov—Tepelná ochrana, Bratislava, Vydavateľstvo JAGA, 1999. ISBN 80-88905-11-7.

Šála, J. Machatka, M.: Zateplováni v praxi, Praha: Grada, 2002, ISBN 80-247-0224-X.

Witzany, J. Čejka, T. Pašek, J. Zigler, R.: Průzkum vad a poruch povrchové úpravy kontaktního zateplovacího systému na vybraných panelových objektech, In: Setkání kateder a ústavů pozemního stavitelství ČR a SR. Praha: Fakulta stavební ČVUT v Praze, 2005, s. 1–8.

Advances and Trends in Engineering Sciences and Technologies II – Al Ali & Platko (Eds)
© *2017 Taylor & Francis Group, ISBN 978-1-138-03224-8*

Technical measurement of the green roof in a winter season

R. Baláž & S. Tóth
Department of Architectural Engineering, Civil Engineering Faculty, Institute of Architectural Engineering, Technical University of Košice, Košice, Slovakia

ABSTRACT: The saying "a good roof is priceless" is not say for nothing. However, it may lead to discussions, which roof is good, because there are lots of requirements and criteria for the functional characterization. It must be understood, that the roof structure defines durability of the building as a unit, thus it defines the lifetime of the building's other parts and, also, the function of space, which is covered by the roof. Therefore, it is very important to pay a particular attention to the design, as well as the realization of the roof structure. The aim of this publication is to judge the physical and technical parameters in the design of the roof cladding module in a climate chamber.

1 IMPLEMENTATION OF THE PANEL TO THE EXPERIMENTAL CHAMBERS

The presented article is treated into two parts. The first part involves the measurements consisting of the roof structure measurement in the climate chamber, where the only difference in the roof structure was in the addition of a vapor barrier into the part of the roof structure S2. Therefore, the results were evaluated, and we conclude after their investigation, that the roof structure must be modernized due to the excessively high temperatures on the waterproofing layer in the summer. Modernization of the roof structure was made by the green roof layer installation to the roof structure S2 (Figure 1). We wanted to eliminate rising surface temperatures on the waterproofing layer in a summer time, and we wanted to demonstrate it on this roof structure module. The study of demonstration is listed in the second part of this article.

The roof skin composition is the result of a detailed analysis. The roof module, with dimensions 1700 (mm) × 2200 (mm), was created thanks to a support centre. It has 2 pieces of the identical roof panels, which are overlapping. However, they are two distinct parts of

Figure 1. Situation of the experimental measuring chambers.

the roof module. The structure of the two roof elements differs. The left side of the module is equipped with the thermal insulation manufactured by NOBASIL with a vapour barrier from JUTAFOL N; the drywall ceiling is anchored to the support above it with using "U" profile. The left side is equipped with NOBASIL 125 (mm) thermal insulation, although a vapour barrier is excluded from it. The drywall ceiling is similarly anchored as the one on the right side. The surface ceiling is left without a finishing treatment. The roof module, with proportions 1700 (mm) × 2200 (mm), was supported and separated by the wooden beam 125 (mm) × 125 (mm) with known thermal properties.

While the module's left half up to centre beam axis contained JUTAFOL N, the right half deliberately had a vapour barrier omitted. The sample of the roof deck above the rafters is the same for the both halves of the roof module. Consequently, they will fit into the roof module. The samples consist of following: OSB particle board with increased load with thickness15 (mm) is attached to the top of the rafter. The second layer—Prominent, is an anchoring layer with a micro-ventilation for the polyurethane insulation. Metal sheet is used

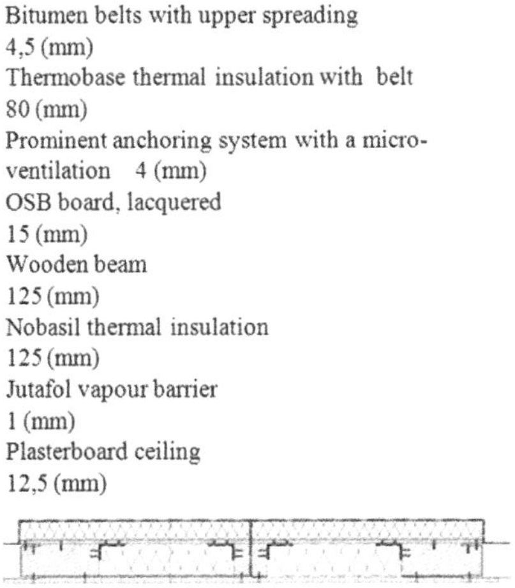

Bitumen belts with upper spreading
4,5 (mm)
Thermobase thermal insulation with belt
80 (mm)
Prominent anchoring system with a micro-ventilation   4 (mm)
OSB board, lacquered
15 (mm)
Wooden beam
125 (mm)
Nobasil thermal insulation
125 (mm)
Jutafol vapour barrier
1 (mm)
Plasterboard ceiling
12,5 (mm)

Figure 2.   Previous roof layers composition.

Soil, thermal insulation along the edges
80 (mm)
Drainage layer
Bitumen belts with upper spreading
4,5 (mm)
Thermobase insulation with belt
80 (mm)
Prominent anchoring system with a micro-ventilation   4 (mm)
OSB board, lacquered
15 (mm)
Wooden beam
125 (mm)
Nobasil thermal insulation
125 (mm)
Jutafol vapour barrier
1 (mm)
Plasterboard ceiling
2,5 (mm)

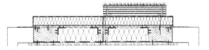

Figure 3.   Current roof layers composition.

as an edge on the bottom and on the top side, and it is made from a galvanized sheet with thickness 2 (mm). It is attached to the OSB particle board and, thereupon the anchoring layer Prominent was affixed. The following panel is modified by a gas burner. It snapped the insulation Thermobase/polyurethane thermal insulation on the top of the sarking felt/with thickness 80 (mm). These manufactured roof panels were fitted into the roof module. The final layer consisted a bitumen, from the company Index mineral, with thickness 4.5 (mm), and its colour is brown. It overlay the whole roof structure. (Katunský, [et al.] 2014).

Possibility of the roof modernisation or thermal properties improvement has to be taken into the account, regarding the evaluation of the testing measurements made in the winter and summer season. The measurements are described and evaluated in the dissertation thesis of R. Baláž. We have reached the decision after the analysis. The addition of a vapour barrier to the roof structure S2 is necessary, and it would be possible to add the green roof layers onto the roof construction for thermal properties improvement (Figure 3).

2   PLACEMENT AND TYPE OF THE SENSOR IN THE EXPERIMENTAL SAMPLE OF THE ROOF CLADDING

The construction of the roof experimental module was divided into the construction S1 and the construction S2. S1 contains the vapour barrier JUTAFOL N in contrast to S2, what is displayed in the Figure1. The sensors were placed to the individual layers of the roof construction and they mirror each other in distance 300 (mm) from the centre, which is defined by the load bearing beam. The first temperature sensor A3.7 was installed under the bitumen layer of the construction S1, and the sensor A3.8 was same way installed to the construction S2. The second layer is the polyurethane thermal insulation with the sensors A3.9 and A3.10, which were installed into the middle of the insulation/40 (mm)/. The connecting layer is the construction's third layer, and it is placed on the OSB particle board, where the sensors A3.11 and A3.12 were installed. The sensors A3.7, A3.8, A3.9, A3.10, A3.11, and A3.12 are temperature sensors on a resistance base without a tip, with a temperature range from −50 to +125 (°C). The internal insulation is from a mineral wool, thickness 125 (mm), and the sensors I1/6, I1/2 are in it with a mirrored distance between them. The sensors measure air temperature, relative humidity, dew point and absolute humidity. The sensors D1/13 and D1/14 were installed as the last ones on the internal surface of the internal drywall. They measure surface temperature in two points. The sensor D1/41 measuring the surface temperature directly below the vegetation pad was mounted into the construction S2 (green roof construction) (Baláž, R., 2013).

The measured parameters are external and internal temperature, relative humidity, temperature and relative humidity inside the roof construction, surface temperature, and heat flow.

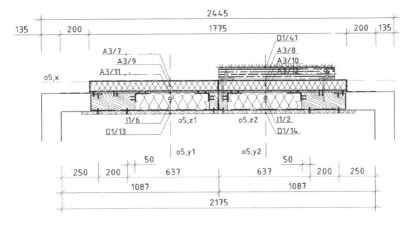

Figure 4.   Sensor's placement.

The temperature sensors on a resistance base without a tip, type NTC, made from NiCr, are used due to purpose of the measurements. The weather station measures a wind direction, and its average and maximum speed, atmospheric pressure, air temperature, relative humidity, and rainfalls. We, also, monitored a global solar radiation by a pyranometer (Flimel M., 2013, Chuchma, L., Kalousek, M., 2014).

3 THE RESULTS OF THE MEASUREMENT IN THE CLIMATE CHAMBER MODULE

All measured data from the experimental roof module, except data obtained from the weather station, were recorded in one minute intervals during four months in the winter at the Civil Engineering Faculty of Technical University in Košice. The weather station assembled in the given climate chamber module supplied the actual time data, also, external air temperature, wind speed, intensity and duration of global solar radiation. The values of external temperature used in this publication were obtained from the weather station and processed by the licensed software. Only three values measured by the sensors are processed in the graphs to avoid overlapping of the individual curves.

3.1 The temperature oscillation phase shift in the roof structures S1 and S2 in the quasi-stationary conditions of the internal environment and the variable external environment

Comparison of the phase of the temperature oscillation was assessed in October 2012, November 2012, January 2013 and February 2013. The internal surface temperature was measured on the plasterboard by the sensor D/13 in the roof structure S1 and by the sensor D/14 in the roof structure S2. These values were used for the calculation. Minimum and maximum values of external surface temperature on the roof structures S1 and S2 were selected in the individual months of the measurement. Subsequently, these values were compared. The full assessment of the temperature oscillation phase shift is displayed in the Table 1.

Further comparison of the temperature oscillation phase was assessed in October, November and December 2014, and in January and February 2015, continuing in October

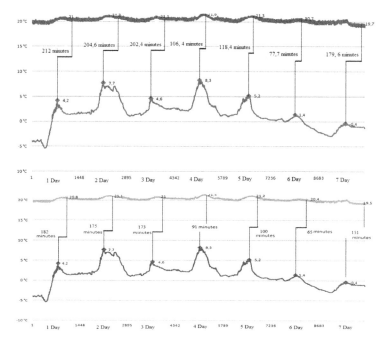

Figure 5.   Daily phase shift in the winter period of November 2015.

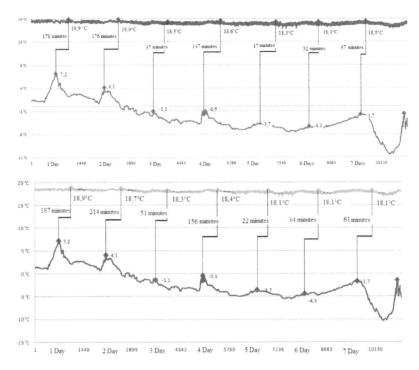

Figure 6.   Daily phase shift in the winter period of February 2016.

Table 1.   Total comparison of the temperature oscillation phase shift in the structures S1a S2.

The measurement 1		The roof structure S1/ vapour barrier/	The roof structure S2
Daily temperature oscillation phase shift in the winter season	October 2012	ψ = 81.59 minutes	ψ = 81.55 minutes
	November 2012	ψ = 118 minutes	ψ = 98.42 minutes
	December 2012	ψ = 104 minutes	ψ = 88.66 minutes
	January 2013	ψ = 103.5 minutes	ψ = 90.57 minutes
	February 2013	ψ = 133.8 minutes	ψ = 124.7 minutes
The average value ($\bar{x}$)		ψ = 108.2 minutes	ψ = 96.7 minutes
Difference		ψ = 11.42 minutes	
			The roof structure S2 green roof
The measurement 2		The roof structure S1	
Daily temperature oscillation phase shift in the winter season	October 2014	ψ = 93.02 minutes	ψ = 109.1 minutes
	November 2014	ψ = 98.3 minutes	ψ = 115.1 minutes
	December 2014	ψ = 120 minutes	ψ = 142.3 minutes
	January 2015	ψ = 128 minutes	ψ = 153.6 minutes
	February 2015	ψ = 80.3 minutes	ψ = 94.1 minutes
Average value ($\bar{x}$)		ψ = 103.9 minutes	ψ = 122.8 minutes
Difference		ψ = 18.97 minutes	
			The roof structure S2 green roof
The measurement 3		The roof structure S1	
Daily temperature oscillation phase shift in the winter season	October 2015	ψ = 80.14 minutes	ψ = 100.2 minutes
	November 2015	ψ = 90.9 minutes	ψ = 98.5 minutes
	December 2015	ψ = 100 minutes	ψ = 134.6 minutes
	January 2016	ψ = 115 minutes	ψ = 142.5 minutes
	February 2016	ψ = 70.7 minutes	ψ = 94.5 minutes
Average value ($\bar{x}$)		ψ = 91.3 minutes	ψ = 103.3 minutes
Difference		ψ = 24.52 minutes	

and December 2015, and January and February 2016. The internal surface temperature value from the sensor D/13 of the roof structure S1, and the sensor D/14 of the roof structure S2 is used for the calculation. The sensors measured surface temperature on the plasterboard. Minimum and maximum values of external surface temperature on the roof structures S1 and S2 were selected in the individual months of the measurement. Subsequently, these values were compared.

The full assessment of the temperature oscillation phase shift of the roof constructions S1 and S2 (the green roof) is displayed in the Table 1.

The measurements of surface temperature on the drywall of the roof S2 (green roof), surface temperature on the drywall of the roof S1.

4 CONCLUSION

The vapour barrier JUTAFOL N had an impact on the roof structure S1 in the aspect of the temperature oscillation phase shift. This statement is substantiated by the comparison of the measurements implemented in the climate chamber module in quasi-stationary conditions of the internal environment. Confrontation of the measurements about the impact is following:

– Daily temperature oscillation phase shift in the roof structure S1 with the vapour barrier JUTAFOL N in the winter was 11.42 (minutes) in contrast to the structure S2 without a vapour barrier. These measurements were conducted in the quasi-stationary conditions of the internal environment in the climate chamber (Baláž, R., 2013).

Our green roof layer had an impact on the structure S2 in the aspect of the temperature oscillation phase shift. This statement is substantiated by the comparison of the measurements implemented in the climate chamber module of the structures S1 and S2 in quasi-stationary conditions of the internal environment and the variable external environment. Records and confrontation from the measurements about the impact are following:

– The roof structure S2 thanks to the green roof layer has more favorable temperature oscillation phase shift in the winter season /2014–2015/ than the roof structure S2 in the same conditions. Its value is 18.97 (minutes) and its measurements were implemented in quasi-stationary conditions of the internal environment in the climate chamber.
– The roof structure S2 due to the green roof layer usage shows 24.52 (minutes) daily temperature oscillation phase shift in the winter /2015–2016/, which is more favourable than the roof structure S1 had. These measurements were realised in quasi-stationary conditions of the internal environment in the climate chamber.

ACKNOWLEDGMENT

This paper was created thanks to the financial support of VEGA 2/0117/14 Research of envelope construction influences on the luminous and thermal environment in the attic spaces.

REFERENCES

Baláž, R. 2013, Posúdenie fyzikálno-technických parametrov strešného plášťa v module klimatickej komory, *Dizertačná práca*.
Chuchma, L., Kalousek, M., 2014, Electricity storage in passive house in Centra Europe region. *Advanced Materials Research*, 2014, Vol. 899, pp. 213–217.
Flimel, M., 2013, Differences Ug—values of glazing measured in situ with the influ ence factors of the internal environment, 2013, *In: Advanced Materials research*. Vol. 649, pp. 61–64.
Katunsky, [et al.] 2014. Numerical Analysis and Measurement Results of a Window Sill. *Advanced Materials Research*, 899, pp. 147–150.

Assessment of ground water infiltration to the basement garage spaces

R. Baláž & S. Tóth
Department of Architectural Engineering, Civil Engineering Faculty, Institute of Architectural Engineering, Technical University of Košice, Slovakia

ABSTRACT: Currently, in the construction of new multifunctional residential houses at places where, mostly industrial buildings were in the past, land appropriateness can be underestimated with it's hidden pitfalls and traps in different forms (for example groundwater level, etc.). A similar case happened in a building in Košice.

1 CASE STUDY

The task is to develop an expert assessment of leaking into non-apartment spaces on 2nd basement floor in the building in Košice.

- Assessment of real condition.
- Assessment of defect of leaking by its influence on the building structures (foundation and exterior constructions).
- Assessment of defect of leaking by a life span of built-in structural materials; by building statics—possibility of damage occurrence (during long term influence).
- Assessment of appropriateness of built-in and used materials and technological processes.
- Assessment of defect of leaking by removal point of view—whether the defect is permanent or it can be removed. Design a way how to remove it, if it's possible.

In this article we will discuss the first, the second, third and the fifth of customer's questions.

2 ASSESSMENT OF REAL CONDITION

Slovak law system defines these basic requirements for structures:

- mechanical resistance and structure stability;
- structure fire safety;
- sanitation and health and environment protection;
- building safety during usage;
- noise and vibration protection;
- energy-savings and building warmth protection.

To fulfill these basic requirements, it is necessary to ensure the water-proofing ability of exterior structures. The water-proofing function, as one of other isolations, has to co-act during assurance of building structure for that function during complete life-span.

Good water-proofing protection depends on thorough knowledge of projection team (architect—designer, static surveyor, geologist, …), experiences and skills of the realization team (implementing workers, foremen, site managers, supervision).

Water isolation system, as an object of this assessment, is localized in lower structure's part of the construction object. It consists of:

– Horizontal isolations, which protect object's structures against water and humidity creeping from subsoil.
– Vertical isolations, which protect object's structures against water and humidity in surrounding soil.

This real condition is characterized by lower water leaching through water isolation system, which is part of the exterior building constructions in lower building of assessed object (2nd basement floor—underground parking garages). This phenomenon appears during the year, especially in the time of increasing atmospheric precipitation. These conditions, also, get help by high level of underground water, which is caused by appearance of underground springs (fount) with high spring-discharge in a relatively close distance to the incriminated structure object despite the fact, that there is permanently functioning well of underground water near the object (near chimney) which perpetually decreases a level of underground water. (Horniaková, L. et.al. 1997).

3 ASSESSMENT OF DEFECT OF LEAKING BY ITS INFLUENCE ON BUILDING CONSTRUCTIONS (FOUNDATION AND EXTERIOR WALLS)

Leaching of lower water causes partial floods of underground floor spaces, it corrupts trouble-free functioning of those spaces for the purpose of what they were designed for. Structures in the underground buildings are from materials, which are more or possibly less porous. It concerns concrete and ferroconcrete constructions, which are mostly used in lower building areas. These types of building constructions are relatively the best resistant against humidity and clear water, but by themselves they has no satisfactory water-proofing function. Damages caused by humidity or water leaching:

– Appearance of efflorescence on building constructions surface (leaching of salts).
– Warmth insulating reduction of used materials in building's constructions by humidity increment.
– Humidity also spreads by help of materials capillarity beyond the leaking places.
– Inner nominal air humidity φ_{ai} increment and inner surface temperature θsi reduction of building constructions and inner air temperature θsi, reduction cause increased risk of fungus formation in inner spaces.
– Small material parts sequentially releasing and leaching from constructions and increasing joints from where water is leaching in long term period (Sedláková, A. et.al. 2015).

4 ASSESSMENT OF DEFECT OF LEAKING BY LIFE SPAN OF BUILT-IN CONSTRUCTION MATERIALS; BY BUILDING STATICS—POSSIBILITY OF DAMAGE OCCURRENCE (DURING LONG TERM INFLUENCE)

This answer follows the above written answer. Clear underground water very rarely occurs in common practice. Water usually occurs containment of other admixture and that's why it has different chemical characteristics as clear (distilled) water. Change of these characteristics strongly affects the risk of other damages occurrence. Mainly in case of existence of aggressive underground waters. These waters are divided in the following categories:

1. Hungry (extract) waters—water hardness is ≤ 6°dH (German degrees)
2. Acid waters—water acidity is < 7 pH (potentiahydrogenii);
3. Sulphate water—contains sulphated anion SO_4^{2-};
4. Carbonate waters—contains carbonate anion CO_3^{2-};
5. Magnesium waters—contains magnesium cation Mg^{2+};

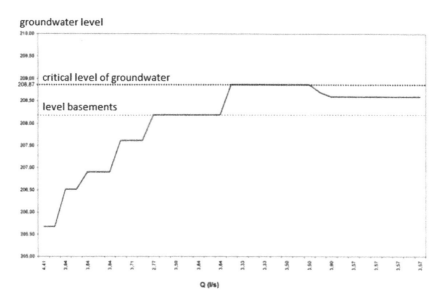

Figure 1. Graph displays the level of the underground water in the well near the chimney, depending on the rainfall (period: November–December 2015).

Aggressiveness can also cause occurrence of these substances: hydrosulfide H_2S, ammonia NH_3, humic acids $C_{349}H_{401}N_{26}O_{173}$ and eventually other substances.

During the assessment production we did not have chemical analysis of underground water at disposal, which is present close to the object, which it is attacked. From maps, which we had at disposal, Košice district is classified as an area with occurrence of the slightly to mid aggressive, carbonate and acid waters.

These waters have a negative influence on standard concrete and ferroconcrete constructions from time view point. It causes pH reduction of alkaline environment of both types of constructions, what has mainly negative influence on occurring steel armature. It can, sequentially, corrode by influence of higher acidity environment until sequential reduction of armature in building construction happens, which can cause static damages.

The so called pressurized water can cause problems with fulfilling constructions function. We talk about pressurized water when level of underground water acts on foundation gap or horizontal water-proofing with pressure at least 0,02 MPa. Pressure value of 0,02 MPa is reached in underground water level in minimum depth at 2 meters over foundation gap or horizontal water-proofing. Based on hydrogeological examination, in the presented case is the highest so called critical underground water at the level of 208,87 m.a.s.l. and decreased underground water at the level to 208,60 m.a.s.l., when pump starts working in shaft near chimney. The underground water level does not rise above this level. Level of foundation gap is in 207,30 m.a.s.l. and level of water-proofing is 208,15 m.a.s.l.

Based on knowledge of these high levels, it is possible to determine the pressure of underground water by the following:

– Hydrostatic pressure in water-proofing level (208,15 m.a.s.l.) with critical underground water level (208,87 m.a.s.l.) is at 7,2 kPa ergo 0,0072 MPa.
– Hydrostatic pressure in water-proofing level (208,15 m.a.s.l.) with decreased underground water level (208,60 m.a.s.l.) is at 4,5 kPa ergo 0,0045 MPa.
– Hydrostatic pressure is in foundation gap level (207,30 m.a.s.l.) with critical underground water level (208,87 m.a.s.l.) is at 15,7 kPa ergo 0,0157 MPa.
– Hydrostatic pressure in foundation gap level (207,30 m.a.s.l.) with decreased underground water level (208,60 m.a.s.l.) is at 13,0 kPa ergo 0,0130 MPa.

In this case it is not pressure water. Underground water is in the height that does not create demands on water-proofing against pressurized water.

It is very problematic to specify, at the given object, exactly when from the time view point, to extend from quantitative and to what level of damage from qualitative point of view static damages may occur; due to the large amount of marginal factors and more detailed pieces of information which co-act and were not accessible to us (exact type of concrete, type and high of reinforcement cover, quality of work performance, state of structures in regard and similar). (The hydrogeological report, 2013).

5 ASSESS DISORDERS LEAKS FROM THE POINT OF VIEW OF REMOVAL— WHETHER IT'S THE FAULT PERSISTS OR CAN BE REMOVED. IF CAN BE REMOVED, DESIGN A METHOD FOR REMOVING

In the view of the fact of the disorder which we find, age of the building, building purpose we believe that the disorder could be eliminated. The options how eliminate this disorder are few. From the passive method, through the active, or by the combination of these two methods.

The aim of the active method is to decrease, respectively completely exclude the impact of the ground water to the assessed object. It could be realised by erection of: underground wells with pumps and drainage channels, pipes drainage system, by the acquisition clips, waterproofing curtain, or by their combination.

To removal of disorder by this purpose is presented in the conclusion of hydrogeological research, which was available. This method is in consideration of the surrounding and adapted terrain relatively complicated.

Passive method consists in use of special chemical waterproofing and reprofiled membrane and barrier system. Material which occurs the water penetration is reinforced concrete, or concrete. Concrete is, though to a lesser extent, the porous material. That is the reason why it has capillaries, which allows water transport (or other liquids) which are situated inside. To prevent this transport from the any direction on the diffusion of the reactive chemical substances, which migrate in concrete capillaries and use its humidity, the chemical components operate just above the membrane, and such a chemical barrier systems. This process

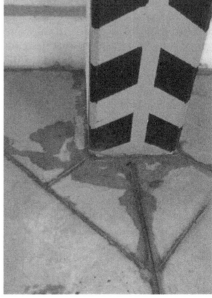

Figure 2.   Intersection of underground water by technological and construction gaps in structures.

Figure 3. Intersection of underground water by technological and construction gaps next to a wall. Rising water level.

accelerates chemical reaction between them, minority products of cement hydration and humidity, where the new insoluble crystalline structures are created. These new structures fill capillaries and create water resistant concrete and also protect reinforcement inside. This chemical reaction could be renewable in case, when the humidity will occur in concrete. The one of the main advantages is possibility to application from the passive side of building structure (mainly in the case of the disorder removal) as also in this case.

These materials are also resistant against aggressive water and another chemical substances—knows permanently protect concrete against chemical substances in source 3,0–11,0 (pH) in state contact.

Also have resistance against hydrostatic pressure (also from the negative side) because after crystallization becomes a direct part of the concrete structure. Some products are able to resist to hydrostatic pressure to the 13 MPa.

On the market are also available products, which are resistant against oil products, which in the case of the basement storey—garage, which is exactly what happens in this disorder is very necessary. The products are not toxic.

Could be applied:

– Across the board of the structure, which have to report waterproofing ability (if necessary in several layers);
– Continuous in the connection of individual structures, technological joints, expansion joints;
– Point at transfer points wiring and acute leaks.

It is this passive system which is primarily by our most convenient way to delete the disorder and its implementation should take place as soon as possible without creating more damage on object.

5.1 *Photo documentation*

There is a photo documentation attached as proof of leakage of underground water which stays in building's garages. Every picture is described. This description contains information as where is water leaking in building's structures.

Figure 4. Intersection of underground water by technological and construction gaps next to pillar. Rising water level.

6 CONCLUSION

Damages made by lower water leaking to the object in Košice, excluding other potential negatives, make difficulties or temporary interfere the proper function of spaces, which they were designed for. In regard to the facts, which are described above in detail, it is necessary to do the object remediation in the measures proposed above. The answers to the questions which were given to us, we analyzed given state and we proposed a system of solution to this defect. For its successful elimination, it is necessary to find established company with at least several years of experience in the given area with appropriate references, following that to propose materials, technology and precise implementation of remediation measures.

ACKNOWLEDGEMENT

This paper was created thanks to financial support VEGA 2/0117/14 Research of envelope construction influences on luminous and thermal environment in attic spaces.

REFERENCES

The hydrogeological report hg. ratios below the object in Kosice.
Horniaková, L. et.al. 1997. Civil engineering works, *JAGA Group*, Bratislava.
Fendeková, M., Roháčiková, A. 2004. Aggressive characteristics of groundwater in Slovakia. *Groundwater X*/2004 No. 2.
Sedláková, A. et.al. 2015. Analysis of material solutions for design of construction details of foundation, wall and floor for energy and environmental impacts, *In: Clean Technologies and Environmental Policy.* Vol. 17, No. 5 (2015), pp. 1323–1332.

Advances and Trends in Engineering Sciences and Technologies II – Al Ali & Platko (Eds)
© 2017 Taylor & Francis Group, ISBN 978-1-138-03224-8

Monitoring the effectiveness of sorbent materials for removing selected metals from water

R. Biela, T. Kučera & J. Konečný
Faculty of Civil Engineering, Institute of Municipal Water Management, Brno University of Technology, Brno, Czech Republic

ABSTRACT: This article is focused on a laboratory experiment of the removal of selected metals from water. The removal efficiency of iron, manganese, cooper and arsenic from modeling water by filtration through four sorption materials was monitored in the laboratory of Institute of municipal water management. These were materials with the tradenames CFH 0818, Bayoxide E33, GEH and DMI-65, the last one is not yet used in the Czech Republic. Removal of the individual elements was monitored at regular intervals and compared with the limit values for these elements given by Decree no. 252/2004 Coll., which is valid for drinking water in the Czech Republic. From the results of water analysis it can be said that all the materials are able to remove selected metals from water, each of the materials, however, removes these metals with different efficiency.

1 INTRODUCTION

Almost all metals and metalloids are naturally contained in water, at least in trace quantities, depending on the geological conditions. The enrichment of water occurs through contact with rocks and soil. The main anthropogenic sources of metals and metalloids are wastewater from mining and processing ores, from ironworks, rolling mills, from the surface treatment of metals, as well as from the photographic, textile and leather industries. Another source may be agrochemicals, algicidal preparations and the leaching of sludge dumping sites. In addition, atmospheric water contaminated by exhaust fumes from the burning of fossil fuels and motor vehicle exhaust can be an important source of metals and metalloids in surface waters (Pitter 2009).

In drinking water with trace metals such as arsenic, nickel and lead they may occur only in units up to tens of micrograms per litter, which in turn creates the need for more complicated water treatment technologies. Higher concentrations of these metals are hazardous to health concerns and some of them may have carcinogenic effects.

There are a lot of ways to remove heavy metals from water. One option is sorption on a granular media mostly based on iron oxides and hydroxides. It is a selective, easy, economically reasonable and highly efficient method of removing metals from water.

2 DESCRIPTION OF SORPTION MATERIALS SELECTED FOR THE EXPERIMENT

Through experimental measurement the efficiency of the removal of selected metals from water by filtration through four sorption materials was monitored, namely CFH 0818, Bayoxide E33, GEH and DMI-65. The material DMI-65 has a chemical composition based on silicon oxide, the other three materials are based on iron hydroxide.

The material CFH is produced in two variants, namely CFH 12 and CFH 0818 that differ from each other in grain size (Tab. 1). They serve as a sorption filling for filters for the removal

Table 1.   Granularity of filtration materials Kemira CFH. (Kemwater ProChemie s.r.o. 2012).

Kemira CFH 12		Kemira CFH 0818	
Dispersion [mm]	Presence [%]	Dispersion [mm]	Presence [%]
2–0.85	92.7	2–0.5	97.6
<0.85	5.9	<0.5	2.4
>2	1.4	>2	0

of arsenic, selenium, phosphorus, silver, nickel, lead, molybdenum, silicon, vanadium, copper and other metals from water. The advantage of these filter materials is their ease of use and the simple pre-treatment of raw water. The Finnish manufacturer, the company Kemira, recommends the inclusion of two filtration units in a row to increase efficiency. The material is put on the market in 25 kg paper bags, which is supplied by the company Kemwater ProChemie s.r.o. (Kemwater ProChemie s.r.o. 2012).

Bayoxide is a dry crystalline medium developed by the company Severn Trent in cooperation with Bayer AG and is manufactured by LANXESS Deutschland GmbH, Leverkusen in Germany. The material is designed for the removal of arsenic, antimony and other metals from water, such as iron and manganese. The material is able to remove As^{III} and As^{V} below the value of 4 $\mu g.l^{-1}$. The lifetime of the material is dependent on the quality of the treated water. The filter material is commercially available in two forms, Bayoxide E33 as a granular filtration material and Bayoxide E33P in the form of tablets (Ilavský & Barloková 2008), (De Nora Water Technologies 2015).

The filter material GEH is a high performance adsorbent on the basis of granulated iron hydroxide. (Ilavský et al. 2015) There is a version of GEH 101, which focuses on removing heavy metals such as chromium, uranium, copper, lead and toxic substances including arsenic, antimony, vanadium, molybdenum and selenium. The GEH 102 version specializes directly in removing arsenic from water. The material was developed at the University of Berlin at the Department of Water Quality Control to remove these compounds from water, without affecting the quality of the treated water. However, it effectively removes iron, manganese and other metals. The adsorption capacity of GEH is dependent on the pH and on the quality of raw water. It is produced by the German firm GEH Wasserchemie GmbH (GEH Wasserchemie GmbH 2011), (Biela et al. 2013).

DMI-65 is a filter medium based on silica sand used for removing mainly arsenic, iron and manganese from water, but also mercury, aluminum, or cyanide. The presence of NaClO is required to activate this medium. This filter material is not yet used in the Czech Republic. It has been used for 20 years in a wide range of applications around the world. The DMI-65 material is manufactured by Quantum Filtration Medium in Australia. DMI-65 obtained a certificate of safety according to NSF/ANSI 61 and can therefore be used for drinking water (K2O Kabana Water Consulting 2012).

3   THE EXPERIMENTAL REMOVAL OF SELECTED METALS FROM WATER

Measurements were performed at the Faculty of Civil Engineering in Brno in the laboratory of the Institute of Municipal Water Management. The water model simulated groundwater with an increased content of iron, manganese, copper, arsenic (Tab. 2) and was prepared from water from the municipal water supply with the addition of chemical solutions of these components in concentrations above the limit for drinking water. The aim of the experiment was to assess the effectiveness of selected sorption materials for removing the chosen metals from water.

The individual sorption materials in the laboratory were put into cylindrical columns, which were formed by glass tubes with an internal diameter of 4.4 cm. The lower part of each glass tube was fitted with a plastic knee and a ball valve intended for the offtake of water. A drainage layer of small stones was formed in the lower part of the tubes, above those

Table 2. Analysis of the raw water with the simulated pollution.

t [min]	pH [−]	Temperature [°C]	Turbidity [ZF]	c(Fe) [mg.l⁻¹]	c(Mn) [mg.l⁻¹]	c(Cu) [mg.l⁻¹]	c(As) [µg.l⁻¹]
0	7.53	13.6	11.40	2.770	0.501	2.040	62.300

Table 3. Analysis of water after filtration through the sorption material CFH 0818.

t [min]	pH [−]	Temperature [°C]	Turbidity [ZF]	c(Fe) [mg.l⁻¹]	c(Mn) [mg.l⁻¹]	c(Cu) [mg.l⁻¹]	c(As) [µg.l⁻¹]
0.5	7.91	17.5	1.84	0.607	0.035	0.180	2.600
1	7.95	17.8	1.41	0.354	0.003	0.097	1.600
2	7.88	17.3	0.83	0.237	0.000	0.066	1.000
3	7.91	17.2	0.81	0.234	0.000	0.069	<1
5	7.89	17.1	0.75	0.116	0.000	0.047	<1

Table 4. Analysis of water after filtration through the sorption material Bayoxide E33.

t [min]	pH [−]	Temperature [°C]	Turbidity [ZF]	c(Fe) [mg.l⁻¹]	c(Mn) [mg.l⁻¹]	c(Cu) [mg.l⁻¹]	c(As) [µg.l⁻¹]
0.5	7.77	18.2	1.37	0.176	0.016	0.063	1.200
1	7.74	18.0	1.03	0.145	0.013	0.055	1.000
2	7.68	17.9	0.97	0.118	0.011	0.045	<1
3	7.64	17.9	0.94	0.112	0.008	0.044	<1
5	7.64	17.9	0.74	0.096	0.005	0.039	<1

Table 5. Analysis of water after filtration through the sorption material GEH.

t [min]	pH [−]	Temperature [°C]	Turbidity [ZF]	c(Fe) [mg.l⁻¹]	c(Mn) [mg.l⁻¹]	c(Cu) [mg.l⁻¹]	c(As) [µg.l⁻¹]
0.5	7.62	18.5	2.80	0.324	0.045	0.105	2.100
1	7.58	18.6	1.98	0.299	0.044	0.101	1.800
2	7.57	18.5	1.42	0.240	0.042	0.082	1.700
3	7.52	18.2	1.32	0.230	0.025	0.074	1.400
5	7.48	18.2	1.13	0.164	0.023	0.073	1.200

Table 6. Analysis of water after filtration through the sorption material DMI-65.

t [min]	pH [−]	Temperature [°C]	Turbidity [ZF]	c(Fe) [mg.l⁻¹]	c(Mn) [mg.l⁻¹]	c(Cu) [mg.l⁻¹]	c(As) [µg.l⁻¹]
0.5	7.63	19.2	7.93	0.180	0.025	2.920	<1
1	7.54	19.0	3.48	0.062	0.021	0.735	<1
2	7.53	19.1	2.40	0.041	0.020	0.588	<1
3	7.52	18.9	1.98	0.026	0.019	0.051	<1
5	7.50	18.6	1.13	0.022	0.012	0.029	<1

there were a layer of glass beads with a diameter of 2–4 mm, so that the leaching of the filter material from the column to the pipe did not occur during filtration. Over the drainage layer the filter filling is located with a thickness of 0.55 to 0.7 m. The remainder of the tube was filled with water up to its upper edge. The upper portion of the tube was then sealed with a plastic closure with a ball valve. The entire filter system consisted of a 30 liter container for raw water, the pump unit, the filter columns, the flow meter and the inlet, outlet, and wash piping.

343

During filtration, the raw water was pumped using a pump from the vessel through a flow meter into the filter column. The quantity of flowing water was regulated on the flow meter through a throttling nozzle, while the filter flow rate was set at 30 l.h^{-1}. During filtration, water samples were taken at the water outlet from the filter column at the predetermined times of 0.5, 1, 2, 3, and 5 minutes. Temperature, pH, turbidity, iron and manganese were measured in the samples. In addition, water samples were taken for the determination of copper and arsenic, which were measured in the accredited laboratory of the National Public Health Institute in Brno. The results of the analysis of water after filtration through individual sorption materials are given in Tables 3 to 6.

4   EVALUATION OF THE MEASURED VALUES

The results of the analysis of the individual water samples can reliably say that all materials are able to effectively remove iron, manganese, copper and arsenic. The material DMI-65 showed excellent effects in the removal of iron (Fig. 1). The value of iron concentration was already below the limit of 0.2 mg.l^{-1} laid down in Decree no. 252/2004 Coll. after 0.5 minutes, while only Bayoxide E33 achieved similar values. During the experiment, the material DMI-65 maintained the concentration of iron at the lowest value of all the materials tested. For the materials CFH 0818 and GEH the values of the concentration of iron in the filtered water were under the limit value for drinking water after about 3.5 minutes.

Manganese was removed from the water by all materials within 0.5 minutes of measurement below the limit value of 0.05 mg.l^{-1} laid down by Decree for drinking water (Fig. 2). The material GEH in this case showed the weakest performance, the material DMI-65 led to the removal of manganese slightly better than the material GEH. However, the best results in the removal of manganese from water were achieved by the material CFH 0818 that after 2 minutes of measurement removed manganese from the water completely.

Copper was removed from the material CFH 0818, Bayoxide E33 and GEH after 0.5 minutes of measurement under the limit of 1 mg.l^{-1} under the Decree for drinking water (Fig. 3). The material DMI-65 reached below the limit value after less than a minute. However, all materials held copper concentration values to very low values. The material DMI-65, showed the lowest values of copper concentration in the filtered water from all the materials investigated, after the third minute of measurement.

Although the concentration of arsenic in the model water was up to 62.3 mg.l^{-1}, it was removed well below the limit value for drinking water which is 10 mg.l^{-1} (Fig. 4) after filtra-

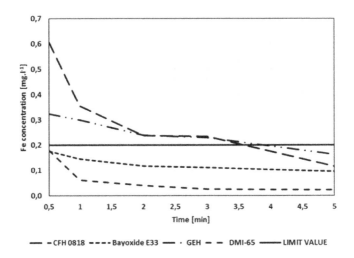

Figure 1.   The course of the removal of Fe from water using the selected filter materials.

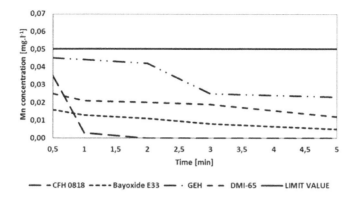

Figure 2. The course of the removal of Mn from water using the selected filter materials.

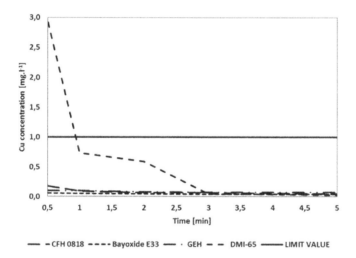

Figure 3. The course of the removal of Cu from water using the selected filter materials.

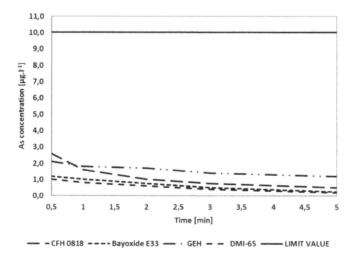

Figure 4. The course of the removal of As from water using the selected filter materials.

tion through the selected materials. However, the material DMI-65 worked best, where after half a minute the concentrations fell below 1 mg.l^{-1}, which was the lowest limit of measurability of this metal. When comparing all four materials, the highest concentrations of arsenic in the filtered water were exhibited by the material GEH.

At the same time as the monitoring of the concentrations of iron, manganese, copper and arsenic the values of pH, temperature and turbidity were also tracked. The ph values for all materials fell during the measurement process. The temperature of the water during the experiment was greater than the temperature of raw water, which is affected by the temperature in the laboratory. The change had no effect on the experiment. The turbidity declined in all materials during filtration, while with the exception of the material DMI-65, it immediately fell below the limit value of 5 ZF stipulated by Decree no. 252/2004 Coll. The reason for the initial increased turbidity with the material DMI-65 was the flushing away of small particles of material, despite the fact that the filtering took quite a long time.

5 CONCLUSION

From the results of the analysis of individual samples of water it can be said that all the tested filter materials are capable of removing the selected metals such as iron, manganese, copper and arsenic from water. The best results in the removal of iron and arsenic from water were achieved by the material DMI-65, which reduced the concentration of these metals to the lowest values of all the tested materials. The best results in the removal of manganese from water were achieved by the material CFH 0818 that after 2 minutes of measurement removed manganese from the water completely. The material Bayoxide E33 did very well in removing manganese from water, which after 0.5 minutes of measurement reduced the concentration of manganese to the lowest of all materials, from 0.501 mg.l^{-1} up to 0.016 mg.l^{-1}. This material was very effective in removing copper from the water, where after 0.5 minutes it reduced its concentration in the water from 2.04 mg.l^{-1} to 0.063 mg.l^{-1}.

ACKNOWLEDGEMENT

The paper was prepared under the solution of the grant project on special research at BUT in Brno No. FAST-S-15-2701.

REFERENCES

Biela, R. et al. 2013. *Účinnost sorpčních materiálů na odstranění niklu i jiných kovů z vody*. TZB—info, roč. 15., č. 26, s. 1–5. Praha, ISSN 1801-4399.

De Nora Water Technologies 2015. Bayoxide—Model E33—Arsenic Removal Media. http://www.environmental-expert.com/products/bayoxide-model-e33-arsenic-removal-media-15343.

GEH Wasserchemie GmbH 2011. GEH 102 Arsenicentfernung. GEH Wasserchemie GmbH, Osnabrük. http://www.geh-wasserchemie.de/files/datenblatt_geh102_de_web.pdf.

Ilavský, J. & Barloková, D. 2008. *Nové sorpčné materiály v odstraňovaní kovov z vody.* In: Sborník konference Pitná voda 2008. České Budějovice: W & ET Team, 2008, s. 195–200. ISBN 978-80-254-2034-8.

Ilavský, J. et al. 2015. *READ-As and GEH sorption materials for the removal of antimony from water*. Water Science and Technology: Water Supply, 15 (3), pp. 525–532.

Kemwater ProChemie s.r.o. 2012. Kemira CFH12, CFH0818. Kemwater ProChemie s.r.o., Kosmonosy.

K2O Kabana Water Consulting 2012. Leaflet materials for the company K2O Kabana Water Consulting. Quantum DMI-65 Media. For Iron and Manganese Removal. USA.

K2O Kabana Water Consulting 2012. Leaflet materials for the company K2O Kabana Water Consulting. Quantum DMI-65 Media. For Arsenic Removal. USA.

Pitter, P. 2009. Hydrochemie. 4th edition. VŠCHT Praha. Praha. ISBN 978-80-7080-701-9.

Advances and Trends in Engineering Sciences and Technologies II – Al Ali & Platko (Eds)
© 2017 Taylor & Francis Group, ISBN 978-1-138-03224-8

Technology of waterproofing membranes based on mPVC for flat roof

M. Božík, M. Ivanko, A. Mečiar & M. Petro
Faculty of Civil Engineering, Slovak Technical University, Bratislava, Slovakia

ABSTRACT: Presently we often come across with dysfunctional leaking flat roofs, where the Damp Proof Course (aka DPC) is of poor quality or was damaged. Until recently, the most widely used material for damp-proofing of flat roofs—bitumen, is being replaced with plastic-based membranes. These plastic membranes, especially mPVC-based, are used not only for new buildings, but also for reconstruction and restoration of old roof structures. The use of these membranes comes both with advantages as well as with problems that occur during their installation. The technology of roof construction using this type of membrane is essential for their quality. Therefore, the aim of this paper is to explain basic attributes of mPVC-based waterproofing membrane´s joints and how the technology of construction affects them.

1   ISSUES WITH WATERPROOFING MEMBRANE INSTALLATION

Plastic membranes, especially mPVC-based ones, are not new in the construction market. However, we can still find faults during their application. For the sake of comparison, bitumen strips installation consists of melting the strips onto the roof surface whereas the joints are done by melting the lower side of the strip, while the installation of plastic membranes is more complicated (Šveda, M. 2007). During the installation of plastic membranes there is a higher possibility of making mistakes both when applying the layers as well as joining individual membranes at the joints. The process of membranes installation should begin with meteorological conditions assessment to ascertain if the environmental conditions are appropriate for such a type of plastic membrane in the first place. However, this step is usually left out because of the investors´ pressure on the construction speed. That is why membranes are often installed during drizzles or low temperatures. Although manufacturers declare that their plastic membranes can be welded at up to −10°C they do not specify what risk could such environmental conditions incur.

One should pay all of his/her attention mainly at the process of welding and particular details during the installation. The most common way of joining mPVC-based plastic membranes and TPO is by hot air welding. Joints of PVC products are considered the strongest among all membranes (Björk, F. 1993). Less common but of high-quality are EPDM membranes. They can be pasted on a large area but it has to be a properly constructed solid base. The advantage of the pasted asphalt or EPDM membrane is that if the roof gets damaged, suffusion should occur where the leak is located. If the roof's DPC is stabilized by anchorage or by additional load, for example gravel, water can freely penetrate through the covering layers and the suffusion can appear far from the spot of leakage. Another problem could lay in the so-called T-joints. This kind of joint consists of an additional membrane that overlaps the seam of two already welded membranes whereas at the meeting point of the three membranes a small canal may form. This spot can be a potential source of leakage of water under the roofing. It is therefore appropriate to thoroughly check this joint and reinforce it with a sealing compound. Another problematic detail could be the joints between the mPVC foil and PVC coated metal sheets which are typical on vertical surfaces

(Gränne, F. & Björk, F. 2000). An important factor for proper welding of membranes can be attributed to the welding machine and its settings i.e. temperature, and in case of automatic welder, its speed. Manufacturers prescribe welding guide values for their products but these need to be adapted to the specific environmental conditions on sites.

Welding temperature is influenced by many factors, the main factors are:

- membrane type and manufacturer, the thickness of the membrane;
- welding speed;
- the type of welding device (hand welder or automatic welder);
- base and ambient temperature and humidity;
- wind speed, drizzle, rain, snow etc.

Real life cases have shown that not setting the right temperature is not as damaging for mPVC-based membranes as it is for the TPO-based membranes. PVC membranes can weld sufficiently well even if the welding temperature is fluctuant but only to a certain extent of course. We tried to verify these PVC-based membranes´ properties by testing their joints in a laboratory.

2 LABORATORY EXAMINATION OF WELDS MPVC-BASED WATERPROOFING MEMBRANE

The prerequisite for research of waterproofing membrane joints was to have the quality of their welding, and/or the actual weld itself, affected by the environmental conditions— temperature, humidity, wind, rain or snow. On construction sites today we get to see foil welding at very low temperatures, during light rain and even snowfall. The producers them- selves recommend that it is possible to weld membranes at temperatures down to about −5 to −10°C, whereas such unsatisfactory conditions have a major impact on the quality of welds. Unsuitable climatic conditions, especially ambient air temperature can be corrected by adjusting the temperature on the welding machine. In order to determine the correct temperature of the welding machine, trials should be performed and then by pulling the two membranes apart we can check if the temperature is correct (if there is a visible grid over the entire width of the torn joint, the temperature is correct). However, in most cases a trial such as this is not performed. Thus, the temperature used in welding has a major impact on the quality of joints.

For large dimension flat roofs modern automatic welding machines are used, which greatly speeds up the whole process of waterproofing membrane welding. The automatic welding machine can have its welding temperature as well as welding speed adjusted. In an effort to speed up the work workers often set the automatic speed higher than what is specified by the manufacturer. This might also affect the final quality of the welds.

Another factor is how clean the membranes are at the point of their subsequent joint— weld. Provided that they are properly stored and all the procedures for installation are followed membranes should not get unclean on the site. However, due to the fact that construction sites have ongoing construction work done with dust produced, even the waterproofing mem- branes are exposed to various contaminants. Producers of membranes recommend inspection of the purity of the welds before welding. If the membranes are dirty, dirt should be removed. For this purpose, water is used, but it is not very suitable. If it is dried insufficiently, the membrane may welded be badly. That is why so called cleaners—diluents-based cleaners— are produced especially for this purpose. As it was mentioned before, workers often do not remove the dirt from the membranes, either because of financial or time-saving reasons.

Based on the above-mentioned assumptions, following methods of research were determined:

1. Welding the membranes with automatic welding machine set at various temperatures.
2. Welding joints of different level of cleanliness.
3. Different working speeds of the automatic welding machine.

2.1 Preparation of materials and samples

Before the actual welding of waterproofing membranes it was necessary to prepare material samples. Samples were obtained from the membranes in accordance with the standard STN 72 7652 (EN 13416), which defines the rules for obtaining samples from waterproofing membranes based on PVC. Standard waterproofing PVC-based membranes with thickness of 1.5 mm intended for installation of the single layer mechanically anchored waterproofing system were used for the research. These membranes are produced by modern technology of the two-sided application on supporting impregnated polyester plate by extrusion.

An undamaged waterproofing membrane was selected and taken from a warehouse where it was stored for the last 24 hours at a temperature of about 20°C (standard requirements are 20 ± 10°C). Subsequently, the roll was unfolded on a flat surface and pieces—samples—which did not contain any visible imperfections were cut. Next, these pieces were stored for over 20 hours in a room with a temperature of about 22°C (the standard requirement s are 23 ± 2°C). Each membrane was checked once again before welding as it is required by the standard STN EN 1850-2 (72 7642). Selected samples of membranes underwent visual inspection from both sides for potential blisters, cracks, holes, scratches or sags. The presence of bubbles or any dirt was checked on the section of each sample. After the inspection, the membranes were marked in the corner according to the method of research used.

2.2 Production of welded samples

An automatic welding machine Leister UNIROOF E was used for sampling. It had a welding nozzle attached capable of making 40mm-wide welds. The process of sampling was as follows: at first, the membranes intended for welding were unfolded and spread on the welding surface and were left there for about 30 minutes. Subsequently, those surfaces of membranes intended for welding were cleaned using the specialized cleaner depending on the research method used. After the cleaner had dried out, the process of welding started. The following sets of samples were produced this way:

1. Welding of membranes where the variable factor was temperature
   Samples had been cleaned by the cleaner prior to welding and the working speed of the welding machine was set to 2m/min. Tthis is how the sets of samples were produced:
   1. A at + 560°C
   1. B at + 540°C
   1. C at + 520°C
   1. D at + 500°C
   1. E at + 480°C
   1. F at + 420°C
2. Welding of membranes where the variable factor was the purity of joint
   Samples were produced by welding at a temperature of 520°C and the working speed of the welding machine was set at 2m/min. This is how the sets of samples were produced:
   2. A—joint uncleaned, the membrane straight off the production line
   2. B—joint cleaned with water, then dried
   2. C—joint cleaned with cleaner (similar to 1.C)
   2. D—joint cleaned with water, not dried (moist membrane)
3. Welding of membranes where the variable factor was the working speed of the automatic welding machine
   Samples had been cleaned by the cleaner prior to welding and welded at a temperature of 520°C. Speed range of this particular automatic welding machine is 2 to 5 m/min. This is how the sets of samples were produced:
   3. A—working speed 2m /min. (similar to 1.C)
   3. B—working speed 3.5 m/min.
   3. C—working speed 5m/min.

The starting temperature of 520°C was selected because this temperature results in proper welding of two membranes. The actual welding was carried out in a closed hall with ambient air temperature of 20°C. After the welding, welded membranes which were produced in the above mentioned way were left to sit for approximately 30 minutes. Subsequently, the process of testing welded samples was performed. These samples were subsequently tested to determine the shear resistance and the peel resistance of joints.

2.3 *Sampling and testing*

10 individual strips with width of 5cm were taken from each type of welded and settled membrane which had been labeled according to its method of production. The strips were cut out with a knife with the help of a stainless template. The length of the samples was chosen to make it possible to grip the sample into the pliers of the tensile testing machine in a way that the membrane covers the entire surface of the clamps of pliers. At the same time, special attention was paid to make sure that the weld is right in the middle of the cut sample.

The welds of waterproofing membranes can be assessed on the basis of two laboratory tests regulated by the standard. It is the test for determining the shear resistance of joints—STN EN 12317-2 and the test for determining the peel resistance of joints—STN EN 12316-2.

2.3.1 *Determination of shear resistance of joints*

Shear resistance is the maximum tensile force that can act on a sample before it is torn or separated on the joint. In this test, the samples (a total of 5 samples were tested in one set) were gripped into the pliers of the tensile testing device that operates by steadily pulling the sample and at the same time recording the progress of the acting force, until the failure of the sample occurs. It is important to ensure that the weld is right in the middle between the two clamps of the pliers. The distance between the clamps has to be 3-4 times the width of the joint, in this case 120 mm. The scheme of the device with the clamped sample is shown in Figure 1.

The device continuously records the pulling force up to the time of tearing or separation of the tested sample. The result is recorded in the form of a diagram which marks the course of acting of the force during the testing. After testing each individual sample a maximum tensile force—Fmax and the way of failure of the strip were recorded into the register. The average value was calculated from the five values of Fmax and the standard deviation was indicated.

2.3.2 *Determination of peel resistance of joints*

Peel resistance test is fundamentally the same as the shear resistance test differing only in gripping the sample into the tensile testing device in regard to the made weld. This test is

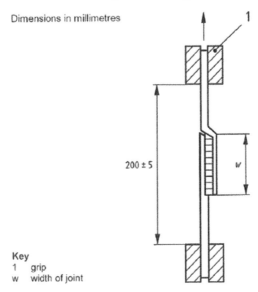

Key
1   grip
w   width of joint

Figure 1.   Test of shear resistance of joints, source: STN EN 12317-2 (72 7639).

stricter than the tensile test. The sample must be made in a way to make it possible to grip the sample into clamps of the testing device as shown in Figure 2.

The evaluation of the results is different as well, since there may be three different types of failure to the sample: a) membranes can be separated without breaking their surfaces; b) one of the strip is torn outside the joint; c) the sample is damaged, there is a visible grid in the joint (Figure 3). The evaluation of such damaged samples has to be carried out according to the standard STN EN 12316-2.

2.4 Evaluation of measured results of shear resistance of joints and peel resistance

As it is indicated by the measured results, mPVC-based membranes have a certain range of temperatures at which it is possible to make a weld of a good quality (see Table 1). At the

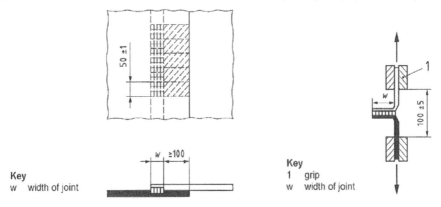

Key
w   width of joint

Key
1   grip
w   width of joint

Figure 2.   Preparation of the test strip and testing of peel resistance of joints, source: STN EN 12316-2 (72 7638).

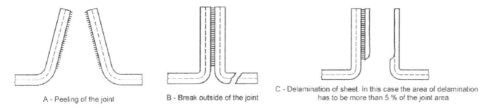

A - Peeling of the joint   B - Break outside of the joint   C - Delamination of sheet. In this case the area of delamination has to be more than 5 % of the joint area

Figure 3.   The modes of sample failure in peel resistance test, source: STN EN 12317-2 (72 7639).

Table 1.   Evaluation of measured results shear resistance of joints and peell resistance.

Norm	STN EN 12316-2		STN EN 12316-2	Failure modes samples	
Resistance	$F_{avg.}$ [N/50mm]	F_{max} [N/50mm]	F_{max} [N/50mm]	STN EN 12316-2	STN EN 12316-2
1.A	–	527	1022	C > 5%	TOJ
1.B	–	485	995	C > 5%	TOJ
1.C	–	561	1070	C > 5%	TOJ
1.D	–	575	1068	C > 5%	TOJ
1.E	–	626	1092	C > 5%	TOJ
1.F	235	–	1063	C < 5%	TOJ
2.A	–	572	1088	C > 5%	TOJ
2.B	–	578	1030	C > 5%	TOJ
2.D	–	645	1017	C > 5%	TOJ
3.B	171	–	1032	A = 100%	TOJ
3.C	85	–	1044	A = 100%	TOJ

Note: TOJ (Tear out of joint).

temperature of 420°C (1.F) the values of strength of joints were not satisfactory in terms of peel resistance. It is also evident from the above-mentioned statistics that those joints which showed low values in the peel resistance test, had values for shear resistance test unchanged in comparison to joints of correctly welded membranes. It means that the joint which is under the shear stress can withstand more than the same joint under the peel stress. Next, it was proven that if workers increase the workings speed of an automatic welding machine, it has a negative impact on the quality of the joint. A joint made at a higher speed does not have to withstand the stress to which the membrane is exposed on the roof structure. Cleaning the surface of the welded membranes did not affect the quality of the weld. An uncleaned joint showed the same values as a joint cleaned with the cleaner. However, we have to bear in mind that this joint was not exposed to a potentially polluted environment of an actual construction site.

3   CONCLUSION

This article discusses the issue of mPVC-based waterproofing membranes and describes an experiment which aimed at determining the quality of joints made with various factors in laboratory conditions. The assumption from the practice that mPVC-based membranes are made of a material which is easy to work with in all environmental conditions was also confirmed. The quality of the joints of mPVC-based membranes tends to be good even if welding takes place in non-ideal conditions, such as bad environmental conditions or human errors but of course only up to a certain extent. This does not mean that we can fully rely on the quality of the material of mPVC membranes. It is necessary to constantly monitor the accuracy of the technology and installation processes given by the manufacturer. The quality and functionality of a building really depends on quality of the roof structure. That is why special attention has to be paid to the technological discipline during the installation process and individual steps have to be checked on a regular basis.

REFERENCES

Šveda, M. 2007. *Hydroizolácie na báze asfaltov, Slovenská technická Univerzita, Bratislava: 2007.* 56–72 p. ISBN 978-80-227-2596-5.
Björk, F. 1993. Single-ply roof coverings on flat roofs, *Geotechnical engineering for the preservation of monuments and historical sites; Proc. intern. symp.*, Volume 7, Number 4, 249–251 p.
Gränne, F. & Björk, F. 2000. Joints between roofing felt and sheet metal flashings, short-and long-term tests, *Construction and Building Materials 14*: 21–35.

Advances and Trends in Engineering Sciences and Technologies II – Al Ali & Platko (Eds)
© *2017 Taylor & Francis Group, ISBN 978-1-138-03224-8*

Importance of aerodynamic quantification in the determination of heat losses in buildings

I. Bullová
Faculty of Civil Engineering, Technical University of Kosice, Kosice, Slovakia

ABSTRACT: To determine the ventilation heat losses it is necessary to accept the air permeability of the building envelope, because it significantly affects pressure conditions in the interior. The present article investigates the effect of wall porosity, ratio of openings A(+)/A(−) and opening location on internal pressure in buildings through the openings. In the article were analyzed and quantified the results for selected factors—the impact of urban form and the layout, the deployment and the size of the openings in the façade—influencing the air pressure difference that causes ventilation heat losses. The article found that the internal airflow field has a significant impact on values of the internal pressure, which is not uniform for openings located on adjacent walls.

1 INTRODUCTION

Air flow in buildings is complex, time dependent and multi-directional. The understanding of air flow through and within buildings assumes that wind forces, thermal effects (stack action) and air movement associated with mechanical cooling, heating and other ventilation systems are the dominant factors relating to air pressure relationships.

Many problems associated with pollutant transfer, smoke and fire spread, moisture effects such as microbial contamination, corrosion and biological decay cannot be explained by cross assembly (one dimensional) air flow.

Even the analysis of energy consumption and comfort within buildings needs to be considered in terms of multi-directional air flow.

2 THE AIR PRESSURE DIFFERENCE

The air pressure difference is created at the interface between the internal and external environment. In real environments the interface is formed by buildings and their façade. Air flow through and within buildings has been based on the requirement for continuity of mass and momentum caused by wind forces and thermal effects. The total air pressure difference $\Delta p_C = \Delta p_\theta + \Delta p_v$ ultimately causes the air movement through joints of the façade (infiltration and exfiltration).

The formation of the air pressure difference between the two environments is caused:

1. by different external and internal temperatures of environment Δp_θ (Pa) expressed:

$$\Delta p_\theta = \pm h.g.\left(\rho_e - \rho_i\right) \tag{1}$$

where h = height from the Neutral Pressure Plane—NPP; ρ_e and ρ_i = exterior and interior air density.

2. by wind effects Δp_v (Pa) expressed:

$$\Delta p_v = C_p \frac{v^2 \cdot \rho_e}{2}$$ (2)

where C_p = total aerodynamic coefficient; v = air velocity; ρ_e = outdoor air density.

3   AERODYNAMIC QUANTIFICATION

Aerodynamic quantification takes into consideration the complex wind effects and particular building's parameters and we have to know:

– outdoor climatic parameters—wind speed and wind direction, outside air density,
– aerodynamic parameters of the building—the total—overall aerodynamic coefficient consisting of aerodynamic coefficients of the external and internal pressures.

3.1   *The parameters of the external climate*

Outdoor climatic parameters—wind speed, wind direction and outdoor air density are obtained from the measurements delivered by hydro-meteorological stations and are processed in the test reference years at hourly intervals.

Wind speed is measured at the height of 10 m above the open terrain. Data on wind speed measured at hydro-meteorological stations are not always identical with the actual speed characteristics of a particular site or urban form.

Wind speed depends on factors that operate in the boundary layer. The factors are:

1. change of height above ground level—expressed mostly in the form of power law:

$$v_z = v_{10} \cdot \left(\frac{z}{z_{10}} \right)^b$$ (3)

where v_{10} = wind speed measured at a reference height of 10 meters as; v_z = wind speed at the height of z meters above the ground; b = exponent characterizes the locality—roughness of the terrain.

2. terrain roughness—expressed by the vertical gradient $b = 1/4$–$1/7 = 0.25 - 0.14$, depending on the type of locality (open terrain, urban form).

The speed of the airflow varies depending on the height above the terrain and urban form. Change of height above the ground level—terrain is expressed in the form:

$$v_x = k \cdot v_{10,meteo}$$ (4)

where $v_{10,meteo}$ = wind speed measured at hydro-meteorological stations at 10 m height; k = the coefficient of variation—indicating the impact of urban fabric—built up area and the height above the ground level.

3.2   *Aerodynamic parameters*

When considering the exterior field, the boundary layer at the building envelope surface is typically of primary significance in building analysis.

The aerodynamic effects on buildings are various and depend on the shape and size of the building, the location of buildings in urban sites and on the orientation of the main wind flows. A building—its shape, surface, air permeability—affect the airflow in its immediate adjoining surroundings and the change of the pressure conditions in the building's interior.

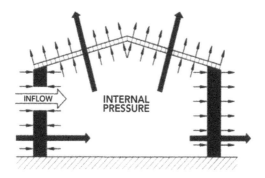

Figure 1. Pressure distribution on the building and effect of openings in a partially-enclosed building.

Wind forces are effective on all buildings, typically creating a positive pressure on the windward wall and negative (suction) pressures on the lateral and leeward walls. This pressure (suction) is expressed in external aerodynamic coefficient C_{pe} (−), depending mainly on the geometric shape of the building and wind direction. Values of C_{pe} (−) are obtained by measuring in the wind tunnels for so-called "Solid model". Several researchers conducted a series of measurements on the models of rectangular shape and investigated the impact of aspect ratio on the external aerodynamic coefficient (Amin & Ahuja 2013).

Values of this pressure coefficient for simple buildings—with a rectangular ground plan, the ratio height/width h/b = 3, the height /length h/l = 2 and vertical exterior walls are: C_{pe} = +0.7 to 0.8 on the windward side and C_{pe} = - 0.1 to −0.5 on the leeward and lateral side (Bielek et al. 1990, STN EN 1991-1-4:2007).

Façade shows a certain degree of the air permeability, which causes the changes of external and internal pressure—Figure 1. Therefore, when dealing with the wind impacts, we must consider the size and dimension of the internal pressure coefficient C_{pi} which acts on the other side of the surface (Stathopoulos et al. 1979, Holmes 1979).

For simple buildings is the internal pressure coefficient C_{pi} a function of the h/b and the ratio of the openings for each wind direction expressed as a function:

$$C_{pi} = f(a) = f\left(\frac{A_{(+)}}{A_{(-)}}\right) \tag{5}$$

where a = the ratio of openings; $A_{(+)}$ = real surface of the openings on the windward wall of a building; $A_{(-)}$ = real surface of the openings on the leeward and lateral sides of a building.

Values a shall be determined for the different wind directions, variable size and layout of the openings and subsequently C_{pi} can be determined utilizing the graphical C_{pi} = f(a) (Bielek et al. 1990, Bielek & Bielek 2005).

Internal and external pressures shall be considered to act at the same time.

4 VENTILATION HEAT LOSS IN THE REFERENCE BUILDING

4.1 Description and basic classification of the reference buildings

A selected reference building with 6 floors (design height of the floor is 3 m) is rectangular and its ground-plan has these dimensions: length: l = 45 m, width b = 13 m, height h = 18 m.

The aerodynamic coefficients C_{pi} were determined for a reference building with windows on all four sides of the building, for different wind directions, the variable size and location of openings (Bullová 2013) utilizing the graph for C_{pi} = f (a) (Bielek & Bielek 2005).

355

Basic classification of the reference buildings and the values of external and internal aerodynamic coefficients for the reference room oriented on the windward and the side walls, towards the wind direction 3600 (on the longer side of the building) and 90⁰ (on the shorter side of the building), for different ratio of openings on each side of the building envelope are given in Table 1.

4.2 Determination of the ventilation heat losses

Based on the more accurate quantification of aerodynamic coefficients were processed values of the ventilation heat losses—expressed (Chmúrny 2003):

$$\phi_v = 1200.\Sigma\left(i_{l,v}.l\right).\Delta pc^m.\left(\theta_{ai} - \theta_e\right) \tag{6}$$

$$\phi_v = 0,33.n.V_m.\left(\theta_{ai} - \theta_e\right) \tag{7}$$

$$n = 3600.\frac{V_{inf}}{V_m} = 3600.\frac{\left[\Sigma\left(i_{l,v}.l\right).\Delta pc^m\right]}{V_m} \tag{8}$$

The parameters of the external climate and the values of air pressure difference depending on wind, on a particular day—January 17th 2016, are shown in Figures 2 and 3. Values of exterior climate parameters are: air temperature and wind speed measured at hydro-meteorological stations and at different heights above the ground—10, 15 and 20 m in the city center.

Table 1. Basic classification of the reference building.

15 m < h < 50 m	The middle high building
0,5 ≤ h/b = 1,38 ≤ 1,5	Spatial proportionality
1,5 ≤ l/b = 3,4 ≤ 4,0	Surface area proportionality

Table 2. Values of external and internal aerodynamic coefficients for different ratio of openings.

	Wind direction on the longer side				Wind direction on the shorter side			
		C_{pi}				C_{pi}		
	C_{pe}	2:1	3:1	4:1	C_{pe}	2:1	3:1	4:1
Windward wall	+0.7	−0.2	−0.15	−0.1	+0.8	−0.6	−0.8	−1.0
Lateral wall	−0.5	−0.2	−0.15	−0.1	−0.5	−0.6	−0.8	−1.0
Leeward wall	−0.3	−0.2	−0.15	−0.1	−0.1	−0.6	−0.8	−1.0

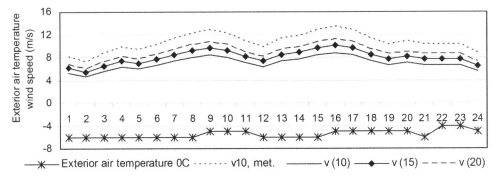

Figure 2. Values of exterior air temperature and wind speed measured at hydro-meteorological stations and at different heights above the ground in the city center on January 17th 2016 in Kosice.

Redistribution of the air pressure inside the building, depending upon the size and proportion of openings on each side of the building, is affected by the location of the reference room towards the wind direction. The effect of the layout and dimensions of openings on each side of the building is shown graphically in Figure 3.

The values of air pressure difference are given for the reference building:

- for windward and side walls,
- wind speed at a height of 10 meters above the ground—open terrain,
- with considering and without considering the impact of openings,
- ratio of openings on each side of the building façade (2:1, 3:1 and 4:1).

Figure 4 shows graphically the values of ventilation heat losses Φ_v on a particular day—January 17th 2016.

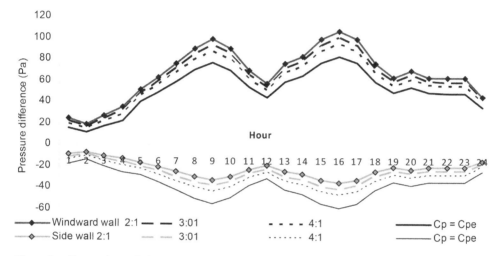

Figure 3. Comparison of air pressure difference for the selected day—on January 17th 2016 in Kosice with considering and without considering the impact of openings for windward and side wall.

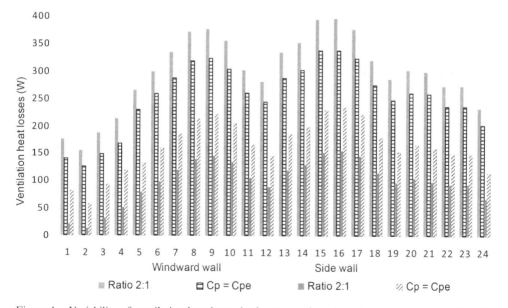

Figure 4. Variability of ventilation heat losses in the room oriented to windward and side walls, with and without considering the impact of openings.

Values have been processed in reference room with volume V = 50 m^3, i$_{1,v}$ = 0.4.10^{-4} m^3/ (m.s.Pa^{0,67}) and the joint length of 14 m, in open terrain at a height of 10 meters above the ground. Values of ventilation heat losses are given for reference room located alternative to windward and side walls, with considering and without considering the impact of openings.

Figures 3 and 4 shows the influence of openings for a room located on different sides of the building. A higher ratio of openings on the windward side increases the effect of the openings, which means a decrease in pressure difference and ventilation heat losses and vice versa.

5 CONCLUSIONS

The air permeability of the building envelope significantly affects pressure conditions in the interior. It had been confirmed in several literature resources (Holmes 1979, Bielek & Bielek 2005).

The exact results of the pressure difference and the ventilation heat losses shows, that acceptance of the air permeability of the façade affecting the pressure conditions in the interior, plays an important role. Therefore it is necessary to calculate the overall aerodynamic coefficient $C_p = C_{pe} - C_{pi}$. The results are significantly influenced by the layout and orientation of openings towards the direction of the applied wind, as well as their size and mutual ratios on every side of the building.

Understanding the complex interactions of the building pressure field—the air pressure regime within and surrounding the building—helps explain the increasingly common failures related to:

- operating costs—energy consumption, maintenance;
- indoor air quality, sick building syndrome;
- smoke and fire spread;
- condensation, corrosion, decay and mold;
- comfort—temperature, humidity and odors;
- in the design of modern double transparent façade.

ACKNOWLEDGMENTS

This article was written as a project VEGA 1/0835/14.

REFERENCES

Amin, J.A. & Ahuja, A.K. 2013. Effects of side ratio on wind-induced pressure distribution on rectangular buildings. *Journal of Structures.*

Bielek, M. at al. 1990. *Aerodynamika budov*, Alfa, Bratislava.

Bielek, M. & Bielek, B. 2005. Interakcia budova—vietor v teórii energetickej potreby budov, In: *6. vedecká konferencia Budova a energia 2005*: 64–69.

Bullová, I. 2013. Aerodynamic quantification—factor influencing heat losses caused by ventilation In: *CzechStav 2013—Sborník příspěvků z mezinárodní vědecké konference: Vol. 4,* Hradec Králové.— Magnanimitas: 40–46.

Holmes, J.D. 1979. Mean and fluctuating internal pressures induced by wind. *Proceedings of 5th international wind engineering conference.* Colorado: 435–440.

Chmúrny, I. 2003. *Tepelná ochrana budov*, Jaga Bratislava.

Stathopoulos, T., Surry, D., Davenport, A.G. 1979. Internal pressure characteristics of low-rise buildings due to wind action. *Proceedings of the 5th International Wind Engineering Conference 1*, Colorado, USA: 451–463.

STN EN 1991-1-4:2007. Eurokód 1. Zaťaženie konštrukcií. Časť 1-4. Všeobecné zaťaženia. Zaťaženie vetrom. Bratislava: Slovenský ústav technickej normalizácie: 2007.

Advances and Trends in Engineering Sciences and Technologies II – Al Ali & Platko (Eds)
© 2017 Taylor & Francis Group, ISBN 978-1-138-03224-8

New methods of optimizing building maintenance

Z. Cúciková, M. Hanko, V. Somorová & R. Matúšek
Faculty of Civil Engineering, Slovak University of Technology in Bratislava (STU), Bratislava, Slovak Republic

ABSTRACT: Recent time is characterized by an increasing trend in the application of sustainable development in civil engineering, including building maintenance. There exist several methods of optimizing maintenance. One of them is the application of state-of-the-art information technologies in the building maintenance process. A detailed analysis of all processes related to the building maintenance right in the design process can affect future maintenance of buildings—whether economic or energy intensity of the maintenance itself. Identification of all maintenance activities of individual building components and technical and technological installations in the proposed service life can determine their financial scope, but also time needs. A suitable version may be selected by a designer according to the future user needs. According to their requirements and resources, this version will be the best for the given group of users within the life cycle process.

1 BUILDING MAINTENANCE

In the legislation system of the Slovak Republic (SR), maintenance is regulated in the Building Act as follows: "A building owner shall be obliged to keep the building in good condition, in compliance with the documentation verified by a building authority and with a building authority's decision (a building permission, a an occupation permit), to prevent the occurrence of any fire and/or sanitary failure, to prevent its devaluation and/or endangering its appearance and to extend its usability to the maximum".

Current status of building maintenance in Slovakia is represented by the following facts:

– investors, designers, owners and tenants do not pay sufficient attention to the maintenance, in most cases no maintenance system with provisional maintenance costs assessment is developed after putting the building into service;
– in newly built buildings, financed from the state budget, an investor or constructor respectively shall be obliged to create a Building Utilization Plane (a building maintenance plan is a part thereof) prior to putting the building into service.

A system step to improve given status is the facility management application, with the building maintenance as an integral part thereof. A Building Utilization Manual is an important aid to help facility managers to manage maintenance processes. It's purpose is to define requirements for failure-free building utilization, avoiding any failures. A part of the Manual is also the maintenance rules determination. By its application, there may be achieved the building effective operation during its operation period and costs and human resources optimization necessary for the maintenance performance (Somorová 2014).

Quality of the Building Utilization Manual layout is directly proportional to its maintenance management quality. The maintenance management quality goal is to minimize costs and maintenance work expenditure, i.e. to optimize the building maintenance. This goal may be achieved through the application of an evaluation methodology for the maintenance of different material versions of building structures in the phase of a design.

The methodology is an important tool of designers in the phase of building design. Using it we can achieve and compare data on the costs and time demands of the execution and maintenance for different versions of particular material design of the construction. The Methodology layout system is presented for the building external constructions, which form percentage highest portion of the building construction from both, the costs as well as the time point of view.

2 EVALUATION METHODOLOGY FOR BUILDING EXTERNAL CONSTRUCTION MAINTENANCE

Within the designing stage, the designer shall decide on the material design of the external constructions by choosing the version from a variety of versions. To make a decision, it is necessary to define for each version the costs and time duration of the execution and maintenance of given version.

The methodology principles including its main goal, is described in the Figure 1.

2.1 Sequence chain in the methodology creation

1. Creation of versions for the external construction material design:

The number of versions in the examined group is optional and depends on the involved parties requirements. External construction version material compositions shall be defined for the selected building. The basic requirement is that all examined external constructions meet the standard "Thermal Protection of Buildings—Thermal and Technical Properties of Building Constructions and Buildings, Part 2: Functional Requirements". The standard defines mandatory values of thermal resistance of external constructions. Every version shall have created a thermal and technical assessment.

2. External construction lifetime determination:

Each of the layers of the assessed external construction versions has its own lifetime. The load-bearing part of the external construction has the longest lifetime—masonry and reinforced concrete skeleton.

3. Calculation table:

For the assessment of the particular external construction versions, in the calculation table there shall be defined:

– execution costs indicator;
– maintenance costs indicator;
– execution work expenditure indicator;
– maintenance work expenditure indicator.

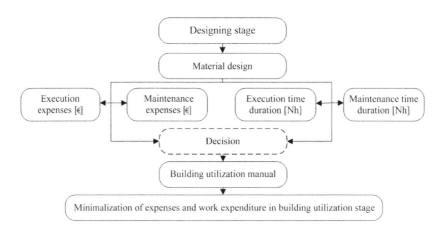

Figure 1.   Evaluation methodology of external construction maintenance.

Execution costs:

For the selected building costs shall be defined for the execution of external construction for particular versions, including the costs for opening fillings (windows, doors). The execution cost indicator to 1 m² for particular external construction versions $C_{C,i}$ [eur/m²] shall be determined by a ratio of the total costs and the building area.

Maintenance costs:

Each of the external case versions shall have exactly determined type of its maintenance. The methodology is based on the assumption that all maintenance activities are carried out in the scope of normal maintenance, characterized by early exchange of components and following prescribed cycle periods with preventive examinations of components and determining their replacement.

The Periodicity of activities and replacement of components with a shorter lifetime is specified by the manufacturer. The periodicity of the activities, such as routine and additional processes for the maintenance of windows and doors, visual inspection and replacement of components with a shorter lifetime shall be met. The periodicity of the cleaning and renewal of external and internal paintings may be adapted according to the designer or investor's requirements respectively.

Costs for the evaluation of all considered maintenance activities is acquired from price offers from companies offering required services and their subsequent averaging to avoid acquired data distortion. The final price will be obtained by multiplying the price for a unit of measure and the number of units of measure belonging to the particular maintenance activity.

Each cost assessed item of the maintenance is a price expressed on the area of 1 m² of the external casing and the selected building object. It is determined its periodicity, i.e. repeating in given intervals. The maintenance cost resulting value in the given year is attained as the sum of all activities belonging to the appropriate year. Considering the fact that the periodicity is different, the resulting value for each year is also different.

The calculation table is created so that in horizontal direction there will be years of the external case component with the longest lifetime. In the vertical direction, in the first line there will be costs for maintenance activities for all periodicities ($c_{m,i}$), whereas in each year of the whole life-time there will be this value variable. In the second, there will be considered the costs for the replacement of components with shorter life-time in the appropriate year of the replacement ($c_{e,i}$) expressed on 1 m² of the external construction. In the third line there will be sum of these values $c_{s,i} = c_{m,i} + c_{e,i}$—maintenance cost indicator.

The values $c_{m,i}$, $c_{e,i}$ and $c_{s,i}$ will be expressed in the current value of money. The value $c_{s,nor,i}$ represents the value $c_{s,i}$ of the appropriate year, which will be increased by the percentage increase (decrease) of process in the civil engineering sector, which may be based on the Statistical Office publications.

From the calculation table gives an indication on the maintenance costs in the whole lifetime of the building (Somorová, 2007).

Table 1. Maintenance activities and their periodicity.

Maintenance activities	Periodicity years
1. Routine maintenance of windows and doors	1×1
2. Additional maintenance windows and doors	$1 \times 5 - 1 \times 10$
3. Visual inspection of external case	$1 \times 5 - 1 \times 10$
4. Cleaning of windows and doors	min. 1×1
5. Cleaning of external case	$1 \times 5 - 1 \times 10$
6. Painting of external surface of external case	$1 \times 10 - 1 \times 15$
7. Painting of internal surface of external case	$1 \times 5 - 1 \times 10$
8. Replacement of components with shorter lifetime	specific

To calculate the line with the value $c_{r,i}$ we need to define the following values:

$$I_d(t) = \frac{1}{(1+k)^{t-1}}$$ (1) (Hromníková, 2004)

where $I_d(t)$ = discount factor; k = interest rate; and t = value of considered calculated year.

The discount factor defines the current value of the future costs as per the end of the year, in which the effect of time distance to the change of values is reflected. To calculate I_d (t) it is necessary to calculate the value of the interest rate k:

$$k = u_B + u_r + d$$ (2) (Ilavský et al. 2012)

where k = interest rate; u_B = basic interest rate of the bank; u_r = global risk rate and d = income tax burden

$$u_r = u_{riz} + u_{PU} + u_{pol}$$ (3) (Ilavský et al. 2012)

where u_r = global risk rate, u_{riz} = risk rate expression; u_{PU} = banking institutions commercial additional charge and u_{pol} = locality effect expression

$$d = \left(u_B + u_r\right) * \frac{100}{100-D} - \left(u_B + u_r\right)$$ (4) (Ilavský et al. 2012)

where d = income tax burden, u_B = basic interest rate of the bank; u_r = global risk rate and D = the absolute value of the tax burden in accordance with the Income Tax Act

the value c_r in the fourth line, representing the future value of the current money, is expressed as follows:

$$c_r = c_{s,nor.} * I_d(t)$$ (5) (Ilavský et al. 2012)

where c_r = future money current value, $I_d(t)$ = discounting factor

The sum of the values $c_{r,i}$ from all years, considered in the calculation, gives us the current value of the maintenance costs in the life cycle of given external case being assessed ($c_{m, \text{life cycle}}$).

Execution work expenditure:

Each from the versions will be equipped with a complete schedule in the appropriate program. There will be defined the area of the building external construction, including openings fillings. The datum on the average work expenditure of the execution of 1 m² of the external construction $P_{R,i}$ [Nh]—execution work expenditure indicator will be obtained by the aggregation of the particular schedule items.

Maintenance work expenditure:

For each version there will be defined all maintenance activities, which will be the same as on the point 4. From the time data of companies providing such services will be acquired nec-

Table 2. Calculation table.

Costs/years	1	2	3	4	5	(90–100)
$c_{m,i}$	–	–	–	–	–	–
$c_{e,i}$	–	–	–	–	–	–
$c_{s,i}$	–	–	–	–	–	–
$c_{s,\,nor.,i}$	–	–	–	–	–	–
$c_{r,i}$	–	–	–	–	–	–
$c_{m,\text{life cycle}}$						

362

Table 3. Output table.

Version	1	2	3
Execution costs C_C	–	–	–
Maintenance costs C_M	–	–	–
Execution work expenditure P_R	–	–	–
Maintenance work expenditure P_M	–	–	–
$c_{m,\text{life cycle}}$	–	–	–

essary standard hours for each maintenance activity. There will be prepared the schedule for the appropriate year and the average work expenditure of the execution of 1 m² of the external construction $P_{M,i}$ [Nh] in the given year, which will depend on the activities belonging to the maintenance—maintenance work expenditure indicator by the aggregation of the items.

For the assessment of the particular external construction versions there will be the following indicators in the output table:

The values representing the costs for the execution and the execution work expenditure of 1 m² of the external construction being assessed are represented only by one value.

The values representing the maintenance costs and the maintenance work expenditure of 1 m² of the external construction being assessed are expressed by several values, which characterize different maintenance activities running in the appropriate year.

From the table there may be found out also the financial cost value for the whole life cycle period of the external construction, which is represented in the current value of money.

The stated resulting data may be used separately in the process deciding on the selection of the most suitable version, but this proposal may be applied also in other areas, especially in the area of modelling in building industry.

2.2 Methodology development parties

Several participants share the use of proposed methodology. Fits of all it is a designer, who makes decisions on material composition of the assessed versions from construction point of view. A budget and schedule shall be prepared by a building preparer. Another important segment is also the facility manager, who is, thanks to his experience in building management, an essential person in the decision-making process. Equally important parts are future owners, whose financial resources and expectations will affect the future result.

2.3 Methodology application period

Versions of the external constructions are prepared according to the methodology in the building procurement stage—in the project preparation. The version assessment should be done during creation of a Building Utilization Manual, as the results of the assessment with subsequent application of the selected external construction version will affect the quality of the maintenance process management, what will affect the building maintenance optimization. The defined maintenance activities may be used for its smooth operation during the building operation itself.

3 BUILDING INFORMATION MODELING

Incorporation of the building construction maintenance evaluation methodology into the BIM (building information modeling) process helps designers and facilitates decisions on the selection of the optimal material version. This is one of the ways the building maintenance optimization.

BIM technology represents basically a process for creation and processing of virtual information on the building, not only on the material and construction part, but also on the technical and technological facilities installed in the building. A result of the process is the model, which is identical to the actually made building in a virtual space (Eastman et al.

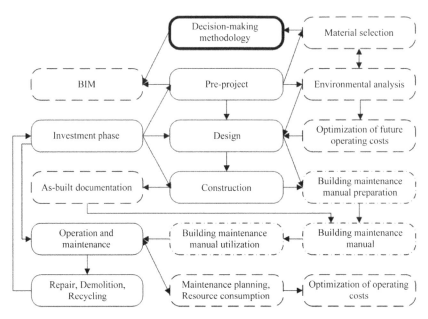

Figure 2. Incorporation of the methodology of maintenance evaluation into the BIM process during the building life cycle.

2011). The **BIM** application facilitates several time-consuming toilsome analyses. Up-to-date technology allows the building materials be assigned with the information on their properties, prices, life-time, and their execution work expenditure. For the analysis of the maintenance version for the building construction material design, the **BIM** technology is an important base from the perspective of time-consumption of their evaluation.

Figure no. 2 shows the way of the methodology incorporation into **BIM** during the building life cycle.

4 CONCLUSIONS

The Methodology of the building construction maintenance evaluation serves as an aid for designers decision making process for material versions. With its application, it gets an overview of the costs and time-consumption of the execution and maintenance of the building constructions. Results of this methodology may be consulted with the investor, a future facility manager or future owners. Therefore, in the final draft version there may be reflected the requirements of all parties taking into consideration their requirements. The methodology gives primary data on the maintenance activities and their financial evaluations also for the Building Utilization Manual. The methodology may be applied also in the **BIM** process, thus providing more accurate analysis of output data. The proposed application reflects and meets the basic concept of sustainable development, which is an integral part of today's building industry.

REFERENCES

Act 50/1976 Zb on Territorial Planning and Building Code (Building Act) as amended.
Eastman, Ch. & Teicholz, P. & Sacks, R. & Liston, K. 2011. BIM Handbook: A guide to building information modeling for owners, managers, designers, engineers and contractors (second edition). New Jersey: Wiley.
Hromníková, M. 2004. Renewal of housing stock. Bratislava: STUV.
Ilavský, M. & Nič M. & Majdúch D. 2012. Real Estate Evaluation. Bratislava: STUV.
Somorová, V. 2007. Construction Object Cost Management Optimization by Facility Management Method. Bratislava: STU
Somorová, V. 2014. Facility management. Prague: Professional Publishing.

Advances and Trends in Engineering Sciences and Technologies II – Al Ali & Platko (Eds)
© 2017 Taylor & Francis Group, ISBN 978-1-138-03224-8

The inclusion of a water feature to the interior and its impact on internal microclimate

K. Čákyová & Z. Vranayová
Civil Engineering Faculty, Technical University of Košice, Košice, Slovakia

ABSTRACT: Nowadays it is increasingly places emphasis on designs of works in harmony with nature and natural elements. And not only within placement of the building into the urban environment, but also within the implementation of those elements into the interior of the building. This trend arise in particular on the basis that people spend time in an artificial environment on average 80–95%. In connection with this facts we speak more often about the quality of indoor air and subsequently about syndrome of sick buildings. This article focuses on the location of water features into the internal environment and its impact on the basic parameters of the internal environment, temperature and humidity of air, and the influence of this parameter on people and their work.

1   INTRODUCTION

Humans in developed countries have, in the past few millennia, advanced from depending on rock shelters, caves, and rude huts to protect themselves from the elements to modem single- and multifamily dwellings and other buildings that provide amenities and conveniences far beyond the basic needs of shelter—conveniences that ensure comfort whatever the vagaries of weather and climate.

Our world is one of the structures that shelter our many activities: the small to grand shells that house a myriad of industrial processes and activities; institutional buildings such as schools, universities, hospitals, and government buildings; automobiles, trains, planes, and ships that provide transportation as well as shelter; shopping malls and office complexes where we trade goods and services; and cinemas, theaters, museums, and grand stadia that provide venues for entertainment. In addition to functional aspects, built environments reflect human aspirations and creativity. They also reflect more fundamental factors, such as the diversity and availability of construction materials, climate, cultural tastes, and human foibles.

The built environments of man are fragile artifacts. They are in constant peril from forces by which the earth renders all things unto itself. Just as water, ice, and wind level the mountains with time, so too do they act to level what man has built. We attempt to keep rain, snow, and wind out of our indoor environments; provide and maintain warm thermal conditions in seasonally cold climates; provide cooler and more acceptable conditions in hot climates; and mechanically ventilate our larger buildings to reduce odors and discomfort associated with human bioeffluents. Our ability to control thermal comfort and other aspects of indoor environments requires the application of a variety of climate-control technologies and a commitment to operate them properly.

2   BIOPHILIC DESIGN

2.1   *The use of natural feature in the interior*

Biophilic design can reduce stress, improve cognitive function and creativity, improve our well-being and expedite healing; as the world population continues to urbanize, these qualities

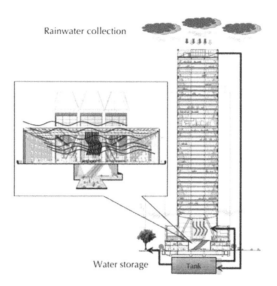

Figure 1. Using of rainwater to create a water feature in the building Hearst Tower, New York (Foster + Partners, 2006).

are ever more important. Given how quickly an experience of nature can elicit a restorative response, and the fact that U.S. businesses squander billions of dollars each year on lost productivity due to stress-related illnesses, design that reconnects us with nature—biophilic design is essential for providing people opportunities to live and work in healthy places and spaces with less stress and greater overall health and well-being. Biophilia is humankind's innate biological connection with nature. It helps explain why crackling fires and crashing waves captivate us; why a garden view can enhance our creativity; why shadows and heights instill fascination and fear; and why animal companionship and strolling through a park have restorative, healing effects. Biphilia may also help explain why some urban parks and buildings are preferred over others. A space with a good presence of water condition feels compelling and captivating. Fluidity, sound, lighting, proximity and accessibility each contribute to whether a space is stimulating, calming, or both (Browning, et al., 2014).

2.2 Hearst tower, NY—example of water features in the interior

Harvested rainwater from the roof is fed into a central tank and used for irrigation and to feed the water feature. This, alongside water efficient fixtures and fittings, has led to a 30% reduction in water usage compared to a typical building. At the base of the tower, there is a large atrium space for the building users. A water feature helps enhance the microclimate by thermally tempering the space, providing acoustic dampening and humidity control. The waterfall cools the space in the summer and humidifies it in the winter (Foster + Partners, 2006).

The biophilia hypothesis contends that natural elements are calming for people today because of the linkage to survival in the past (just as common fears such as snakes, spiders, and heights are rooted in the past and related to survival): "Throughout human existence, human biology has been embedded in the natural environment" (Wilson E.O., 1984).

The example shows that the natural element (water feature), which is part of the interior may have also affect the physical parameters of the internal environment.

3 INDOOR AIR QUALITY

One of the most important goals for building design is to satisfy the people occupying the building. This satisfaction is related to the thermal comfort, Indoor Air Quality (IAQ),

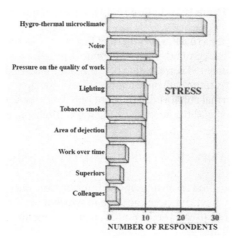

Figure 2.   Environmental factors (stress) that people are feeling especially in the interior of the building (Jokl, 2011).

acoustical and visual comfort. Among the factors influencing perceived IAQ and human comfort, indoor Relative Humidity (RH) is one important parameter. High moisture levels can damage construction and have an adverse effect on health (Bornehag, et al., 2001). High humidity harms materials, especially in case of condensation and it helps mould development increasing allergic risks. Microbial growth, encouraged by high relative humidity, leads to poor IAQ and building deterioration. Poor IAQ results in discomfort, health problems and could ultimately lower productivity and spawn lawsuits. Many people are concerned about wood floors and wood furniture being damaged if humidifiers are not installed. Besides, static electricity can be greatly diminished by maintaining the relative humidity of air in this level because dry air can also cause static electricity in an environment (Maalouf, at al., 2016)

3.1   Impact of indoor environment on performance and user satisfaction

The American Lung Association estimates that most people spend 90% of their time indoors. This confirms that the IAQ is a very important factor in maintaining healthy lives. Finally, it is noted that air quality affects productivity in offices. The performance of simulated office work is estimated to increase on average by 1.5% for every 10% decrease in the percentage of persons dissatisfied with the air quality (Maalouf, at al., 2016).

According to a survey carried out by the departments of banks and insurance companies in Germany (the German Trade Union, Bank and Insurances, HBV), almost a third (27.1%) of employees in the interior of buildings complained about hygro-thermal microclimate, 13.5% more noise, 10.6% for lighting, 10.2% of the tobacco smoke, and 9.9% of the area of dejection, i.e. area accounts for over 70% (71.3%) of discomfort at work (Fig. 1 1). At least they are making people worry about work over time (8.9%), with their superiors (4.0%) and colleagues (2.9%) (Weber, 1995).

Proximity and availability of the natural environment can foster many desired outcomes, even if the employee does not spend a great amount of time in the natural setting. Already view from the window to nature can contribute to improving the working environment to improve the work effectiveness. While providing windows at work may not be a simple matter, other ways to increase contact with nature may provide a low-cost, high-gain approach to employee well-being and effectiveness (Kaplan R., 1993).

3.2   Basic parameters of indoor climate

Indoor climate (its physical condition) describe the various climatic factors indoor environment—composition of air, air temperature and other heat properties, humidity air,

air velocity, vibration of air (sound), radiation (light), electrical and magnetic properties of the environment.

Of these factors appear as the crucial hygro-thermal factors. If not secured the desired level, other factors relegate to the background. Thermal comfort is defined as a feeling of satisfaction with the thermal environment. Thermal state of the indoor environment affects four basic parameters—ambient air temperature, effective temperature of the surrounding areas, air velocity and relative humidity. Thermal comfort depends not only from those stated parameters, but also on other related on person and the purpose of the room (Székóva, et al., 2004).

4 AIR TEMPERATURE AND RELATIVE HUMIDITY

The first condition of the satisfaction of man in the interior from thermal state of the internal environment is the thermal neutrality, ie the state in which a person does not feel heat or cold. Placement water features affects the basic parameter of the internal climate, and that temperature and humidity of air.

4.1 *Air temperature*

The optimum air temperature depends on the purpose room—internal heat production, the type of clothing but also on the person's age, gender, etc. The Central European conditions can be considered the most suitable temperature for an average sitting man dressed up in the interior that does not physically work, performs intellectual work or resting, the air temperature 18–22°C in winter and 23–25°C in summer (Székóva, et al., 2004).

Higher temperatures in the summer substantiates that man is easier to wear, and thus to remove the same amount of heat needed for the same amount of body surface temperature ambient. It is particularly important vertically uneven temperature distribution in the heating, respectively. air-conditioned rooms. This vertical distribution arise in particular due to uneven heat input and uneven cooling of individual walls of the room. Vertical unevenness is a particular problem with convection heating method, which is at a higher air circulation to promote an increase in temperature-dependent height of the space (Székóva, et al., 2004).

4.2 *Relative humidity*

The relative air humidity is an important parameter to express the thermal comfort in interior, but also an important parameter from a hygiene point of view. In assessing the thermal environment with air temperature, the humidity can be disregarded only if the relative humidity is in the summer less than 60%. A person in a room at 20°C and almost does not perceive the

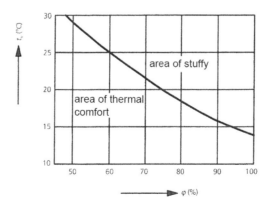

Figure 3. Dependence of air temperature and relative humidity on the overall well-being (Székóva, et al., 2004).

difference between the values of humidity of 35%–70%. These values are considered to be beyond the allowable humidity (Székóvá, et al., 2004).

4.3 *Low relative humidity of air*

At low relative air humidity respiratory mucous membranes drying out, reduces the formation of mucus and the cilia activity of the nasal mucosa, thereby weakening the human defense mechanism against the penetration of microorganisms, including aerosols and allergens in the human body. Microscope can be seen that the cilia (cilia) in the mucosa are in constant motion, what pre-vented the accumulation of dust. Slime formation depends mainly on the relative humidity of the inhaled air, if it falls below 30%, its formation decreases rapidly and the movement of cilia as well. Bacteria and viruses are the optimal conditions for their development (Jokl, 1993). Low relative air humidity will adversely reflected on the skin and eyes, too. The human body has no sensors to moisture, but is very sensitive to changes in humidity. A decline from 25% to 15% is able to perceive as negative (Jokl, 2011).

The effect of 15% relative humidity on feelings of the workers in office buildings has been studied in Sweden (540 in the office building 160, 4943 respondents). 54% of subjects reported that the air was too dry. The most common were indicated dry mouth and throat (31%), also on the face (44%), dry eye (36%), lips (38%), and also a cold (46%) (Jokl, 2011).

4.4 *High relative humidity of air*

According to the Mollier diagram it is clear that the lower the temperature, the higher the humidity. It is therefore necessary to take into account the ambient temperature. It is a direct effect of high RH to respiratory tract infection. The indirect effect is the impact on the growth of fungi that can cause infections and allergic reactions. Mold appear on the inner surface of the outer walls when the RH in the space for a longer period of time exceeds 70%. Lime plaster is absorbed about 40% of the nascent humidity in interior. Absorption takes about 30 minutes, but for re-release is required several hour. Adults often suffer from nausea, vomiting, shortness of breath, constipation, back pain, fever and neurological difficulties. The number and severity of these problems grow proportionately with moisture living rooms. Relative humidity indoors should not exceed 70% if it is to be prevented the proliferation of mold and mites. Failure to comply with this value then determines the required air exchange. Research in Denmark showed that the mite in Denmark are the main source of allergies—are the cause of allergies in 200,000 inhabitants. On following place are coat of dogs and cats, mold and cigarette smoke. Humidity is not affected by smells, while high humidity reduces the perception of odors: RH over 80% of already completely eliminates the perception of both pleasant and unpleasant odors (Jokl, 2011).

4.5 *The changes of state of air at its contact with wet surfaces (water level)*

In contact air with a wet surface (water drops, water running down by webbing solid porch, water level and so on.), some water molecules break away from the surface and penetrate into the air, while other water molecules that move linear and high speed, hit the surface that swallows them up. The frequency of phase transformations depends on kinetic energy of the molecules. Depending on which direction (from the air on a wet surface or vice versa) resulting power flow prevails, we observed precipitation by water vapor or evaporation of water. Changes of state in the air over a wet surface produces phenomena that take place at the phase boundary. General can have a layer of air-saturated above wet surface a different temperature, or another specific humidity as surrounding air. Between the air in this layer and high above it then there exists temperature, possibly moisture potential. Thermal potential to cause transfer of sensible heat and moisture potential transfer of moisture. At a lower surface temperature than the air temperature, set wet cooling air. While if the surface temperature higher than the dew temperature of the air, in addition to cooling the air and humidify. If the temperature of surface is lower, water vapor from the air in it condensing and air in addition to cooling also dehumidify. During cooling air passes therefrom on a wet surface heat flux by convection, which causes a decrease air temperature and evaporation of water (Széková, et al., 2004).

5 CONCLUSION

Microclimate conditions, especially temperature and humidity, extremely influence performance when they are in extreme values. Extremely high temperatures and humidity, as well as extremely low temperatures and humidity, adversely affect the performance and its quality. To ensure a suitable working environment, working conditions and being disturbed over all mental functions in work activity must be set optimum values of each microclimate factors and their mutual links and conditionality. Keeping the indoor RH in a desired level are necessary and essential to improve building performance in the terms of IAQ, energy performance and durability of building materials.

Use of natural elements in the interior may have a positive effect on human labor, its performance and overall well-being. Biophlic design and design in harmony with nature and natural elements are becoming increasingly popular. One approach to bring a natural element in the interior is the useage of water features. However active water element (water running down the glass) is not only an aesthetic element, based on the above examples and the theory it is expected to influence the internal microclimate, especially humidity and temperature. This influence can have positive or negative effects in terms of hygiene requirements for internal microclimate. We should also say that I for more details and the extent of this, the affected carry out a further investigation, because it is a complicated issue which impinges on several scientific disciplines.

ACKNOWLEDGEMENTS

This work was supported by: VEGA 1/0202/15 Bezpečné a udržateľné hospodárenie s vodou v budovách tretieho milénia/ Sustainable and Safe Water Management in Buildings of the 3rd. Millennium.

REFERENCES

Bornehag, C. G., G. Blomquist, F. Gyntelberg, B. Jarvholm, P. Malmberg, L. Nordvall, A. Nielsen, G. Pershagen, and J. Sundell. 2001. Dampness in Buildings and Health. Nordic Interdisciplinary Review of the Scientific Evidence on Associations Between Exposure to 'Dampness' in Buildings and Health Effects (NORDDAMP). IndoorAir 11 (2): 72–86.

Browning, W.D., Ryan, C.O., Clancy, J.O. 2014. 14 Patterns of Biophilic Design. New York: Terrapin Bright Green, LLC.

Foster + Partners, 2006. "Heast tower". [online] [available] Retrieved from http://www.fosterandpartners.com/projects/hearst-tower/

Jokl, M.V. 1993. The Theory of Microenvironment. In Czech. CTU Publishing House, Prague 1993, p. 263.

Jokl, M. V. TEORIE VNITŘNÍHO PROSTŘEDÍ BUDOV. 2011. [online] [available] Retrieved from http://www.ib.cvut.cz/sites/default/files/Studijni_materialy/TVPB/Teorie_vnitrniho_prostredi.pdf

Kaplan, R. 1993. The role of nature in the context of the workplace, Landscapeand Urban Planning, pp. 193–201.

Maalouf Ch., Tran Le A. D., Douzane O., Promis G., Ton Hoang Mai & Thierry Langlet. 2016. Impact of combined moisture buffering capacity of a hemp concrete building envelope and interior objects on the hygrothermal performance in a room. Journal of Building Performance Simulation.

Széková, M., Ferstl, K., Nový, R. 2004. VERTANIE A KLIMATIZÁCIA. JAGA: Bratislava.

Weber, J.H. 1995. Sick building syndrome-dangerous game with spread characters". Air Infiltration Review 16, 3:12–13.

Wilson, E.O. 1984, Biophilia: the human bond with other species. Cambridge (MA): Harvard University Press.

Advances and Trends in Engineering Sciences and Technologies II – Al Ali & Platko (Eds)
© 2017 Taylor & Francis Group, ISBN 978-1-138-03224-8

A human induced landslide in the centre of a town

W. Dąbrowki, M. Zielina & J. Bąk
Institute of Water Supply and Environmental Protection, Cracow University of Technology, Cracow, Poland

ABSTRACT: The potential reasons for a dangerous landslide are discussed. A valley historically drained by a seasonally flowing stream was filled with soil, levelled out and used for housing development. The estuary of the stream was crossed by a busy avenue constructed on a dyke. Forty years after the development, the road suddenly slid down, demolishing the underground infrastructure and several garages. The local waterworks company has been accused for causing the slide because of operation of two drinking water pipelines. However, the consideration of several potential causes for the accident concludes in the statement that the blocking of the groundwater outflow from the former valley could be the main or even the sole cause. Long term leaking of the water pipelines and storm water sewerage could also have additionally contributed to the accident, but a possible pipe break or disconnection has been excluded as a potential cause of the disaster.

1 INTRODUCTION

Human life and health are the highest and most important values in a society, therefore, all efforts should be taken to avoid disasters in the urban space, which is also the living environment of man. This should concern, especially, the architecture and technical infrastructure. Even in ancient times, the Roman architect Marcus Vitruvius Pollio (who lived in the 1st century BCE) in his work 'De Architectura libri decem' (Vitruvius, 1956) stated that "architecture is to keep to three principles: durability, utility and beauty". This principle has numerous references in contemporary technical literature. As examples can be cited works such as Gronostajska 2009, Kosiński 2009 and Marcianiak 2009. The aforementioned 'durability' in this aspect can be understood as safety. All buildings, including public roads, should be durable and safe for all users at all times.

All activities during the design process should aim at avoiding disasters as well as their prevention. The risk of accidents and incidents should be minimised and safety maximised. For these reasons, it is extremely important to examine in detail the disaster that took place and to discover their causes in order to avoid similar situations in the future, or better cope with their consequences.

Operational problems with roads happen relatively frequently. There are various events including small problems, such as cracks in the road and small potholes, and bigger problems, such as large landslides. Among the major disasters related to roads in recent years, some well-known ones are listed here. In 2014, a roadway collapsed in Manhattan. A bus with 11 children fell into the formed hole. Large temperature differences may have contributed to this crash. According to the New York road builders, the hole in the road was created by the malfunction of an aqueduct (cf. www.polskieradio.pl, 2014). In 2015, in Brazil, a roadway collapsed under a bus; the vehicle fell into the crater created and was then carried away by rushing flood water (cf. www.reuters.com, 2015). In 2016, in the town of Guiyang in China, a bus was suddenly swallowed up by a very large hole that formed in the road (cf. www.dailymail.co.uk, 2016). It should be noted that these events are only a few examples from a long list of disasters of this type.

The causes of each aforementioned disaster are different so it is important to properly recognise them and, if there was human error, to eliminate them. We try to limit and monitor effects of dangerous situations caused by natural factors. We can also try to prevent and counteract with danger arising with the participation of a human factor. In order to take such action is necessary knowledge of the possible causes of such phenomena. In this work, the potential reasons of disaster in the centre of a town are discussed.

2 DESCRIPTION OF THE PROBLEM

A valley drained by a stream has been filled in with soil. The surface of the soil has been profiled horizontally and supported by gravity and cantilever walls, without drainage. The embankment, eight metres high, covered the bridge over the stream. The bridge was removed from the maps in the 1960s and forgotten about by the living population. In general the changes in the water table elevation during a hydrological year are such, that the highest groundwater levels occur usually in January for oceanic climate, while for continental climate sometime between April and May (Rethati 1983). The area of interest is located nearby the boundary between oceanic and continental climates. Just before a midnight, on 23rd December 2012, a sudden landslide occurred, breaking up a busy road and the underground infrastructure including: two drinking water mines, old storm sewers, warm water pipelines and electricity and telephone cables. The landslide demolished two rows of garages but fortunately no-one lost their lives. The main causes of the landslide were officially recognised as: liquefaction of the soil and poor soil mitigation, including the retaining walls. The local water supply and wastewater disposal company was accused of being responsible for operating 55-year-old cast iron pipelines, whose breakage was supposed to have caused the soil liquefaction.

A different opinion is presented here, as a sudden break of the pipeline could be a trigger, but definitely not the cause of the problem. At this stage it is necessary to realise a

Figure 1.   A view of the location of the landslide from the level of the road.
Source: Photograph taken by the Ostrowiec Świętokrzyski waterworks company.

difference between these two terms. The cause of a disaster is the process responsible for an accident—tending to the loss of strength of the soil finally resulting in a collapse of the retaining walls, but not necessarily at this specific point in time. In contrast, the trigger is a single event causing a landslide to occur at a particular time, but not responsible for this specific disaster in the sense that, even if not at this time, the landslide would not have been avoidable in the future.

A view of the location of the landslide from the level of the road is presented in Figure 1. The vertical surface of the culvert wall, presented in Figure 2 below, created an ideal opportunity for a sudden soil movement. Existing fragments of an old and non functioning storm water sewerage system were totally clogged by fine sandy-clay. The critical shear stresses initiating sediment erosion processes in sewers and open channels strongly depend on void ratio (Partheniades 2007), the pipe diameter and the depth of flow (Nalluri, Dąbrowski 1994). The circular sewers totally clogged by sediments are the evidences of the soil erosion process in the catchments of the storm water sewerage.

From the underground infrastructure, only a steel pipe remained undisturbed after the accident. It is well known that welded steel pipe connections are more resistant to unstable soil than any ductile iron pipe fittings, including even a variety of blocked gasket fittings. Maybe the earthquake-resistant ductile iron pipes are of a similar durability under the stress induced by sudden soil movement, but the mechanism of protection is the opposite. In the case of steel welded pipes they resist the soil movement, while earthquake-resistant ductile iron pipes change their position, without leakages from joints. In the area of the disaster, old-fashioned grey iron pipes were used. In spite of the weaknesses of their connections, perhaps they supported the asphalt road surface layer over a short period of time after the slide before the pipelines collapsed. This supposition is based on the observation documented in Figure 3. A part of the road surface has not been displaced in the direction of the soil slide, but instead failed vertically downwards.

Figure 2.   A vertical culvert wall being a relatively smooth boundary of the soil slide.
Source: Own photograph.

Figure 3.   A large piece of the road surface at the bottom of the hole.
Source: Photograph taken by the Ostrowiec Świętokrzyski waterworks company.

3   A SUDDEN OR A LONG STANDING PROCESS?

During a storm the infiltration velocity into the soil is more rapid at first and then decreases to a constant value depending on the soil type. Several approaches have been used for modelling this process (Rossman 2010, Turner 2006), and the Horton and Green-and-Ampt methods are the most common. In the first method, the infiltration rate into the soil is calculated as the sum of the minimum infiltration value for the soil saturated with water and the second component. This component is the product of the difference between the initial and the minimum value of the infiltration rates multiplied by "e" raised to the power "–kt", where t is time and k the decay coefficient, expressed in 1/s. For a long period of time, the infiltration rate is asymptotic to the minimum value, which for loam and silt loam is recorded usually in the range of between 3.8 and 7.6 mm/h. In the case of a sudden break of a ductile iron pipe, the infiltration around the place of leakage is governed by a high pressure but, at a distance of a metre or more in silt loam the pressure is much lower.

The flow through the saturated zone far away from a pipe leakage can be recognised as being governed by Darcy's law. The filtration coefficient for compacted silt loam and loam soils takes values somewhere between $0.9 + 10^{-6}$ and $4.6 + 10^{-6}$ m/s (Davis S,N, 1969). Assuming a pressure of 5 bars at the place of the break and the atmospheric pressure at the outflow from the soil, it is evident that it would take at least two days to cover the distance of 20 metres if the flow would be bounded to a constant cross section in only one direction. In the case of a pipe break or disconnection, the water flows in all directions, so the liquefaction of the soil at a distance of twenty metres would take at least several days. In spite of the inadequacy of infiltration and Darcy's flow models to the situation of water flow from a broken pipe through, at the beginning, unsaturated soil and then saturated soil in all directions, it is evident that liquefaction of the soil would take such a long period of time that the outflow through the ground surface would be both well visible and alarming. Such an outflow would release the high pressure of the water in soil, drastically decreasing the filtration velocity. In consequence, a pipe break could be the trigger, but not the cause of the landslide.

4 LIFETIME SERVICE

The cast iron gasket pipes were connected using rope and lead sealing. Such sealing was flexible but not reliable as small leakages could occur after any movement of soil and the deformed lead ring would never go back to its previous shape. The pipelines of low diameters should be constructed below pavements, preferentially one metre far from the kerbside. However, because of the road broadening, one of the cast iron pipe-lines was finally located below the roadway. Shocks from the traffic could contribute additionally to small leakages and, as a consequence, to liquefaction of the soil. It is questionable how long such pipelines should remain in operation before renovation or replacement. According to the DIPRA report (Bonds et al. 2005), more than 600 utilities in the United States of America and Canada keep cast iron pipes older than 100 years in operation, and 20 utilities above 150 years. Usually this time is much shorter but the frequency of failures is usually one of the key factors regarding the decision of replacement or renovation. In Japan rigorous regulations were established requiring the replacement of cast iron pipes after fifty years of operation. However, because of the higher mortality than birth rate, even the Japanese economy is not able to cover the expenses of all required replacements, and more that 10% of grey iron pipes are over 50 years old. In spite of this, water losses in the supply systems are estimated as being below 1.7 percent across the whole country. The life period of modern ductile iron pipes produced by Kubota is expected to reach 100 years in moderate soil-water conditions.

5 CONCLUSIONS

The analysis of the available materials enables the authors to put forward the following conclusions:

– neglecting of the natural formation of the valley drained by a stream, filling this valley in with soil and protecting the embankment by retaining walls without drainage could be one of the main, or even the only, cause for the landslide;
– a long wall previously supporting the bridge was one of the slide planes;
– a sudden break of cast iron pipes is unlikely but can happen from time to time. It is not possible to predict the place of such a break in advance, because in most cases such an accident requires both a material defect and corrosion destruction;
– such a break could be one of possible landslide triggers, but not the cause;
– the lifetime of cast iron pipes depends on the soil corrosivity, protection methods and the cast iron's quality. In corrosive soils, the pipes are usually replaced after fifty to sixty years, but there are many utilities keeping cast iron pipes older than 100, 120 and some even older than 150 years in operation. Such cases are the exception, and can be explained by much thicker pipe walls produced in old casting technologies, as well as low soil and transporting water corrosivity;
– small leakages from cast iron pipes connections or small leakages through the pipes' walls because of graphitisation could be an additional likely, but unproven, cause of the landslide. The decision about replacement or renovation of cast iron pipes is taken usually based on the experience of the failure frequencies;
– in hilly areas the law should protect land and buildings against so-called Best Management Practice, using drainage and storage of underground storm water (Dąbrowski 2011a, Dąbrowski 2011b);
– the safety of the people should be one of the key aspects in the design phase, modification or erection of buildings and their subsequent operation;
– it is important to provide designers with access to all the important documents, including historical maps, of a location;
– some research carried out in the field is crucial in the design process.

There are disasters caused by natural factors and disasters caused by the so-called human factor. There are also events that are the result of both natural processes and human negligence.

It is extremely important that, if possible, eliminate any human errors. Knowledge of the causes of such disasters and experience is needed for such activities. The required cognizance and information can be obtained from the analysis of the reasons of other disasters.

It seems necessary that the results of studies to identify the triggers and the causes of disasters show for creators. This knowledge will help them to protect the highest value in society—human life.

It is crucial that designers have always had in mind one of the pillars of the Vitruvian triad, mentioned previously durability of created objects. Durability is in fact inextricably linked to the safety of people who use buildings and objects of infrastructure.

REFERENCES

Bonds, R.W., Bernard, L.M., Horton, A.M. & Oliver, G.L. 2005. Corrosion and corrosion control of iron pipe: 75 years of research, *Journal AWWA*, 97(6): 88–98.

Davis, S.N. 1969. *Flow through porous media*, Academic Press, New York–London.

Dąbrowski, W. 2011a. Storm water of the neighbour, part I, The best practice of governing, *Instal*, 11: 50–54.

Dąbrowski, W. 2011b. Storm water of the neighbour, part II, Our theoretical privileges, *Instal* 12.

Gronostajska, B. 2009. Vitruvius' theory in contemporary architecture of XXI century. *Czasopismo Techniczne. Architektura*. 106 (1-A): 45–49.

Kosiński, W. 2009. The Vitruvian message—seriousness, classicism, minimalism. *Czasopismo Techniczne, Architektura*. 106 (1-A): 60–67.

Nalluri, C., Dąbrowski, W. 1994. Need for new standards to prevent deposition in wastewater sewers, Journal of Environmental Engineering, *ASCE*, 120 (5): 1032–104.

Marciniak, P. 2009. Topicality of Vitruvius' theory and the change of the architectural paradigm. *Czasopismo Techniczne. Architektura*. 106 (1-A): 392–396.

Partheniades, E. 2007. *Engineering Properties and Hydraulic Behavior of Cohesive Sediments*, CRC Press, Taylor & Francis Group, Boca Raton, London, New York.

Réthàti, L. 1983. *Groundwater in Civil Engineering*, Akadèmiai Kiadó, Budapest.

Rossman, L.A. 2010. *Storm Water Management Model user's manual, version 5.0, U.S.* Environmental Protection Agency, report EPA/600/R-05/040, July.

Turner, E.R. 2006. *Comparison of infiltration equations and their field validation by rainfall simulation*, University of Maryland, Master of Science thesis, 2006.

Vitruvius, M.P. 1956. De architectura libri decem–Ten books about architecture, Państwowe Wydawnictwo Naukowe, Warszawa.

http://www.dailymail.co.uk/news/article-3485971/Now-don-t-Shocking-moment-bus-disappears-sinkhole-China-matter-seconds.html, access 14.07.2016.

http://www.polskieradio.pl/5/3/Artykul/1062056,Na-Manhattanie-zapadla-sie-jezdnia-W-dziure-wpadl-szkolny-autobus-wideo, access 14.07.2016.

http://www.reuters.com/video/2015/03/25/brazilian-bus-falls-through-hole?videoId=363619110, access 14.07.2016.

Advances and Trends in Engineering Sciences and Technologies II – Al Ali & Platko (Eds)
© 2017 Taylor & Francis Group, ISBN 978-1-138-03224-8

Implementing effective Supply Chain Management (SCM) strategies

N. Daneshjo & M. Kravec
Economical University in Bratislava, Košice, Slovakia

A.A. Samer & J. Krivosudská
Faculty of Mechanical Engineering, Technical University of Košice, Slovakia

ABSTRACT: Supply chain management is a major issue in many industries as firms realize the importance of creating an integrated relationship with their suppliers and customers. Managing the supply chain has become a way of improving competitiveness by reducing uncertainty and enhancing customer service. The role of planning and coordination in complex integrated systems and information technology to synchronize the supply chain is described in a framework that creates the appropriate structure and installs proper controls in the enterprise and other constituents in the chain.

1 INTRODUCTION

The primary purpose in establishing supply chains is to minimize the flow of raw materials and finished products at every point in the pipeline in order to enhance productivity and cost savings. Successful supply chain ventures manage some critical elements for parts such as individual business unit in the entire supply chain. The strategy covered in different aspects contributes to the overall performance.

Supply chains encompass the end-to-end flow of information, products, and money. For that reason, the way they are managed strongly affects an organization's competitiveness in such areas as product cost, working capital requirements, speed to market, and service perception, among others. In this context, the proper alignment of the supply chain with business strategy is essential to ensure a high level of business performance.

Leading ISCM practitioners display certain common characteristics. For one, they focus intensely on actual customer demand. Instead of forcing a product into the market that may or may not sell quickly (and thereby inviting high warehousing costs), they react to actual customer demand. And by doing so, these supply chain leaders minimize the flow of raw materials, finished products, and packaging materials at every point in the pipeline.

2 ESTABLISH SUPPLIER RELATIONSHIPS, INCREASE CUSTOMER RESPONSIVENESS

It is important to establish strategic partnerships with suppliers for a successful supply chain. Corporations have started to limit the number of suppliers they do business with by implementing vendor review programs. These programs strive to find suppliers with operational excellence, so the customer can determine which supplier is serving well. The ability to have a closer customer or supplier relationship is very important because these suppliers are easier to work with.

With the evolution toward a sole supplier relationship, firms need information such as financial performance, gain-sharing strategies, and plans for jointly designed work. They may establish a comparable culture and also implement compatible forecasting and information technology systems. This is because their suppliers must be able to link electronically into the customer's system to obtain shipping details, production schedules and any other necessary information.

Seen from 3PL companies in inbound processes, building relationships to suppliers plays important roles in providing high quality service in satisfying both supplier and manufacturers.

To remain competitive, firms focus on improved supply chain efforts to enhance customer service through increased frequency of reliable product deliveries. Increasing demands on customer service levels are driving partnerships between customers and suppliers. The ability to serve their customers with higher levels of quality service, including speedier delivery of products, is vital efforts. Having a successful relationship with a supplier results in trust and the ability to be customer driven, customer intimate and customer focused.

3 BUILD A COMPETITIVE ADVANTAGE FOR THE PRODUCT ORIENTED CHANNEL, INTRODUCE SCM SOLUTIONS AND ENABLING INFORMATION TECHNOLOGY

Achieving and maintaining a competitive advantage in an industry is not an easy undertaking for a firm. Many competitive pressures force a firm to remain efficient. Supply chain management is seen as a competitive advantage for firms that employ the resources to implement the process. It also serves to increase the influence on the channel because these firms are recognized as the leading edge and are treated with respect.

A competitive advantage can be supply chain management, which would help firms to implement better processes. Attaining a competitive advantage in the channel comes with top management support for decreased costs, waste management, and enhanced profits. Many firms want to push costs back to their supplier and take labour costs out of the system. These cost reducing tactics tend to increase the competitive efficiency of the entire supply chain.

Firms have become more market channel focused. They are observing how the entire channel's activities affect the system operation. In recent times, the channel power has shifted to the retailer. Retailer channel power in the distribution channel is driven by the shift to some large retail firms, such as Wal-Mart, Kmart, and Target. The large size of these retailers gives them the power to dictate exactly how they want their suppliers to do business with them. The uses of point of sales data and increased efficiency of distribution also have been instrumental in improving channel power and competitive advantage.

Information is vital to effectively operate the supply chain. The communication capability of an enterprise is enhanced by an information technology system. However, information system compatibility among trading partners can limit their capability to exchange information. An improved information technology system where partners in the channel have access to common databases that are updated in real-time is needed. The stages of Supply Chain management include:

The fundamentals. Basic tools such as spreadsheets are used to help ensure delivery of quality goods at a reasonable, predictable cost. Management focus is internal, built around independent departments, with a Premium placed on achieving repeatable results through a standard operating procedures and automation. Companies see their management mission in supply chain management only as controlling finished goods, transportation, and warehousing. They emphasize expediting today's workload above all.

Cross-functional teams. Management focus is on consolidation at the operational level, bringing together people from manufacturing, logistics, and customer service to solve problems. l *Integrated enterprise.* Management focus moves from consolidated operations to an integrated supply chain, at least internally. All enterprise functions, ranging from finance to logistics, as well as related business units composing the internal supply chain, are integrated to form the foundation of a unified business system. Achieving this objective requires not only the close synchronization of all daily operational and planning processes, but also the removal of departmental biases and the establishment of strategic congruence and consensus.

Extended supply chain. Companies recognize the competitive advantage and the potential for profitable growth in extending integration to trading partners. As much as possible, functions throughout the supply chain are integrated to form the foundation of a unified value chain. In addition, mass customization of goods and services and finer segmentation of customer groups becomes the norm. To make this possible, there is interoperability within customer and supplier systems, even though they may involve different computing architectures, operating systems, applications, data definitions, and performance metrics. The extended supply chain stage requires organizations to develop mutual, well-defined objectives, advanced negotiating skills, knowledge of critical business processes, and the ability to work and make decisions that span the operations of multiple channel partners. In addition, this stage requires organizations to create and empower effective intra-channel and inter-channel process teams. These process teams assume fundamental agreement of operational objectives both within individual organizations and between channel members.

Supply chain communities. The capabilities and capacity for innovation found among individual organizations included in a supply chain are fused into a single competitive entity. Networks of preferred suppliers are created in this stage. When this occurs, a synchronized planning solution is in place along with a collaborative capability to connect operations with ever-changing sets of trading partners.

4 SCM—A FRAMEWORK FOR ANALYSIS

We will begin with the analysis of the supply chain framework, in order to gain a clear understanding of what a supply chain is about, its basic tenets of linking structural strategies with prescriptive strategies.

Many strategists agree that firms may not be able to rely either on a price leadership role or on a differentiation strategy alone to guarantee sustained market strength. To sustain long-term growth, however, combinations of both strategies are typically needed to operate effectively within constraints imposed by the environment. Such is also the case for a supply chain of products and services offered by a firm. However, since a number of autonomous business entities belong to the supply chain network, it becomes imperative to develop a common mission, goals, and objectives for the group as a whole, while pursuing independent policies at individual members' level. This scenario offers opportunities for design, modelling, and implementation of supply chain networks for maximum effectiveness, efficiency, and productivity in a dynamic environment.

As noted, a supply chain network, depicted in Fig. 1, can be a complex web of systems, sub-systems, operations, activities, and their relationships to one another, belonging to its various members, namely, suppliers, carriers, manufacturing plants, distribution centres, retailers, and consumers. The design, modelling and implementation of such a system, therefore, can be difficult, unless various parts of it are cohesively tied to the whole. The motivation in proposing a framework to manage a supply chain system is to facilitate the integration of its various components through a common set of principles, strategies, policies, and performance metrics throughout its developmental life cycle.

An example of a manufacturing supply chain network, depicted in Fig. 2, captures the essence of the proposed framework. It has been derived from the general architecture of a supply chain network depicted in Fig. 1.

This supply chain is made up of a manufacturer and a two-level hierarchy of suppliers. Each sub-system in the supply chain network incurs costs that are to be monitored and controlled.

At each level in the supply chain, delay due to procurement activity is incurred, which has the potential of imposing waste, and thus incurring additional costs in the system.

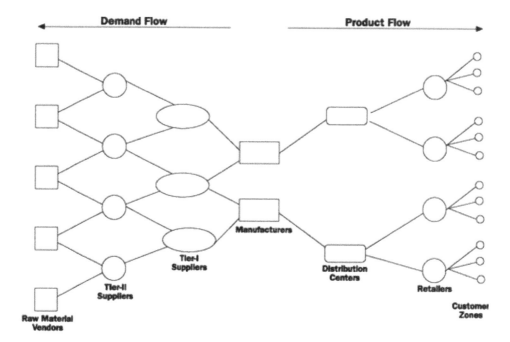

Figure 1.   A supply chain network.

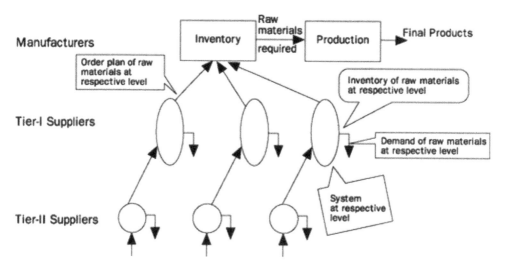

Figure 2.   A manufacturing supply chain network.

5   THE SUPPLY CHAIN OBJECTIVES AND PRINCIPLES

5.1   *Objectives*

Supply chain objectives directly support its stated goals; such as a common manufacturing supply chain goal can enhance revenue through eliminating or reducing bottleneck operations in the system. Supply chain objectives that directly support this goal can be identified as:

1. Increase throughput.
2. Reduce cycle time.

3. Reduce inventory at different stages (Raw materials—work-in-process—finished goods).
4. Reduce overall capital tied up.
5. Postponed management.

It is easy to realize that these objectives are complementary to each other. For example, a primary objective of increased throughput in the supply chain must be supported by a secondary objective to reduce cycle time. A reduction in processing time and set-up time will allow smaller batches to be processed faster, thereby lessening congestion in the system and registering shorter cycle time. This will also create increased throughput, and consequently, a higher revenue stream in the supply chain. As a result of this improvement in the supply chain, the tertiary objective of reduced inventory at different stages, which supports both the primary and secondary objectives, can be realized, since inventory at different stages will not have to wait for the availability of operations for further processing.

Objectives can be set both at the group level for the supply chain, and at the member level for individual members. However, the two sets of objectives ought to be coordinated in order to be effective performance measures for the supply chain. This may require tuning individual objectives of members so that common supply chain objectives can be met.

5.2 *Modelling principles*

In general, the principles support objectives for manufacturing the supply chain. By applying these principles, out-of-control processes, inefficient logistics, and inefficiencies that are inherently present in any system, can be developed. These principles are:

1. Reducing the influence of inventory variability at different stages and locations in the supply chain.
2. Reducing the influence of inventory variability at different stages and locations in the supply chain.
3. Reducing the influence of batching effects variability in the productive system.
4. Reducing the influence of variability due to bottleneck operations in the supply chain.

5.3 *Developing coordinated strategies*

The supply chain management perspective enables a developing interaction between production and marketing policies in the supply process of raw materials and the production of finished products. The element of coordination in developing effective strategies for a manufacturing supply chain is built by incorporating planning and control function as the integration unit.

The coordination of the end product is effected with the help of a common model that performs the planning and controlling functions of the supply chain.

In this manner, common policies agreed to among various members of the supply chain are implemented. For example, it may be possible to enforce common quotas for capacities, mutually agreeing to price and cost structures, as well as production schedules, etc.

5.4 *Implementation*

Thus, while the focus of the single product supply chain is different, similarities in approaches to design goals and objectives and model various principles should enable the developing of structural solutions to problem solving for diverse industry environments. These principles shall be applied individually.

6 CONCLUSION

Supply chain management is a major issue in many industries as firms realize the importance of creating an integrated relationship with their suppliers and customers. Managing the supply chain has become a way of improving competitiveness by reducing uncertainty and enhancing customer service. The role of planning and coordination in complex integrated systems and information technology to synchronize the supply chain is described in a framework that creates the appropriate structure and installs proper controls in the enterprise and other constituents in the chain.

During the past few years, supply chain excellence, optimisation, and integration have become the focus and goal of many organizations worldwide. Strengthening the supply chain management is perceived by many firms as the way to enhancing customer satisfaction and enabling profitable growth.

REFERENCES

Charu C. & Sameer K. 2000. Supply chain management in theory and practice: a passing fad or a fundamental change. *Industrial Management & Data Systems*, Vol. 100 Iss: 3, ISSN: 0263–5577. http://www.imanet.org/docs/default-source/thought_leadership/operations-process-management-innovation/implementing

Coopern, M. C. & Lisa, M. E.1993. Characteristics of Supply Chain Management and the Implications for Purchasing and Logistics Strategy. *The International Journal of Logistics Management.*

Coyle, John J. & Edward J. Bardi, & C. John Langley, Jr. 1996. The Management of Business Logistics. *6th ed. St. Paul, MN: West Publishing Company*.

Daneshjo, N. & Štollmann, V. 2013. Logistics systems and supply chain management. *In: Interdisciplinarity in theory and practice—Journal for Presentation of Interdisciplinary Approaches in Various Fileds.ITPB, Arad- Romania*: Nr: 2-year: 2013, pp. 73–76, ISSN: 2344–2409, ISSN-L: 2344–2409.

Daneshjo, N. 2014: Podniková logistika. *Petit s. r. o, Zimna 3/A, Košice-Slovakia.* p. 267—ISBN 978-80-971555-3-7

Joan Magretta 1998. What Management Is How it works and why it's ever yone's business ISBN 978-1-78125-147-8

Swaminathan, J.M. & S.F. Smith & N.M. Sadeh.1996 "A Multi-Agent Framework for Modeling Supply Chain Dynamics", *Proceedings NSF Research Planning Workshop on Artificial Intelligence and Manufacturing*, Albequerque, NM, June.

Tzafestas, S. & Kapsiotis, V.G. 1994. "Coordinated control of manufacturing supply chains using multilevel techniques", *Computer-integrated manufacturing systems*, 7(3), pp. 206–212.

Willis, J. (1995). Recursive, reflective instructional design model based on constructivist-interpretist theory. *Educational Technology, 35* (6).

Detecting fibers in the cross sections of steel fiber-reinforced concrete

M. Ďubek, P. Makýš & M. Petro
Slovak University of Technology, Bratislava, Slovakia

P. Briatka
COLAS Slovakia, Košice, Slovakia

ABSTRACT: Control of fibers´ dispersion in hardened fiber-reinforced composite, especially their spatial orientation and uniformity of distribution is very important for the final
characteristics of the fiber-reinforced composite, which is foreseen in its proposal. The aim of
the article is the application of the results of research into the systematic evaluation of sections of fiber-reinforced structures using image processing software. A mathematical expression of dispersion of fibers in fiber-reinforced concrete is incorporated into the evaluation
software, which detects fibers, their angles of rotation and evaluates ellipse areas identified
on the detection surface. Image processing and fiber detecting algorithm was developed specifically for research purposes. The resulting parameter of fiber dispersion reflects the distribution of printed fabrics and theoretically estimated amount of fibers. The detection area
is randomly inserted via fiber-plane structure. Methodologies to be used for detection and
evaluation forms the basis for determining the spatial homogeneity of the dispersion of fibers
with a 3-D scanning devices.

1 INTRODUCTION

Steel fiber-Reinforced Concrete (SFRC) represents a composite material consisting of concrete (composite itself) and fibers. The fibers in the concrete matrix are supposed to meet all
performances which are used in static design of SFRC structures. The fibers, however, can
meet the designed performances only if they are evenly distributed. The definition of homogeneity says that in each particular section of the structure, an equal number of fibers are
always found. From a mechanical point of view, not only distribution, but also orientation
around three main axes plays role. The definition of homogeneity has already been a subject
of recent research. SFRC has been subjected to measurements of both metal detectors and
X-rays. The purpose was to define the homogeneity. The research dedicated to the influence
of filling the form with fresh SFRC on homogeneity and orientation of the fibers was led by
(Ferrara et al., 2008; Stahli et al., 2008). There, the samples have been tested also for mechanical properties. The tested samples were also checked by X-rays. In detail, it is described in
(Ferrara et al., 2011; Ozyurt et al., 2009). Some research have proved an influence of procedure of filling the form with SFRC on ultimate position of the fibers. A shape of the structure influences its filling with SFRC and position of the fibers.
    According to (Duppond, 2003), the orientation factor expresses the prevailing orientation
of the fibers. Therefore, the dimensions of the prisms (samples) are crucial for the selection
of fiber lengths. The orientation factor was investigated by (Gettu et al., 2005) and indicates
the direction and spacing of the fibers. Ensuring fibers homogeneity in concrete is a difficult
task. There is plenty of aspects affecting the distribution and orientation of the fibers in
whole procedure of construction of the SFRC. To maximize their potential in structure, it is

desired to manage their distribution in mass of the structure and manage their orientation (longitudinal to tensile stresses).

2 EXPERIMENTAL PART

2.1 Tested samples

The samples were manufactured of prismatic shape with dimensions $150 \times 150 \times 550$ mm (Fig. 1). These were cut transversally, creating samples of thickness 100 mm. Cut surfaces (150×150 mm) were photographed and processes by software specially developed for automatic identification of the fibers. These pictures were used for testing of the software. For assessment of construction of SFRC structures on fibers' distribution, the core drills from original SFRC structures were used. These represented the methodology of check of homogeneity of SFRC by conventional/standardized methods. The developed software was supposed to identify the fibers, determine their number, determine their orientation in 3D and of course distribution in cross section. The dimensions of core drills were, 180 mm in diameter and 250 mm in length.

2.2 Used materials

The samples were manufactured using cement CEM II/B-M (S-V-LL) 32,5 R, natural gravel and steel fibers of length 60 mm and diameter 1 mm. The recipe of concrete was in accordance with STN EN 14845: 2007(water to cement ratio 0,55). The dosage of cement was kept at 320 kg/m³; dose of water at 176 kg/m³; aggregate 0/4 mm and aggregate 8/16 mm was kept at 952 kg/m³, both. The dosage of steel fibers was kept as recommended for industrial floors –30 kg/m³.

The SFRC industrial floor, out of which the core drills were taken, was constructed from concrete of similar recipe. The materials were identical, however the construction was performed in accordance with relevant procedures (Motyčka et al., 2005).

2.3 Pictures for assessment

Each cut of the samples was performed on high-frequency saw. The cuts were subsequently photographed. These pictures shown all steel fibers crossing the cross section. This represents the distribution of steel fibers in matrix of concrete.

All samples were photographed in the same manner. The distance between surface of the cut and camera's sensor was set to 1100 mm (based on previous empirical experience). For identification of the steel fibers in the picture, the software identified bright areas, representing

Figure 1.   Freshly poured prismatic samples curing in laboratory conditions.

polished cross sections of the fibers which reflect the ambient daylight. The cross sections were not treated in any manner.

3 HOMOGEITY—DISTRIBUTION

Homogeneous SFRC must contain SF evenly distributed throughout its volume. Random distribution should represent fibers oriented around all three axes.

3.1 *Spatial distribution of the fibers*

The fibers in concrete are distributed randomly. The spatial distribution as well as orientation are affected by used combination and proportion of the materials. Their position in space (Zhu Yt et al., 1997) can be determined by method of e.g. magnetic induction or X-ray (Rudzki et al., 2013; Ponikiewski et al., 2012). Direction of filling and flow of concrete has significant effect on direction of the fibers (Stähli, 2008; Ferrara et al., 2008; Yong-Zhi, 2000) in hardened concrete.

Management of distribution and orientation of fibers has impact mostly in production of precast concrete structures. Figure 2 shows randomly inserted plane in space. It is surface of one cut of samples. Cut fibers are present in this cross section in form of either a circle or ellipse. An inclination of cut fiber says about orientation of the fiber (Stroeven, 2008, 2009; Yong-Zhi, 2000). Therefore, the developed software expressed the ellipse as a ratio of its axes. For purposes of this project, it was essential to develop and tune-up automatic indication of projection of cut fiber. Then, based on prevailing angle, the fibers were attributed to one of three axes. If length of one axe of ellipse is more than 1,311 multiple of shorter axis of the diameter/projection of the fiber (expressed in pixels), then the length is attributed to axis "z" or "y". This multiple comes from imaginary plane under the angle 45°.

3.2 *Expression of distribution of the fibers*

Estimated number of cut fibers is based on probability of finding the fiber in certain volume. It comes out of dimensions of the fibers and conversion of dosage per 1 m³ of concrete. The problematics of derivation of spatial projection through 2D projection was handled with basics of stereography and morphometry. It expresses number of points per unit of volume, spatial orientation, density of the points or volume density of the structure. The characteristics of the fibers are known or identifiable. For estimation of detected fibers we came out of their dimensions and dosage of the fibers + with 3rd law of stereometry, formula 1 (Kalousová, 2009).

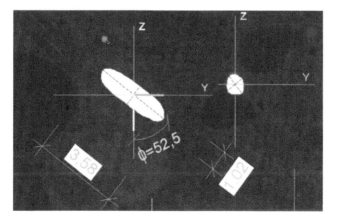

Figure 2.   Analysis of cut fiber.

$$E(N) = \frac{Ls}{2} \tag{1}$$

where $E(N)$ = mean value of number of the intersections (ks); L = total length of the fibers (m); s = area inserted to the volume, randomly (m²).

Length of the fibers "L" we express total length of added fibers (based on chosen dosage of the fibers). After the detection of cut fibers in cross section, we are able to count number of the fibers. This enables us to derive a stochastic number (concentration) of fibers of SFRC structure. If we set Lv = L and PA denote as number of intersections per unit of area (1 m²), we obtain a stereological formula for length concentration (Kalousová, 2009).

$$[L_V] = 2P_A \tag{2}$$

where LV = cumulated length of all found fibers in cross section (m); PA = number of found fibers (–).

For purposes of stereological calculation, the thickness of the fiber is negligible when consider its length. For application of the formula it is necessary to know the fibers' length and their amount per unit of volume. Each cross section should find fibers oriented in prevailing direction of each of three main axes of space. In case of homogeneous orientation, the percented of orientation of hte fibers in all three directions would be equal. The fibers perpendicular to cutting plane (cross section) appear circular, whereas all other orientations cause appearance of elliptical shape.

For purposes of expression of homogeneity of distribution of the fibers in cross section, it is needed to express index of homogeneity. It can be done by number and size of squares assigned to individual fibers detected in cross section. The edge "a" of the squares is a square root of area of cross section divided by number of identified fibers. In case of homogeneous distribution, there should be one fiber in center of gravity of each square. These squares should be tightly close to each other without any overlap. The cumulative area of these squares divided by area of cross section gives index of homogeneity (formula 5)—a number in interval (0–1).

$$a = \sqrt{\frac{P_S}{V}} \tag{3}$$

where a = side of square assigned to each fiber (mm); P_S = Area of cross section—(e.g. 150 mm × 150 mm (mm²)); V = number of fibers identified by software (-); P_V = Area of square assigned to each fiber (mm²).

$$P_V = a^2 \tag{4}$$

Since spacing of the fibers is not homogeneous, the squares assigned to each fiber overlap. That reduces cumulative area of these squares and this way reduces also index of homogeneity. The lower the index is, the more probably fibers are not evenly distributed over all area of cross section.

$$i_r = \frac{P_S}{\sum P_V} \tag{5}$$

where ir = index of homogeneity (-); $\sum Pv$ = cumulative area of squares assigned to the fibers (mm²).

4 ANALYZING SOFTWARE

For detection of the fibers, special software application has been developed (fig. 3) in C++. The architecture of the application is based on external libraries OpenCV, which is developed for image processing in real time (Bradski, 2000). The environment of the application was created in library Microsoft Foundation Class Library (MFC), what enables compatibility of this application with most of the platforms of Windows.

The uploaded image is processed by application of several filters of Library OpenCV. If the input picture is too large and contain no relevant areas (sample) on the sides, the application recognizes it and cut them off. Subsequently, the contours are detected whereas the defects of convexity are merged so that one solid object is created. The result of all this is cut of concrete sample representing cross section without any other digital information on the picture. Next, this picture is processed by morphopogical reconstruction to eliminate local peaks. Subsequently, the erosion and dilatation methods of the bright areas are applied. In combination with binary thresholding, the application finds bright areas (steel fibers) and analyze them in terms of their position (coordinates) and appeared shape.

5 RESULTS

The main goal of the topic described in this paper was to identify the fibers in SFRC and define scatter. It was concluded that the index of homogeneity meets this. In real tests it reached an average value 0,48 (far away from homogeneous level). The number of fibers in these tests counted 51 per cross section, representing a dosage of approximately 32,8 kg/m³. The number of fibers from core drills reached 24 per cross section, representing a dosage of approximately 15,2 kg/m³ (Table 1). In both types of samples the same dosage of fibers was used: 30 kg/m³.

Table 1. Index of homogeneity and number of fibers in samples.

Cross section	1	2	3	4	Average
Prism 1 i_r	0.437656	0.447398	0.46013	0.478764	0.455998
Fibre	44	28	55	55	45.5
Prism 2 i_r	0.46266	0.375694	0.471637	0.436333	0.436581
Fibre	52	87	53	35	56.75
Drill 1 i_r	0.40518	0.396268	0.507002	0.546124	0.46364
Fibre	32	25	27	18	25.5
Drill 2 i_r	0.590236	0.493774	0.553189	0.554792	0.4799
Fibre	28	20	16	31	23.75

 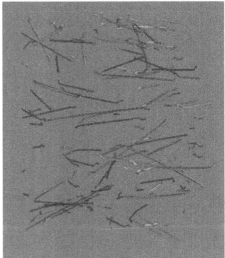

Figure 3. Tomography of the sample.

Rotation of the fibers and their affiliation to the axes are calculated in order to determine bearing capacity and transmission of the load in these directions.

The ratio of the bearing capacity in the direction of main axes (defined by the pre mentioned application), which represents the orientation of the fibers during LAB tests was recorded x:y:z in average 52:37:44. That confirmed the influence of filling of the form on final position and orientation of the fibers. In core drills taken from inductrial floor the ratios were as follows, x:y:z 38:33:37. In reality the fibers are oriented homogeneously, in a isotropic manner, in all directions. The fact however is that despite the ideal orientation of the fibers in real construction, it lacks approximately half the fibers which were supposed to be present. This might be caused by the sequence of mixing the fibers in the truck. Definitely not affected by pouring and compaction.

Important information is that data obtained by this software application were confirmed by microfocal X-ray tomography (Fig. 3). Areal interpretation of the data from X-ray tomography is now under development.

REFERENCES

Bradski, G. 2000. The OpenCV Library Dr. Dobb's Journal of Software Tools.

Duppond, D. 2003. Modelling and experimental validation of the constitutive law (σ-ε) and cracking behaviour of steel fibre reinforced concrete. Dissertation thesis. Department of Civil Engineering, University of Lueven.

Ferrara, L., Park, Y.D. & Shah, S.P. 2008. Correlation among fresh state behaviour, fiber dispersion and toughness properties of SFRCs. ASCE J. Mat. Civ. Eng. 20 (7). 493–501.

Ferrara, L., Ozyurt, N. & Prisco, M. 2011. High mechanical performance of fibre reinforced cementitious composites: the role of "casting-flow induced" fibre orientation. RILEM, Materials and Structures.

Gettu, R., Gardner, D.R., Saldivar, H. & Barrgán, B.E. 2005. Study of the distribution and orientation of fibers in SFRC specimens. RILEM, Materials and Structures 38.

Kalousová, A. 2009. Počátky geometrické pravděpodobnosti a stereologie, Praha. [online]. Available at: http://math.feld.cvut.cz/0rese/kolokvia/prezentace/MK_240409_kalousova.pdf.

Motyčka, V., Hrazdil, V., Dočkal, K., Lízal, P. & Maršál, P. 2005. Technologie staveb I. Technologie stavebních procesů, Část 2 Hrubá vrchní stavba, ISBN 80-214-2873-2. Akademické nakladatelství CERM, s. r. o. Brno. Brno.

Ozyurt, N., Tregger, N., Ferrara, L., Sedan, I. & Shah, S.P. 2009. Adapting fresh state properties of fiber reinforced cementitious material for high performance thin-section elements. Proc. Rheo-Iceland, O. Wallevik et al. eds., 313–321.

Ponikiewski, T. & Golaszewski, J. 2012. The new approach to the study of random distribution of fibres in high performance self-compacting concrete. Cem Wapno Beton 17(3):165.

Rudzki, M., Bugdol, M. & Ponikiewski, T. 2013. Determination of steel fibers orientation in SCC using computed tomography and digital image analysis methods. In Cem Wapno Beton 80:257–263.

Stähli, P. 2008. Ultra-Fluid oriented Hybrid-Fibre-Concrete. Dissertation thesis. Swiss Federal Institute of Technology Zurich. Switzerland, ISBN 978-3-9523454-0-5.

Stähli, P., Custer, R. & Van Mier, J.G.M. 2008. On flow properties, fibre distribution, fibre orientation and flexural behaviour of FRC. Mat. & Struct., 41 (1). pp. 189–196.

Stroeven, P. & Guo, Z. 2008. Distribution and Orientation of Fibers in the Perspective of Concrete's Mechanical Properties. Fibre reinforced Concrete: Design and Applications, Proceedings of the 7th RILEM International Symposium. France, pp. 145–154.

STN EN 14889-1: 2006. Vlákna do betónu. Časť 1: Oceľové vlákna. Definície, špecifikácie a zhoda.

STN EN 14651+A1: 2008. Skúšobnémetódy na betónvystuženýkovovýmivláknami. Meraniepevnosti v ťahupriohybe (medzaúmernosti (LOP), zostatkovápevnosť). (Konsolidovanýtext).

STN EN 14721+A1: 2008. Skúšobnémetódy na betónvystuženýkovovýmivláknami. Meranieobsahuvláken v čerstvom a zatvrdnutombetóne. (Konsolidovanýtext).

STN EN 14845-1: 2007. Skúšobnémetódyprevlákna v betóne. Časť 1: Porovnávané betony.

Yong-Zhi, L. 2000. DeutscherAusschussfürStahlbeton im DIN DeutschesInstitutfürNormunge.V. 494DAfStb-Heft 494. Tragverhalten von Stahlfaserbeton, BeuthVerlag GmbH. ISBN 9783410656944.

Zhu Yt, Blumenthal wr, LOWE TC 1997. Determination of non-symmetric 3-D fiber orientation distribution and average fiber length in short-fiber composites. J Compos Mater 31(13):1287–1301.

Advances and Trends in Engineering Sciences and Technologies II – Al Ali & Platko (Eds)
© 2017 Taylor & Francis Group, ISBN 978-1-138-03224-8

Waste water treatment in North Moravia and Silesia, from the past to the present

T. Dvorský & V. Václavík
Institute of Environmental Engineering, Faculty of Mining and Geology, VSB-Technical University of Ostrava, Ostrava, Czech Republic

P. Hluštík
Institute of Municipal Water Management, Faculty of Civil Engineering, Brno University of Technology, Brno, Czech Republic

ABSTRACT: The requirement for wastewater discharge was based on the knowledge of its negative impact on human health as well as the negative development of the society. This article describes the historical development of waste water treatment from the global perspective, and it is also focused on the development of this sector in the Czech lands. This article also describes the development of urban centres in Northern Moravia and in Silesia and the associated requirement for the discharge and sanitation of waste water. This article describes the development of sanitation technologies in the cities of Ostrava, Frýdek—Místek, Opava, Třinec and other cities, and it provides information on the current state of waste water treatment in these cities.

1   INTRODUCTION

The efforts of the human race focused on securing sustainable economic growth, while maintaining or even improving the living conditions on the Earth, is known under the name of "sustainable development". At the same time, the development of cities and densely populated areas relies on more and more intensive use of the resources, which also includes water. There has been an increasing requirement to discharge higher and higher amounts of waste water, which goes hand in hand with the development of cities, and is not possible without the reconstruction of sewerage systems, which often date back to the early 1920s and have already lost their ability to hydraulically and hygienically safely transport the required amount of produced wastewater. At the same time, the amount of sewage has been growing, especially together with the amount of storm waste water which, as a result of the extension of the areas with low infiltration and evapo-transpiration capabilities in urban drainage basins, keep filling the sewerage systems with increasing amounts of water and overload the often outdated wastewater treatment plants.

2   HISTORY OF WASTE WATER TREATMENT

2.1   *The world history of waste water treatment*

From a historical perspective, the use of sewerage systems for the drainage of waste water from residential areas is nothing new. From the time of the first modern human beings, when the population started to grow and the hunters and gatherers became farmers, the people began to struggle with the increasing production of waste materials and they returned them back to the countryside. The consequences of such activities soon became evident in urban densely populated areas in the form of serious health problems. It was already at that time, when people understood that to eliminate these unwanted health problems, it is not sufficient to take steps to obtain high-quality drinking water, but it is also necessary to deal with the drainage of waste water (Vuorinen 2007, Larsen 2008).

The first confirmed records related to the solution of the problem with the drainage of waste water in the form of discharge of sewage from houses and the construction of latrines with cesspits dates back to the period of 2500–3000 BC from the Mesopotamian Empire. Although this system worked well, the streets were full of garbage and faeces. (Lens, Zeeman, Lettinga, 2001) Further references can be found in the period from 2500 to 26 BC in various areas. Ruins of pits were discovered in the Indus valley. They were connected to houses, where solid impurities built up, and the pre-treated wastewater was then drained into the streets. The remains of latrines and bathrooms, from which waste water was discharged into pits or directly into the desert, were found in Egypt. (Lofrano, Brown, 2010).

Another significant civilization, which is also referred to as the predecessor of the modern sewer systems, isthe Greek civilization. In the period of 1700–2000 BC, we found records of a separate sewer system in the royal palace at Knossos or other systems on the island of Crete. These were primarily used to drain storm water, but they were also used to discharge sewage water (Angelakis 2005; Wolfe, 1999).

We can move in time to the period dominated by the Roman Empire, where the famous Cloaca Maxima was built around the year 800 BC. It is interesting that it was built 300 years earlier than the first aqueduct Aqua Apia. Sewage at the time was discharged by a system of sewers into the Tiber River. (Hodge 2002, De Feo et al. 2010).

The period covering the Middle Ages up to the Industrial Revolution is referred to as the Dark Ages. Water was bad, the level of hygiene of the population was decreasing, and dirt was covered with perfumes, wigs, etc. There were widespread myths about water saying that it causes the opening of skin pores, thereby exposing humans to disease hazards. The first fragments of possible sanitation in Europe are dated back to the times of epidemic diseases, more precisely to 1539, when France began to build the first pits for collecting waste water from the newly constructed houses (Lens, Zeeman, Lettinga, 2001).

2.2 *The history of drainage in the Kingdom of Bohemia*

The first records of waste water or just faecal waste drainage in the Czech lands date back to the Middle Ages. The nobility tried using specific devices to drain faeces from castles at least to the walls, where they made the advance of attackers during the sieges of the castle more difficult. The equipment found at Czech castles reminds an earth-closet and it was called "garderobe", or there are also kitchen pits, from which the waste was discharged on the walls as well. As far as the residential cities at that time are concerned, there were no sanitary facilities similar to sewer systems there, and all the sewage and waste ended up on the streets. Later, when the streets were paved, there were the first open ditches, where liquid sewage was flowing and spreading foul odor and diseases around. The spreading of epidemics forced the population of the Czech lands to a gradual solution of the situation with sewage, but everything was influenced by the social status and the development of sewer systems was very slow. There was a boom of sewer systems in the Kingdom of Bohemia in the 19th century and after the foundation of Czechoslovakia, followed by stagnation of this development during the economic crisis and World War II. (Broncová 2002).

2.3 *The history of drainage in North Moravia and in Silesia*

The first records of today's cities in the North-Moravian Region date back to the 13th–15th centuries, but they were mostly only small residential areas of the size of a settlement. A big boom in the development of these settlements occurred with the discovery of coal in the Ostrava region, but the major expansion came in the 19th century. Coal mining has led to the development of other industries, such as the chemical industry or the metallurgical industry. All these industries required manpower for their expansion and the city began to grow, not only in terms of the factory complexes but also in terms of civic buildings. All this urban development in the area of Ostrava was hampered by inconvenient and sometimes very primitive waste water drainage. The only facilities were often open races flowing into the adjacent streams.

The problems with drainage were first addressed by the city of Ostrava in 1855, when a directive of the district office ordering a construction of a cesspit for each residential house was had already been valid for 5 years. The construction of sewer systems in the individual municipalities, which are parts of Ostrava now, began in 1890. The construction work in these municipalities was uncoordinated and without any concept, and it did not reflect the development of the settlements and the sewer systems flew into streams. The only exceptions were the cities of Opava and Vítkovice, where the sewer systems were built conceptually and the waste waters were treated in urban wastewater treatment plants, which were equipped with settling tanks and biological filters. Later, up to 1932,the constructions of double-storied septic tanks with a biological filter began in other cities such as Třinec, Orlová, Fulnek, and Bruntál. (Broncová 202).

By 1935, Ostrava managed to eliminate all races and, after the merger of the cities in 1924, it began an intensive construction of main sewers and built a network of 115 kilometres of sewers. At that time, Ostrava had a population of 120,000, but the main sewers were not connected to any wastewater treatment plants yet. A boom in the construction of wastewater treatment plants started after World War II, around 1950, when a huge intensification of mining and metallurgical industry took place as well, resulting in an enormous increase in civic, cultural and other constructions. It was the period of the beginning of the construction of huge housing estates of Ostrava - Zábřeh, Karviná, Havířov, and Poruba. Under the weight of these circumstances, there were projects of independent wastewater treatment plants, but without any experience with wastewater treatment plants of similar size and with minimum information about the application of the processes of biological wastewater treatment. In 1953, there were the first projects for wastewater treatment plants for 11 to 20,000 inhabitants.

3 THE DEVELOPMENT OF SANITATION TECHNOLOGIES IN NORTH MORAVIA AND SILESIA

3.1 *Ostrava*

Ostrava, as one of the largest cities in the former Czechoslovakia and the largest city in North Moravia and Silesia, completed one of the first wastewater treatment plants with a capacity of over 10,000 Population Equivalents (PE) in the region in 1957. It was the wastewater treatment plant of Ostrava—Zábřeh and it was a mechanical—biological wastewater treatment plant with bio-filters and separate mesophilic digestion of sludge and sludge fields.

1951 was the year of the beginning of the construction of Poruba housing estate near Svinov, which was included in the Porubka river basin. This stream had very little water and ran through a built-up area, and it was necessary to quickly resolve the discharge of waste waters from the housing estate to avoid situations with more sewage than water flowing in the river. These were the reasons why a temporary wastewater treatment plant with bio-filters and mesophilic digestion of sludge for 25,000 population equivalents was built in 1955. The continuing construction of the housing estate and the introduction of a new technology of compressed air production led to the construction of a new wastewater treatment plant for Poruba in Ostrava—Třebovice in 1960. The size of the housing estate according to the plan was 120,000 inhabitants and this area also became the location of light industry in the form of meat processing plants together with dairy works and a bakery. The wastewater treatment plant was designed to treat waste water of 137,500 population equivalents.

The drainage of large housing estates of Ostrava had been solved, and that is why the execution of a continuous sewer system for the entire city was commenced in 1959. The concept was based on a division of the city into individual sewer districts, where sub-collectors were to be used to drain waste water into the main sewer and then to a wastewater treatment plant. This concept was maintained and the central wastewater treatment plant in Ostrava—Přívoz was launched in 1969. It was designed for 252,000 population equivalents, including 43,000 population equivalents of pollution from the industry. The central wastewater treatment plant was gradually becoming outdated and, together with the changing requirements of the city, the effects of mining activity and long-term overloading, it was decided to build a new central wastewater treatment plant with a designed capacity to 638,850 population equivalents. Along with the commencement of the construction of the central wastewater treatment plant in 1987, a construction of the D main sewer started as well. The central wastewater treatment plant was completed in 1996 and, after a trial operation, the wastewaters directed to the wastewater treatment plant in Ostrava—Třebovice were connected to the D main sewer in 1997 and they were diverted to the central wastewater treatment plant. The design parameters of the central wastewater treatment plant of Ostrava are presented in Table 1.

The central wastewater treatment plant of Ostrava was affected by major floods in 1997, when the water reached up to 5 m and put the wastewater treatment plantout of operation. The overall restoration of operation took 4 months, and the first operational part was the mechanical stage followed by the biological stage.

The current capacity of the central wastewater treatment plant is designed for the flow rate of 254,666 m^3/d and the pollution of 47,975 kg BOD5/d (biochemical oxygen demand in 5 days—BOD5), which corresponds to a capacity of about 800,000 population equivalents. This capacity is to be achieved in 2030.

The hydraulic plant capacity utilization is 54% and the material utilization is about 45%, however, from the perspective of today's limits, which do not take into account only the organic pollution within the BOD5 indicator but also the removal of nitrogen from the wastewaters, it is near the capacity limit, which is about 340–350,000 population equivalents.

3.2 Frýdek—Místek

The 19th century was the time of the foundation of textile factories of J. Munk and sons, Landsberger, the Neumann brothers, Lembergerin Frýdek—Místek,and the history of Charles Ironworks in Lískovec, currently known as Arcelor Mittal Ostrava a.s., with the rolling mill plant in Frýdek—Místek, started in 1833. The development of these industrial enterprises caused aninflow of new inhabitants and increased demands with regard to wastewater treatment. However, the construction of the first wastewater treatment plants for Frýdek—Místek started many years later, more precisely in 1967.

The original mechanical-biological wastewater treatment plant was built in Sviadnov. It was intended for the treatment of wastewaters from the inhabitants of F-M, as well as the neighboring municipalities of Sviadnov, Dobrá, Staré Město and others. The wastewater treatment plant was also intended to treat wastewaters from Frýdek—Místek industrial plants. The original wastewater treatment plant from 1967 was soon insufficient with respect to its capacity and was overloaded, which lead to the preparation of a new wastewater treatment plant project, the execution of which began in 1990. The only parts left from the original wastewater treatment plant include the rainwater basin and part of the sludge treatment and disposal.

The actual construction of the new plant started in July1990 but in the 1990s, however, there were major changes in the quantity and quality of waste waters due to the restructuring of the local industry and it was necessary to modify the original project.

Table 1. Design parameters of the central wastewater treatment plant of Ostrava.

Design parameters		
Flow-rate 24h Q_{24}	[m^3/day]	184372
BOD_5	[kg/day]	38331
CHSK	[kg/day]	76662
N_{celk}	[kg/day]	5015
NH_4	[kg/day]	6446
NL	[kg/day]	32265
P_{celk}	[kg/day]	1596

The new wastewater treatment plant was put into operation in 1995, it is of mechanical-biological type with nitrification and pre-denitrification, chemical coagulation of phosphorus, entire sludge treatment and disposal stage, and energy production in cogeneration units.

3.3 Třinec

Thanks to the discovery of large reserves of iron ore and the foundation of Třinecké Ironworks in 1839, in the middle of the 19th century this farming village became an important seat of North Moravia. Another important milestone was the commencement of the construction of the Košice—Bohumín railway, which connected Třinec with new ore deposits in Slovakia, allowed the transport of coal from the Ostrava—Karviná mines, and opened the way for the development of the whole area. The development of the Ostrava region was connected with the beginning of the construction of Lyžbice housing estate with a planned capacity of 20,000 inhabitants in Třinec in 1956–1977. In 1965, this expansion forced the city to build a large wastewater treatment plant, which was intended to purify not only the sewages but also the industrial phenyl-ammonia water from Třinecké Ironworks. There were 2 wastewater treatment plants in Třinec in late 70s-Ropice and Sosna, which were overloaded and could not meet the stricter limits of the removal of nutrients, which is why a new wastewater treatment plant was launched in 1994. At present, both the sewage and the phenyl-ammonia waters are treated together in a mechanical—biological cleaning cycle, where the biological treatment is carried out in three-chamber activation with an anaerobic selector, pre-gentrification and nitrification, and two settling tanks.

3.4 Opava

There were several mainly textile companies in operation in Opava during the 19th century. A brewery was established in 1825, and two sugar refineries were opened by the middle of the 19th century. A significant recovery of the industry, however, did not come until the opening of th Ferdinand Northern Railway in 1855. (The Archive of the city of Opava).

Opava built one of the first wastewater treatment plants in North Moravia in the early 20th century. It was a system of lay-by and settling tanks from which the deposited crude sludge was drained to sludge fields. In 1967, the wastewater treatment plant underwent a complete overhaul, including a construction of a biological cleaning stage with an activation tank and a capacity of 18,144 m3/day. However, the gradual connection of new industrial plants and the population growth required further adjustments and, in particular, the intensification of the biological part of the facility. Despite all the modifications and modernizations, the wastewater treatment plant was constantly overloaded during the 1980s. These were the reasons for a refurbishment and extension of the treatment plant to the capacity of 22,276 m^3/day and 183,600 population equivalents. The wastewater treatment plant was in operation in this mode until 1997, when the extended and refurbished wastewater treatment plant with double-corridor cascade activation, with pre-denitrification and nitrification, biological and chemical phosphorous removal, with the production of electricity from surplus biogas, and with a complete automation of the operation management was launched. The designed capacity of the wastewater treatment plant is 149,000 population equivalents, with the wastewater volume of 33,500 m^3/day, and it ensures the treatment of both municipal wastewater from the city of Opava and the waste waters from major industrial and food companies from the city surroundings.

3.5 Český Těšín

The history of Český Těšín began in 1920 when it was decided to split the territory of Silesian Těšín between the two newly created states, Czechoslovakia and Poland. Olše has become the border, and this time marked the beginning of the development of the independent city of Těšín, which was until then only a suburb of the old city of Těšín. The development of the city and the building of the missing urban infrastructure attracted new residents and industrial enterprises.

Sewage was drained by means of a sewage system flowing into the adjacent recipient—the Olše River. The first wastewater treatment plant was built in Český Těšín in 1972 as a mechanical—biological plant with full sludge treatment and disposal, and it was designed to 9,245 m^3/day. This wastewater treatment plant became outdated and unable to meet the stricter limits for nutrient removal, which resulted in several renovations and technological changes in 2001. It included the introduction of the technology of long-term activation with nitrification and gentrification and separate sludge regeneration. The extension of the technology included a dosing of ferric sulphate solution at the outlet of the activation tanks, which facilitated a reduction of the phosphorus concentration at the outflow of the wastewater treatment plant.

4 CONCLUSION

Water management should be aimed at the least possible pollution level. The process of wastewater discharge should be understood as a complex assessment of the environmental impact, with a special focus on surface and groundwater. This assessment should include the most important parts, i.e. sewage system, wastewater treatment plant and recipient. The operators of the sewer networks in the aforementioned cities and densely populated areas (Ostrava, Opava, Frýdek—Místek, Třinec) are taking such steps to make sure the wastewater disposal is closely connected with the monitoring of surface and groundwater and the environment as a whole. When suggesting a solution of the problem of waste water discharge, we should take into account and assess the mutual links of the individual parts of the system to make sure that waste water management complies with the concept of sustainable development, and is not chaotic and without a clear concept as it was in the past, for example in the territory of the city of Ostrava. At the same time, the building of new wastewater treatment plants or the extension of the sewerage systems shall take into account the environmental effects of the individual facilities and they shall be designed in such a way to achieve the highest possible efficiency, while trying to minimize their energy consumption. These environmental effects and energy management involve all the recent or planned renovations of the wastewater treatment plants in Ostrava, Opava, Český Těšín and other smaller towns of the Moravian-Silesian Region. The effort to make the wastewater drainage and treatment in the Moravian-Silesian Region more efficient represents a permanent effort and is supported by the governing bodies of the Region.

REFERENCES

Aiello A.E. et al. 2008. *Hidden heroes of the health revolution. Sanitation and personal hygiene*. American Journal of Infection Control 36 (2008): 128–151.
Angelakis A.N., Koutsoyiannis D., Tchobanoglous G. 2005. Urban wastewater and storm water technologies in ancient Greece. *Water Research* 39 (2005): 210–220.
Broncova, D. (ed.) 2002. *The History of Sewerages—The History of the Draining and Treatment of Waste Water in the Czech Lands*. Prague: Milpo media.
De Feo, G. et al. 2010. Water and waste water management technologies through the centuries. *Water Science and Technology: Water Supply* 10 (3):337–349.
Hodge, A.T. 2002. *Roman aqueducts & water supply (2nd Edition)*. London: Gerald Duckworth & Co. Ltd.
Kucerova, R., Fecko, P., Lyckova, B. 2010. *Water treatment and waste water treatment*. Ostrava: VSB-TU Ostrava.
Larsen, O. 2008. The history of public health in the Ancient World. *International Encyclopedia of Public Health*. (2008):404–409.
Lens P., Zeeman G., Lettinga G. (Eds.). 2001. *Decentralised sanitation and reuse: concepts, systems and implementation*. London: IWA Publishing.
Lofrano, G. Brown, J. 2010. Over view of waste water management through the ages: A history of mankind. *Science of the Total Environment* 408 (22):5254–5264.
Vuorinen, H.S. et al. 2007. History of water and Health from ancient civilizations to modern times. *Water Science and Technology: Water Supply* 7 (1): 49–57.
Wolfe, P. 1999. History of waste water. World of Water 2000—The Past, Present and Future:24–36.

Advances and Trends in Engineering Sciences and Technologies II – Al Ali & Platko (Eds)
© 2017 Taylor & Francis Group, ISBN 978-1-138-03224-8

Model house with modifiable disposition

J. Gašparík, J. Piatka, P. Šimko, M. Šmotlák & P. Orosi
Faculty of Civil Engineering, STU in Bratislava, Bratislava, Slovakia

ABSTRACT: The space requirements of families are constantly changing and evolving during their life cycle. Young people have lower demands on living space compared to families with children, whose demands decrease again once their children have grown. Thus the family space demands have a sinusoidal function. The possibility to modify the disposition of a living space seems there for to be the most effective. A designed model house has only 30 square meters of floor space and is growing according to family demands. The aim of this article is to design a method of assembling various stages of the house according to the family life cycle.

1 INTRODUCTION

Accommodation, together with nutrition and clothing, is one of the three main existential needs of the humans. Nobody should be devoid of it and it should be satisfied on a scale adequate to the overall grade of the socio-economic development of the society. For this reason, accommodation is among the basic social rights with a special character. According to certain studies every fourth marriage doesn't last because of problems linked to their accommodation. It is a sad fact that nowadays many "wiser" people exploit this basic need of ordinary people to have a roof over their head. Young people live in constant uncertainty, they live in the fear of losing employment, and thus also from the loss of a steady income. They say a mortgage loan haunts you throughout the whole life, and thus many young people and young families rather chooses to live in rental flats—the state doesn't offer them a helping hand. They have no choice but to fill the pockets of others. Our times and situations simply aren't favorable to people wishing to buy their own house. Not that they wouldn't want to, they simply don't have the resources. That's why the term "model house" or "model flat" becomes more and more important part of the discourse of architects, reality markets and in the society at large. In short, it is the first option for young people after the high school or university to find accommodation.

This paper will attempt to bring a new view of typology and construction solutions of the space of a model family house, whose aim is to provide the middle reasonable compromise between the aspects of quality, culture and affordability of accommodation. It deals with the question on how to build the cheapest possible house to accommodate young people. The idea of such an accommodation, which doesn't require young people to indebt themselves for 30 years, was verified on a particular conception, based on a sociological survey published by Šimko (2015) (where I've gained an overview about how young people wish to live, using data gained by questionnaires from 330 young people). The idea is to have the house grow with the family together.

The data of our sociological survey among 330 respondents, gained by means of a questionnaire with 14 questions, show the following demands of young people for their first house:

- for the most time they are away from home, where they mostly need to only sleep,
- to save money, they don't require a working table at home,
- to save money, they prefer a shower enclosure to a traditional bathtub,

- to save money, they would rather have a sleeping corner accessible per stairs from the living room than a sofa bed in the living room,
- they prefer the living room connected with the kitchen to the enclosed kitchen,
- to save money, they wouldn't mind having an integrated WC in the bathroom,
- to save money, they would accept the usable height of 2.2 m in the short vestibule, WC and the bathroom,
- to save money, they would accept the usable height of 2.1 m in the sleeping corner.

2   LAYOUT OF A MODEL FAMILY HOUSE (1ST PHASE)

Our projected house has only 30 m2 (7.5 × 4 m) and contains one multifunctional living space, integrating the living room, the kitchen and the sleeping corner (Figure 1). Kitchen units are connected to the living room. From there it is possible to get to the sleeping corner by stairs. Under the sleeping corner there is the bathroom with the vestibule. The boiler is built in the vestibule. According to our sociological survey, the most important reason why young people lower their demands for accommodation is to pay as little as possible. Also for that reason they accept the mentioned functions of accommodation, which minimize their costs. The shape of first phase construction (i.e. living room, sleeping corner, bathroom and vestibule) isn't random. It is based on minimal usable heights as provided by our question-naire. In the vestibule and the bathroom it is 2.2 m, in the sleeping corner, the highest point over the stairway, the height is 2.1 m.

3   LAYOUT OF A MODEL FAMILY HOUSE WITH THE FAMILY MODULE (1ST + 2ND PHASE)

The family house grows with the family. It is projected in such a way, that when the child is born, an additional module is built: the dining room, two further separate rooms and a sleeping room for the parents. In the beginning we expect to lay two "fundaments" of con-crete. One will bear the house, and the other, as long as the residents don't have resources to construct the additional module, can serve as a parking lot for two cars. The construction is

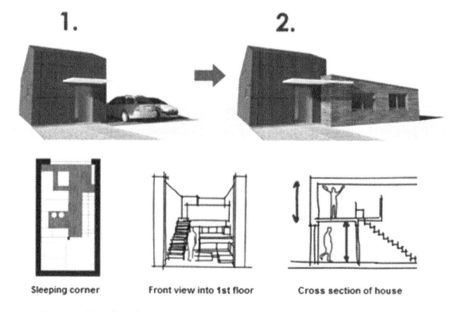

Figure 1.   Layout of the first phase of the model house.

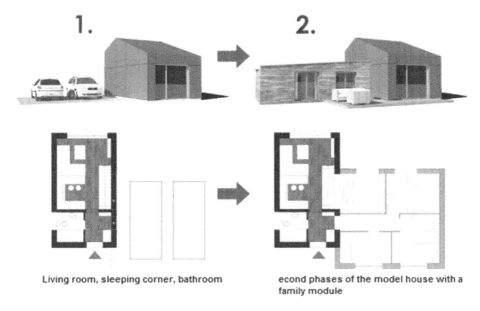

| Living room, sleeping corner, bathroom | econd phases of the model house with a family module |

Figure 2.   Layout of the first and second phases of the model house with a family module.

projected with two assembly holes to connect it with the future family module. This concept offers a new way of construction of model family houses, splitting it to two phases. In the first phase, we build the main module with the living room, sleeping corner, bathroom and vestibule—the minimal needs of two young people—and in the second we build the family module according to their growing requirements—two rooms, sleeping room, dining room. The sleeping corner then may become a library or a leisure gallery.

4   MATERIAL AND STRUCTURAL BASIS OF THE MODEL HOUSE

Our analysis of various construction solutions showed that the variant (from two possible alternatives) using wood would be the most effective. Not only is it less costly, its realization is faster than traditional masonry work, as all wet processes are eliminated. We've projected the vertical structure as a wooden framework fitted with OSB panels, with thermal isolation fitted between them. Contact points between the OSB panels will be covered with sealing tapes. The horizontal structure over the sleeping corner and the roof is projected with beams and rafters, which remain visible. The roof is designed with waterproofing membrane is joined by hot air welding. The roof edging will be made from a foil welded on a metal sheets.

We propose to replace the outer wall plaster with less expensive Cement bonded particleboards ("cetris") with a surface treatment. Cetris will be fitted on the underlying wooden pallet. The pallet can be filled with additional thermal isolation, but our cost estimation doesn't contain this isolation. To lower the heating costs, it is also possible (at an additional costs) to add isolation to the roof coating. The family module is covered by spruce panels as a sign that the family has earned money. The spruce covering beautifully contrasts with the "cheap" cetris. We tried to minimize the filling structures, and thus they are composed of one glass wall, an entrance door and a small window in the bathroom. We've projected the surface treatment of the walls and roofs to contain only the paint on the OSB panels.

The wall with assembly holes contains a steel frame, which will be uncovered during the construction of the family module and connected to the steel frame of the new part of the house with screws. The framework and material basis of the family module is similar to that of the first part of the house. The individual frames are anchored simply by screws, containing a threaded rod and nuts (Figure 3).

397

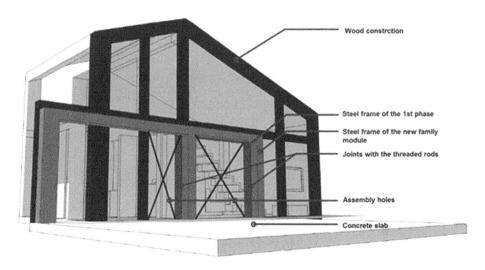

Figure 3. Anchorage scheme of the steel frame of the first and second phase of the model family house.

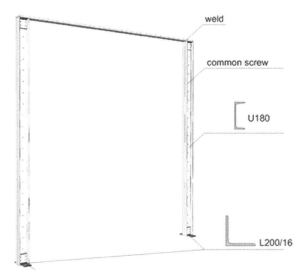

Figure 4. Principal details of the frame composition of both the first and the second phase of the model family house.

The steel frame of both the first and the second phase is composed by pillars with a U 180 profile 180, anchored to the concrete slab using L profile L200/16. The frame beam will be made of steel strips welded onto the L profiles. Such a frame anchored to the concrete slab will be pre-built and then brought to the construction site and integrated to the first phase of the house. The firmness of the whole house after the second phase will be provided by the same frame from the second phase and screws for the pre-drilled holes. The principally detailed scheme of frames can be seen in the Figure 4.

5 PRICE OF THE MODEL HOUSE

We've used the price database CENEKON 2013 for the second half-year, used in the Cenkrosplus software by the KROS Company, to estimate the costs of the first phase of our

Table 1.   Costs of foundation.

Code	Description	Costs [EUR]
HSV	Main building production—Foundation	3 515.993
1	Earthworks	319.542
2	Foundation. reinforcement	2 903.690
3	Vertical and complex structures	292.761
	Total	3 515.993

Table 2.   Costs of own construction.

Code	Description	Costs [EUR]
HSV	Main building production—Own construction	620.191
6	Adaptation of surfaces. floor and placing of fillings to openings	620.191
PSV	Associated or Auxiliary building production	15 747.736
711	Waterproofing. humidity and gas insulation	1 238.424
712	Built-up roofing	795.043
713	Heat insulation	2 756.352
725	Sanitary—technical installations	175.928
762	Timber work	5 154.941
763	Wood construction. plasterboards	2 544.507
764	External plumbing	516.688
766	Joiner's construction	1 728.276
771	Tile floor	113.268
775	Timber floor (parquet floor. deal floor. etc.)	105.705
781	Finishes work—ceramic facing	326.930
784	Finishes work—painting	291.674
N00	Undefined	876.560
N01	Undefined works	876.560
	Total	17 244.487

model house. The final sum was € 27 thousand with VAT, calculated for only the first phase, i.e. the living room, sleeping corner, bathroom, vestibule and the concrete steel base for both phases. Connections to local infrastructure aren't a part of this estimate, as it depends much on the location of the construction site. The price is only for the orientation; people using the mentioned database for evaluation of objects would know, that this house can be built for a 10–20% lower cost. A self-build construction could be realized even for even half that price, as the material costs are around € 10 thousand. The house can be built in a short time and its layout is very variable. The foundation costs are in the first table, the cost of own construction in the second one.

6   CONCLUSION

To have one's own family house is for many people merely a dream. It is no wonder. for one's financial situation simply doesn't allow it for the masses. while the life-long yoke of a high mortgage scares off many others. leaving the wish for a house to stay in the realm of dreams. But now we have an opportunity to make it to a house without being overwhelmed by the costs. For example. we can find suitable lots south of Bratislava with a price around € 40 thousand. When we add the house for € 27.000 we are close to the price for a new studio apartment (garçonniere) or a one-room flat in its original state in Bratislava. However. when the family comes. such flats become too cramped. The flat size simply can't be adjusted. On the other hand. the described model house. with the same price as the smaller flats. foresees the future needs of the family. It has a perspective. for the basic structure can be expanded by an additional module anytime. thus satisfying the needs for living space. To live like in a family house. yet not indebt oneself for the lifetime—that's the philosophy of the model house. which grows together with the family.

REFERENCES

Šimko. P. 2013. Requirements of young people for their starting flats. In 15th International Conference of PhD Students: 27–30. Brno: University of technology. Faculty of Civil Engineering.

Šimko. P. 2013. The Social housing—The term exact or not. In 15th International Conference of PhD Students: 30–33. Brno: University of technology. Faculty of Civil Engineering.

Advances and Trends in Engineering Sciences and Technologies II – Al Ali & Platko (Eds)
© 2017 Taylor & Francis Group, ISBN 978-1-138-03224-8

Optimizing method of building machine group selection implemented into soil processes with software support

J. Gašparík, S. Szalayová, Z. Hulínová, L. Paulovičová & V. Laco
Slovak University of Technology in Bratislava, Slovakia

ABSTRACT: The optimal selection of a machine or machine group for building processes is a very important role of the building planner during the process of building planning. In our contribution the ability of machines to realize the designed building process (quality aspect), duration of the mechanized process (time aspect) and minimizing of energy consumption (cost and environmental aspect) are analyzed. We have developed the present state of knowledge of the purpose of the machines and machine groups for building processes. The optimization method was proposed and implemented into soil processes and selected building group of machines (excavators and trucks), which are often used in construction. In our contribution will be presented key mathematical models for model example solution and software built in Java, which has been created as a support for method described in this contribution. Application of this method and software will increase the effectiveness of building machine selection.

1   ANALYSIS OF PROBLEM

The optimal selection of a machine or machine group for building processes is very important role of building planner during the process of building planning. During the process of building planning the planner must analyze suitable selection of building machines and its group for the effective proposal of mechanized building processes. There are several criteria for selection of building machines. In our contribution the ability of machines to realize the designed building process (quality aspect), duration of the mechanized process (time aspect) and minimizing of energy consumption (cost and environmental aspect) are analyzed. Many researches are interested in earthwork operations and most of them use optimization and simulation as the methodologies that can be used for analyzing soil processes. CYCLONE and STROBOSCOPE are commonly used simulation tools specified for construction (Zhang, 2008). These tools for construction modeling, such as STROBOSCOPE enable for an accurate and detailed modeling of any complex situation but these tools demand a high level of training (Martinez, 1996). In the context of STROBOSCOPE Martinez developed an Earth-Mover, which is a discrete-event special-purpose simulation modeling tool for earthwork planning. This tool includes STROBOSCOPE as a simulation engine, Visio for the graphical and interactive model definition, Excel for tabular and graphical output and Proof Animation for dynamic output (Martinez, 1998). Halphin developed CYCLONE methodology for modeling and simulating repetitive construction processes (Halphin, 1977). Shi and AbouRizk introduced the Resource-Based Modeling (RBM) methodology in order to automate the modeling process and by using this methodology the project manager can construct a simulation model for a project in a few minutes, but it consisted of only eight basic atomic models and is connected only with earthmoving operations (Shi & AbouRizk, 1998). Marzouk and Moselhi analyzed earthmoving operations by combining a Genetic Algorithm (GA) with CYCLONE and other simulation techniques. Their simulation and optimization considered multi-objectives for selecting a near-optimal fleet configuration for earthmoving processes, but could not select any potential combination of various type of equipment which are in the fleet (Marzouk & Moselhi, 2004). The work of Struková, Z., Kozlovská, M., Kováčová, B. (2012) and Zhang formed a framework of multi-objective simulation-optimization for optimizing equipment-configurations of earthmoving operations and it is proposed by integrating an activity object-oriented simulation, multiple attribute utility theory, a statistical approach like the two-stage ranking and selection procedure and a particle swarm optimization algorithm. His procedure is equipped to help compare the alternatives that have random performances and thus reduce an unnecessary number of simulation replications. It can speed up the evaluation process, but this integrated framework is still developed (Zhang, 2008). In this study a computational example is provided to justify our selected scientific methods and theories like theory of systems, multi-objective optimizing method, queuing theory and method of scientific analysis and synthesis. These methods were implemented into soil processes and building machines and its group and will be presented in proposed mathematical model by software which was developed in JAVA. Applications of these methods and software will increase effectiveness of building machines selection from the point of key criteria of optimizing: quality, time and fuel consumption, thus speeding up whole process and avoiding exhaustive calculations and experiments.

2 MACHINE SELECTION OPTIMIZING METHOD CHARACTERISTICS

By suggesting the "Machine Selection Optimizing Method" (MSO Method) we have developed the present state of knowledge of the purpose of the machines and machine groups for building processes (Gasparik, 2013) and also of the information which has been obtained by a study of the theory of systems (Štach, 1983) and optimization theory of the process (Niederliňski, 1983). The "MSO Method" consists of the three phases (Figure 1)—entry, decision and optimization.

An analysis of all these phases except the introductory is examined:

- the input universe of the system: that is the set of the machines submitted for analysis in the given phase,
- the criterion, according to it is the input universe of the system of given phase analyzed,
- the procedural steps being necessary to realize the appreciation of the input universe of the system according to the criterion of the given phase,
- the output universe of the system: that is the set of the machines fulfilling the criterion of the given phase.

3 MSO METHOD APPLICATION

This "MSO method" was applied into the selection of machine group for the excavation and the removal of the earth at the given distance from the above mentioned criteria.

With regard to the great number of model variables and the extent of the work this paper is considering the decision and optimizing phases. Basic input data:

- final product of building process—building pit: width—50 m, length—90 m, depth—3,5 m,
- soil type and class—sandy soil, the 2nd class of cohesion (according to Slovak National Standard STN 733050),
- required work capacity $V_r = 15\,750$ m³,
- transport distance $L = 4$ km,
- required time of duration of works $T_p = 14\,400$ min. (30 shifts),
- season of year of realization of works—April, May,
- kind of road surface—mastic asphalt, plane on the whole length.

Comment: presupposition of approximate identical operation of machines during shifts, time for lunch and inspection of machines at the beginning and the end of shift have not being included in time of shift duration. The input universe of the system of the decision

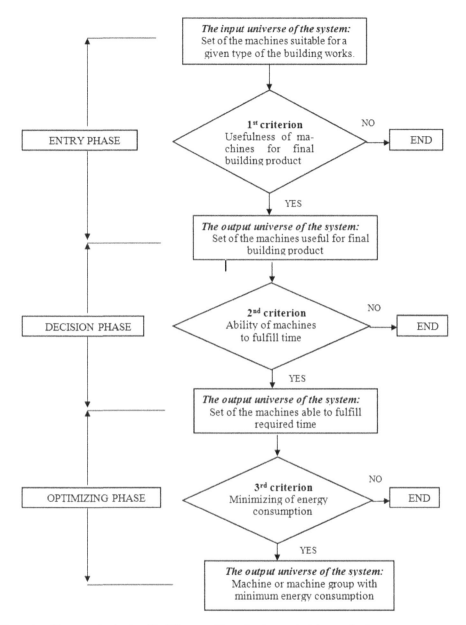

Figure 1. Phases and criteria of building machine selection optimizing method.

phase is being created by 3 types of depth shovel excavators: DH 411, DH 621, Cat 225 and 3 types of folding transport means: T 148 S1, T 815 S3, S 706 MTSP 24. The same transport means were applied to every type of the excavator. There are 9 variants of the excavator machine group together with the transport means and in every variant we used 1 to 13 pieces (pcs) of transport means.

For the evaluation of the machine groups in the decision and optimization phase the concept of queuing theory is applied. In our contribution the final mathematical models (1, 2) of the decision and optimizing phases are shown. All data necessary for equations 1 and 2 can be found in Gasparik (2013)—see references.

The mathematical model of the 3rd eliminating criterion of the decision phase is in the form:

$$T_r = V_r \cdot t_{caj} \cdot (V_{naj} \cdot k_{caj} \cdot k_{kaj} \cdot k_{daj} \cdot k_o \cdot N_{aj})^{-1} \text{ (min)}$$

$$\text{for } j = 1, 2, 3; \; N_a = 1, 2, \ldots, 13, \tag{1}$$

Where T_r = duration of work of machine group by earthworks of required volume (min); V_r = required volume of earthworks (m³); t_{ca} = duration of duty cycle of transport mean (min); V_{na} = volume of earth removed by transport mean in loosened state (m³); k_{ca} = plant factor of transport mean (–); k_{ka} = coefficient of influence of operation of transport mean at its capacity (–); k_{da} = coefficient of influence of transport distance at capacity of transport mean (–); k_o = coefficient of calculation of soil in loosened state at volume of soil in natural state (–); N_a = number of transport means in machine group (pcs).

The output universe of the system of the decision phase follows from graphical interpretation in Figure 2, where suitable variants of machine groups are placed under line representing required time of duration of works T_p. The suitable variants of the machine group of the decision phase are being evaluated in the optimizing phase from the point of view of the minimal F.C. (Diesel oil).

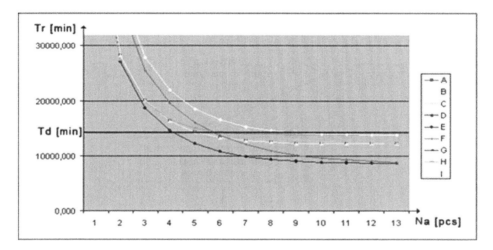

Figure 2. Dependence of actual duration of earthwork Tr (min) on number of vehicles (pcs) of machine group variants (excavator + vehicles) by required volume of earthworks Vr = 15 750 m3. Td = 14 400 min. (required time) for variants A-I.

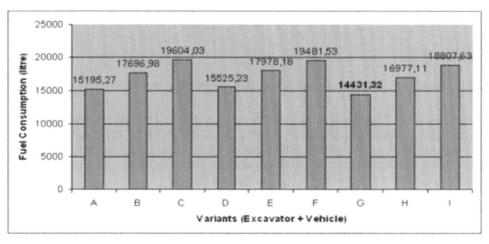

Figure 3. Machine group (excavator-trucks) variants evaluation from the point of minimum consumption of fuel for variants A-I.

The mathematical model of the optimizing criterion is in form as follows:

$$MS = T_r \cdot T_{ps}^{-1} \cdot V_r^{-1} \cdot [T_{mri} \cdot S_{mri} + T_{pri} \cdot S_{pri} + (T_{caj} \cdot S_{caj} + L_{naj} \cdot S_{naj} + L_{paj} \cdot S_{paj}) \cdot N_{aj}] \; (l.m^{-3})$$
$$\text{for } i = 1,2,3; \, j = 1, 2, 3; \, N_a = 1,2, \dots, 13, \tag{2}$$

where MS = specific F.C. (Fuel Consumption) of machine group excavator + transport means by the required volume of the works $(l.m^{-3})$; T_{ps} = duration of operation of machines during a shift $(min.shift^{-1})$; T_{mr} = time of excavator maneuver $(min.shift^{-1})$; S_{mr} = fuel consumption of excavator at maneuvering $(l.min^{-1})$; T_{pr} = duration of work regime of excavator except time of maneuvering $(min.shift^{-1})$, S_{pr} = fuel consumption of excavator in operating regime $(l.min^{-1})$; T_{ca} = duration of waiting regime of transport mean during running engine $(min.shift^{-1})$; S_{ca} = fuel consumption by waiting regime of transport mean $(l.min^{-1})$; L_{na} = length of road covering by transport mean with a load, from place of loading to place of unloading $(km.shift^{-1})$; L_{pa} = length of road covering by transport mean without of load, from place of unloading to place of loading $(km.shift^{-1})$; S_{na} = fuel consumption of transport mean by driving with a load $(l.km^{-1})$;

S_{pa} = fuel consumption of transport mean by driving without a load $(l.km^{-1})$.

Input data concerning the consumption of fuel were given by producers of excavator and transport means. The best energy saving machine groups of each kind are being compared in Figure 3. The most advantageous solution for the realization of output and removal of earth at given distance from the point of view of minimizing of fuel consumption is at analyzed model example a choice of the machine group Cat 225 + 6 pcs of T 148 S1.

4 MSO METHOD SOFTWARE SUPPORT

As a software support for method described in this contribution, the Machine Selection software was created. The software called the Machine Selection is a desktop application, which was built in Java. Therefore, it is runnable on all operating systems that support Java Virtual Machine. Introduction screen contains panels to enter input variables. User can choose number of selected excavator and selected vehicle types. For both—one as minimum and three as maximum. Clicking "Check Inputs" button provides control of input variables values. Wrong values are marked as red, acceptable as green. Button called "Calculate" leads to result screen, which is divided into these sections: Optimal Solutions(s), Complete Work Time Table, Complete Fuel Consumptions Table. For seeing the results, "Optimal Solution(s)" contains of a list, which is displaying all variants of excavator and vehicle(s) able to solve the task in desired time and the required volume of work. This part of result screen is displayed on Figure 4.

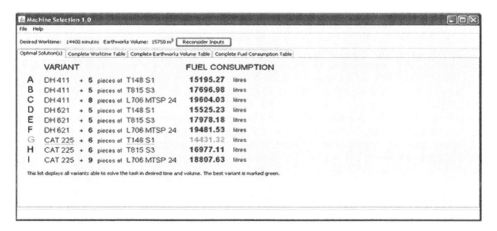

Figure 4.   Result screen, Optimal Solution(s) section.

Best variant is marked as green. In this example, the best variant is marked as a G, it means excavator CAT 225 and 6 pieces of transport means Tatra T148S1. The final fuel consumption is 14431,32 liters, according to formula 1. It is also possible, that task in desired volume with desired work time is not solvable with maximum number of vehicles 13. In this case, fuel consumption of variant is not calculated and this variant is marked as "out of range" error. To overcome the problem of comparing the chosen machines a mathematical modeling approach leading to multi-criteria optimization was adopted to make the step wise decision. To overcome the problem of comparing the chosen machines a mathematical modeling approach leading to multi-criteria optimization was adopted to make the step wise decision. The proposed software and methodology gives a key mathematical models, by which we can find the right type of excavators and transport means.

5 CONCLUSION

Machine selection optimizing method was during our research work implemented into earthwork processes and selected building group of machines (excavators and trucks), which are very often used in construction and mining processes. In our contribution, the selected mathematical models for model example solution and optimization was presented. For calculating the selected criteria of optimality we used a software, which has been created in Java as a support for method described in this contribution. Application of this method and software will increase the effectiveness of building machine selection from the point of key criteria of optimizing: quality, time and energy consumption. For a practical application of the proposed MSO method it is necessary to improve the quality of input data, especially energy use information. The volume of savings of the operating expenses possible to be obtained already in the preparation phase of buildings by this method are not negligible, vice versa, it shows the disclosure of reserves that are available in the choice of machines for building processes. This MSO method will find a full application only when these reservations will be removed.

REFERENCES

Gašparík, J. (2013). *Automated system of optimal machine group selection implemented into soil processes*. Brno, TRIBUN EU, ISBN 978-80-263-0542-2.
Halphin, D.W. (1977). *CYCLONE: A method for modeling job site processes*, Journal of Costruction Division, ASCE, 103 (3), pp. 489–99.
Martinez, J. C. 1996. *STROBOSCOPE: State and resource based simulation of construction operations*. Doctoral Dissertation. University of Michigan. Retrieved from: http://www.lib.gan, Ann Arbor, MI.
Martinez, J. (1998). *Earthmover-Simulation Tool for Earthwork Planning*. In: Proceedings of the 1998 Winter Simulation Conference, pp. 1263–1271.
Marzouk, M., Moselhi, O. (2004). *Multiobjective Optimization of Earthmoving Operations*. In: Journal of construction engineering and management, pp. 105–113.
Niederliňski, A. (1983). *Numerical systems of control technologic processes II*. Prague, CZ: SNTL.
Shi, J., AbouRizk, S. (1998). *An Automated modeling System for Simulating Earthmoving Operations*. In: Journal of Computer-Aided Civil and Infrastructure Engineering, pp. 121–130.
Štach, J. (1983*). Bases theory of systems*. Prague, CZ: SNTL.
Struková, Z., Kozlovská, M., Kováčová, B. (2012). *Determination of optimal supporting system of deep foundation pit based on multi-criteria decision analysis (conference paper)*. 12th international multidisciplinary scientific geoconference and expo, SGEM 2012; Varna; Bulgaria; 17 June 2012 through 23 June 2012; code 101586.
Zhang, H. (2008). *Multi-objective simulation-optimization for earthmoving operations*. In: Automation in Construction, 18, pp. 79–86.

Advances and Trends in Engineering Sciences and Technologies II – Al Ali & Platko (Eds)
© *2017 Taylor & Francis Group, ISBN 978-1-138-03224-8*

Spatio-temporal analysis of soil properties for the eastern border of the European Union

E. Glowienka
Faculty of Environmental, Geomatic and Energy Engineering, Kielce University of Technology, Poland

K. Michalowska
Faculty of Environmental Engineering and Land Surveying, University of Agriculture in Krakow, Poland

A. Pekala
Department of Geodesy and Geotechnics, Rzeszow University of Technology, Poland

ABSTRACT: The goal of this paper is to provide a spatio-temporal analysis of the distribution of strontium (Sr) in to soils of eastern Poland, and an investigation of abnormal concentrations of that element, depending on various environmental factors. The study takes into account type of geological structure, land inclination, type and soil reactions, its sorption capability and radioactivity. Using a specialized GIS software, relevant interpolation procedures have been performed to enable spatial visualization of the geochemical data obtained from the Institute of Soil Science and Plant Cultivation (IUNG) in Pulawy (Poland) in the form of raster and vector maps. In addition, the application of GIS spatial analyses has made it possible to study correlations between the content of the examined element in the soil and the aforementioned factors that affect the soil properties.

1 INTRODUCTION

1.1 *Geochemical soil properties*

Strontium belongs to the group of oxyphilic and diadochically dispersed elements. It can be found in particular in plagioclases, gypsums and carbonate minerals, i.e. calcite and dolomite. Strontium released from rock-forming minerals during the process of weathering, penetrates into the solution and migrates in surface water in the form of bicarbonate, chloride, sulfate, and also possibly as complex compounds (Migaszewski & Gałuszka 2007). The average Sr concentrations in the continental crust amount to 320 mg/kg, while in the world soil 240 mg/kg (Koljonen 1992, Rudnick & Gao 2004). Research conducted in the area of Europe proved, that concentrations of strontium were higher in the lower parts of soil (95.0 mg/kg) than in the upper parts (89.0 mg/kg) (Pekala 2012, Foregs 2016). Increased contents of strontium are identified in those areas, where the presence of carbonate and salt rocks has been discovered (Foregs 2016).

In organisms of animals and humans strontium takes part in the metabolism of calcium and phosphorus. The excess of that element causes decalcification and deformation of bones. The geochemical anomalies in the soil may lead to increase of the content of toxic elements in plants, ground water and surface water. Mapping of that type of pollutions requires collecting proper samples in the field.

1.2 *Geographic Information System (GIS) using in environment monitoring*

Presently, traditional methods of mapping of that sort of phenomena are more and more often replaced with GIS tools, which enable research, analysis and presentation of spatio-temporal

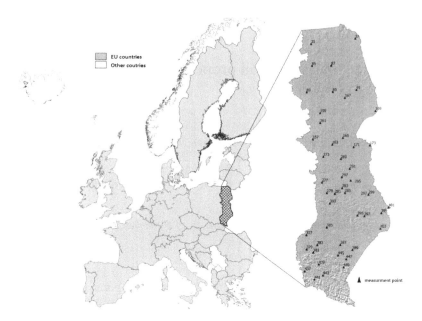

Figure 1.   The localization of area of interest and spatial distribution of measurement points.

distribution of concentrations of elements in the soil. Even open-source GIS applications make it possible to perform spatio-temporal analyses of data gathered within the framework of nature monitoring in the form, which so far could only be attained with the use of dedicated discipline systems (Neteler & Mitasova 2002, Jonesa et al. 2014).

Geographic Information System broadly applied in many fields of science, including but not limited to geomorphology, ecology or geochemistry, support the performance of spatio-temporal analyses and examination of causes and effects of spatial distribution of phenomena as well as modelling and projecting them (Jabbar 2003, Zhang & McGrath 2004, Michalowska & Glowienka 2008, Machiwal 2015, Michalowska et al. 2016).

During the research works on the analysis of spatial distribution of strontium a number of spatio-temporal analyses have been conducted in consecutive periods relying on thematic maps that have been generated based on an attribute database.

1.3   Research area

Research works have been conducted in relation to the buffer zone of land along the eastern border of the European Union, ca. 150 km wide (Figure 1). The area encompasses eastern provinces of Poland. The soil environment of this area was molded by exogenous processes. In the Pleistocene analyzed area was covered by ice ages (from the north to the Outer Flysch Carpathians). Therefore the overwhelming part of sediments which have developed, tested soils from the Quaternary. Moreover, this is a poorly urbanized area, with point pollution sources. The major threat to the natural environment is posed by cement mill, which is the biggest plant of that type in the central and eastern Europe, as well as a brown coal mine and a nitrogen plant situated in the central part of the research area.

2   DATA AND METHODOLOGY

2.1   The research materials

Data gathered by the Institute of Soil Science and Plant Cultivation (IUNG) in Pulawy (Poland) in the form of an attribute database provided the basis for studying variability of: strontium, replaceable calcium, radioactivity level and sorption capability of soil. The data were gathered within the framework of the programme of "Monitoring the Chemism of Arable Soils in Poland" financed from resources granted by the National Fund for Environmental Protection and Water Management (access on website)(GIOS 2016). The research materials were collected from 216 selected measurement points located all over Poland at the depth of 0–20 cm and from the areas of ca. 100 m^2. For the purpose hereof, 48 points located in the area under consideration were selected (Figure 1). Research works were performed in 1995, 2000, 2005 and in 2010.

Moreover, the SRTM (Shuttle Radar Topography Mission) digital terrain model was used (NASA 2016). According to Guth's research results (Guth 2006), SRTM data are well fit for spatial analyses (that is e.g. to generate maps of land slopes, average elevations, height differences, etc.) at average land slopes exceeding 5%. Data for the examined area were acquired, free of charge, from the USGS (United States Geological Survey) server, in the form of an *.hgt files. The raster digital terrain model that was utilized had ground resolution of 3″ (NASA 2016). Analyses were performed only for the territory confined within the boundaries of the area of interest (Figure 1).

2.2 Research methods

Using wide possibilities offered by GIS technologies and the potential of the dynamically developing open source-type applications, geochemical data have been applied to present spatio-temporal variability of soil environment. All operations relating to the process of mapping and analysis of spatial distribution of geochemical parameters of soil were performed with the use of the open-source QGIS/GRASS system.

Based on coordinates of sample collection points, a numerical grid of measurement points was prepared. To perform a detailed topographic and situation analysis, the grid of 48 measurement points (Figure 1) has been superimposed on the base map of a topographic map and orthophotomap that were available as Web Map Services.

Next, using the RST interpolation algorithm, which has been implemented into the GRASS system, map compositions of the strontium concentration in the soil for particular annual data sets (1995, 2000, 2005, 2010) were made. The RST method uses an interpolating function, which is the aggregate of the tendency function and local variability (Mitasova & Hofierka 1993). In this way raster maps have been obtained, constituting two-dimensional models of strontium concentration in the examined area (Figure 2). The raster maps, constituting data of a continuous nature, were then basis for generating isolines of strontium concentration, further utilized in the analysis of the impact of land profile upon the possibility of translocation of the examined element in the soil. On the basis of SRTM altitude data, the exposure of particular measurement points was investigated. This has made it possible to check the potential migration of pollutants, which might have been caused by land height differences.

3 SPATIO-TEMPORAL ANALYSIS OF STRONTIUM (SR) AND RADIOACTIVITY

The spatio-temporal analysis has demonstrated that the median concentration of strontium in the examined area in the period of 1995–2010 amounted to ca. 10.0 mg/kg. That value was lower than the level of content of that element in topsoil (humus) layers demonstrated in studies conducted for the whole Europe area, where the median value was 17.4 mg/kg (Foregs 2016). Within the area under examination, one can notice three characteristic and diverse zones of strontium concentration levels: A—zone of low Sr content, B—zone, in which Sr content approximates to results of European studies, C—zone of extremely high Sr content (Figure 2). With reference to zone A areas, values from the range of 3.0–10.0 mg/kg have been recorded. These are areas, in which luvisols and podzol soils occur. In the south and north parts of the examined area (zone B), the values were close to the European median value, and amounted to 18–32 mg/kg. Following the spatial analysis, it has been noticed that the range of zone B coincided with the occurrence of soils classified as brown soils in accordance with

USDA (United States Department of Agriculture) classification system. In addition, spatial distribution of the content of strontium clearly shows that there exists an extreme anomaly in the central part of the examined area that is marked as zone C (Figure 2).

Strontium concentration values recorded in that area for the particular years of 1995, 2000, 2005, and 2010 amounted to 560.0, 490.0, 517.5, and 453.4 mg/kg, respectively. Values that have been measured considerably exceed values of strontium concentration for the areas of Poland and Europe (Pekala 2012). Soils in the examined region have been defined as rendzina limestone soils (NSRI 2011), while according to the USDA classification (Baldwin et al. 1938), they are silt loams. These are soils rich in divalent calcium and magnesium cations, and poor in iron content. Their acidity measured in KCl ranges from 6.8 to 7.7, and permits to classify them as neutral soils. In the area under examination this is the only sample of soil of that type. The remaining soils were largely dusts and sandy dusts. A thorough analysis of that area as regards other determined soil features allows to notice certain relationships between those features and the concentration of strontium.

High content of strontium in the examined soils may be attributed mainly to natural origin and types of soils. Limestone soils are formed on substrata that are built of carbonate and silica rocks. Examinations of such formations confirm that the content of strontium in opoka-rocks may reach maximum values amounting to 396 ppm, irrespective of the anthropogenic factors. However, in the case under examination one cannot exclude industrial pollutions, resulting from the immission of dusts from the adjacent cement mill. In many studies, scientists focus particularly on the cement mill dust, which is composed mainly of calcium oxide that not only is alkaline but also highly reactive with water. Harley (1966, quoted by Farmer 2004) pointed out that cement dust solutions not only demonstrated a high alkaline reaction (pH 12) but also contained considerable admixtures of heavy metals in the form of sulfides.

The analysis of the IUNG attribute data (GIOS 2016) shows that sorption capabilities of soil that has a high strontium content have become reduced within 15 years from 40.3 down to 22.9 cmol $(+)*kg^{-1}$. Together with decreasing of soil sorption capabilities, its ability to accumulate strontium is decreased. Also the content of divalent calcium cations is changed. In the period under consideration, the content of those cations was reduced almost by half

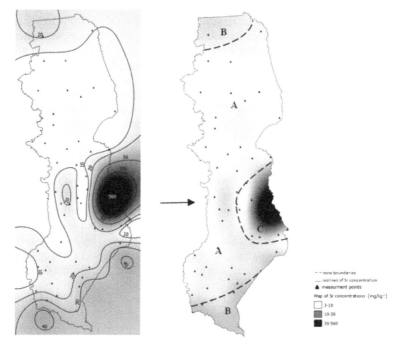

Figure 2.   Map of the spatial distribution of strontium content in the analyzed area—the result of interpolation RST (left). The demarcated zones of strontium concentration (right).

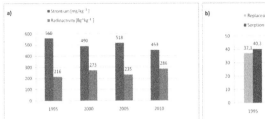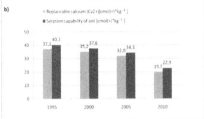

Figure 3. Relationship between: a) strontium content and radioactivity level b) replaceable calcium and sorption capability of soil in 1995, 2000, 2005, and 2010.

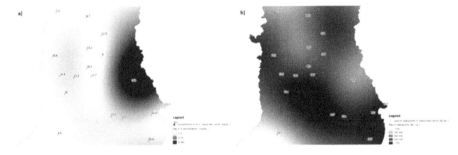

Figure 4. Comparison of strontium and radioactivity levels on basis of maps showing the spatial distribution: a) Sr c.oncentration, b) radioactivity level.

from 37.1 to 19.7 cmol (+)*kg^{-1} (Figure 3). The pace of translocation of pollutants depends mainly on mineral composition of soil and its physical and chemical properties. Within a relatively short period of time (15 years), the specification of soil as regards its parameters subjected to analysis has undergone diametrical changes. In particular, the sorption capability of soil has decreased.

Moreover, on the basis of analysis of maps (zone C), a relationship between soil radioactivity level and strontium concentration has been noticed (Figure 2, Figure 4). One can state that together with the increase in Sr content, the level of soil radioactivity is decreased (Figure 4b). Considering in detail the relationship between strontium content and radioactivity, an analysis of distribution of those parameters in relation to land profile has been performed. The analysis and isolines maps discovered connections between land profile and spatial distribution of radioactivity. The analysis demonstrated that the level of radioactivity was much higher in upland areas than in lowland areas (north part of zone C). Values of radioactivity in upland areas in the period of 1995–2010 fell within the range of 430–860 Bg* kg^{-1}, whereas lowland areas exhibited values ranging from 140 to 430 Bg* kg^{-1}.

4 CONCLUSION

During the study the following factors were considered: type of geological structure, DTM, soil type and pH, soil sorption capability, and radioactivity. The specialized GIS software was used to prepare proper interpolation of data obtained from the Institute of Soil Science and Plant Cultivation (IUNG). Appropriate procedures were carried out to enable the spatial visualization of geochemical data from the Institute in the form of raster and vector maps. Furthermore, application of the GIS spatial analysis enabled checking the correlation between the content of the element in the soil, and the aforementioned factors affecting properties of the soil. The increased concentrations of strontium, which are regional anomalies, can mainly be seen in areas where the geological structure is dominated by carbonate rocks, as in the case of test area (zone C). Strontium concentration values recorded in that area for the particular years of 1995, 2000, 2005, and 2010 amounted to 560.0, 490.0, 517.5 and 453.4 mg/kg, respectively. Values that have been measured considerably exceed values of strontium concentration for the areas of Poland and Europe.

On the basis of analysis of maps for zone C, a relationship between soil radioactivity level and strontium concentration has been noticed. One can state that together with the increase in Sr content, the level of soil radioactivity is decreased.

The study also revealed connections between land profile and spatial distribution of radioactivity. The analysis demonstrated that the level of radioactivity was much higher in upland areas than in lowland areas (north part of zone C).

REFERENCES

Baldwin, M., Kellogg, C., Thorp, J. 1938. *Soil Classification*. Soils and Men: Yearbook of Agriculture 1938. U.S. Government Printing Office. Washington D.C.: 979–1001.

Farmer, A. 2004. Wpływ zanieczyszczeń pyłowych. In Bell J.N.B., Treshow, M. (ed.). *Zanieczyszczenie powietrza a życie roślin.* WarszawaWNT: 209–222.

FOREGS, 2016. *Geochemical Atlas of Europe*. Electronic version (951-690-913-2) http://weppi.gtk.fi/publ/foregsatlas/index.php.

GIOS 2016. http://www.gios.gov.pl/chemizm_gleb/index.php?mod=pomiary.

Guth, P.L. 2006. Geomorphometry from SRTM: Comparison to NED. *Photogrammetric Engineering and Remote Sensing* 72(3): 269–277.

Jabbar, M.T. 2003. Application of GIS to estimate soil erosion using RUSLE. *Geo-spatial Information Science* 6(1): 34–37.

Jonesa, W.R., Spencea, M.J., Bowmanb, A.W., Eversb, L., Molinari, D.A. 2014. A software tool for the spatiotemporal analysis and reporting of groundwater monitoring data. *Environmental Modelling & Software*55 (ISSN: 1364-8152): 242–249.

Kabata—Pendias, A. 2011.Trace Elements In Soils and Plants. *4th ed. CRC Press Taylor and Francis Grup* (978-4200-9368-1).

Koljonen, T. 1992. The geochemical atlas of Finland. Geol. Survey of Finland, Espoo. *I Environ.Qual.* 32:2230–2237.

Machiwal, D., Katara, P., Mittal, H. 2015. Estimation of Soil Erosion and Identification of Critical Areas for Soil Conservation Measures using RS and GIS-based Universal Soil Loss Equation. *Agricultural Research* 4(2): 183–195.

Michalowska, K., Glowienka, E. 2008. Multi-temporal data integration for the changeability detection of the unique Slowinski National Park landscape. *The International Archives of the Photogrammetry, Remote Sensing and Spatial Information Sciences*, Vol. XXXVII, part B7, WG VII/5: 1017–1020.

Michalowska, K., Glowienka, E., Pekala, A. 2016. Spatial-temporal detection of changes on the southern coast of the Baltic sea based on multitemporal aerial photographs. *Int. Arch. Photogramm. Remote Sens. Spatial Inf. Sci.*, XLI-B2, 49–53, doi:10.5194/isprs-archives-XLI-B2-49-2016.

Migaszewski, Z.M., Gałuszka, A. 2007. *Podstawy geochemii środowiska* (978–83-204–3223–7) WNT Warszawa.

Mitasova, H., Hofierka, J. 1993. Interpolation by Regularized Spline with Tension: II. *Application to Terrain Modeling and Surface Geometry Analysis. Mathematical Geology* 25: 657–67.

NASA, 2016. http://www2.jpl.nasa.gov/srtm/

Neteler, M., Mitasova H. 2002. Open Source GIS. *The Kluwer International Series in Engineering and Computer Science* 689 of the series: 1–5.

NSRI, 2011. Soil classification system of England and Wales. *Cranfield University, National Soil Resources Institute.*UK.

Pękala, A. 2012. Mineralogical—geochemical study of the transitional rocks from the Mesozoic—Neogen contact zone in the "Bełchatów" lignite deposit. *Górnictwo i Geologia* 7(2): 187–205.

Rickwood, P.C. 1983. Crustal abodance distribution, and crystal chemistry of the elements. InGovet G.J.S. (ed.). *Hendbook of Exploration Geochemistry* 3, *Rock Geochemistry in Mineral Exploration.* Amsterdam, Elsevier Scientific Publishing Co:347–387.

Rudnick, R.L., Gao, S. 2004. Composition of the Continental Crust. *Treatise on Geochemistry*. Holland, H.D. & Turekian, K.K. (eds.), Elsevier 3, Amsterdam: 1–64.

Zhang, Ch., McGrath, D. 2004. Geostatistical and GIS analyses on soil organic carbon concentrations in grassland of southeastern Ireland from two different periods. *Geoderma* 119(3–4): 261–275.

Advances and Trends in Engineering Sciences and Technologies II – Al Ali & Platko (Eds)
© 2017 Taylor & Francis Group, ISBN 978-1-138-03224-8

Use of the program SWMM to simulate rainfall runoff from urbanized areas

P. Hluštík & M. Úterský
Faculty of Civil Engineering, Institute of Municipal Water Management, Brno University of Technology, Brno, Czech Republic

V. Václavík & T. Dvorský
Faculty of Mining and Geology, Institute of Environmental Engineering, VSB-Technical University of Ostrava, Ostrava, Czech Republic

ABSTRACT: This document describes the program SWMM (Storm Water Management Model) developed U.S.EPA Agency (Agency for Environmental Protection of the United States). SWMM program is a dynamic rainfall-runoff simulation model that allows you to simulate rainfall-runoff processes from an urbanized area and is used around the world. The benefits of this program are: free download, simple user interface, a wide range of inputs and outputs. Small drawbacks include: failure to insert a dwg. format. Incompatibility with GIS. Runoff process is based on a sub-catchment where rain falls and where there is surface runoff and pollution. SWMM monitors the quantity and quality of wastewater within each area. It also monitors the flow, depth and quality of water in different parts of the pipes and drains during the simulation, consisting of several time steps.

1 INTRODUCTION

1.1 Storm water management model

The program SWMM (Storm Water Management Model) is a program developed by the U.S. agency EPA (the United States Environmental Protection Agency). The program SWMM is used around the world for planning, analyses and designs related rainfall runoff, combined and separate sewerage systems and other drainage systems in urban areas. It is a dynamic, rain-fall run-off simulation model used for single events or for long-term simulations of run-off volumes and quality of water from largely urbanized areas.

2 ALTERNATIVE SIMULATION PROGRAMS

2.1 MOUSE pipe flow

This model is the computation tool for simulation of non-stationary flows in the pipeline network enabling to apply both free level and pressure flow conditions for computation. The computation is based on the implicit final distribution of the numerical solution of the basic I-D in accordance with the equation (Saint Venant). The implemented algorithm renders the effective and precise solution concerning quantity of drains and branched pipelines.

2.2 Flow 3D

FLOW-3D is specialized CFD software of the company Flow Science Inc., solution of the tasks with free level flow being its core strength. When solving tasks of this type for steady and unsteady flows, it is possible to define one liquid only (e.g. water) and it is not inevitable to simulate two liquids (e.g. water + air) as in other CFD programs. When modeling free level, the VOF (Volume of Fluid) method can be extended by the method named TruVOF®.

2.3 Mike urban

Mike Urban consists of several modules, such as the model for the sewerage network MOUSE model for rainwater Mike and SWMM models for water distribution, including HYPRESS for the calculation of water hammer and the module Mike NET (ODULA) for water supply systems. The Drain program uses Manning equations for the calculation of water flow in circular or non-circular sewage pipes. The program supports the use of civil structures and profiles of all sizes and any roughness.

2.4 Drain URB

The program has been created by the Brno University of Technology, Faculty of Civil Engineering, Institute of Municipal Water Management.

3 OBJECTIVES OF THE WMM PROJECT

3.1 Input data

As a basis for the simulation we obtained a geodetic survey of the sewer system with the necessary data, such as altitude of manhole covers, manhole depth and length of pipe sections in the assessed area. Furthermore, hydraulic data of the sewer network such as the pipe size, sewer materials etc. was available. Based on expert estimates we determined the roughness coefficient for each material, which is necessary for calculating the flow velocity. Hydrological data of the area, total rainfall and rainfall duration were then determined on the basis of the nearest rain-gauging station.

The developed project is a set of predefined objects where it is necessary to complete corresponding data. The quantity and type of input data required by the programme depends on the outputs required by the investor. In case it concerns the assessment of sewer system capacity, the following data is required: geodetic survey of the sewerage (necessary data is manhole cover altitude and manhole depth) in the assessed area, hydrological data for the area (rainfall) and hydraulic data of the sewerage (pipe size, sewerage material, estimated roughness coefficient. Units of the input and output (calculated) parameters can be entered in the SI metric system or in the US format.

3.2 Data import to the SWMM programme

Data imported to the SWMM programme may be primarily addressed in the form of an BMP image, data import may also be performed based on entering object coordinates (e.g., manholes, outlets, CSOs). Rain data may also be entered in the form of a text file to be entered as: RainGauge → FileName and then you must select the path to the text file containing data for the required simulation. If no data is available (coordinates, map data in image format), the layout may be plotted directly in the SWMM programme as the visual form of the layout is irrelevant. All necessary lengths, surfaces etc are entered in the tables of the relevant objects and sections.

3.3 Outputs

The main output is the assessment of the capacity of sewerage in smaller agglomerations using the software program. The objective aim is to describe the area using the so-called sewerage model and assess the behaviors of rainwater in the sewer system at various rainfall loads. To develop the mathematic model of the sewerage system it is necessary to develop a detailed data collection of the sewerage, identify the technical condition of sewerage-related structures and pipes and collect necessary information about the sewerage.

3.4 Data export from the SWMM programme

Data exported to the SWMM programme may be addressed in two ways offered by the programme. All outputs are addressed via File → Export. The first export option is Map Export, where the programme offers three output format, the most commonly used being DXF, which can be opened in AutoCAD. AutoCAD can be used to plot various manholes and sections between manholes. Thus, the sewerage of the assessed area may be obtained. The most frequent output form is Status and Summary Report, where the programme generate results of the entire simulation in a RPT text file that can be opened in a notepad.

3.5 Design rains

To address the sewerage systems, short-term local heavy rains are essential as these are characterized by high intensity, low surface area and short duration. Rainfall forms the most essential input data for determining the flow rate in the sewerage.Determining the design rainfall is understood as determining the intensity, duration, frequency and type of rain event. For the mathematical model we identified synthetic design storm by Sifaldas the most relevant. This synthetic Sifald rain captures the rainfall time course better than block rain. Sifald's rain is divided into three parts, i.e. pre-rain, main rain and post-rain. Of the total rainfall, pre-rain amounts to 15%, main rain to 55% and post-rain to 30%. The volume of rainfall should equal the volume of block rain. The intensity of the design rain was determined using Trupl's tables (VUV Prague, 1958).

To assess the capacity of the sewerage system, use was made of the following 6 design rains. For each of them we provide outputs showing the relevant part of the sewerage system. Use was made of periodicity $p = 1$, $p = 0.5$ and rain duration of 5, 15, 120 minutes. The main design rain was rain with the periodicity of 1, duration of 15 minutes and intensity of 129 l/s/ha. Other types of rain were chosen in order to expand the simulation and make comparison of the effects on sewerage.

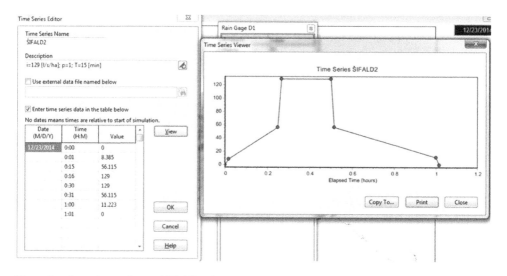

Figure 1.   Intensity setting and Sifald's rain graph, i = 129 l/s/ha.

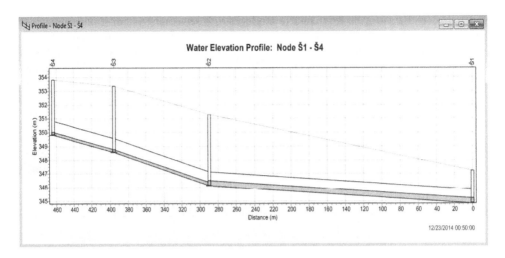

Figure 2.   Flow rate simulation in the end section of the sewer system, i = 129 l/s/ha.

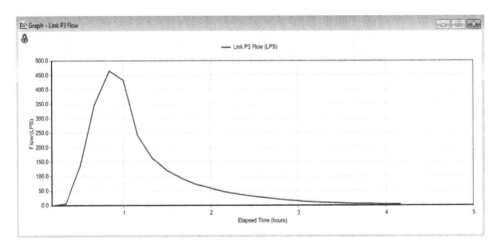

Figure 3.   Graphic output of simulation in the form of flow-rate in pipe in dependence overtime.

Link	Type	Maximum |Flow| LPS	Day of Maximum Flow	Hour of Maximum Flow	Maximum |Velocity| m/sec	Max / Full Flow	Max / Full Depth
P1	CONDUIT	618.62	0	00:35	2.17	0.35	0.41
P2	CONDUIT	589.69	0	00:44	3.62	0.15	0.26
P3	CONDUIT	572.66	0	00:43	3.32	0.16	0.27
P4	CONDUIT	569.28	0	00:43	4.49	0.10	0.22
P5	CONDUIT	502.65	0	00:43	2.84	0.10	0.21
P6	CONDUIT	503.60	0	00:43	2.93	0.10	0.21
P7	CONDUIT	503.56	0	00:43	2.53	0.12	0.23
P8	CONDUIT	503.96	0	00:43	2.70	0.11	0.22
P9	CONDUIT	492.52	0	00:43	3.10	0.09	0.20
P10	CONDUIT	493.06	0	00:43	3.10	0.09	0.20
P11	CONDUIT	494.48	0	00:42	3.20	0.08	0.19
P14	CONDUIT	490.85	0	00:42	1.13	0.68	0.59

Figure 4.   Summary table—simulation results. i = 129 l/s/ha.

3.6 Sewerage assessment

As part of the simulation we assessed the entire sewer system and, in particular, the endsection of the sewer with respect to the design rain. The attached longitudinal section (Figure 2) shows pipe sections P1 to P3 between manholes S1-S4 during the maximum load for the relevant rain. The attached table (Figure 4) compares parameters such as flow rate, flow speed and filling during peak loads in the given pipe section.

The entire sewer system was simulated using six design rains and, in all cases, the results were evaluated both in graphic and tabular formats. The summary results of the design rains are presented in Table 1. The advantage during the simulation is the possibility to adjust the time steps at which we wish to follow the progress of the simulation. The result is then a simulation of sewer filling, e.g. in each minute of the rain event at the selected rainfall intensity. In each minute we can therefore evaluate whether the sewer has been overfilled or not. The selection of the time step can be accelerated or decelerated for any potential discussion about the sewer filling with rainwater.

The simulation of flow rate in the sewerage can either be monitored in the selected longitudinal profiles that are assessed or in the entire considered area. To do so, use is made of the main work environment of the program where, using the "Map" function, we can select a color to distinguish between the filling heights in the individual time steps. This makes it easy to recognize when a particular section of the sewer changes into the color corresponding to the percentage of filling.

Figure 4 shows that in the P1–P3 sections the maximum flow rates are reached in time 35 minutes–44 minutes after the rainfall start. In terms of capacity, the pipe in the end sections is filled up at ca. 35%. It can therefore be said that the pipe capacity is adequate for our main design rain.

4 SEWERAGE ASSESSMENT

4.1 Identified results

Table 1 shows the evaluation of the end section of sewer according to the individual design rain events, both in terms of maximum flow rate, velocity, capacity and filling height of the circular pipe profile.

Our task was to assess whether the end section of the sewerage upstream the inlet to rainwater separator meets the size requirements. It was detected that even the end section meets the size requirements given all selected design rainfalls. For main design rain $p = 1$; $t = 15$ min; $i = 129$ l/s/ha it was identified that the end section of the sewer is oversized.

The SWMM program established that the end section could be one order of magnitude smaller i.e. DN 800. If the program was employed when designing the sewer network, it would be determined that the proposed profile DN 1000 was oversized. During the construction of the relevant sewerage section the initial capital costs were increased (pipe profile, earthworks and others).

Table 1.   Summary table of design rain assessment.

Rain	Max. flow [l/s]	Time [min]	Velocity [m/s]	Capacity (flow rate) [%]	Capacity (height) [%]
$p = 1$; $t = 15$	618,62	35	2,17	35	41
$p = 1$; $t = 5$	1601,42	15	2,58	91	75
$p = 1$; $t = 120$	34,23	265	0,91	2	10
$p = 0,5$; $t = 15$	652,78	35	2,21	37	42
$p = 0,5$; $t = 5$	1880,01	15	2,66	100	91
$p = 0,5$; $t = 120$	52,89	263	1,03	3	12

5 CONCLUSIONS

The final output was the assessment of the sewer system capacity. As a basis for entering the input parameters into the program we used the municipality layout showing land survey and hydraulic data, aerial images of the municipality to specify the hydrotechnical situation of the area and Trupl's tables (VUV Praha, 1958). The applied data were sufficient to carry out rainfall runoff simulation. Based on the outcomes we concluded that the capacity of the sewerage system was sufficient in all design conditions simulated in the SWMM program. The simulation of the rainfall runoff processes was also performed in other programs which, similarly to SWMM, assessed the capacity of the pipe profile as satisfactory. None of the simulations resulted in any major result deviations.

Advantages of the SWMM program: freely downloadable program, easy to install, user environment, work in the program, program stability, wide range of inputs and outputs.

Modeling capabilities of SWMM: time variable rainfall, evaporation of stagnant surface water, snow accumulation and melting, rainwater infiltration into unsaturated soil layers, percolation into groundwater layers, leakage of groundwater into sewer systems, rainwater retention and storage, etc.

Disadvantages of the SWMM program: inability to enter data in the DWG format., it does not work with GIS compared to competing program, it cannot perform calculations if the pipe slope is reversed, absence of "back" button, not too frequent program updates (latest version 5.1 from September 2015).

ACKNOWLEDGMENT

This paper has been worked out under the project No. LO1408 "AdMaS UP—Advanced Materials, Structures and Technologies", supported by Ministry of Education, Youth and Sports under the „National Sustainability Programme I".

REFERENCES

Hlavínek, Petr. *Stokování a čištěníodpadníchvod*. Vyd. 1. Brno: Akademickénakladatelství CERM, 2003, 253 s. ISBN 80-214-2535-0.

Rossman, Lewis. U.S. EPA. *STORM WATER MANAGEMENT MODEL USER'S MANUAL*. 2010.

Starý, Miloš. *Hydrologie*. Brno: Vysokéučenítechnické, Fakultastavební, 2005, 368 s. ISBN Hydrologie.

Trupl, Josef. Intensity krátkodobých dešťů v povodích Labe, Odry a Moravy.Praha, 1958, p. 76.

Advances and Trends in Engineering Sciences and Technologies II – Al Ali & Platko (Eds)
© 2017 Taylor & Francis Group, ISBN 978-1-138-03224-8

Causes and impacts of dropping water consumption on a wastewater treatment plant

P. Hluštík
Faculty of Civil Engineering, Institute of Municipal Water Management, Brno University of Technology, Brno, Czech Republic

V. Václavík & T. Dvorský
Faculty of Mining and Geology, Institute of Environmental Engineering, VSB-Technical University of Ostrava, Ostrava, Czech Republic

ABSTRACT: This paper identifies factors that influence the process of dropping water consumption and their impacts on the owners and operators of wastewater treatment plants. The impacts of decreasing water consumption on the WWTP are mainly economic aspects for the population, i.e. higher water and wastewater tariffs affected by the contractual terms related to the granted subsidies and sustainability, oversized civil structures at the wastewater treatment plants with a high buffer capability, unused capacity of equipment at wastewater treatment plants, lower energy intensity of various production plants and local factors. Currently, sewage treatment plants are not operated to the maximum performance of the WWTP and are not fully utilized.

1 INTRODUCTION

Drinking water supply, wastewater collection and treatment form an important part of living a quality life in modern society. These services also condition economic and social development at the local, regional, national, and to some extent, international level. Water consumption is characterised by a continuous year-on-year drop. The drop in water consumption results in insufficient water-related structure capacity utilisation and oversized water pipes where water is often transported over tens of kilometres with a several-day retention.

The main factors affecting the dropping consumption of drinking water are primarily the economic increase in water tariffs (construction or reconstruction of water management infrastructure), development of individual water supply systems, decreased water consumption in agriculture and industries, environmental aspects and other factors (e.g. efficient appliances, rainwater and grey water harvesting, etc.).

2 WATER MANAGEMENT INFRASTRUCTURE REFURBISHMENT

Reliability of the infrastructure contributes to low operating costs. Achieving optimal service-life and reliability of the water supply infrastructure is ensured via all necessary conditions in the process of planning, design, preparation and implementation, and operation and maintenance of the infrastructure. The service life of the infrastructure is the result of the interaction of a number of factors, proposed concepts, design, materials and technologies used, the implementation method and the overall quality of construction and installation, correct operation, including compliance with design parameters and regular and high-quality maintenance and repairs of the infrastructure. Every part of this chain can have a significant impact on the service life of the resulting whole as well as the individual parts. A newly refurbished wastewater treatment plant can reliably keep all limits and standards set by the European Union and conserve the environment. Given the terms and conditions of the provided subsidies and recommendations made by the State Environmental Fund the sewage tariffs necessarily increase in order to ensure the sustainability of this project.

2.1 Self-funding and cost recovery

In European Union countries, the principle of self-financing and cost recovery applies, for wastewater treatment plants average recovery is about 85%. With respect to infrastructure operation it is generally stated that cost recovery is achieved at 100% in relation to the operation, however, this decreases rapidly if new investments and reconstruction of the infrastructure is included, to about 10–15%.

The funds to operate, reconstruct and develop water management facilities must be generated from the water and sewage tariffs, i.e. without subsidising operating and other costs. This gradual increase has been set as one of the basic conditions for the sustainability of the project of reconstruction and construction of new wastewater treatment plants and a failure to follow these conditions could result in financial sanctions.

2.2 Tool to calculate sustainability of water management facilities

The State Environmental Fund has published a tool to calculate the sustainability of water management projects submitted under Priority Axis 1 of the Operational Programme Environment (herein under OPE) for 2014–2020.

This software tool for calculating the sustainability is applicable to water management projects co-financed from the Operational Programme Environment for the programming period of 2014–2020 under Priority Axis 1 Improvement of water management infrastructure and reduction of flood risks, Specific objective 1.1 Reduction in the amount of pollution discharged into surface water and groundwater from municipal sources and infiltration of pollutants into groundwater and surface water and Specific objective 1.2 Provision of drinking water supplies in adequate quality and quantity. The published Sustainability tool will set the minimum amount of funds to be generated by the owner (applicant) over the reference period and to be re-invested in the reconstruction of infrastructure facilities. It will be mandatory to submit the filled-in tool along with the documents needed for the decision to award the grant.

The generation of funds to ensure the refurbishment of infrastructure facilities under OPE 2014–2020 projects means that the owner (applicant) of the infrastructure facilities should generate such a minimum amount of funds within 30 years from the start of the project implementation to cover the renewal of infrastructure facilities (in the form of lease payment/ depreciation/profit/repairs/funds for refurbishment directly specified in accordance with the financing plan for water supply or sewerage systems, etc.) in order to guarantee financial sustainability of the project in terms of the minimum funds that should be reinvested in the infrastructure of the owner (applicant/beneficiary). These funds should be generated in such an amount so that at the end of the 30-year reference period the so-called "full depreciation" has been reached, with the full depreciation defined as the value of the existing and newly constructed infrastructure divided by the average economic service life of the assets (15 years for technology, 40 years for water pipelines, 60 years for sewerage and 40 years for other structures).

Subsequently, the funds should be used for the management, rehabilitation and extension of water management infrastructure of the owner (applicant/beneficiary). The beneficiary should submit the completed sustainability tool at latest as part of the supporting documents for the grant decision, the applicant has committed itself to generate such minimum funds and this will be audited during the subsequent project monitoring for a period of 10 years from the commencement of the project operation. The sustainability tool is submitted together with the draft financing plan related to the rehabilitation of water mains or sewers, which must be developed by every owner of the infrastructure for a min. of 10 years in accordance with Annex No. 18 to the Decree of the Ministry of Agriculture No. 48/2014 implementing Act No. 274/2001 Sb., on water supply and sewerage systems for public use and amendments to certain acts (the water supply and sewerage systems act).

It must also be mentioned the fact that the mechanism of how to suitably accumulate and keep funds is not clear (this is due to changes in municipal councils, low interest rates on deposits and, on the contrary, high interest rates on loans, etc.). In this respect it is interesting that the required fulfilment of the OPE terms and conditions for OPE eligibility for grants from the European Community does not provide for a solution to the situation.

3 CAUSES WATER CONSUMPTION ON WWTPS IN THE CZECH REPUBLIC

The development of specific consumption has been different from the expectations after 1990. Historically, water consumption was increasing with water mains construction in municipalities. The highest water consumption was in the socialist era, when the price was determined by the planned economy and did not reflect the actual costs of water treatment plants. The price was negligible and, as a result there was huge waste of water. In the period after 1990, when the price of water was set on the basis of actual costs, the specific water consumption began to drop again. We expect a continuous drop in specific water consumption as we are pressed to do so by the ever-increasing water and sewage tariffs. The current specific household water consumption in 2015 totalled 89 l/inhab./ day, the total consumption was 130 l/inhab./day and it is one of the lowest in Europe. The price of water in the whole country increased by almost 85% over the past decade, which is twice the average in the European Union. The current average water and sewerage tariff is CZK 85/m3.

3.1 Causes and reasons for dropping water consumption

3.1.1 Technical-economic reasons

The dropping water consumption is mainly due to economic reasons. Every year, the water tariff in the Czech Republic depends on a price assessment issued by the Ministry of Finance of the CR. The price assessment contains regulations that specify costs that may be included into eligible costs, and costs that cannot. Eligible costs represent about 40% of the total price of water. They cover depreciation used to reconstruct sewers, water mains, etc.

The price of water also includes stands repairs, chemicals including laboratory research and energy consumption, wages, various services, purchase of water from available sources and fees for wastewater discharges. These cost items keep increasing. The profit along with depreciation is invested in the refurbishment and development of the water management systems. This cost has a major share that cannot be significantly reduced despite the dropping water production and consumption. The final price of water is also determined by the VAT rate, in every locality it also depends on the condition of sewers, water mains, on the quantity of water consumed and on water availability. The water and sewage tariffs are also influenced by illegal abstracting. A developed questionnaire has determined aspects that affect the decreasing water consumption. The questionnaire has been drawn up for the general public for 2,000 respondents, equally distributed in agglomerations of various sizes and age structure. The aspects of dropping water consumption can be included in the following points.

Savings population
– In-house drinking water resources
    The number of inhabitants who own drinking water sources is 13%. These are wells utilising water for direct use and for further use. The conducted survey shows that nearly 90% of the wells were non-compliant in at least one indicator. As regards to the fitness to drink, ca. 60–70% of wells meet this parameter.
– Energy-saving appliances
    Nowadays, appliances come with various energy categories. For new washing machines, dishwashers and toilets, the manufacturers not only reduce energy consumption, but also

water consumption. Economical appliances are owned by 86% of people. The remaining 14% are students living in university dormitories and hostels.
– Rainwater use for irrigation and technical maintenance

The use of rainwater can save up to 50% of drinking water. The most common ways of using rainwater instead of drinking water relate to green area watering—79%, toilet flushing—5%, technical maintenance—4% (car washing, swimming pools, etc.) and washing—1%.
– Recycled water (grey water) harvesting

Recycled water is the dominant solution mainly small sources and localities. In the Czech Republic the use of grey water is in its infancy, recycling is used in new houses by 2% of the inhabitants.

Charges for groundwater and surface water abstraction
In 2013, according to the CSO data, a total of 601.7 mil. m3 of drinking water was produced, of which 50.3% from groundwater resources. The state revenues from water companies for groundwater abstraction thus totalled around CZK 623 mil and half of this amount was received by the State Environmental Fund of the Czech Republic and the other half by local regions. The collected funds for groundwater abstraction are primarily intended to protect and develop groundwater resources which is the responsibility of the state; however, the funds are in fact used in other areas. As regards public resources, it is not possible to replace the raw water resource, which leads to higher water and sewage tariffs. The lower payment limit is at abstraction of 500 m3/month, which practically means that this is another reason to get disconnected from the collective distribution drinking water systems as the individual abstraction is not subject to a charge (domestic wells).

Groundwater abstraction can be made only on the basis of time-limited permits issued by water authorities, which must be accompanied by hydrogeological surveys and long-term measurements must be provided for larger abstractions. These measures ensure compliance with the balancing system in the relevant river basin.

Design stage
The primary design problem may occur already in the preparatory stage if not all supporting data is collected, with respect to civil-technical amenities, business activities in the municipality, technical condition of the sewer network, ballast water, current data on design rain events. Due to the constant weather changes (alternating periods of drought and heavy rainfall), it is necessary to make the data more accurate. Statistical evaluation has been simplified as a result of the unavailability of high-performance IT equipment in the past years. However, rainfall observations have not been updated to determine more accurate design rainfall values.

WWTP technologies
Selecting the correct technology is critical to the operating and capital costs and machinery at the WWTPs. With a higher number of PE the technology changes along with the values of permissible emission standards, tank volumes and mechanical equipment at the WWTP.

Inadequate capex
At the preparatory, design documentation stage, costs inadequate for the municipalities resulting in debts incurred by the municipalities are expended with respect to the terms of Council Directive no. 91/271/EEC, and to the provision of the required quality of discharged water. The prepared WWTP projects are often oversized, designed to for higher mass loads and quantity of wastewater. These are mainly investments in the previously approved projects implemented at present.

3.1.2 *Ecological reasons*
Another reason is the lack of water coupled with water saving, which can be classified as ecological reasons for the population. In the Czech Republic, the condition of the environment deteriorates which is reflected in the dropping water yield of water resources, disrupted subsurface runoff, increased flow-rate fluctuation in water courses and deteriorated quality

of groundwater. The worst situation of extremely dry soil with water immediately running off and the soil unable to retain water is in the South Moravian Region. These causes rank amongst ecological reasons.

Drought with respect to water-related structures
Changes in rainfall are highly uncertain; nevertheless, most climate models agree on the stagnation of annual precipitation and changes in its distribution during the year, specifically, the drop in summer rainfall and rise in winter rainfall. This indicates an increased risk of adverse hydrological balance in the summer period, both in terms of providing water for human consumption and for food production, and in terms of the ecological status of bodies of water.

Environmental limits of localities
Although the quality of surface and groundwater in the Czech Republic has greatly improved in recent years, the condition of water courses cannot be considered as satisfactory. Problems are mainly related to streams with lower water yield and high accumulation of point and diffuse sources of pollution.

At the preparatory stage of the design documentation it is necessary to analyse the risk rate of surface water in terms of quantity (abstraction vs. primary runoff from the body of water) and in terms of the chemical condition (infiltration of substances from diffuse sources from agriculture—nitrogen, pesticides, atrazine, impacts of point sources of pollution—old environmental pollution). Water eutrophication increases the costs of treatment, which is reflected in the water tariff. To evaluate groundwater according to the environmental indicators, use is made of abiotic indicators, biotic components (phytoplankton, phytobenthos, macrophytes and fish), but also pollutants such as BOD, COD, SS, Nt and Pt.

More stringent concentration limits
In the forthcoming Amendment to Water Act No. 274/2001 Sb. there are more stringent limits set for Pc and AOX. These limits can not be met anymore by simply increasing the dosage of chemicals that are not negligible, but in practice it will be necessary to keep the limits through investments—construction of tertiary wastewater treatment stages. Increased annual consumption will also require more complicated tender documentation in competitive bidding (above-limit for public procurement). Reduction in the Pt indicator means payments by municipalities with WWTPs from the size of 400 EO and municipalities with polluted wastewater even from about 260 EO. The more stringent weight limit AOX will affect more WWTPs over 5,000 EO. To keep the limits, it is necessary to modify the technology and sorbent dosing—e.g. activated coal, again with an impact on operating and capital costs. There are no new current methods in technically and economically acceptable conditions available.

4   IMPACTS OF DROPPING WATER CONSUMPTION ON WWTPS IN THE CZECH REPUBLIC

Due to the constantly decreasing water consumption it is important to correctly determine the volume and mass load, i.e. determination of specific water consumption for the equivalent determination of the volumes of civil structures and mechanical-technological equipment.

Storage capacity of the tanks
At the oversized existing wastewater treatment plants it is possible to use the free capacity of the tanks for connecting new sewer sections and localities. Wastewater treatment plants with a large storage capacity are able to withstand rain events and thus prevent unwanted biological stage flooding.

Technology and equipment modifications
Currently, more and more modern technologies and mechanical-technological equipment is used to increase efficiency while minimising the requirements for operating and energy costs. By selecting the proper option in planning investments or reconstruction it is possible to reduce energy intensity, for example, frequency converters in pumps, blowers, WWTP control systems.

Capital and operating costs

Operating costs related to wastewater treatment are usually lower compared to higher capital costs of civil structures in terms of the reduction in volumetric quantity of wastewater.

Water course yield

In the summer, as a result of drought there are more frequent cases of surface water drying off. The main problem related to drought is scarcity of water and reduced water quality. Besides water scarcity for water supplies, the drought manifests itself in the lack of water for economic purposes, or in the restricted or interrupted use of water for power generation purposes.

5 CONCLUSIONS

Due to inadequate management, fragmented competence and lack of communication between authorities a fully functional system ensuring sustainability of water supply and sewerage systems is still missing in the Czech Republic. Furthermore, the Czech Republic has not fulfilled the conditions of the European Union concerning the treatment of wastewater and is threatened with penalties which have not been imposed yet. When refurbishing the existing wastewater treatment plants and constructing new plants it is important to identify the input data (current information, monitoring, analyses) and to propose suitable wastewater treatment technologies.

ACKNOWLEDGMENT

This paper/contribution/abstract has been worked out under project No. LO1408 "AdMaS UP—Advanced Materials, Structures and Technologies", supported by Ministry of Education, Youth and Sports under the "National Sustainability Programme I".

REFERENCES

Evropské vodohospodářské společenství. Směrnice Rady o čištění městských odpadních vod. In *Úřední věštník* č. L 135. 1991, 40, 91/271/EHS.

Nástroj pro výpočet udržitelnosti vodohospodářských projektů v programovém období 2014—2020. *EVROPSKÁ UNIE: Evropské strukturální a investiční fondy Operační program životního prostředí* [online]. 2015 [cit. 2016-04-26]. Dostupné z: http://www.opzp.cz/o-programu/aktuality-a-tiskove-zpravy/nastroj-pro-vypocet-udrzitelnosti-vodohospodarskych-projektu-v-programovem-obdobi-2014-2020.

Zákon č. *254/2001 Sb., o vodách a o změně některých zákonů (vodní zákon)*. In: 2001.

Český statistický úřad [online]. 2016 [cit. 2016-04-27]. Dostupné z: https://www.czso.cz/.

Životnost a obnova vodohospodářské infrastruktury. CzWA: *The Czech Water Association*[online]. 2015 [cit. 2016-04-26]. Dostupné z: http://www.czwa.cz/os/oszovi.html.

Zákon č. *274/2001 Sb. o vodovodech a kanalizacích pro veřejnou potřebu a o změně některých zákonů (zákon o vodovodech a kanalizacích)*, In: 2001.

Effectiveness of costs incurred for labor protection

Z. Hulínová, T. Funtík, J. Madová & A. Bisták
Faculty of Civil Engineering, Slovak University of Technology in Bratislava, Bratislava, Slovakia

ABSTRACT: The Occupational Safety And Health (OSH) approach is very different for various organizations. It is particularly evident in the attitude to compliance with safety regulation. Despite the high risk of building production there are organizations that do not comply with these regulations at all or only under the threat of penalties and fines, but also those which approach the OSH obligations beyond the legal requirements. The voluntary principle in the OSH approach needs to be encouraged through resources to help organizations navigate when deciding to invest in prevention. An appropriate means of obtaining specific indicators is a mathematical model that expresses the dependence between costs of prevention and accident costs in relation to attitudes towards OSH.

1 INTRODUCTION

Construction is a high risk industry. New technologies are emerging with more intense development of science and technology, and with them new types of risk and new methods of realization. New work situations can also result in unexpected health and safety risks that would require some approaches in order to ensure a high level of occupational safety and health (OSH) (Struková & Kozlovská, 2014).

The OSH system is enforced through state legislation. Compliance with the laws and others regulations to ensure occupational safety and health falls under the preventive activities of the organization and can bring benefits both for employers and employees. Non-compliance with safety rules results in corrective actions and may lead to losses, especially on life and health in the form of occupational accidents or to financial losses in the form of penalties or fines.

Both of these areas, preventive and corrective, require certain resources and time to achieve them. These funds must be incurred by the employer. Due to their activity and approach to OSH they constantly come into the process of deciding whether it is more effective to spend resources for preventive measures or the correction of adverse events, and whether they will release adequate resources and time to protect workers, or they will not adequately protect workers, and thus risk adverse impacts. However, we must also consider the possibility that no accident happens, and so these resources are saved.

Some authors think that the most important incentive for businesses to invest in occupational safety and health is the fact that the costs of disaster recovery are very high (Laufer, 1987; Levitt & Parker, 1976; Simmonds & Grimaldi, 1963; as cited in Ibarrondo-Dávila et al., 2015).

Besides the state when safety regulations are fully complied with, or are followed in part or almost not at all, the employer also has the possibility of exceeding these requirements beyond the legislative framework. Even in this case it is important for them to know the limit to which the inserted means for protecting workers will still be effective.

The effectiveness of spending on occupational safety and health is therefore essential for the employer. The costs of prevention and the costs of accidents can be determined via mathematical modeling of individual items and based on the statistical research. One of the important statistical indicators that need to be introduced into the model is the incidence of occupational accidents.

Authors Chua & Goh, 2005 considered the incidence of occupational injuries to be random events that successfully approach the Poisson distribution values defined by a certain mean value of occupational accident frequency.

By comparing the costs of prevention and correction with regard to the production unit, we receive information about the advantages and disadvantages of these approaches to OSH, or even the threshold when it is still effective to invest in OSH.

2 METHODOLOGY

2.1 *Approach to OSH*

The level of knowledge in the field of health and safety in the industry is unsatisfactory and security is considered an unnecessary expense. More specifically, in order for organizations to remain economically competitive, sustainable, and also to achieve maximum profit, many perform only basic precautions and ignore many important programs of prevention and professional training during construction (Cheng et al., 2010; as cited in Gurcanli et al., 2015).

The legislation prescribes minimum requirements to the employer that must be respected in the workplace in the context of occupational health and safety. In the event of non-compliance with legislative requirements there are costs that adversely affect the economics of the organization, create a bad name for the company, but particularly endanger the lives and health of employees with this attitude.

The employer may voluntarily expand occupational protection, but it has meaning for them only if the incurred resources are still effective. But the organization's management needs to have a good motive to increase occupational protection over the legislative framework. Such a motive can be a zero accident rate. If the organization wants to achieve zero accident rate in OSH, occupational safety and health of workers must become a top priority, and it is therefore necessary that all participants are motivated to cooperate. Gains from such a management system appear in the form of higher productivity, better culture of the organization, but also the valuation and recognition by external collaborators and a better position in the labor market.

Given the maturity of the OSH management we can classify organizations into four stages:

- organizations with partial compliance with legislation: organizations which have poor OSH management and expertise as well as response to problems at the moment when an accident happens or when an employee of the labor inspectorate comes for an inspection;
- organizations complying with the legislative framework: organizations that comply with the legislation, carry out regular risk assessments, actively plan and set priorities for troubleshooting, and perform planned control activities;
- organization with active exceeding of legislation: organizations that implement and maintain OSH as a management system pay continuous structural attention to OSH organized before starting new activities;
- organizations with the inclusion of OSH in other management levels: organizations that incorporate the OSH management into other management systems and/or their business processes, management focus is on continuous improvement and collective learning (Kiviniemi et al., 2011).

The role of each employer is to design their system of protection of workers so that its efficiency is in accordance with the efficiency of expenditure on ensuring the OSH.

The proposal to eliminate or remove the risk has a higher priority than simply controlling the risk or protecting workers from risks (Manuele, 1997; as cited in Gambatese et al., 2008).

Tam & Fung, 1998; as cited in Esmaeili et al., 2015, studied the relationship between standard safety management strategies and their accident rates with the use of multiple regression analysis. They found that four variables are important in determining safety performance, namely: (1) investigation after the accident; (2) the proportion of work sub-contracts; (3) safety awards; (4) safety training. In particular the effects of informal approaches and practices in the field of education are significantly higher than the effects of formal approaches (Kozlovská & Struková, 2011).

At present, there are more and more emerging views that investing in prevention is more profitable than repairing the damage, and if its claim is substantiated, the investor will voluntarily put money into preventive activities (Brody et al., 1990; as cited in Feng, 2013). When examining the relationship between security investments and safe operation, authors found (Tang et al., 1997; as cited in Feng, 2013) a weak correlation coefficient of 0.25 in construction projects in Hong Kong. They speculated that this may be a difference in the safety culture of the different companies.

Creating a safe working environment requires proper preparation, a systematic approach, but also specific expenditure to ensure measures to eliminate or minimize risk. When designing measures, it should be noted that there is a limit where the insertion of other resources for protection does not bring its increase.

2.2 *Methods*

The procedure of mathematical model creation is outlined in this paper. This procedure allows to determine the efficiency of investments in OHS. Base for this procedure is detailed knowledge of OHS field with application of statistical and mathematical means, especially Poisson´s models. For the verification of the model it is necessary to obtain sufficient number of components of costs for prevention and costs of accidents. This is possible to obtain via questionnaires filled by civil engineering organizations as well as statistical information from OHS supervising authorities.

3 THEORY

The basis for the OSH approach is the attitude of the organization towards compliance with obligations arising from safety regulations. The decision whether to invest resources in the prevention or correction of unwanted damage arises based on this approach. The relationship between the costs of prevention and accident costs should be expressed by a mathematical model.

3.1 *Structure of costs to ensure OSH*

The indicative cost structure associated with ensuring OSH includes the following items: personal protective equipment (PPE), collective security, education and training, control, qualification development, services such as occupational health service and safety-technical service etc.

Structure of costs associated with non-compliance with legislative requirements for ensuring OSH may include: fines, sanctions, payment of sickness absence, compensation, damage repair, psychological damage, loss of reputation etc.

According to Feng, 2013, the security investments consist of all types of costs incurred in organizing the project by the contractor (including subcontractors).

These safety investments—basic safety investments ratio (BSIR) and voluntary safety investments (VSIR) were used to compare the level of total safety investments (TSIR) between projects of different sizes.

The study provided new insights into the relationship between investment safety and safety of the operation:

1. investments in voluntary security measures are more effective in preventing accidents than investments in basic security measures
2. the effect of the basic safety investment in accident prevention is tempered by safety culture and the level of projected danger of the construction projects

3. basic safety investments have a stronger positive impact on accident prevention on the basis of a higher level of safety culture and the projected level of danger
4. the effect of the basic security investment in injury prevention may not be positive, if the projected level of hazard and the level of project safety culture are low

López-Alonso et al., 2013, divide safety costs into several groups: safety costs, cost of non-safety and other extraordinary costs. Safety cost is even further divided into: prevention costs as well as evaluation and monitoring costs. Costs of prevention are those incurred in order to comply with legal requirements relating to accident prevention.

Cost of non-safety is divided into: tangible costs of accidents (can be estimated or calculated using traditional cost accounting methods) and intangible costs of accidents (according to Gosselin, 2004; as cited in López-Alonso et al., 2013 are unmeasurable costs, such as the deterioration of the company image, low staff morale, labor disputes or losses on the labor market).

The authors also present other extraordinary costs, which are all losses caused by events that cannot be prevented by technical or human resources available on constructions, or which are absolutely necessary, such as natural disasters.

Intangible costs or extra costs cannot be included in a structured calculation model, because they generally are not calculable or estimable in the building project, and also because they are uncontrollable.

3.2 *Mathematical model*

Mathematical model describes the method by which it is possible to compare the cost of prevention (C_P) and the costs of accidents (C_A).

Cost of prevention (C_P)

Given the organization's approach to compliance with the OSH rules prescribed by legislation, we can split organizations into three groups after a certain simplification by analyzing the different types of behavior (Table 1):

– Group I: ensuring of protection is almost zero;
– Group II: basic protection of workers determined by the legislation is ensured;
– Group III: above standard protection of workers beyond the legal requirements is ensured.

The activities the organization is performing to ensure occupational safety and health belong to the area of prevention activities. In relation to the three groups of access to OSH we distinguish three prevention costs groups:

– for Group I costs $C_Z = 0$, since the organization risks and does not protect workers;
– for Group II costs C_B [€], these are costs in terms of legal requirements;
– for Group III costs C_{BR} [€], these are costs for above standard work protection beyond the scope of legislative requirements.

Costs of accidents (C_A):

Occupational accidents tend to be the adverse consequences of unsafe activities at construction sites. Chua & Goh, 2005, considered the incidence of occupational injuries to be random events that successfully approach the Poisson distribution values defined by a mean

Table 1. Input data for the model.

Definition Groups of Protection	I	II	III
Protection of workers	(almost) ZERO	BASIC	BEYOND the REQUIREMENTS
Prevention costs (C_P)	C_Z	C_B	C_{BR}
Number of accidents	Poisson [λ_z]	Poisson [λ_B]	Poisson [λ_{BR}]
Cost of accidents (C_A)	EXP [μ_Z]	EXP [μ_B]	EXP [μ_{BR}]

value of work accident frequency. Even if no other model is available, the Poisson distribution is sufficient for the initial estimate.

Poisson distribution can also be easily applied in practice, since λ can be interpreted as the average number of incidents per 50,000 hours of human labor, known as the accident frequency rate ~ AFR. This means that the required information for the Poisson distribution is readily available (Chua & Goh, 2005).

In accordance with the results obtained with this distribution, the number of accidents varies in direct proportion depending on the total number of workers, the average of subcontractors and the budget on health and safety, and varies indirectly with the costs of prevention.

The application of the Poisson model allows us to predict the number of accidents that happen on the construction site. Once we estimate the number of accidents on the construction site, we can also estimate their costs and thus the total cost of workplace safety to be added to the cost of prevention.

For each monitored group, we define the incidence of occupational accidents as discrete variable (X) using Poisson distribution with parameter λ:

– for Group I number of occupational accidents (λ_Z);
– for Group II number of occupational accidents (λ_B);
– for Group III number of occupational accidents   (λ_{BR}).

Hypothesis 1: The incidence of occupational accidents will diminish with increasing investment in prevention, i.e. $(\lambda_Z) > (\lambda_B) > (\lambda_{BR})$.

The costs of rectifying can be seen as a continuous random variable (Y) and expressed using exponential random distribution with parameter μ:

– for Group I cost of occupational accidents (μ_Z);
– for Group II cost of occupational accidents (μ_B);
– for Group III cost of occupational accidents (μ_{BR}).

Hypothesis 2: Cost of occupational accidents will diminish with increasing investment in prevention, i.e. $(\mu_Z) > (\mu_B) > (\mu_{BR})$.

The resulting equation is the sum of the prevention costs and costs of accidents for each study group. The X and Y variables are independent and uncorrelated, so the mean total cost value can be expressed as:

$$E\left(\text{Total costs}\right) = C + E\left(X \times Y\right) = C + E\left(X\right) \times E\left(Y\right) \tag{1}$$

Total costs for individual groups shall be expressed as follows:
for Group I:

$$E\left(\text{Total costs}\right) \approx 0 + \frac{\lambda_z}{\mu_z} = \frac{\lambda_z}{\mu_z} \tag{2}$$

for Group II:

$$E\left(\text{Total costs}\right) \approx C_B + \frac{\lambda_B}{\mu_B} \tag{3}$$

for Group III:

$$E\left(\text{Total costs}\right) \approx C_{BR} + \frac{\lambda_{BR}}{\mu_{BR}} \tag{4}$$

If we combine the one-dimensional random variables X and Y and solve them as multidimensional statistical problems, we can model the marginal division separated by domes.

The present mathematical model allows us to compare the costs arising from the different approaches to OSH, i.e. to compliance or exceeding of legislative duties. Validation of the model requires a set of statistical cost data for individual activities that will be individual for each project.

4 DISCUSSION

In the OSH there are several hypotheses that need to be acknowledged and supported with specific outputs of practical research. The paper presented several hypotheses in which it is assumed that certain causes will accompany the expected consequences. Some of the hypotheses are essential not only for the content of this paper, but also for the whole OSH, such as:

Hypothesis 1: The probability of occupational accidents decreases with an increasing degree of protection for workers.

Hypothesis 2: The higher the investment in prevention, the lower the costs of accidents.

Hypothesis 3: Investment in security is directly related to workplace safety.

Hypothesis 4: There is a certain limit at which the investments entered into the OSH do not increase protection for workers anymore.

Many research teams tried to confirm or refute these assumptions in their studies. For example, Teo & Feng, 2010; as cited in Feng, 2013, reported in their study in connection with Hypothesis 3 that there is still no evidence that an increase in security investment also brings improvement in operational safety. This fact supports Hypothesis 4, claiming that there is a certain threshold beyond which the investment put into OSH does not increase protection for workers anymore. This may be the human factor that plays a major role in all causes of occupational accidents.

The assumption about the return of prevention activities and investment in OSH requires the support of mathematical models and their verification in practice.

5 CONCLUSION

The paper's theme is now very topical. OSH has become a very important aspect of building production worldwide. Legislation in this area is encouraging the trend to reduce occupational accidents to zero, through the change of status of OSH in the organization to the level of priority number 1.

Organizations decide with their attitudes towards OSH on the amount of investment in this area. Although the prevention costs are higher in the case of occupational protection beyond the legal requirements, the benefits of such protection are larger as well. These benefits can be tangible, such as higher labor productivity, but also intangible, such as better image and culture in the organization. If the organization fails to comply with safety regulations, these investments can be spared, but the penalties in the event of an accident may even exceed the prevention costs.

An adverse result of an improper approach to OSH tends to include work accidents in which the lives and health of workers are endangered. The cost of these injuries will vary according to the mentioned approach to OSH. A large share of the costs of accidents is primarily represented by penalties and fines, which are usually very highly unsecured in the event of not ensuring the appropriate protection of workers.

Currently, organizations comply with safety regulations, especially under the threat of penalties. The principle of voluntary in the approach to OSH can be supported by specific indicators obtained based on relevant mathematical models and statistical verification.

Using mathematical modeling we can determine the limit at which the investment in preventive measures is more effective compared to the cost of damage repair. The model presented in the paper will be verified in practice in the next step on specific construction activities, and the obtained statistical data will be evaluated in order to identify generally applicable dependances between the costs of prevention and accident costs, which need to be incurred to repair the damage.

REFERENCES

Chua, D.K.H. & Goh, Y.M. 2005. Poisson model of construction incident occurrence. *Journal of Construction Engineering and Management* 131(6): 715–722.

Esmaeili, B., Hallowell, M.R. & Rajagopalan, B. 2015. Attribute-Based Safety Risk Assessment. II: Predicting Safety Outcomes Using Generalized Linear Models. *Journal of Construction Engineering and Management* 141(8).

Feng, Y. 2013. Effect of safety investments on safety performance of building projects. *Safety Science* 59: 28–45.

Gambatese, J.A., Behm, M. & Rajendran, S. 2008. Design's role in construction accident causality and prevention: Perspectives from an expert panel. *Safety Science* 46(4): 675–691.

Gurcanli, G.E., Bilir, S. & Sevim, M. 2015. Activity based risk assessment and safety cost estimation for residential building construction projects. *Safety Science* 80: 1–12.

Ibarrondo-Dávila, M.P., López-Alonso, M. & Rubio-Gámez, M.C. 2015. Managerial accounting for safety management. The case of a Spanish construction company. *Safety Science* 79: 116–125.

Kiviniemi, M., Sulankivi, K., Kähkönen, K., Mäkelä, T. & Merivirta, M. 2011. BIM-based safety management and communication for building construction. *VTT Tiedotteita-Valtion Teknillinen Tutkimuskeskus* (2597): 1–123.

Kozlovská, M. & Struková, Z. 2011. Environmental and safety education in building industry through unconventional teaching techniques. In *11th International Multidisciplinary Scientific Geoconference and EXPO—Modern Management of Mine Producing, Geology and Environmental Protection.* SGEM.

López-Alonso, M., Ibarrondo-Dávila, M.P., Rubio-Gámez, M.C. & Munoz, T.G. 2013. The impact of health and safety investment on construction company costs. *Safety Science* 60: 151–159.

Struková, Z. & Kozlovská, M. 2014. The green construction technologies from safety and health point of view. In *International Multidisciplinary Scientific GeoConference Surveying Geology and Mining Ecology Management.* SGEM.

Advances and Trends in Engineering Sciences and Technologies II – Al Ali & Platko (Eds)
© 2017 Taylor & Francis Group, ISBN 978-1-138-03224-8

Fiber optic temperature measurement of mass concrete at an early age – case study

N. Humar
Institute for Water of the Republic of Slovenia, Ljubljana, Slovenia

D. Zupan & A. Kryžanowski
Faculty of Civil and Geodetic Engineering, University of Ljubljana, Ljubljana, Slovenia

ABSTRACT: Large dams are among the most frequently monitored structures. This is understandable, having in mind the consequences of dam failure or equipment malfunction. Over the last decades we have witnessed great progress of measurement techniques and methods, due to the rapid development of materials and equipment. Distributed fiber-optic measurements are among the very promising methods for monitoring dams. The method solves some of the weaknesses of the commonly used measuring methods, e.g. installation and density of the network, which is useful when measuring parameters, such as temperature field changes, formation of cracks, or leaking. In this paper we want to present the results of measurements of the release of temperature measured using fiber optic cables in an early age concrete directly after the placing of mass concrete in the spillway section, the advantages of the presented method for monitoring and controlling the process of hydration, and the release of temperature increment.

1 INTRODUCTION

1.1 *Phenomenon in early age concrete*

In early age concrete, temperature is one of the most important parameters—temperature changes indicate the development of the hydration process and thus the rate of hardening of the concrete. Thermal processes caused by the hydration of cement generate unavoidable early and very-early age deformation that is the origin of premature cracking of concrete. The constant monitoring of the temperature field in the early stages of hydration of mass concrete blocks allows for estimating their characteristics at different stages of hardening of concrete, as well as the prediction of the final material characteristics, when the concrete reaches the adequate strength and the process is completed.

Due to the particularities of the successive (progressive) construction used during the construction of mass concrete blocks in dam construction, inducing the effects which occur on the contact between the new blocks and the one in which the process of hydration already started, the construction of new blocks influences the temperature field in the pre-built concrete blocks. Thus, with the technology in which we proceed gradually, we change the stress state of the concrete blocks which can lead to undesired processes such as the formation of cracks.

In order to avoid the undesired unfavourable stress states, constant control and adaptation of work dynamics to achieve better results are needed. The measurements of temperature changes inside the concrete block and on the surface of it, during the process of hydration and concrete hardening, proved to be the most efficient method/tool in the current engineering practice (Glišić et al, 2000). By monitoring the temperature rise during the hydration period, we can observe or detect excessive temperature rise early enough to intervene with adequate treatment of concrete or with implementing technological changes in the construction process (Inaudi et al, 2005).

1.2 *Temperature field measurements*

Temperature field changes in concrete structures are normally monitored using temperature sensors installed at various points of the construction. This means that each sensor captures the data of the measurement range at the point at which it is placed. The disadvantage of this method is the problems related to sensor installation and the fact that we capture the point only, which therefore hardly gives a detailed overall picture of the thermal field (Nikolovski et al, 2014), which is important when analyzing the transfer of heat to the surroundings and the stress conditions at the edge of the block. From an overview of the tests and measurements where fiber optic cables have been used to monitor the early age concrete, one can see that many measurements have been performed under laboratory or other controlled conditions; however, one can hardly find any data or analyses that include (field) monitoring of early-age concrete during construction, particularly the construction of dams.

In 2014, the construction of **HPP Brežice**, the fifth in the row of six hydropower plants on the Lower Sava River, started; this provided a rare opportunity to test the distributed fiber-optic system in the field and observe the influences of progressive construction. The investor gave us the opportunity to install a high-resolution Fiber-Optic System (FOS) in the impact basin of one of the overflow sections of the dam before pouring the concrete. From these measurements, we were able to observe and analyse the behaviour of the early-age concrete under the influences of progressive construction in the natural environment.

2 MEASUREMENTS

2.1 *Description of the test field location*

The FOS grid was installed in the stilling basin base plate of the first overflow section, one day prior to the beginning of the concrete works. The test field of a size of $17.7 \times 17 \times 1.5$ m was part of the stilling basin base plate, which consists of 4 concrete blocks (Figure 1). The test field is in the upper part of the base plate, just below the spillway chute. This location was selected on purpose; firstly, to include ambient influences in the measurements and, secondly, due to the proximity of the inspection gallery where the measuring equipment can be installed. The size of the concrete block was determined based on the experience with dam construction on the Sava River, and generation capacities.

A specific hydrotechnical concrete was used for the placement, with a low heat of hydration. A concrete mixture was prepared by using blast furnace cement of strength class 32.5 and ordinary early strength (N), with Low heat of Hydration (LH), type: **CEM III/B 32.5 N-LH**, which is in accordance with EN 197–1. The aggregate was obtained by separation of the

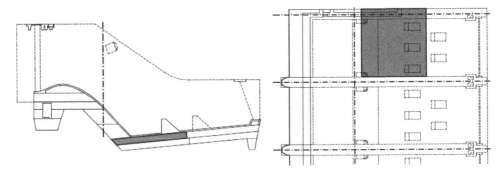

Figure 1.   Location of the test field at the Brežice dam spillway.

natural crushed gravel from the alluvial quaternary filling up of the Sava River at the site. Fractions 0–4, 4–8, 8–16 and 16–45 mm were used. The concrete was prepared at the construction site concrete mixing plant, and placed by pumping. A hyper plasticizer was added to allow for an appropriate workability of the fresh concrete mix during placement, and air entrainer was added to increase frost resistance, which also favourably affects the workability of concrete during placement. The average value of compressive strength at the age of 28 days was 34.6 MPa.

2.2 Installation of the FOS

The FOS grid was installed in the form of seven loops aligned continuously one after another, from the left to the right side of the stilling basin (in the direction of flow). Each loop was installed in 3 height levels. The choice of their position was based on the position of the main reinforcements in the concrete block: at the bottom, in the middle, below the surface of the concrete block. The edges of the side loops as well as the ends of the loops were installed along the border surfaces of the concrete block, thereby ensuring the registering of temperature at the edges. The installation of the measuring cable and the distribution density of the points allow the registration of temperature in any section of the concrete block.

The optic fibre was mounted on the reinforcement bars. The cable had length markings on it while, prior to pouring of concrete, the positions of the markings were precisely surveyed using a 3D laser scanner. The effective length of the active part of the cable installed inside the concrete block (measuring field) was 798 m and consisted of a total of 2,793 measuring points. The optical cable was georeferenced using the scanner manufacturer's software, by determining the line of the cable within the concrete block. The cable markings allowed for the positioning of measuring points at a distance of 25 cm, with an accuracy of ±<1 cm (in the flat part), presenting the temperature reading points. Each measuring point was georeferenced according to the local coordinate system with a reference position in the left upstream angle of the overflow section with attribute data of the location: main loop, height position, location of the loop, and status of location: whether the point was located in the flat part (inside the field—status 1) or in the uneven part (at the edge of the field—status 2). This allowed the spatial determination of each measuring point (Figures 2 and 3).

2.3 System and measuring equipment

The FOS was designed in a way to guarantee precision, the possibility of control of the collected data, control of operation of the system, and the possibility of subsequent use of the dam surveillance system after dam completion. The system consisted of sheathed optical fibre, a central measuring unit with a storage unit for data recording and acquisition (SIL-IXA XT-DTS which allows measurement of temperature ranging from −40°C to + 65°C),

Figure 2.   Coordinate system positioning.

435

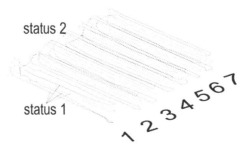

Figure 3.   Positioning of the cable in the test plot.

and an external unit for controlling the operation of the system (Silixa, 2014). In addition, thermometers positioned on the surface of the concrete (in shadow) for control of the collected data were foreseen, as well as a climatological station for acquisition of the data on climate conditions such as ambient temperature, humidity, etc.

The point data along the cable were to be recorded every 25 cm. The lowest measurement time selected was 5 seconds. The measurements were performed with a 15-minute time step. The chosen distribution of measuring points and the time interval was identified as the best compromise for optimal acquisition of spatial and temporal distribution of temperature changes. Notably, the accuracy of data could be increased by densifying the distribution of the measuring points, but a finer distribution of measurement points goes at the expense of the longer time interval, or vice versa. Nevertheless, the scope of technical feasibility would be disproportionate to the accuracy of results. As the analysis of measurements showed later, the assumed density of measuring points was completely adequate for the purposes of this study. To compare and control the results acquired on the ground, during the pouring of the overflow section, an adiabatic test was conducted in laboratory on a standard specimen made of the same concrete as that placed in the overflow section.

3   MEASUREMENTS AND ANALYSIS

We began with measurements simultaneously with the beginning of concrete works in June 2015. The measurements were taken for a total of 21 days and captured the data on temperature changes in the first 21 days after the start of pouring of concrete. In parallel to temperature of concrete recording, meteorological data (temperature, precipitation, humidity, wind speed) were monitored using an automatic weather station, and the ambient temperature in proximity of the concrete block on the shady side by the right column of the overflow section was measured. The ambient temperature was recorded directly using a free cable not embedded into the concrete block and a classical thermometer on the weather station. We present the preliminary analysis of the measurement recordings intended for a qualitative interpretation of the field work. The measurements were conducted in the first week after the placement of concrete, when temperature changes in the concrete block were the greatest. Throughout measurements, intensive wet curing of concrete was conducted with an additional protection of the concrete surface with felt and PVC foil.

The measurements were carried out in dry weather conditions. The average daytime maximum temperature ranged between 22°C and 24°C, and the night time minimum was below 11°C. The daytime maximum in the period was 30°C (Figure 4). During the measurements, some short measuring equipment outages occurred due to power supply interruptions, which resulted in minor losses in the data set. Fortunately all interruption intervals happened at a time when no works were underway on the structure; furthermore, they were short enough to allow linear interpolation of the missing data. Regardless of power failures, the measuring equipment provided enough information to calculate the temperature curve characteristic for concrete hardening. To compare and control the results acquired on the ground, during the pouring of the overflow section, an adiabatic test was conducted in laboratory on a standard specimen made of the same concrete as that placed in the overflow section (Kryžanowski et al, 2015).

The measurements at the measuring points in the central part of the concrete block in the bottom "layer" were closest to the measurements in adiabatic conditions in the laboratory. The temperature rise in the concrete block was considerably faster than that in the reference laboratory test (Figure 5). The reason for this was probably that the hydration process in the concrete block was underway at considerably higher external temperatures than the test conducted under laboratory adiabatic conditions. The comparison of final temperatures showed that the differences of final temperatures were negligible: the maximum temperature in the mass block was 43.8°C after 48 hours after the pouring, while the final temperature under adiabatic conditions was 42.9°C, reached approx. after 100 hours, e.g. 48 hours after the pouring the concrete reached 37.7°C. The most notable temperature rise was registered in the centre of the concrete block, which considerably decreased towards the edges of the block. When comparing the lateral temperature progression at the bottom of the block, the most notable temperature rise was registered in the centre of the concrete block, which considerably decreased towards the edges of the block (Figure 6).

Figure 7 shows the evolution of the temperature field over time inside the block in the first five days after the placement of concrete. This presentation shows the measurements of a half of the block (loops 4 to 7). The spatial presentation of the evolution of the temperature field was based on the interpolation of results of point measurements, for various levels between the individual branches, and spatially between the loops. If the measurements are analysed in the same way as they have been until now, i.e. based on a "limited" number of points or sections selected in advance, these findings confirm the temperature rise distribution known so

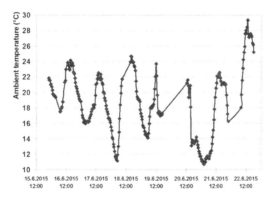

Figure 4.   Ambient temperature.

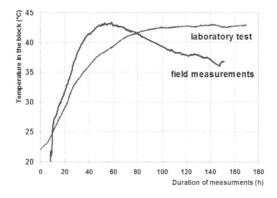

Figure 5.   Temperature rise in the concrete block and the reference laboratory test.

far, where the most notable temperature rise was registered in the centre of the concrete block and considerably decreased towards the edges of the block (Figure 6). A larger number of measuring points allows for a considerably better detection of locally limited temperature differences than could be expected according to the results of the measuring techniques known so far. The interpretation of measurements of temperature rises shows a slightly different picture of the spatial distribution of temperature gradients than the expected one.

The temperature field is significantly different than the expected graded distribution of temperature gradients from the inside towards the edge of the concrete block. Although the distribution of temperature gradients inside the concrete block is within the expected values this changes significantly in the impact area of adjacent blocks. This is clearly visible at the left side of the block (Figure 7) where there is a pronounced impact on the temperature field in the area where the plate is mounted in the side column of the spillway or in the areas where impact blocks were envisaged (the upper part of the temperature field in Figure 7). Conventional analyses of the evolution of the temperature field over time fail to detect such anomalies. More importantly, the spatial measurement data acquisition helped us to register local

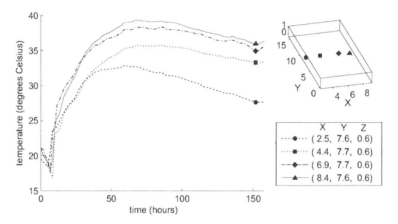

Figure 6.   Comparison of lateral temperature progression at the bottom of the block.

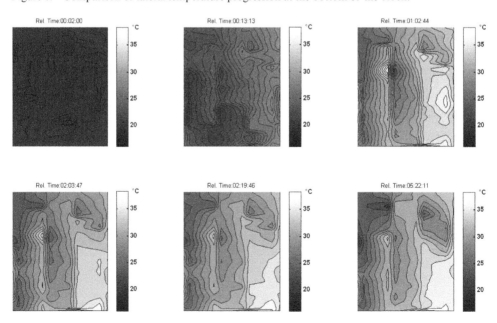

Figure 7.   Evolution of the temperature field over time inside the block.

temperature gradients that significantly exceed the expected values and can cause critical stress conditions in unexpected places. Determining the significance of the impact of local changes on boundary conditions will guide and direct us in the future modelling of temperature conditions in mass concrete and finally influence the final characteristic of the concrete by adopting the proper treatment or adjusting the speed of pouring early enough.

4 CONCLUSIONS

The test clearly demonstrated the great applicability of the optical cable system for monitoring the processes in concrete during construction, and pointed out to the possibility of using such a system for controlling hydration and hardening of mass concrete, particularly when heat release is of key importance in ensuring appropriate quality and safety of a structure.

The main advantages of the precise optical system of monitoring include the simplicity of installation and the quantity of measuring points, which can be selected freely. The size of the optical cable is much smaller than that of most other measuring instruments, which makes the intervention in the structure negligible, while, on the other hand, it is relatively insensitive to external influences and, due to the simplicity of construction, is not susceptible to defects and compatibility with more advanced instruments and, hence, its service life can be extremely long. All of this allows for an efficient control of the state of the structure also at later stages (e.g. filling of the reservoir, exploitation).

The results of thermal measurements in concrete were as expected in terms of monitoring the temperature rise during hydration inside the concrete block. An advantage of the spatial data acquisition lies in its ability to check the thermal condition at any point or section of a structure. In the future, the results of this study will be used to develop a mathematical model to analyse the temperature field in mass concrete. The optical cable system will remain a component of the operating monitoring system based on which the mathematical model will be validated. The key objective of this study is to develop a mathematical model for monitoring the temperature condition in mass concrete to support the provision of dam safety policy, and a basis for decision-making in the planning and construction of mass concrete in other dams on the Sava River.

REFERENCES

Glišić, B, Simon, N (2000). Monitoring of concrete at very early age using stiff SOFO sensor, *Cement & Concrete Composites.* Issue 22, pp. 115–119.
Inaudi, D, Glišić, B (2005). Field Application of Fiber Optic Strain and Temperature Monitoring Systems. *Opto-electronic Sensor-based monitoring in Geo-engineering, Nanjing, P.R.China*, November 2005.
Kryžanowski, A, Vidmar, A, Zupan, D, Brilly, M, Humar, N (2015). Analysis of measurements of temperature changes in early age massive concrete at HPP Brežice—preliminary research study, University of Ljubljana, Faculty of Civil and Geodetic Engineering.
Nikolovski, Z, Slavenski, M (2014). *Monitoring of Sveta Petka dam during different phases of dam's life*, 34th Dam days 2014.
Silixa (2014). XT-DTS—Software manual, Silixa Ltd, Elstree, UK.

Advances and Trends in Engineering Sciences and Technologies II – Al Ali & Platko (Eds)
© 2017 Taylor & Francis Group, ISBN 978-1-138-03224-8

The effect of distance and noise barrier at the dispersion of pollutants in the vicinity of roads

D. Jandacka & D. Durcanska
Department of Highway Engineering, Faculty of Civil Engineering, University of Žilina, Žilina, Slovakia

ABSTRACT: Monitoring the status of air quality is the set of activities leading to the discovery of the environment and monitoring the evolution of air pollution in time. Knowledge of the environment is not the primary objective of monitoring, but a means of predicting its further development and designing sustainability measures. A clear statement of the objectives of monitoring is a vital precondition for the correctness of the decision about monitoring pollutants, how and where they should be monitored and the accuracy and precision required. Many factors impact air quality that determines its development and changes. Air pollution is subject to the especially primary source of pollution and consequently secondary influences which have a different impact on current concentrations of pollutants in the air. This contribution deals with problems of the monitoring of air pollution from road traffic and the influence of the location of the monitoring station from the primary source of pollution on detected concentrations of selected pollutants.

1 INTRODUCTION

The dispersion of pollutants in the atmosphere is a complex process that is not only subject to different yield sources producing this pollution. Of course, the source of various pollutants is decisive for a number of substances that get into the ambient air. Physical factors influence the spread of emissions from the sources and determine the dispersion of pollutants into the environment. They are mainly meteorological parameters, the stability of the atmosphere and diversity of the surrounding terrain. Several studies have processed the dispersion of pollutants in the vicinity of roads, which to some extent confirmed different levels of concentration of pollution given the distance of the monitoring station from the intended source (Sharma et al., 2009, Jung, et al., 2011). A generally acknowledged relationship is the greater the distance from the road = lower concentrations of pollutants. However, this may be different for different fragmentation of space around the road by natural or artificial barriers or specific process of dispersion of pollutants. No less important is the nature of stratification of pollution in the vertical direction, which shows a decrease with a greater height above the ground (Adamec et al., 2006, Wu et al., 2002, Morawska et al., 1999, Hitchnins et al., 2000, Roorda—Knape et al., 1998). So we decided to perform measurements that would confirm or refute the facts known to the spread of pollutants. We decided to focus these measurements on the influence of measurement distance from the road to the changes of concentration of pollutants and the impact of noise barriers at the dispersion of pollution from the road.

2 METHODOLOGY OF MEASUREMENT

The experimental measurements presented in this paper took place during the two measurement cycles from 11. 1. 2014 to 16. 1. 2014 on the monitoring station at the D1 highway and from 18. 3. 2014 to 23. 3. 2014 on monitoring station near a feeder road D3 Žilina, Strážov.

Experimental measurements near the D1 highway were carried out in order to determine the effect of distance monitoring station from road to the changes of pollutants concentration in the atmosphere. The selection of monitoring station at a feeder road Žilina, Strážov was developed to detect the influence of Noise Barrier (NB) on the dissemination of air pollution from road transport.

Mobile air quality measuring station University of Žilina (MAQMS ŽU) was placed at a distance of about 7.0 meters from the edge of the highway. Mobile Air Quality Measuring Station Envitech (MAQMS ENV) was placed from MAQMS ŽU about 30 m away from the D1 motorway (37 m from the border line of the road). This area can be described as a flat terrain surrounded by hilly topography forming the river Váh valley in the direction of southwest ↔ northeast.

A monitoring station near a feeder road D3 Žilina, Strážov was placed on the right side in the direction to Žilina. MAQMS ŽU was placed in front of the company KABE as a monitoring station without NB about 35 m from the edge of the road. MAQMS ENV was placed in front of the company TECHPLASTY as a monitoring station with NB with a height of about 4.5 meters at a distance from the edge of the road about 15 m. The distance between the monitoring stations is approximately 100 m.

The road is oriented in the direction of northwest—southeast. The road is surrounded by several family houses, railways and the river Váh from the northern side. From the west side performs moderate hilly terrain, while creating a barrier to the flow of wind from this side.

This area is very rugged, not only in terms of terrain, but also by the occurrence of diverse artificial obstacles and barriers that particularly affect wind flow and natural ventilation zone.

There were measured pollutants NO, NO_2, NO_x and PM_{10} and used the following measurement methods on measuring stations: NO_x, NO_2, NO—Standard chemiluminescence method of measuring the concentration of nitrogen dioxide and nitrogen monoxide, PM_{10}—Method of beta radiation attenuation. There were also surveyed meteorological parameters (temperature, humidity, wind speed and direction, pressure) and traffic volume on the measuring stations.

3 RESULTS OF MEASUREMENTS

3.1 *The effect of measuring station distance from the source on measured concentrations of pollutants*

One of the objectives of the experimental measurements was to determine the effect of measuring station distance from the road on the concentrations of pollutants. For this purpose, experimental measurements were carried out in the vicinity of D1 motorway.

Some facts resulted from realized experimental measurements which substantially affect the adequate assessment of air quality in selected locations. The average concentration of PM_{10} identified by MAQMS ENV (32 µg/m^3) was about 8 µg/m^3 lower than the average concentration of PM_{10} by MAQMS ŽU (40 µg/m^3) during the reporting period. Reduction of the average PM_{10} concentration is 20% at a greater distance during the reporting period. The highest observed daily mean concentration of PM_{10} was on monitoring station MAQMS ŽU 75 µg/m^3 and MAQMS ENV 58 µg/m^3 during the reporting period (Figure 2). Decrease of PM_{10} concentrations was found at a greater distance from the primary source of pollution (road transport) despite of the weather conditions during the period in which the prevailing winds were oriented in a direction along the highway respectively in the direction from the monitoring stations to the highway.

The decrease of the nitrogen oxides was detected at a greater distance from road as well. The lower concentration of NO µg/m^3 was detected at MAQMS ENV, decrease was representing by the average value 13.8 µg/m^3 i.e. 29% decrease of concentration in greater distance from the road. Also the concentrations of NO_2 and NO_x were measured lower in greater distance from road—MAQMS ENV. The average decrease of NO_2 is 4.9 µg/m^3 i.e. 13% and NO_x is 25.8 µg/m^3 i.e. 23% over the measuring period (Figure 1).

Regression and correlation analyses of measurement data were performed for investigation interdependence of particulate matter and nitrogen oxides measured at 7 m and 37 m from the motorway. Linear regression model demonstrated the possibility to describe one variable (PM_{10}, NO_x—7 m) the other one (PM_{10}, NO_x—37 m). The coefficient of determination and correlation coefficient for the PM_{10} reached $R^2 = 0.85$, $r = 0.92$ (Figure 3) and for the NO_x $R^2 = 0.93$, $r = 0.97$ (Figure 4). The measured particulate matter PM_{10} and NO_x in 37 m distance from the motorway and 7 meters from the motorway are intertwined and it is the same spread of pollutants from the highway.

3.2 Impact of noise barrier on the dispersion of pollutants

The second aim of the measurements was to determine the impact of NB at the dispersion of air pollution from road transport. Measuring station was located near the feeder road Žilina, Strážov on the right side of road in the direction to Žilina.

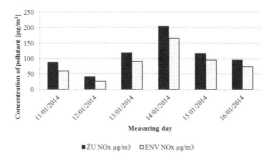

Figure 1. Comparison of the average daily concentrations of NO_x at the measuring station in the vicinity of highway D1 in a distance of 7 m (ŽU) and 37 m (ENV) from the border line of road.

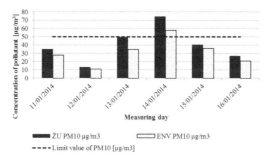

Figure 2. Comparison of the average daily concentrations of PM_{10} at the measuring station in the vicinity of highway D1 in a distance of 7 m (ŽU) and 37 m (ENV) from the border line of road.

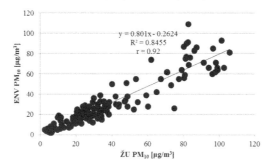

Figure 3. Regression and correlation analysis of measured PM_{10} concentrations on measuring station in the vicinity of highway D1 in a distance of 7 m (ŽU) and 37 m (ENV) from the border line of road.

Expected impact of NB at reducing concentrations of pollutants is noticeably manifested only for the PM_{10}. The average concentration of PM_{10} identified by MAQMS ENV (18 µg/m³) was about 5 µg/m³ lower than average concentration of PM_{10} identified by MAQMS ŽU (23 µg/m³) during the reporting period. Reduction of average PM_{10} concentration behind NB is 22% compared to a place without NB during the reporting period. The highest observed daily mean concentration of PM_{10} was at monitoring station MAQMS ŽU 27 µg/m³ and MAQMS ENV 20 µg/m³ during the reporting period (Figure 6). Only minimal differences between concentrations of NO, NO_2 and NO_x at monitoring station with NB and without NB were found. Impact of NB on measured concentrations of gaseous pollutants was not practically detected (Figure 5).

The results of the regression analysis showed that measured concentrations of particulate matter PM_{10} at the measuring station without NB and with NB are not largely influenced by each other. The coefficient of determination for PM_{10} model is $R^2 = 0.39$. The model describes

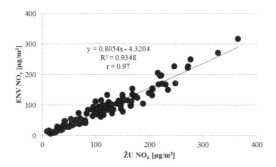

Figure 4.   Regression and correlation analysis of measured NO_x concentrations on measuring station in the vicinity of highway D1 in a distance of 7 m (ŽU) and 37 m (ENV) from the border line of road.

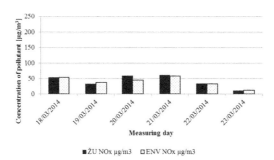

Figure 5.   Comparison of the average daily concentrations of NO_x at the measuring station near the feeder road Žilina, Strážov without PHS (ŽU) a with PHS (ENV).

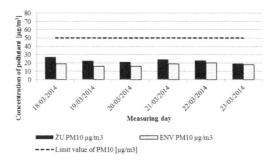

Figure 6.   Comparison of the average daily concentrations of PM_{10} at the measuring station near the feeder road Žilina, Strážov without PHS (ŽU) a with PHS (ENV).

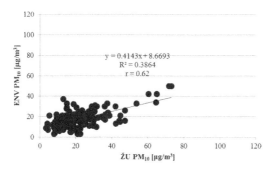

Figure 7. Regression and correlation analysis of measured PM_{10} concentrations at the place with noise barrier (ENV) and without noise barrier (ŽU).

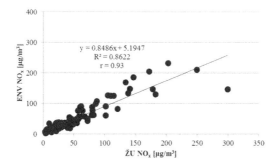

Figure 8. Regression and correlation analysis of measured NO_x concentrations at the place with noise barrier (ENV) and without noise barrier (ŽU).

only 39% of the variability of the original data (Figure 7). Measured concentrations of nitrogen oxides NO_x at the measuring station with NB and without NB are strongly linked. The coefficient of determination for NO_x is $R^2 = 0.86$ (Figure 8).

4 CONCLUSIONS

The experimental measurements were carried out on two diametrically different locations. In the vicinity of highway D1 is an open, flat area without any artificial barriers that will affect and disturb the natural dispersion of pollution subject only to the production of pollution at its source and meteorological factors. Near the feeder road Žilina, Strážov is a site with a great articulation of the surrounding terrain, a significant area of family and corporate buildings, and of course it is a stretch of road with extensive noise barriers where these components of environments present significant barriers to the natural dispersion of pollution. The results of the measurements show difference dispersions of pollutants. A significant decrease of concentrations of measured pollutants was confirmed in the vicinity of highway D1 due to different monitoring station distances from the road. The effect of noise barrier at concentrations of particulate matter was determined near the feeder road Žilina, Strážov by reducing the concentration of particulate matter behind noise barrier. A significant impact noise barrier at reducing the concentrations of nitrogen oxides was not found. The fact could play an important role that MAQMS ENV (with NB) was placed closer to the road as MAQMS ŽU (without NB), what could reduce the impact of NB expressions effect on the concentration of nitrogen oxides. Sensitively and precisely selected place and time of monitoring is imperative to adequately preparing air quality monitoring, since even a slight deviation of monitoring environmental conditions can significantly affect the results, as demonstrated by these experimental measurements.

445

ACKNOWLEDGEMENTS

The paper originated as being supported by means of a grant for the scientific research task VEGA 1/0557/14.

REFERENCES

Adamec, V. et al.: Traffic related dust and its effect on particulate matter air pollution level. Annual Report of Project 1F54H/098/520 in 2006. CDV Brno. (In Czech).

Ďurčanská, D. a kol. 2014. Analysis of evaluation methods of traffic related air pollution: research task. Žilina: University of Zilina, 2014. 84 s. (In Slovak).

Hitchnins, J., Morawska, L., Wolff, R., Gilbert, D.: Concentrations of submicrometre particles from vehicle near a major roads. *Atmos. Environ.* 34. 2000, pp. 51–59. ISSN: 1352–2310.

Jung, K. H., Artigas, F., Shin J. Y.: Seasonal Gradient Patterns of Polycyclic Aromatic Hydrocarbons and PM Concenrations near a Highway. *Atmosphere*, 2011, vol.2, pp. 533–552. ISSN: 2073–4433.

Morawska, L., Thomas, S., Gilbert, D., Greenaway, C., Rijnders, E.: A study of the horizontal and vertical profile of submicrometer particles in reletion to a busy road. *Atmos. Environm.* 33. 1999, pp. 1261–1274. ISSN: 1352-2310.

Roorda-Knape, M. C., Janssen, N. A. H., De Harthog, J. J., Van Vliet, P. H. N., Harssema, H., Brunekreef, B.: Air pollution from traffic in city districts near major motorways. *Atmos. Environ.* 32. 1998, pp. 1921–1930. ISSN: 1352-2310.

Sharma, A., Massey, D. D., Taneja, A.: Horizontal gradients of traffic related air pollutants near a major highways in Agra, India. *Indian Journal of Radio a Space Physics*. Vol. 38, 2009, pp. 338–346. ISSN: 0975-105X.

Wu, Y. Hao, J., Fu, L., Wang, Z., Tang, U.: Vertical and horizontal profiles of airbone particulate matter near major roads in Macao. China. *Atmos. Environ.*, 36, 2002, pp. 4907–4918. ISSN: 1352-2310.

Advances and Trends in Engineering Sciences and Technologies II – Al Ali & Platko (Eds)
© *2017 Taylor & Francis Group, ISBN 978-1-138-03224-8*

The effective use of renewable energy in the design of construction equipment

E. Jankovichová, M. Nguyen Tien & S. Hajduchová
Slovak University of Technology in Bratislava, Slovakia

ABSTRACT: Construction site equipment is still part of the construction of all civil structures. The correct design of the construction equipment objects and their usage plays an important role in the overall efficiency of the construction with regard to the execution time, the resulting price, performance of workers on the construction site and use of deployment mechanisms. The current trend in construction is not only in the proposal itself, but also the possibility of using natural resources. Their proper design and appropriate combination of renewable energy sources can contribute to the overall efficiency of constructions.

1 INTRODUCTION

Currently humanity solves one of the main challenges, namely to ensure the sufficient supply for maintaining the continuous development of society without negatively affecting the creation and protection of the environment. Construction, still requires stricter conditions for the energy efficiency in buildings, not only in their design, but also from the beginning of the construction of buildings (Bielek et al. 2014). Implementation of alternative sources into operation proves that we can cooperate with nature. And by reducing CO_2 production, essential consumer safety it provides for construction workers—(improving productivity) and last but not least opportunity to get a **BREEAM** (Building Research Establishment's Environmental Assessment Method) certificate.

Solar energy falling on the earth's surface is the richest available source of energy. In view of its huge potential in the natural state and its quality properties is also from an economic side an alternative inexhaustible energy source (Dušička et al. 2014). Wind energy has its origin in solar energy. Earth's surface is unevenly heating by solar radiation. As a result, there is a temperature difference of the ambient air, which form conditioning factors of differential pressures. Warmer air heated from the earth's surface rises up and cooler air is pushed to the surface. This results in motion—the air mass flow from the higher pressure area to a lower pressure area that forms the subject of wind. These two systems are complementary (Mackay 2008).

2 MATERIAL AND METHODOLOGY

2.1 *Research material*

The paper presents an application of selected renewable energy sources in the construction site during the implementation of the multifunctional residential building. The residential building is located in Bratislava—Ružinov. The multifunctional building has 11 floors and one underground. Outdoor parking is designed on the south side of the project. This place will be used during the construction for location of the construction equipment. The planned locations of renewable sources such as wind and collector systems are contemplated by over the cabins. The construction period is expected 15 months. Renewable energy for the construction site equipment are designed to cover the need electricity for the purposes of construction.

2.2 Research methodology

2.2.1 Proposal of solar collectors

The most important step to determine the optimal collector area and the number of collectors that provide the required of the energy cover in the time period according to the timetable, given the size of the structure is given in the calculation of solar energy systems. For application we choose for a given area collectors A_k (m²) heat balance for a certain period and determine the ratio of energy consumption covered by solar energy and additional resources (Petráš et al. 2009).The total collector area A_k is calculated from the formula:

$$A_k = \frac{Q_n}{Q_k}$$

(1)

where Q_n = the thermal energy needed for the day or month(kWh); Q_k = energy captured by the collectors for the day or month (KWh/m²),by the time of the equipment operation.

$$Q_k = \eta_k \times Q_s$$

(2)

where η_k = effectiveness of collectors shall be determined by calculation or it is specified by manufacturer; Q_s = solar energy falling on the panels (kWh/m²).

Considering the accurate calculation Q_k: the sunlit surface makes the overall intensity of solar radiation continuously at the time of clear days, the intensity of diffuse radiation acts only at cloudy sky. Time rotation of clear and cloudy sky during a certain period is usually unknown and for climate data and is expressed as the average monthly relative sunshine (Figure 1).

$$S_m = \frac{T_{real}}{T_{teor}}$$

(3)

where S_m = average monthly sunshine (–); T_{real} = actual time interval determine the energy for the day sunlit area 1 m² (10–14 hod); T_{teor} = theoretical time interval determine the energy for the day sunlit area 1 m² (6h–18h).

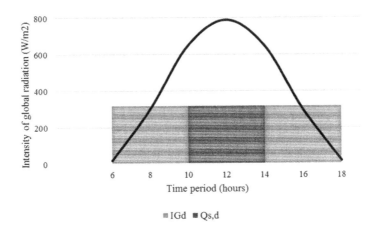

Figure 1. Graphical method of determining energy $H_{s,day,teor}$ energy falling on a day sunlit area 1 m² (Petráš et al. 2009).

448

After taking into account factors that can affect the performance of the collectors we have created a table with the following values. The collectors will be placed on the southern side of the construction site in 45° degrees.

In Table 1 are measured the average values of solar radiation energy Hs the month (kWh/m²) for selected part—Bratislava.

2.2.2 Wind turbines

To determine the type of wind turbine preferred to the building, it is necessary to know some of the main parameters of the building. One of the most important parameters is the average wind speed in the surroundings, the height of the proposed building, or a building in an built or unbuilt territory, and around the building are not above the higher buildings or whether it is not located in the protection zone and etc. In cooperation with the Slovak Hydrometeorological Institute was established a graph for the period 2010–2015 for this locality (Nguyen Tien et al. 2014).

Accordance with the previous figure 2, it can be said that the wind speed is over the last three years stable and not changing. The average wind speed for this location is about 3.6 m/s.

For the calculation of energy needs it is necessary to know the length of the workday, which is 10 hours and apparent input power for the construction site equipment. Specifically for offices lighting, washrooms, changing rooms and construction sites (Makýš 2003). To calculate the apparent input power for construction site equipment is used this formula:

$$S' = 1.1 \times \sqrt{\left(0.5 \times P_1 + 0.8 \times P_2 + P_3\right)^2 + \left(0.7 \times P_1\right)^2} \tag{4}$$

where S' = apparent input power (kVA); 1.1 = the contingency reserve coefficient of increase input power; P_1 = installed capacity of the electric motors on the construction site (kW); P_2 = installed capacity of the interior lighting power (kW); P_3 = installed capacity of the outdoor lighting(kW).

The aim is to cover P_2 and P_3.

Determination of the input power P_2

Table 1. Average value of solar energy (Petráš et al. 2009).

Month	Jan	Feb	Mar	Apr	May	June	July	Aug	Sep	Oct	Nov	Dec
Energy per day (kWh/m²) in 45°	3.26	5.27	6.73	7.88	9.57	9.64	9.57	7.88	6.73	5.27	3.26	2.69
Average sunrise S_m		0.25	0.31	0.42	0.53	0.57	0.61	0.63	0.63	0.58	0.44	0.24	0.21

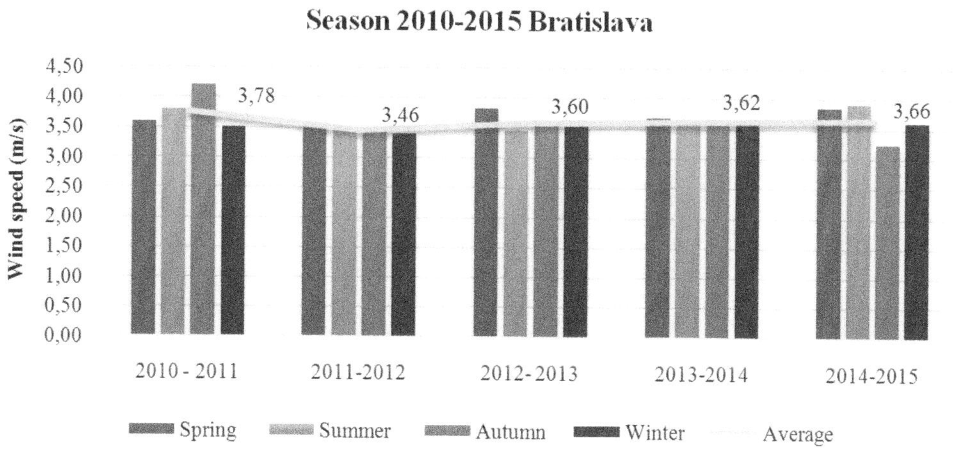

Figure 2. Graph of wind flow velocities at different periods (Nguyen Tien et al. 2014).

449

- Lighting offices—20 W/m², 91.7 m² × 20 W/m² = 1.834 kW
- Lighting washrooms, changing rooms—10 W/m², 170.3 m² × 10 W/m² = 1.703 kW
- Lighting store—8 W/m², 50 m² × 8 W/m² = 0.4 kW

Determination of the input power P_3

- Lighting construction site 0.5 km × 5 kW = 2.5 kW
  Total input power   6.446 kW

The selected part of the electricity consumption to be covered will be 32.30 kW/per day, when it is considered with 5 hours of continuous use of all equipment simultaneously.

The total free floor area of the proposed cabins is 86.4 m².

3 RESULTS AND DISCUSSION

For multifunctional residential building is designed:

- For wind system type: Zephyr Airdolphin Mark-Zero Z-1000-24 with specific data such as Blade diameter: 1.8 m, Weight: 17.5 kg power system: NdFeB permanent magnet three-phase synchronous motor is mounted, Winds from the starting power—2 m/s, Material of blades—carbon fiber surface, body material—cast aluminum, body structure—Japanese traditional craft screw less card Tenon. It is considered to cover by the turbine at least 2.4 kWh/per day. The manufacturer declares that the turbine at a wind speed of 3.6 m/s can produce 1.21 kWh/day. The required area to install the turbines are 2 m². The price of one wind turbine is about 1,300 euros. For construction site equipment are designed two turbines. Details are provided in table 2.
- The type of solar system: FV modul Amerisolar 250 Wp, poly-crystalline photovoltaic modules with aluminum frame with 16.9% efficiency. With dimensions 1.64 m × 0.99 m × 0.04 m (Figure 3). The total weight of construction is 18.5 kg. Considering all the factors that

Table 2. Zephyr Airdolphin power capacity (Saturn Power Co. Ltd. 2015).

Wind speed (m/s)	Every time (Wh)	Wh per day	kWh per month	kWh per year	Wind speed (m/s)	Every time (Wh)	Wh per day	kWh per month	kWh per year
2	6	144	4	53	9	566	13.584	408	4.958
3	28	672	20	254	10	662	15.888	477	5.799
4	73	1.752	53	639	11	738	17.712	531	6.465
5	140	3.36	101	1.226	12	795	19.8	572	6.964
6	231	5.544	166	2.024	13	834	20.016	600	7.306
7	339	8.136	244	2.97	14	858	20.592	618	7.516
8	455	10.92	328	3.986	15	871	20.904	627	7.63

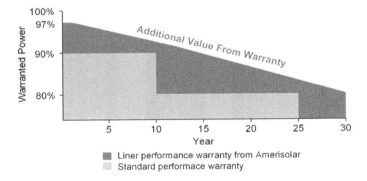

Figure 3. Special warranty for this type of solar panel (SVP SOLAR 2015).

can have a major impact on the design, figures are presented in table3.The dimensioned number of photovoltaic panels for the construction site equipment is 34 units.

The scheme is illustrative and shows a way of involving the combination of solar panels and wind turbines (Figure 4). Diagram shows also the other necessary components for the smooth functioning of the system, including lead-acid gel batteries with a capacity 5.8 or16 kWh. An approximate selling price of one batter is 3,600euros by the manufacturer.

The implementation of these alternative equipment have many advantages at the beginning of construction. Not only the reduction of CO_2 production during construction, as well as independent power to ensure continuous operation cabins and also it allows the investor a better chance to get world **BREEAM** rating(Building Research Establishment's Environmental Assessment Method). **BREEAM** is the world's foremost environmental assessment method and rating system for buildings. A BREEAM assessment uses recognized measures of performance, which are set against established benchmarks, to evaluate a building's specification, design, construction and use. The measures used represent a broad range of categories and criteria from energy to ecology. They include aspects related to energy and water use, the internal environment (health and well-being), pollution, transport, materials, waste, ecology and management processes (**BREEAM**® 2016).

3.1 *Economic assessment*

From the above data for the assessment example it is considered with the estimated total initial investment for the realization around 13,570 €. When we calculate the return on investments, we consider the current price of electricity from power plants ZSE, i.e. 0.1377 €/kWh

Table 3. The calculation of the required number of photovoltaic panels.

Features	
Average energy	3.28 kWh/m²/day
Effectiveness	0.17%
Area for 1 panel	1.6 m²
Real power	0.89 kWh/panel
Energy need to be covered	29.88 kWh/day
Needed solar number	33.69 panels
Proposal solar number	34 panels
Approximately the cost of 1 panel	panel 205 €
Available are upper cabin	86.4 m²

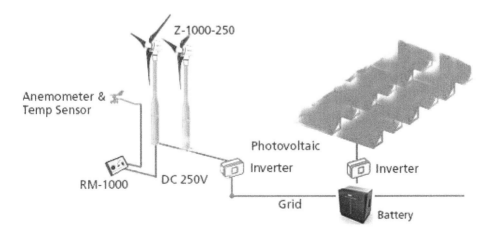

Figure 4. An illustrative connection diagram of wind and solar system (Saturn power Co. Ltd. 2015).

451

in the first quarter of the year 2016. The return on investments for these units is expected about 8.47 years. The payback period may vary depending on the specific conditions of construction equipment, specific conditions of each building, the development of electricity prices, inflation and the overall development of micro and macroeconomic indicators.

4 CONCLUSION

This article is directed to a method of supplying construction site equipment with electricity using alternative sources for interior lighting buildings and lighting the construction site equipment. Complete supply system is located so as to ensure minimum electricity losses. The whole system can be compiled according to the requirements of each construction site equipment. Life time semi-crystalline solar panels is 20–25 years and the wind turbine is 20–23 years. Another advantage is that these systems are not difficult to install. All the panels and turbines at the construction site can be used again and again in other buildings. The most efficient and appropriate conditions for these alternative sources are places when the construction site is situated outside the town and the contractor needed supplies energy construction site equipment before supply points of electricity are realized. The introduction of renewable energy sources fulfilled one of the requirements (In: chapter: Responsible construction practices—Environmentally Aware—Alternative energy sources have been considered) allowing for the achievement of higher global assessment BREEAM. With this proposal we declare that nature is a part of us all, last but not least, the solution can help Slovakia to a higher percentage of use BREEAM.

REFERENCES

Bielek, B., Bielek, M., Vranay, F., Lukášík, D., Vranayová, Z., Vilčeková, S., Ehrenwald, P., Hríše, J., Majsniar, V., Mikušová, M. 2014 *Low Energy, Green, Sustainable Building-Climate-Energy*: 16–25. Bratislava: Faculty of Civil Engineering STU.

BREEAM® 2016, [online] available at: http://www.breeam.com

Dušička, P., Hutňan, M., Janíček, F., Kutiš, V., Murín, J., Paulech, J., Perný, M., Šály, V., Šulc, I., Šulek, P., Šurina, I.2014, *Renewable Energy Sources II: Biomass-Sun-Water,139*–142. Bratislava: STU v Bratislave

Mackay, D. J.C. 2008, *Sustainable Energy-Without the Hot Air*: 22–150.UIT Cambridge.

Makýš, O. & Makýš, P. 2003, *Technological Project. Construction Site Traffic and Construction Site Equipment.* Bratislava: Slovak University of Technology in Bratislava, ISBN 80-227-1847-5.

Nguyen Tien, M., Ingeli, R., Jankovichová, E., Čekon, M. 2014, *Integration of Small Wind Energy Source for Optimization of Energy Efficiency in Residential Building.* In Advanced Materials Research Vol. 1041, CD ROM, SCOPUS, s. 162–166. ISSN 1022-6680(P).

Petráš, D., Lulkovičová, O., Takács, J., Furi, B. 2009, *Renewable Energy Sources for Low Temperature Systems:* 21–35. Bratislava: JAGA GROUP.

Saturn Power Co. Ltd. 2015, *Wind Power*, [online] available at: http://www.saturnpowerltd.com/danfoss_a_en.php?cid1 = 2&cid = 14

SVP SOLAR 2015, *Photovoltaic Panels*, [online] available at: http://www.solar-eshop.cz/p/fv-modul-amerisolar-250wp/

Advances and Trends in Engineering Sciences and Technologies II – Al Ali & Platko (Eds)
© 2017 Taylor & Francis Group, ISBN 978-1-138-03224-8

Properties of autoclaved aerated concretes produced with different types of fibers

C. Karakurt
Bilecik Seyh Edebali University, Bilecik, Turkey

Ö. Korkmaz
Anadolu University, Eskisehir, Turkey

ABSTRACT: Autoclaved Aerated Concrete (AAC) is a well-known Lightweight Concrete (LWC) and consists of a mixture of sand, lime, cement, gypsum, water and an expanding agent. The AAC is generally used in structures for the insulation properties against heat and sound. However, the mechanical properties of this material is comparatively lower. The aim of this study is to increase the mechanical properties of AAC by adding different fibers. The experimental studies, were carried on polypropylene, steel and glass fiber. Firstly, aerated concrete have been prepared in the laboratory and placed in the molds of $70 \times 70 \times 70$ mm dimension. Then, AAC specimens have been pre-cured in the furnace at 40°C for 4 hrs. After that, specimens have been cured in an autoclave for 8 hrs. Finally, AAC specimens were subjected to compressive strength and unit weight tests. Test results showed that fiber usage has beneficial effect on mechanical properties of AAC.

1 INTRODUCTION

Autoclaved Aerated Concrete (AAC) is a well-known LWC and consists of a mixture of sand, lime, cement, gypsum, water and an expanding agent. AAC can be molded and cut into precisely dimensioned units and cured in an autoclave. During the production process, the ingredients combine to form the calcium silicate hydrate gels that establish the special properties of the finished product. Air is entrapped artificially by the addition of metallic powders, such as Al and Zn, or foaming agents (Murdock, 1979). The chemical reaction caused by adding of aluminum makes the mixture expand to about twice its volume, resulting in a highly porous structure. The thing that provides the feature to be high thermal insulation and the feature of being the most lightweight material is dry air tucked into these tiny pores (Pehlivanlı et al., 2016). Curing it in an autoclave under pressure considerably reduces drying shrinkage and water movements. Consequently the final products with an average compressive strength of 2.5–7.5 MPa and an oven-dry unit weight of 400–600 kg/m³ (G2–G6 according to TS EN 771-4) offer considerable advantages over other construction materials, such as improved high thermal and sound insulation, excellent fire resistance, high resource and energy efficiency and outstanding structural performance (TS EN 771-4, 2015; Karakurt et al., 2010).

Moreover, lightweight concrete provides resistance against fires that might occur during an earthquake. During a fire the crystal water present in the AAC plays an important role in reducing the temperature. As a result of the porous structure of AAC the steam runs without causing dispersion or leaving broken parts. During a fire, temperature in AAC is much lower than that of the ordinary concrete. Temperature is effective on only one face and only about 5 mm deep, on the other side of the AAC temperature is not effective. As a result AAC is able to protect the reinforcement materials in it very well (Demir et al., 2010). The typical production step of AAC in a plant is given in Figure 1.

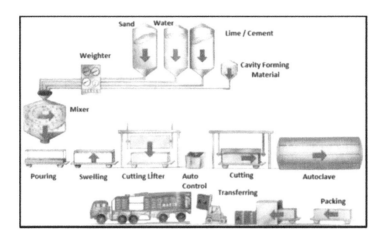

Figure 1.   Production steps of AAC.

Fibers are generally used to increase the tensile strength of traditional concrete. Another beneficial effect of fiber is the reduction of shrinkage cracks in concrete composites. For this purpose, some studies have investigated the effects of fiber on lightweight concrete and AAC (Gül et al., 2007; Pehlivanlı et al., 2015; Tanyıldızı, 2008). The fibers used in concrete are steel fibers, polypropylene fibers, glass fibers and fibre. In this study, the effect of fiber type and size on the production of AAC on compressive strength has been investigated. Polypropylene fiber (PP), Glass Fibers (GF), Steel Fibers (SF), and Fibre (F) was used as the fiber in the AAC production. Physical and mechanical characterization of fiber-reinforced AAC was experimentally investigated.

2   MATERIALS AND METHOD

2.1   *Materials*

The AAC mixtures are prepared by using limestone, sand, aluminum powder and water. In addition to these main materials different fibers are used as additive (polypropylene fiber, steel fiber, glass fiber and fibre) in fresh AAC slurry. The ordinary Portland cement CEM I 42.5 R was supplied from Bilecik SANCIM cement plant. Limestone, quartz sand and aluminum powder are supplied from YTONG A.S. aerated concrete factory. Aluminum powder is a very important material for AAC production. Because it generates hydrogen gas with the reaction between limestone and expands the volume of the fresh AAC mixture. The fibers used for reinforcing the AAC (Ploypropylene fiber, steel fiber, glass fiber and fibre) are provided from the Anadolu University Structure Division Laboratory. The fibers used in the AAC mixtures are given in Figure 2.

2.2   *Method*

In this study, it is investigated that the effect of different fibers (polypropylene, steel fibers, glass fibers and fibre) on the physical and mechanical properties of AAC. For this purpose these fibers are used as additive in AAC mixtures with appropriate proportions due to the manufacturer's recommendations. The mix design of AAC specimens are given in Table 1. After mixing process the prepared slurry was molded in pre-heated 70 × 70 × 70 mm steel molds. The dimensions of the cubic specimens have been selected as the most appropriate due to the limited capacity of the autoclave apparatus. These samples were pre-cured in humidity saturated oven at 40°C for 2 hrs.

During this pre-curing period, the slurry in the mold increases its volume with the reaction of Al powder. Figure 3 shows the samples which were taken from curing oven after

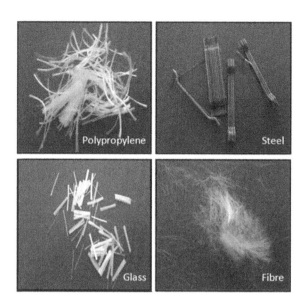

Figure 2.   Fibers used in AAC specimens.

Table 1.   Mix proportions of AAC specimens for 1 m³.

Specimen type	Cement (gr)	Water (gr)	Sand (gr)	Lime (gr)	Alumina (gr)	Fiber (gr)
Reference 1	100	140	535	30	0.75	–
Reference 2	150	210	800	45	1.5	–
Polypropylene	150	270	800	45	1.5	1.6
Steel Fiber	150	270	800	45	1.5	1.6
Glass Fiber	150	270	800	45	1.5	1.6
Fibre	150	270	800	45	1.5	5.0

Figure 3.   Specimens after pre-curing process.

pre-curing process. Then, the autoclaving process was performed on cubic specimens under 12 MPa pressure at 180°C for 8 hrs. At the end of the autoclaving process the hardened AAC specimens are removed from the autoclave after a cooling period for overcoming the thermal shock effect on heated specimens. The physical and mechanical properties of the AAC specimens are determined by unit weight and compressive strength tests respectively.

3 EXPERIMENTAL STUDY

In this study, the effect of fiber type and size in the production of AAC on compressive strength has been investigated. G2 classes of AAC were produced. Polypropylene (PP), Glass Fibers (GF), Steel Fibers (SF), and Fibre (F) was used as the fiber in the AAC production. These productions are also supported by experiments. Unit weight and compressive strength experiments were performed. These results are obtained, respectively.

3.1 *Unit weight test*

This test is carried out on all AAC specimens which were dried in the oven after autoclave curing. The weights of the specimens are divided by the volume of the cubic specimens. For this purpose all specimen dimensions are measured by a sensitive caliper. The unit weight test results of specimens are shown in Figure 4. As can be seen from the test results, the unit weight of fiber used AAC specimens are increased except for glass fiber type. This behavior can be attributed to the distribution of the air bubbles generated by the Al powder in the

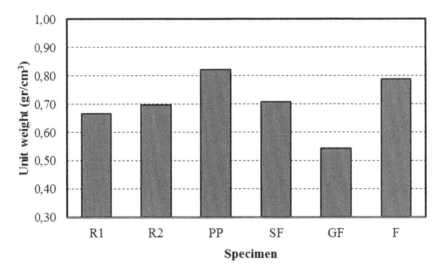

Figure 4.   Unit weight results of AAC specimens.

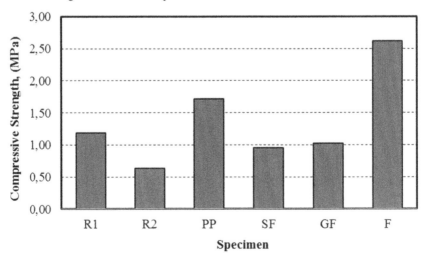

Figure 5.   Compressive strength of AAC specimens.

AAC slurry. Especially the polypropylene fiber and fibre increased the cohesion of the slurry and reduced the expansion of the slurry volume. Thus the adequate expansion of these specimens is limited by the fibers used in the AAC mixtures. The unit weight of AAC mixtures are varied between 0.665–0.822 gr/cm³ due to the type of the mixture.

3.2 *Compressive strength test*

Compressive strength of AAC specimens are determined by uniaxial compression test. The experiment is carried out on smooth bilateral surfaces of the cubic AAC specimens by the compressive test apparatus. The compressive strength test results are presented in Figure 5.

The ultimate load is divided to the surface area of the specimen in order to find the compressive strength. As seen from Figure 5 utilization of polypropylene and fibre in the production of AAC increases the compressive strength value, while the steel fibers and the glass fibers decrease. The reason for that is the distribution homogeneity of the fibers in the mixing and placing processes of the fresh AAC mixtures. Steel fibers couldn't mix homogeneously in the production. However steel fibers are placed at the bottom of the mold depending on their higher density. According to test results the compressive strength of AAC specimens are varied between 0.633–2,623 MPa. The strength values showed that fiber additive in AAC increase the strength of the hardened composite. However the workability of the fresh AAC is reduced with the increased cohesion of the slurry by the additive of fibers in the AAC mix design.

4 CONCLUSION

In this study, the effect of different type of fibers on the main physical and mechanical properties of AAC is investigated by experimental studies. The outcomes of this study can be defined as follows.

– Curing is very important to gain strengthening on the aerated concrete production. The moisture effect on the curing is more important on AAC production for the volume stability and hydration products of the cementitious composite.
– It is very important to provide the right temperature of water in the AAC slurry to perform the reaction between alumina powder and lime for the hydrogen gas generation. The molds for AAC production should be heat resistant or pre-heated in order to overcome the temperature decrease of the AAC slurry rapidly.
– Of course, the main purpose of the AAC is not to resist against external loads but increasing of the ultimate strength of AAC will help to reduce the dead load of the structure to become safer.
– Fibre is the most effective on the compressive strength of AAC, while it is the heaviest concrete in the mix proportions. In addition to this, glass fiber has the minimum unit weight with the higher compressive strength value than the R2 specimen. According to this result glass fiber showed the best weight / strength performance than the other fiber types.

REFERENCES

Demir, A., Karakurt, C. & Topcu, I. B. 2010. Utilization of Crushed Autoclaved Aerated Concrete as Aggregate in Concrete. Second International Conference on Sustainable Materials and Technologies, 28–30 June 2010.
Gül, R., Okuyucu, E., Türkmen, I. & Aydin, A. 2007. Thermo-mechanical properties of fiber rein force draw perlite concrete. *Materials Letters,* 61:5145–5149.
Karakurt, C., Kurama, H. & Topcu, I. B. 2010. Utilization of natural zeolite in aerated concrete production, *Cement & Concrete Composites,* 32(1):1–8.
Kartal O. 2014. *Materials of Construction Design Thesis,* Eskisehir: Anadolu University Turkey.
Murdock, L.V. & Brook, K.M. 1979. *Concrete Materials and Practice,* British Council.

Pehlivanlı, Z. O. & Uzun I. & Demir I. 2015. Mechanical and microstructural features of autoclaved aerated concrete reinforced with autoclaved polypropylene, carbon, basalt and glass fiber. *Construction and Building Materials*, 96:428–433.

Pehlivanlı, Z. O. & Uzun I. & Demir I. 2016. The effect of different fiber reinforcement on the thermal and mechanical properties of autoclaved aerated concrete. *Construction and Building Materials*, 112:325–330.

Tanyıldızı, H. 2008. Effect of temperature, carbon fibers, and silica fume on the mechanical properties of light weight concretes. *New Carbon Materials*. 23: 339–344.

TS EN 771-4. 2015. *Specification for masonry units—Part 4: Autoclaved aerated concrete masonry units*. Ankara: Turkish Standarts Institutition.

Selected problems of temperature conditions in the interior of large area of single store halls

J. Katunská, D. Katunský & S. Tóth
Faculty of Civil Engineering, Institute of Architectural Engineering, Technical University of Kosice, Slovakia

ABSTRACT: The present contribution shows the temperature conditions in various single-store halls in the reflection on the formation of the envelope structures. In several halls in eastern Slovakia were taken measurements of temperature and humidity microclimate. It can be noted that presently the traditional methods is not possible to create optimal indoor conditions that would be acceptable by multiple users. Problems associated with the formation of envelope constructions (as external walls, roofs and transparent parts), with heating and ventilation. The paper deals with selected problems that were found during the measurement directly in real terms of building use. The selected building is a single-store hall, which is located in a small town in Eastern Slovakia in Rožňava.

1 INTRODUCTION

1.1 *Measuring time*

In winter period, they have been carried out experimental measurements in production buildings Wang et al. (2016). There were also realized measurements in several manufacturing plants in eastern Slovakia Katunská et al. (2002)—Katunský et al. (2014). Contribution shows the results of the two-week measurement from January 1 to 12 last year (selected 12 days).

1.2 *Description of selected hall*

Description of selected buildings is as follows. Hall has one nave, steel (steel shell) with a span of 16 m in the longitudinal direction of the warp modular 11×4.5 meters i.e. length of about 50 meters (over $16{,}150 \times 49.80$ m).

The columns are of steel rolled profiles and welded elements. Span halls to bridge truss are mounted on steel columns with height of 1.8 meters.

Figure 1 and 2.   Front and side view of the selected hall.

Table 1. Measuring points in the selected one store hall.

Setting	Position
6911	Under the soffit
6985	Left wall inside surface temperature
6979	The temperature at 1,5 m high from the floor
6980	Right wall surface temperature of the inner
7640	Right wall surface temperature of the outer

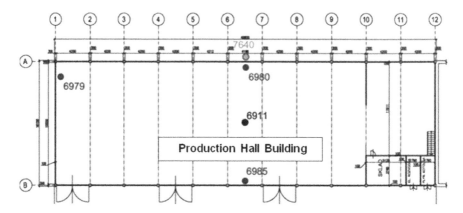

Figure 3. Plan view of a single-store hall.

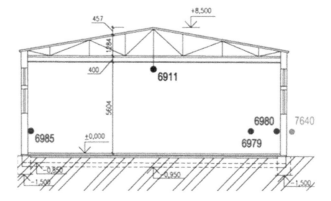

Figure 4. Cross-section of a single-store hall.

The roof is made of corrugated and insulation is with polyurethane foam. Waterproofing cover is formed from metal sheet. The walls are stacked sandwich (sheet metal, polyurethane foam and sheet metal) of the total thickness of 150 mm. The windows are double with insulating double glazing and frames are made of steel. Doors and gates are metal. The whole design was implemented for the hall design temperature 12–16°C which results from the nature of the work (mechanical workshop). Hall height is 8 m and the total usable area of the house is 785 m².

2 EXPERIMENTAL MEASUREMENTS IN SITU

2.1 *External and internal temperatures*

Measurement of outdoor temperature in the location of hall was realized by logger. Indoor air temperatures were recorded at hourly intervals in 1.5 meters above the floor, below the ceiling, under the roof. Outdoor temperatures had significantly decreasing and increasing character. Descent was from the maximum temperature of +6.0°C reaches the minimum of nearly −14.00°C. This was achieved on the night of January 7 to 8. The internal temperature of the external temperature copied out of phase 6–12 hours, depending on the rate of change of external temperature.

Maximum daily outdoor temperature was +6.0°C at 12.00 on January 2. Maximum of daily indoor temperature will only take effect for several hours shortly. This temperature change will take effect in the interior area between 4:00 p.m. to 6:00 p.m. in the afternoon. In view of the heat storage capability to the envelope structures, but a large volume of air in the indoor environment is a time difference of temperature changes of about 4–6 hours.

2.2 *Measuring of horizontal slabs and roof*

Measuring the temperature and the horizontal level of the roof was carried out in the points which are shown in figure 4—schematic cross-section of hall. The maximum daily air temperature in the hall was measured about 16°C, which is reflected on January 7 and 12 at afternoon, with the highest surface temperature of horizontal external structures have

Figure 5. Internal views in hall.

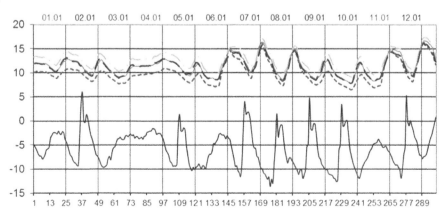

Figure 6. External and internal air temperatures.

461

been reported early in the morning on January 8. It is because of its low heat accumulation properties. High surface temperature is always recorded on the surface of the ceiling. The lowest is the interior surface of the roof. This is the sub-cooling from the outside. Indoor air temperatures replicate the outdoor air temperature with a phase delay. This has not been mentioned. As might be expected, the lowest temperatures are just above the floor and the highest surface temperature of the ceiling truss. The difference between these temperatures reaches a value of 5–6°C, which is in this case quite a lot.

2.3 Measuring of vertical structures

Surface temperature measurement of vertical envelope structures—e.g. curtain wall was realized on the internal surface of about 1.5 meters above the floor. Measurement was carried out on the wall of the left side (red squares—course on the chart) and the wall of the right side (blue diamond's). Surface temperatures were also measured on the inner surface of the glass of the windows. From the waveform it can be seen that the highest surface temperature of the internal surface of windows and the greater the wall of the left (facing south)

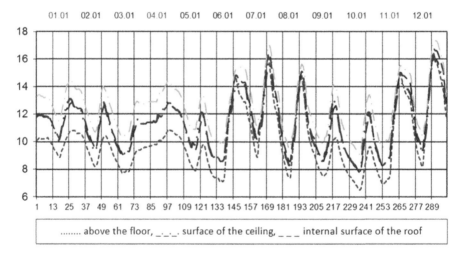

Figure 7.   Internal surface temperatures of horizontal structures.

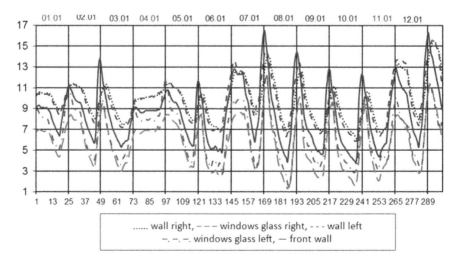

Figure 8.   Internal surface temperatures of vertical structures.

than the right wall. Left and right wall is a symbolic indication—clear from looking at the entrance to the hall. It also suggests a cross section of the measurement points. Sectional visible beams and walls (left, right). Interior surface temperatures of external walls replicate the outdoor temperature with a phase delay (6–12 hours). Surface temperatures on the windows are almost identical to the air temperature outdoors. The maximum surface temperatures were at January 7, 12. These temperatures are close to the value of 17°C. Slightly lower by about two degrees are temperatures January 2, 6, 8, and 11. Their value is about 11°C. The lowest temperatures are from January 8 to 11. They are located on the right side of the window. Because wall is oriented north. Difference high and low temperature reach 6 to 7°C. The lowest surface temperature below the dew point, resulting in the occurrence of water drops on the glass.

3 DISCUSSION

Measuring the temperature in vertical and the horizontal level of the roof was carried out in the points which are shown in figure—schematic sectional hall. Measurements were also carried out in the indoor environment. The measurement showed that the internal environment is very cold. Thermal performance envelope structures are inappropriate, Cuadrado (2015).

In the paper it is not evaluated relative humidity, which is the well-being of employees is very necessary. When employees used hall for a long time (shift work), it may be required specific relative humidity. It can only be achieved by mechanical treatment of the air. For that purpose it is particularly suitable hot air heating, which is always put into operation at night, Brinks, P. et al (2015). Before starting working hours and during working hours is mechanical treatment of the air off. Therefore, the internal temperature of the air dropped very rapidly during the working time of the already mentioned low values of the inside air temperature of 6–9°C for hot air heating weaning. Due to this fact, thermal comforts, as well as working conditions are worse, because they are very low temperature indoor air Yan et al (2014).

Such method of heating the production hall in terms of micro-climatic conditions in the workplace as well as in terms of energy efficiency is not appropriate and is not efficient. By evaluation of measurements inside the single-store factory building was found that the temperature of the desired temperature conditions has not been fulfilled. It is characterized by the fact that the measured values have not been met, such as requirements for thermal comfort of workers and the technological process, Katunský et al (2011, 2012, and 2013).

4 CONCLUSIONS

Mandatory values of heat-moisture conditions are required for internal microclimate in the type halls. But it is not a sufficient condition for thermal comfort. It cannot be achieved. The required environmental conditions inside buildings cannot be achieved through thermal properties only envelope of hall. Buildings shall be provided with technical equipment that allows delivering heat or cold, for example with air conditioning and air heating—Šikula et al (2013), Streckiene & Polonis (2015), Grossmann et al (2016).

There may be a link between them with a common goal to provide the required state of the internal environment of the hall. This was also confirmed in the present case, only the radiators which are the (convection heating) at some time during the measurement, and at certain times of the hot air generators are turned on. Measured variables—temperature differences in the two methods of heating were great. Their evaluation indicates that the inadequacy of these two types of heating in the building lobby for the operation has resulted in a dramatic change in temperature.

Thermal performance of the building envelope, especially glass surfaces, shows a large heat loss, depending on the climate conditions and orientation to the cardinal, Lopusniak & Katunský (2006). It had been confirmed in several literature resources by Abbasi (2014) to Wang (2016).

Recent research in this area, which is mainly engaged in the solution heat gain and the development of new building envelope industry, is in references by Abbasi (2014), Subelzu & Alvarez (2015) and Grossmann et al. (2016).

ACKNOWLEDGMENT

This contribution is a sequel to the grant project solution VEGA 1/0835/14 "*Experimental research on the physical properties of fragments and construction details of the building envelope in non-stationary heat-air-moisture conditions*" supported from Grand Agency of Ministry of Education and Slovak Academy of Science.

REFERENCES

Abbasi, N. 2014. Industrial Design and Production of Buildings in: *Journal of Applied Environmental and Biological Sciences* Vol. 4, no. 8, p. 169–175 ISSN: 2090–4274.

Brinks, P. & Kornadt, O. & Oly, R. 2015. Air infiltration assessment for industrial buildings in: *Energy and Buildings* Vol. 86, No. 1, p. 663–676.

Cuadrado, J. et al. 2015. Sustainability-related decision-making in industrial buildings: an AHP analysis In: *Mathematical Problems in Engineering* Vol. 19, no. 10, p. 1–14.

Grossmann, D. et al. 2016. A methodology for estimating rebound effects in non-residential public service buildings: Case study of four buildings in Germany in: *Energy and Buildings* Vol. 111.

Katunská, J. & Katunský, D. 2002. Evaluation of the Quality of the Internal Microclimate in Large Industrial Halls, *International Journal for Restoration of Buildings and Monuments*. Vol 8, No. 4 p. 371–378.

Katunská, J. & Katunský, D. 2008. Diagnostic of a Selected Industrial Building and Design for its Thermal Refurbishment, *Selected Scientific Paper / Journal of Civil Engineering*, Vol.3, 1-, pgs 37–44.

Katunský, D., Korjenic, A., Katunská, J., Lopušniak, M. 2011. Evaluation of energy consumption for heating of industrial building in-situ in: *Engineering*, 3, p. 470–477.

Katunský, D., Lopušniak, M., Bagoňa, M., Dolníková, E., Katunská, J., Vertaľ, M. 2012. Simulations and measurements in industrial building research In: *Journal of Theoretical and Applied Information Technology*. Vol. 44, No. 1, p. 40–50.

Katunský, D., Zozulák, M., Vertaľ, M., Kondáš, K., Baláž, R. 2013. Measuring Methodology and Results of Heat-Air-Moisture Performances at Building Envelope Levels, In: *Advanced Materials Research*, Vol. 649, pp. 85–88.

Katunský, D., Zozulák, M., Kondáš, K., Šimíček, J., 2014. Numerical Analysis and Measurement Results of a Window Sill, In: *Advanced Materials Research*, Vol. 899, pp. 147–150.

Lopušniak, M. & Katunský, D. 2006. Interaction of selected parameters within design of suitable working environment. In: *Healthy Buildings June 2006 Lisboa*, ISIAQ, p. 147–152.

Sikula, O. & Plasek, J. & Hirš, J. 2013. The Effect of Shielding Barriers on Solar Air Collector Gains In: *Energy Procedia*, Vol. 36, p. 1070–1075.

Streckiene, G. & Polonis, E. 2015. Skirtingos konfiguracijos maža energiu daugiabuciu pastatu energios poreikiu tyrimas, In: *Mokslas—Lietuvos, Ateitis Science—Future of Lithuania: Environmental Protection Engineering*, Vol. 6, No. 4, s. 414–420.

Subelzu, S. &. Alvarez, R. 2015. Urban planning and industry in Spain: A novel methodology for calculating industrial carbon footprints In: *Energy Policy* Vol. 83, no. 1, p. 57–68 ISSN: 0301-4215.

Yan, B. et al. 2014. Influence of user behavior on unsatisfactory indoor thermal environment, In: *Energy Conversion and Management*, Volume 86, pages 1–7.

Wang, Y. et al. 2016. Measurement and evaluation of indoor thermal environment in a naturally ventilated industrial building with high temperature heat sources In: *Building and Environment* Vol. 96, pages. 35–45.

Advances and Trends in Engineering Sciences and Technologies II – Al Ali & Platko (Eds)
© 2017 Taylor & Francis Group, ISBN 978-1-138-03224-8

The efficiency of green roofs to mitigate urban heat island effect in Rio de Janeiro

S. Konasova
Department of Economics and Management in Civil Engineering, Czech Technical University in Prague, Prague, Czech Republic

ABSTRACT: Nowadays, a frequent topic of discussion is deterioration of the quality of the urban environment. Given that the majority of people live in urban areas, it is necessary to address this issue. The buildings, streets, and other infrastructure that comprise urban environment typically make cities hotter than surrounding rural areas. The integration of vegetation in the form of green roofs into the urban fabric is particularly important to reduce this temperature discrepancy. This paper attempts to evaluate the efficiency of green roofs to mitigate the urban heat island effect through monitoring temperatures in Rio de Janeiro. The 1.82°C of the difference in average air temperature between the green roof and the reinforced concrete flat roof was found within this study. This work provides useful information for policy-makers and decision-makers to utilize one of the green strategies for lowering urban temperatures, saving energy, and improving the quality of life.

1 INTRODUCTION

According to the report of the United Nations (2014), the world population is increasing rapidly and in the next few years, more than half of the population will live in urban areas. While the major cities offer many opportunities; their habitants are exposed to the pollution, noise, higher temperatures, and the remoteness from nature. By integrating the vegetation in the built urban environment, these issues could be minimized.

Integration of vegetation into the urban fabric offers many benefits for habitants, mainly enhances the quality of their lives and connects the buildings with their unique local environment. One of the possibilities how to incorporate the vegetation in the building envelope is in the form of green roof. The green roof provides shade and insulation, reduces energy consumption, mitigates the urban heat island effect, and protects against the wind and temperature extremes, thunderstorms, hail, and ultraviolet radiation. Based on the study of Peck et al. (1999) substrate depth, shade from plant material and transpiration can reduce solar energy gain by up to 90% in comparison with non-shaded buildings. Therefore, green roofs can reduce indoor temperatures from 3 to 4°C when outdoor temperatures are between 25°C and 30°C. It was also calculated by Peck et al. (1999) that 5.5°C reduction in the temperature immediately outside of a building can reduce the amount of energy needed for air-conditioning by 50% to 70%. Concerning these characteristics, green roofs have a great potential in urban areas, especially in tropical cities where the solar radiation is very intensive.

2 URBANIZATION AND THE LOSS OF GREEN SPACES IN RIO DE JANEIRO

Cities occupy a small proportion of the Earth, but they are the most concentrated areas of population in the world. The percentage of the population living in cities continues to increase, as the rural population decreases. Most of this growth occurs in the economically undeveloped areas, especially in tropical areas, where urban infrastructure has to be built yet.

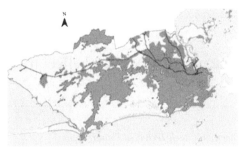 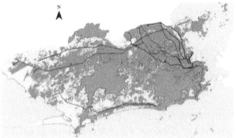

Figure 1,   Rio de Janeiro 1940 (left) and 2009 (right),
Source: Rio em Mapas, http://portalgeo.rio.rj.gov.br.

Thus, in the near future, the largest proportion of humanity will live in places where the local environment has been profoundly modified by the cities, which are already constructed or will be constructed.

The urbanization process in Brazil took on a major dynamism in the late nineteenth century, when the cities acquired the growing importance in the territorial organization of the country. The process of industrialization began in the first half of the twentieth century, and it has always been strongly linked to urbanization, with a direct influence on the structure and formation of the urban network. This urban network currently consists of regional systems located mainly along the coast, mostly in the south and southeast, such as Rio de Janeiro, Sao Paulo, and Salvador. The strong relationship between urbanization and industrialization has been characterized the territorial, demographic, and economic dynamics in Brazil since the 1950s.

Rio de Janeiro, along with many cities in Brazil, has undergone a process of expansion during last five decades. Through this process of urban development, the trees were cut out, and the vegetation was largely replaced by asphalt or concrete surfaces. Despite the fact that Rio de Janeiro has the largest urban forest in the world, the built-up areas have been gradually changing the city landscape. During these last 50 years, the city lost at least thousands hectares of vegetation (Platonow, 2012). The urban growth of Rio de Janeiro between 1940 and 2009 is shown in Figures 1 and 2. From the figures, it can be seen an increment in urban areas, when urban development occupied almost every non-built-up area. The loss of public green spaces has a significant effect on the urban heat island in Rio de Janeiro, especially in the city center.

3   URBAN HEAT ISLAND EFFECT AS CLIMATE PHENOMENA OF CITIES

The urban microclimate is influenced by urban form and surfaces. Cities are characterized by impervious surfaces with a high concentration of anthropogenic activities leading to significant increases in the air temperature and the surface temperature, which are higher than the temperature of countryside. Such effect is known as a phenomenon, is called "Urban Heat Island" (UHI) (Oke, 1987), which its magnitude depends primarily on the size of the city and local climatic conditions.

The urban heat island effect occurs in cities all around the world, and it is a result of the different thermal properties of surfaces in urban areas. In Rio de Janeiro, especially during the hottest period of the year, up to 25°C differences can be observed in comparison to rural areas of the city. Some heat layers occur on the surface while others occur a few meters above the surface. The majority of studies on urban heat islands in Rio de Janeiro are focused on surface temperatures rather than focusing on air temperatures. There are several types of heat islands depending on the climate and topography. Cities at higher latitudes face the urban heat island effect at night when the stored heat within the buildings and the asphalt street is released. In contrast, cities in tropical latitudes, such as Rio de Janeiro, face this effect during the day (Reynolds, 2015).

4 MITIGATION OF URBAN HEAT ISLAND EFFECT THROUGH GREEN ROOFS

Green roofs have a positive impact on urban heat island effect for their surrounding climates for two main reasons: (i) protecting buildings from solar radiation by the physical act of shading; (ii) causing the process of evapotranspiration and photosynthesis (Niachou et al., 2001). The plants and the substrate prevent the sunlight from reaching the membrane of the roof and reduce the surface temperature below them. In summer, usually 70% to 90% of the sun's energy is absorbed by the leaves for photosynthesis and the rest is reflected back into the atmosphere (EPA, 2008). The evapotranspiration occurs through processes of evaporation and transpiration, which is important in urban climates for assessing environmental benefits including attenuation of stormwater runoff and mitigation of urban heat islands.

The densification of existing built-up areas is responsible for the decreasing vegetation, which results in the lack of evapotranspiration to cool the air. Such decreasing vegetation causes urban heat islands, which follows high temperature during days in the major cities, like Sao Paulo or Rio de Janeiro. Moreover, roofs and pavements comprise more than 60% of urban surfaces in most cities. Since both roof and pavement have a very low albedo, within 0.15 and 0.25, they absorb a lot of sunlight. Besides that, concrete and asphalt release their heat slowly during the night and day (Akbari et at., 2009). Several authors have shown that natural and permeable surfaces, as in the case of green roofs, can play an important role in mitigating this negative climate phenomenon and providing higher energy efficiency for the building, leading to energy savings.

5 STUDY AREA

The city of Rio de Janeiro (22°54'10''S 43°12'28''W) is a Brazilian city, located in the southeast of the country, at an altitude of about 10 meters in relation to the mean sea level. Rio de Janeiro is the capital of the state of Rio de Janeiro, the third most populous state in Brazil. The city's metropolitan area is the second most populous metropolitan area in Brazil and seventh most populous in the Americas, which has 1197 km² surface area. In 2014, according to Brazilian Institute of Geography and Statistics (IBGE, 2014) the population of Rio de Janeiro was 6,453,682.

The diverse landscape of the metropolitan area of Rio de Janeiro, which is surrounded by the Atlantic Ocean and Guanabara Bay in the east, contributes to the significant variability of weather elements such as temperature, precipitation, humidity, wind, cloud cover, and evaporation. Rio de Janeiro is located around the Tijuca Massif, the Gericinó-Mendanha Massif in the north and the Pedra Branca Massif in the west, all affecting such mentioned climate.

According to the Köppen-Geiger climate classification system, climatic condition of Rio de Janeiro is Atlantic tropical with an average air temperature of 16°C throughout the year and a dry season in which the average monthly precipitation is less than 60 mm for at least one month of the year. The average annual temperature varies between 23°C and 24°C, where the highest monthly average occurs during summers in February (28.7°C), and the lowest monthly average occurs during winters in July (21.3°C) (INMET, 2015).

5.1 *Case study: Copacabana*

Copacabana is a neighborhood located in the south of the city of Rio de Janeiro. It is known for its beach that is one of the most famous in the world. In 1892, the inauguration of the tunnel Velho, which connected neighborhood Botafogo with Copacabana, allowed to integrate Copacabana with rest of the city (KAZ, 2010). Due to the demand to live in Copacabana by growing middle class, the neighborhood began to expand vertically up to thirteen-story. Nowadays, residential buildings with ten to twelve-story built next to each other dominate the neighborhood. Houses of two or three-story are rare. According to IBGE (2014) total population of Copacabana was 147,021 inhabitants in 2014. The lack of green spaces inside of the densely built-up area of Copacabana creates many heat islands. For this reason, Copacabana was chosen as a study area to demonstrate the possibility to mitigate the urban heat island effect.

The program EVNI was used in order to determine the roof systems applied in Copacabana. Three systems of conventional roofs were defined through satellite images; flat concrete roof, pitched roofs with ceramic tile, and tin roof. Three representative areas of the district have been subjected to a proportional analysis of construction systems, considering the color and shape of the roofs. This analysis resulted in a conclusion that estimated that 85% of roofs are flat concrete roofs, 13% of pitched roofs with ceramic tiles, and 2% of a tin roof.

The green roof was not found through Google Earth in any of the randomly selected areas. In some cases, vegetation in the form of flowers or palm trees was possible to find on rooftops. In the process of searching a green roof for this case study, one was found through the local news, on the building of supermarket Zona Sul. Since the installation was finished recently and satellite images probably weren't actualized, it was not possible to find it by Google Earth.

6 METHODS

The mitigation effect of the green roof on urban heat islands was assessed using meteorological measurements, which were collected using two automatic weather stations that were installed on two types of roofs; an extensive green and a conventional roof in Copacabana, during 1st of February to 7th of February of 2016.

The selected study area is a densely urbanized area of Copacabana neighborhood. This area was selected based on the only green roof that was found in the neighborhood on the supermarket Zona Sul. Finding green roof in Rio de Janeiro is challenging, as the matter of fact, just a few green roofs are installed in whole city. The conventional roof of reinforced concrete slab was found in the same area as green roof to undertake a fair comparison analysis. The distance between the two buildings is around 200 meters. Both buildings have similar geometric characteristics; the number of floors, side of street, environment, and sun path direction.

The three-story supermarket Zona Sul has an extensive green roof covered by grasses and low plants. The supermarket Zona Sul is located at 29 Dias da Rocha Street. The green roof system consists of drainage system leading to rainwater harvesting tanks to keep the water for reuse in the irrigation system. The building is surrounded by lower building in the Northwest and the taller building in the Southeast. The nearest obstruction around the green roof is just twelve floors building in the Southeast, but the direction of this building prevents the creation of shadow during day. The weather station was installed one meter above the roof surface and was located three meters away from the Northeast and Northwest edge of the green roof.

Second two-story building with a conventional roof is located at 572 Barata Ribeiro Street, is a residential building situated on corner of Barata Ribeiro Street and Raimundo Corrêa Street. The surrounding neighbor buildings are located in the Northwest and Southwest. Seven-story building located in the Northwest in relation to selected conventional roof creates a shadow at the end of afternoons. The nearest vegetation is in the form of trees on the street. The roof consists of reinforced concrete slab and concrete tiles. The weather station is installed one meter above the roof surface and is located three meters away from the Southwest and Southeast edge of the roof.

The two weather stations were borrowed from the metrology laboratory of UFRJ. The metrological stations are comprised of perforated plywood shelter, tripod, thermometer, and thermocouple. Waterproof HOBO Pro Series-onset sensor was chosen for measuring relative air humidity and air temperature. HOBO sensor has an integrated data logger, which is fully automatic. Data logger was programmed to measure air temperature and relative humidity every 10 minutes. The data logger with the full battery could record data for 151 days, and then it is important to download all the data through the Box Car software and program the data logger again. Thermocouple wire with data logger Log Box Novus was used to measure surface temperature. The sensor at the end of the wire was calibrated based on the method using hot and cold water with the utilization of another thermometer. The data logger was programmed to measure surface temperature also every 10 minutes. The data logger with the full battery could record data for 227 days, then it is important to download all the data through the Log Chart II software and program again because all previous data will be removed. Both data loggers are placed inside the shelter to be in shadow, only the thermo-couple is fixed on the surface of the roof.

7 RESULTS

The results point out that the average air temperature of the green roof is cooler than the conventional roof during the first week of February. The average air temperature results for the entire period demonstrate that the average air temperature of the green roof of supermarket Zona Sul was 28.20°C and the conventional roof of the residential building was 29.83°C. The difference in average air temperature between two samples is 1.82°C (Figure 2). It can be seen from the Figure 2, the air temperature of the green roof is below that of the conventional roof almost all the time. The factors behind this result are evapotranspirational cooling and shading. The greatest reduction in air temperature occurred between 00:00 and 10:00.

In February, the relative humidity varies between 60% and 90% in Rio de Janeiro (Leal, 2013). Based on collected data, Figure 3 shows the relative humidity of green roof and conventional roof. The results reveal that the average relative humidity of the green roof was 70.13% and the conventional roof was 65.21%. The difference in average relative humidity between the two sites is 4.93%. The maximum differences occurred between 1:00 and 11:00 and reached up to 13.7% at 10:42.

Figure 4 demonstrates the surface temperatures that were obtained from data logger of thermocouple wire during the same period of the summer. As can be seen from the graph, the average surface temperature of green roof was 32.14°C and the conventional roof was 38.49°C. The difference in average surface temperature is 6.35°C. The maximum differences occurred between 19:00 and 7:00 and reached up to 14°C.

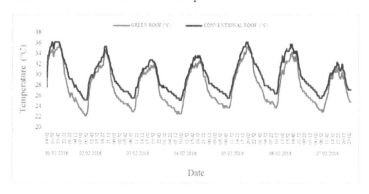

Figure 2.   The air temperature.
Source: Author, 2016.

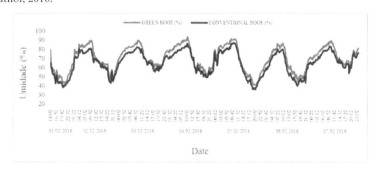

Figure 3.   The relative humidity of air.
Source: Author, 2016.

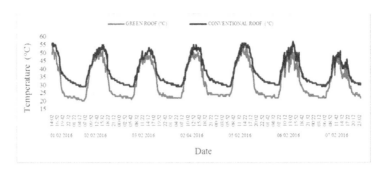

Figure 4.   The surface temperature.
Source: Author, 2016.

8   CONCLUSION

The urban heat island effect, as one of the ecological consequences of urbanization, has a considerable impact on the quality of life in cities. The replacement of vegetation by impervious surfaces and anthropogenic activities tends to be the main factors responsible for the temperature increases in the urban area of Rio de Janeiro. The objective of this paper is to demonstrate that green roof is an effective tool for reducing the air and surface temperatures and increasing relative humidity to mitigate urban heat island effect. To achieve this goal a case study was defined and observed. The case study is located in the built-up area, where the most urban heat islands occur, and mitigation is most needed. The measurements were obtained simultaneously from two buildings, close to each other, with the different type of roofs in the selected area to make the comparison legitimate as much as possible. The results prove that green roofs are efficient to improve its environment, specifically by reducing the temperature and increasing humidity. Moreover, the observed data highlights the positive effect of green roofs on the reduction of energy usage for cooling, based on decreasing temperature.

REFERENCES

Akbari, H.; Menon, S. & Rosenfeld, A. 2009. Climatic Change, 94. *Global cooling: increasing world-wide urban albedos to offset CO$_2$*, pp. 275–286.
EPA, Environment Protection Agency. 2008. *Reducing Urban Heat Islands: Compendium of Strategies: Green Roofs, EPA Report*, http://www.epa.gov/heatisld/resources/pdf/GreenRoofsCompendium.pdf.
IBGE. 2010. IBGE Report. *População residente em 2000 e população residente em 2010*.
INMET, National Institute of Meteorology. 2015. *Weather information for the Olympic and Paralympics Game s in Rio de Janeiro 2016*, RIO 2016.
Kaz, S. 2010. Tese de Doutorado, *Um jeito Copacabana de ser: o discurso do mito em O Cruzeiro e Sombra*, Departamento de Artes e Design da PUC-Rio, p. 249.
Leal, M. 2013. INMET Report. *Boletim meteorológico para o Rio de Janeiro, Instituto Nacional de Meteorologia*. p. 2.
Niachou, A.; Papakonstantiou, K.; Santamouris, M.; Tsangrassoulis, A. & Mihalakakou, G. 2001. Energy and Buildings, 33. *Analysis of the green roof thermal properties and investigation of it energy performance*, pp. 719–729.
Oke, T. R. 1987. *Boundary Layer Climates*, London: Routledge, ISBN 0-203-71545-4.
Peck, S.W., C. Callaghan, M.E. Kuhn, and B. Bass. 1999. *Greenbacks from green roofs: Forging a new industry in Canada*. Canada Mortgage and Housing Corporation. Ottawa, Canada.
Platonow, V. 2012. *Área verde por habitante cai 26% no Rio com avanço de favelas e especulação imobiliária*, Meio Ambiente, Agência Brasil.
Reynolds, L. 2015. Sustainability. *Rio de Janeiro's Urban Heat Islands: A Primer, International Observers, Research & Analysis*, Solutions, Understanding Rio.
United Nations. 2014. World Urbanization Prospects. *World's population increasingly urban with more than half living in urban areas*, Published by the United Nations, ISBN 978-92-1-151517-6.

Thermal mass of building construction versus operative temperature

K. Kovacova & M. Kovac

Technical University of Kosice, Civil Engineering Faculty, Kosice, Slovakia

ABSTRACT: The content of paper is energy analysis where the impact of accumulated effect of building construction on operative temperature trends in analysed room is evaluated. In this case study of classroom with sloping roof and roof windows was selecting. There are compared two types of building constructions, so called lightweight and heavyweight construction. There is taken account with natural and mechanical ventilation in analysed room for booth construction types. The results from energy simulation show positive impact of heavyweight construction on max operative temperature in room for booth ventilation regimes. The results are output from dynamic energy simulation was elaborated by tool DesignBuilder. This simulation programme uses calculating tool EnergyPlus.

1 INTRODUCTION

Thermal mass is defined as the ability of a material to store (charge) heat, and later on release (discharge) it when activated mechanically or naturally through one or a combination of conduction, convection and radiation (Workie, D.M. 2015, Clark, J. 2001). The most important parameters that affect the performance of thermal mass are material thermal properties, location of thermal mass, ventilation and occupancy too (Balaras, C.A. 1995). When a building is naturally ventilated, thermal mass can be used to regulate indoor air temperatures (Yam, J. at al. 2003). The content of paper is evaluating the impact of thermal mass in building constructions on operative temperature trend in summer period. Further the impact of ventilation mode on operative temperature is monitored in case of both types of construction.

2 THE SUBJECT OF ENERGY ANALYSIS

2.1 Model geometric

The subject of energy analysis is existing loft space used as university classrooms. In order to gain relevant results the model geometric of loft space was created in simulation tool Design-Builder. Existing loft space is built from wooden framed structure filled in mineral wool as thermal insulation what we can mark as lightweight construction in terms of thermal mass. The model of two classrooms and corridor is shown in the following figure (Figure 1). The presented results from energy analysis are valid for classroom with windows in external wall oriented South-Westerly (classroom SW). There were used the climate data of Test Reference Year (TRY) for Kosice from DesignBuilder database.

Floor area of classroom is 66 m². Full height by gradient of roof plain 23° is range from 1.9 to 4.5 m. The total internal volume of room is 210 m³. The external windows area is 7.65 m², roof windows area is 10.4 m². Max capacity of classroom is 30 people per room. The technical parameters of building constructions are in Table 1. In order to eliminate solar heat load there are used light internal shade roll on roof windows (solar transmittance = 0.4).

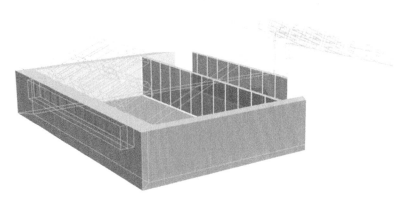

Figure 1. Model geometric of loft space in DesignBuilder (analyzed classroom oriented south-westerly is highlighted).

Table 1. Parameters of building constructions (STN 73 0540-3. 2013).

Building construction	Coefficient of thermal transmittance U [W/(m²/K)]	Thermal properties SHGC	Optical properties τ
External wall	0.29	–	–
External windows	3.00	0.72	0.78
Roof	0.20	–	–
Roof windows	2.00	0.60	0.77
Internal wall	0.68	–	–
Internal windows	2.90	0.75	0.82

2.2 Thermal mass of building constructions

This analysis compares the impact of lightweight and heavyweight building construction on operative temperature in classroom during summer period. Existing loft space is built from wooden framed structure filled in mineral wool as thermal insulation what we can mark as lightweight construction in terms of thermal mass. As alternative is used heavyweight construction from reinforced concrete.

In case of lightweight construction the thermal insulation thickness 180 mm with plasterboard lining on inside surface is used for roof construction. The floor construction is built from wooden grid filled in thermal insulation thickness 60 mm. The wear layer is from OSB board thickness 24 mm. The heavyweight construction could be built from ferro-concrete slab thickness 150 mm with thermal insulation thickness 180 mm in part of sloping roof. On the floor there is thinking with using the concrete layer thickness 40 mm placing on thermal insulation thickness 60 mm.

2.3 Internal environment conditions and ventilation modes

The analyzed classroom is used every workday in time from 7.30 a.m. till 6.20 p.m. The teaching time is 1.5 hours with pause running 10 minutes or 70 minutes during lunch time. The maximum density of classroom is 30 students (0.45 people/m²). There was under consideration with this operating mode of classroom and with metabolic rate from students at value 108 W/person in energy simulation. The classroom has no ventilation or cooling equipment for air treatment in summer period nowadays. It is possible to use only natural ventilation by external windows.

The ventilation is important parameter affecting the performance of thermal mass too. The other partial result from energy analysis is comparison of natural and mechanical ventilation

Table 2. Thermal properties of used building materials (STN 73 0540-3. 2013).

Material	Thermal conductivity λ [W/(m.K)]	Specific heat capacity c [J/(kg.K)]	Density ρ [kg/m³]
Mineral wool	0.038	840	140
OSB board	0.13	1600	600
Plasteboard	0.15	1060	750
Ferro-concrete slab	1.58	1020	2400
Concrete layer	1.30	1020	2200

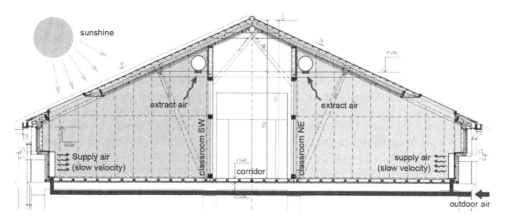

Figure 2. Scheme of displacement mechanical ventilation of classrooms in loft space.

mode and their impact on operative temperature trend during the summer. In case of natural ventilation of classroom there is under consideration that 25% area of external windows are opened. Roof windows with shade roll are closed in order to eliminate heat load from solar radiation. In case of mechanical ventilation of classroom there is modelled displacement ventilation with supply outdoor air and exhaust of extract air by scheme in the Figure 2.

The required outdoor airflow for classroom was calculated by valid standard requirements (STN EN 15251. 2007). There was thinking the basic ventilation rates for emissions from people and for building emissions pollutions in category I by equation (1).

$$q_{tot} = n.q_p + A.q_B \qquad (1)$$

where q_{tot} = breathing zone outdoor airflow rate in l/s; n = number of people in room; q_p = airflow rate per person in l/(s. person); A = zone floor area in m² and q_B = airflow rate for buildings emissions pollutions in l/(s.m²).

The resulted value of breathing zone outdoor airflow rate for analyzed classroom is 366 l/s. This value corresponds to the outdoor airflow rate per person 12.2 l/s and to ventilation rate of classroom at value 6.0 ac/h. Exterior windows and roof windows are closed during mechanical ventilation.

3 RESULTS FROM ENERGY ANALYSIS

In order to analyse indoor environment state in classroom there were modelled 8 variants total. These variants are mainly different in type of building construction. One is lightweight building construction with low accumulated effect, and other is heavyweight building construction with high accumulated effect.

The second parameter whose impact was monitored is ventilation mode. There were modelled two operation regimes for natural ventilation and for mechanical ventilation too that are different in ventilation time. At first there is under consideration operating time of ventilation only in time of teaching (from 7.30 a.m. till 6.20 p.m.). The second operation regime calculates for 24 hours ventilation (so called night precooling).

The energy simulation of the all variants was made from 1st till 15th May. The results for period from 13th till 15th May are presented in the graphs below. From results of all variants it is clear that thermal mass of buildings constructions have positive impact on operative temperature in room.

3.1 Thermal mass of building construction and natural ventilation

The first two graphs below (Figure 3) show simulation results for natural ventilation mode. On left side the results for natural ventilation only in time of teaching (from 7.30 a.m. till 6.20 p.m.) are presented. The graph on right side displays the results for 24 hours natural ventilation (night precooling).

In case of natural ventilation only in time of teaching the operative temperature achieves max values 31–32°C in room with lightweight construction (low accumulated effect). If we apply heavyweight construction (high accumulated effect) the max operative temperature decreases at value 27–28°C, the drop by to 4 K. If we want to use the night ventilation for room precooling (24 hours ventilation), the impact on operative temperature trend during teaching is only insignificant in case of lightweight construction (max operative temperature is 30–31°C). The more effective is the combination of heavyweight construction and night ventilation (room precooling) where the max operative temperature decreases at value 24–25°C (drop by to 6 K).

In case of natural ventilation there is under consideration that 25% of the external windows area is open when the air temperature in classroom is higher than outside air temperature. The ventilation rate (Total fresh air [ac/h]) is calculated based on the pressure difference across the opening calculated from wind and stack pressure effects. The resultant ventilation rate for natural ventilation during teaching time and for 24 hours natural ventilation are shown in the Figure 4.

3.2 Thermal mass of building construction and mechanical ventilation

Another two graphs below (Figure 5) show simulation results for mechanical ventilation modes. On left side the results for mechanical ventilation only in time of teaching (from 7.30

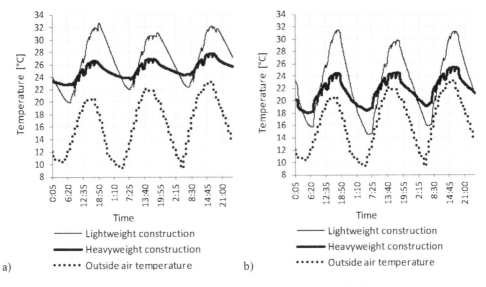

Figure 3.   Operative temperature profiles in the analysed classroom (May 13–15) for natural ventilation modes, a) ventilation only in time of teaching, b) 24 hours ventilation.

474

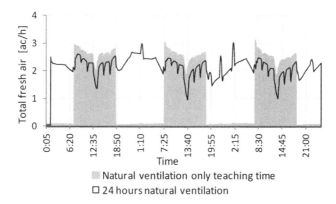

Figure 4.   The resultant ventilation rate in the analysed classroom (May 13–15) for natural ventilation modes.

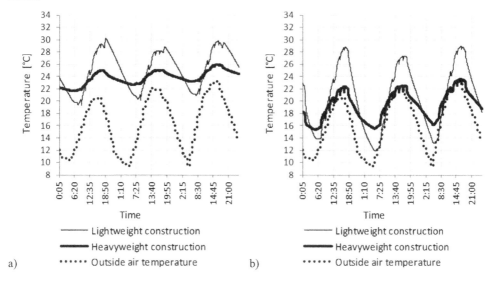

a)   b)

Figure 5.   Operative temperature profiles in the analysed classroom (May 13–15) for mechanical ventilation modes, a) ventilation only in time of teaching, b) 24 hours ventilation.

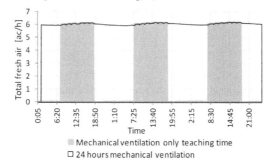

Figure 6.   The resultant ventilation rate in the analysed classroom (May 13–15) for mechanical ventilation modes.

a.m. till 6.20 p.m.) are presented. The graph on right side displays the results for 24 hours mechanical ventilation (night precooling).

In case of mechanical ventilation only in time of teaching the operative temperature achieves max values 28–30°C in room with lightweight construction (low accumulated effect). If we apply heavyweight construction (high accumulated effect) the max operative temperature decreases at value 25–26°C, the drop by to 3–4 K. The combination of night mechanical ventilation for room precooling (24 hours ventilation) and lightweight construction is not very effective (max operative temperature 27–29°C). The results are similar as in case of natural ventilation mode. The more effective is the combination of heavyweight construction and night ventilation (room precooling) where the max operative temperature decreases at value 22–24°C (drop by to 6 K). By application of mechanical ventilation we can keep the required air change in room. In our case it is the air change at value 6.0 ac/h (Figure 6). This is in accordance with standard requirements STN EN 15251 for type of space and category I.

4 CONCLUSIONS

The aim of energy simulation was to evaluate the accumulated effect of building construction on operative temperature trends in analysed room by using the natural or mechanical ventilation.

The results from energy simulation show positive impact of heavyweight construction on max operative temperature in room in comparison with lightweight construction. In analysed room there was possible to achieve the decrease of max operative temperature about 4–6 K in case of natural ventilation and the decrease about 3–5 K in case of mechanical ventilation. In order to eliminate of heat load in analysed room during summer period the most effective is the combination of heavyweight construction with 24 hours ventilation for room precooling.

Providing a natural overnight ventilation has encountered with difficulties. It is needed to leave windows open during night what is non-permissible in terms of security. Mechanical ventilation has more benefits. By its application we can keep the required air change in room. In our case it is the air change at value 6.0 ac/h what is 12 l/s per person. From analysis results is clear that the max operative temperature in room can drop 1–2 K if we double air change rate from 3.0 to 6.0 ac/h.

ACKNOWLEDGEMENT

This paper is the result of the projects implementation: University Science Park TECH-NICOM for Innovation Applications Supported by Knowledge Technology, ITMS: 26220220182, supported by the Research & Development Operational Programme funded by the ERDF.

REFERENCES

Balaras, C.A. 1995. The role of thermal mass on the cooling load of buildings. An overview of computational methods. *Energy and Buildings* 24(1996): 1–10.
Clark, J. 2001. *Energy Simulation in Building Design*. Oxford: Butterworth-Heinemann Ltd.
STN 73 0540-3. 2013. *Thermal protection of buildings. Thermal performance of buildings and components. Part 3: Properties of environments and building products*. Bratislava: TSUS Bratislava.
STN EN 15251. 2007. *Indoor environmental input parameters for design and assessment of energy performance of buildings—adressing indoor air quality, thermal environment, lighting and acoustics*, Bratislava, SÚTN Bratislava.
Workie, D.M. 2015. *Thermal Mass Modelling in Whole Building Simulation Tool, for Model Predictive Control of Peak Demand Limiting and Peak Load Shifting*. Glasgow: University of Strathclyde Engineering.
Yam, J. at al. 2003. Nonlinear coupling between thermal mass and natural ventilation in buildings. *International Journal of Heat and Mass Transfer* 46(2003): 1251–1264.

Advances and Trends in Engineering Sciences and Technologies II – Al Ali & Platko (Eds)
© 2017 Taylor & Francis Group, ISBN 978-1-138-03224-8

Effect of admixtures on probabilistic indexes of concrete quality with consideration of environmental impact

S. Koval, M.J. Ciak & N. Ciak
University of Warmia and Mazury in Olsztyn, Olsztyn, Poland

ABSTRACT: The effect of admixtures was evaluated based on the average values of concrete parameters (e.g. strength) and numeric distribution characteristics of these parameters, including probabilistic indexes used for finding guaranteed material properties. In the described study, the empirical distributions of parameters were characterized by parametric values, including a mean variability index and asymmetry index as well as with non-parametric indexes (e.g. quantile) which are determined based on repeated experiments. It has been demonstrated that the technical measures taken in order to increase the mean values may differ from the measures aiming at improvement of performance, with the more the distribution deviates from normal, the more profound the difference is.

1 INTRODUCTION

One of the main directions defining the progress of concrete at the present stage is the use of multifunctional admixtures—modifiers—which make it possible to solve complex issues related to the quality and durability of constructions. Effective shaping of the structure and obtaining concrete of desired properties and increased durability is possible due to the application of multi-component admixtures.

Increasing requirements with regard to the quality and reliability of constructions (striving to avoid or reduce economic and other types of losses resulting from the implementation of less-than-optimal engineering solutions) are a driving factor favouring the improvement of methodologies for the prediction of the impact of admixtures on the technological and practical properties of concrete (Neville 2011, Koval et al. 2013). The durability of concrete is predicted based on a variety of physiochemical models which make it possible to explain certain aspects of the mechanism of destruction processes. However, using such models in order to establish the optimal composition of a concrete mix is hindered by the need to make various assumptions and accept limitations, to take into account constant changes in the structure of cement paste over time, and the impact of a range of technological and utilization factors. In the case of concrete used in various climatic conditions and technological environments the application of admixtures should be properly substantiated (Neville 2011).

In the quantitative analysis of multifunctional admixtures accompanied by the physiochemical approach, it is also effective to use experimental-statistical (ES) models—multifactorial regression equations calculated from results obtained during the realization of optimal experiment plans (Voznesensky & Lyashenko 1998, Montgomery 2012).

2 PROBABILISTIC INDICES OF EVALUATION OF CONCRETE QUALITY

The criteria for the assessment of the impact of admixtures on the durability of concrete include not only average marks of properties $\bar{Y}$ but also probabilistic indices—Y_α as numerical characteristics of the spread of results which need to be taken into account while designing concrete (Montgomery & Runger 2012).

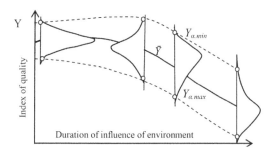

Figure 1. A hypothetical diagram of the impact of influence time of an aggressive environment on the change in the distribution of the quality index.

Experience has shown that as the influence increases, the distribution of results may undergo various transformations (Figure 1), thus reflecting changes in the vulnerability of material to defects (deficiency). In the initial phase of influence exerted by the environment, processes related to the shaping of the structure (nucleation) may continue in one elementary space while irreversible processes may be accumulated in another, which, with the mean value of $\bar{Y}$ remaining unchanged, will result in increased dispersion (spread) of the quality index.

While the theoretical basis of distribution probability is in most cases unknown, the probabilistic indices of Y_α are determined based on numerous repetitions of the experiment or by using computer statistical methods to generate parameters of properties (Voznesensky & Lyashenko 1998, Montgomery 2012). In these tests the values of Y_α were established on the basis of the results of experiments in which the number of parallel tests (repetitions) was high enough ($N \geq 50$). Based on ES-models describing the impact of factors X_i on mean values $\bar{Y}$ and probabilistic indices Y_α, guaranteed technological solutions were accepted both for quality indices and for formulation- and technology-related factors. In such a case it is possible not only to assess the contribution of individual admixtures to the increased reliability of concrete, but also to enhance the knowledge of manners in which to shape the distribution of its properties during utilization (Voznesensky & Lyashenko 1998, Koval 1997).

3 EVALUATION OF ADDITIVE INFLUENCE ON PROBABILITY INDEXES OF CONCRETE QUALITY DURING DESTRUCTION ENVIRONMENT EXPOSURE

Histograms presented in figure 3 reflect changes in distribution of tensile strength with bending f_i for two fine-grained concretes (C/P = 1/2.5; W/C = 0.48)—without admixtures (curve 1) and modified with a complex additive "superplasticizer + microsilica" (SP+Mk) (curve 2).

The test analysed at least 150 concrete samples undergoing cyclic wetting in a NaCl solution and drying. The empirical distributions are described using the following parametric numeral characteristics: $\bar{f}$—mean value, MPa; $v\{f\}$—the coefficient of variation,%; A^*—asymmetry; E^*—kurtosis and other factors, as well as non-parametric indices as a part of the distribution, differ from the normal one. This includes in particular quantile f_{05} of the minimal acceptable strength (confidence interval $\alpha = 0.05$) with "relative boundary" $\gamma_\alpha = 1 - \bar{f}/f_\alpha$ (Lyashenko & Voznesensky 1998), which were used to assess the critical values (in the tail of the distribution) of empirical distributions.

Comparative analysis of probabilistic indices of two types of concrete before testing shows, in the desired confidence interval of strength, an increased guarantee of concrete modified with the same mean values of f. After 600 cycles the average strength of concrete with complex modifier "SP+Mk" slightly changed, and the strength of concrete without admixtures even increased slightly.

However, the "dispersion" of distribution (Figure 3, Table 1) indicates not only a strengthening of the least defective elements of the structure, but also an increased tendency to degrade the most defective, "weak" microareas. A significant increase in the spread of the f_i

478

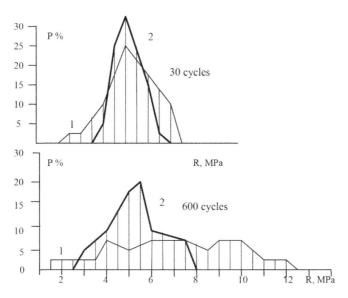

Figure 3. Changes in the characteristics of distribution f_i of two fine-grained concretes as influenced by the environment (1—concrete without admixtures; 2—concrete with admixtures).

Table 1. Statistical parameters of concrete strength after specific number of cycles.

| Cycles | Concrete | Distribution characteristics f_α | | | | | |
		f_{tm} MPa	v %	A^*	E^*	f_{05} MPa	γ_5 %
30	"1"	4.9	20	+0.4	−0.4	3.3	33
	"2"	4.9	14	−0.5	±0.0	3.9	21
300	"1"	9.9	22	+1.0	+0.5	6.3	37
	"2"	5.5	22	+0.4	−0.6	3.7	33
600	"1"	7.0	33	±0.0	−0.8	2.9	59
	"2"	4.6	26	±0.0	−0.7	2.5	45

value (judging by variance $v\{f_i\}$), the lengthening of the critical distribution tail (judging by the value of coefficient γ_α) and its increased asymmetry (based on values A^*), as compared to the original state, indicate quick structural changes in concrete without admixtures.

The use of admixtures decreases the mark spread as the so-called "worse" elements of modified concrete influenced by the environment are degraded less intensively than in the case of concrete without admixtures.

4 MODELING THE INFLUENCE OF ADDITIVES DURING CYCLES MOISTENING AND DRUING

4.1 *Probabilistic strength indices*

In order to assess the possibility of controlling probabilistic indices in tests with fine-grained concrete from mixtures of identical workability (C/P = 1/3) according to plan B_2, the content of aerating admixture (DN = X_1 = 0.02 ± 0.02%) and naphthalene-formaldehyde superplasticizer (SP = X_2 = 0,3 ± 0.3%) was changed. In point $x_1 = x_2 = -1$ of the schedule concrete without admixtures was tested. In other points of the schedule information on the action of particular admixtures was collected. Destruction of concrete was caused by repeated moistening with water and drying (t = 70°C).

479

A two-factor ES-model (1) in the form of an incomplete fourth-degree polynomial describes the effect of admixtures on the average strength at bending f_{tm} of fine-grained concrete after 28 days of "regular" hardening:

$$f_{tm} = 3,86 - 1.28x_1 + 0.55x_2 + 1.13x_1^2 + 1.02x_2^2 + 0.91x_1^2x_2^2 + 1.16x_1x_2^2 \qquad (1)$$

An experimental and statistic model contains only significant amounts of indices; with $s = 0.10$ MPa and $t_{0.05} = 1.645$ the values of coefficients b_{12} and b_{112} are minute. The mean strength of $\bar{f}_{tm}$ along the vector on the graph (Figure 4a) changes 1.8 times and reaches a value of 6.5 MPa with superplasticizer added in the amount of 0.42% of the cement mass.

Judging by the change of probabilistic parameters (v, A^*, E^* and γ_{05}), a complex admixture not only contributes to a change of the properties' assessment, but also causes changes in the form of the distribution curve of the results, wherein, as the analysis of the models shows, the distribution may differ from the normal one in a significant part of the factorial space. Controlling the amount of superplasticizer leads to the decrease of the coefficient of variation to 7.3% (Figure 4b); with optimal combinations of X_1 and X_2 in the factorial space a zone of increased homogeneity f is created.

The index of relative boundary of acceptable strength γ_{05} indicates that destructive processes influence first of all the reference concrete ($X_1 = X_2 = 0\%$). Critical distribution values of $\gamma_{05}\{f_{tm}\}$ (Figure 4c) are mainly the function of controlling the amount of superplasticizer. However, after 300 cycles the impact of admixtures changes significantly (Figure 4d) and $\gamma_{05}\{f_i\}$ is minimized only by adding a complex admixture (DN = 0.03%, SP = 0.77%).

4.2 Evaluation of additives influence on water absorption

Activities aimed at increasing the resistance of concrete in liquid environments require knowledge and information on the impact of modifiers on capillary and porous structures,

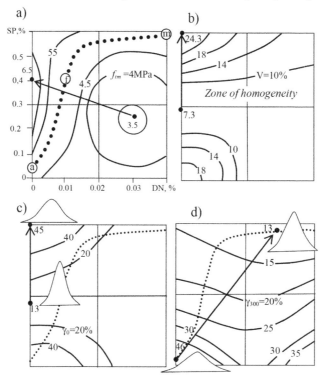

Figure 4. Impact of admixtures on the initial average strength f_i (a), its variability index (b) and "relative boundary" γ_{05} before testing (c) and after 300 cycles of tests (d).

determined particularly during testing water absorption parameters W with concrete. A reasonably composed admixture helps obtain more homogeneous pores which actually participate in the absorption process (Neville 2011).

An analysis of experimental and statistical models showed a substantial impact of the modifier, not only on average values W_m, but also on probabilistic indices W_α, which may characterize the homogeneity of pores' structure. Increasing homogeneity translates into a 2.4-time decrease of standard deviation $s\{W\}$, a 2.5-times decrease of the coefficient of variation $v\{W\}$, limit water absorption (a 95% probability)—quintile W_{95} and "critical" positive asymmetry of distribution A^* and elimination of a significant pointed kurtosis E^*.

The assessment of the impact of admixtures on probabilistic indices of absorption was performed in accordance with the method of isoparametric analysis (Voznesensky & Lyashenko 1998.). Going along contour line $f_i = 5 \pm 0.08$ MPa (marked on Figure 4a-d with dots) probabilistic properties are calculated for many types of concrete which have the same strength but vary in the composition of the admixture. The data in the table 2 show that for composition "f" (with an average amount of admixture) as compared to composition "a" (without admixture), the variation of strength $v\{f_i\}$ decreased and the total average absorption of water W within 24 hours and variability index $v\{W\}$ decreased. The structure of modified types of concrete with better statistical parameters can be described as more reliable.

Changes in distribution $p\{W\}$ were assessed on the basis of changes of quintile W_{95} and relative maximum possible limit (with a 95% probability) for water absorption $\gamma_{95}\{W\} = W_{95}/W$, which are characteristic of the lengthening of critical distribution tails as a result of local

Table 2. Characteristics of different types of concrete before exposure to the influence of the environment.

Characteristics of compositions and statistical indices	Composition index		
	a	f	m
Water and cement index C/W	0.57	0.46	0.40
Aeration, %	8.00	14.00	21.00
Dosing DN, %	0.00	0.012	0.04
Dosing SP, %	0.04	0.46	0.54
Deviation of strength at bending, f_c, %	23.00	12.50	13.40
Average strength at compression, f_c, MPa	20.50	16.70	16.00
Deviation of strength at compression, f_c, %	13.50	13.00	20.80
Average water absorption, W, %	8.50	7.80	6.90
Deviation of absorption	13,50	5.80	7.80

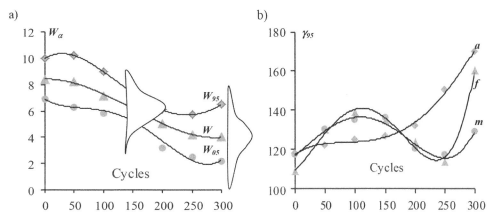

Figure 5. Change of absorption quintiles W_α of concrete "f" (a) and a relative limit of critical absorption $\gamma_{95}\{W\}$ for three types of concrete of similar strength (b).

defects (large pores, losses, fistulas, etc.). This is particularly significant in the case of thin-walled structures.

For types of concrete of similar strengths assessments of structure in homogeneity on the basis of changes W_α (Figure 5a) and limit $\gamma_{95}\{W\}$ (Figure 5b) may change with an increasing number of cycles, according to the third-degree curve: the stage of "adjustment" of the structure to the influence of the external environment where the dispersion (scatter) of indices may increase, the stage of increasing homogeneity and the stage of the "loosening" of the structure or its fatigue. Between 200 and 300 of cycles degradation processes begin to prevail, which is reflected in the increase of average value $\bar{W}$ and in the widening of interval $W_{95} \div W_{05}$ accompanied by an increase in critical spread tails.

After 300 cycles of saturation and drying of concrete, the index of $\gamma_{95}\{W\}$ increased between 5 and 9 times (depending on the composition of the complex admixture) as compared to the original state, which indicates a decrease in the homogeneity of the structure and, consequently, the development of defects and an increased probability of local defects and damage. Nevertheless, the confidence interval of absorption decreases 1.5–2 times with the introduction of admixture composed of **SP** (0.06%) and **DN** (0.03–0.035%), which allows the obtaining of concrete of a guaranteed level of quality in use.

5 CONCLUSIONS

ES-modelling of probabilistic indices makes it possible to establish areas of recipe and technological solutions, thanks to which sufficient reliability of the material in construction can be provided, naturally taking into account the acceptable level of risk. The testing proved that reasonable control of the composition of complex modifiers makes it possible to improve probabilistic structural and mechanical indices of properties of concrete, which helps increase its reliability in use.

REFERENCES

Koval S. 1997. Zur Auswahl optimaler Rezepturen bei der Steuerung rheologischer Eigenschaften von Morteln. *Proc. 13 Int. Baustofftagung (Ibausil)*. Weimar Bauhaus-Univer (T. 2): 489–494.
Koval S. 2012. Design and optimization of composition and properties of the modified concrete, Odessa: Astroprint.
Koval, S., Ciak M.J. & Sitarski M. 2013. Searching of self-compacting concrete composition with the statistical models. *Cement wapno beton* 4: 197–206.
Lyashenko T. & Voznesensky V. 1998. Modeling the influence of composition on probabilistic indices of building polymer composites quality. *Proc. Int. Congr. on Polymer in Concrete*. Bologna: 201–208.
Montgomery D.C. & Runger G.C. 2012. *Applied Statistics and Probability for Engineers*. 5th edition, New York: Wiley & Sons.
Montgomery, D.C. 2012. *Design and analysis of experiments*. 8th edition. New York: Wiley & Sons.
Neville, A.M. 2011. *Properties of concrete*. 5th ed. Harlow: Pearson Education.
Voznesensky, V.A. & Lyashenko T.V. 1998. *Experimental-statistical modeling in computational materials science*. Odessa: Astroprint.

Advances and Trends in Engineering Sciences and Technologies II – Al Ali & Platko (Eds)
© 2017 Taylor & Francis Group, ISBN 978-1-138-03224-8

The possible areas to use 3D printers in building constructions

B. Kovářová
Faculty of Civil Engineering, Institute of Technology, Mechanization and Construction Management,
Brno University of Technology, Brno, Czech Republic

ABSTRACT: The article deals with the history of 3D printing and its possible use in the design of civil construction overall. The author of the article will try to clarify the current status and the level of use in a specific of civil constructions. In the article are shown examples of current application and prepared projects for near future especially from abroad

1 INTRODUCTION

Projects taking advantage of digitization and computing are often criticized for their frivolous or indulgent nature. On the other hand, there has been an emergence of work that exemplifies the most optimistic of this "Third Industrial Revolution"—an architecture that appropriates new technology and computation for the collective good of our cities and people (Oh, 2016).

2 EVOLUTION TO 3D PRINT IN BUILDING CONSTRUCTIONS

Computer-Aided Design (CAD) is the use of computer systems to aid in the creation, modification, analysis, or optimization of a design. CAD software is used to increase the productivity of the designer, improve the quality of design, improve communications through documentation, and to create a database for manufacturing (Horák, 2016).

CAD is an important industrial art extensively used in many applications, including automotive, shipbuilding, and aerospace industries, industrial and architectural design, prosthetics, and many more. Because of its enormous economic importance, CAD has been a major driving force for research in computational geometry, computer graphics (both hardware and software), and discrete differential geometry. CAD is just another example of the pervasive effect computers were beginning to have on industry. Current computer-aided design software packages range from 2D vector-based drafting systems to 3D solid and surface modelers. Modern CAD packages can also frequently allow rotations in three dimensions, allowing viewing of a designed object from any desired angle, even from the inside looking out.

CAD and support of IT is a prerequisite for any 3D print application. CAD systems can be divided into general CAD systems (2D or 3D—surface and volume) and specialized. In the area of construction and architecture are especially: AEC (Architecture-Engineering-Construction), BIM (Building Information Model), CAAD (Computer-aided architectural design). BIM technology is increasingly used when designing (Venkrbec, 2016). The designer who decides to switch from designing in general CAD system for the use of BIM systems will initially deal with plotting elements, e.g. the walls. They will be suddenly faced with the decision on whether to draw the wall as usual, i.e. the compositional dimensions, which do not show plaster. Or individual layers to include layers of walls to allow more accurate reporting of the bill of quantities, which, using existing principle of drawing, cannot achieve. For basic CAD systems it is based on the principle of graphical representation of elements by drawing substitution tags where it was not necessary to address this issue. Walls were represented by only two lines, without any relationship and were placed at a distance corresponding to modular dimensions. When creating the design documentation in BIM, it leads to the involvement of other parameters (price, quality, other properties …), so that the model can be used for further processing. The purpose for which this documentation is determined must be therefore carefully considered, and then the level of detail of the model adjusted accordingly. If we compare the way of working in CAD systems, it is as the use of electronic drawing board. When using BIM you can create a complex model that you can use later within life-cycle of the building.

3   BASIC PRINCIPLES OF 3D PRINTING IN BUILDING INDUSTRY

We define 3D printing or additive manufacturing as a process of making three dimensional solid objects from a digital file or digital source in general. The creation of a 3D printed object is achieved using additive processes. In an additive process an object is created by laying down successive layers of material until the entire object is created. Each of these layers can be seen as a thinly sliced horizontal cross-section of the eventual object.

Not all 3D printers use the same technology. There are several ways to print and all those available are additive, differing mainly in the way layers are built to create the final object. Some methods use melting or softening material to produce the layers. Selective Laser Sintering (SLS) and Fused Deposition Modeling (FDM) are the most common technologies using this method of printing (Figure 1). To be more precise: since 2010, the American Society for Testing and Materials (ASTM) group "ASTM F42—Additive Manufacturing", developed a set of standards that classify the Additive Manufacturing processes into 7 categories according to Standard Terminology for Additive Manufacturing Technologies. These seven processes are: Vat Photopolymerisation, Material Jetting, Binder Jetting, Material Extrusion, Powder Bed Fusion, Sheet Lamination, and Directed Energy Deposition. In building industry the most used is Material Extrusion.

We can see an analogy with the Superadobe system (Kovářová, 2012) an industrial way of implementation. Wall, which is built by layering of potato bags, which are filled with mixture of soil and straw. Bags are connected by barbed wire which is also reinforcement for the wall. This barbed wire is freely laid between bags. The walls being constructed with these bags is 50 cm thick. Openings for windows and doors are prepared during the building process. Instead of windows temporary wooden fillings are placed into openings. These wooden frames are later substituted for windows and doors (Figure 2).

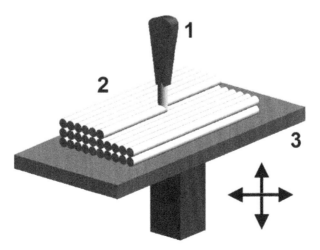

Figure 1.   Fused deposition modelling, 1—nozzle ejecting material, 2—deposited material, 3—controlled table (White, Thompson, 2013).

Figure 2. Superadobe technology—example of implementation (Forumhouse, 2015).

Figure 3. 3D printing of church—Material extrusion technology (Alec, 2015).

In building construction is used in two directions movable nozzle because the table (unlike applications in other industries) is fixed. For 3D print application, there has to be prepared flat area for printing building itself. This area has to be extended for possibility to assembly 3D printer on construction site and is for printing itself. From a logistic point of view it is important to have the connection to a source of material. As material is used concrete based mixture. Print speed/construction depends on the planar size, the layer thickness and consistency of the mixture, since it does not use shuttering. Compared with the traditional construction of reinforced concrete we can see an analogy—you can print directly on site (reinforced concrete), or assemble pre-printed components (prefabricated concrete). Pre-printed parts

Figure 4.   Building based on pre-printed parts—example (Bakir, 2014).

can be produced directly on the site (site-cast units), or in central factories (Figure 4). Both types of print outs are the result of computer-aided design, mostly based on a BIM system.

When printing directly on the construction site, the building creates printing consecutive layers which resemble a technological analogy to "classical" construction based on bricks. In contrast, when using pre-printed components, it leads to their assembly into the finished work.

Components can be assembled by stacking, or may be used in part shaped "sections" of the object (Figure 4). Both options can of course be combined. The actual construction should use all possibilities of BIM, e.g. simultaneous incorporation of piping and cabling, which increases the efficiency and reduces the price demands of finishing work on the subject.

4   FUTURE OF 3D PRINTING IN BUILDING INDUSTRY

3D printers gradually penetrate into all industries. This is true also for the area of construction. Using a 3D printer is an increase in productivity, reduces waste due to improved accuracy of production, even if the diversity of requirements—3D printing is easing requirements on modular coordination.

In comparison with the "classical" methods of construction use of 3D printing is another alternative in technology. Outputs of 3D printing show improved mechanical and thermal properties compared to the current monolithic and prefabricated structures. Another highly positive benefit is less impact on the environment due to reduction in waste and lower environmental demands during the construction process.

Netherlands within the EU supports the most advanced 3D technology. Let the occasion of the EU Presidency to propose a mobile pavilion Building Europe for meetings with politicians' entrance facade from the studio DUS Architects. Attractive shape concept enabled 3D printed parts made from fully recyclable plastic (Figure 5).

Of course we should mention the disadvantages of this method of construction. The main disadvantage is current material base of paste for printing. The technology is relatively new, materials are subject to research and development. Another disadvantage of printing on the construction site is relatively complicated installing of very large printer on site to achieve the required accuracy of printing, thus resulting accuracy of the works.

Figure 5. Building with implementation of pre-printed facade parts (Scott, 2016).

Figure 6. Bridge over the water in the center of Amsterdam—plans (Molitch-Hou, 2015).

It could be use metal as "printing" material as well. Printing an intricate, ornate metal bridge for a special location is the ultimate test for robots and software, engineers, craftsmen and designers. The bridge by designer Joris Laarman will be ready in 2017. The design process using new Autodesk software (AutoCAD Civil 3D) is a research itself, synchronized with the technical development and taking into account the location. The project is a collaboration between MX3D, software giant Autodesk, construction company Heijmans and many others (Figure 6).

3D printing is not limited only to Earth. Setting up a lunar base could be made much simpler by using a 3D printer to build it from local materials. Foster + Partners devised a weight-bearing 'catenary' dome design with a cellular structured wall to shield against micrometeoroids and space radiation, incorporating a pressurized inflatable to shelter astronauts. 3D printing offers a potential means of facilitating lunar settlement with reduced logistics from Earth. The new possibilities this work opens up can then be considered by international space agencies as part of the current development of a common exploration strategy. The UK's Monolite supplied the D-Shape printer, with a mobile printing array of nozzles on a 6 m frame to spray a binding solution onto a sand-like building material. First, here is needed to mix the simulated lunar material with magnesium oxide. Then for structural 'ink' is applied a binding salt which converts material to a stone-like solid (ESA, 2013).

REFERENCES

Alec, 2015. Andrey Rudenko plans to 3D print a full sized fantasy-style concrete village. Available at: http://www.3ders.org/articles/20150129-andrey-rudenko-plans-to-3d-print-a-full-sized-fantasy-style-concrete-village.html

Bakir, D., 2014. Dieses Haus kommt aus dem 3D-Drucker. Available at: http://www.stern.de/wirtschaft/immobilien/hausbau-in-24-stunden-dieses-haus-kommt-aus-dem-3d-drucker-3686824.html

European Space Agency (ESA), 2013. Building a lunar base with 3D printing. Available at http://www.esa.int/Our_Activities/Space_Engineering_Technology/Building_a_lunar_base_with_3D_printing

Forumhouse, 2015. Available at: https://www.forumhouse.ru/articles/house/5985

Horák, J., 2016. Projektování TZB v Revitu: Problémy a úskalí. Available at http://www.cad.cz/stavebnictvi/79-stavebnictvi/6894-projektovani-tzb-v-revitu-problemy-a-uskali.html

Kovářová, B., 2012. Buildings built by Superadobe technology and other possibilities use of clay plaster. Available at http://www.scientific.net/AMR.649.227

Molitch-Hou, M., 2015. Construction of World's 1st 3D Printed Bridge Begins in Amsterdam. Available at: https://3dprintingindustry.com/news/construction-of-worlds-1st-3d-printed-bridge-begins-in-amsterdam-60110/

Oh, E., 2016. 7 Futuristic Fabrications Leading Us Towards a Newer Architecture. Available at http://www.archdaily.com

Scott, C., 2016. Netherlands Leads the Council of the European Union from a 3D Printed, Recyclable Building. Available at: https://3dprint.com/114533/netherlands-europe-house/6/

Scott, R., 2014. Chinese Company Showcases Ten 3D-Printed Houses. Available at http://www.archdaily.com/543518/chinese-company-showcases-ten-3d-printed-houses

Scott, R., 2015. Chinese Company Constructs the World's Tallest 3D Printed Building. Available at http://www.archdaily.com/591331/chinese-company-creates-the-world-s-tallest-3d-printed-building

Venkrbec, V., 2016. Optimization of Re-Mixing Recyclated Concrete Aggregates. In *Applied Mechanics and Materials*. Switzerland: Trans Tech Publications, pp. 116–123. Available at: http://www.scientific.net/AMM.824.116.

Winsun, 2011. Company web pages. Available at http://www.yhbm.com

White, C. & Thompson, A., 2013. How the uPrint Rapidprotyping Machine Works. Available at: http://www.newhaven.edu/518887.pdf

Advances and Trends in Engineering Sciences and Technologies II – Al Ali & Platko (Eds)
© 2017 Taylor & Francis Group, ISBN 978-1-138-03224-8

Knowledge database of modern methods of construction

M. Kozlovska, D. Mackova & M. Spisakova
Faculty of Civil Engineering, Technical University of Kosice, Kosice, Slovakia

ABSTRACT: Construction industry is under great pressure to increase construction efficiency. Currently, all over the world, there are increasingly widespread innovative methods of construction also called Modern Methods of Construction (MMC). MMC are synonymous with off-site manufacturing and prefabrication of building components and modules in factory settings, including complete buildings. The modern method of construction has a great potential to improve the efficiency of construction production, quality, customer satisfaction, environmental impact, sustainability and predictability of construction design delivery in particular terms. MMC is focused on looking at the possibility of improving the performance of construction. The aim of this paper is to form a structure of knowledge database of modern methods of construction as a tool to aid in decision-making processes of investors. The result of research is data collection aimed to MMC material base, description of MMC systems and their components, which provides information for processing of compendium of MMC in Slovakia and ICT tool for selection of MMC by investors in Slovakia.

1 INTRODUCTION

The priority themes of construction not only in Europe but also globally include sustainable construction and integrated design and delivery solutions of each project. One of the key assumptions for the solution of sustainable construction and integrated design and delivery solutions is research focused on more effective methods of design and realization of construction projects used in Modern Methods of Construction (MMC) and increasing the performance as a sustainable construction, as well as integrated design and delivery solutions. Authors Chen et al. (2010) argue that MMC in the construction industry has enhanced productivity and improved quality as well as several benefits as shortened construction time, lowered overall construction cost, improved quality, enhanced durability, improved architectural appearance, enhanced occupational health and safety, material conservation, reduced construction site waste, reduced environmental emissions, and energy and water consumption. MMC are about better products and processes. They aim to improve business efficiency, quality customer satisfaction, environmental performance, sustainability and the predictability of delivery timescales. MMC are, therefore, more broadly based than a particular focus on product. They engage people and process to seek improvement in the delivery and performance of construction. The authors (Burwood & Jess, 2005) defined MMC as those which provide an efficient product management process to provide more products of better quality in less time. It can be classified in various ways and may involve key services (e.g.) plumbing, key items (e.g. foundations) inner shell (walls etc), external walls, or any combination of these elements. It can also be classified by material (timber, steel, concrete and masonry). According to Warszawski (1999) MMC are defined as a set of elements or component which are inter-related towards helping the implementation of construction works activities. He also expounded that MMC are an investment in equipment, facilities, and technology with the objective of maximizing production output, minimizing labour resources, and improving quality. Trikha (1999) defined MMC as a system in which concrete components prefabricated at site or in factory are assembly to form the structure with minimum in situ construction.

Many authors mentioned various classifications of modern methods of construction. Warszawski (1999) classified MMC according to geometrical configuration of their main framing components which are (i) linear or skeleton (beam and column system), (ii) planer of panel system or (iii) dimensional and box system.

Housing Corporation (2000) divides MMC into construction (i) volumetric, (ii) panellised, (iii) hybrid, (iv) sub-assemblies and components (v) inovative traditional methods of construction. In Malaysia, CIDB (2003) classified MMC into five categories which are (i) pre-cast concrete framing panel and box system, (ii) steel formwork systems, (iii) steel frame system, (iv) timber frame system and (v) block work system.

According the study "Current practices and future potential in modern methods of construction" processed in the UK (2007), MMC can be divided into following categories (i) volumetric modules, (ii) wood or steel frame constructions, (iii) kitchen and bathroom pods, (iv) composite insulated non load-bearing or load-bearing sandwich panels, (v) light steel frame systems, (vi) prefabricated panels, (vii) prefabricated lightweighted ceiling and roof panels, (viii) structural insulated panels, (ix) prefabricated cladding systems, (x) insulated concrete formworks.

According to many authors modern methods of construction have a great potential to improve the efficiency of construction production, quality, customer satisfaction, environmental impact, sustainability and predictability of construction design delivery in particular terms.

Adopting MMC has the following main benefits to the practitioner when compared to the conventional construction method (Warszawski, 1999, Barlow, 2000, Pasquire & Connoly, 2002, Thanoon, 2003, Engstrom, 2009, Baldwin et al., 2009, Lovell & Smith, 2010). However, despite these undisputed benefits some stakeholders think, that MMC has not yet had the impact they expected, or hoped for and they are little bit skeptical on using MMC (Engstrom, 2009). The same authors who described benefits of using MMC, pointed also to barriers that prevent the spread of MMC in many countries. Summary of MMC benefits and barriers is shown in Table1.

The MMC are increasingly becoming more popular because of their undisputed benefits in technological (Hulinova, 2014, Gasparik & Gasparik, 2012), economical (Mesaros & Mandicak, 2015), environmental (Kozlovska et al., 2014, Lesniak & Zima, 2015) and social areas and gradually find their place in the market in the Slovak republic, where are primarily used in residential buildings.

Figure 1.   a) Volumetric modules (www.pcko.co.uk), b) Panellised construction (hhcontractor.com).

Figure 3.   Insulated concrete formwork (beaver-vuconstruction.com).

Table 1. Summary of MMC benefits and barriers.

Benefits	Minimization/reduction of time Minimization on-site operations Reduced cost Increase quality	Fewer people on-site Less environmental impact Improved safety on-site Reduced defects (better control)
Barriers	Higher cost compared to the traditional methods Client resistance Negative image of prefabrication	Not locally available Insufficient worker skills Inflexible for design changes

Each country has its own specifics and investors have different decision factors for building type selection. The Institute of Construction Technology and Management at the Faculty of Civil Engineering, Technical University in Kosice have conducted research which was conducted questionnaire survey in Slovakia in the second half 2015 with the purpose of identification of decision factors order. By e-mail were interviewed 884 respondents from all regions of Slovakia and 213 returned with a response rate 24%. Respondents (not experts in construction, as a potential investor) were asked to determine the order of selected factors (type of construction system, complexity of construction system, material base and construction method) for the residential buildings. The results of this survey pointed to the fact that for the majority of respondents (41.8%) was the most imported factor "material base". Because of this fact, the aim of paper is to design the knowledge classification of modern methods of construction through MMC database based on construction material decomposition.

2 MATERIAL AND METHODOLOGY

The aim of paper was a creation of knowledge database structure for modern methods of construction as a tool to aid in decision-making processes of investors. The research material presents information about the MMC focused mainly on:

– material platform of MMC,
– MMC system,
– components of MMC systems.

Basic division can be understood as three decomposition levels of MMC classification. This information presents inputs into the forthcoming decision-making model for selection of suitable MMC according the requirements of potential investors. Besides these three basic types of information, the knowledge system is extended by construction composition of MMC components, benefits and barriers of MMC related to design, production and transport of MMC, construction process and use of building, construction limits of MMC, machine requirements.

Data collection was processing through analysis of available knowledge about various Slovak and foreign MMC. Data concerning to investment costs, costs for energy and construction time are also important for future investors in their decision-making processes. Therefore these items were calculated based on the analysis of particular MMC. Investment costs and costs for energy were expressed per unit of measurement—euro per 1 cubic meter of building volume. The costs were estimated through the detailed quantity take off processing and assigned the unit costs for all items in software Cenkros Plus (the most widely used software for a cost estimation and management of building production in Slovakia) and construction time was processed in programme MS Projects.

3 RESULTS AND DISCUSSION

Each MMC has been assigned by the number within numbering of MMC classification. The numbering is divided into 3 levels with serial number and has a form:

$$m - s - c_1 c_2 c_3 c_4 - n \qquad (1)$$

where $m = $ 1st level of numbering; $s = $ 2nd level of numbering; $c_1 c_2 c_3 c_4 = $ 3rd level of numbering; and $N = $ serial number of MMC

According the results of questionnaire survey, the investors identified the material base of load-bearing construction as a key decision factor for selection of MMC. Considering this, the first level of numbering, as well as first level of classification of MMC, is material base of load-bearing construction. The first numbering figure "m" indicates the relevant material base of MMC, the second numbering figure "s" indicates the MMC system and the third numbering figure "$c_1 c_2 c_3 c_4$" indicates the existence of particular component in MMC system (Table 2). Number "1" on particular c-th figure means that the given MMC system has that component; and number "0" on particular c-th figure means that the given MMC system does not have that component. The last numbering figure "n" specifies the serial number of MMC in particular type of MMC. For instance, the code of MMC—Structural Insulated Panels (SIPs) is 3-3-1101-1, ie. the material base of load bearing construction is wood, it is panelised MMC system containing horizontal and vertical load-bearing components and roof components, the serial number 1.

Determination of material base was processed according the analysis of MMC material composition. The proposal of MMC system and its components classification were adjusted by authors. MMC systems vary considerably, the nuances of which include modular, framed, panelised and volumetric variants which involves modular, framed, panelised and volumetric approaches. For instance, framed systems present load-bearing construction and is closed by panels. Panelised systems are produced in a factory and assembled on-site to produce a three dimensional structure. Conversely, volumetric construction involves the production of three-dimensional modular units in factory conditions prior to transport to site.

The better arrangement of the necessary information has been achieved through "MMC summary", which provides comprehensive information about a particular MMC. Comprehensive information about all available MMC (in MMC summary) will be processed in Compendium of MMC in Slovakia. Besides classification of MMC according to the above mentioned criteria, each MMC was characterized by benefits and barriers in the related to design, production and transport of MMC, construction process and use of building, construction limitation of MMC, machine requirements. The list of these additional criteria was adapted to Slovak construction conditions and Slovak investors' requirements by authors (Table 3).

Specification of additional criteria enhances the knowledge base of MMC and provides important information for design the decision model to select the MMC by investors. Of course, each investor needs to know the investment costs, costs for energy and construction time. Database contains the investment costs, costs for energy and construction time calculated per 1 cubic meter of building volume.

The result of knowledge classification is processing in tabular form according the criteria—material base, MMC system and components of MMC system. Each MMC has assigned his code number in prescribed structure.

Table 2.   Numbering of MMC classification.

Classification of Modern Methods of Construction (MMC)		
1st level—material base of load-bearing construction—m	2nd level—MMC system—s	3rd level—component of MMC system—c
1 - brick	1 - modular	c_1 - horizontal load-bearing components
2 - concrete	2 - framed	c_2 - vertical load-bearing components
3 - wood	3 - panelised	c_3 - staircase
4 - steel	4 - volumetric	c_4 - roof
5 - other		

Table 3. Additional criteria of modern methods of construction.

Benefits	Barriers
Related to design of MMC	
Standardized components	Lack of components variability
Customization	Lack of standards
Availability of components	Standardized designs
Shorter design time	Longer design time
Related to production of MMC	
Standardized production processes	Waste reduction
Minimal material wastage	Higher investment cost of the production line
Controlled production environment	Lack of production skills
Higher product quality	Limited number of producers
Reduction of components failures	
Related to transport of MMC	
Transport within one delivery cycle	Need of larger vehicles
Reduction of vehicles	Need of lifting machinery
	Need of parking area for vehicles on sites
Related to construction process of MMC	
Shorter construction	Need of lifting machinery
Reduction of operations number	Need of assembly area on sites
Increased quality of construction	Larger storage area on sites
Construction process in spatial limitation	
Waste reduction	
Safer construction sites	
Related to use of building	
Lower costs for energy	Potential problems with component maintenance
Longer lifetime of building components	Failure/exchange construction distribution systems
Construction limitations	
Flexibility to change components, variability of shapes and types of components	
Need of storage area on sites, need of assembly area on sites, need of lifting machinery	

4 CONCLUSION

Modern methods of construction should not be perceived as a threat to conventional methods. The modern methods of construction present the latest trends of construction sector to ensure sustainability through pre-fab construction, permanent modular construction, energy efficiency, single-design model and materials. Modern methods of constructions are not new in developed countries in Europe and the USA. Currently, the MMC are increasingly becoming more popular because of their undisputed benefits in technological, economical, environmental and social areas and gradually find their place in the market in Slovakia where they are primarily used in residential buildings.

The increased use of MMC can be achieved by creating tools (knowledge models) for selecting the MMC type according the investors requirements. An initial step for the creation of knowledge model is to collect data aimed at MMC material base, description of MMC systems and their components. Based on this data is a database of MMC available in Slovakia is processed which is arranged by the submitted knowledge classification and numbering according to the mentioned system. Each MMC is characterized by additional information (benefits and barriers related to design, production and transport of MMC, construction process and use of building, construction limitation of MMC, machine requirements) and investment costs, costs for energy and construction time calculated per 1 cubic meter of building volume. The knowledge classification provides information for processing the compendium of MMC in Slovakia and ICT tools for the selection of MMC by investors in Slovakia. The result of paper presents an inputs for development of MMC technology in Slovakia.

ACKNOWLEDGEMENTS

The article presents a partial research result of project VEGA—1/0677/14 "Research of construction efficiency improvement through MMC technologies".

REFERENCES

Baldwin, et al. 2009. Designing out waste in high-rise residential buildings: Analysis of precasting methods and traditional construction, *Renewable Energy* 34: 2067–2073.

Barlow, J. 2000. Innovation and learning in complex offshore construction projects, *Research Policy* 29: 973–989.

Burwood, S. & Jess P. 2005. Modern Methods of Construction. Evolution or Revolution? A BURA Steering and Development Forum Report. Available on http://www.buildicf.co.uk/pdfs/1%20 mmc%20evolution%20or%20revolution%20%20paper.pdf

Chen, Y. et al. 2010. Sustainable performance criteria for construction method selection in concrete buildings. *Automation in Construction* 19: 235–244.

Engstrom, S. et al. 2009. Competitive impact of industrialised building—in search for explanations to the current state. In: *Procs 25th Annual ARCOM Conference*, 7–9 September 2009, Nottingham, UK, Association of Researchers in Construction Management, pp. 413–424.

Engstrom, S. et al. 2009. Competitive impact of industrialised building—in search for explanations to the current state. In: *Procs 25th Annual ARCOM Conference*, UK, Association of Researchers in Construction Management, pp. 413–424.

Gasparik, J. & Gasparik, M. 2012. Automated quality excellence evaluation, *Proceedings of the 29th International Symposium of Automation and Robotics in Construction*, Eindhoven, the Netherlands

Housing Corporation. 2000. NHBC Foundation: A guide to modern methods of construction. Available on <http://www.nhbcfoundation.org/Publications/Guide/A-guide-to-modern-methods-of-construction-NF1>.

Hulinova, Z., 2014. Need of change in the attitude to OS, *Construction, Technology and Management 2014*, Bratislava, Slovakia, pp. 209–217.

IBS Survey. 2003, Construction Industry Development Board Malaysia (CIDB), Kuala Lumpur

Insulated concrete formwork. Available on: http://beaver-vuconstruction.com/wp-content/uploads/2013/11/ICF-Example-3.jpg

Kozlovska, M., Krajnak, M., Baskova, R. & Tazikova, A. 2014. Sustainability of construction from perspective of applications Technologies for green buildings. *14th International Multidisciplinary Scientific Geoconference*, Albena, Bulgaria, pp. 323–330.

Lesniak, A. & Zima, K. 2015. Comparison of traditional and ecological wall systems using the AHP method, *15th International Multidisciplinary Scientific Geoconference*, Albena, Bulgaria, pp. 157–164.

Lowell, H. & Smith, S. J. 2010. Agencement in housing markets: The case of the UK construction industry, Geoforum, pp. 457–468.

Mesaros, P. & Mandicak, T. 2015. Information systems for material flow management in construction processe, *IOP Conference series*, Tomsk, Russia, vol. 71, pp. 1–5.

Pannelised construction. Available on: http://hhcontractor.com/wp-content/uploads/2014/11/Panel.jpg

Pasquire, Ch. & Connolly, G. 2002. Leaner construction through off-site manufacturing. *Procceding IGLC*. Available on <http://www.iglc.net/Papers/Details/201/pdf>.

Thanoon, W. A. M., Peng, L. W., Abdul Kadir, M. R., Jaafar, M.S. & Salit, M.S. 2003, The Experiences of Malaysia and Other Countries in Industrialised Building System in Malaysia, Proceeding *on IBS Seminar*, UPM, Malaysia.

Trikha, D.N. 1999. Industrialised Building System: Prospect in Malaysia. *Proceeding of World Engineering Congress*. Kuala Lumpur.

Volumetric system example. Available on: http://www.pcko.co.uk/word/wp-content/gallery/barling-court/volumetric_system_example2.jpg

Warszawski, A. 1999. Industrialized and Automated Building System. Technion-Israel Institute of Technology. E & FN Spoon.

Waste and resource programe (2007). Current practices and future potential in modern methods of construction. WAS003-001: Summary Final Report. Available on http://www.wrap.org.uk/sites/files/wrap/Modern%20Methods%20of%20Construction%20-%20Summmary.pdf

Advances and Trends in Engineering Sciences and Technologies II – Al Ali & Platko (Eds)
© *2017 Taylor & Francis Group, ISBN 978-1-138-03224-8*

Impact of construction activities on GDP in the Czech Republic and its regions

L. Kozumplíková, J. Korytárová & B. Puchýř
Brno University of Technology, Brno, Czech Republic

ABSTRACT: An important link between the volume of investment activity and economic growth are represented by the structure of the implemented investments. The research has investigated a correlation between performance in the construction activities and GDP. The analysis has been performed based on data collected during 1998–2014 period in terms of current and constant prices. Regression analysis and correlation coefficients have been applied in order to determine the dependence between regional GDP and civil engineering production. Research results have determined the highest value of the correlation for GDP and construction activities in total values for the whole Czech Republic in a current price dataset, the medium value of this coeficient in constant price dataset. This dependence was not confirmed at the regional level. It can be concluded that construction activities has a very strong influence on GDP, but in case of smaller areas, the impact of public investments shows a significant time delay.

1 INTRODUCTION

The original idea of the harmonious development of all regions has been replaced by the idea of the equality of opportunity and creation of common financial funds for financing regional policy. The requirement for efficiency in invested funds comes with a new implemented mechanism. Funded projects have to address local issues, but at the same time they should represent a character of growth effect. Strongly developing regions should provide significant economic opportunities for business entities from the weaker regions. Economic growth is getting to the foreground while the indispensable determinant of growth becomes the increase in the competitiveness of municipalities, regions and states. Gross domestic product, unemployment rate, sectoral unemployment rate, range and investment sectors in the region, tax revenue and others represent the major indicators of the area development. These aspects enter into the evaluation of the area in terms of security and stability of the environment for investment projects as well. Long-term monitoring clearly shows that the construction industry and especially the investment construction activities contribute to the performance of the state by a relatively high percentage.

2 LITERATURE REVIEW

The national economy of each state goes through an economic cycle, which is defined as a swing in total national output, income and employment usually lasting for a period from 2 to 10 years (Samuelson, 2007). Most authors distinguish two basic phases of the economic cycle which are expansion and recession. Within recession falls depression, which is a really considerable and long decline in economy's performance and subsequent crisis, when this decline lasts for more than 24 months and its quite significant (Kučerová, 2009). Besides the partial phases, two turning points of the economic cycle exist—peaks and troughs. According to the length of a cycle, three types of economic cycles are considered: short term (Kitchin) cycles, midterm (Juglar) cycles and long term (Kondratiev) cycles (Groligová, 2004). Most authors use the same names for partial phases of the economic cycle, but the authors differ in determination of the reasons of the cycle (Konečný, 2013). The authors' approaches can be divided into two groups—neo-classics and neo-keynesians (Holman, 2010). Similarly to national economy, its partial sectors go through their market life cycles as well. But in this case, terms for its partial phases are not generally used, because there are more models of market life cycle. The market life cycle includes phases of growth, maturity and decline (Lu, 2000). Other authors distinguish phases of introduction, growth, maturity and decline (Liang et al, 2009; Wong, 1997).

Economic growth and welfare depend on productive capital, infrastructure, human capital, knowledge, total factor productivity and a quality of institutions (Guide to CBA, 2008). Construction projects are important because economic growth produces constant pressure on the renewal and expansion of the existing infrastructure. However, growth may not be uniform, both intensive and extensive growths may occur. Intensive growth is associated with the growth in labour productivity based on the application of new knowledge in the field of science and technology. Extensive growth is associated with extended contributions of labour and capital at the set technical level. The net intensive and extensive growth currently does not appear, so economic growth can be marked within a certain time period either as mostly intensive or as mostly extensive (Wokoun, 2008). The potential GDP growth rate is determined by supply-side factors, including labour, capital and Total Factor Productivity (TFP). In a growth accounting equation, holding constant labour force participation rate and the natural unemployment rate (i.e. the Non-Accelerated Inflation Rate of Unemployment (NAIRU)) and reduction in the working age population will directly reduce the potential GDP growth rate (Cai, 2013).

With rapid social development as well as large-scale construction of infrastructure, construction projects have become one of the driving forces of the national economy which energy consumption, environmental emissions, and social impacts are significant (Chang, 2011).

Economic impacts of construction projects are quite diverse. They are usually grouped into direct or indirect economic impacts. Direct economic impacts are directly related to the projects in which the savings in transportation costs usually form the most important impacts. Conversely, indirect economic impacts or economic impacts result from direct impacts, e.g. increasing the productivity of companies and distributive effects (Hromádka et al, 2011). Due to the economy of scale effect, service efficiency, e.g. utilization rate of the stock is usually high for services provided by the formal economy, as compared to people's use of their own stock of durable goods (Nørgård, 2006).

The environmental impact e.g. pollution or noise (Thanos et al, 2011; Browne, 2011), the benefits of the natural environment (Quintana et al, 2010), impacts on health (Oxman et al, 2009), cohesion of settlement, accessibility, urban planning, urban area and habitat fragmentation (Thomopoulos, 2013; Santos et al, 2010) are also included among the socio-economic impacts of transportation projects. There are a variety of modelling studies which deal with the changes in climate, policy and markets impact on land use. (Agarwal et al, 2002; Aspinall et al, 2008; Irwin et al, 2001; Koomen et al, 2007; Lambin et al, 2006; Schaldach et al, 2011).

3 METHODOLOGY

The research worked with two views for the surveyed problems—in current and constant prices. Current research examined 15 regions, data of the whole Czech Republic and data of 14 regions (local government unit, NUTS 3) in which the investment activities were assessed during 1998–2014 period. The results of the empirical part of the research were processed in the following steps. Firstly, linear regression analysis was performed in order to determine the relationship between a particular variable (GDP value) and the regressor (construction output). Regression analysis has the potential to express the relationship between examined variables, while the slope of regress curves may describe the significance of the GDP with respect to the investment activities.

Secondly, the correlation analysis was carried out in order to determine the existence of dependence between variables as well as its strength (the intensity level of dependence is expressed by Pearson's correlation coefficientρ). The size of dependence of both variables is determined by the correlation coefficient of the two time lines. Generally the values of the correlation coefficient ρ are assessed in terms of dependence level of the observed variables as follows (Budíková et al, 2007): ρ < 0,3—weak dependence to independence; 0,3 ≤ ρ < 0,5—moderate dependence; 0,5 ≤ ρ < 0,7—considerable dependence 0,7 ≤ ρ < 0,9—strong dependence, 0,9 ≤ ρ ≤ 1,0—very strong dependence.

Finally, both the results (in current and constatnt prices) for the Czech Republic area and for local government units have been evaluated.

4 RESULTS AND DISCUSSION

The dependency rate of the performance of the entire economy of the Czech Republic on the construction production is considerable (ρ = 0.6180). This development is represented in the constant prices relative to 2010. The equation of the regression line in this case can be described by:

$$y = 0,2998x + 1,8988 \tag{1}$$

However, the dependency rate of the performance of the entire economy of the Czech Republic on the construction in current prices is very strong dependence (ρ = 0.9173). GDP tends to lead the construction flow, not vice versa (Tse, 1997). The equation of the regression line in current prices can be described by:

$$y = 7,3605x + 339673 \tag{2}$$

The following charts in Figure 1 and Figure 2 show the development of GDP and the construction works in the Czech Republic.

Chart 1 shows the development of the construction activity and the Czech GDP in the period from 1998 to 2014. The development is represented in constant prices relative to 2010. Chart 2 shows the development of the construction activity and the Czech GDP in the period from 1995 to 2014. The development is represented in the current prices relative to 1995.

The chart 1 shows not only the obvious growth (GDP or investment activities), but also a drop in performance reflecting the financial crises. Construction industry tends to be the first affected sector in the crisis period, and vice versa. Improvement in outcomes of construction industry indicates an improvement of the entire economy of the country. A slowdown in construction investments brings an immediate reflection in the country's economy because construction industry possesses a high multiplier effect, a large number of manufacturers and suppliers from other industries are related to it and the completion of objects brings additional employment.

From a regional perspective, however, the dependence is not so clear. Dynamics of construction activity in the regions is so intensive that the relationship between GDP and construction

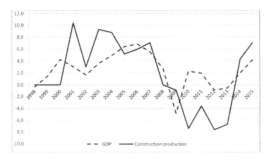

Figure 1.   GDP and construction activities development during the 1998–2015 period (constant prices, average of 2010 = 100; source: In-house processing using data from Czech Statistical Office).

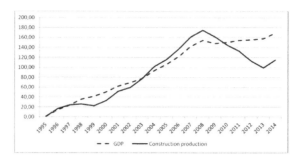

Figure 2.   GDP and construction activities development during the 1995–2014 period (current prices, average of 1995 = 100; source: In-house processing using data from Czech Statistical Office).

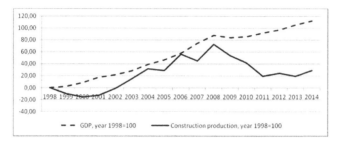

Figure 3.   South Moravian GDP and construction activities development during the 1998–2014 period (current prices, average of 1998 = 100; source: In-house processing using data from Czech Statistical Office).

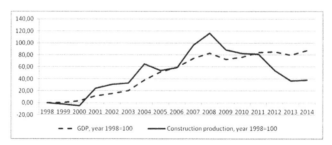

Figure 4.   Moravian-Silesian GDP and construction activities development during the 1998–2014 period (current prices, average of 1998 = 100; source: In-house processing using data from Czech Statistical Office).

activity cannot be deduced. In relation to the size of the budgets of individual regions, any major investment project financed from national or international sources can cause an enormous fluctuation of the curve of construction activities. Another important influence on these smaller territorial units is the focus of the investment. The research clearly shows that it is crucial whether the building activities are directed to investments in public or private (business) sector. At the regional level, the differences in investment focus play a crucial role, as it is evident from a comparison of data of the South Moravian and Moravian-Silesian regions.

Within the investment activities in the South Moravian Region in the researched period, especially public construction investments were realized, which will display in the performance of the territorial unit in the long term horizon (the transport infrastructure, the new passenger terminal of the Brno-Tuřany airport, the project completion and reconstruction of the Znojmo Hospital, the Technology Incubator II in Brno, the Biotechnology Incubator INBIT in Brno-Bohunice and new department of radiological and clinical oncology of Znojmo Hospital. From this list it is clear that they are mainly public investments whose output mainly affects the society as a whole. These investments have no immediate (short-term) effect on the economic performance as it is in the case of private investments. Their impact on the economic performance, measured by GDP, is delayed into subsequent years. Their main objective is to provide public service and public infrastructure for the operation of all functions of the national economy.

In contrast, in the Moravian-Silesian region in the researched period, private investments into the industrial infrastructure are significantly represented. Private investments pursue just one goal which is profit. Performance is therefore more rapidly reflected in the GDP. Both graphs show rises and falls of the curve in the same periods (Figure 4)

Moravian-Silesian region has a strong industrial base and tradition, but it is monoculturally targeted, mainly in metallurgy and automobile industries, and it is linked to the dominant large enterprises that are owned mainly by multinational corporations (construction investments in engineering and automotive industries, Shimano; Bang & Olufsen; Dura Automotive and a production plant Hyundai in Nošovice. Another important investment in the Moravian region was the construction of the D 47 highway.

The impacts of investments into the regional economy are already reflected in the following two years, when there was a significant increase in GDP.

5 CONCLUSION

From a nationwide perspective, it can be confirmed that the relationship between the performance of the national economy and construction activity is evident. The curve of construction activity varies slightly around the curve of the gross domestic product.

Dependence between the performance of the economy, characterized by the GDP and construction activity was confirmed by regression and correlation analyses. The strength of dependence has been proved on a certain number of observations. In the case of constant prices, there were 14 monitoring periods (data from 2001 to 2014). A considerable dependence can be confirmed with a 99% probability. In the case of current prices, there were 20 monitoring periods for the Czech Republic and 17 monitoring periods for the regions. The very strong dependence can be confirmed by a 98% probability. Correlation coefficient describing the strength of the dependence in the Czech Republic has been 0.6180 for constant prices and 0.9173 for current prices; therefore it is possible to consider there to be a considerable dependence.

Conversely, in individual regions this dependence has not been confirmed. One of the reasons is the fact that in each region the investments are focused in another direction. While in one region they are investments in the public sector, in another region it may be largely private sector investments. From the general public and private investment perspective, it is obvious that the effect is not always the same. While private sector investments concentrate on mainly the financial aspect (profit), the public sector primarily pursues the common welfare and societal effects. This was demonstrated in two regions—South Moravian and Moravia-Silesian regions. While in the South Moravian region the structure of investments was mainly focused on social infrastructure (health care), in the Moravian-Silesian region it was the industrial infrastructure. Investments in the South Moravian Region demonstrated themselves in the performance of the region with some delay, and therefore, although since 2008 there is a decline in construction investment activities, the performance of the region has not been affected by this decrease. Conversely, in the Moravian-Silesian region, the link between construction investment activities and economy performance is much closer. The research results refer to the fact that mainly the investment focus plays a crucial role in determining the relationship between GDP and the investment activity at the territorial unit level NUTS 3.

ACKNOWLEDGMENTS

This paper has been worked out under the project no. LO1408 "AdMaS UP—Advanced Materials, Structures and Technologies", supported by Ministry of Education, Youth and Sports under the "National Sustainability Programme I"

REFERENCES

Agarwal, C.; Green, G. M.; Grove, J. M.; Evans, T. P. & Schweik, C. M. 2002. A Review and Assessment of Land-Use Change Models. Dynamics of Space, Time, and Human Choice. *CIPEC Collaborative Report Series No.1.*

Aspinall, R. J. & Hill, M. J. (Eds.), 2008. *Land Use Change: Science, Policy and Management*, CRC Press, Boca Raton.

Browne, D. & Ryan, L. 2011. Comparative analysis of evaluation techniques for transport policies. *Environmental Impact Assessment Review*. Volume 31, Issue 3, pp. 226–233.

Cai, F. & Lu, Y. 2013. Population Change and Resulting Slowdown in Potential GDP Growth in China. *China & World Economy*, 1–14, Vol. 21, No. 2.

Chang, Y. & Ries, R. J.; Wang, Y. October 2011. The quantification of the embodied impacts of construction projects on energy, environment, and society based on I-O LCA. *Energy Policy*. Volume 39, Issue 10, pp. 6321–6330.

Groligová, I. & Mandelík, P. *Makroekonomie*. 2004. CERM, Brno, 130 pp.

Holman, R. 2010. *Makroekonomie. Středně pokročilý kurz*. 2nd ed., C. H. Beck, Praha, pp. 424.

Hromádka, V.; Korytárová, J.; Kozumplíková, L.; Bártů, D.; Špiroch, M. & Adlofová, P. 2015. *Risk of Megaprojects in Transport Infrastructure*. In Advances in Civil Engineering and Building Materials IV. Hongkong: Taylor and Francis Group, London, UK, s. 223–227. ISBN: 978-1-315-69049-0.

Irwin, E. & Geoghegan, J. 2001. Theory, data, methods: developing spatially-explicit economic models of land use change. *Agriculture, Ecosystem and Environment*, 85, pp. 7–24.

Konečný, Z. 2013. Structure of owner's risk rewards depending on the sector sensitivity to the economic cycle. *Trends Economics and Management*, Volume VII, Issue 15.

Koomen, E.; Stillwell, J.; Bakema, A. & Scholten, H. J. 2007. *Modelling Land-use Change; Progress and Applications*, vol. 90, Geojournal Library, Springer, Dordrecht.

Kučerová, V. 2009. Fuzzy modelování hospodářských recesí. *Trendy ekonomiky a managementu*, 3(5), pp. 29–36.

Lambin, E. F. & Geist, H. J. 2006. *Land Use and land Cover Change. Local Processes and Global Impacts*, Springer Verlag, Berlin/Heidelberg/New York.

Liang, T. P.; Czaplewski, A. J.; Klein, G. & Jiang, J. J. 2009. Leveraging first-mover advantages in internet-based consumer services. *Communications of the ACM*, 52(6), pp. 146–148.

Lu, J. & Wu, Ch. 2000. Cost and benefits models for logic and memory BIST. *ACM*, pp. 710–714.

Nørgård, J. S. 15 April 2006. Consumer efficiency in conflict with GDP growth. *Ecological Economics*. Volume 57, Issue 1, pp. 15–29.

Oxman, A., Lavis, J., Lewin, S. & Fretheim, A. 2009. Support tools for evidence informed health policymaking (STP) 10: Taking equity into consideration when assessing the findings of a systematic review, *Health Research Policy and Systems*, 7 (Suppl. 1), S10, 1–9.

Quintana, M. S., Ramos, M. B., Martinez, C. M. & Pastor, O. I. 2010. A model for assessing habitat fragmentation caused by new infrastructures in extensive territories—Evaluation of the impact of the Spanish strategic infrastructure and transport plan, *Journal of Environmental Management*, 91(5), 1087–1096.

Samuelson, P. A. & Nordhaus, W. D. 2007. *Ekonomie*. 18nd ed., NS Svoboda, Praha, pp. 775

Santos, G., Behrendt, H., Maconi, L., Shirvani, T., Teytelboym, A. 2010. Part I: Externalities and economic policies in road transport. *Research in Transportation Economics. Road Transport Externalities, Economic Policies And Other Instruments For Sustainable Road Transport*, 2–45.

Schaldach, R. & Priess, J. A. 2008. *Integrated Models of the Land System: A Review of Modelling Approaches on the Regional to Global Scale*. URL: www.livingreviews.org/lrlr-2008-1[11.5.2011].

Taylor, R. 1990. Interpretation of the Correlation Coefficient: A Basic Review. *Journal of Diagnostic Medical Sonography*, January, 6: 35–39.

Thanos, S., Wardman, M. & Bristow, A. 2011. Valuing aircraft noise; Stated Choice experiments reflecting inter-temporal noise changes from airport relocation, *Environmental and Resource Economics*, 50(4), 559–583.

Thomopoulos, N. & Grant-Muller, S. 2013. Incorporating equity as part of the wider impacts in transport infrastructure assessment: an application of the SUMINI approach. *Transportation*, 40 (2), 315–345.

Tse, R.Y.C. & Ganesan, S. 1997. Causal Relationship between Construction Flows and GDP: Evidence from Hong Kong. *Construction Management and Economics* 15 (4), pp. 371–376.

Wokoun, R. 2008. *Regionální rozvoj: (východiska regionálního rozvoje, regionální politika, teorie, strategie a programování)*. Praha: Linde, 475 s. ISBN 978-80-7201-699-0.

Wong, Y. Y. & Maher, T. E. 1997. New key success factors for China's growing market. *Business Horizons*, pp. 43–52.

Advances and Trends in Engineering Sciences and Technologies II – Al Ali & Platko (Eds)
© 2017 Taylor & Francis Group, ISBN 978-1-138-03224-8

Wooden structures: Traditional and yet forgotten

I.H. Krčmář
VŠB-TU Ostrava, The Czech Republic

ABSTRACT: Examples of traditional construction procedures, their details and technologies are stated in the paper. Despite being well verified by time, these methods face marginalization due to new approaches, which are very often short-sighted or too complicated. The author also deals with particular examples of traditional roofing, gutters, edges and ground anchoring.

1   INTRODUCTION

Over the course of time wood lost its role as the basic material used for constructions, tools, and furniture making. As a result, there has been a significant decrease in experience concerning woodwork in the field of architecture and construction. The notion that it is possible to imitate wood with the help of other materials often results in ridiculous situations.

The shift away from wood in the Austro-Hungarian Empire became particularly significant with Maria Theresa's introduction of new construction rules, which resulted in the majority of wooden buildings being reconstructed or demolished. Cities on fire during the Second World War damaged the reputation of wood and its substitution for steel took place nearly everywhere, despite steel's worse characteristics concerning fire-safety.

Wooden buildings have recently experienced a revival. The loss of continuity and interrupted cultural traditions, however, resulted in the introduction of confusing designs and a return back in time through timber framing to log huts. At present, logs are protruded beyond roof overhangs and the only reason rot has not afflicted these buildings yet is their relatively low age. Metal plating applied to the logs tops this nonsensical approach.

The influence of wooden buildings is much more significant than we are willing to accept, due to the traditional folk's house model, stone imitations of wood details in Classical antiquity and masonry in classicism. Old buildings have proven the traditional construction methods, from the facade layout to profiling of bevelled edges, making the structure look thinner and smoothing out the splinters, which results in a reduced area needing for painting and the ability to absorb humidity. Despite this, the well verified and time tested parameters face oblivion (Krčmář, 2008).

2   BASE

The best seal for wood is a hole, which is most often positioned in the base or in the roof edge; it is able to take away precipitation as well as condensed water. Anchoring points are already equipped with drains, but horizontal masonry constructions are not allowed to have hollows leaving the wood lying in water, either.

Non-sanded felt on a concrete base can also be encountered when trying to protect wood, in which case the wooden wall plate creates pressure and a pool is made, especially if the case when the higher layer blocks humidity and water goes down right from the ridge. It is amusing to watch how old barns, hay sheds, granaries and mills put on flat stones under corners and frame joints can stand for such a long time without upkeeping.

3   STRUCTURAL STABILITY

There is a saying that a wooden building can stand everything except a structural analysis following the latest norms. Sagging beams make an issue for the structural stabilty analyst even innew constructions, despite the fact that lots of old buildings serve well even with dramatic sagging.

Delicate jointscan be treated by putting in laminate or composite plates, whichsolves sagging and buckling andinterrupts the thermal bridge at the same time. Dowel joints with no metal reinforcement, thus having similar properties, have been the best solution since time immemorial. Carpenters with experience in this field are able to perform partial replacement of wood constructions using these joints.

4   INFLAMMATION

Flammability of wooden structures aret one of the limiting factors for wooden buildings. Besides having to stick to lots of mandatory rules, property owners tried to separate the buildings from fire hazards in their own interest.

Even when wooden smoke fluesare banned in our countries, scorching-treated wooden chimney cowls are still in use in the Alps. Scorching makes the wood more fire-resistant and in combination with moisture control it also helps to prevent rot.

Proper treatment of massive wooden members includes e.g. beam-ends hammering, which brings about the ability of 3 cm thick wood to withstand the equivalent of 30 cm thickness in the case of fire. Smooth wood finishing also results in a reduced surface exposed to humidity and paint. Should the wooden building not be properly designed and treated, a high pressure fire engine will be necessary to extinguish the fire (Krčmář 2011).

5   EAVES EDGES

The drip edge nose makes another significant part of wooden structures, its recommended treatment being homogenization of shingle or hardwood slats using hammering or scorching and finally grinding-wheel polishing.

Should the previous treatment not be followed, the moisture remains in the wood, which will consequently be destroyed by rot or wood-decay fungi. This has to be taken into account especially concerning moisture barriers, airtight insulations as well as at crossing points of reciprocally-supporting horizontal constructions.

Overhanging structure makes the best protection also for protected masonry or vertical wood constructions. Hardwood slats and shingles are in preferenceto metal flashing, which is -in majority of cases—unsuitable for historical buildings (Krčmář 2013).

6   SURFACE FINISHING

Suitable surface finish—impregnation in the past, pressure treatment used at present—is another important factor for wooden buildings to survive. The surface is required to be breathable and slightly greasy = water repelling. It is of importance to bear in mind that pure lime causes wood decomposition in a similar way to waste oil. Modern materials are also available in a wide range of colors and only 150 years ago painting shingles red was quite frequent, as it imitated more expensive ceramics.

Certain paints (including modern in tumescent coating) or cladding with non combustible plaster work go against their purpose when used for a wooden construct protection as they increase weight, change aesthetics and make the surface airtight. Finishing with the use of wooden planks or shingles has better parameters than half-timbered framing using split log filling.

Half-timbered constructions using wainscoting or cladding are questionable due to the cladding instability, humidity problems and the never ending necessity to repair cracks arising between different materials. The troubles have not been solved so far, as there is no putty elastic enough and not changing the facade colour. Despite being the most suitable surface treatment, scorching and hammering are not promoted due to having no exclusive manufacturer (Krčmář, 2008).

7 ACCESSORIES OF WOODEN CONSTRUCTIONS

For a credible impression of historical wooden buildings it is necessary to come back to traditional accessories, e.g. poles made from hard and flexible wood used mainly as wooden gutter holders.

Thatched roofing represents the development of materials still in use today, whether straw or reed is used. Despite better slate shingle qualities, wooden shingle remains the most frequently used material for traditional structures. Asphalt felt put on triangular slats is also acceptable for use in wooden village buildings originating from the turn of 19th and 20th

Photo 1.   Gerstein House, Author I. Krčmář 2013.

century. In this case it is necessary to bear in mind the possible condensation to be found on the bottom part.

Should you have a look at log cabin constructions, they are running out of folklorefeaturing metal on edges of a modern roofing, with lots of metal flashings reaching up to the ridge.

Vents, windows, and masoned fillings of rotten parts introduce problems with structural stability and condensation, made even worse by excessive heating in winter period and by putting thermoinsulating foam in the chinks and vents.

Log cabins constructed in the traditional way,having minimal gap in the centre and slate or plaster filling,featured more stable humidity and temperature and they did not suffer from micro-cracks occurence.

Diversity of colors on wooden members is one of the region specialities. Carpatian area and Jurkovičovy Pustevny or Luhačovice make an extreme case of polychrome wooden structures in our conditions. There are many buildings, however, whose coloured finishing would look like an imitation.

Historical building do not use just green, yellow, red, brown, and white colors, but gray as well. Sun-faded shades of gray are not so traditional in our country as is the case in Switzerland or Germany, we consider it to be acceptable only on forgotten storage houses and huts. Despite this opinion it is necessary to accept it as a fully-fledged color finishing.

Imitations of wood and other materials, e.g.foils or plastic cladding, are alien to the environment, feature a short lifetime and the volatile substances can—in the long term—be potentially hazardous to human health. Popularity of the pleasant wooden finishing results in illogical extremes, where even metal constructions are painted with brown color.

When a new building is designed, it is significant to consider the interaction between traditional structures and materials. This concerns especially herritage sites, as any modification performed influences surrounding structures and materials.

After abolishing private companies in the Central Europe in 1948, the knack of old masters got lost as well, because the craftsmen were degraded into temporary workers. The impact is—unfortunatelly—visible all around us. It can, however, be gradually rectified with paragons of good realizations.

It is necessary to at least discuss the wooden construct interlaid with wedges and logs extending far beyond roof, a roof that looks like it should be placed on a completely different house.

Photo 2.   Concrete Roof of Hrčava, Author I. Krčmář 2015.

Photo 3.   House of Píšť—Silesia, Author I.Krčmář 2016.

8   CONCLUSION

Forests of Central Europe represent a natural recreational environment. They also serve as a significant source of construction material, serving both for new buildings and restoration of previously builtones. Earlier destroyed bypollution created by heavy industry, monoculture forests are nowadays damaged by hurricanes and particularly swindlers. This results in the degradation of trees,whether it comes from covered bio-factors orwell noticeable mistletoe.

According to Erwin Thoma, there are three main criteria concerning woodcutting. It is of utmost importance to cut the wood during the winter (September to January or February), when the sap circulation is limited. Harvesting the woodin the second moon phase is also critical due to its higher quality in comparison with wood cut at full moon. Zodiac oriented rule states the best months for wood harvest are those of Capricorn, Virgo and Taurus (Thoma 2006).

Differing quality, way of treatment and color of wood are the main determinating factor for its use in constructions and we should realize that so called "modern" layered lamela does not represent a cure-all. Despite having a "limited" lifespan, wood is a synonym of quality and excellent insulating, optical and tactile characteristics (Titscher2002).

The author presents a list of simple, but time-proven rules for treating wood as a construction material:

1. Cutting construction wood after Christmass is the best time due to low humidity. Downhill storing will allow the humidity to gather in the tree top. Consequential removing of the top will make it substantially easier to dry the wood while minimizing the amount of cracks.
2. The span of building season should last from Saint Joseph to Saint Venceslas, which allows for wood to dry up and will therefore increase its lifespan.
3. Wood should never be put in puddles.
4. The fact that a hole makes the best seal for wood has been well-proven in time.
5. Fighting fire with fire is a good method, as scorching represents a traditional and effective way of protection against fire. The use of a heavy, fire-resistant drywall devastates the constructions, while the protection time from fire spans from 30 to 45 minutes only.
6. Permeability is an important factor in insulation, because when constructions breathe, they are dry.
7. Flashing has no place in roofing (with the exception of chimney roof connections) and can be easily replaced by hardwood, examples of which are still visible today.

8. Unplastered wall fights humidity far better than a wall with any additional layer.
9. Not painting the wood is the best solution, as the quality treated wood does not need any paint. In the case of extreme weather conditons it is possible to use transparent wax.
10. The form of a wooden structure has to matchits contents, for which there are certain architectural and technological rules to be followed. Panelling masonry structure with wood or a wooden structure with drywall does not represent a good example.

REFERENCES

Krčmář, I.: Dřevo jako tradiční stavební materiál z pohledu architekta, in. Sanace dřevěných konstrukcí staveb, Proceedings of Společnost pro technologie ochrany památek, Národní Muzeum Praha, 15. května 2008, s.4–16.

Krčmář, I.: Zapomenuté mechanické úpravy dřevěných prvků, in. Proceedings of international conference - Termosanace, Church St. Martin, 10. -11. 8. 2011 Bratislava / Pressburg.

Krčmář, I.: Pomalu a tradičně, in. Vodní mlýn Wesselsky v Loučkách nad Odrou, Vade Mecum Bohemiae, Loučky 2013str. 66–73.

Thoma, E.: …viděl jsem tě růst: o prastarém a novém životě se dřevem, lesem a Měsícem. Ústí nad Labem, 2006. Translated by Jiří Fiala.

Titscher, F.: Die Baukunde / Tradice c.k.stavebnictví, Praha 2002. Translated by prof. Josef Michálek a doc. Luděk Novák.

Advances and Trends in Engineering Sciences and Technologies II – Al Ali & Platko (Eds)
© 2017 Taylor & Francis Group, ISBN 978-1-138-03224-8

Microorganisms growing on insulated façades

M. Kubečková, V. Kučeriková & D. Kubečková
Faculty of Civil Engineering, VŠB-Technical University of Ostrava, Ostrava, Czech Republic

ABSTRACT: Occurrence of microorganisms on façades is a common problem,mostly on prefabricated houses which were additionally insulated. There are two main reasons of the insulation: lowering energy costs and new look of the facade. Algae, mould and bacteria can grow on almost every surface, but not all insulated façades are attacked by them. Suitable conditions for growth of organisms on facades are certain ranges of temperature and high moisture level, but also the surface structure, nutrient availability, pH, etc. Different organisms have different demands that decides if an organism can grow in a certain environment. At the present time there is no material that can resist microorganisms for long time. This problem depends on many factors which are described in the article in more detail.

1 INTRODUCTION

In the construction industry, from the second half of the preceding century the implementation of the strategy to improve energy performance of buildings has led to relatively extensive remediation of residential houses and buildings of urban civic amenities in terms of improving the energy concept of construction, which leads to savings. Thus, the energy performance of a building is currently one of the most important and also most monitored characteristics of all buildings.

Since 2000, we have witnessed a relatively high rise in the revitalisation of prefabricated concrete panel housing complexes. The present day's increase in energy prices and thereby the awareness of the inevitability to cut down energy consumption are the reasons why prefabricated panel housing complexes started to be fitted mostly with contact thermal insulation systems. For prefabricated panel housing complexes to meet the new requirements pursuant to the implementation of new thermo-technical and energy-related legislation, the available, most affordable and most widely used insulation method is the application of the contact-type External Thermal Insulating Composite System, hereinafter referred to as "*ETICS*".

The advantages of this kind of thermal insulation are beyond a doubt. First of all, it eliminates thermal loss, increases the lifetime of building structures, especially in the area of external building envelope, and creates a new aesthetically favourable appearance of prefabricated panel buildings. However, in the course of time deficiencies also emergedin the area of contact-type ETICS additional thermal insulation. This also includes the presence of microorganisms on the surface of façades of prefabricated panel housing complexes. Façades having ETICS show green to black coatings indicating that the façade has been afflicted with microbiological cultures.

2 REDUCING ENERGY CONSUMPTION FOR HEATING—THERMAL PROTECTION OF BUILDINGS

In association with reducing energy consumption for heating, the Czech Republic issued a set of standards (ČSN 73 0540-1, 2005), (ČSN 73 0540-2, 2007), (ČSN 73 0540-3, 2006), (ČSN 73 0540-4, 2005),dealing with the thermal protection of buildings. The primary impetus for their publication was the United Nation's programme document titled Agenda 21.

Subsequently, there has been an adjustment of standards in thermo-technical requirements and energy regulations. Figure 1 illustrates the development of requirements for thermal protection of external walls in the period of 1962–2016 in the Czech Republic. The indicator here is the heat transfer coefficient U_N [W/(m².K)]. The coefficient reflects how much heat is dissipated through a structure having a surface area of 1 m² when the difference between the temperatures of the surfaces is 1 K.

It is evident that external structures of prefabricated panel buildings are not compliant with the standard requirements (ČSN 73 0540-2, 2007) of the current legislation in the field of heat transfer coefficient U_N[W/(m².K)]. The requirement, for the external wall is at the required value of $U_N = 0.3$ W/(m².K) (for heated residential buildings). The best way to achieve these requirements the in order to improve the situation is to apply the ETICS additional contact thermal insulation system.

2.1 The thickness of the thermal insulation of external envelope

The growing thermo-technical requirements lead to an increase in the thickness of the thermal insulation of external building envelopes in the ETICS configuration. In the model example we assessed a panel external envelope fitted with ETICS having a thickness of the insulation of 80–200 mm thick.

Prefabricated panel housing complex in type series: T 08B.

Configuration S1:

The external wall made ofsandwich construction (Gable230 mm thick).(Steel-reinforced concrete panels—90 mm thick, thermal insulation made of polystyrene foam—80 mm thick, protective layer of steel-reinforced concrete—60 mm thick).

Configuration S2:

The external wall made ofsandwich construction (Frontage290 mm thick).(Steel-reinforced concrete panels—145 mm thick, thermal insulation made of polystyrene foam—80 mm thick, protective layer of steel-reinforced concrete—65 mm thick).

The calculation (Software Heat 2014) was used for thermal and technical evaluation of the two different circumferential structural designs of prefabricated housing buildings for configurations S1 and S2.

It is logical that as the thickness of thermal insulation grows, the thermo-technical parameters of the panel external envelope improve provably, the thermal resistance R [(m².K)/W]

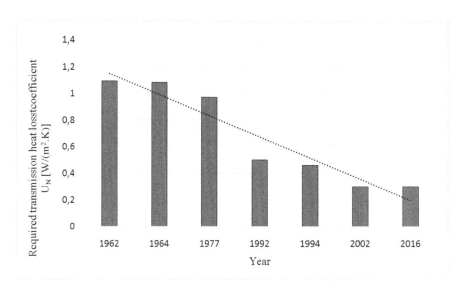

Figure 1.   Requirements for heat transfer coefficient U_N[W/(m².K)] in the Czech Republic in 1962–2016 for external walls (Ateliér DEK, 2007).

rises linearly, and so the heat transfer coefficient U_N [(W/m^2.K)] decreases. The temperature on the inner surface in the interior increases to values amounting to 19.8 °C with thermal insulation of about 180 mm, refer to Table 1 and 2. The temperature under the thermal insulation increases as well.

Thermal insulation on prefabricated panel walls with EPS polystyrene foam boards having a minimum thickness of 80 mm in the ETICS configuration reduces heat loss significantly. Thermal insulation maintains a required temperature inside a building and the insulation will not let any heat escape to the exterior. This phenomenon can be *inter alia* demonstrated using the values calculated, refer to Table 1 and 2, where the temperature on the inner surface rises with increasing thickness of the thermal insulation.

2.2 *The disadvantage of the thermal insulation in respect of microbiotic attack*

The consequence of the application of thermal insulation having large thickness is that the outer surface of the façade receives less thermal energy. This means that the outer surface of the insulated façade has got lower surface temperature than in the previous situation.

The reduction of the façade's surface temperature is the first step to its hypothermia. This phenomenon is generally defined in terms of construction physics as follows: if the temperature of an object, which is surrounded by air of a certain temperature and relative humidity, is lower than the dew point's temperature, then water vapour is condensed on the object's surface. Most frequently, this phenomenon occurs at night and in winter(Büchli&Raschle, 2011). Due to hypothermia, an aqueous film is formed on the façade, and the surface becomes damp. External condensation may also occur on façades with no thermal insulation. However, water may penetrate the structure (porous materials) and therefore it does not accumulate at the

Table 1. Evaluation of heat transfer coefficient with an increase in thermal insulation for the S1 configuration, (Software Heat 2014).

Thickness EPS	U [W/(m^2.K)]	T_1 [°C]*	T_2 [°C]**
No insulation	0.524	17.5	–
80	0.232	19.3	6.4
100	0.204	19.4	8.1
120	0.181	19.6	9.5
150	0.156	19.7	11.1
180	0.136	19.8	12.2
200	0.126	19.9	12.9

*Temperature on the inner surface of the structure.
**Temperature under the thermal insulation.

Table 2. Evaluation of heat transfer coefficient with an increase in thermal insulation for the S2 configuration, (Software Heat 2014).

Thickness EPS	U [W/(m^2.K)]	T_1 [°C]*	T_2 [°C]**
No insulation	0.521	17.5	–
80	0.230	19.2	6.2
100	0.202	19.4	7.9
120	0.180	19.5	9.3
150	0.155	19.7	10.9
180	0.136	19.8	12.1
200	0.125	19.9	12.7

*Temperature on the inner surface of the structure.
**Temperature under the thermal insulation.

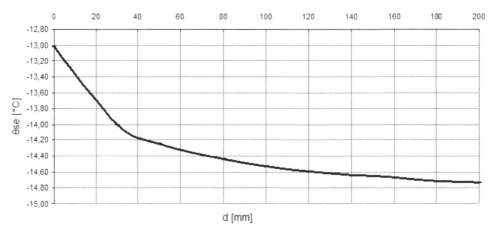

Figure 2. The relation of the thickness of thermal insulation in the ETICS system d [mm] and the temperature of the outer surface of the façade θ_{se} [°C], (Software Heat 2014).

surface and thanks to the dissipation of heat from the interior to the exterior it is gradually evaporated from the wall.

For the time being, the software support available in the area of thermal engineering may not be used to model and evaluate condensation on exterior surfaces. What is possible at least is making a model using the Heat 2014software to calculate the façade's external surface temperature.

Figure 2 shows the relation of the thermal insulation's thickness and the temperature of the façade's surface in standard model conditions of $\theta_e = -15°C$, $\theta_i = 21°C$. The temperature of the façade's surface before applying additional thermal insulation is $\theta_{se}(0) = -13.0°C$. After applying thermal insulation having a thickness of merely 20 mm there is a major temperature drop of 0.7°C. After the application of ETICS having 200 mm thick insulation the temperature decreases to $\theta_{se}(200) = -14.7°C$, i.e. a difference of 1.7°C.

3   THE OCCURRENCE OF MICROORGANISMS ON THE FAÇADES OF THERMALLY INSULATED BUILDINGS

Façades of thermally insulated prefabricated panel buildings attacked by microorganisms may be understood as a natural, but negative phenomenon, which is caused by physical and natural phenomena and influences. The basic requirements for the life of microorganisms include moisture, temperature, nutrients, light, pH value and oxygen. There are those kinds of microorganisms growing on the surface the façades of prefabricated panel housing complexes for which the climate and local conditions are best suited to.

The available scientific literatures report that the microorganisms which occur on the surfaces of insulated façades (they are mostly fungi, algae, cyanobacteria), are also found in the interiors of prefabricated panel housing complexes, for example, as a result of poor regime of ventilation, insufficient heating control, and as a result the natural migration of microorganisms. It is therefore due to the natural phenomenon of migration of microorganisms in the given environment, and, consequently, their colonisation on the ETICS surfaces, because most of them are borne in the air.

Additional ETICS thermal insulation contributes to the deterioration of the situation to a certain degree. The cultivation of micro-organisms on the façades of prefabricated housing complexes is evident especially in shaded places, or sides oriented to the North or Northwest, where the façade is damp most time of the day.

A big role is also played by the time duration of the action of humidity. However, as mentioned above, for non-insulated wall moisture may migrate towards dryer parts of the wall (interior) and does not accumulate on the surface. In winter, during the heating season, heat may accumulate throughout the entire thickness of the external wall. Thus the warming of external building envelope prevented the action of moisture also on the exterior side in a longer term. The thermal insulation board prevents the penetration of moisture which the accumulated on the exterior side, where there is only a thin layer of plaster. The difference in the distribution of the relative humidity is shown using an example of a prefabricated panel complex's wall in Figure 3.

3.1 *Impact of surface finish of ETICS on attacking insulated façades*

For surface finish of ETICS, the today's construction market offers us to choose from acrylic, silicone, silicate, mineral or mosaic plasters, and their properties are determined by the type of the binding material. The diffusion resistance factor μ [-] of these plasters varies in the range from 20 to 120 (ČSN 73 0540-3, 2006).The diffusion resistance factor reflects the ability of a material or a substance to interrupt water vapour. It specifies how many times the resistance of the material/substance is greater than that of an equally thick layer of air at the same temperature (ČSN 73 0540-3, 2006). This property defines how much the plaster is permeable to vapours and capable of "breathing". Due to the presence of micro-organisms those plasters are preferable which have high vapour permeability and do not trap potential moisture.

Acrylic plasters are bonded by synthetic resins and are more susceptible to the deposition of dust particles, which can be a cultivating substrate for the vegetation of algae. Silicate plasters are bonded with potassium water glass, which has a pH of up 12 and forms a natural protection against the growth of algae. It is necessary to add a note that the pH value of a plaster reduces over time, and thereby also the natural protection. The binding material of silicone plasters are silicone resin, which have hydrophobic effects. Contaminants are washed from these by rain water. For preventive purposes, it is possible to add algistatic and fungistatic ingredients to all types of plaster (Rovnaníková, 2002).

Given the generally known context, these measures do not have a long-term effect, because the active ingredients are washed away by rain over time. The structure of the plaster, grain size and direction of application also affect the capability of the organisms and dust particles to stick on the surface. On smooth surfaces, both dust particles and spores of microorganisms may be washed away by rain water. While on the surfaces of plaster having large grain size and for example horizontal deposition of dirt easily sticks, and drops of water are hitting grains of the aggregate.

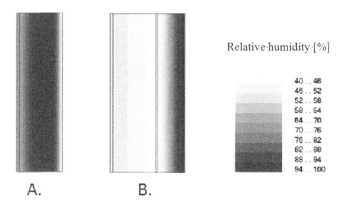

Figure 3.   Progress of relative humidity—External building envelope made of single porous concrete panels T08B without ETICS and wall with ETICS[%]—A. wall without ETICS, B. wall with ETICS (Software Area 2014).

4 CONCLUSION

Nowadays we can say that the contact-type thermal insulation system (ETICS) fully meets the requirements to eliminate heat dissipation through external walls. Compared to the state without insulation, the heat consumption is significantly reduced. At the same time, ETICS also protects the façade from weather and ensures the aesthetic function of the entire façade.

A definitive solution to prevent the occurrence of microorganisms on insulated façades has not been found yet. To some extent, the occurrence may be postponed by an appropriate type and structure of plaster and remediation coatings. Some new plasters already contain biocidal substances having a preventive purpose. For example, in heavily burdened environment such as industrial sites with high concentrations of air-borne dust PM_{10}, with air containing SO_2, NO_2, CO, the prevention using biocidal substances in plasters has got only a small or limited effect. In industrial sites with polluted air having a concentration of $PM_{10}>$ 180 as measured 1 hr. in $\mu g/m^3$, there is repetitive biodegradation growth on the façades of buildings, and biocidal substances in plaster materials do not provide sufficient resistance to microorganisms. Therefore, a system of regular prophylactic care and maintenance of the external building envelopes becomes inevitable. It must be remembered that the application of ETICS will not provide for a maintenance-free system for façades, and any façade fitted with ETICS requires regular care.

REFERENCES

Ateliér DEK. 2007. *Development and thermal binding technical requirements.*Information on http://atelier-dek.cz/
Büchli, R. &Raschle, P. 2011. *Algae and fungi on the facades*. Ostrava: Missions.
ČSN 73 0540-1. 2005. *Thermal Protection of Buildings—Part 1*: Terminology.Prague: Czech Standards Institute
ČSN 73 0540-2. 2007. *Thermal Protection of Buildings—Part 2*: Requirements. Prague: Czech Standards Institute, includingchanges made in April 2012.
ČSN 73 0540-3. 2006. *Thermal Protection of Buildings—Part 3*: Design valuequantitles. Prague: Czech Standards Institute
ČSN 73 0540-4. 2005. *Thermal Protection of Buildings—Part 4*: Calculationmethods. Prague: Czech Standards Institute
Rovnaníková, P. 2002. *Plasters, chemical and technological properties*. Prague: The Society for Protection of Monuments of technology, ISBN 80-866657-00-0.
Software Building physics, *Heat 2014*. 2014. Svoboda Software
Software Building physics, *Area 2014*. 2014. Svoboda Software

Advances and Trends in Engineering Sciences and Technologies II – Al Ali & Platko (Eds)
© 2017 Taylor & Francis Group, ISBN 978-1-138-03224-8

Changes in housing demands and their influence on overall user comfort for a block of flats in the Czech Republic

D. Kuta & J. Ceselsky
Department of Urban Engineering, Faculty of Civil Engineering, VŠB-TUO, Ostrava, Czech Republic

M. Zubkova
Department of Economics and Management of Building, Faculty of Civil Engineering, STU, Bratislava, Slovakia

ABSTRACT: After 1948, housing construction was socially-oriented. The main emphasis was placed on the quantity of flats built and on improving technical level of housing (this phrase does not fit). All efforts were aimed at construction restoration and the removal of war damage. New possibilities were sought with regard to the acceleration and economization of construction, frequently at the expense of long-term quality and operational efficiency. This paper deals with the overall user comfort of residential houses, which is, among others, influenced by the existence, spatial efficiency and usability of facilities in residential houses. This paper is focused on the very facilities of residential houses and their influence on overall user comfort, in as much as this issue has up to now primarily been solved in the terms of the dwelling area, while the other areas of residential houses have been neglected.

1 INTRODUCTION

Housing constructions form the basis of the built-up environment, be it whether the significance of this statement is viewed from the perspective of the quantity of constructions built or their impact on every individual's needs.

Housing is often regarded as a relatively conservative sphere of life whereas, as a construction task, it may easily evoke the impression of commonness, perhaps even banality, particularly with regard to collective housing. Whether ordered by public administration, corporations or enterprises, owing to their repeatability, it is often bound by numerous regulations, unwritten rules and cultural restrictions from which present society is still not able to disengage (Kohout & Tichý & Tittl, 2015).

For the field of collective housing, detailed knowledge of the issues is usually crucial by reason of maneuverability in a small space. This includes awareness of various circumstances and concurrences determining the character and methods of utilization of the individual rooms of residential houses used for collective dwelling.

Interest in the issues of collective housing in typological terms, irrespective of special areas such as housing for seniors, housing for the handicapped, starting flats, etc., is not very typical of our cultural environment even from a historical perspective.

2 HOUSING IN THE CZECH REPUBLIC AFTER 1948

After 1948, the period of socialist lifestyle began in the Czech Republic, based on egalitarianism and the socialist building industry, the general obligation to work and general right to abode; nevertheless, one should point out, with a very long waiting period. Obviously, the general right to abode meant the necessity of quick construction of new flats, which resulted in overall construction of housing estates according to five-year plans and the technical-economic indicator (Act on the First/Sixth Five-Year Plan No. 241/1948 Coll.). With regard to their characteristics, these housing estates are regarded as social housing today.

At that time, the issue of social housing was not a breakthrough idea at all; the history of social housing dates back to the turn of the 20th century, when the labour force began to flow into cities; therefore, the first social housing buildings appeared in the early 20th century; for example, construction of the Jubilee Settlement in Ostrava from 1921 to 1932 is worth mentioning. Housing of this type was provided for a specific group of the population, as housing for workers and employees or, on the contrary, housing, for instance, for the old, the sick, young families with children and the unemployed. Within this primary social construction, the floor area of every dwelling was about 80 sqm. After 1921, the Czechoslovak Republic placed great emphasis on social issues and further developed social housing; laws on building activities enabled the foundation of housing associations and building activities of municipalities as a counterbalance to private enterprise. Social housing became an important subject of prominent architects; there was a perceptible distinctive pursuit of efficient ground plan layout and reduction of architectural expression. Another important milestone in the area of social housing occurred in the year 1927, when the "building activity act" was passed, involving preference for ground plan quality and granting of support to buildings in which even fixtures were lit and ventilated, with flats above the ground level. Furthermore, emphasis was placed on facilities such as a common laundry and dryer, often located on the ground floor or basement, which are today considered of much value, even in the parterre area. As an example of this period, the small-flat settlements in Obřany and Komárov can be cited.

The demand for cheap, efficient and undemanding construction predetermined building standardization. Brick residential houses of various types were implemented, differing in their layout and size. The original, functionally undifferentiated rooms of the same dimensions were eventually replaced by rooms with closer differentiation of functions; over the course of time, the floor area standard was also increased. They further included minimum sanitary installations and a small balcony. In the 1960s, panel buildings arrived. The first residential houses implemented by the method of prefabrication appeared in 1955; their construction continued until 1992, after which it was terminated. Emphasis was placed on rationalization and standardization. The civil-engineering design, low standard of sanitary installations and low variability of dwellings soon failed to conform to the ever-developing needs of dwelling users.

In 1989, the recurrent subject of social housing appeared again as a result of differentiation in society. Once again, the subject was topical and appealing to architects; towns and municipalities were motivated by a more favourable housing policy. Architects were encouraged by the high architectural level of social housing in many Western European countries (Austria, the Netherlands, Spain and Germany). However, the standard of living as well as housing demands also increased; the first shortcomings of the panel building housing estates were revealed in support of suburban development as a trend which, paradoxically, was to be curbed according to the original idea.

The construction of residential houses started to decline quickly. There was sparse construction of individual residential houses.

In the 1990s and at the beginning of the 21st century, humanization of the original panel housing estates took place, being primarily restricted to heat cladding, roof extensions and replacement of windows; the internal space of residential houses was totally ignored, or changes in the actual dwelling proportions were considered regardless of the space of the entire internal environment of the residential house including all of its rooms belonging to the dwelling, such as the house facilities.

3   CONTEMPORARY HOUSING DEMANDS

Contemporary people's housing demands go hand-in-hand with the housing function. In general, housing is considered one of the basic necessities of human life (UDoHR, 1948).

The right of abode is defined in the Universal Declaration of Human Rights. The most fundamental function of housing is specified as protection against adverse external effects; at present, the functions of relaxation and fulfilment of personal and life needs are somewhat preferred. One can state, therefore, that housing is one of the essential factors of development of the living standard.

The traditional content of housing, or rather its actual function, undoubtedly involves the common life of the family; with the increasing societal living standard, this will mean rising demands for the housing standard, particularly for the general standard, above all in the central living room. Moreover, privacy of the individual family members is of similar significance; even here, society is raising its requirements, posing claims for a separate workroom for every household member living in the given dwelling. Other interference in the layout of present-day dwellings features higher demands for storage areas, pantries and dressing rooms (Kuta & Kuda, 2005).

Up to now, only the demands for the actual housing function have been analysed, with their influence on changes to the layout as well as user comfort. Within the solution to the user's comfort of residential houses, however, it is also necessary to deal with the issue of facilities. As a whole, traditionally conceived rooms, such as laundries, drying and ironing rooms, areas for bicycles and prams as well as cellars, present functions, many of which were gradually transferred to the dwelling in the past; the remaining part (such as the storage of prams and bicycles) has become highly problematic, for safety reasons in particular.

Whereas during the period of dwelling housing construction, it was common for house facilities to be designed as well as frequently used in every new house, these rooms rarely perform their originally intended functions at the present time. Before the house and dwelling layout is designed, demands for operational relations, functional and spatial requirements should be clarified. Therefore, one should particularly consider the usability and spatial efficiency of selected rooms of house facilities, particularly the storage room for prams, bicycles and wheelchairs, storage rooms for objects, unless included in the dwelling, which the Czech National Standard No. 73 4301, concerned with the issue of residential buildings in the Czech Republic, defines as mandatory to ensure economic and technical operation of the residential building.

Generally speaking, house facilities form an important part of the residential house—they may have a great influence on the housing comfort in the object. At the same time, implementation of the rooms for facilitie location adds to the construction costs, which in turn have an unfavourable effect on dwelling prices. Moreover, one should realize that the developing companies primarily determine the contents of house facilities in residential houses with regard to the highest possible profit from their investments (Kuta & Ceselsky, 2015). The area intended for the dwelling can be utilized for various layout solutions—they can differ in the size and shape of the individual rooms, position of doors and windows, etc. The user's wishes should be respected and the layout adapted to them. Similarly to the dwelling area, the house facility area can employ a certain variability in their layout solution; thus over the years, when the facilities have ceased to perform their basic function, they can be transformed to a different functional use by simple modification, in which way a defunct room can be revived again. This would have to be considered as early as during the design and actual construction of the residential houses. For the existing residential houses, which mainly involve panel buildings, possibilities should be considered of rediscovering the options for changes in the layout or functions of the existing layout.

4   RESEARCH ON USABILITY AND SPATIAL EFFICIENCY OF FACILITIES OF RESIDENTIAL HOUSES TO DETERMINE THE USER COMFORT

The crucial requirement of the present is increasing sustainability of new projects and existing buildings as well as increasing their future benefit. This condition primarily applies to the housing and dwelling fund group; there are rising demands for user comfort and spatial efficiency of residential houses. In the methodology of the national tool for building quality certification with regard to the sustainability principles (SBToolCZ), the weight of the user comfort criterion amounts to 9% and that of the spatial efficiency criterion to 7% (Vonka, 2012).

With regard to the absence of any statistical materials concerning the real usability and efficiency of the house facilities in residential houses, a direct survey of the real state has been conducted in the form of questionnaires within the university project SP2015/127.

The survey aimed either to confirm or to rebut the hypothesis that the house facilities are not really utilized sufficiently and, pursuant to this result, to propose subsequent forms of amendments of legal and technical standards governing the subject issues (e.g., by partly transferring the areas to the areas for dwelling accessories—pantry, cellar room, etc.).

Within the period of May-October 2015, a questionnaire entitled Survey of Usability and Spatial Efficiency of House Facilities of Residential Houses was sent off and made publicly available on the Netquest portal (intended for the creation and publication of surveys in the form of questionnaires) to obtain a broad spectrum of respondents. Among others, one can state on the basis of the answers that the highest number of respondents (62%) live in dwellings located in panel buildings.

This is almost double in comparison with dwellings in brickwork estates. The questionnaire survey further revealed that the highest number of respondents, 35% or 52 persons, live in dwellings sized 3 + 1 (three-room apartment with kitchen). According to the size, the other most frequent dwellings inhabited by the respondents include 2 + 1 (two-room apartment with kitchen) with 19%, 2 + kitchenette (two-room apartment with kitchenette) with 14% and 3 + kitchenette (three-room apartment with kitchenette) with 12%. Dwellings of other sizes make up about 20% in total; they mostly include smaller-size dwellings; 4 + kitchenette (four-room apartment with kitchenette) dwellings only make up 3% of the total. Inhabitants of larger dwellings did not participate in the survey. Further, the questionnaire survey revealed that house facilities of some types mostly do not fulfil the needs of the residential house inhabitants. These primarily include pram and bicycle rooms, cellar cubicles, parking areas and flat roof adaptations (which, however. only exist in 4% of the respondents' residential houses). Furthermore, house facilities such as letter box, waste disposal, cellar cubicle, pantry, parking areas and assembly rooms, if present in the residential house, are really utilized at 100% or almost 100%. On the contrary, facilities such as heating and maintenance rooms, laundries, rolling presses and rooms for carpet beating, if present in the residential house, are

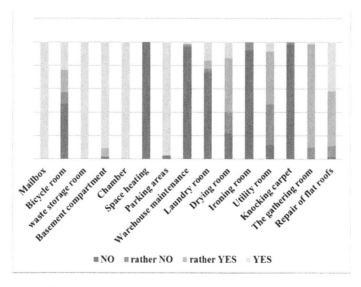

Figure 1.   The real usefulness of home furnishings as stated by the respondents (in %) (Source: author's archive).

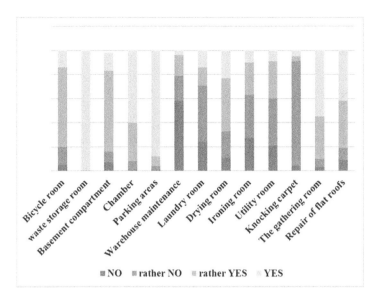

Figure 2.   The need for home furnishings in an apartment building according to the respondents (in%) (Source: author's archive).

hardly ever used. The other house facilities are used variously; for example, in dependence on the locality, respondents' dwelling size, or the structure of the residential house's inhabitants. In the respondents' opinion, the original purpose of house facilities, for which they were designed, is today only fulfilled by the letter box, waste disposal room, heating rooms and parking areas; largely, by cellar cubicles as well. If the respondents' residential houses included (provided they do not include) rooms such as a pram and bicycle room, waste disposal rooms, cellar cubicles, pantry, parking areas, dryers, assembly rooms and an adapted roof area, then these rooms would be utilized. Next, the respondents responded to the question of "Where in your house would you like to have a room to store objects outside your dwelling?" with 34% of the respondents wished to have a room for the storage of objects on the ground floor at their dwelling, or in the form of a cellar cubicle in the house basement (20%), at the main entrance on the ground floor of the house (17%), or in separate rooms in the immediate vicinity of the residential house as well (15%). And it is this very answer that might indicate the demands made for the rooms by the present-day user. To the final question of "How in your opinion can the situation of house facilities be improved in your residential house?" respondents replied in almost the same way regardless of their different permanent residence, dwelling size or usability and spatial efficiency of house facilities. Namely, they requested more storage rooms in the dwellings, or even outside the dwelling, including their larger area, more parking places; about 70% of respondents considered safety to be of the utmost importance. Safety was mentioned in relation to the house facilities, the entrance door as well as the surroundings of residential houses.

4.1   *Conclusions of the questionnaire survey*

If present in the residential house, the following are used often:

– cellar cubicles and pantries (insufficient for the inhabitants' needs),
– parking areas (insufficient for the inhabitants' needs),
– assembly rooms (insufficient for the inhabitants' needs).

If present in the residential house, the following are seldom used:

– laundries (empty room, storage),
– rolling presses (empty room, storage).

The other house facilities (such as pram and bicycle rooms—insufficient for the inhabitants' needs) are used in various ways; for example, depending on the locality, dwelling size, or structure of the residential house inhabitants.

5 CONCLUSION

Only understanding the significance of collective housing and the influence exerted by an environment built in this way on the quality of its users' lives could lead to achieving a solution of this issue as well as preventing any innovations from being merely "morphological"; instead, they should rather be based on a well-considered concept of housing in its social, cultural, as well as economic and technical context.

There are considerable differences between the dwelling areas and the area for house facilities in residential houses. This fact also influences the usability and spatial efficiency, and subsequently the overall user comfort of residential houses.

In the Czech republic sphere, relatively little attention is paid to housing and almost none to house facilities. This fact further means that even though there are various other areas directly related to housing, such as building and administrative legislation, landscape planning, public housing policy, housing market and new construction, as well as consumer culture, demographic changes and behavioural patterns, changing over time, they are often unrelated in public documents.

Finally, it should also be said that it is appropriate to focus consistently on the organization of rooms within the layout, as well as their proportional relations. Material and aesthetic qualities should be rigorously applied at the moment of creation. Provision of efficiency, user, material and aesthetic quality produces the desirable user comfort.

ACKNOWLEDGEMENTS

The paper has been supported from the funds of the Student Grant Competition of VŠB—Technical University of Ostrava. The project registration number is SP2016/55.

REFERENCES

Kohout, M. & Tichy D. & Tittl F. 2015. Collective Housing: a Spatial Typology. Prague: The Czech Technical University.

Kuta, D. & Ceselsky, J. 2015. Use and Spatial efficiency of Housing Facilities and its Influence on the Overall User Comfort in Blocks of Flats in the Czech Republic. In WIT Transactions on The Built Environment: Sustainable Development and Planning VIII. Turkey: WIT Press, Wessex institute.

Kuta V. & Kuda F. 2005. Housing—Component of Life Quality, its Functions and Changes. In Urbanism and Landscape Planning VIII(1). Prague: Ministry for Regional Development of the Czech Republic.

The Universal Declaration of human rights [online]. 1948 [quoted 2010-04-20]. Available at: <http://www.ohchr.org/EN/UDHR/Pages/Language.aspx?sLangID = czc>., Art. 25.

Vonka, M. & et. 2012. The SBToolCZ Methodology, Assessment of Residential Buildings. User Comfort. Prague: The Czech Technical University.

Advances and Trends in Engineering Sciences and Technologies II – Al Ali & Platko (Eds)
© 2017 Taylor & Francis Group, ISBN 978-1-138-03224-8

Contactless method of measuring displacement in the study of brick prisms

D. Łątka & M. Tekieli
Cracow University of Technology (CUT), Cracow, Poland

ABSTRACT: The article compares results of strain measurements performed by using two methods: contactless and traditional. Two destructive tests were conducted, each on brick prism cut from brick wall made in the laboratory. The Ist method was based on traditional Linear Variable Differential Transformers (LVDT), while the IInd method was based on optical measurement (aimed to extend the scope of information on the behaviour of the tested fragment of the structure). The optical measurement was based on the Digital Image Correlation Method (DIC) and taken at 196 points on the two side surfaces of each of the test prisms. The results obtained by the optical measurement were first compared with those obtained using the traditional measurement, and shown to be compatible. Then, maps of displacements and deformations (presenting the behaviour of the structure during loading) were generated.

1 INTRODUCTION

In order to evaluate the homogeneity of mechanical parameters of existing masonry, numerous Non-Destructive Tests (NDT) are usually performed, such as Schmidt hammer (Matysek & Łątka 2012; Łątka 2014), ThermoVision, sonic pulse velocity, ultrasonic or radar tests (Niker 2010). The Minor-Destructive Techniques (MDT) includes mainly: the hole-drilling method (Onsiteformasonry 2005), single and double flat-jack test (Łątka 2015) and pull-off test. Unfortunately, the main drawback of the above mentioned tests is that they allow to determine the mechanical parameters for the outer layer of the wall only. The most popular Destructive Tests (DT) that allows the assessment of the wall is taking into account the simultaneous existence of bricks and mortar are: large core testing and prism testing. In terms of adequacy of the obtained results, tests conducted on prisms cut from the construction are the most desirable. They allow simultaneous determination of the cracking stress level, compressive strength and material characteristics such as Poisson's ratio and modulus of elasticity (Matysek & Witkowski 2013).

Normally, in tests determining the strength and elastic characteristics of masonry prisms, the LVDT sensors are used. Unfortunately, they often require installation directly on the tested surface. Other weaknesses of this method of measurement is that the LVDT sensor measures displacement only in one direction, and that the sensor measuring range (maximum lengthening or shortening/extending) and the base length need to be determined even before the test. Also to be emphasized, important limitations of this approach are: the long time required to install the LVDT sensors, their high price as well as the need to involve a qualified personnel. During the test, the sensors veil a part of the surface, thus making it difficult to observe the occurrence and development of cracks. They also tend to fall off at higher levels of deformation. The alternative is to use the contactless displacement measurement method, based on optical recording using a high resolution camera or camcorder. The purpose of this article is the juxtaposition of the results obtained using both of these methods.

2 METHODOLOGY OF EXPERIMENT

2.1 *Test stands description*

The set I, based on the LVDT sensors is shown in Figure 1a. Used were 2 types of sensors WA/20mm and WA/50mm; the displacement was measured on 3 different lengths of the bases: 26 cm and 42 cm (vertical), and 25 cm (horizontal).

The set II for contactless measurement is shown in Figure 1b. The applied lens allowed the registration of the change of all markers position placed on the tested prism's surface. Used were paper markers, with dimensions of 3 x 10 mm and high contrast, which allowed increased reliability of the measured displacement, as compared with the measurement based on the texture of the brick itself. Other ways to use the visual method can be found in the literature (Cucchi 2012; Tekieli 2014).

In the case of the method I, measurements were made on all four side surfaces of the prism (a total of 16 LVDT sensors), while in method II, due to the limited access, the results were recorded for two side surfaces (a total of 196 markers). A detailed distribution of sensors for both methods is shown in Figure 2.

In order to carry out this comparative study, 2 prisms were cut out of brick walls previously made in the laboratory. These prisms were then prepared for testing by aligning the top and bottom surfaces with cement mortar and installing sensors and markers. The test was performed with the constant increase of compressive force until the prism's destruction by means of collapse. In order to reduce the effect of friction between the bearing plates of the testing machine and the upper and lower surfaces of the prism, used were: PE layer and

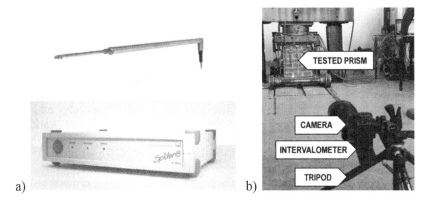

Figure 1.   a) HBM's measurement kit (inductive displacement transducer and amplifier), b) The optical method test stand.

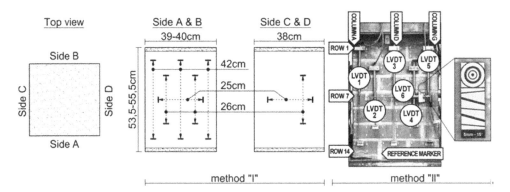

Figure 2.   Distribution of sensors and markers in method I and II.

synthetic oil. Data from the LVDT sensors was registered using the software HBM Catman Professional. Cameras used in method II were set within a distance of a few meters as shown in Figure 1b. Photos were taken simultaneously by two cameras with a fixed time interval.

2.2 Basis of optical method

Optical measurements can be performed at any points on the sample surface, provided the surface is appropriately prepared and is not too uniform. In the case of masonry, due to the uniform outer texture of bricks, it is not possible to use its natural surface in measurements. To make the measurement possible, specially prepared markers were attached on the surface of the bricks, forming a measuring grid composed of 7 columns (A-G) and 14 rows (1–14). Additional markers were attached at the terminal points of the bases of the LVDT sensors. Arrangement and appearance of each marker are shown in Figure 2. To determine the displacement of the marker, two photos from the previous and the current measurement moment are used iteratively. The total displacement of the marker is defined as the sum of displacements determined on the basis of two consecutive images. The idea of the method of tracking the pattern, which is a fragment of the sample surface is presented in Figure 3.

For pattern tracking on the surface of the sample, the digital image correlation method was used. This method is based on finding the maximum value of the correlation coefficient between two matrices containing total values describing the pixel intensity, which are included in the corresponding image subsets. The designated value is equal to the displacement of one image subset in relation to the corresponding subset of the second image (Peters et al. 1982; Bremand et al. 2010).

To calculate the correlation coefficient between the reference subset "f" and the target subset "g", whose dimensions are equal and are MxN pixels, the zero mean normalized cross correlation method was used, as described by Equation 1 below:

$$CC^{ZMN} = \frac{\sum_{i=1}^{M}\sum_{j=1}^{N}((f(i,j)-\mu_f)\times(g(i,j)-\mu_g))}{\sqrt{\sum_{i=1}^{M}\sum_{j=1}^{N}(f(i,j)-\mu_f)^2}\times\sqrt{\sum_{i=1}^{M}\sum_{j=1}^{N}(g(i,j)-\mu_g)^2}} \tag{1}$$

where μ_f means luminosity of the reference subset and μ_g means luminosity of target subset.

To determine the displacement with the accuracy greater than one pixel without increasing the image resolution, using the sub-pixel measurement is required. Photos collected during the measurements are first converted to grayscale, so that each pixel discretizing the image is described by an integer value from 0 (black) to 255 (white). Here, the most frequently used method of interpolation of intermediate values is the bi-cubic spline interpolation (Rastogi 2015; Xiong et al. 2011; Luu 2011).

By using sub-pixel measurement, it is possible to increase the measurement resolution to the level of 1/10 or even 1/50 of the integer pixel unit. A vision system with such a resolution of the measurement can be used to study deformation of masonry structures. By using a

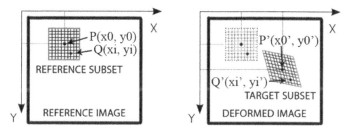

Figure 3.   General principle of operation of the digital image correlation method.

suitable resolution of the optical sensor and high quality of the lens, it is possible to obtain measurement resolution similar to that provided by the conventional LVDT sensors.

The vision system was equipped with two Digital Single-Lens Reflex cameras (DSLR) with a resolution of 16 Mpx and 24 Mpx. Both cameras were set on a stable tripod, approx. 2 m from the plane of the sample. The cameras were equipped with digital intervalometers to take photos automatically, at regular intervals. In the case of the first prism, photos were taken every 5 sec., while in the case of the second prism, every 10 seconds. The photos were taken using the manual settings of the aperture, shutter speed and ISO value, without image compression. RAW files were processed using an optical measurement system called CivEng Vision, developed at the CUT.

3 RESULTS

3.1 LVDT sensors

Based on the recorded data, deformation was averaged for each side of prism 1 and 2. The resulting curves are shown in Figure 4. The curves located under the deformation axis correspond to the horizontal sensors, and the other curves, to the vertical sensors.

The parameters of elasticity were determined for both prisms at the load equal to 440 kN, which corresponded to the stress of 2.97 MPa and was equal to 1/3 of the destructive value in the weaker prism. The resulting values of the destructive force, as well as the modulus of elasticity and Poisson's ratio are significantly higher for prism 1. The measured values are summarized in Table 1.

3.2 Optical measurements

The first step of result analysis was to compare the results with those obtained with conventional LVDT sensors. The correlation between the stress and the deformation shown in Figure 5.

Based on these results, the material constants were calculated. Young modulus for the results from the sensor no. 11 on the first prism was 2.56 GPa for the optical method, and

Table 1. The summary of measured parameters for prisms 1 and 2.

The measured parameter	Prism 1	Prism 2
Destructive force	1666 kN	1322 kN
The modulus of elasticity	3.27 GPa	1.91 GPa
Poisson's ratio	0.27 [-]	0.18 [-]

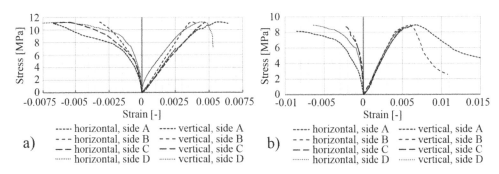

Figure 4. σ–ε relation for every side of the prism: a) Ist specimen, b) IInd specimen.

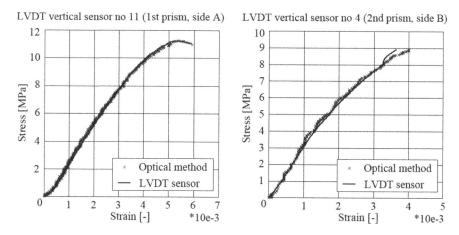

Figure 5. Comparison of the results of the optical method with the results from the selected sensors.

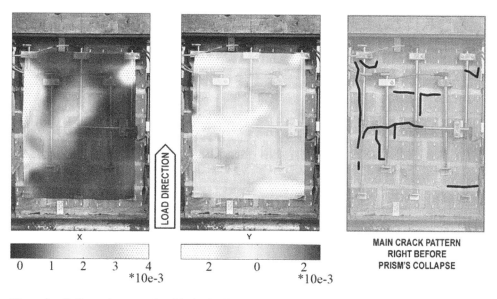

Figure 6. Deformation maps for side A of prism 2.

2.43 GPa for conventional measurements. For sensor no. 4 on the second prism, these values were respectively 3.32 GPa and 3.09 GPa.

Similar compliance was achieved for the other measuring points corresponding to the LVDT sensors, which allowed generating maps of displacements and deformations, describing the behaviour of the sample during the test. Figure 6 presents a map of deformation in the direction X and Y, together with a schematic sketch of cracks occurring in prism 2 just before it is destroyed.

4 CONCLUSION

The results of the correlation 6-ε and the modulus of elasticity, which were presented in this article, as determined by the classical method based on LVDT sensors and visual methods, are similar. The difference in displacement, recorded with the two methods, ranged between 3 and 8%. A great advantage of the method II (contactless) proved the recording of the resultant of the markers' displacement, and not only their component, as in the method I.

This allowed to determine the location of cracks even before it can be seen with naked eye; this is especially useful when the crack occurs at the bonding of brick with the vertical seam. The multi-pattern measurement also enabled generating a deformation map, both vertical and horizontal. Due to the still insufficient number of publications in the field of contactless measurement in masonry structures, it is recommended to use the analysed measurement system as complementary to the classical method of measurement. The usage of optical measurement system can help to reduce the number of the masonry prisms to be extracted for a reliable assessment of mechanical parameters of the existing structure.

REFERENCES

Bremand, F. et al. 2010. Monitoring of civil engineering structures using digital image correlation technique. In: *EPJ Web of Conferences*. Volume 6., EDP Sciences, 2010.

Cucchi, M. & Tiraboschi, C. & Antico, M. & Binda, L. 2012. Optical system for real-time measurement of the absolute displacements applied to flat jack test. Proceedings of the International Conference on Structural Analysis of Historical Constructions: 2528–2535. Wrocław, 15-17 October 2012. Wrocław: Dolnośląskie Wydawnictwo Edukacyjne.

Luu L. 2011. Accuracy enhancement of digital image correlation with B-spline interpolation, *Opt Lett*. 2011 Aug 15;36(16):3070-2.

Łątka, D. 2015. Zastosowanie poduszek ciśnieniowych typu flat-jack do określania stanu naprężenia w konstrukcji murowej. Wiedza i eksperymenty w budownictwie: prace naukowe doktorantów: praca zbiorowa pod red. Joanny Bzówki, volume 558: 225–232. Gliwice: Wydawnictwo Politechniki Śląskiej.

Łątka, D. 2014. Możliwości wykorzystania sklerometru elektronicznego do wstępnej oceny konstrukcji murowej. Współczesny stan wiedzy w inżynierii lądowej: prace naukowe doktorantów: praca zbiorowa pod red. Joanny Bzówki, volume 519: 327–334. Gliwice: Wydawnictwo Politechniki Śląskiej.

Matysek, P. & Łątka, D. 2012. Comments on the application of the sclerometric method in the diagnosis of brick masonry. Proceedings of the International Conference on Structural Analysis of Historical Constructions: 2471–2479. Wrocław, 15–17 October 2012. Wrocław: Dolnośląskie Wydawnictwo Edukacyjne.

Matysek, P. & Witkowski, M. 2013. Badania wytrzymałości i odkształcalności XIX-wiecznych murów ceglanych. Conference proceedings. Awarie Budowlane: zapobieganie, diagnostyka, naprawy, rekonstrukcje: XXVI Konferencja Naukowo-Techniczna: 183–190. Szczecin-Międzyzdroje, 21–24 may 2013. Szczecin: Wydawnictwo Uczelniane Zachodniopomorskiego Uniwersytetu Technologicznego.

Niker: New integrated knowledge based approaches to the protection of cultural heritage from earthquake-induced risk. Research project no. 244123 final report, 2010, Padova.

Onsiteformasonry: On-site investigation techniques for the structural evaluation of historic masonry buildings. Recommendations for the end users. Research project no. EVK4-2001-00091 final report, 2005, Florence.

Rastogi P. 2015. Digital Optical Measurement Techniques and Applications, Artech House.

Tekieli M., Słoński M. 2013. Computer vision based method for real time material and structure parameters estimation using digital image correlation, particle filtering and finite element method. In *Artificial Intelligence and Soft Computing*, volume 7894 of Lecture Notes in Computer Science, pages 624–633. Springer, 2013.

Tekieli, M. & Słoński, M. 2014. Particle filtering for computer vision-based identification of frame model parameters. *Computer Assisted Methods in Engineering and Science volume 21* (online access: http://cames.ippt.pan.pl/pdf/CAMES_21_1_5.pdf): 39–48.

Xiong L. et al. 2011. Evaluation of sub-pixel displacement measurement algorithms in digital image correlation, *Mechatronic Science, Electric Engineering and Computer (MEC)*.

Advances and Trends in Engineering Sciences and Technologies II – Al Ali & Platko (Eds)
© 2017 Taylor & Francis Group, ISBN 978-1-138-03224-8

Evaluating adhesion of weatherproofing sealant joints: Price vs. quality

P. Liška, B. Nečasová & J. Šlanhof
Faculty of Civil Engineering, Brno University of Technology, Czech Republic

ABSTRACT: Joint sealing in facades of buildings is a technologically very complicated process that cannot be solved without connection to other constructions. To ensure proper functionality of the joint it is necessary to use a suitable sealant. The core idea of this experiment was general conviction that the more expensive material or product is used the better properties it should demonstrate. For presented experimental measurements only polyurethane sealants that are intended for usage in exterior were selected. The selection of tested sealants was based on previous experiences as well as on some previous laboratory experiments. The performed experiments involved verification of adhesive and cohesive sealant properties to the selected substrate. Even though the obtained data demonstrate that high quality sealants were selected, certain differences between cheap and expensive test specimens had appeared.

1 INTRODUCTION

Sealing joints in facades of structures subjected to climatic effects such as weathering is a very demanding process, as stated by (Novotný 2016) that cannot be solved without connection to other constructions. We adopted the idea for testing real sealed joint from previous research project that focused on joint sealing in the building envelopes. The core idea of the experiment was general conviction that the more expensive material or product is used the better properties it should demonstrate. Our team decided to confirm or disprove the hypothesis. That is why we chose to test two sealants with a maximum difference in their price. Producers often declare application of the sealant in the façade but they do not state under what conditions. Only sealing without any connection to other circumstances such as preparation of the base leads to defects in the future. Sealants in joint should provide a permanent sealing (Pocius 2012, Klosowski and Wolf 2016, Bull and Lucas 2015). Without ensuring sufficient adhesion of the sealant to underlay materials and without possibility of expansion moves damage can be caused. On the market, there are plenty of combinations of tile materials, sealants and conditions of applicability. That is the reason why producers cannot include the entire area.

While choosing cheaper sealants, it was requested that the producer allowed for the application without improving coatings. In case of the application of a more expensive sealant, an improving coating was added to enhance adhesion of the sealant to the cladding. This fact not only increases the price difference of used sealants, but also the assumption of more significant differences in the results.

As we wanted to verify actual suitability, a whole range of tests in the scope of several tens of samples had to be carried out. Performed tests, i.e. the method of testing, or the shape of test bodies are not in compliance with requirements of the standards, by which sealants are tested. Testing methodology is adopted from sets of tests that are used for testing bonded façades. The authors of the article selected the shape and size of the test bodies based on previous experience. The authors of the article are convinced that testing methodologies used characterize better the conditions that the sealants are actually exposed to. Another conclusive advantage of the selected methodology is utilization of own test equipment.

2 EXPERIMENTAL

2.1 *Materials*

Two polyurethane sealants of different producers were chosen for experimental testing. They were selected based on experience from previous research. Application of sealants that are commonly sold on the Czech market was emphasized. Gutta PU 40, sealant made by GUTTA ČR—Praha spol. s r.o., was selected as the representative of cheap sealants. This product is about half the price of the second representative. SikaFlex 11 FC+ sealant with improving coating Sika Primer-3 N produced by Sika CZ, s.r.o. was selected as the representative of expensive sealants. Basic technical parameters are stated in producers' data sheets (Gutta ČR—Praha spol. s r.o. and Sika CZ, s.r.o.).

In both cases, one-component bonding and sealing materials on polyurethane basis hardening by air humidity were concerned. Selected sealants are suitable for exterior and interior application. The basic version of cement-bonded wood particleboard "Cetris" without any surface treatment was selected as the tile material. Such application is not quite usual because cement-bonded wood particleboards with surface treatment are used primarily. On the other side, producers allow it. Cutting into requested shapes was done using woodworking saws in compliance with producers' procedures. Basic technical parameters are stated in producer's data sheet (CIDEM Hranice, a.s.).

2.2 *Test of adhesion and cohesion of samples*

The aim of this case study was to measure a maximum elongation at break of selected sealants and compare the recorded data. Two test methods, described in the text below, were selected for this purpose. The normal as well as shear extension of test samples was executed. The principle of both methods was normal and shear stressing of test samples by its extension until it was damaged. For each temperature and test method 6 test samples, i.e. total of 72, were made.

2.2.1 *Manufacturing and curing of test samples*

For production of selected shape of sample, it was first necessary to produce a special mould that would allow shaping of the sealant. Dimensions of the test sample that consists of two items is 40 × 160 mm. To create a joint the test sealant is inserted in-between the items. The shape and dimensions comprise a practicably sealed joint. The width of joint is 10 mm, length of 160 mm and depth of 6 to 8 mm are considered for the purposes of testing. Depth of the joint depends on its width. Foam filling "sealing strand—lite foam polyethylene" was used in production of samples in order to avoid three-point stressing of the sealant. Production of samples was fully in compliance with technical procedures of the sealant producers (Gutta ČR—Praha spol. s r.o. and Sika CZ, s.r.o.).

Treatment of sealed surfaces depended on the requirements of sealant producers. Therefore, in case of cheap sealant application, the surface was not treated only dust was removed mechanically.

However, in case of expensive sealant application, tile surfaces were first cleared of dust. Then underlay improving coating "primer", which required ventilation for 30 minutes, was applied by a small brush. While handling, the treated surfaces of samples had to be protected against possible impurities.

Test samples cured in a dry and clean environment at average air temperature $(20 \pm 3)°C$ with relative humidity $(50 \pm 5)\%$ as required by the standard (Czech Standards Institute 2009).

2.2.2 *Conditioning of test samples*

The environment as well as aging has a crucial impact on the efficiency and durability of sealed joints (Wolf 1999, Chew et al. 2002). For testing the samples were conditioned so that they characterize the weather conditions in the place of building in as close as possible.

In the first set of tests, the samples were not conditioned at all. Samples started to be tested after curing. The next set of samples was conditioned for sudden temperature changes

(Czech Standards Institute 1983). The principle of conditioning was alternating of warming the samples by infra-red lamps and subsequent cooling in water bath in 25 cycles. A construction for keeping the samples that allows tilting into water bath needed to be made for testing. Warming was ensured by a free standing halogen heating. The samples were warmed to the temperature of $(70 \pm 3)°C$ and subsequently cooled to $(20 \pm 2)°C$. Warming was carried out by reaching the requested temperature in 30 to 40 minutes, cooling was reached in 5 to 10 minutes. Warming and subsequent cooling formed one cycle. Surface temperature of the samples was measured by a contact thermometer. Measurement was carried out for 30 seconds after warming was completed. While cooling down the temperature was measured for 15 seconds after taking out of the water bath (Liška et al. 2014a). The last set of samples was conditioned to frost resistance (Czech Standards Institute 1981). The principle of conditioning was alternate freezing and defrosting of test samples in 15 cycles. Test samples were immersed into water bath of the temperature of $(20 \pm 3)°C$ for 6 hours. Subsequently they were taken away and placed, in vertical position, into freezing area of the temperature of $(-20 \pm 2)°C$ for 18 hours. Time of placing in water and freezing area, i.e. 24 hours, formed 1 freezing cycle. After requested 15 cycles were carried out, the samples were let dry in upright position at air temperature of $(20 \pm 3)°C$ with relative humidity of $(55 \pm 10)\%$ (Liška et al. 2014b) for at least 4 days. Then the tests were performed.

2.3 *Experiment execution*

After producing the test samples, curing and possibly conditioning them, the samples were inserted into testing equipment to be tested. The equipment that was used for testing is not shown intentionally, which is due to ongoing patent proceedings. It is a mechanical equipment for testing of sealants in real joint allowing extension of sealed joint until it is damaged. Test sample of the length of 160 mm consists of two small boards 40 mm high that are interconnected at the upper edge using a sealant. The sample is inserted into the testing equipment that allows moving the underlay materials away at shear extension, during which the defects of adhesion and cohesion of the sealant with underlays are monitored. After the sample was fixed in the equipment, extension by 100% of the original joint width in the speed of c. (5.5 ± 0.7) mm/min was carried out. Extension was done manually through screws without using electrical equipment. In case the sample was not damaged during the first extension, the samples remained extended for 24 hours. After that the extension continued until the sample was damaged. While extending, behaviour of the sealant and small underlay boards was monitored in detail, as can be seen in Figure 1. The width of extending the joint was measured by sliding digital gauge. Six pieces of equipment for normal stressing and 6 pieces of equipment for shear extension were produced for testing in order to speed it up maximally.

3 ANALYSIS AND DISCUSSION

The obtained results evidently imply that used tiles can be sealed. Moreover both sealants are suitable for application in the exterior, which was tested here. As the presented results of conditioning to sudden temperature changes and frost resistance that represent external weather conditions show, adhesive and cohesive properties of tested materials do not decrease significantly unless the maximum elongation is achieved. Cohesive and adhesive failure of the sealant occurs at extension exceeding 350% of the original width of the joint, see Figure 1. In real situation such expansion move does not occur, not even due to temperature or humidity expandability, despite the fact it is useful to extend the samples until they damage.

In Figure 2 and Figure 3 are presented measured average values of samples maximum extension. The presented data show that differences between cheap and expensive representative are almost negligible. Therefore the assumption that cheaper product has to have worse properties was disproved.

At extension by 100% of the original width of the joint adhesion and cohesion did not appear, even not after 24 hours when sealant remained extended. This testifies high quality of used

Figure 1. Example of adhesive failure—normal extension.

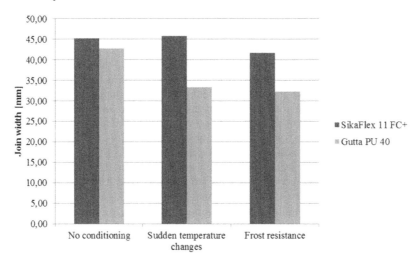

Figure 2. Test results—normal extension.

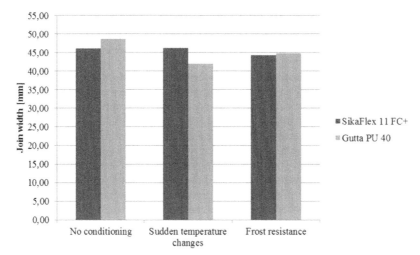

Figure 3. Test results—shear extension.

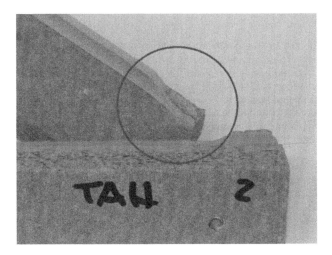

Figure 4. Defect of tile material—normal extension.

sealants and tile material. Recording the course, i.e. where and how the defect occurs, is what is considered in particular in this research case. In several cases defect of tile material, e.g. due to insufficient treatment of tile edges, occurred, see Figure 4. It was found that if slight defect of tile material occurs, subsequently the entire joining is gradually damaged regardless the used sealant.

Contrary to existing methodology of testing sealants according to standards, the method we chose provides essential findings regarding the defects of adhesion and cohesion that are decisive in the cases when the sealing function of sealant is principal.

4 CONCLUSION

Based on the experimental measurement, it is evident that achieved results of cheaper sealant without improving coating are worse than results of the sealant with applied coating. On the other side, the results are not as different as the authors expected at the beginning of the experiment. In case of a defect of sealant cohesion, both sealants are at quite comparable level. The sealant exposed to weather effects has a relatively short durability in terms of adhesion as has been described previously (Šlanhof et al. 2014). That is why there is the endeavour for maximum efficiency of the joining, thus also longer lasting durability. Costs are connected with durability of the joining. That is why the following question should be asked: "To what extent will application of an expensive sealant with better adhesion increase durability, thus decrease costs of possible renewal?". The answer should be given in the next research.

Although 6 samples were used for each test, the total number from all tests, however, is 36 per one sealant. Such number is representative and can be handled statistically.

Finally, it should be noted that it is not appropriate to "abandon" or "recommend" either of the sealants tested. Further tests where the testing methodology, i.e. production of samples, their conditioning, testing and subsequent evaluation of results will be fully in compliance with technical standards, need to be carried out to confirm or disprove the results. The authors of the article are aware that it is only the beginning of studying this issue and further verifications of this methodology appropriateness are required.

ACKNOWLEDGEMENT

This research was carried out as a part of the FAST-S-16-3434 project—Experimental Verification of Technological Design of the Construction of Sustainable Mobile Building with Emphasis on its Construction-physical Behaviour.

REFERENCES

Bull, E. D., Lucas, G. M. 30 Year Outdoor Weathering Study of Construction Sealants. In *Durability of Building and Construction Sealants and Adhesives: 5th Volume*. DOI: 10.1520/STP1583-EB; 2015.

CIDEM Hranice, a.s. *Documentation—Cetris: Basic*. Available at: http://www.cetris.cz/pagedata/boards/technical-data-sheet-basic-cz.pdf?1433418712 [Accessed 2016].

Chew M Y L, Zhou X. Enhanced resistance of polyurethane sealants against cohesive failure under prolonged combination of water and heat. *Journal of Polymer Testing*. Amsterdam: Elsevier BV; 2002 pp. 188–193. Available from: http://linkinghub.elsevier.com/retrieve/pii/S014294180100068X.

Czech Standards Institute. 1981. *ČSN 73 2579—Test for frost resistance of surface finish of building structures*. Prague.

Czech Standards Institute. 1983. *ČSN 73 2581—Test for resistance or surface finish of building structures to temperature variations*. Prague.

Czech Standards Institute. 1995. *ČSN ISO 10365—Adhesives. Designation of main failure patterns*. Prague.

Czech Standards Institute. 2009. *ČSN EN ISO 291—Plastics—Standard atmospheres for conditioning and testing*. Prague.

Gutta ČR—Praha spol. s r. o. *Tmely PU—Gutta Czech*. Available at: https://www.gutta.com/html/cz/produkty/stavebni-chemie-gutta/tmely-a-lepidla/tmely-pu/ [Accessed 2002].

Klosowski J., Wolf A. T. Sealants in Construction. Taylor and Francis Group: ISBN 9781574447170; 2016.

Liška, P., Šlanhof, J. & Nečasová, B. 2014a. Testing of Bonded Facade Joints in Sudden Temperature Changes. In *Construction Technology and Management CTM 2014 International Scientific Conference Bratislava (Slovakia): proceedings*. Bratislava: Slovak University of Technology in Bratislava. 8.

Liška, P., Šlanhof, J. & Nečasová, B. 2014b. Frost Resistance of Bonded Facade Joints. In *CRRB—16th International Conference of Rehabilitation and Reconstruction of Buildings: proceedings*. Brno: Scientific-technical association for remediation of buildings and care of monuments WTA CZ. 4.

Novotný, M. 2016. The applicability of existing methods for testing the base layers of ETICS. In *Applied Mechanics and Materials*. Switzerland: Trans Tech Publications. 8.

Pocius A V. Adhesives and sealants. *Polymer Science: A comprehensive Reference*. Amsterdam: Elsevier BV; 2012, pp. 305–324.

Sika CZ, s.r.o. *Documentation—Sikaflex®-11 FC+*. Available at: http://cze.sika.com/dms/getdocument.get/55cbf88f-a5a6–3b00–89eb-db6d4a022cbc/Sikaflex-11%20FC+_i_cure.pdf [Accessed 2014].

Sika CZ, s.r.o. *Documentation—Sika® Primer-3 N*, Available at: http://cze.sika.com/dms/getdocument.get/f93df705–20b3–3fcd-9c5b-41c08a519980/Sika%20Primer%203 N.pdf [Accessed 2015].

Šlanhof, J., Liška, P., Šimáčková, M. & Nečasová, B. 2014. Sealing Possibilities for Wood Elements. In *People, Buildings and Environment*. Kroměříž: Brno University of Technology, Faculty of Civil Engineering. 9.

Wolf A. T. *Durability of building sealants*. Bagneux: RILEM Publications s.a.r.l.; 1999.

Advances and Trends in Engineering Sciences and Technologies II – Al Ali & Platko (Eds)
© 2017 Taylor & Francis Group, ISBN 978-1-138-03224-8

Vertical distribution of compact air jets in the hall of an ice arena

E.A. Loktionova, D.Y. Malyshevsky, D.I. Schemelinin & D.D. Zaborova
Peter the Great St. Petersburg Polytechnic University, St. Petersburg, Russia

ABSTRACT: In recent years, a lot of ice arenas and skating rinks have been built in Russia. Such constructions have to work under specific air conditions. Therefore, one of the most important technological parts is to optimize the microclimate in halls of ice complexes. This article identifies a problem of the existing air distribution scheme of a given hall. The simulation model built in the program ANSYS Fluent, showed problems of design and choice of equipment for the ice hall, and also exceeding the recommended standards for air temperature and relative humidity. According to the resulting data, adjustments to the existing scheme were made. It allowed us to obtain the optimal values of temperature and humidity in ice arena zone and tribunes, corresponding to the standards.

1 INTRODUCTION

The use of simplified balancing methods, laws of jet-stream formation, and etc. for consistent description of air flow behavior in "bowl-shaped" ice rinks is quite difficult and, in most cases, practically impossible. This is conditioned by such characteristics as:

- the essential non-iso-thermality of air flow along the space (in high and in horizontal sections);
- the interaction of forced air flows (supply air jet from nozzles, diffusers) and intensive free convective flows (warm air flow rising from viewers array);
- the necessity of taking into account the radiation component on a large part of the surface, participating in the heat transfer (the surface of ice and roof) (Samarin, 2011).

These characteristics lead to the necessity of using computational fluid dynamics for analysis of air distribution of ice arenas and corresponding adjustments in their design. In other words, attraction methods that would be based on the numerical solution of combining three-dimensional differential equations of Navier-Stokes are required (Klimaregelung, 1984).

The most problematic design issue is related to air feeding in the ice arena (Pozin, 2012). Important problem lies in the choice of equipment, for example, muzzles. Engineers usually set the temperature in the exit from muzzles and in the working zone, in this situation near the ice arena, as a metric for initial program selection. However, in the case with ice arenas with sitting areas for viewers, it is not correct (Chujkin, 2014; Vatin 2014; Strelets 2014).

The purpose of this work is to analyze the project scheme for air distribution of the hall of ice arenas by temperature, speed, humidity fields computational simulation in 2D position (summer-time) (Gaas, 2014; Klimaregelung, 1984; Pozin, 2012; Samarin, 2011; Vatin, 2014; Petrichenko, 2015).

2 PROBLEM STATEMENT AND LAYOUT GEOMETRY

Temperature of the ice surface, energy of heat capacity from viewers, humidity uptake intensity, the power split of heat radiation from heat radiation from light system, heat input from roof of the building are given. Ingress of air and wall heat gain is neglected. Four groups of air conditioners for the tribune zone and ice zone are working. System performances, characteristics of supply air, types of supply device and schemes of air distribution are given. Position and productivity of exhaust air terminal are given in Figure 1 and Table 1, 2 (Greene, 2003; Kraft 1973; Petrichenko 2006; Petrichenko 2008).

– Tribunes. Heat liberation from viewers 60 W/m², water generation 7.2 g/m²*hr;
– Ice surface. Temperature –6°C;
– Bridge seating. Heat liberation (light, building envelop) 1 W/m²;

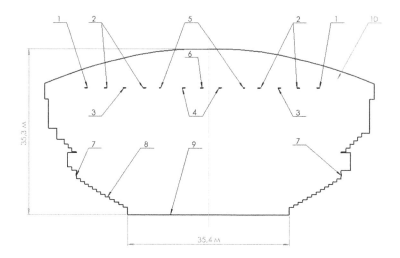

Figure 1. Cross-section of ice arena hall and boundary conditions.

Table 1. Boundary conditions, project results.

No	Name	Characteristic dimension mm	Spending m³/hr	Temperature °C	Humidity g/kg
1	Tributary	630	4800 × 2	21	10
2	Exhaust	630	3750 × 4	–	–
3	Tributary	630	5000 × 2	21	10
4	Exhaust	630	3900 × 2	–	–
5	Tributary	315	750 × 2	21	10
6	Tributary	630	5000	21	10
7	Exhaust	550	1650 × 2	–	–

Table 2. Boundary conditions, corrected results.

No	Name	Characteristic dimension mm	Spending m³/hr	Temperature °C	Humidity g/kg
1	Tributary	630	4800 × 2	21	10
2	Exhaust	630	3750 × 4	–	–
3	Tributary	630	3150 × 2	21	10
4	Exhaust	630	3900 × 2	–	–
5	Tributary	315	–	–	–
6	Tributary	630	10200	14	4
7	Exhaust	550	1650 × 2	–	–

The values of surface thermal emissivity was not taken into account in this calculation. RMS residual levels are equal 1E^{-4}.

Computation is carried out in ANSYS Fluent software package in stationary position with structured mesh—81 111 elements. Energy equations, equations of motion and mass transfer of vapor under action of gravity are solved. Turbulence Model RNG is used, scheme SIMPLE, 430 iterations (Sotnikov, 2012).

3   RESULTS OF COMPUTATIONAL SIMULATION

Results of simulation are shown in Figures 2, 3, 4.

For clarity of presentation, the limit value of maximum speed is 1 m/s, and relative degree of humidity—to 70%.

It is possible to observe the following:

1. Dead zone is being formed in ice arena (Hakan, 2011);
2. The maximum lowering altitude of air jet from specified 27 m over ice arena is about 16 m;
3. Air temperature in ice arena zone and sitting area exceeds the recommended values: 6°C and 22°C, respectively;
4. The value of relative degree of humidity over ice arena and sitting area exceeds the recommended value 55% (Bonecrusher, 2006; Melkumov, 2008;Petrichenko, 2014);
5. In this air distribution scheme it seems unreasonable to accommodate the air flow 5 with 750 m³/h spending—supply air jets do not reach ice arena zone and sitting area, stay in bridge surface, leading to unreasonable expenses of air circuits.

To correct the problem with the existing scheme, the following alternative processes are proposed:

– Increase spending in the central air inlet 6, so that air flow could reach the ice arena zone. Guide spending value could be defined as (Flynn, 2006; Kharkov, 2014;Petrichenko, 2014; Nemova, 2014):

$$Z^{\mathrm{max}} = 0.58 \cdot H_j \tag{1}$$

where H_j—maximum of lowering the jet, m; $Z^{\mathrm{max}} = 27$ m—the height of the air inlets over the ice arena. Proceeding from the position that the initial speed of the jet must be equal:

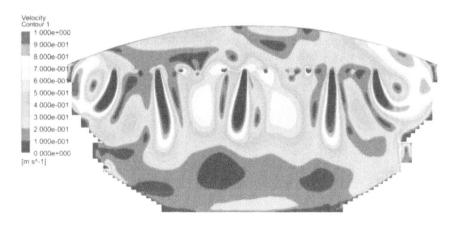

Figure 2.   Velocity contour diagram.

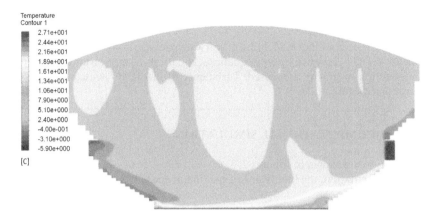

Figure 3.   Temperature contour diagram.

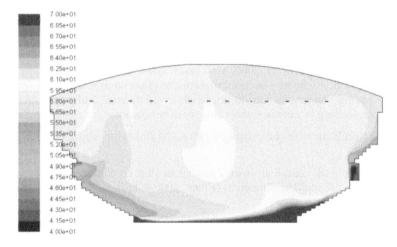

Figure 4.   Relative degree of humidity diagram.

$$U_0 = H_j \cdot \frac{\sqrt{n \cdot dt}}{5.45} \cdot m \cdot \sqrt[4]{F_0} = 8.8 m/s \qquad (2)$$

where m = 6.4; n = 4.8—kinematic and heat characteristic of diffuser; dt = 5°C—initial temperature difference between supply and ambient air; F_0 = 0.32 m²—sectional area of diffuser.
– Reduce the temperature and humidity in air inlet No 6 to values 14°C and 4 g/kg, respectively, creating climate control by air-conditioning and dehumidification.

Corrected targets are shown in Figures 5, 6, 7.

4   RESULTS

The application of the above-mentioned adjustments to the existing scheme of air distribution allowed:

1. To dispose of hold up spots in ice arena area;
2. To get temperature values in the ice arena zone and tribunes within the limits 6–22°C;
3. To provide the relative degree of humidity on tribunes near the optimum value—55%.

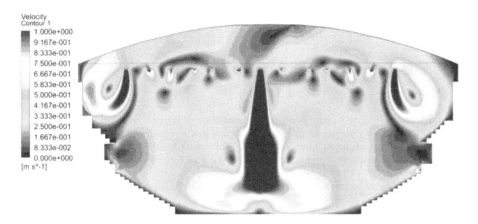

Figure 5.   Velocity contour diagram.

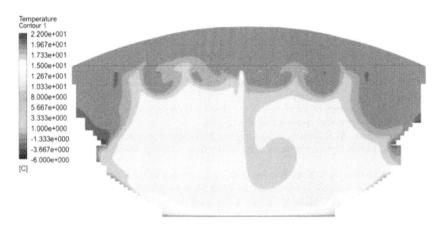

Figure 6.   Temperature contour diagram.

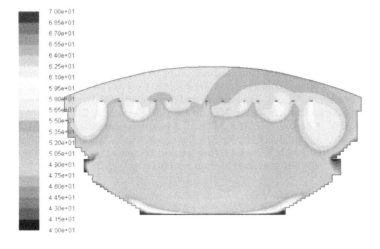

Figure 7.   Relative degree of humidity diagram.

This calculation is conventionally shown by a comparison of two methods of air distribution for the 2D model. More accurate results can be obtained when modeling point sources for the 3D model.

REFERENCES

Bonecrusher, I.V., Kutushev, A.G. 2006. Numerical investigation of a free convection indoor air with a heat source. *Teplofizikaiaeromehanika [Thermophysics and Aeromechanics]* 3: 425–434.

Chujkin, S.V., Dubrovskaya, E.E., Proskuryakov, A.E., Ryabtsev, O.A. 2014. Modern ways of creating of the microclimate of covered ice arenas and skating rinks. *Voronezh State University of Architecture and Civil Engineering*: 25–37.

Flynn, M.R. 2006. Natural ventilation in interconnected chambers. *Journal of Fluid Mechanics* 564: 139–158.

Gaas, I. A., Startsev, S.A., Kharkov, N.S., Shuravina, D.S. 2014. Numerical modeling of stationary heat exchange panel building OD4 series. *Construction of Unique Buildings and Structures* 1(16): 23–35.

Greene, G. 2003. Heat Transfer. Encyclopedia of Physical Science and Technology (Third Edition). San Diego: Academic Press.

Hakan, O. F., 2011. Natural convection in wavy enclosures with volumetric heat sources. *International Journal of Thermal Sciences* 4: 502–514.

Kharkov, N.S., Chamkina, M.V., Petrichenko, M.R. 2011. Whether the recovery of a pressure in the cylindrical twirled flow is possible. *Magazine of Civil Engineering* 1(19): 12–16.

Kharkov N.S., Ermak O., Aver'yanova O. 2014. Numerical simulation of the centrifugal separator for oil-water emulsion. *Advanced Materials Research* 945–949: 944–950.

Klimaregelung, J.B. 1984. *GrundlagenPrax is der Projektierung*. München.

Kraft, G. 1973. *Lehrbuch der Heizungs. Laftungs und Klimatechnik*. Dresden: Verlag Theodor Stimkopff.

Melkumov, V.H. 2008. Dynamics of formation of air flow and temperature fields in the premises. *News of higher educational institutions construction* 4: 172–178.

Nemova, D.V., Petrichenko, M.R., Vatin, N.I., Kharkov, N.S., Staritcyna, A.A. 2014. EOR (Oil Recovery Enhancement) technology using shock wave in the fluid. *Applied Mechanics and Materials* 627: 297–303.

Petrichenko, M.R., Kharkov, N.S. 2006. Extreme properties of spin motion of viscous fluid. *Proceedings of the Fourth Russian National Conference on Heat Transfer RNKT* 4(2): 230–232.

Petrichenko, M.R., Kharkov, N.S. 2008. Hydraulic losses over the main stretch of a cylindrical channel at a low degree of twisting. *St. Petersburg State Polytechnic University Journal* 4(63): 237–242.

Petrichenko, M.R., Kharkov, N.S. 2014. Boundary problems for the Crocco equation in the transport theory. *St. Petersburg State Polytechnic University Journal. Physics and Mathematics* 3(201): 47–56.

Petritchenko, M.R., Kharkov, N.S., Nemova, D.V. 2014. Hydraulic version of the Business model of free convective motion in vertical slits. *St. Petersburg State Polytechnic University Journal. Energetics* 4(207): 26–40.

Petrichenko M., Nemova D., Reich E., Subbotina S., Khayrutdinova F. 2015. Propulsive force and average velocity of free-convective motion in a vertical channel. *Procedia Engineering* 1: 127–1134.

Pozin, G.M., Ulyasheva, V.M. 2012. Distribution of air parameters in premises with heat release sources. *Magazine of Civil Engineering* 6: 42–47.

Samarin, O.D., Azivskaya, S.S. 2011.On numerical calculation of an assimilation factor of variable heat ingress at automation of micro-climate systems. *Magazine of Civil Engineering* 23(5): 31–33.

Sotnikov, A.G., Borovitskiy, A.A. 2012. Theoretically and experimental validation of the air changes in industrial ventilation optimization method. *Magazine of Civil Engineering* 2: 32–38.

Vatin, N.I., Nemova, D.V., Kharkov, N.S. 2014. Method of numerical simulation of a centrifugal separator for cleaning petroleum products. *Applied Mechanics and Materials* 680: 354–358.

Vatin, N.I., Strelets, K.I., Kharkov, N.S. 2014. Gas dynamics in a counter flow cyclone with conical nozzles on the exhaust pipe. *Applied Mechanics and Materials* 635–637: 17–21.

Advances and Trends in Engineering Sciences and Technologies II – Al Ali & Platko (Eds)
© *2017 Taylor & Francis Group, ISBN 978-1-138-03224-8*

Theoretical and methodological principles of impact assessment of strategic documents on environment and health

M. Majerník, N. Daneshjo & G. Sančiová
Economical University in Bratislava, Košice, Slovakia

ABSTRACT: Formally, Environmental Impact Assessment (EIA) and SEA are structured approaches for obtaining and evaluating environmental information prior to its use in decision-making in the development process. This information consists basically of predictions of how the environment is expected to change if certain alternative actions are implemented and advice on how best to manage environmental changes if one alternative is selected and implemented. Whilst EIA focuses on proposed physical developments such as highways, power stations, water resource projects and large-scale industrial facilities, SEA focuses on proposed actions at a "higher" level such as new or amended laws, policies, programmes and plans. Often, physical developments and projects are the result of implementation of a policy or plan, for example an extended highway network may be an outcome of a new transport policy.

1 INTRODUCTION

In order to meet the present and future needs and principles of balanced socio-economic and environmental development, i.e. sustainable development, preference is given especially to preventive tools such as Impact Assessment of projects and plans in general (strategies, plans, programs, documents or projects) on environment and human health. Methods, processes as well as legislatives of such assessments are constantly improved and new effective methodologies developed. An example of legislation and its global coordination (particularly in relation to cross-border impacts and public awareness) is the EIA process. Also methodologies for Health Impact Assessment are currently well developed for instance by World Health Organization.

The problem is that the Environmental Impact Assessment and health impact assessment processes are formally separated. Although it is now undeniable that population health and environmental quality are "communicating vessels". The health and well-being of a population depend on many different factors such as environmental, individual, social and economic aspects. Improving the health status of a population and reducing ill health pose challenges to national and local governments in multi-sectoral decision-making.

2 METHODOLOGICAL PROCESS OF IMPACT ASSESSMENT OF STRATEGIC DOCUMENTS

The particular steps of the impact assessment are intertwined and all participants of the Strategic Environmental Assessment (SEA) process access them virtually from the initial phase (Figure 1). For the application of the law, particularly in terms of guaranteeing the time flow, it is particularly important that each step of the assessment (except for step—the preparation of the report on the evaluation of strategic document) are designed binding deadlines which have to be respected. This consistency creates the preconditions for good communication between participants in the SEA process and mutual respect of rights and obligations arising from the law. Deadlines for each step are designed to allow a thorough assessment of the proposed strategic document without undue delay.

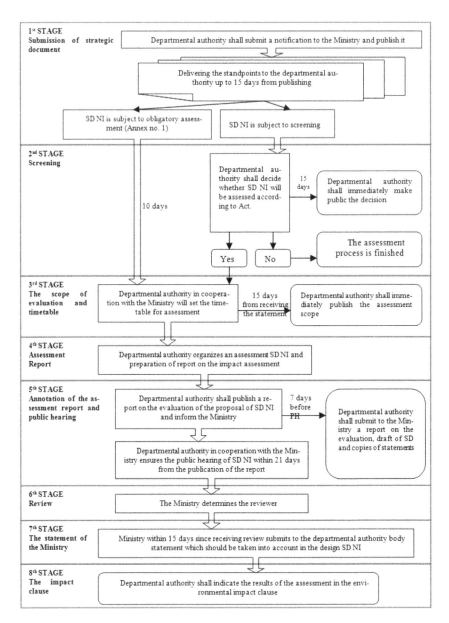

Figure 1.   The sequence of steps for assessing strategic documents of national importance, SD NI-strategic document with national importance, Ministry—Ministry of environment of Slovak Republic, Departmental authority—relevant ministry to ensure the development of SD NI, PH—Public Hearing.

3   SUBMISSION OF STRATEGIC DOCUMENT

Submission of the announcement of strategic document (next referred as the announcement) is the first step in the SEA process. At this stage, by an announcement is submitted the first official information on the considered strategic document, its nature, objectives, content, relation to other documents, as well as its environmental impacts, including impacts on human health and proposes measures to eliminate, reduce or compensation of the anticipated negative impacts. The contracting authority which is responsible for preparation of the strategic document, whether it has developed this document itself or entrusted it to an expert, submits the announcement to the competent authority. In the case of the strategic document of national importance, the announcement is submitted to the departmental authority. An announcement is a basic material in the process of the assessment of strategic documents to decide whether the assessed strategic document, respectively the change in strategic document will be assessed according to the Act no. 24/2006. If the strategic document is the subject of an impact assessment, this step also serves to determine the scope of the assessment of the strategic document and, if applicable timetable.

When strategic documents are subject to mandatory impact assessment under the Act it is recommended to create a team for the preparation of the strategic document and involve them in the development of the Announcement of the strategic document. The announcement of a strategic document, including land planning documentation must be prepared in accordance with Annex no. 2, which determines the content and structure of the Announcement of strategic document.

One of the most important requirements for the announcement is the basic data on the expected impacts of the strategic document on the environment, including impacts on human health for each proposed variant of the Strategic document. The description of these impacts is of great importance in relation to the further course of the assessment process. The sequence of steps of Announcement of strategic document is carried out according to the Act no. 24/2006. This process is summarized in Figure 2.

4   SCREENING PROCESS OF ASSESSING THE STRATEGIC DOCUMENT

The screening stage is applied in the case of optional (voluntary) nature of the process and consists of two parts. The first is to submit and comment on the announcement, for which the requirements and procedures are the same as in the compulsory nature of the process. The second part of the decision is whether the strategic document will or will not be assessed according to Act no. 24/2006. In the case of a decision of the competent (departmental) authority that the strategic document will not be assessed according to the law, the SEA process is finished. If relevant the (departmental) authority decides whether the strategic document will be assessed according to the law, the SEA process continues to the next step of the assessment process.

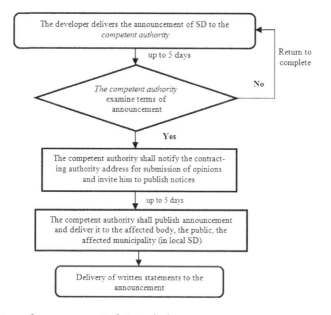

Figure 2.   The steps of announcement of strategic document.

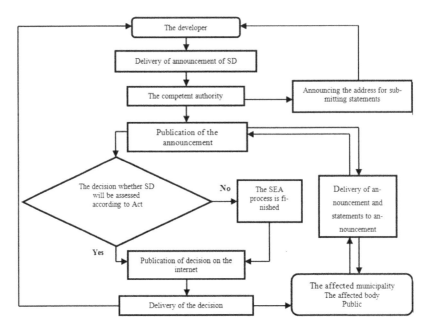

Figure 3.   The steps of screening.

The criteria of screening, also covering health issues are governed by Annex no. 3 of the Act and in deciding whether the draft of strategic document will or will not be assessed by the law takes into consideration the received written statements of the participants in the process. The sequence of steps of screening is shown in Figure 3.

5   THE SCOPE OF EVALUATION AND TIMETABLE

This step is important to ensure an appropriate assessment of the strategic document, whether or not the impact on the environment or health is anticipated. At the same time it paves the way for effective communication between the relevant (departmental) bodies, authorities and other entities of the assessment process (written statements) before the actual drafting of the assessment report.

This step should ensure that in the next step, i.e. in the assessment report of the strategic document, a comprehensive finding was conducted, with a description and evaluation of expected impacts on the environment and human health of the strategic document. Whereas individual assessed strategic documents are different, it is impossible to carry out the assessment the same way. In determining the scope of assessment it is important to point out these facts that are closely related to the impact of the strategic document on the environment and human health and is necessary to take into account the opinions and requests received from participants in the process. The scope of the assessment determine in particular:

a. Which alternatives for the proposed strategy document need to be further developed and evaluated.
b. Which points of the content and structure of the assessment report of the strategic document is necessary in the assessment report pay a special attention.
c. The number of copies of the assessment report.

The timetable is determined the time sequence and, if necessary, the sequence of steps of the assessing strategic document. This step is important for the timing of the assessment process because it creates the opportunity for the public to get more information between the initial information about the strategic document and a public hearing on the assessment of the strategic document.

6 THE REPORT ON THE STRATEGY DOCUMENT AND A PUBLIC HEARING

The main objective of the assessment report is to summarize the impact of assessed strategic document on the environment, including the health of the population in the territory where the strategic document will be implemented, and the area to be affected by implementation of the strategic document. The assessment report, together with the draft of strategic document is submitted to the competent (the departmental) board authority, its development is not affected by any deadline. In case of land-use planning documentation, the developer submits the concept of urban study together with an assessment report to the competent authority or body.

It is appropriate, but not mandatory, that the report is prepared by qualified persons because it contains information from different areas and places high demands on expertise. Assembled team of processors assumes to contain the specialists in different areas of environment and health, which must have knowledge on the assessment of the activities (transport, energy, agriculture, etc.). Within the team a representative should be also a wide range of views on size and forms of protection of the environment and health, as well as local experts who are familiar with the area in question. The evaluation report of the strategic document is required to include information that corresponds to current knowledge and methods of assessment. Degree of specificity of information is directly proportional to the degree of specificity of assessed strategic documents and the expertise and experience of the processor. The lowest degree of specificity is in the generally defined strategic documents. In this case there remains only the possibility of strategic environmental impact assessment in general terms. Quantification is not always applicable; in this occasion, it is advisable to use a qualitative forecast. For example, the effects can be expressed by verbal valuation. The prognosis can be supported by facts as references for research, peer discussions or consultations, which are also the basis for the achieve findings of the strategic assessment. The strategic document containing proposals for specific projects evaluated at a sufficient level allows foreseeing precisely their impact on the environment and health. Quantitative impact assessment, for example determination of noise emission, emissions of pollutants into the atmosphere, it is usually possible only in the assessment of buildings, facilities and activities (EIA).

As the report on the evaluation will benefit many people and many of them are not familiar with the report on the subject, this report must be prepared in a form which can be understood by non-specialists, too. The report generally contains also comprehensible final summary. The public hearing on the assessment report and the draft of strategic document is provided by developer in cooperation with the competent authorities (in case of the strategic documents of national importance also departmental body and the Ministry of Environment). This is the space for the developer where they can provide the public with information on strategic document via

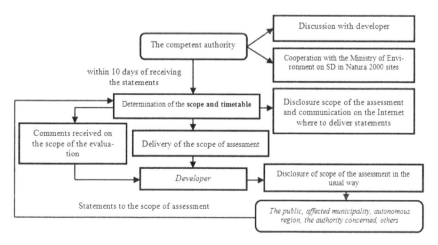

Figure 4. The steps of scoping of assessment and timetable.

different forms and types of promotions. Public hearing is a key stage of public participation in the assessment process, since it allows assessment of the objectivity of the conclusions in the assessment report and expressed public support for the draft of strategic material. An expert review of the strategic document can be only developed by professionally qualified natural or legal persons entered on the list of professionally qualified persons. In elaboration of the expert review, people who have participated in the elaboration of a draft of strategic document or the assessment report of the strategic document cannot participate. The expert reviewer is determined by the competent authority, having regard to the statements to the assessment report and the record of the public hearing. If it is a strategic document with national importance, the expert reviewer is always set by the Ministry of Environment. For developer, it implies the obligation to provide expert reviewer at his/her request additional information necessary for external expertise. In the expert review shall be evaluated in particular:

a. The completeness of the assessment report of the strategic document.
b. Statements to the assessment report of strategic document.
c. The completeness of assessment of the positive and negative impacts of the strategic document on the environment, including their interaction.
d. The evaluation of methods used and completeness of the input information.
e. Alternatives for the implementation of the strategic document.
f. Suggestions of measures.

7   CONCLUSION

The current stage of development of production and consumption activities, processes and services is strongly marked by their environmental aspect and compliance with the manners and the principles of sustainable development. Impact assessment of herein mentioned activities on components of environment and human health is undoubtedly the most efficient and effective tool for sustainable development. Experience and knowledge of the assessment processes (EIA / SEA / HIA) of concrete activities and strategic documents highlight the need for improvement and elimination of subjective (often lobbying and political) approaches and attitudes. The way to such aims is quality legislation, corresponding environmental awareness of society as a whole and efficient mathematically supported procedures of quantitative evaluation and assessment of impacts on the health of the population resulting from changed quality of the environment. It seems necessary to assess impacts of the forthcoming activities and development strategies on environmental components and population health interact with globally recognized, monitored and assessed indicators of sustainable consumption, production, green growth and sustainable socio-economic development in the future.

ACKNOWLEDGEMENT

This paper was written in frame of the work on the projects VEGA 1/0936/15.

REFERENCES

Ministry of Environment. 2008. Environmental Impact Assessment in the Slovak Republic: *General Guide.* 63 pp. ISBN 978-80-88850-79-3.
Dahlgren, G. &Whitehead, M. 2007. Policies and strategies to promote social equity in health. ISBN 978-91-85619-18-4.
Majerník, M. 2007. Posudzovanie vplyvov činností na životné prostredie. *Skalica: SEVŠ v Skalic*i. 276 s. ISBN 978-80-969700-1-8.
Majerník, M. & Szanková, E. 2011. Integrované environmentálno-zdravotné posudzovanie vplyvov činností a stratcgických dokumentov. *Metodická príručka. Košice*. ISBN 978-80-89284-80-1.
Majerník, M. & Husková, V. & Bosák, M. & Chovancová, J. 2008: Metodika posudzovania vplyvov na životné prostredie. *Edícia študijnej literatúry Košice*. 212 s. ISBN 978-80-8073-947-8.

Advances and Trends in Engineering Sciences and Technologies II – Al Ali & Platko (Eds)
© 2017 Taylor & Francis Group, ISBN 978-1-138-03224-8

Modelling of integrated impact assessments of development strategies

M. Majerník, G. Sančiová & N. Daneshjo
Economical University in Bratislava, Košice, Slovakia

J. Krivosudská
Faculty of Mechanical Engineering, Technical University of Košice, Slovakia

ABSTRACT: Development policies and strategies derived there from aimed at satisfying socio-economic needs and quality of life of society are sustainable only if their preventive assessment in the face of their impact on the elements of the environment and human health. The authors in this paper present a concept of a model for integrated environmental-health-safety assessment of strategy documents. The proposed model is based on the implementation of the management of health risks according to methodology HIA into standardized SEA process and its specific stages.

1 INTRODUCTION

Socio-economic development activities aimed at improving the quality of life favoring the socio-economic dimension at the expense of environmental-safety are unsustainable in the future. The balance of the development of individual areas must be ensured at the stage of preparing a development strategy. Safety and health risks that occur in meeting social and economic objectives as a result of reduced quality components of the environment, must so be managed in preparation, and also after adopting the strategy documents, and derived concrete actions, plans and projects there from. These risks need to be identified during the first stage of preparation of future activities, i.e. in the strategic development documents and to assess them from the first step of assessing impacts on the environment and human health in the face of expected economic and social objectives.

The identification of the potential risk by revealing its source and subsequent impact area, through a risk analysis that includes, inter alia, finding measure of the hazards and threats for risk assessment and management in which it is possible to identify or it is possible to measure the level of threat, and by managing evaluation of feedback adopted measures are included in a comprehensive risk management process.

2 THE CONCEPT OF A MODEL FOR INTEGRATED IMPACT ASSESSMENT

If such risk management becomes a part of the impact assessment process of strategic documents, the SEA process will include not only the priority environmental assessment, but also an assessment of effects on elements of the environment, air, water, soil, fauna and flora as well as health assessment, which includes an assessment of health risks arising from strategic documents focused primarily on the socio-economic improvement, thus the environmental-safety-health impact assessment of strategic development documents.

A model of integrated environmental-safety-health assessment of strategic development documents as a result of our scientific and technical studies is shown in Figure1.

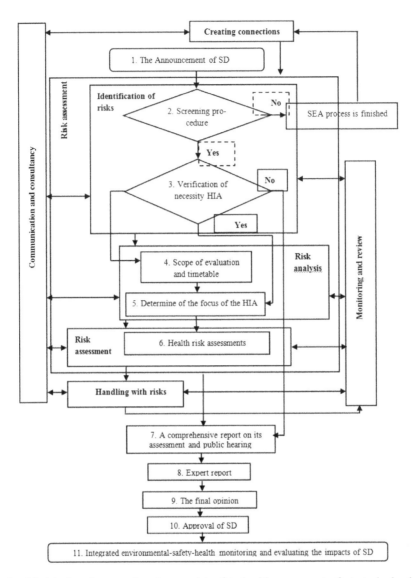

Figure 1. Model of an integrated environmental-safety-health assessment of strategic development documents (SD—Strategic document).

3 THE SEQUENCE OF STEPS IN AN INTEGRATED ENVIRONMENTAL-SAFETY-HEALTH ASSESSMENT OF THE STRATEGIC DEVELOPMENT DOCUMENTS

3.1 *The announcement of strategic document*

The first step in an integrating environmental-safety-health assessment is as well the assessment of strategic documents: „The announcement of strategic document". The strategic document authority prepares it under current globally coordinated legislation and conventions in Slovakia by Act no. 24/2006 Coll., which has six amendments. The announcement of the strategic document is a central and essential document in the process of assessment in determining if the strategic development document will be assessed under the law through all eight phases or the stage of the screening procedure is sufficient. Therefore the announcement of a strategic document should include indicators relating to environmental, safety and health issues, in the face of the expected economic and social benefits which makes it possible to proceed better, more detailed and demonstrable in the stage of the screening procedure.

Screening procedure, verification of the necessity of Health Impact Assessment (HIA) and identification of risks—The next step in a standard SEA process is a screening procedure with the verification of the necessity to assess the strategic development document in terms of its effects on the health of the population. This stage consists of two parts. The first is the presentation and subsequent comments to the announcement by the stakeholders. The second one is a decision whether the strategic document will or not be comprehensively assessed under Act no. 24/2006 Coll. as amended by its amendments.

Based on the announcement of the strategic document and the decision of the investigation procedure, it is necessary to determine the need for assessing the impact on public health, i.e. examining the necessity of HIA. This screening should be done in cooperation with experts from a public health authority. If the strategy document is a subject to compulsory assessment under the Act, the stage of screening procedure and verification of necessity of HIA is not acted, the process continues to the next step. If the health impact assessment is decided on the need to be assessed in step verification necessity of HIA, the process continues to the next step, if verified that there is no need to carry out HIA, because the impact on health is minimal and its assessment would unnecessarily delay the SEA process, the process of HIA does not end, but found facts, arguments, and results are quantified given in the Assessment report, which will be discussed with the public in the affected area. When identifying risks in its management, a detection, accurate identification and specification of all risks are performed, it is an important stage, because if risks will be not demonstrably identified, they cannot be analyzed in more detail. The results of preliminary risk assessment must be documented.

The scope of assessment and timetable, determination the focus of the HIA and risk analysis—The task of the next step of an integrated assessment—Scope of assessment and timetable—is to ensure that impacts assessment of strategic development documents have been organized with high measure of efficiency. In terms of health assessment, the focus is set of a health impact assessment of the population in this step. At this stage, it provides a space to refine and analyze the risks arising from the strategic document on the basis of the risks identified in the previous step—from drawn up register. The scope of the assessment is determined mainly:

- Which variants of the strategic document (always drawn up with alternatives) are necessary to elaborate and assess (in terms of risk, health and safety issues) between options and compared to business as usual (status if the document is not implemented).
- Which parts of the Assessment report of the strategic document are necessary to take into account detailed in the assessment report (better health impacts assessment of the population and the risks in terms of a comprehensive security strategy document).
- The number of copies of Assessment report.

In this stage of process of assessment, it is appropriate to perform risk analysis based on identifying risks, i.e. the process of risk identification and risk assessment for the concerned population, objects, concerned environment and other examined objects. Risk analysis is usually a subjective process that results in not only quantitative indicators, but also to compromise solutions, expert assessment and so on. The risk analysis identifies the likelihood and the measure of the consequences of the strategic development document, and reveals the measure, or the degree of risk.

Health risk assessments and risk assessment—As a part of an integrated environmental-safety-health assessment it is a very important step of risk assessment. Health risk can be assessed only when data on specific activities resulting from the strategy document and their impacts on the environment and health are available. The statement of the assessment of strategic documents that the health risk assessment will be the scope of concrete derived projects within the EIA process is partly logical because in the strategic development documents and programs, it is not always possible to determine what health risks will arise from specific projects within the strategic document. These impacts can be only assumed.

3.2 Risk assessment

When assessing a risk measure (degree) of risk of danger, which have been analyzed for the protection of the health of the population, material values and others, associated with implementing the development strategy is determined. Mathematical expression of risk:

$$R = P \times D \qquad (4)$$

$R = (1: 10) \times 10 = 1 \ m^2$ contaminated area for one year. Where:

R—degree of the risk (arising from the expected adverse effects of the strategic document),

P—probability of risk occurrence during the preparation and after the approval of the strategic document

D—degree of severity of the risk

Risk = (likelihood of risk occurrence of strategy document) x (severity of risk of strategy document)

$$R = f(P,D) \qquad (2)$$

$$R = \sum_{i}^{n} P_i \times \sum_{j}^{m} D_j \qquad (3)$$

where i, j—are the indexes relating to the possible threat to the i-th and j-th result

$$R = P \times D^s \qquad (4)$$

where S—expresses different types of consequences (based on their weight, $S \geq 1$).

Risk assessment based on risk analysis assesses the severity of the estimated measure (risk assessment) and assesses the necessity of its reduction (acceptability of the risk). Identified risks are placed in so-called risk matrix. The place where the risk is placed in the risk matrix is determined by the estimated probability of occurrence and the probable financial impact of risks on a project/strategy document. A view of the specific risks assumed in the risk matrix then will provide a picture of the acceptability or non-acceptability of this risk and it will allow for a comparison of individual risks. The algorithm of risks management arising from the implementation of the proposed strategic document, as the consequences of its negative environmental, health and safety impacts, is shown in Figure 5.

When estimating risk of strategy document *the risk* should be *handled*, i.e. determine, if it is necessary to reduce risk of negative impacts of SD further or if the required safety has already been achieved. If it is necessary to reduce the risk, preventive and corrective safety measures are chosen, and the whole process will be repeated. If new measures are taken, they need to be checked for any critical or other hazards from the proposed strategy document and to determine preventative measures. When managing the existing risk of the strategic document, an appropriate timing of interaction linkages of the stages of the environmental assessment (SEA) process and steps safety-health risk assessment (HIA) is important and intervention linkages and in communication, consulting, monitoring, review and creation of connections.

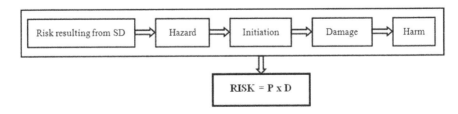

Figure 2.   Causal dependence of risk assessment.

546

3.3 The comprehensive report on the assessment of strategic document and a public hearing

The aim of a comprehensive impacts assessment report is to summarize the environmental-safety-health impacts assessment of the strategic development document in the place where the strategy document will be implemented and in the place that will be affected by the implementation of the strategic document (national, regional, local range of character of territorial planning documentation). The terms and the structure of the report are given by legislation and the results of the investigation procedure.

Expert judgement on the Assessment report of the strategic document- A professionally qualified body, a person or an organization which is registered in the list of professionally qualified bodies (Decree no. 113/2006), will elaborate an expert judgement based on a comprehensive report on environmental-safety-health assessment of the strategic development document, the results of public hearing and the views of stakeholders.

3.4 The final judgement on the assessment of the strategic document

It will be determined in the final judgement whether to recommend or not to recommend an approval of the strategic document or under what conditions, also and the extent of monitoring and assessment of the actual impacts after implementation of the strategy.

The final judgement of the assessment process of the strategic document has a character of an expert judgement in terms of its environmental-safety-health impact. It is issued by the competent authority based its proposal in the elaborated judgement.

3.5 Approving of the strategic document

The decision on the approval of the strategy document is a result of the process of integrated environmental-safety-health assessment. For this decision, the approval of strategic document is absolutely necessary final opinion from the assessment. Requirements of the process are set out in the Act 24/2006 Coll. and its amendments.

Integrated environmental-safety-health monitoring and assessment of the real impact of SD—The role of the final stage in the integrated assessment is to determine methodology for identifying and monitoring the real environmental-safety-health impacts of the strategic document. This is a comparison of the predicted (assessment) impacts with the real operational impacts of individual actions arising from the strategic document. When worse impacts are found, it is necessary (under the threat of stopping concrete actions) to implement measures to minimize or to eliminate them.

Figure 3. Schematic illustration of risk parameters.

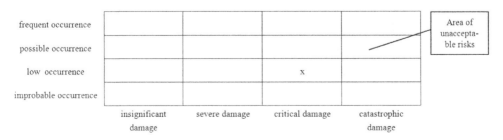

Figure 4. Risk matrix for the implementation of the strategic document.

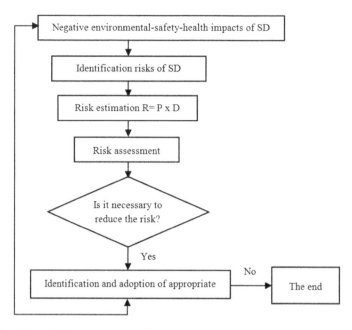

Figure 5.   Algorithm of risk management of strategic development document.

4   CONCLUSION

The proposed model of integrated environmental-safety-health assessment of strategic development documents has demonstrated a broad applicability in preventive risk management of documents of national, regional and local range in its decomposition within the standard process of the SEA. The identified and assessed risks (environmental, health, safety) in the form of an integrated registry should be a part of "Assessment report of the SD". Such a report will be then thorough, accurate from professional perspective. It will help the process of making comments.

It will add requirements and uncertainties arising from the positions of participants in the assessment process and during the public hearing significantly. It demonstrates that not only the assessment of the strategic document by competent persons will be improved by application of the proposed methodology but also the actual decision-making process of approving a strategy for implementation.

ACKNOWLEDGEMENT

This paper was written within the work on the project VEGA 1/0936/15

REFERENCES

Chovancová, J. Environmental risk management. *University of Prešov in Prešov*, 2015, ISBN 978-80-555-1394-2.
Šimák, L. & Hollá, K. & Ristvej, J. Crisis Management in Public Administration. Crisis Management I. Theory of Crisis Management. EDIS—*vydavateľstvo ŽU*. 2013, 213s., ISBN 978-80-554-0651-0.
Majerník, M. & Chovancová, J. & Daneshjo, N. & Danishjoo, E. Environmental and Health Impact Assessment of Development Strategies. *Albersdruck GmbH & Co. KG, Düsseldorf, GERMANY*, 2015, 148s. ISBN 978-3-00-050052-7.
Zvijaková, L. & Zeleňáková, M. Riziková analýza v procese posudzovania vplyvov objektov protipovodňovej ochrany na životné prostredie. *Leges Praha*. 2015, 255s., ISBN 978-80-7502-062-8.

Advances and Trends in Engineering Sciences and Technologies II – Al Ali & Platko (Eds)
© 2017 Taylor & Francis Group, ISBN 978-1-138-03224-8

Fighting moisture in baroque buildings – case study

O. Makýš & D. Šrobárová
Department of Construction Technology, Civil Building Engineering Faculty, Architectural Heritage Conservation Technology Centre, Slovak University of Technology in Bratislava, Slovakia

ABSTRACT: Fighting moisture in historic masonry, made of mixed construction material of bricks and stones, is always very complicated. Besides the source of moisture, which such cases usually lies in the basement of the buildings, also the quality of construction material, which was used for constructing the walls originally, plays a major role. Unfortunately the exact character of the masonry is usually not known and only in some cases is it possible to get this information. Therefore the fighting moisture efforts are in many cases based on experiences of the expert's team. In the case on the church of Piarists´s in Nitra the internationally approved method of continues steps was introduced. In the following article results of the first three years of introducing it are described.

1 INTRODUCTION

In the year of 2010 the Slovak University of Technology was asked by the order of the Piarists in the town of Nitra to assist with the problem of fighting moisture in wet masonry of their church and sacristy. Besides elaborating a concept for fighting the moisture, regular measurement of the development of moisture level in the wet masonry was undertaken. The elaborated concept was mostly followed and that is why it was possible to evaluate the real influence of proposed measurements. However—unfortunately, besides the influence of university experts, also other influences on Piarists, undertaken by other practicing persons was the case. As a result of such conflicts only a partial success, but also a partial non-success occurred.

2 MATERIAL AND METHODOLOGY

The church of St. Ladislav in Nitra stands under law protection (Act No. 49/2002) as an architectural heritage building with the No. 18/1. The church was built in the period of early baroque, based on a rectangular floor plan and it is situated in the centre of a monastery building. The building itself was built using a ship plan, accented by two huge towers, situated at both sides of the main entrance. In the interior the ship divided into three vaulted fields. The whole interior is richly decorated by the use of stucco. In close contact with the presbytery is a building of the sacristy. The main altar, constructed in 1750, is constructed with six columns, made of marble.

The church itself was constructed between 1742 and 1789 on the place of an older basilica, which was fully dilapidated. Constructing the church took an enormous amount of time—during the construction process the building was damaged by three (!) large fires.

There is not any documentation, i.e. plans, from the time of construction and re-reconstructions of the church. Even from construction repair activities, which can be dated into 20th century. During the 40-ies there were marble panels installed on the side walls, partly reaching up to the vaults. The marble was gathered from Italy. The aim of this installation was probably in an effort to hide the surface damage on the walls—plasters, caused by rising humidity, combined with the crystallizing of water soluble salts. It can not be excluding, that exactly during this time also the ventilation openings, coming out from the churches crypt were closed. The last minor damage, which the church achieved, took place during the WWII. This damage was repaired only in the year of 2002.

In 2010 the representatives of the order of Piarists in the town of Nitra decided to solve the problems with the moisture in the wet masonry of their church and sacristy possibly definitively and in a right way. Due to this decision experts from the Architectural Heritage Centre, established in 2004 besides the Department of Construction Technology at the Civil Building Engineering Faculty in Bratislava were asked to solve these problems, as mentioned above.

2.1 *Research methodology*

For dealing with the problem of unwanted moisture in the masonry of the church of Piarists in Nitra, the implementation of the: Method of Sequential Steps was recommended with technologies, coming out of it. This method advanced the realising of repair interventions in a step by step order. Between the steps, there should be always time organised for evaluation of the influence / results of the previous steps. During these steps the masonry has some time to react on changed situation. There is always the possibility to adapt the following interventions, according to changing situation.

Figure 1. The church of Piarists in Nitra. Figure 2. Interior of the church of Piarists in Nitra.

Figures 3 and 4. The interior of the crypt—after it was cleaned of the remains of human bodies.

The main disadvantage of this way to fight the moisture is, that it consumes a lot of time. The advantage lies in the fact, that it can save quite a significant amount of money in total and that the achieved success can be a longer lasting. But this method has anyway a significant advantage, seeing it from the point of view of architectural heritage conservation—in first phase of its introduction, it is based on less invasive and more reversible technologies. More invasive and less reversible technologies are to be used only in the case that previous group of technologies has not enough influence.

2.2 *Air flow inside the crypt*

The first step, which was to recommend was the opening of the original ventilation windows in the crypt, which were closed (maybe) during the 40-ties of 20th century. Thanks to this action the quality of the interior air of the crypt raised dramatically and also the contents of relative moisture in the masonry was lowered—even not enough.

As a part of the ventilation technologies also the main entrance doors to the crypt were arranged in a more likely ventilation function. The function of the ventilation was afterwards proofed by using the flame of a candle.

2.3 *Dealing with marble panelling*

The next important step, which had to lead to the reduction of the moisture level, was the removal of the non diffusive panelling in the interior of the church, made of marble. The panels were, until now, removed only partially, but in the part of the church where they were removed, the moisture went down.

2.4 *Rainwater transporting pipes*

As the, by accident organised control in autumn of 2013 showed, there was not enough attention given to cleaning the pipes. During the following control in July of 2014 the situation was much better.

Figures 5 and 6. A ventilation window, inside the crypt. Before it was opened and after it.

Figures 7 and 8. Arrangement of the crypts ventilation window, friendly towards ventilation and proofing the ventilation activity by the use of a candle.

2.5 *Dealing with moisture in the basement of the church*

In our original proposal a possibility to introduce the side insulation of basements was also given. Unfortunately this recommendation was not understood properly. A strange construction

Figures 9,10, 11 and 12.   Original and partially removed marble panels in the church.

Figures 13,14 and 15.   Results of the controls in autumn 2013 and during the summer 2014.

Figures 16 and 17.   Scheme, originally proposed and the reality on site.

Figures 18 and 19.　Interior of the "ventilation" channel—proofing of the air flow in it by the use of a candle—there is no air movement in it.

Table 1.　Data gathered in the crypt of the church.

Date	9.3.	6.10.	21.6.	5.10.	19.6.	4.12.	12.4.	30.10.	10.12.	2.7.
Place	2010	2010	2011	2011	2012	2012	2013	2013	2013	2014
No.	%	%	%	%	%	%	%	%	%	%
45	17,4	13,2	13,0	11,5	10,1	9,4	8,4	8,3	8,8	9,1
48	13,7	13,2	13,1	12,5	12,7	12,4	12,2	9,5	9,1	9,1
51	14,0	13,9	14,0	13,5	13,2	11,8	11,0	11,4	11,2	11,6
54	14,3	13,4	14,1	13,7	13,6	13,4	13,4	12,5	12,6	12,0
59	14,8	14,9	15,2	15,6	13,5	13,7	13,2	13,0	13,7	13,3

Table 2.　Data gathered on the facade of the church.

Date	9.3.	6.10.	21.6.	5.10.	19.6.	4.12.	12.4.	30.10.	10.12.	2.7.
Place	2010	2010	2011	2011	2012	2012	2013	2013	2013	2014
No.	%	%	%	%	%	%	%	%	%	%
4	8,0	7,5	7,2	7,5	6,3	6,6	6,3	6,0	6,2	6,1
5	11,5	11,2	10,8	10,7	10,5	10,6	10,6	9,5	8,6	8,0
9a	9,6	9,1	9,0	9,0	9,0	9,8	10,0	9,2	8,3	8,0
15	7,0	7,2	8,0	6,9	7,2	7,7	7,3	7,5	7,6	7,1
18	8,0	7,5	7,2	6,3	6,6	6,3	6,0	6,2	6,1	7,0
22	6,0	5,0	4,9	4,5	4,1	4,9	4,4	6,0	6,1	7,5
27	8,4	7,9	7,2	6,6	5,2	4,0	4,0	5,7	6,1	8,6
29	5,3	5,0	4,9	4,5	4,6	4,0	4,0	7,7	8,5	9,1
32	8,6	9,1	9,3	9,6	9,7	8,2	7,3	4,2	4,7	5,9
33	4,6	4,6	4,6	4,2	3,8	3,5	3,0	3,5	3,7	4,7

Table 3.　Data gathered from the interior of the church.

Date	9.3.	6.10.	21.6.	5.10.	19.6.	3.12.	12.4.	30.10.	10.12.	2.7.
Place	2010	2010	2011	2011	2012	2012	2013	2013	2013	2014
No	%	%	%	%	%	%	%	%	%	%
34b	13,8	13,5	13,2	11,7	9,7	7,4	5,1	4,9	4,8	4,9
36a	12,0	10,9	10,3	9,8	8,3	4,1	3,9	3,8	3,6	3,6
38	6,7	6,0	5,8	4,6	3,5	3,7	4,0	4,3	3,6	3,9
40b	2,7	4,1	4,6	5,3	4,5	4,2	4,0	3,9	3,6	3,5
43	10,5	10,6	9,1	7,8	6,0	4,9	5,0	5,8	5,7	5,3

Table 4. Evaluation of moisture according to ČSN Standard.

Degree of moisture	Moisture (uM) [%]	Increased moisture	5,0–7,5
Very low moisture	<3,0	High moisture	7,5–10
Low moisture	3,0–5,0	Very high moisture	>10

company was chosen to introduce this technology, which had too little experiences with such work. Therefore it resulted into a strange proposal, which we were asked to comment—we did not recommend introducing it. Anyway—our opinions were not accepted and a "ventilation" channel was built, even in a more worse way, as it was originally proposed.

3 RESULTS AND DISCUSSION

Due to fix (or even correct) our statements, given towards the ventilation technologies for fighting the moisture, a line of measuring were undertaken with following results [measuring was done with the help of the GANN UNI—1 machine]. Due to lack of space only a selected part of gathered data are published here—just to illustrate the article.

The above mentioned figures give a very clear depiction on the real influence of ventilation technologies, used for drying out of historical masonry. It is clear, that they have some positive effect, but this is very limited. In general the lowering of the moisture level in the masonry of the Piarists church in Nitra, caused thanks to ventilation technologies, was only about 3–5%, which is in a case of heavily moistened masonry definitively not enough. Also the construction of the air tunnel plays an important role in it. As it is visible—the concrete channel, introduced out of our expertise, made the situation even worse.

The best results were achieved at the figures No 34a and 34b—the basement of the altar, but in this case an additional injection technology of liquid sealing barrier was used. This is definitively one of such technologies, which can really help with moisture in historic masonry.

REFERENCES

Ashurst, J. & Ashurst, N. (1989): Practical Building Conservation.—Vol. 2. Brick, Terracotta and Earth, Hants: English Heritage Technical Handbook, Gower Technical Press.
Balík, M.—Starý, J. (2003): Sklepy—opravy a rekonstrukce. Praha: Grada Publishing, 110 pp. ISBN 80-247-0221-5.
Branson, G. (2003): Home Water and Moisture Problems: Prevention and Solutions, Richmond Hill: Firefly Books, 144 pg. ISBN 1552978354.
Lebeda, J. et al. (1988): Sanace zavlhlého zdiva. Praha: SNTL.
Makýš, O. (2004): Technologie renovace budov, Bratislava: Jaga group, ISBN801-8076-006-3, pp. 262.

Advances and Trends in Engineering Sciences and Technologies II – Al Ali & Platko (Eds)
© 2017 Taylor & Francis Group, ISBN 978-1-138-03224-8

Comparison of wind pressure on high-rise building obtained by numerical analysis and experimental measurements

S. Medvecka, O. Ivankova & L. Konecna
Department of Structural Mechanics, Faculty of Civil Engineering, Slovak University of Technology in Bratislava, Bratislava, Slovakia

ABSTRACT: In this article the resulting effect of wind pressure on a high-rise building obtained by an experimental measurement in the wind tunnel is compared with results of a numerical method performed according to the valid standards. The high-rise building with a circular floor plan in two variants—skeletal structural system with vertical columns and with columns with the slope was analyzed. In the calculation a load in the construction stage, where one can find the most unfavorable conditions for horizontal load, was considered. Maximum horizontal displacements of these buildings for horizontal wind load defined by the experimental and numerical method have been evaluated. The difference of reserve for numerical methods compared with the measured values in the wind tunnel has been formulated.

1 INTRODUCTION

High-rise buildings must securely transmit the load through the bearing system. Categories of load are: horizontal and vertical. The most unfavorable load for high-rise buildings is a horizontal load represented by the wind and seismic effects. Essentially the wind is a large scale horizontal movement of free air, Stathopoulos & Baniotopoulos (1997). It plays an important role in the design of tall structures because it exerts loads on the buildings. In this article, models loaded by wind are calculated according to EN 1994-1-4 and analyzed experimentally in

VARIANT I   VARIANT II

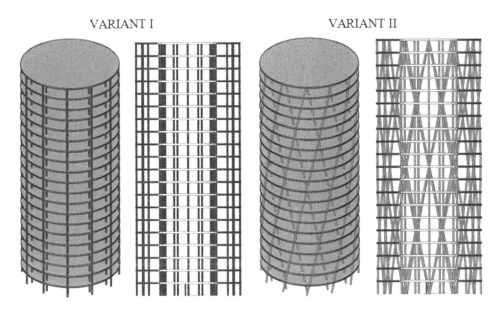

Figure 1. Axonometry and view of two variants of a high-rise building.

the Boundary Layer Wind Tunnel (BLWT) of Slovak University of Technology in Bratislava. Subsequently, the resultant horizontal displacement of the high-rise building has been compared. This building was also analyzed for the Serviceability Limit State (SLS).

2   BASIC PARAMETERS OF BUILDING

Two variants of high-rise building were modeled in Scia Engineer software. The shape of building is a cylinder of height 70 meters and its diameter is 30 meters. A high-rise building with 20 floors was considered, where construction height of each floor is 3.5 meters. Plates are 250 mm thick and they are made of reinforced concrete C20/25. Structural system of the whole building is skeletal system with columns having diameter 400 millimetres. Columns are made of reinforced concrete C30/37.

Variant I is a building with vertical columns. Peripheral columns in variant II are inclined. Columns with inclination are located from the centre of outermost peripheral ß = 8,37°, Figure 2. Columns located in the central area are not inclined. The model, which was measured in the tunnel, was in a scale 1:360.

3   LOAD OF THE BUILDING

The high-rise building was loaded by two types of load: its self weight, and wind load. Self weight was generated automatically by using the program. Wind load was calculated according to the standard and obtained by experimental measurement in the tunnel. We will specify the calculation procedure of an external wind pressure coefficient on the cylinder in next chapters. The terrain roughness III (area with regular cover of vegetation of buildings or with isolated obstacles with separations of maximum 20 obstacle heights, EN 1991-1-4) was used in both calculations. Wind speed $v_{b0} = 26$ m/s, has been considered, which is given in the standard for a location of Bratislava. The wind loads have been applied as a liner load acting on the slabs, through which the external cladding transmits horizontal wind loads into the structure.

3.1   *Calculation of external pressure c_{pe} according to EN 1991-1-4*

We have proceeded according to Equation 1.in the calculation of external pressures c_{pe} on a building of a cylindrical shape

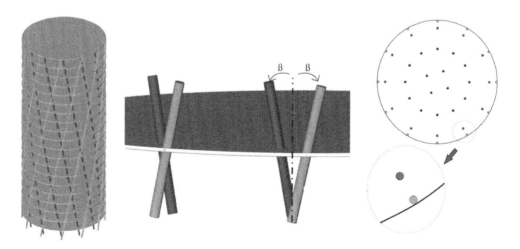

Figure 2.   Principle of columns' inclination.

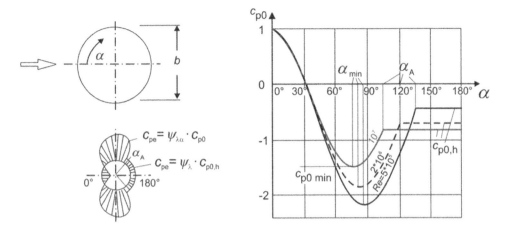

Figure 3. Pressure distribution for cylinder and for different Reynolds number ranges and without end-effects, EN 1991-1-4.

$$c_{pe} = c_{p0} \cdot \psi_{\lambda\alpha} \tag{1}$$

where, c_{p0} is the external pressure coefficient without free-end flow and $\psi_{\lambda\alpha}$ (Equation 2) is end-effect factor.

$$\psi_{\lambda\alpha} = \psi_\alpha + (1 - \psi_\lambda) \cdot \cos\left\{\frac{\pi}{2} \cdot \left(\frac{\alpha - \alpha_{min}}{\alpha_A - \alpha_{min}}\right)\right\} \tag{2}$$

The value of c_{p0} is given in Figure 3 as a function of angle α, where for analyzed model the Reynold's number 1×10^7 was calculated. The resulting values and direction of external pressure coefficients c_{pe} for analyzed circular cylinders are shown in Figure 4.

3.2 External pressure c_{pe} calculated by experimental measurement

There are two operating spaces in BLWT STU wind tunnel. In the Front Operating Space (FOS), the uniform wind flow with low deviations from the mean wind velocity is generated. It is suitable for testing of the larger models and for the determination of extreme distribution of pressures on examined objects. In the Rear Operating Space (ROS), turbulent wind flow is developed by the boundary layer placed in front of ROS. This wind flow simulates the roughness of the real terrain. The model scale is in the range from 1:360 up to 1:380. It satisfies all required particulars of the model. The minimum value of the Reynold's number is equalled to 1×10^4, in order to the similarity of the circumfluence around the structure has been ensured, (ACSE Manuals and Reports on Engineering Practice, Hubova 2016).

The value of pressure p(z) on model was determined by experimental measurement made in the Rear Operating Space (ROS). From this pressure, we have calculated the exposure factor c_e according to Equation 3, while we have obtained basic velocity pressure q_b on the basis of air density ρ and reference wind velocity v_{ref} in tunnel (Franek et al. 2016)

$$c_e = \frac{p(z)}{q_b} = \frac{p(z)}{1/2 \cdot \rho \cdot v^2_{ref}} \tag{3}$$

In the next step, we have expressed the value of external pressure coefficients c_{peM} according to Equation 4, where c_{pe} is the value the external pressure coefficients of analyzed circular cylinders from the Eurocode (Figure 4).

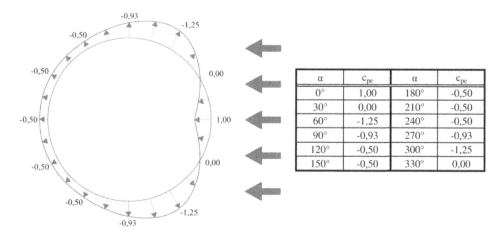

Figure 4. Value of external pressure coefficients c_{pe} for analyzed high-rise building according to the Eurocode.

α	c_{pe}	α	c_{pe}
0°	1,00	180°	-0,50
30°	0,00	210°	-0,50
60°	-1,25	240°	-0,50
90°	-0,93	270°	-0,93
120°	-0,50	300°	-1,25
150°	-0,50	330°	0,00

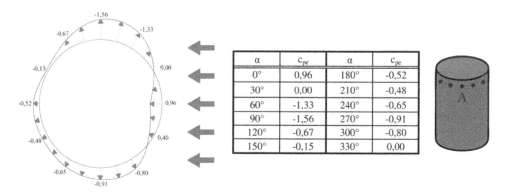

Figure 5. Values of external pressure coefficients c_{pe} for model from experiment—level A.

α	c_{pe}	α	c_{pe}
0°	0,96	180°	-0,52
30°	0,00	210°	-0,48
60°	-1,33	240°	-0,65
90°	-1,56	270°	-0,91
120°	-0,67	300°	-0,80
150°	-0,15	330°	0,00

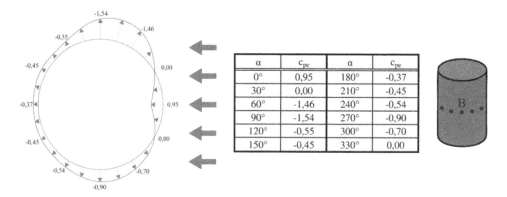

Figure 6. Values of external pressure coefficients c_{pe} for model from experiment—level B.

α	c_{pe}	α	c_{pe}
0°	0,95	180°	-0,37
30°	0,00	210°	-0,45
60°	-1,46	240°	-0,54
90°	-1,54	270°	-0,90
120°	-0,55	300°	-0,70
150°	-0,45	330°	0,00

$$c_{peM} = c_e \cdot c_{pe} \qquad (4)$$

We have applied this procedure at three levels of model, that means at the upper edge (A), in the middle (B) and at the lower edge (C).

4 CONCLUSION

For high-rise buildings the impact of horizontal forces is represented by wind-caused horizontal displacement. For a proper design it is necessary to know the external wind pressure coefficients c_{pe} that are given in the Eurocode. We have compared values of c_{pe} with the values measured experimentally in the wind tunnel, (measured in turbulent wind flow), thereby we wanted to express the real effect of the wind on the building. Comparison of different pressures is shown in Figure 8. It is evident from the graph that results of measurements are very close to the Eurocode, taking minor variations into consideration.

The comparison of differences between the external pressure coefficients c_{pe} measured by experiments and according to the maximum horizontal displacement of high-rise building can be seen in Figure 9. Maximum horizontal displacements due to the wind load obtained by the experiment did not exceed values of displacements calculated according to the Eurocode. Thus, we can review that the Eurocode provides safe values of external pressure coefficients.

Due to the conformity of results compared to the standard, we can venture into more experimental measurements of external pressure coefficients for the shapes of buildings that are not mentioned in the standard.

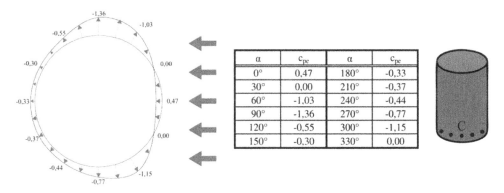

α	c_{pe}	α	c_{pe}
0°	0,47	180°	-0,33
30°	0,00	210°	-0,37
60°	-1,03	240°	-0,44
90°	-1,36	270°	-0,77
120°	-0,55	300°	-1,15
150°	-0,30	330°	0,00

Figure 7.   Values of external pressure coefficients c_{pe} for model from experiment—level C.

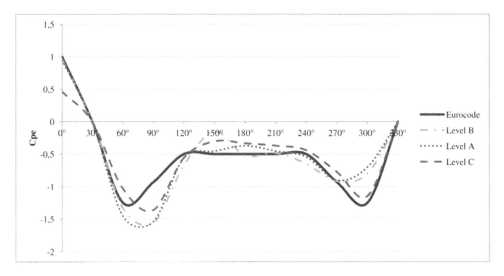

Figure 8.   Comparison of external pressure coefficients c_{pe} from the Eurocode and from experiments.

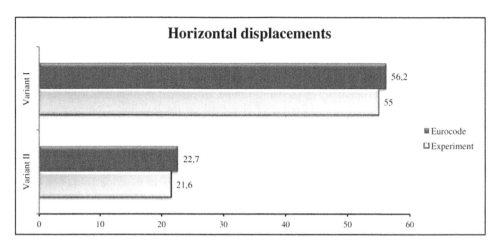

Figure 9. Comparison of horizontal displacements due to wind load from the Eurocode and from experiments.

ACKNOWLEDGEMENTS

This paper were created with the support of the Ministry of Education, Science, Research and Sport of the Slovak Republic within the Research and Development Operational Programme for the project "University Science Park of STU Bratislava", ITMS 26240220084 and the Scientific Grant Agency—VEGA grants No. 1/0544/15. Authors thank to the Slovak University of Technology for the support—financial source No. 1619. This paper was created with the support of the TU1304 COST action "WINERCOST".

REFERENCES

ACSE Manuals and Reports on Engineering Practice, no.67, Wind Tunnel studies of buildings.
EN 1991-1-4 Eurocode 1: Actions on structures. Part 1–4: General action, 2005.
M. Franek, L. Konecna, O. Hubova, J. Zilinsky, 2016, Experimental Pressure Measurement on Elliptic Cylinder, *Applied Mechanics and Materials: Advanced Architectural Design and Construction*. Vol. 820, 332–337.
O. Hubova, L. Konecna, 2016, The Influence of the Wind Flow Around the Free Ends of High-Rise Building on the Values of External Wind Pressure Coefficients, *Applied Mechanics and Materials:* Vol. 837, 203–208, Switzerland.
Stathopoulos T. & Baniotopoulos Ch. C., 2007, Wind Effects on Buildings and Design od Wind-Sensitive Structures, Udine.

Semi-automated budgeting of structures included in modern methods of construction based on 3D model

P. Mesároš, J. Talian & M. Kozlovská
Faculty of Civil Engineering, Institute of Construction Technology and Management, Technical University of Kosice, Slovakia

ABSTRACT: Budgeting from the 3D model is dependent on the cooperation between 3D design applications and the budget application. For an automated budgeting system from the 3D model to the budget application, it is necessary to ensure the mutual interoperability of both applications. In Slovakia, separate applications are used and completing data transfer is complicated. System of data transmission using data standards IFCxml was used, but data transmission is dependent on price-list database TSKP (Classification of building structures and works). Therefore it is not possible to automate the process of estimating for these structures. For this reason, semi-automated approach for budgeting from the 3D CAD application to the budget applications was developed. The paper explains own semi-automatic approach to budgeting from 3D CAD applications with demonstration of the cost estimating procedure on specific elements included in modern methods of construction that are not in the price-list database.

1 INTRODUCTION

1.1 *Automated and semi-automated budgeting*

In construction projects, significant cost overruns and schedule delays are now the norm rather than the exception (Venkatarman and Pinto, 2008). The success of any construction project depends on accurate estimates of the cost. To achieve the lowest difference between the cost of construction of the design and implementation phase you need to get as much information as possible in the initial phase of the project. Lack of information in the early stages of the construction project has mostly resulted in higher costs during implementation (Harrison and Lock, 2001). Automated budgeting in an integrated design approach is based on a complex 3D model data. Any design data model is proposed parametrically. In addition to basic geometric parameters structure contains parametric data defining not only the basis of material structure itself, but also other parameters such price may be the structure or identity elements.

In order to access to integrated planning, the software environment that communicates directly with each other should be used. The basis of direct communication between CAD systems and cost estimate and budget applications is the use of appropriate transmission format that will provide the most data (Zhang, 2010). For this purposes, data exchange format known as IFCxml was designed primarily for data transmission. The concept of modern methods of construction is most often seen as an innovative concept for prefabrication or off-site construction (Kozlovská & Spišáková, 2013; Mačková & Mandičák, 2015). Modern construction methods mainly include the production of building structures in manufacturing (Tažiková, 2014, Kyjaková et al., 2014). Budget creation for construction belong to the modern methods of construction in the case where the price-list database is missing, is dependent on the creation of new price-list items. Automated budgeting of these items is not possible and it is necessary to manually create items. To accelerate and simplify budgeting, combined approach between automated and manual budgeting has been developed. The combination of both approaches allows to create a semi-automated system of budgeting. The main requirement for semi-automated budgeting is to create an assistant database of design elements in CAD application. The new assistant database will build on existing database structures and prices. The following section classify the breakdown of the individual structures which are a part of modern methods of construction with regard to existing price database "Classification of building structures and works" (TSKP).

2 INTEROPERABILITY BETWEEN CAD AND COST ESTIMATING SOFTWARE

Interoperability between CAD software application and estimating software is basis for automated and semi-automated estimating. In order to automated and semi-automated budgeting, it is necessary to create a data model in CAD applications from the structures containing the identification data. The identification data are for correct processing of the transmitted data. In the next part of the paper we describe how to process imported data.

Processing of imported data in the budget application consists of several steps:

– data analysis;
– data processing;
– assignment price to the appropriations;
– creating construction budget;
– construction budget control.

As a model of interoperability between CAD systems and budget in a semi-automated system, the wall system NUDURA ICF 96-6was selected. This is a sandwich wall structure consisting of three layers with a total thickness of 286 mm. The first and the third layer is a layer of the polystyrene panel with a thickness of 67 mm. The middle layer is the bearing part of wall system consisting of reinforced concrete with thickness of 150 mm. In Allplan CAD application, the assistant for NUDURA ICF wall constructions was created. Using this assistant, a wall with the door from the material NUDURA ICF 96-6was created. Subsequently, this construction was exported to a format called IFCxml model_1.ifcxml. This data file was imported and processed in the budget application Kalkulus.

2.1 Data analysis

The entire budget processing begins with import of data from CAD applications IFCxml format and then analyzing them. This analysis is carried out in the budget as a hidden application process during the data import. The result of the analysis is the precise identification of data for subsequent processing. The basic task of data analysis is the distribution of the transmitted data to the identified elements and unidentified elements.

Identified data made up of individual items are separated by a predefined structure construction budget and database of structures and works (TSKP). Distributed data are divided into the groups (structures), which imply information about the work necessary for the construction of the structure (with or without material specifications), quantity, unit of measure and identification code. The following process after identification is creating the summary list of identified items to be entered into the budget. Summary list contains identified items generated from the 3D model. Each item contains the basic data for processing the budget like description, unit of measure, number and identification code.

Automated unidentified data are subject to semi-automated processing of the budget. Unidentified data may be the case if the transmitted data are not complete, but only fragmentary.

2.2 Data processing

Data processing follows after analyzing all transmitted data that is entered in the summary list. The basic task of data processing is the correct identification of items based on the identification element item and provides additional parameters necessary for processing construction budget as a unit of measure and quantity.

2.3 *Item ID*

All transmitted data include identification items. Identification of items based on the parameters established for identification. In previous research we looked at creating a new identification parameter for correct identification of the transmitted data. New identification parameter consists of two codes TSKP and CodeText.

After examining the structure of IFCxml, it was found that TSKP code is written to the part ifcPropertySingleValue. Entered parameter is assigned a chosen parameter name in subpart "Name" in our case "TSKP". The value of the parameter is registered in the part NominalValueandsubpartIfcInteger-Wraper. In our model example wall is value consists of partial TSKP code "311351". Entire entry identifier TSKP code in XML code is shown in Fig.1.

Given the fact that the ID item is not a complete TSKP code for proper material identification in semi-automated system design includes CodeText. This code is written in subpart "Name" in our case "CodeText". The value of the parameter is registered in the part NominalValueandsubpartIfcInteger-Wraper. In our example is value of text code "Wall_NUDURA ICF_96-6". The combination of both codes will allow us to properly identify and link the transmitted data.

2.4 *Unit of measure*

One of the main parts of the itemized budget is a unit of measurement specification in bill-off quantities. Unit of measure given in what kind of units is indicated the amount of construction work and materials. This parameter is defined for each construction in the parameter menu structures called "unit" (see Figure 2 Unit registered in IFCxml structure). The units can be m, m2, m3, kg or piece. The choice of units shall be based on the units of measure used in database TSKP or database ZOCM for that construction work and materials. In our example, the model parameter measurement units when exporting to IFCxml was entered into the structure of ifcPropertySingleValue assigned an ID in this case id = "i1825". Registered parameter application automatically assigned parameter name in subpart "Name" in our case, the application was in the Czech language localization and therefore, the name was "Jednotka". The value of the parameter is registered in the part NominalValue in Subpart IfcInteger-Wraper. Unit of measure of our model example is given in cubic meters "m3". Writing unit of measurement in IFCxml format is shown in Figure 2.

2.5 *Amount*

The most important information by the data for calculating the indicative budget price is the amount of calculated structure, construction work or building element. The creation of the bill-off quantities of individual items is the most time-consuming activity of the estimating. This was a major reason for the research of an automated production system of construction budget from the CAD applications. For each item of construction budget (labor and material)

```
<IfcPropertySingleValue id = "i1817">
  <Name> TSKP </ name>
  <NominalValue>

  <IfcInteger-wrapper>311351</IfcInteger-wrapper>
```

Figure 1. TSKP code written in IFCxml structure; own source.

```
<Name>Jednotka</name>
    <NominalValue>
        <IfcDescriptiveMeasure-wrapper>m3
</ IfcDescriptiveMeasure-wrapper>
```

Figure 2. Unit of measure entered in IFCxml structure; own source.

563

it is defined in what units will indicate the amount. Next step is then calculating the actual amount. Calculate the amount in CAD applications executed automatically with defining the actual geometry of the structural elements. During the examination of the exported data file format IFCxml it became known that the program writes to a file more data about the volume and surfaces. Wall construction volume has been written to the two parts GrosseVolume and NetVolume. First value "GrosseVolume" indicates the amount of wall construction, including wall openings. The second value called "NetVolume" is the net amount of the wall without opening constructions.

Our model included wall construction door openings to make it possible to the accuracy of calculation of the net amount. For correct calculation, the statement of area budget items transmitted data with net amount must be involved. The calculated amount is written in the section IfcQuantityVolume specific id = i1786 under name (NetVolume) in the amount registered in the "VolumeValue" concrete 4.92 m3. This value is the exact value of wall volume after deduction of the door opening (see Fig. 3).

The accuracy of the calculation depends on the geometrical accuracy of the calculations of each CAD application. In this case, the accuracy was 1×10^{-5} mm registered in the part "IfcGeometricRepresentationContext"

2.2.1 *Assigning the prices to items*

Assigning a unit price of a specific item from a database of prices is basic condition of estimating. Assigning of unit prices can be automated, semi-automated or manual.

The main criterion for automated approach of budgeting is a complete code TSKP for finding an item in the database prices. Manually assigning of unit price for items is in the case if the item is not in the database. In case of the price which is not in the database or program cannot match the correct item from the database, you must manually assign items in the budget and manually obtain the unit price. Manual creation of new items to the correct section of the budget extends the budgeting time. New semi-automated approach to budgeting was developto simplify budgeting and time savings.

Semi-automated approach of creating a construction budget is an intermediate step between automated and manual access. Semi-automated system occurs in two cases. The first case is if the program cannot assign a unit price because unit price is not in the database. In this case, the estimator has to create a new entry with all parameters (name, unit of measure, unit price).

The second case is when a program cannot match the unit price from the database because identification code is not compatible with database, or incomplete code. Deficiencies in the code may be the case where the material or work is not in the database price. Despite non-existent entries in the database, it is appropriate to assign sub-code all structures (material, labor) that do not contain the full code. Any design in a semi-automated system must include at least the first three codes for correct classification of items in the estimating section.

The process of linking prices of items in a semi-automated system based on a price database of section in which the items are adding by partial TSKP code. The program allows you to select from the section of the database. The second option is an Internet database with the

```
</ IfcQuantityVolume>
  <IfcQuantityVolume id = "i1786">
    <Name>NetVolume</ name>
    <Unit>
      <IfcSIUnitxsi: nil = "true" ref = "I1566" />
    </ Unit>
    <VolumeValue> 4.92 </ VolumeValue>
  </ IfcQuantityVolume>
```

Figure 3.   The volume of the wall minus the hole in IFCxml structure; own resource.

assistant search. The last option is to create new items. In this case, the user must create a new budget item and the unit price of the item must calculate. Calculating of new items will be done using the calculation formula or the unit price is found based on information from the manufacturer or another supplier. Advantage of semi-automated system in case of a manual bidding is that the program can find similar items from the same section.

Assigning unit prices to all items (automatically, semi-automatically, manually) which were found with budget software are generates a partial budget. Partial budget is not complete budget, and should be supplemented with items that are not included in the 3D model. Examples of items that model does not include is supplement for stickiness. The second case of items that partial budget does not include is the cost of auxiliary structures, for example shuttering in concrete work or scaffolding. The result of this in whole process of adding an additional fee and auxiliary structures is the full budget. This budget should be checked.

Different approaches to assigning price (automatic, semi-automatic, manual) the items and the total budget of the process is shown in Figure 4.

2.2.2 *Checking the construction budget*

After importing data from CAD applications automatically or manually, the second most important step in creating the construction budget completeness check and then the resulting control of the budget. Everything that is modeled in the CAD application, if it contains the identification data is transmitted to the budgetary applications as partial budget either manually or automatically. Created partial budget is not complete budget, but it is necessary to process control, and then adds missing data to the budget. The reasons for the partial control of the budget are several:

- Incompletely processed 3D project AEC CAD applications.
- Processing errors in 3D model AEC CAD applications.
- 3D model does not include ancillary and supplementary materials, or construction.
- Indirect costs are not included in the 3D model.
- The additional budget costs are not included in the 3D model.
- The completeness of the budget after adding the missing parts of a partial budget.

As a first step of control the budget is checked for completeness and accuracy of processing the 3D model from CAD applications. A virtual 3D model of the building must include all structures which will also include the actual building and all must be modeled from data assistant in order to implement an automated budget. Control is determined by the structure, if virtual model does not contain a structures, they must be supplemented.

The second step is addition of the auxiliary and additional materials and structures because a virtual model in most cases contains only the basic structure.

3 CONCLUSION

Creating a construction budget with its complexity is similarly complicated than the mere creation of project documentation, except that the data on prices are changing. Frequent changes in the price levels of individual structures and works, was the reason not to transfer parameter of prices in CAD applications, butt keep the assignment of prices in the budget application. The proposed approach to data transfer from CAD applications in budget applications for computer aided design in construction budget has several advantages but also disadvantages. The major advantage of automated budgeting is automatically generated indicative budget for the construction of structures containing TSKP code. In other cases, the semi-automated budgeting has to be applied. It is based on creating drawings of structures containing an identification parameter for correct data import and processing. Paper demonstrated in detail the process of semi-automated budgeting based on the assignment of sub-identifiers TSKP with minimum 3 digit code and with identifiers CodeTex. Assigning the code to ensure partial assignment of the imported data in a particular section of the budget will reduce the demands on time and the cost of budgeting.

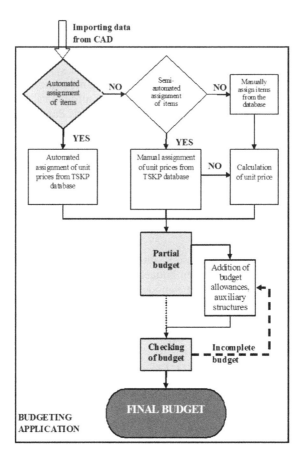

Figure 4.   The treatment process of the budget; own source.

ACKNOWLEDGEMENTS

The paper presents partial research result of project "VEGA—1/0677/14 Research of construction efficiency improvement through MMC technologies".

REFERENCES

Harrison, F., Lock, D. Advanced Procect Management, A Structured Approach, 4th Rvised edition, In: Gower Publishing Ltd., 2001.

Kempton, J: Modern methods of construction: Maintenance issues in the registered social landlord sector, In: Dainty, A.R.J. (Ed) Procs 25th Annual ARCOM Conference 2009 Nottingham.

Kozlovská, M., Spišáková, M.: New construction technology in terms of waste reduction In: Waste Forum, Vol. 2013, no. 2.

Kyjaková, L., Mandičák T., Mesároš, P.: Modern methods of constructions and their components. In: Journal of Engineering and Architecture. Vol.2, no. 1, 2014.

Mačková, D., Mandičák T.: Acceptance theories of innovation and modern methods in construction industry. In: Open Journal of Business Model Innovation. In Press (2015), p. 1–8.

Tažiková, A.: Cost analysis of modern methods of construction. In: Improving the efficiency of construction by means of MMC Technology, Košice, 2014.

Venkataraman, R., Pinto, J. Cost and Value management in projects: In: John Wiley & Sons, Inc., New Jersey USA, 2008.

Zhang, J. et al. (2010). Improving the usability of standard schemas In: Information Systems.Elsevier, 2010.

Advances and Trends in Engineering Sciences and Technologies II – Al Ali & Platko (Eds)
© 2017 Taylor & Francis Group, ISBN 978-1-138-03224-8

Scheduling of tower cranes on construction sites

V. Motyčka & L. Klempa
Faculty of Civil Engineering, Brno University of Technology, Brno, Czech Republic

ABSTRACT: Tower cranes are often crucial mechanisms in the construction of buildings. The choice between cranes and the manner of their deployment has an important influence on the course and costs of the construction process. Despite this, construction investors and contractors pay little attention to the proposal of cranes for construction work. The currently-used procedures for the proposal of cranes are based on various input indicators which only characterize the requirements for the transport of materials very roughly. The results achieved on the basis of such procedures are of varying quality, but mostly only very approximate. The capacity of the proposed cranes to supply all construction processes in the required time is not being evaluated sufficiently. The following contribution describes new, more precise ways of dealing with this task, i.e. a proposal for an evaluation procedure to determine whether proposed secondary transport using tower cranes corresponds to the transport requirements set out in the schedule for continuous construction sub-processes.

1 SCHEDULING OF TOWER CRANES

1.1 *A more accurate evaluation of the scheduling of tower cranes at construction sites*

Tower cranes are some of the most widely used construction machines, and their correct deployment and use has a significant influence on the smoothness and financial costs of the construction work. Even though this fact is generally known, no one is systematically dealing with the development of a more exact way of evaluating the scheduling these machines. Even today, only approximate methods are used for the scheduling of cranes. Such methods are based mainly on experience and estimates (Lizal et al. 2003). More accurate determination of the service time can be calculated according to Prof. Girmscheid thesis from Zurich (Girmscheid et al. 2002), very simply, without the particular technical parameters of the applied crane and applying the spatial and time links on the particular site. Other authors, who have dealt with the issue, were Prof. Bottcher from Wiesbaden (Bottcher et al. 1954) and Prof. Zhang (Zhang 2008). Their models use mathematical statistics for the analysis of monitored parameters of tower cranes or groups of machines at sites and their optimization.

As a result, several years ago at Brno University of Technology, Faculty of Civil Engineering, Institute of Technology, Mechanization and Building Management, we decided that we would try to contribute towards a more accurate way of dealing with this issue.

1.2 *Possible approaches to the task*

There are several options one can take in investigating this issue. One of the possible options was a use of the queueing theory. The result was the production of what is known as the extensive method (Motycka 2007), which is based on the concept that it is impossible to predict all of the movements of a crane during its deployment as part of the realization of a structure. We therefore also took the element of chance into consideration during calculations. A mathematical model was developed utilizing the queueing theory system according to Kendall in which individual construction sub-processes are circulating elements of a system which require service via a queueing system. The service line, or lines, is a crane or a set of cranes. In order to deal with the task in this way, it is necessary to introduce simplifying prerequisites into the calculation that are able to affect the overall results of the evaluation of the time requirements of the construction sub-processes needing crane operation. We have therefore also approached the investigated issue in other ways which could lead to more exact results.

One of these ways is a method which utilizes a simulation model. This method simulates the expected movements of a construction site tower crane which are the results of the time requirements of individual construction sub-processes needing the use of a crane in connection with the binding time schedule of the construction project.

Another possible solution to this issue is to use the control process method. This method tries to simplify the whole process of evaluating the workload over time by choosing only one control process from the evaluated construction project to which all other construction processes are temporally linked. This control activity is subjected to thorough temporal analysis as regards to the utilization of the tower crane. Other construction sub-processes, which have time requirements tied to this control construction sub-process, are gradually linked to it with regard to their material supply requirements. However, the control process always has, when served by the crane, priority over other construction sub-processes. The result of this solution is, again, an evaluation of whether the total time requirements on the tower crane (or a set of cranes) are in accordance with the binding time schedule of the construction project (Motycka et al. 2014).

Currently, we are primarily focused on the creation of a simulation model which is described in the following section in greater detail.

2 SIMULATION MODEL FOR THE WORK OF TOWER CRANES

2.1 *Input prerequisites*

A simulation model is an abstractly created system which describes a certain group of phenomena observed in the real world as realistically as possible. We replace these real phenomena with the created model.

The basic input prerequisite is the same as in the case of other methods. We start from the assumption that the construction schedule, which is confirmed on the basis of the contract between the client and the contractor, is binding. When creating the simulation model for the evaluation of crane workload, we model the work of the crane as if it were working exactly according to the aforementioned assumption, i.e. according to the prepared and approved binding construction schedule. When signing the project contract, the participating parties also assume that the project deadline will be complied with. We are therefore looking for a simulation model for the work of a crane or set of cranes which will describe the expected and required crane workload with regard to compliance with the contractually approved deadline for the construction of the whole structure.

If we assume that the construction work really does take place according to the plan, i.e. in an ideal manner (this is what is considered during the construction preparation stage, though in reality the construction process will not be like this), then the service requirements placed on cranes are not random but are repeated at regular intervals (Funtik et al. 2014). The duration of service then also does not have a random character—each activity will occupy the crane for an exactly calculated time which is governed by the nature of the task. The model is thus strictly based on selected individual specific processes and sorts them into sequences according to real crane service requirements (Klempa et al. 2013). It thus differs from the approach mentioned earlier, which utilizes the queuing theory.

In order to create a functional simulation model, it is necessary to define a group of phenomena (the course of each monitored construction sub-process) which we observe in the real world (at construction sites) and which we want to describe using the model.

For every evaluated construction sub-process (construction activity), a model track is determined for the suspended load which is to be transported from the place of storage to the place of unloading. We will consider the suspended load as a mass point for the time being. The model track is a trajectory which the mass point will move along when travelling from the centre of the surface on which it was placed, with vertical movement upwards to the transport height, movement tracing a curve in the horizontal plane above the centre of the surface where it will be placed, vertical movement downwards to the place of unloading, and then movement back to the centre of the storage area.

On the basis of this track and the speed of the individual movements of the suspended load, we will calculate the duration of load movement in one model working cycle. By taking into consideration the time needed for hanging (loading) and disconnecting (unloading) the load from the crane's hook, we will obtain the duration of one model work cycle.

Next, we will calculate the number of cycles necessary to supply the construction sub-processes within a given limited time interval. Here, we consider the planned quantity of material to be transported by the tower crane. It is assumed that the material is transported in a constant unit form (pallets, containers, pieces, etc.). The number of cycles for the transport of material in the construction sub-process is determined as the ratio of the total volume of material to be transported within a specified time interval to the unit quantity of this material. The number of all crane working cycles in the evaluated time period is then the sum of all cycles for all individual transported materials.

If we assume that during all cycles involving the supply of one type of building material, an item of identical volume or unit form will be transported each time, we are able to determine the length of the period needed to transfer the material based on the performance standards for the specific number of workers who are processing this material. It specifies the periodicity of service requirements involving the use of a crane to supply the material in question. If a requirement is announced to a crane and the crane is servicing another process at that moment, the requirement has to wait to be dealt with. This disturbs the continuity of construction work, causing delays in the time schedule.

2.2 *Phenomena and their characteristics*

For the creation of a simulation model, it is necessary to clearly define a group of phenomena which are the specific monitored construction processes that take place at the construction site and which we wish to describe via the model. A phenomenon is thus the course of a construction sub-activity such as the execution of formwork, reinforcement or concreting, mould release work or, for example, masonry work. These individual phenomena are defined by a trio of basic characteristics: the time period for which the crane serves, the time spent processing the given material without the assistance of the crane, and the number of working cycles during the evaluated time interval (e.g. one work shift). The basic time elements are known as steps (more details can be found in Part 3.3: The principle of the simulation model).

3 MODEL OUTLINE

3.1 *Basic properties of the model*

Based on the facts described above, we are creating a model which combines all of the periodic, continuous and single requirements of construction sub-processes for crane service in gradual steps. It will adapt the duration of the individual building process cycles to include the time a process spends waiting in a queue for service, and it will determine the expected time which is needed for all processes to be supplied within an allocated period (we will usually evaluate one work shift).

The order in which the construction sub-processes are serviced can be decided before calculation begins. It can be, for example, the order in which they join the queue, or can depend on whether or not they are on a critical route or are performing activities which have been set to be served with priority.

One of the properties of the model is also the option to input elements—work cycles of construction sub-processes—which can be pre-stocked. The model allocates these work cycles to periods when work queues do not form.

The model is also able to consider several service lines (sets of tower cranes) operating at once. Part of the work performed by each crane is only done to supply its own assigned processes (or supply processes in a designated area of the construction site), while some processes are shared with other lines (if in an area serviced by several cranes). More details can be found in Part 4: Sets of cranes.

3.2 Initial conditions for modelling

Before modelling the work performed by a crane, initial conditions need to be defined on whose basis the model will be rendered generally applicable.

- the model is linked to the construction schedule,
- the schedule determines the time period covered by the model (e.g. one work shift),
- the total volume of work for individual construction sub-processes is known,
- the concurrences of construction sub-processes are known,
- the crane always carries the same unit quantity of material,
- all construction sub-processes are involved in the system at the same time, or the time shift of their start is known,
- the volume of work in the individual construction sub-processes in the individual evaluated work cycles—elements of the system—is known (if it is not known for the individual time units exactly),
- the positions where material is stored and deposited with regard to the position of the crane are known,
- the exact type of the crane is known,
- the model does not consider random influences.

3.3 The principle of the simulation model

First, one work cycle needs to be defined for all monitored construction sub-processes, i.e. its duration needs to be determined.

This cycle consists of a part where the construction sub-process is served by a tower crane and a part where the supplied material is being processed without requiring service from the crane. The service period is marked as t_c, and the material processing period as t_p, as is shown in Figure 1. The measure unit is the second. The work cycles must be defined for all construction sub-processes which occur in the evaluated time interval.

Once all processes are defined in this way, we can start the modelling itself. The mathematical model consists of gradual steps, as is shown in Figure 2. A gradual step is generally defined as a sub-section of the whole evaluated time interval of the deployment of the crane, which starts at the moment when the first activity is served and ends at the moment when one of the activities served in this step returns to the system with a repeated requirement to be served. The first and the last step of the whole evaluated time interval are exceptions.

In the very first step, all activities which are to be served by the crane from the beginning gradually connect to the system. Construction sub-processes A, B and C are shown in Figure 2. The decision to have them join gradually was made because some of the activities

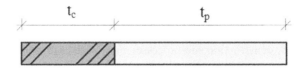

Figure 1. One work cycle of a construction sub-process, tc—time of the tower crane work, tp—the supplied material processing period.

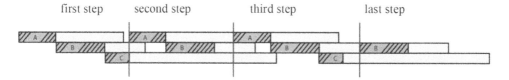

first step   second step   third step   last step

Figure 2.   Gradual incorporation of system elements A, B a C (construction sub-processes) into the simulation model in individual steps.

may not get the opportunity to join the system immediately in the initial stage. In such a case, a construction site work team would be ready to work on their allocated construction sub-process but would not have any material to perform their task. The system must therefore be started in such a way that this situation cannot occur.

The very last step is specific in that no new requirement for repeated service arrives.

During modelling, two types of construction sub-processes can be distinguished: those with or without priority. Prioritised activities are those which are important for the smooth operation of the construction site. In the mathematical model system, when activities which are given priority start queuing with a requirement for service, they skip all the waiting activities that are without priority and appear in the first place of this queue, or behind any other waiting prioritised activities.

The simulation model also considers a situation when activities occur within the evaluated time interval that connect to the system later, i.e. not from the very beginning. In the case of such activities, the length of the time interval which passes from the launch of the system until the moment when this activity first asks for service needs to be known. At this moment, it will join the queue with its requirement and start behaving like the other activities involved in the system.

Each construction sub-process has a finite number of cycles in the evaluated time interval. If this number is exhausted, the process will be excluded from the system and further service requirements no longer arise for it.

Some construction sub—processes require a break for technical reasons after several cycles. If this is the case, the number of cycles before the break needs to be known in advance, as well as how long the break will last. When the break ends, the requirement will return to the system again and start behaving like other activities.

We know that some processes can be pre-stocked. The system will make use of this character-istic in moments when the crane is idle and, simultaneously, the length of this idle time is longer than the period required to service the process we wish to pre-stock. The use of this property has an advantage in that the pre-stocked activity doesn't place any service requirements and thus does not increase the length of the queue of elements which are waiting for service.

3.4   *Evaluation of the simulation model*

The resultant period needed for the deployment of a tower crane at a construction site within the evaluated time span equals the sum of the durations of all individual steps, or equals the time which elapses from when servicing of the first activity begins in the first step until servic-ing of the last activity ends in the last step. In this way, the work done by a crane in a given time interval can be evaluated.

We may sometimes wish to evaluate the whole length of the work shift in the given time interval, i.e. including the processing of the supplied material. If so, after the period of crane utilisation ends, it is necessary to add the period which will elapse from the time the servic-ing of the last activity in the last step ends until the processing of the final, longest-lasting construction process is complete.

This require creation of the user-friendly computer software to support workers in the preparation of construction projects, as well as the site managers, who will commonly use this software to evaluate proposals for the deployment of tower cranes at the construction sites. Conceiving of the simply software for this issue is next task. The principle will be based on the simulation model.

4 SETS OF CRANES

As has already been mentioned, two or more tower cranes are often deployed at larger construction sites. They usually serve one part of the construction site independently, while working on another part of the site together with another crane. Such sets of cranes are currently placed at construction sites on the basis of the experience of staff involved in construction preparations and are not based on any empirical design or calculation. Their deployment in groups is thus even less effective and causes further economic losses to building companies.

In the case of the servicing of a shared area by several cranes simultaneously, the simulation model handles this situation in such a way that cranes are allocated not only the activity which they serve but also the area in which they perform this activity (construction sub-process). If a requirement for servicing arises for an activity which is served by several cranes in a shared area, this requirement is placed in a queue for all the cranes which are serving it together. The crane which is first to have free capacity will deal with this requirement, which is then cancelled for the other cranes, allowing the other requirements waiting in the queue to move forwards. This means that when there is a service requirement for an activity which is serviced by more cranes, the construction sub-processes are also classified according to the section (e.g. Section 1, Section 2 and Section 3) in which the service is required (activities can be labelled A1, A2, A3, B1, B2, B3, etc.). It is thus obvious which activities are being served by several cranes working together in a shared area.

5 CONCLUSION

The approach describes options for the efficient evaluation of tower crane capacity and the satisfactory supply of several concurrent construction processes at once in accordance with a construction project time schedule.

The results achieved by its application can make a significant contribution when checking designed construction schedules, selecting the most suitable machinery for a given construction project, or planning the deployment of sets of machines.

REFERENCES

Böttcher, P. D. N. & Neuenhagen, H. 1954. Baustelleneinrichtung. Bauverlag GmbH. Berlin.
Funtik, T. & Makys, P. 2014. BIM based tower crane placement optimization. *CTM 2014—Construction Technology and Management, proceedings International scientific conference*. Bratislava: Slovak University of Technology in Bratislava.
Girmscheid, G. 2002. Leistungsermittlung fur Baumaschinen und Bauprozesse. Springer, Zurich.
Klempa, L. & Motycka, V. 2013. Modelování práce věžových jeřábů. *Stavitel*. Praha: Economia a. s.
Lizal, P. & Marsal, P. & Musil, F. &Henkova, S. &Kantova, R. &Vlckova, J. 2003, Technologie stavebních procesů pozemních staveb. *Technologie stavebních procesů pozemních staveb*. Brno: CERM s.r.o.
Motycka, V. 2007. Věžové jeřáby v pozemním stavitelství. *Věžové jeřáby v pozemním stavitelství*. Brno: CERMs.r.o.
Motycka, V. & Klempa, L. 2014. Simulation model of tower crane work. *CTM 2014—Construction Technology and Management, proceedings International scientific conference*. Bratislava: Slovak University of Technology in Bratislava.
Zhang, H. 2008. Multi-objective simulation-optimization for earthmoving operations. Automation in Construction, 18, 2008.

Research summary: Analysis of selected adhesive systems intended for façade bonding

B. Nečasová, P. Liška & J. Šlanhof
Faculty of Civil Engineering, Brno University of Technology, Czech Republic

ABSTRACT: This contribution presents an overall summary of several years of research. The focus was placed on an experimental testing of solid wood and wood-based composite materials used as a sheathing of vented façade systems with bonded joints. The potential of this arrangement was verified through experimental measurements. Two different test methods, defined by relevant European and Czech standards, were adopted and modified. Moreover, the frost resistance of some test specimens was determined by subjecting them to a cyclic action of freezing and thawing. Another group of specimens was subjected to conditioning with sudden temperature changes and finally, the substructure bonding retention to the cladding and strength in shear in tension stress were tested, analyzed and discussed. The obtained data revealed that the crucial factor for the bonded joint is not its strength but the quality and stiffness of the cladding material or of the substructure. Additionally, it was verified that any violation against technological discipline may weaken the functionality of the entire façade system.

1 INTRODUCTION

Bonded systems are not an uncommon technique. It is a technology that was first studied about 70 years ago (Goland & Reissner 1944). Although many years have passed since the first application, it has not yet become an established technological option for anchoring of façade cladding (Nhamoinesu & Overend 2012). In the Czech Republic, the percentage of implemented bonded systems is almost negligible. In recent years, this method has found application mainly in renovated buildings, where the emphasis is on pure aesthetic and architectural expression of the building. The vast majority of the implementations use aluminum supporting substructures to which facing material of artificial stone, aluminum composites or cement based materials is attached. Wood-based materials or pure solid timber are used occasionally (Kovářová 2016). The main reason might be a fact that wooden structures and wood elements used in exterior can experience a series of chemical and physical changes that spoil its aesthetic appeal, durability and service life, if a proper and regular maintenance is neglected. On the contrary this material originate from renewable resources, wood is considered as environmentally friendly, provides good insulation and good weather resistance.

The potential of bonded joints in combination with wood has not been studied sufficiently yet, even though this technique of anchoring is able considerably decrease stress concentration and fatigue in areas where the cladding material is connected to the load-bearing supporting structure. In the case of failure of a facade plate the adhesive joint is able to hold large fragments in place (Gobakken & Vestøl 2012) for a necessary period because the elastic adhesive technique allows an invisible back-panel fixing without any weakening of the cross-section of the plate. Due to an extensive bonding area along the edges of the panels, the loads resulting from wind, temperature and the self-weight of the plate are carried out uniformly.

Generally, the conception and construction of the façade is crucial, not only for the external appearance of the building, but also for the serviceability, durability, costs and energy consumption of the entire building.

2 MATERIALS AND EXPERIMENT DESIGN

2.1 *Materials*

The wrong combination of selected materials can cause a substantial reduction in the durability of the entire façade system and in particular can lead to a remarkable increase of requirements concerning maintenance. Thus the selection of appropriate facade components is a significant phase in the design of everyfaçade.

Cement-bonded particleboard (hereinafter 'Cetris'), Siberian larch and wooden plastic composite, (hereinafter 'WPC') were selected as a sheeting material for the purpose of this research. A good knowledge of mechanical properties is crucial for every scientific work, the comparison of material properties can be found in Table 1.

Spruce joists were selected as a material for the load-bearing substructure. This material is the most commonly used material for this purpose. The only demand on this façade element is the use of a specific strength class, specifically for this case it was class C22, and the surface of all bonded profiles have to be roughened e. g. by sandpaper, more details can be seen in Table 1.

2.2 *Adhesives*

To compare more adhesive systems, two different groups of high-strength structural adhesives were tested, see Table 2. The first group of selected adhesives were one-component polyurethanes. The second chosen group of the adhesive system were one—component structural adhesives based on modified silyl polymer.

2.3 *Test sample matrix*

The presented series of test methods were modified to accomplish requirements given by European and Czech standards. First applied test method was used to determine the adhesion

Table 1. Mechanical properties of tested adherents.

Material	Spruce C22	Cetris	Siberian larch	WPC
Density	$410.0 \ kg \cdot m^{-3}$	$1350.0 \ kg \cdot m^{-3}$	$660.0 \ kg \cdot m^{-3}$	$890.0 \ kg \cdot m^{-3}$
Shear st. parallel to the grain	$3.8 \ N \cdot mm^{-2}$	$1.8 \ N \cdot mm^{-2}$	–	$8.0 \ N \cdot mm^{-2}$
Shear st. perpendicular to the grain*	–	–	$1.5 \ N \cdot mm^{-2}$	–
Tensile st. perpendicular to grain (or board plane)	$0.4 \ N \cdot mm^{-2}$	$0.63 \ N \cdot mm^{-2}$	$1.5 \ N \cdot mm^{-2}$	$5.0 \ N \cdot mm^{-2}$
Thickness of the tested material	19.0 mm	20.0 mm	19.0 mm	9.0 mm

*This direction was tested here.

Table 2. Mechanical properties of selected adhesive systems.

Adhesive	Polyurethane		Modified Silyl Polymer	
System	PU-I	PU-II	MS-I	MS-II
Ultimate tensile strength	$4.0 \ N \cdot mm^{-2}$	$9.0 \ N \cdot mm^{-2}$	$2.3 \ N \cdot mm^{-2}$	$1.8 \ N \cdot mm^{-2}$
Ultimate shear strength	$2.5 \ N \cdot mm^{-2}$	$5.5 \ N \cdot mm^{-2}$	$2.0 \ N \cdot mm^{-2}$	$2.25 \ N \cdot mm^{-2}$
Temperature resistance	−40°C to +90°C	−40°C to +100°C	−40°C till 100°C	
Application temperature	5°C to 30°C		−5°C till 30°C	
Skinforming	approx. 20 min		approx. 15 min	
Curing speed	3 mm/24 hrs at 23°C/RH 50%		3 mm/24 hrs at 20°C/ RH 50%	

*Adhesive marked MS-II is designed specifically for a facade bonding of wood and wooden materials. The other selected adhesive, marked MS-I is not initially intended for vented façade systems. PU-I and PU-II are also intended for façade bonding.

of the surface finish to the substructure(Czech-Standards 1981a).The maximal force that would be able to tear off the given surface finish area from the substrate while applying the perpendicular tension was measured.

The second used method, the test of the tensile lap—shear of bonded assemblies (Czech-Standards 2009) was modified because the intentions of the presented experimental research were to test the whole façade system. Therefore all test samples were tested in the same position as they are implemented in the real façade system.

The tensile tests were performed with an FP 10/1 universal test tearing machine. The tests were carried out at room temperature. The displacement rate was set at 8.00 mm·min^{-1}.The grips were designed to be adjustable so that the loading was put centric on the adhesive bond.

The production of test samples itself involved several steps. Each adhesive required a different modification to be made to the bonded surfaces depending on the requirements of individual manufacturer, these modifications were precisely described in previous articles (Nečasová 2015a, b). Moreover, series of tests included testing of the frost resistance of specimens which was determined by subjecting them to a cyclic action of freezing and thawing (Czech-Standards 1981b). The second conditioning method included conditioning of test specimens with sudden temperature changes (Czech-Standards 1983).

3 RESULTS AND DISCUSSION

The bonding of the surface finish to the substrate and the determination of shear strength during stress was calculated according to the equation, see Eq. 1. The values, which are presented in Table 3. and Table 4., were calculated from the maximal force F_{max} required for debonding of the test sample in N and from the bonded area of samples, see Figure 1.

Table 3. Test results of the adhesion of the surface finish to the substructure.

Material	Cetris		Siberian larch		WPC	
Adhesive system	Average σ_{adh} [N·mm^{-2}]	Elongation at break [mm]	Average σ_{adh} [N·mm^{-2}]	Elongation at break [mm]	Average σ_{adh} [N·mm^{-2}]	Elongation at break [mm]
PU-I	0.778	1.152	1.245	3.949	0.891	2.267
PU-II	0.740	1.115	1.315	2.388	0.751	1.379
MS-I	0.688	1.028	1.275	2.742	0.195	0.480
MS-II	0.747	1.522	1.027	2.788	0.080	0.205
MS-II*	–	–	–	–	0.686	1.509

*The bonded surface of the adherend was roughened with sand paper.

Table 4. Test results of the tensile lap—shear of bonded assemblies.

Material	Cetris		Siberian larch		WPC	
Adhesive system	Average τ [N·mm^{-2}]	Elongation at break [mm]	Average τ [N·mm^{-2}]	Elongation at break [mm]	Average τ [N·mm^{-2}]	Elongation at break [mm]
PU-I	1.327	7.717	1.637	11.766	1.561	10.840
PU-II	1.460	5.468	1.651	6.938	3.134	9.291
MS-I	1.107	4.321	1.153	5.772	0.235	0.891
MS-II	1.450	5.453	1.455	5.545	0.000	0.000
MS-II*	–	–	–	–	1.857	7.525

*The bonded surface of the adhered was roughened with sand paper.

$$\sigma_{adh}(\tau) = \frac{F_{\max}}{A_{ef}} = \frac{F}{l_{ef}b_{ef}}. \qquad (1)$$

where σ_{adh} = calculated adhesion of the surface finish to the substructure in $N \cdot mm^{-2}$; τ = calculated shear strength of the single lap joint in $N \cdot mm^{-2}$; F = the force required for debonding of test samples in N; A_{ef} = the effective area of bonding or over lapping in mm^2; l_{ef} = the length of bonded joint in mm; and b_{ef} = the breadth of bonded joint in mm.

Based on the resulting values it can be noted that in the case of the 'Cetris' cladding material, the material was deteriorated in almost all tested cases and the wooden substructure with the adhesive layer was pulled out from the cladding, see Table 5 and Figure 2. The limit strength of any of tested adhesives was not reached which demonstrates a high quality of used adhesives and a low resistance of the cladding material, see Table 3. As can be seen in Table 3., the highest strength determined on the basis of the adhesion test in combination with Cetris was achieved in the case of the PU-I adhesive system, where the average adhesion measured is 0.778 $N \cdot mm^{-2}$. In contrast, in the case of the tests for the determination of shear strength, test samples with the PU-II system showed the highest strength, the average shear strength of samples is 1.460 $N \cdot mm^{-2}$, see Table 4. However, the obtained results are very similar, which only supports the fact of the low resistance of the selected cladding material.

The combination of the adhesive system and a wooden façade cladding, here Siberian larch, has proved to be a very useful option for monitoring of heterogeneity violations during testing. Three disruption modes appeared. Firstly the deterioration of the adhesion between the cladding and the adhesive system The second observed disruption mode relates to cohesion, namely deteriorated integrity of the bond—i.e. cohesive failure, was very frequent in the case of the MS-I system. This is an ideal way of failure. The last observed way of disruption is the disruption of the wood substructure or wood cladding, see Figure 2.

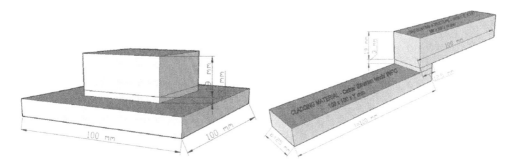

Figure 1. Sample geometry: (left) adhesion of the surface finish; (right) single-lap shear sample assembly.

Table 5. Evaluation of the most common failure mode of test samples.

Material	Cetris		Siberian larch		WPC	
Adhesive system	Failure mode	Percentage failure rate	Failure mode	Percentage failure rate	Failure mode	Percentage failure rate
PU-I	FTF	100%	SF	56%	AF	89%
PU-II	FTF	94%	SF	67%	AF	100%
MS-I	FTF	89%	CF/SF	39%/33%	AF	100%
MS-II	FTF	67%	SF	61%	AF	100%
MS-II*	–	–	–	–	AF	94%
Total	FTF	88%	SF	54%	AF	97%

AF = adhesive failure; CF = cohesion failure; FTF = fiber—tear failure; SF = substrate (adherend) failure.

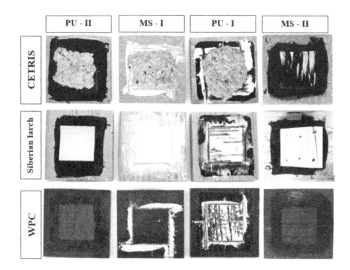

Figure 2. Failure surfaces after adhesion test: (top) fiber-tear failure; (middle) cohesive failure and delamination (substrate) failure on the left; (bottom) adhesive failure and thin-layer cohesive failure on the right.

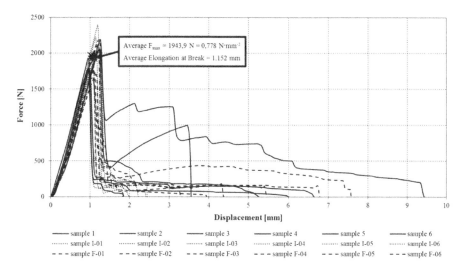

Figure 3. An example of stress-strain diagram for PU-I and Cetris: Test of adhesion of the surface finish to the substructure—solid lines are representing samples which were tested immediately after curing, I and dotty lines are samples tested after conditioning to sudden temperature changes, F and dash lines are samples tested after conditioning to the frost resistance.

The last tested version with the 'WPC' cladding has proved wholly inadequate when combined with adhesive systems based on modified silyl polymers, see Table 3. And 4. In almost all cases the cladding peeled off from the substructure, resulting in deteriorated adhesion between the cladding and the adhesive system. It can be assumed that compatibility between MS polymers and materials containing polyethylene is low. Based on the diversity of obtained results, it can be also stated that the selected cladding is not suitable for anchoring by adhesive bonding. Although the values measured in combination with PU-I, PU-II and MS-II* are satisfactory, the adhesion between the cladding and the adhesive system was deteriorated in all cases. Moreover, in two cases, in combination with PU-I, the adhesion was deteriorated already during conditioning to resistance of surface finish to sudden temperature changes.

4 CONCLUSION

Based on the presented measurements carried out and described above it was verified that a wooden substructure is more than a suitable alternative in the construction of ventilated facades. The mean value of tensile strength perpendicular to grain in radial direction for spruce is only around 0.4 N·mm^{-2}. The direction was actually tested here and in all tested cases the obtained results were two or more times higher. This fact can be put down to improving properties of adhesive systems because the tensile forces were distributed equally. Moreover, the mean value of tensile strength perpendicular to grain (or to board plane) of 'Cetris' is 0.63 N·mm^{-2}, however, in all tests the measured average adhesion to 'Cetris' was higher than 0.7 N·mm^{-2}, which again supports the idea of good compatibility of tested materials with adhesive system. In test cases with Siberian Larch the measured values were also very close to the presented mean value of the material. Similar results were obtained in tests of shear strength.

The adhesion testing also showed that the variety of selected cladding materials revealed several failure modes of the test samples. A series of tests showed that in some cases the adhesive properties of the tested cladding material, i.e. WPC, will be one area of weakness for bonded joints.

It was also demonstrated that different cladding materials are suitable for use in combination with bonded joints but that certain technical principles must be adhered to. It is mainly necessary to obtain information from the manufacturer of the given bonding system as to whether the system is suitable for use c. g. in combination with wood plastic composite. The results of this research might be a suitable foundation for alternative scientific work that might be focused on adhesive systems intended for façades.

ACKNOWLEDGEMENT

This research was carried out as a part of the FAST-S-16-3434 project and also as a part of a project FAST-J-15-2728.

REFERENCES

Czech-Standards 1981a. *ČSN 73 2577—Test for surface finish adhesion of building structures to the base*. Prague: Czech Standards Institute.
Czech-Standards 1981b. *ČSN 73 2579—Test for frost resistance of surface finish of building structures*. Prague: Czech Standards Institute.
Czech-Standards 2009. *ČSN EN 1465—Determination of tensile lap-shear strength of bonded assemblies*. Prague: Czech Standards Institute.
Czech-Standards 1983. *ČSN 73 2581: Test for resistance or surface finish of building structures to temperature variations*. Prague: Czech Standards Institute.
Gobakken, L. R. & Vestol, G. I. 2012. Surface mould and blue stain fungi on coated Norway spruce cladding. *International Biodeterioration and Biodegradation*, November, pp. 181–186.
Goland, M. & Reissner, E. 1944. The stresses in cemented joints. *Journal of Applied Mechanics*, Issue 11, pp. A17–A27.
CIDEM Hranice, C. 2014. *Basic Properties of CETRIS*, Hranice, pp. 149.
Kovářová, B. 2016. *Construction philosophy "light building"*. High Tatras Mountains; Slovakia, CRC Press/Balkema, pp. 293–298.
Nečasová, B., Liška, P. & Šlanhof, J. 2015a. Wooden facade with Bonded Joints - Experimental tests. *Advanced Materials Research*, Issue 1122, pp. 23–27.
Nečasová, B.,Liška, P. & Šlanhof, J.2015b. Determination of Bonding Properties of Wood Plastic Composite Façade Cladding. *Scientific journal: News in Engineering*, 30 05, 1(2), pp. 1–9.
Nhamoinesu, S. & Overend, M. 2012. The Mechanical Performance of Adhesives for Steel-Glass Composite Facade System. Delf, Louter Nijssem Veer.
Somani, K. P. et al. 2003. Castor oil based polyurethane adhesives for wood - to - bonding. *International Journal of Adhesion and Adhesives*, 4(23), pp. 269–275.
Weber, J., Hugues, T. & Steiger, L. 2004. *Timber Construction: Details, Products, Case Studies*. Birkhauser GmbH.

Advances and Trends in Engineering Sciences and Technologies II – Al Ali & Platko (Eds)
© 2017 Taylor & Francis Group, ISBN 978-1-138-03224-8

The impact of green roofs on thermal protection and the energy efficiency of buildings

D.V. Nemova, A.K. Bogomolova & A.I. Kopylova
Peter the Great St. Petersburg Polytechnic University, Saint-Petersburg, Russia

ABSTRACT: The problem of reducing energy demand of buildings has become a prerequisite for the improvement of energy saving technologies. One of the possible solutions of energy saving of buildings and structures is the application of technology for green roofs. Energy efficiency of buildings is realized by enhancing the heat transfer through the external covering. The improving thermal properties of enclosing structures are the most rational way to achieve this goal. In this regard, the main indicator of covering is the heat transmission resistance which depends on the amount of lost heat. The aim of this work is to assess the influence of heat transmission resistance of green roofs to improve energy efficiency of the building. The present paper provides the calculation of heat transfer resistance of the multilayer roof structure. That it is the possibility to significantly improve the energy efficiency of the building if the roof structure is correctly selected.

1 INTRODUCTION

The problem of reducing energy consumption of operating buildings has become a prerequisite for the improvement of energy saving technologies. Due to its thermal and economic benefits, innovative green roof technologies are getting a lot of attention. Increasing the thermal characteristic of building envelopes is probably the best way of achieving energy efficiency and cost cutting. In this case, heat transmission resistance is the main indicator of roof structure. The amount of energy lost heat depends on heat transmission resistance.

The objective of the research is the construction of a green roof representing a covering for two buildings with different ratio of the area of the covering to the area of enclosure structures for to the climatic conditions of the city of St. Petersburg. Thermotechnical calculations of a typical roof structure are carried out to estimate the energy efficiency of the green roof.

The calculation was carried out by the technique specified in National Standard 50.13330.2012 'Thermal performance of the buildings'. Thermotechnical calculations include: calculation of heat transmission resistance of the roof structure and calculation of the thermal energy consumption of the building.

2 OVERVIEW

Significant contribution to the development of this subject made the following specialists: Karachalioua, Santamouris, Pangalou (2016), Coma, Pérez, Castell (2016), Virk, Jansz, Mavrogianni, Mylona, Stocker, Davies (2015), Sopian, Salleh, Lim, Riffat, Saadatian (2013).

The paper (Karachaliou et al. 2016) is devoted to the research of thermal performance and energy efficiency of intensive green roofs with planting. The impact of the roof on energy saving is investigated using experimental and theoretical methods. According to the research results, the use of this type of planting reduces the average temperature in the room at 0.7 K that can significantly reduce the annual cost of heating and cooling the building.

Coma, Pérez, Solé, Castell, Cabeza (2016) jointly conducted experimental work to assess energy consumption and thermal behavior for three identical houses with different roofing designs. The results of the experiment design of green roofs show reduced power consumption in the warm period and a higher level of energy consumption in the heating period compared to a conventional roof.

The study of the technology of green roofs in Russia was engaged in the following authors: E.E. Semenova (2015), A.A. Shamarina (2013), N.B. Kondratyeva (2011) and others.

Currently, there is a large amount of publications related to the review of the technology green roofs, many of which does not focus on the problem of the influence of these technologies on the energy efficiency of buildings. In this regard, the study green roofs is quite promising direction, which requires further development.

3 GOALS AND OBJECTIVES

The goal is to assess the influence of heat transmission resistance of green roofs to improve energy efficiency of the building. The main objectives of the study are as follows:

1. to produce thermotechnical calculation with application of typical roof structure;
2. to produce thermotechnical calculation with application of green roof structure;
3. to assess the impact of the green roof technology on energy efficiency of buildings.

4 THERMOTECHNICAL CALCULATION FOR A BUILDING WITH A SMALL RATIO OF THE AREA OF COVERING TO THE AREA ENCLOSURE STRUCTURES

The first an investigate building is a residential seven-storey building with a small ratio of the area of the covering to the area of enclosure structures. The second building is a two-storey school with a large ratio of the area of the square to the area of enclosure structures.

The structure of external walls: brick laying of ceramic solid bricks (120 mm); air gap between the layers (3 mm); air heat insulation material ISOVER (80 mm); gas-concrete blocks (200 mm).

The windows are triple-pane windows from ordinary glass with the distance between the panes is 6 mm with heat-insulated window profile.

Climatic characteristics:

- design temperature of internal air for a residential building, $t_{int} = 20°C$;
- design temperature of external air during the cold period, $t_{ext} = -26°C$;
- duration of heating season for residential building, $z_{ht} = 220$ days;
- average outdoor temperature during the heating period, $t_{ext}^{av} = -1.8°C$;
- heating degree day, $D_d = 4796°C·day/year$.

The calculated heat transmission resistance of enclosure structure should be not less than the prescribed regulated value, the maximum value of which depends on the number of

Table 1. Area of external enclosure structures of residential building.

Type of enclosure structure	Parameter identifier	Square, (m²)
External walls	A_W	4558.74
Windows	A_F	1558.1
External doors	A_{ed}	190.21
Covering	A_c	1581.5
Coverings over passages	A_{c2}	219.0
Floors on the ground	A_{fl}	1564.9

Table 2. Thermal performance of the individual layers of standard roof structure.

Material	Layer thickness δ, m	Material density ρ, kg/m^3	Coefficient of heat conductivity λ_B, W/(m·°C)	Vapor permeability coefficient μ, mg/(m·h·Pa)
Gypsum neat plaster	0.005	1000	0.35	0.11
Reinforced concrete roof plank	0.2	2500	2.04	0.03
Heat insulation material URSA	0.14	45	0.043	0.62
Expanded clay gravel	0.04	600	0.26	0.26
Sand cement screed	0.025	1800	0.93	0.09
Four layers of ruberoid on bitumen mastic	0.01	600	0.17	0.00136
Layer of gravel treated with bitumen mastic	0.01	600	0.2	0.23

Table 3. Required and calculated value of heat transmission resistance.

Type of enclosure structure	Parameter identifier	Required heat trans mission resistance R_i^{req}, (m^2·°C)/W	Calculated heat transmission resistance R_i, (m^2·°C)/W
External walls	R_w	1.94	3.78
Covering	R_c	3.68	3.8
Coverings over passages	R_{c2}	3.68	3.8
Windows	R_F	0.49	0.51
External doors	R_{ed}	0.79	1.1
Floors on the ground	R_f	3.25	3.3

degree-days of heating season area of construction. They were taken according to National Standart 50.13330.2012. The following results of thermal resistance were obtained (refer to: Table 3).

The calculation of the transmission losses of thermal energy for heating, (kWh)/year, considering a residential seven-storey building during the heating period will be produce according to equation (1):

$$Q_{h1}^y = 0.024 \cdot D_d \cdot \sum_i \frac{A_i}{R_i} \cdot n \tag{1}$$

where 0.024 = conversion coefficient of thermal energy through external enclosure structures from W·day to kWh; D_d = heating degree day, 4796°C·day/year; A_i = area of i-th type enclosure structures (walls, windows, e.g.) for the considered building, m^2; R_i = reduced total thermal resistance of i-th type enclosure structures, (m^2·°C)/W; n = coefficient taking into account the dependence of the outer surface position enclosure structures (walls, windows, e.g.). For external walls, windows and balcony doors, combined coverings over passages windows n is equal to 1.

Calculate according to the equation (1) for the considered residential seven-storey building thermal energy losses through the external enclosure structures for the heating period:

$$Q_{h1}^y = 0.024 \cdot 4796 \cdot \left(\frac{4558.74}{3.78} + \frac{1558.1}{0.51} + \frac{190.21}{1.1} + \frac{1581.5}{3.8} + \frac{219}{3.8} + \frac{1564.9}{3.3} \right) = \tag{2}$$
$$= 619495.6 \,(\text{kWh})/\text{year}$$

5 RESULTS OF THE THERMOTECHNICAL CALCULATION WITH APPLICATION OF GREEN ROOF TECHNOLOGY

Similar thermotechnical calculation was carried out for described above residential building using a green roof instead of a standard roof, the composition of which is given in Table 5.

The following results were obtained: thermal resistance of the roof structure, $R_c^* = 5.64$ $(m^2 \cdot °C)/W$; annual consumption of thermal energy, $Q_{h2}^y = 603867.2$ (kWh/year).

6 THERMOTECHNICAL CALCULATION FOR A BUILDING WITH A LARGE RATIO OF THE AREA OF THE COVERING TO THE AREA OF ENCLOSURE STRUCTURES

For a more accurate assessment of the impact of green roofs on the annual heat consumption, consider the school building, which has a large area of covering commensurate with the area of the external walls.

Thermal performance of external enclosure structures of the school coincide with the characteristics previously considered residential building. The calculation of the annual consumption of thermal energy through the external enclosure structures was carried out for two variants with the use of the model of the typical roof structure and use the green roof structure.

The following results of annual consumption of thermal energy were obtained for school building:

1. using the model of the roof structure (refer to: Table 2), $Q_{h3}^y = 974003.8$ (kWh)/year;
2. using the green roof structure (refer to: Table 4), $Q_{h4}^y = 893358.6$ (kWh)/year.

The calculation's results of the transmission losses of thermal energy in the building during the heating period for the two considered buildings with different covering's type are shown in (refer to: Table 6).

Calculate the difference between the annual costs of thermal energy for residential building with typical and green roof structures:

Table 4. Thermal performance of the individual layers of green roof construction.

Material	Layer thickness δ, m	Material Density ρ, kg/m³	Coefficient of heat conductivity λ_B, W/(m·°C)	Vapor permeability coefficient μ, mg/(m·h·Pa)
Gypsum neat plaster	0.005	1000	0.35	0.11
Reinforced concrete slab	0.2	2500	2.04	0.03
Haydite compression	0.1	1000	0.41	0.14
Sand cement screed	0.02	1800	0.93	0.09
Water isolation	0.008	3.5	0	0.09
Root Barrier WSF 40	0.0008	1400	0.23	0.009
Polystyrene insulation	0.2	12	0.044	0.035
Separation membrane ρ, [kg/m²]	0.005	0.4	0.023	0.002
Floradrain FD 40	0	916	0.26	0.00002
Filter Sheet TG; ρ, [kg/m²]	0.0009	0.13	0.023	0.002
Substrate	0.3	1800	1.16	0.15

Table 5. Area of external enclosure structures of school building.

Type of enclosure structure	Parameter identifier	Square (m²)
External walls	A_W	7008.7
Windows	A_F	936.8
External doors	A_{ed}	50
Covering	A_c	8160.8
Coverings over passages	A_{c2}	444.7
Floors on the ground	A_{fl}	8120.9

$$\Delta Q_{h1}^y = Q_{h1}^y - Q_{h2}^y = 619495.6 - 603867.2 = 15628.4 \, (\text{kWh})/\text{year} \qquad (3)$$

$$\Delta Q_{h2}^y = Q_{h3}^y - Q_{h4}^y = 974003.8 - 893358.6 = 80645.2 \, (\text{kWh})/\text{year} \qquad (4)$$

As it can be seen from the data presented in Table 6, the smallest loss of heat through external enclosure structures of the buildings using the green roof technology. The biggest difference between the annual costs of heat energy is observed for buildings with a large ratio of the area of coverage to the area of the enclosing walls.

7 ECONOMIC COMPARISON OF HEATING COSTS

Compare economic costs of heating the building using standard roof structure and using the green roof structure for the two buildings. In accordance with the Order of the Committee on Tariffs of St. Petersburg from 27.11.2015 No 364-r the cost of 1 Gcal for St. Petersburg is 1295.18 rubles. Multiply this by the annual heat losses, calculated for one heating period (refer to: Table 6). The results are shown in Table 7.

As it can be seen from Table 7, energy-efficient green roof significantly reduce the economic cost of heating the building. It can be concluded that when a large ratio of the area of the covering to the area of the enclosure walls the annual heating costs of the building are less. On the basis of numerical modeling we can build a graph of behavior of the difference of the annual costs of heating the building to the two types of roof depending on the ratio of the area of covering to the area of enclosing walls.

In figure 4: ΔC = the difference of the annual costs of heating the building for the two types of roofing; r = the ratio of the area of covering to the area of enclosing walls

8 SUMMARY

The research presents the analysis of heat losses through the envelope of the building and the comparison of costs for heating of buildings. The study revealed the following results:

1. At a higher level of insulation of enclosing structures of the building heat losses due to heating during the heating season tend to decrease. Consequently, it is necessary to supply less heat to the radiators of the building and costs for heating may be much lower.

Table 6. Annual consumption of heat energy losses through the external enclosure structures.

External enclosure structures	Parameter identifier	Annual consumption of heat energy, Q_{hi}^y		
		(kWh)/year	MJ/year	Gcal/year
Residential building (typical roof)	Q_{h1}^y	619495.6	2230184.2	533.2
Residential building (green roof)	Q_{h2}^y	603867.2	2173921.9	519.3
School (typical roof)	Q_{h3}^y	974003.8	3506413.7	837.6
School (green roof)	Q_{h4}^y	893358.6	3216091.0	768.3

Table 7. Annual costs of thermal energy for heating of buildings.

External enclosure structures of	Annual consumption of heat energy	Costs of thermal energy for heating
	Gcal/year	Rub/year
Residential building (typical roof)	533.2	690590
Residential building (green roof)	519.3	672587
School (typical roof)	837.6	1084843
School (green roof)	768.3	995087

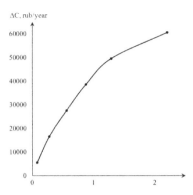

Figure 4. The economic efficiency of the building.

2. Heat losses of the building with a typical roof structure are bigger than heat losses of the building with green roof technologies.
3. Heat losses are less at a higher ratio of the area of the covering to the area of enclosure structures of the building (the school building).
4. Using green roof construction significantly reduces the economic costs of heating the building.

REFERENCES

Averyanova, O.V. 2011. Economic efficiency of energy saving solutions. *Magazine of Civil Engineering* 5: 53–59.
Beskorovaynaya A.V. & Semenova E.E. 2015. *The effectiveness of the exploited roofs. Scientific Herald of the Voronezh State University of Architecture and Civil Engineering. Construction and Architecture. High technology. Ecology* 1: 86–89.
Bilyalova, L.M., Tsymbal, A.A. & Kondratyeva, N.B. 2012. The use of green technologies in architecture. *Actual problems of architecture* 4: 6–11.
Coma, J., Pérez, G., Solé, C., Castell, A. & Cabeza, L.F. 2016. Thermal assessment of extensive green roofs as passive tool for energy savings in buildings. *Renewable Energy* 85: 1106–1115.
Costanzo, V., Evola, G. & Marletta, L. 2016. Energy savings in buildings or UHI mitigation? Comparison between green roofs and cool roofs. *Energy and Buildings* 114: 247–255.
Gorshkov, A.S. 2010. Energy efficiency in buildings. Questions of regulation and energy consumption reduction measures for buildings. *Magazine of Civil Engineering* 1: 9–13.
Hodo-Abalo, S., Banna, M. & Zeghmati B. 2012. Performance analysis of a planted roof as a passive cooling technique in hot-humid tropics. *Renewable Energy* 39(1): 140–148.
Karachaliou, P., Santamouris, M. & Pangalou, H. 2016. Experimental and numerical analysis of the energy performance of a large scale intensive green roof system installed on an office building in Athens. *Energy and Buildings* 114: 256–264.
Kondratyeva, N.B. & Bessmertnaya, E.A. 2011. The relationship of structural elements of buildings and their energy efficiency by the example of the device green roofs. *Actual problems of architecture* 3(1): 28–31.
Nemova, D.V. 2012. Ventilation systems in residential buildings as means of increase of power efficiency. *Construction of Unique Buildings and Structures* 3: 83–86.
Saadatian, O., Sopian, K., Salleh, E., Lim C.H., Saadatian, E., Toudeshki A. & Sulaiman M.Y. 2013. A review of energy aspects of green roofs. *Renewable and Sustainable Energy Reviews* 23: 155–168.
Shamarina. A.A. 2012. The roof as a means of urban design. *Scientific Herald of the Perm National Research Polytechnic University. Urban studies* 4(8): 47–59.
Spala, A., Bagiorgas, H.S., Assimakopoulos, M.N., Kalavrouziotis, J., Matthopoulos D. & Mihalakakou G. 2008. On the green roof system. Selection, state of the art and energy potential investigation of a system installed in an office building in Athens, Greece. *Renewable Energy* 33 (1): 173–177.
Vatin, N.I., Nemova D.V., Rymkevich, P.P. & Gorshkov, A.S. 2012. Influence of building envelope thermal protection on heat loss value in the building. *Magazine of Civil Engineering* 8: 4–14.
Vatin, N.I., Gorshkov, A.S. & Nemova D.V. 2013. Energy efficiency walling during overhaul. *Construction of Unique Buildings and Structures* 3 (8): 1–11.
Virk, G., Jansz A.A., Mavrogianni A., Mylona A., Stocker J. & Davies, M. 2015. Microclimatic effects of green and cool roofs in London and their impacts on energy use for a typical office building. *Energy and Buildings* 88: 214–228.

Advances and Trends in Engineering Sciences and Technologies II – Al Ali & Platko (Eds)
© 2017 Taylor & Francis Group, ISBN 978-1-138-03224-8

Thermal insulation systems (ETICS) – important layers of the system and testing the implementation of non-standard temperatures

M. Novotný, P. Liška, B. Nečasová & J. Šlanhof
Faculty of Civil Engineering, Brno University of Technology, Czech Republic

ABSTRACT: Composite system ETICS is composed of several different layers. The most important of them in terms of their performance and durability is the thermal insulation layer and the basic (reinforcing) layer. The first layer specifies the parameters in terms of heat transmission, and second in terms of overall life of the system. The article discusses the testing material base layer in response to different temperatures. This fact has a significant effect on shortening the life of the system. Shortly we will discuss the advantages and disadvantages of testing methods and their application to the materials.

1 LOW-ENERGY BUILDING CONSTRUCTION

The aim of current designers and architects is to construct mainly low-energy and "passive" houses. These efforts are most limited by the financial means of investors. Such constructions themselves—particularly passive ones—are indeed eco-friendly, needing a minimum of (or no) energy from external sources (coal, gas, electricity, etc.). However, in the majority of cases, private individuals are discouraged from building them by the large sums which need to be spent on the construction of such buildings. It should be said, though, that with current construction methods it is also possible to adapt existing structures to conform to these trends. This means reconstructing the building so that the resultant structure has parameters that fulfil the requirements and standards expected of low-energy or passive buildings. This is primarily achieved via the external thermal insulation of buildings, either via ventilated facades or contact-based thermal non-ventilated ETICS systems. The ETICS contact-based External Thermal Insulation Composite System is the main topic of this article. The quality of ETICS envelopes does not merely derive from their correct thermotechnical and structural design, as the technical aspect is also crucial, i.e. their correct installation. This is often influenced by the climatic conditions present when they are being fixed in place. The article will deal both with thermal insulation types and with parameters which can be tested with regard to base (formerly "reinforcement") layers.

2 THERMAL INSULATION LAYER—WHICH MATERIALS CAN BE USED?

What is the definition of thermal insulation? We can look for definitions in, e.g. the existing standards. The ČSN 732901 Standard—Execution of External Thermal Insulation Composite Systems (ETICS) using thermal insulation (ČSN 732901; 2005)—does not define thermal insulation as such, though a direct definition can be found in Standard 730540—Thermal protection of buildings—Part 1: Terminology (ČSN 730540—p. 1; 2005), where thermal insulation is defined in the following way: "Thermal insulation material—a material which significantly limits the propagation of heat and which has a characteristic thermal conductivity coefficient of max. 0.1 W/mK for the reference temperature and humidity conditions, and the given age." Generally speaking, it is a material which prevents the penetration or propagation of heat inside a structure. It is common that efforts are made to prevent heat from escaping towards exterior spaces, i.e. to prevent what are known as heat losses which increase the cost of heating the building. Use is made either of sandwich structures with insulation inserted within the composition, or by thermally insulating the envelope of the building—either contact-based or in a ventilated form. In this paper, we will mainly deal with the contact-based type of thermal insulation. The following picture shows the most common composition of such systems.

For our purposes, mainly layer 4 is of interest. The ČSN 732901 (ČSN 732901; 2005) standard defines the base layer in the following way: "Screed material for ETICS—a material specified for the system which along with reinforcement forms the base layer of ETICS. The nature of this base depends on the type of binder used: it can be either dispersed, where the prevailing binders are synthetic polymers dispersed in water, or it can be mineral, where the prevailing binder is cement or possibly something different." Our work and research is mainly devoted to cementitious materials, with various parameters being tested and examined. The most important of these for the lifespan of the system is the determination of the absorbability of the material, the tensile strength of the material (including when it is subjected to flexure or compression), impact resistance, adhesion to the base, and frost resistance. The essence of the tests will be detailed below.

As far as thermal insulation materials are concerned, they can be either natural or synthetic. The following table shows the most common current types of thermal insulation, including their thermal conductivity values. In the case of insulation it is of course true that the lower the conductivity value a material has, the better its insulating properties.

It is obvious that these are not the only materials available these days. Others include, e.g. insulation with straw bales, foam glass, sheep's wool, hemp, cellulose fibre or cork. Today, thermal insulation foam (PUR and PIR) is becoming more and more popular. Furthermore, we would also like to mention the most commonly used kinds of insulation for contact-based thermal insulation systems—mineral fibres, polystyrene and PUR/PIR foam:

A. Mineral, basalt and glass fibre insulation materials are basically very similar, differing only in the input raw material from which they are made. According to the input raw material and

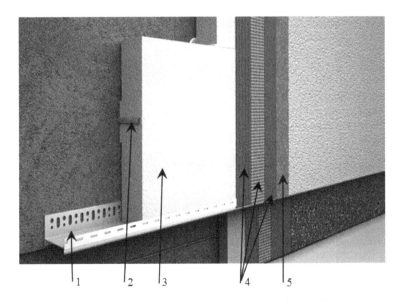

Figure 1. System ETICS—layers. The individual numbers show the layers of the system. The picture shows these layers: 1—Foundation rail; 2—Mechanical anchorage of the insulation; 3—Thermal insulation—polystyrene or other material; 4—Lower and upper part of base layer with embedded mesh; 5—Penetration layer with final plaster layer (optional with color paint). (resource: www pages, STOMIX Inc.).

Table 1. Thermal conductivity of most widely used thermal insulation materials—for ETICS systems.

Material	λ W/m² K
Mineral or glass wool	0.035–0.04
Pressed mineral fibres	0.06
Expanded polystyrene	0.036–0.045
Extruded polystyrene	0.028–0.032
Expanded grey polystyrene with added graphite	0.032
Polyurethane	0.024–0.028

certain differences in the production process, we can distinguish between mineral materials, i.e. stone or basalt wool, and glass wool. In the case of mineral insulation, basalt and silica are the most common raw materials (if basalt itself is used, the end product is basalt wool), with glass supplied only as a component. In the case of glass insulation, the main raw material is glass, which today is mainly recycled. The input raw material for the production process is the mined minerals basalt and silica, or recycled materials such as glass cullet. The main advantage of these materials is their high fire resistance. Another advantage and reason for the use of such insulation is its high water vapour permeability, i.e. the very high diffusion openness of the thermal insulation system. However, such insulation materials are unsuitable for places with higher humidity, where they quickly lose their thermotechnical properties.

There are newer recycled materials which can be used for insulation these days. Their basis as the dismantled facade insulation, for example, can be old cloth and other similar materials. These insulations can be used as a standard insulators as well as insulators for green roofs. My colleagues Martin Mohapl and Petr Selník are interested in this topic deeply. You can read more about this topic in the article "Technological analysis of the orientation of the pitched roof icelandic turf to the cardinal points and their effect on the stability of the vegetative layers" (Mohapl; Selník; Nečadová; 2014). Thermal insulation and modeling of their behavior in relation to other elements of the building envelope is discussed in the article "The influence of the position of the steel element to heat flow in the wall of the industrial freezer" (Mohapl; Vojkůvková; 2014).

B. Polystyrene was invented in Germany in 1949. Today it has many practical uses, of which one of the main ones is as thermal insulation for homes, roofs, facades and floors. It has a low volume weight—30 kg.m⁻³, and very good mechanical, thermotechnical and acoustic properties. Its uses are well known, so we will not analyze it in greater detail, only mentioning that new variants of polystyrene now exist which contain graphite, which has improved properties as far as thermal transmittance is concerned.

C. Insulation from thermal insulation foam (PUR or PIR)—this insulation is suitable for all types of residential, commercial or industrial buildings, including ecological and energy-saving structures. It can be installed all year round without any regard to climatic conditions. The foam consists of a two-component foam insulation material (PUR or PIR) which creates a completely compact, air-tight and non-absorbent thermal insulation layer. This type of insulation is used in, e.g. various types of sandwich structure in which the load-bearing structure is either a wooden or steel frame. Furthermore, brick houses, attics, ceilings, walls, floors and exterior spaces can all be insulated with thermal insulation foam. The material is non-absorbent, and thus is suitable for applications where humidity poses a threat. Particularly with regard to the newly-introduced PIR systems, several verification tests will need to be carried out. However, it is obvious that that it will be possible to replace, e.g. polystyrene in contact systems with boards made of PIR materials. Within the framework of my research, I am preparing a project which will investigate these issues. It will be focused on the resilience and adhesiveness of materials.

As we can see, very different materials are used to provide thermal insulation in the construction industry. However, the use of mineral insulation (wool or boards and polystyrene) still prevails for contact-based facade systems. It is good, though, that some of the other materials mentioned above are also starting to make headway—e.g. PUR sandwich board systems, and recently also PIR systems. In the case of these systems, development and testing are ongoing. One of the most important parameters is the adhesiveness of the insulation to the base, along with the parameters and execution of the base layer. I have also been investigating these issues both in my dissertation thesis and within the framework of research and development at my home institution. Test methods are described below.

3 PARAMETERS OF CEMENT MATERIALS—BASIC METHODS OF TESTING FOR ETICS BASE LAYER

Various tests can be used to determine the basic properties of a material for the execution of the base layer. As far as the evaluation of the quality of the base layer and its lifespan is concerned, the main tests which can be used are those described below. I have also used some of these tests to test materials which I analyzed in my dissertation thesis. The results of these tests can be found within that work. (Novotný; 2013/2014)

As the tested materials do not exhibit the same consistency after their preparation according to their manufacturers' instructions, it was necessary to start by defining a "normal density mash". The materials prepared in this way then have the same properties as far as fluidity and mixture consistency are concerned. The test is standardized and its complete methodology is detailed in the ČSN EN 196–3 Standard—Methods of testing of cement—Part 3: Determination of hardening period and volume stability (ČSN EN 196–3; 2009). The test involves the measurement of the resistance of the ready mixture against penetration by a standardized penetration roller. The penetration value is measured; when the measured value is 6 ± 2 mm from the bottom of a sample of 40 mm in height, the mash has a "normal density". The materials prepared in this way were then subjected to the tests outlined below.

3.1 *Tear-off test*

This test is performed on a joint layer featuring insulation material. With tests executed in such a way it cannot be guaranteed in advance that tearing-off will occur within the screed material. A more common result (which is correct in accordance with the standard) is that tearing-off occurs in the insulation material. This test was executed as part of the author's dissertation regarding the application of the material on an insulator (Novotný; 2013/2014). It was discovered in this way that in the majority of cases, the exact adhesiveness value of a given material cannot be determined due to the fact that the damage occurred within the insulator. I am therefore planning another series of tests where the material will be applied to a firm base (e.g. concrete) in such a way that it is possible to evaluate the influence of low temperatures on the force needed to tear off the screed layer from the base, and to measure this force exactly.

Disadvantages of the method:

1. the preparation of samples is more complex than with other methods, and is more demanding as far as material and time are concerned,
2. it is necessary to apply samples not only to the tested insulator but also to a firm base in order to achieve the exact measurement and comparison of tear-off values. Both base variants have to be non-absorbent so that frost resistance tests can also be performed to assess the possible effect of low temperatures on adhesiveness.

Advantages of the method:

1. the tear-off method is standardized for building envelopes as well as for other materials, i.e. the exact procedure is given and standardized instruments can be used,
2. the method is commonly used in practice for the determination of the parameters of bases and screeds, and thus is generally well-known.

3.2 Tensile test

For this test, the samples were prepared in a way which is closely linked to the execution of the layer when performing real-world construction work, i.e. with a 4–6 mm thickness with inserted mesh in the form of a strip with a 600 × 100 mm pattern. The test consists in grasping the sample in a tensile load press, subjecting it to tensile loading and simultaneously monitoring the development of cracks and the load value at 3% extension and then during tearing. This method is described in the regulations of the Guild for Building Insulation but is not supported by any official document at present. Using this method, the influence of low temperatures during the realization process can be examined—i.e. the influence on various properties of the base layer executed at various temperatures. Due to the breakdown of equipment in the laboratory which is available to me, I have not yet carried out such tests on the materials which I am examining, though their realization is expected in the near future.

Disadvantages of the method:

1. sample preparation is highly demanding as far as both material and time are concerned,
2. sample handling is more difficult, and laying conditions are more demanding due to the thin layers,
3. it is impossible to test the effect of freezing cycles on the adhesiveness of samples to the base.

Advantages of this method:

1. the sample corresponds to the highest possible degree to the execution of such a layer during real construction work—thickness 4–6 mm, planar dimensions prevail, inserted mesh, the execution of the layer as well as the position of the mesh in the layer,
2. the influence of low temperatures on the elasticity modulus can be observed; it is demonstrated via the higher occurrence and growth of cracks during tensile loading of the sample,
3. it is possible to observe the growth of cracks, and how they increase in size in comparison with the growth in the loading force, with the aid of a video extensometer (or visually with the aid of templates and a tensometer).

3.3 Compressive and flexural tensile strength of material

This test was carried out on samples of material which I examined—40 × 40 × 160 mm beams prepared according to the ČSN EN 1015–11 Standard (ČSN EN 1015-11; 2000; Schmid; BI02—M02). These trusses were subjected to destructive flexural tensile stress testing according to the method defined in the given standard. During the test, both the force affecting the sample during fracture and the development of cracks were monitored. The values obtained from the test can again be found in the dissertation thesis (Novotný; 2013/2014). A pressure test was subsequently carried out on one of the two halves of the beams remaining after the destruction of the samples according to the ČSN EN 196-1 (ČSN EN 196-1; 2005) and ČSN EN 1015-11 Standards (ČSN EN 1015-11; 2000). The pieces were formed into 40 × 40 × 40 mm cubes for these tests. The obtained data from the tests allowed flexural tensile and compressive strengths to be determined with the help of a computational formula given by the standards. The advantages or disadvantages of these tests cannot be commented on, as these are commonly-used test methods for the given type of materials.

3.4 Frost resistance of adhesive and screed materials

The second halves of the beams left over from the flexural tensile test were used for this test. The samples for these tests were also shaped into 40 × 40 × 40 mm blocks. The test was mainly carried out according to the ČSN EN 12091 (ČSN EN 12091; 1998). Standard but was adapted to bring it closer to industry practice. The freezing cycles were monitored and the individual samples were weighed and checked during testing. The main monitored parameters were the weight of the sample after a given number of cycles and the loss of material due to the strain imposed upon it by temperature cycles. The sample was also inspected for possible visible damage caused by the deterioration of pieces of material which then fell off. The data concerning material losses during the execution of the test can also be found in the thesis mentioned above (Novotný; 2013/2014).

In the case of this method, the way the samples are produced can be taken as the main advantage. The preparation of the samples from the beams used in previous testing is fast and efficient and removes the duty to produce a whole set of samples right from the beginning, where there is a risk of other entry conditions.

However, the test has also its disadvantages:

1. the size and shape of the samples, which do not correspond exactly to real life applications of the material in structures—40 × 40 × 40 mm vs. a layer of 4–6 mm,
2. the material was not applied to an insulation base layer—the influence of cycles on adhesiveness thus cannot be checked—this should be specified with the aid of other measurements.

4 SUMMARY

It is a fact that within the framework of the development and testing of materials it was found that the methods used have their advantages and disadvantages. In retrospect it can be said that some of the tests could have been performed in a different way so that the results are more conclusive. In the case of the tests whose results were distorted the most by the manner of execution, i.e. for example freezing (a greater number of cycles are needed), efforts are being made to carry out these tests again and verify the results. We know for the future that the foundations of the studied research and its methods—see the tests mentioned—have been laid correctly, and that a significant proportion of the results mentioned in the dissertation thesis are useable in the further development of the test methodology. However, it is necessary to adapt some methods and carry out comparative measurements. Efforts are also being made to carry out all of the previously described tests both on previously tested materials and insulators and also (and mainly) on new insulators—for example, the aforementioned PIR foam is reaching the stage when it will be possible to also use it in contact-based facade systems. It will thus be important to check and test the functionality of the system. It will also perhaps be necessary to test other application options, for example mechanical anchorage, and to determine the technical installation procedure with regard to climatic conditions.

This article is only a tentative outline of a solution to a problem which is constantly being dealt with in practice. Its solution is important, both with regard to climatic conditions and the related situation in the building industry (which is under pressure to extend the construction season), and also in connection with the existence of newly accessible materials and substances. The test methods used are standardized, but as is mentioned in the text, their modification also needs to be considered with regard to new materials, which can have slightly different properties. It is therefore necessary to carry out research and development constantly, not only on a material basis but also with regard to the conditions under which the execution and realization of a given system takes place.

REFERENCES

1998. ČSN EN 12091 Tepelně-izolační výrobky pro použití ve stavebnictví—Stanovení odolnosti při střídavém zmrazování a rozmrazování (in Czech). Prague. Czech Standardization Institute.
2000. ČSN EN 1015-11 Zkušební metody malt pro zdivo—Část 11: Stanovení pevnosti zatvrdlých malt v tahu za ohybu a v tlaku (in Czech). Prague. Czech Standardization Institute.
2005. ČSN 730540 Tepelná ochrana budov—Část 1: Terminologie (in Czech). Prague. Czech Standardization Institute.
2005. ČSN 732901 Provádění vnějších tepelně izolačních kompozitních systémů (ETICS)—(in Czech). Prague. Czech Standardization Institute.

2005. ČSN EN 196-1 Metody zkoušení cementu—Část 1: Stanovení pevnosti (in Czech). Prague. Czech Standardization Institute.

2009. ČSN EN 12808-5 Lepidla a spárovací malty pro keramické obkladové prvky—Část 5 Stanovení nasákavosti. Prague. Czech Standardization Institute.

2009. ČSN EN 196-3 + A1 Metody zkoušení cementu—část 3: Stanovení dob tuhnutí a objemové stálosti. (in Czech). Prague. Czech Office for Standards, Metrology and Testing.

2016. Picture from www pages of *"Stomix- expert na zateplení"*. STOMIX company web. www address — *www.stomix.cz/cz/sortiment/zateplovaci-systemy/stx alfa/stx alfa.html*

Mohapl, M.; Selník, P.; Nečadová, K. 2014. Technological analysis of the orientation of the pitched icelandic turf roof to the cardinal points and their effect on the stability of the vegetative layers. In *Advanced Materials Research, Volume 1041*. Pages 19–22. Trans Tech Publications, Switzerland.

Mohapl, M.; Vojkůvková, P. 2014. The influence of the position of the steel element to heat flow in the wall of the industrial freezer. In *Advanced Materials Research, Volume 941–944*. Pages 2474–2477. Trans Tech Publications, Switzerland.

Novotný, M. 2013/2014. Vliv venkovní teploty na vlastnosti výztužné vrstvy při realizaci ETICS. Brno. Brno University of Technology, Faculty of Civil Engineering, Institute of Technology, mechanization and construction management.

Schmid, P. Year of publication not stated. Zkušebnictví a technologie—Modul BI02–M02—Stavební zkušebnictví (in Czech). Brno. Faculty of Civil Engineering, Brno University of Technology.

Advances and Trends in Engineering Sciences and Technologies II – Al Ali & Platko (Eds)
© 2017 Taylor & Francis Group, ISBN 978-1-138-03224-8

The influence of form of public space around residential complexes on quality of living

D. Orsáková & A. Bílková
VŠB-TUO, Ostrava, Czech Republic

ABSTRACT: Public space is an integral part of urban areas, where it's quality affects their success. The aim of this paper is to define main parameters of quality of public space around residential complexes such as courtyards, with subsequent monitoring of their effects on quality of living and comfort of their users. The output from the paper is to achieve a comprehensive view on the issue and define key parameters which conditionally ensure functionality and usability of public spaces and pointing out the deficiencies identified based on conducted case study.

1 INTRODUCTION

Efforts to raise the standard of living for the greatest possible number of inhabitants after World War II meant for Europe a mass construction of housing estates.

Currently, one third of the population of the Czech Republic live in housing estates, which indicates that the Czech population have become used to this type of housing. However, for example, compared with the period of Czechoslovakia, when life in a housing estate was perceived as a lifestyle of a socialist man, today, especially the more affluent part of the population, replace it with living in detached or terraced houses in the suburbs.

For the competitiveness of this type of housing, the crucial part is played by the quality standard, which is determined not only by the housing unit itself, but also by the surrounding environment in the form of public, half-public, or private space.

This contribution is focused on the condition and quality of the courtyard area.

2 BASIC INDICATORS OF THE QUALITY

Courtyards represent a specific part of the public space, which is an integral part of the living environment. They can be categorized as community spaces, characterized by stronger relations between users.

The main parameters affecting not only the actual quality associated with their use, but also the quality of living, include the following.

2.1 *Security*

One of the basic psychological human needs is the sense of security that gives a person a feeling of security and protection against any accident or crime as described by Gehl (2010).

The main sources of danger for these areas are represented by transport and criminality.

Compared to other types of public spaces, courtyards have a great advantage, since in most cases they are not directly threatened by the dangers related to traffic. The speed on roads passing through these areas is usually limited to 30 km/h, and these roads are used primarily for approaching parking spaces. Nevertheless, it is appropriate to equip roads in this area with deceleration elements—speed bumps or chicanes.

The most common types of criminal activity characteristic of courtyards include vandalism, noise disturbance, drug abuse and attacks with the aim of robbery or rape.

Security can be increased, for example, by providing patrol services and improving maintenance. Furthermore, by ensuring sufficient intensity of lighting, new interior, improving the details (maintenance of trees, street furniture) or total new space solutions (linked walkways, location and schedule of activities) as described by ČSN P CEN/TR (2009).

An important role in the prevention of crime and increasing the safety of these spaces is played by natural surveillance, when the behaviour of individuals is watched by many eyes looking through the windows of individual flats. This is also necessary to ensure by adequate lighting after dark. When designing the layout of these spaces, it is necessary to avoid the pitfalls in the form of walkways lined with mature greenery and dense bushes that may represent a potential hiding place for criminals as described by Towers (2005).

2.2 *Accessibility and passableness*

The notion of public accessibility means accessibility for all, without restriction, regardless of the ownership form and a limited regime of use as described by Melková (2014).

It is, therefore, directly linked with use by the elderly, parents accompanying a child in a stroller and persons with reduced mobility and orientation. For persons with reduced mobility, the important attributes for the possibility of use include, for example, walkway gradients, their condition and width and surface finish. With regard to use by persons with impaired orientation (the partially sighted, blind), it is necessary to create a clear orientation trail through natural and artificial guide lines as described by Orsáková and all (2016).

2.3 *Functionality*

The functionality of courtyard public spaces indicates the number of users that perform different kinds of activities in them, whether of motion or residential nature. What is important is their mutual interconnection, so that they complement and do not restrict each other. An important prerequisite for ensuring their durability and sustainability is multi-functionality. The offer of individual activities should adapt both to the phases of individual days and seasons. Furthermore, it should ensure the satisfaction of all ages and social groups of inhabitants as described by Melková (2014).

Activities of a residential nature include activities such as sitting, standing around and lounging. For these activities, it is necessary to identify appropriate areas based on the possibility of interesting views, reasonable viewing distance, adequate lighting, reduced noise intensity and choice of sunlit and shaded areas.

Various forms of play elements for children, or elements of exercise for adults and seniors inviting activity are intended for motion activities. It is important not to forget the preservation of open grassed areas for optional activities.

Courtyard spaces are characterized by a higher degree of social contact prerequisites of social and societal functions. Social interactions arise there, which can contribute to the actual form of these areas and affect their quality, for example by planting small green vegetation or complementing street furniture at individual entrances. Great interest today is about creating "front gardens" either for plant-growing or purely aesthetic intent.

Less obvious, but equally important are the structural and symbolic functions completing the atmosphere of place and allowing the identification of the user with the place.

An important prerequisite for this type of space is the fact that it is not necessary that just one courtyard offers a large variety of activities. Since individual blocks are often connected to each other, it is desirable to disperse offers of various kinds of activities.

The original functional use of courtyards is being largely replaced with parking areas in connection with the increasing degree of motorization.

2.4 *Pleasure and comfort*

To achieve the proper enjoyment of the use of these facilities, the human scale and attention to detail is crucial. The quality of materials and street furniture is also important. The space as such should offer interesting experiences and sensations, all in architectural unity.

It is essential to use aspects of climate—light/shade, heat/cold, shelter and sensory perception—pleasant views, the rustle of trees, gurgling water and the like

2.5 *Sustainability*

An important prerequisite for sustainability is that this kind of space can be successful, if used. Using high-quality and durable materials with intensive maintenance leads to the sustainability of these areas, as well as their variability and timelessness.

An integral part of the economic point of view that not only includes the cost of realization of adjustments to these spaces, but also ensures their maintenance and management of the required quality.

2.6 *Hygiene*

Environmental hygiene is strictly connected not only with protection against undesirable sensory experience, such as glare, noise, dust, pollution and odour, but also against protection against certain forms of infection.

For courtyards, an important role in respect of environmental hygiene is played by green vegetation. It contributes to the formation of a microclimate by reducing dust and increasing humidity. It is also important in terms of shading (it absorbs 60–80% of solar radiation) and damping noise. Finally, it serves to mitigate temperature fluctuations.

When planting trees, it is necessary to remember the allergens produced, for example, by birch, hazel and alder, and to try to replace them with other, more appropriate species.

Absorption of sound, which is considered one of the main stress factors, for example, affects the breakdown of facades. However, for reducing the intensity of sound, it is crucial to limit its main source—car traffic.

The originator of odour in the immediate vicinity of courtyards is represented by containers for municipal waste. They should be placed away from residential areas and in shaded spaces, which partially prevents the degradation rate in the summer months.

A big problem for courtyards is the contamination of lawns by dog excrement. Polluters are often users themselves due to the lack of rubbish bins or their poor placement.

3 CASE STUDY OF A SELECTED COURTYARD IN PORUBA HOUSING ESTATE

3.1 *History of the Poruba housing estates*

The history of the Poruba housing estate dates back to the early 1950s, when the first plans to build a giant urban complex were prepared with the idea of creating the necessary building capacity for 150,000 inhabitants. Sorel—or architecture of socialist realism shaped both individual blocks of flats in the form of decorative elements (graffiti, reliefs, etc.) and the whole urbanism.

Symmetrical units were created formed by grouping several-floor buildings—blocks that are interconnected by roads. Within the individual blocks, there are recreational and relaxation areas, parks and courtyards. In 2003, the Poruba urban complex in the Sorel style was declared a conservation area.

3.2 *Description of the studied area*

The studied area is located inside a block of buildings between the streets Španielova, Čkalovova and Dětská near Hlavní třída in the city district of Ostrava Poruba. Part of the area of the courtyard is reserved for the kindergarten garden, see Figure 1. The remaining portion is used primarily by the residents of the nearby blocks of flats.

3.3 *Evaluation of the area*

3.3.1 *Security*

Safety related to traffic is very good. Individual roads lead around the perimeter of the blocks rather than through the courtyard. Moreover, they are designed as one-way roads with a speed limit, which leads to increased safety of access to the space.

The feeling of safety in the courtyard itself is enhanced by visual check from the windows of the neighbouring flats and elimination of dense shrubbery along the walkways and entrances to the individual blocks of flats.

Night lighting is sufficient. It runs along a central road leading through the area, to which access roads to the individual entrances are connected.

3.3.2 *Accessibility and passableness*

Accessibility of the area in terms of wheelchair access is quite good thanks to the sufficient width of the footpath running through the central portion; it has an increased curb serving as a natural guide line for people with impaired orientation.

This communication also ensures passableness of the area, which is easy to navigate (entry/exit). Access roads to the individual entrances, however, are inconvenient, especially due to their width and condition. The major problem is the lack of paved areas providing access to various game elements and street furniture, thus limiting the possibility of their use, see Figure 2.

3.3.3 *Functionality*

The solution of the particular space offers sufficient facilities for physical activities, especially for children of preschool age due to several playing elements (playground and sandpit). Part of the courtyard is a kindergarten garden, but that is not freely accessible. It was worth considering making it accessible at least within a limited timeframe, thereby expanding the possibilities of activities for children see Figure 3.

An extensive grassy area with a relatively flat terrain also allows various activities for youth and adults who use it mostly for football.

For passive rest, there are benches near the playground (the busy zone), and one in the quiet zone in the peripheral area.

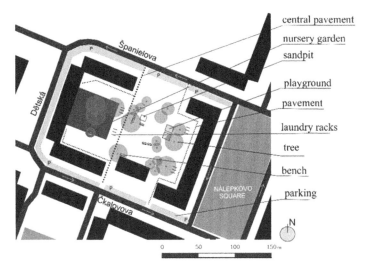

Figure 1. The studied area.

Figure 2.   Absence of footpaths.

Figure 3.   Game elements for children.

Courtyards usually include elements associated with household equipment, such as carpet beating frames or laundry racks. It is also the case here, where they occupy a considerable area, but are never used. Their removal would allow a free space for another activity, such as a workout playground for youngsters.

3.3.4   *Pleasure and comfort*

The pleasant feeling in this space is created by grown and maintained green areas, which at the same time provide plenty of shady places. The surrounding buildings serve as a barrier against the spread of noise, which greatly affects the overall impression.

The furniture used for seating is strategically placed in shady and sunny places both in the quiet zone, where it is possible to observe what is happening around, and in the busier zone with playing elements for children, see Figure 3.

The use of mostly wooden materials creates an attractive natural character and also provides sufficient comfort in physical contact.

The overall impression is spoiled by plastic containers for waste, unused laundry racks and graffiti on the facade of one of the adjacent buildings.

3.3.5 *Sustainability*

The big problem here in terms of hygiene is the amount of dog excrement. A ban on walking dogs in the area, or the delimitation of areas to this purpose could solve this problem.

I evaluate very positively the coverage of children's sandpit with tarpaulin preventing the access of animals and thereby minimizing the risk of infection from their droppings.

As already mentioned in section 3.3.4, there are large amounts of mature vegetation providing ample shade and absorbing dust in the particular area.

The stench of waste containers does not reach this area, because the containers are located outside the courtyard. However, an uncomfortable feeling may be caused by the smell from the kindergarten kitchen due to the orientation of the ventilation vents towards the courtyard.

4 CONCLUSION

This article is based on partial results of the research project Requirements for public space of residential complexes and their impact on the quality of housing, which is focused on identifying and monitoring sub-parameters or indicators that determine the quality of public spaces, especially in the residential environment, followed by determining the extent of their influence on the very quality of living.

It is very important to realize that generally a large number of requirements and the associated quality parameters relate to public spaces. In this paper, various indicators of the quality of public space within housing estates—courtyards are first defined, which are then monitored and evaluated in a case study based on the method of systematic observation and mapping user behaviour.

The weaknesses found are described in detail in other sections of the text of this paper.

Based on the results of the case study of one of the courtyards of the Poruba housing estate, we conclude that it is a highly humane space, its quality is very good, as evidenced, among other things, by the large number of users. This space provides some comfort for the residents of the nearby blocks of flats and improves the potential and value of the site.

ACKNOWLEDGEMENT

The work was supported by the Student Grant Competition VŠB-TUO. Project registration number is SP2016/104.

REFERENCES

ČSN P CEN/TR 14383-2. 2009. Prevention of crime-Urban planning and building design—Part 2: Urban planning. Prague: Czech Office for Standards.

Gehl, J. 2010. Cities for people. Washington: Island.

Melková, Pavla (ed.). 2014. Strategy of development of the public areas of the capital city Prague/proposal, Prague: IPR/SDM/KVP. ISBN 9788087931134.

Orsáková, Diana, Alžběta Bílková, Bohuslav Niemiec a Renata Zdařilová. 2016. Qualitative parameters of a public place and their analysis using a case study of the complex of the former bituminous coal mine of Karolina in Ostrava. Advances in Civil, Architectural, Structural and Constructional Engineering—Kim, Jung & Seo (Eds). 69–72. ISSN 978-1-138-02849-4.

Towers, Graham. 2005. An introduction to urban housing design: at home in the city. Oxford: Elsevier/ Architectural Press. ISBN 0-7506-5902-5.

Advances and Trends in Engineering Sciences and Technologies II – Al Ali & Platko (Eds)
© 2017 Taylor & Francis Group, ISBN 978-1-138-03224-8

Cultural influences conditioning the development of funeral architecture, its current status and its future direction

K.F. Palánová & O. Juračka
VŠB-TUO, Ostrava, Czech Republic

ABSTRACT: Burials of the dead are the necessity of each society through their development. Culture aspects changed the way of burials and dictate the appearance and disappearance of each typological type of funeral architecture. The development of these types or transformations of current realizations of unused types, their current use and future routing need to be specified, clarified and compared with experiences in other European states. This article focuses on a new building type of the 20th Century—crematorium. The development of these buildings was not continuous in all European countries, and it did not take place in the same time period; the most recent development has affected the current situation most. Through comparing, it is possible to find missing ties that have influenced the future of building types of this specific architecture. In the next phase, it is necessary to focus on the regional differences, approaches and facts, as well as to trace the consequences of the ongoing secularization of society

1 INTRODUCTION

As we have written in previous articles, with reference to the work of Markéta Svobodová (Svobodová, 2013), it is possible to divide crematoriums into three epochs of development. The first epoch characterizes crematoriums realized during the interwar period. It was the time of the First republic, the time full of enthusiasm for building a new democratic state, the time of new hope and, finally, its own laws. Mostly old ones from the days of the Austro Hungarian Empire were used, because of instantaneous national division. However, in case the former empire was not reluctant to consider, new laws were adopted. An example of this is the Law Lex Kvapil with two paragraphs, which allowed cremation. The second important thing for development of crematoriums is the process of secularization, which was started as a result of events of the First World War. "It (War) means a fundamental conceptual background—and not just because of the break-up of old Austria. It meant primarily a departure from the old world symbolized by the Christian religion. This departure had two results: People stopped believing in God…people stopped believing in Man…as an image of God, the Man who is a pinnacle of creation or autotelic guarantor of values." (Klaus, 2014)

The second period of the crematorium development can be seen after the Second World War, more after the Second Vatican Council, it means from the 60's of the 20th Century till 1989. The third period progressed during the 90's of the 20th Century. The fourth period in the Czech Republic so far has not occurred. It can be characterized by exhaustion, saturation, keeping of existing buildings, transformation of interwar realization into contemporary needs and conditions and maybe also waiting for the new future of this young building type.

2 VIEWPOINTS AND CULTURAL ASPECTS INFLUENCING DESIGN, FORM AND SYMBOLISM OF THE CREMATORIUMS

Newly-minded society and also the new laws were the signal for the launching of the prepared German crematorium in Liberec in 1918. From that time 14 crematoriums were realized up to 1937. They are mostly inspired by the concept of the Christian church (the ceremonial part—auditorium).

The new building type is looking for its own symbolism which was through centuries anchored in Christian churches and is transposed into the auditorium with slight changes. Instead of a refectory there is a catafalque, which is separated from the mourners by an Arc de Triumph with a slight rise. Ambon is used for the speaker, who comes from the room replacing the original sacristy. The choir compresses the entrance area and premises the explosion of the space of the main funeral hall. Naturally, the building is augmented with the technological part of the crematorium located under or behind the ceremony hall, whose size corresponds to a common church. The typical example is the Pardubice Crematorium designed by Pavel Janák in 1923.

Further development has brought the departure from the composition of a separate presbytery which is still slightly elevated. Axial symmetry with the gradations of catafalque space is still strongly respected. The catafalque becomes a part of the assembly, but it is slightly elevated. The auditorium is still quite spacious. As a result of the fast changes in styles we can get from the rondo-cubism of Pavel Janák, through individualistic modernity, to functionalism.

In several steps crematoriums in all regions of Czech Republic were realized, they are evenly spread, despite the low number of cremations in regions with higher representation of believers. The religiosity of regions has an important impact on location of crematoriums and also their number, even though they were realized in all regions. In the regions with higher percentage of believers, the development of crematoriums was delayed until the secularization of the society comes to these regions.

Dramatic events of the 20th Century, whose consequences are still visible today, had an impact on the population structure. There was 30% of German population in Czechoslovakia, which was displaced from the part of the country called Sudeten in 1945. New settlers in this part of the republic, due to a negative support of religion by communistic regime after 1948, did not have enough time to introduce new traditions. According to the statistics from 2001, regions in this part of the republic have the lowest proportion of the believer population. (Kotrlý, 2008) These are Ústí, Liberec and Karlovy Vary Regions. The highest religiosity is in Moravia and Silesia (Moravian-Silesian Region, South-Moravian Region, a part of Vysočina and Zlín Region), there are only 6 crematoriums in Morava of total 27 cremateries in the

Figure 1.   Even spread of crematoriums in the Czech Republic.

Czech Republic and there is only one in Silesia (Ostrava—Slezská Ostrava), although 30% of population in the republic live in Moravia and 10% in Silesia.

However, it is possible to see the spreading of crematoriums, it means in each region is at least one crematorium, in parts with higher density of population, e.g. Prague or Central Bohemian Region (these are also parts with lower number of believer population), there are more crematoriums.

Even the spreading of crematoriums is not usual in each country in Europe. An example can be the situation in Norway, where there are more cremations in towns and cities (37% of all cases), although in rural regions there are 100% of burials into the grave. Compared with bigger towns, for example with Oslo, there are 71% cremations (the year 2012). The reason is bad availability of crematoriums, higher costs, enough place for burial into the grave and traditional thinking about burial practices, despite the fact that the church never interferes to the choice of burial, and cremation is completely legitimate form of disposal over the last century. Where cremation is available, it is a well-established practice (Hadders, 2013).

Another example can then be Romania, where there is only one functional crematorium for 20 million of inhabitants. The first crematorium was built in 1928 and served as the only one until the realization of the second crematorium in 1994. The first one ceased its service in 2002. Currently, there is the crisis of urban space in Romanian towns from the perspective of enough free places for burials into the graves. Therefore there should be discussion about the possibility to use cremation as a comparable way of the burial. Opinion about cremation is influenced by a strong impact of the Romanian Orthodox Church and in general the lower degree of secularization in Romania. Despite the fact that Romania crematorium was the first from all Balkan countries, the reaction of the church, which issued the prohibition of Christian funerals for cremated dead (it still applies), predestined the stagnation of this new building type and cremation. This trend was not changed during communistic epoch, even though invasive supply of atheism after the year 1948. Presumably thanks to the silent acceptance of the Romanian Orthodox Church. (Rotar, 2015), favourable time for cremation arose in 1980, there was a historical peak in the number of cremations (1,000), nowadays, 850 cremations are carried out per year (compared with the Czech Republic with 84,388 cremations per year with the population of 10 million, it means 80% cremations of all cases in general).

More crematoriums were built in the neighbouring countries, where, in contrast to Romanian regression, three crematoriums were built in Hungary (1951 in Debrecen, 1968 in Budapest, 1985 in Szeged) and the same number in the territory of the former Yugoslavia (1964 Belgrade, 1978 Ljubljana, 1985 Zagreb), where the Catholic church cancelled the prohibition of the cremation. A different situation is in Serbia, where the Orthodox Church has a dominant position. In Bulgaria, there is just one crematorium for 7.3 million inhabitants (2011). (Rotar, 2015; International Cremation Statistics 2014).

In Czechoslovakia, until 1937, the first crematoriums were built in the territory of Bohemia, where secularization of the society is more widespread. Later and to a smaller extent, crematoriums also were built in the territory of Moravia. In Slovakia, and in the period between the wars, crematoriums were not realized at all.

The first crematorium in Czechoslovakia was built in the town of Liberec (1918), in 1921 the crematorium in the original funeral parlour in Olšany Cemetery in Prague was built (it was later cancelled and a new one was built in a different locality), then rondo-cubistic building of crematorium in Pardubice (1923), in Most (1924, later cancelled), Nymburk (1924), original Ostrava crematorium, the only one in the cubistic style (1925, later demolished), in České Budějovice (1925), Plzeň (1926), Brno (1930), Olomouc (1932), Karlovy Vary (1933), in Ústí nad Labem (1934, later cancelled) and finally in 1937 in Semily.

3 THE CHANGE OF THE COMPOSITION IN POST-WAR REALIZATIONS

Physical and later also symbolic approximation of the live and the dead gradually occurs; it includes the removal of the barrier as the Arc of Triumph, the reduction of the elevation until their disappearance in the post-war realizations. The catafalque is located off-axis; also the aisle is not in the middle of the benches anymore. When both worlds (the living and the dead) are getting closer, the ceremoniousness is reduced at the same time: the coffin is the part of the gathering like in a small village church, but the design of the interior does not force the visitors to silence, contemplation, a feeling of exceptionality, to an experience to be remembered for the whole live, vice versa it forces to rush, to deal with the matter quickly and to oblivion.

The view of mourners, previously strictly directed to catafalque with the coffin, is—gradually—directed to the surrounding forest cemetery or to views of the landscape. For example, it is the case of the crematorium in Zlín and Bratislava, in comparison to later realization in Netherlands (in Zoetermeer). The cases demonstrate that views of nature play an important role in crematorium design, and are thought to create an uplifting feeling. This seems based on the notion that the physical environment can influence human emotions and create feeling so well-being. This idea is embraced by designers of health institutions such as hospitals and hospices. (Klaassens and Groote, 2014).

This way of the development can be traced back to post-war realizations, which, due to political situation and due to the people´s thinking, started to be realized in 1954, when a crematorium was built in Prague-Motol. This sophisticated and relatively small building in the middle of a forest cemetery is innovative and important in the development of crematoriums in Czechoslovakia due to its size, as well as location in the landscape and also because of simplification of whole disposition and organization of the process, permeation of the technological part and the ceremonial part. The next realization was finished in 1969 and it was a new crematorium in Ostrava. It initiated the boom of the post-war crematoriums and other 12 buildings were built by 1989.

The expansion of this building type, caused by the demand for cremation, resulted, besides other things, from the Second Vatican Council (1962–65), which accepted the cremation as possible, but not recommended. This aspect probably motivated the construction of crematoriums in Slovakia, where a strong influence of the Catholic Church prevailed despite the post-war totalitarian regime. The first one was built in Bratislava in 1968–69 and it was designed by architect Ferdinand Milučký, another one as late as in the 1980s in Bánská Bystrica (1980) and in Košice (1982).

In the Czech Republic, the next development was as follows: the crematorium in Česká Třebová was built in 1970, the new crematorium in Most in 1974, and in the same year in Blatná, 1976 in Jihlava, the crematorium in the existing forest cemetery in Zlín was built in 1978, in the same year in Šumperk, in 1981 in Jindřichův Hradec, in 1982 in Tábor, in 1985 in Klatovy and Mělník and the new crematorium was built in Ústí nad Labem in 1986. (Kotrlý, 2008)

After the Velvet revolution in 1989 and after the division of Czechoslovakia, the situation developed in different ways in each country. In the Czech Republic the needs for the crematoriums have gradually been fulfilled. In the 1990s, only 4 buildings of this type were realized (Kladno 1992, Jaroměř 1994, Hrušovany u Chomutova 1995 and Hustopeče 1997), no more crematoriums have been built since 1997. In contrast with Slovakia, where the time for this building type came at the beginning of 21st century and crematoriums were built in Nové Zámky (1993), Nitra (2003) and Žilina (2003). Altogether, 31 crematories were built in the Czech Republic, 4 of them were cancelled during the 20th century. 27 of them are functioning and 6 of them are privately owned. In Slovakia, there are 6 crematoriums and 3 of them are privately owned.

4 CURRENT GROPING

During the last century, predominantly design and symbolic function of the crematorium developed; as for the technology, the space more or less conformed to its function and needs. Gradually, both aspects (ceremonial and technical) have been simplified, reduced and they have been freed from religious influence, even though it is still evident that there is a search for symbolism, as it was abandoned in the diversion from faith because of the secularization of the society. The First World War was a strong impetus for this. The war of unprecedented dimension and also duration, the war, which affected each person from the countries which were involved in it. This war changed the society and its consequences were evident in the society during the whole 20th century and they are still present today.

The last realization of a crematorium in Czech Republic was in 1997 (until this time, the building, its continuity to inner dispositions, but also the design had been gradually simplified, … up to the disappearance of the ceremoniousness), consequently we had no space for the fight with secularization, for thinking about new contemporary mourners´ needs, which could be verified, for example, by the realizations in the Netherlands after 2000, marked as post-modern phase(one of the four development steps, similar as in the Czech Republic, where, however, the last step is missing). The crematoriums in the Netherlands after 2000 have reflected the changes of the society, their new needs. Architect Zeenstra (the crematorium in Haarlem, realized in 2002) tried to cancel the rating system, commonly used for visitors until then. This system was never used in our republic. It means an entrance into the object from one side, going through the auditorium during the ceremony into the condolence room and exit in different location so that visitors to the different ceremonies do not meet. This system, according to Klasens and Groote, is like a production line in the factory, and it lacks ceremoniousness and also the respect and there is not enough time for reconciliation. In the new buildings of crematorium, trying to innovate and exclude mere functionality, a new approach is obvious. For example, in the crematorium in Leusden (2003), the architect Arnold Sikkel tried to create multicultural and multifunctional space for different cultural and religious groups. He spoke with their representatives to find out their needs and expectations. The crematorium in Zoethemeer (2006) is extraordinary in its outer design. Architect Martijn de Gier covered the auditorium by a "second skin" as a protective wall. (Klaassens and Groote, 2014).

The above mentioned examples also raise the idea of a possibility to open or close auditorium to the external environment. Open space offers the feeling of belonging to surrounding landscape, reconciliation with the course of life, which leads to its end that the world continues in our followers, finally then the psychological effect of the nature to the soul of mourners. In contrast, an enclosed building offers safety, concentration on the ceremony, the deceased, sadness. The skin of the building forms protective arms, enclosing the gathering in shared sorrow, protects it, but also separates it from the surrounding world and everyday life. It enhanced concentration, allows sadness flow by along its grievous journey, which then cleanses the soul and helps the mourners go on. Interwar crematoriums in Czechoslovakia were situated near the existing cemeteries in their immediate vicinity; the dominant building was accompanied by columbarium and by urn graves; it was situated in a strictly delimited area surrounded by walls. Nearby, there were roads which provided a good traffic availability, as well as parking space. The entrance area to the auditorium leads through monumental staircase or wide plateau, for better awareness of exceptionality of the time and space. The buildings are strictly enclosed, the windows provide mostly diffuse light during the day, but do not allow any view.

In the case of the crematorium in Brno, the light is supplied through the ceiling skylight and the whole ceremony hall is surrounded by accompanying rooms in the layout. In contrast to this example, there are crematoriums from the second half of the 20th century, which are realized more often in the middle of the forest cemeteries; these cemeteries are founded together with the realization of the building (except for the existing forest cemetery in Zlín). The crematorium is no longer a close part of the town, but it is a part of the landscape. There is a pathway from the cemetery leading to the entrance to the building.

5   CONCLUSION

The development of the new building type of funeral architecture was influenced by several aspects which resulted from the development of Czechoslovakia, later the Czech Republic and Slovakia and also from the comparison of the development in the selected countries in Europe. The most important aspect is the influence of the Church. In Czechoslovakia, the secularization of the society increased as a result of the First World War. But the effect of the prohibition of cremation by the Catholic Church is still evident. Its effect was suppressed after the Second World War, due to the communistic regime, nevertheless, it persists in the traditions. Burials in the church are eliminated by the construction of crematorium chapels, which were promoted as a modern space suitable for the funeral ceremony. Even though, a tendency to cremation started after the Catholic Church lifted the ban of cremation in 1960s. This caused the second wave of the realization of crematoriums in Bohemia, Moravia and Silesia, and the construction of the first crematorium in Slovakia (in Bratislava).

A different situation is in Norway, where the Church never had negative attitude to cremation, in contrast to Romania, where the Romanian Orthodox Church has never lifted the ban of cremation.

Other influence, evident in the new Czechoslovak Republic in 1918, but also in other countries, is legislation, which makes the cremation a legitimate form of the disposal.

Last but not least, it is necessary to focus on the situation at the beginning of 21st century. In the countries of Eastern Europe, the influence of the fall of the communistic regime has been manifested, new crematoriums have been established, not only as a state enterprise as before the revolution in 1989; some of them are private.

Slovakia is ready for the establishment of new crematoriums (the reason is the number of 6 crematories for 5.4 million inhabitants), in the Czech Republic, the construction of other buildings is not expected, because of the high number of crematoriums. The existing buildings have been transformed to the current needs and wishes of mourners and to the needs of the technology. The situation in Belgium or the Netherlands is different; a new epoch of the development of this young building type is in progress there, and there are new types realized with the composition corresponding to the current needs. This is the part of the development, which we cannot test in the Czech Republic, but we can be inspired by details in the transformation of the existing buildings.

ACKNOWLEDGEMENT

This work was supported by means of the Institutional development projects in 2016. Project registration number is 20/2016/RPP-TO-1/a.

REFERENCES

Hadders, H. 2013. Cremation in Norway: regulation, changes and challenges.
In Mortality. Vol. 18, No. 2, 195–213. UK: Taylor & Francis.
International Cremation Statistics 2014. In: http://www.effs.eu/cms/fileadmin/members_only/documents/news_articles/Cremation_statistic_2014.pdf.
Kotrlý, T. 2008. K proměnám postavení hrobníka. In: Český lid. Vol. 95, 273–292.
Klaassens M., Groote P.D. 2012. Designing a place for goodbye: The architecture of crematoria in the Netherlands. In Emotion, Identity and Death: Mortality Across Disciplines: 145–159. Ashgate Publishing Ltd.
Klaassens M., Groote P.D. 2014. Postmodern crematoria in the Netherlands: a search for a final sense of place. In Mortality. Vol. 19, No. 1, 1–21. UK: Taylor & Francis.
Rotar, M. 2015. Attitudes towards cremation in contemporary Romania. In Mortality. Vol. 20, No. 2, 145–162. UK: Taylor & Francis.
Svobodová, M. 2013. Krematorium. Praha: Artefaktum.

Advances and Trends in Engineering Sciences and Technologies II – Al Ali & Platko (Eds)
© 2017 Taylor & Francis Group, ISBN 978-1-138-03224-8

Vegetation of the green roof affecting temperature under the foliage

Z. Poórová, F. Vranay & Z. Vranayová
Faculty of Civil Engineering, Technical University of Košice, Slovakia

ABSTRACT: The paper presents the start of an ongoing PhD research, which main focus are green roofs. The experiment explained in this paper is a part of the study. The paper presents existing building in Kosice with green top of the building. Theoretical knowledge about water cycle and its cooling effects in the vegetation are the main ideas of this study and the experiment. Three different types of vegetation creating green top of the building are crucial for this experiment, where the main idea is to picture the temperature differences under different vegetation.

1 INTRODUCTION

The two biggest phenomena of water collide in a problematic of modern urbanism—these are water cycles and water types (ocean/sea water, land water, atmosphere water and water in living organisms).

2 HEAT BALANCE

Water is known in three states. Solid, liquid and gaseous. During the change from one state to another, heat is released or consumed. During the change from solid to liquid or gaseous state, it gains high mobility thanks to which it is capable of quick motion. Water also has the highest specific heat capacity, thus the ability to receive thermal energy from known materials. With its ability to bind and release energy, and transfer skills, reflection and dissipation of energy, water in all its states according to the needs cools or heats the planet. It is keeping it at a temperature that supports life on Earth (Kravčík, 2007, Poórová et al. 2014).

By observation we can differ water balance among day and year phases or diverse areas/ locations. Thus water reduces weather extremes. The more water in the atmosphere, the

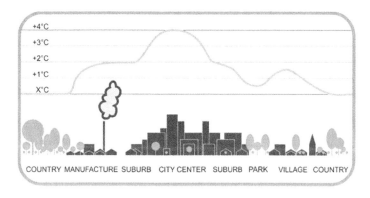

Figure 1. Temperature differences between different areas.

stronger effect of temperature balance. Therefore there are less weather amplitudes. The less water in the atmosphere, the weaker effect of temperature balance. Thus there are more extreme weather amplitudes. Where is the lack of water in the soil and in the atmosphere, extreme temperature conditions usually persist. Water and water vapor affect the climate in the most significant way on the Earth (Kravčík, 2007, Poórová et al. 2014).

3 WATER CYCLE

Water cycle is a complex combined of partial processes (Figure 2): evapotranspiration, evaporation, condensation etc. Water cycle can be understood in two ways. Global and local.

3.1 *Global water cycle*

Global water cycle is an exchange of water between the ocean and the land. About 550000 km³ of water evaporates into the atmosphere each year. From the seas and the oceans around 86% evaporates, from the mainland 14% of the total evaporation from the surface of the Earth. Out of the total atmospheric precipitation, which arise from the evaporation, 74% drops over the seas and the oceans, and 26% drops over the land. The seas and the oceans through the evaporation and precipitation subsidize land with some volume of water. This amount of water by the atmospheric and thermodynamic horizontal flows (Figure 3) is getting through long distances over continents where expires (or falls in the form of snow) (Kravčík, 2007, Poórová, 2014).

3.2 *Local water cycle*

Local water cycle is a closed water cycle in which vaporized water falls in the form of precipitation on the land. Just like local water cycle exists over the land, it exists over the sea

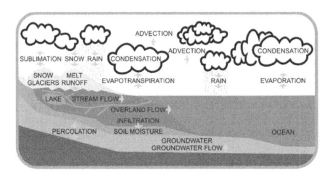

Figure 2. Partial processes of water cycle.

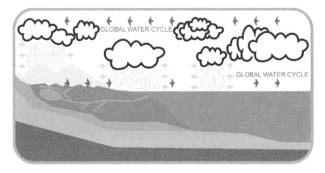

Figure 3. Water cycles—global and local.

or ocean. Between the local water cycles, over the large territories with different morphology and surfaces with varying humidity, ongoing interactions are going on. The local water cycle performs vertical water circulation, but unlike the global water cycle, it is characteristic horizontal (Figure 3) movement for it. Evaporation from neighboring areas with different temperatures can cooperate on the design and conduct of cloud. We can say that local water cycles circulate around the country at the same time. We can say that above the landscape the water is circulating in many local water cycles that are donated by the amount of the large water cycle (Kravčík, 2007, Poórová, 2014).

3.3 *Collapse of water cycle*

If there is a widespread disruption of vegetation cover (deforestation, agricultural activities, urbanization), solar energy hits all the surfaces with low vapor and a part is converted to heat. This is how extreme gives rise to significant variations in temperature and the temperature difference between day and night, or only between sites with a different temperature regimes grow. Air circulation will increases, hot air is drifted away and most of the evaporated water from the country is being lost. Small and frequent rainfall decrease and more powerful and less frequent rainfall from the sea increase. The cycle opens, global water cycle starts to dominate, which is in contrast to the local one characteristic with erosion and washing away of soil and nourish to the sea (Kravčík, 2007, Poórová, 2014).

4 WATER

Water of the seas and oceans covers 70,8% of the Earth's surface and forms the largest part of all water on the Earth. Without rhe seas and the oceans the planet would be suffering from changing extreme temperatures, what would make life as we know it impossible. Among other functions of the seas and the oceans water from seas and oceans supply to precipitation on the land (Figure 2) (Kravčík, 2007, Kravčík, 2007).

4.1 *Land*

Water is often fixated on water in the rivers or natural or artificial lakes. Water in solid form (ice, snow) forms 2,05% of all water on the Earth and shelters up to 70% of the world's freshwater supplies. Visible surface water in rivers and lakes (including salt lakes and inland seas) forms only small amount of all water on the Earth. Groundwater and water forming soil moisture presents besides eccentrically placed glaciers the greatest wealth on the land that exceeds several times the volume of water in all rivers and lakes of the entire world. Water in the soil in terms of the quantity of benefits is more important than water in rivers (Kravčík, 2007, Kravčík, 2007).

4.2 *Atmosphere*

The volume of water in the atmosphere, in all three states is approximately 10 times bigger than the volume of water in all the rivers. If all water in the atmosphere felt in time in the form of precipitation, it would create on the imaginary ground surface 25 mm layer of water. Just like the seas and oceans have key global thermoregulation role on the planet, water in the atmosphere has crucial local thermoregulation function (Kravčík, 2007, Kravčík, 2007).

4.3 *Biota*

Water surrounds us. It is not just around us, but it is inside us. In living organisms, water volume is only about 0,00004% of all water on the Earth, what is the smallest amount of total volume of water, but what is lacking on the volume is the highly balanced in crucial importance of this water for daily individual form of life. Water content in plants varies depending on the species and often is much higher than water content in animal bodies. The volumes of water accumulated in the vegetation cover are not negligible, just like the volumes of water stored in the soil due to the existence of vegetation. The vegetation on the land, among other functions, has in particular the critical role in the regulation of evaporation from the soil. Therefore, on the land greatly aids thermal stability. On the existence and prosperity of vegetation depends consequently all higher life on the Earth (Kravčík, 2007, Kravčík, 2007).

5 COOLING

When the solar radiation hits water well-stocked area, most of the solar energy is consumed for evaporation and only the rest is consumed for sensible heat, heating the soil, reflection, or photosynthesis. When the solar radiation hits the drainage area, most of the solar energy turns into sensible heat, in the year-long sufficiently humid areas, most of the solar energy is consumed for evaporation. Therefore, water areas, soil saturated with water and vegetation have important role in the water cycle on the land. Functional vegetation fulfills the function of the valve between the soil and the atmosphere. It protects the soil from excessive overheating and thus drying out and optimizes the amount of the water evaporation through the transpiration of amount of air channels on the leaves. Vegetation well stocked with water thus has a significant cooling and air conditioning feature. Deforestation, agricultural and urban activities are changing the amount of water in the country (Figure 1) (Poórová, 2014).

6 URBANIZED GREEN

6.1 *Košice green roof*

Green roof is situated in Košice city in Slovakia, Magnezitárska 2/C, 04013. Name of the building is EcoPoint Office Center Košice and one of the mottos of this building is "Where ecology meets economy". The EcoPoint project has been awarded with Silver DGNB Precertificate. That certificate is awarded based on assessment of authorized experts of ÖGNI. EcoPoint is thus the first project of a building in Slovakia that has been awarded with the precertificate. It guarantees to the investor, but especially to tenants, that principles of sustainable construction will be adhered to [7, 8]. The building is using active technologies. Heating and cooling using activation of the concrete core in ceilings while heat pumps will provide environmental comfort without air-conditioning. Other qualities of this project are: possibility of natural airing through opening windows, fire and evacuation radio, air ventilation system in all offices provide fresh, thermally treated air, clear height of offices, double flooring, antistatic carpets, adequate natural light, connection to public data and telecommunication networks, video monitoring system for building protection, high-speed elevators, access to the parking lot and all floors protected by a security card system etc.
The roof construction and all its layers are shown in Table 1.

Table 1. Green roof construction.

Vegetation	–
Medium	100 mm
Filter fabric	–
Water holding drainage layer Dekdren L 60	60 mm
Filter membrane TYPAR SF	–
Waterproof PVC foil Sikaplan SGmA	1,5 mm
Filter fabric PP 300 g/m2	–
Thermal insulation Roofmate SL-A	40 mm
Thermal insulation	160 mm
Vapour barrier Isoroof Plus	–
Interpenetration paint SBS	–
Bearing construction	250 mm

Three significant plants of this roof used for an experiment are: Sedum Diffusum 'Potosinum', Sedum Kamtschaticum and Sedum Spurium 'Voodoo'. Diffusum 'Potosinum' on Figure 4 described in Table 2 has significant yellow/green color, Sedum Kamtschaticum on Figure 5 described in Table 3 is vivid green and Sedum Spurium 'Voodoo' on Figure 6 described in Table 4 has significant burgundy color.

Figure 4.   Sedum Spurium 'Voodoo'.

Table 2.   Sedum Diffusum 'Potosinum'.

Plant type	Perennial, Cactus/Succulent, Groundcover
Height × Spread	15 × 30 cm
Soil needs	Average, Well-Drained, Rich, Rocky, Sandy/Gritty
Water needs	Regular, Occasional, Low, Drought Tolerant
Sun exposure	Sun, Part Sun
Special situation	Pollution Tolerant
Flowers	Yellow
Foliage	Evergreen, Silver/Gray/Blue
Flowering time	Summer

Figure 5.   Sedum Kamtschaticum.

Figure 6.   Sedum Spurium 'Voodoo'.

Table 3. Sedum Kamtschaticum.

Plant type	Cactus/Succulent, Groundcover
Height × Spread	10–15 × 30–60 cm
Soil needs:	Well-Drained, Sandy/Gritty
Water needs:	Regular, Occasional, Drought Tolerant
Sun exposure:	Sun
Special situation:	–
Flowers:	Orange
Foliage:	Green
Flowering time:	Summer

Table 4. Sedum Spurium 'Voodoo'.

Plant type	Perennial, Cactus/Succulent, Groundcover
Height × Spread	10–15 × 45 cm
Soil needs:	Well-Drained, Sandy/Gritty
Water needs:	Occasional, Drought Tolerant
Sun exposure:	Sun
Special situation:	–
Flowers:	Pink, Red
Foliage:	Evergreen, Burgundy/Maroon
Flowering time:	Summer

8 CONCLUSION

Aim of this paper was to present theoretical knowledge about water cycles and types of water, focusing on water in biota. The idea was to picture its importance in urban units in form of green walls. Second part of the paper presents ongoing research on green roof in Kosice. Following the theoretical knowledge about water in vegetation, proving that lowering the temperature under different types of vegetation is in progress in this research.

ACKNOWLEDGEMENTS

This work was supported by: VEGA 1/0202/15 Bezpečné a udržateľné hospodárenie s vodou v budovách tretieho milénia/ Sustainable and Safe Water Management in Buildings of the 3rd. Millennium.

REFERENCES

Kravčík, M. 2007. Voda, Wasser, Water, Woda. Municipalia: Žilina.
M. Kravčík et al. 2007. Water for recovery of the climate-new water paradigm. Municipalia: Žilina.
Kravčík M. et al. 2007. Water for recovery of the climate-new water paradigm. Municipalia: Žilina.
Poórová Z. et al. 2014. Green roofs performance towards cooling heat island and boosting water conservation. In: Městské vody 2014: Sborník přednášek konference s mezinárodní účastí: Velké Bílovice, Brno: Ardec, pp. 205–210.
http://www.ecopoint.sk/en/dgnb-pre-certificate
http://plantlust.com/plants/sedum-diffusum-potosinum/
http://plantlust.com/plants/sedum-kamtschaticum/
http://plantlust.com/plants/sedum-spurium-voodoo/

Advances and Trends in Engineering Sciences and Technologies II – Al Ali & Platko (Eds)
© 2017 Taylor & Francis Group, ISBN 978-1-138-03224-8

Impact of environmental aspect and safety at work in designing of mechanized building processes

K. Prokopčáková, Z. Hulínová, S. Szalayová & L. Prokopčák
Faculty of Civil Engineering, Slovak University of Technology in Bratislava, Bratislava, Slovakia

ABSTRACT: Industrialization, economic as well as scientific and technological development have an impact on the environment and last, but not least, these factors also affect the area of working conditions of people and safety at work (hereinafter only OSH—occupational safety and health). Building production negatively affects environment and it is our obligation to treat these impacts with giving preference to new technologies, new methods and understanding, concurrently aiming at efficient implementation of a building project, building processes. The possible ways—of treatment of issues in the systems of building production—thorough preparation, designing, individual approach to assessment of the internal structure of building processes.

1 INTRODUCTION

Efficiency of building processes, simply said, depends on the efforts exerted to achieve maximum results with minimum of socially needed work. Treatment of issues in the systems of building production, due to their "specificity" is only possible by thorough preparation, designing, individual approach to assessment of the internal structure of building processes in consideration of the environment as well as working conditions of people.

Under the present development within the construction sector, where it is supposed that the amount of the new building realization will be lower and the most important factor will be modernization and reconstruction of existing buildings, the processing of new technological projects acquires new dimensions (Prokopčáková 2008 Brno Tribun EU). We have to anticipate some restrictions which also implies that the process of projects implementation will gradually become more and more complicated. The environment and protection of human life undoubtedly represents such impacts.

The negative impact on the environment over the past 25 years has been so conspicuous and the ecological balance is being overridden to such an extent that nature will not be able, without scientific regulation, to ensure the basic conditions for healthy development of man on the Earth. Due to this, the criterion of workforce protection should have a significant impact on the decision taken in selecting the optimal alternative of the mechanized building process. The decision-making procedure in connection with the mechanical design of the project can be supported by a risk management system. This includes risk assessment as well as proposal of measures for exclusion, or at least, minimization of the risk.

2 DESIGNING OF BUILDING PROCESSES

The essential precondition for efficient building production is thorough and timely pre-production and production preparation of building projects. It establishes conditions for optimization of building projects implementation. Optimization and its tools, including designing, play key roles in finding acceptable solutions to actual tasks. The designing itself represent a set of theoretical and practical analyses of certain processes that in respect to some special applications have the highest possible general applicability.

(Rockstroh 1972 as cited in Technologické projekty), The designing process includes a comprehensive set of:

– material relations (machines, devices), human relations, time-related relations (optimization of design preparation works, optimization of implementation).
Design phases:

– initial phase, definition of the purpose and objective, positioning of the design in the macro-space,
– lead phase, definition of production resources, positioning within the micro-space and time,
– design processing, analysis of the current status and input data, elaboration of alternative solutions,
– phase of optimization, selection of the optimal alternative.

The final objective of designing is not just any technical solution with some economic effects, but rather an optimal overall solution providing for best technical results, favorable working conditions and the highest possible economic effect. In spite of the fact that the economic effect is becoming the dominant requirement, under certain conditions, can be prioritized and other aspects. Each aspect has its position within the system of efficiency assessment of a building process and at a certain point of time it can be the very decisive one.

3   MECHANIZED BUILDING PROCESSES AND DESIGNING

Current development of building production, implementation of specific building processes as well as deployment of equipment implies that thorough preparation of construction projects has to cover some new dimensions. One has to consider a greater number of restricting factors and therefore designing is becoming more complex. Nevertheless, the objective of designing stays the same: achieving the maximum economic effect.

There is developed a set of alternative solutions and in their general assessing it is necessary to determine a level of meeting the set goals. In selecting individual alternatives their technical parameters do not provide sufficient representations about possible consequences. In the majority of cases, therefore, technical units must be transformed to economic ones, so that we can assess the effects of partial measures on the overall solution by demonstrating or comparing the efficiency. Such analyses contribute to achieving the optimal partial as well as final results with the least incurred costs.

To avoid any partial view on the optimal design of implementation of some building process it is necessary to consider the decision-making criteria comprehensively and in case of technically adequate alternatives select the alternative that is economically optimal. For the sake of completeness, it is necessary to add that the optimal alternative is not always the one achieving the maximum economic effect, i.e. with the minimum of implementation costs.

Under specific conditions, in the forefront there can be also other aspects that must be considered and the overall selection must be adapted accordingly. Such aspects can include, according to (Prokopčáková 2008 Brno Tribun EU) the following:

– meeting of needs of a community, market,
– employment issue in the region, issue of a certain social group,
– improving of labor culture and hygiene etc.
– specialized works in the course of the process implementation,
– quality of the resulting product of the mechanized process,
– decreasing of energy demands,
– working environment,
– occupational health and safety,
– environmental protection etc.

The Fig. 1 contains a set of functionally related factors having different mutual connections. A significant factor in selecting the optimal alternative is the possibility of implementation

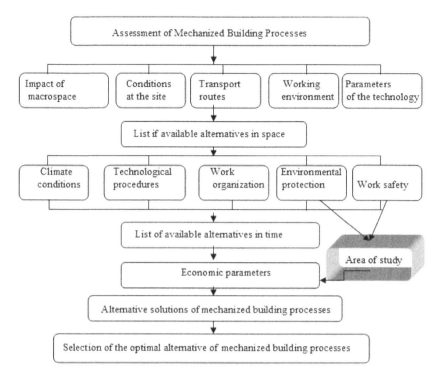

Figure 1. General model of assessment of mechanized building processes.

(Prokopčák, L. 2015 Bratislava, SR, CD-ROM). In case the selected alternative fails to be technically or organizationally feasible, many a time, in order to achieve a suitable solution one selects a combination of two or several alternatives.

To avoid any partial view on the optimal design of implementation of some building process it is necessary to consider the decision-making criteria comprehensively and in case of technically adequate alternatives select the alternative that is economically optimal.

4 HEALTH AND SAFETY AT WORK

Currently the factor of health and safety at work (hereinafter only OSH—occupational safety and health) is becoming a significant factor in connection with all the production as well as non-production activities, in many companies OSH has even been put in the 1st position, above all the other topical activities.

Regardless of the relation to OSH, the mechanized building processes belongs to high-risk activities and therefore it is necessary to consider them minimally from the point of view of legislative obligations. Each piece of equipment, not only individually but also as a constituent part of machinery, brings along also certain risks into the process of building. Due to this, the criterion of workforce protection should have a significant impact on the decision taken in selecting the optimal alternative of the mechanized building process. The decision-making procedure in connection with the mechanical design of the project can be supported by a risk management system. This includes risk assessment as well as proposal of measures for exclusion, or at least, minimization of the risk.

4.1 Risk assessment

Risk assessment comprises two important steps, namely: identification of threat and risk assessment. By identification of the threat we specify all the sources that may endanger workers at work. We are able to characterize these threats from several points of view. In general we analyze threats posed by operation of each device individually, but also as a part of several serially running mechanisms or mechanisms working in parallel.

The biggest threats to workers stem from moving parts of machines or from their actual movement. Connecting these parts into machinery sets piles up the threat due to additional risks created by their mutual interaction.

Upon a thorough analysis of threats, as the next step it is necessary to assess the risk and consequently to assess whether the risk is still acceptable, i.e. to determine:

1. probability (p) of occurrence of damage to health and level of possible consequences (s)
   a. to health of workers,
2. risk assessment on the basis of combination of probability and consequences as $R = p \times s$ (Risk = probability × severity).

The risk assessment, i.e. determination of the level of threat to workers represents the basis for a proposal of suitable measures as well as consideration of further steps in the process.

4.2 *Proposal of measures in risk management*

When drafting the measures we derive from a hierarchy prescribed by safety regulations. The most effective way of elimination or at least minimization of the risk is the use of suitable technical, organizational and methodological measures. In case of mechanized building processes the correct organization of work represents the most significant safety measure.

We shall validate the success rate of the proposed measures during practical operation of construction machines. This stage of risk management also includes continuous monitoring of design works and outputs of the risk analysis.

4.3 *Principles of the design of mechanized building processes*

When designing the mechanized building processes it is also necessary to respect certain principles deriving from OSH requirements. In the decision-making process these principles may significantly affect the choice of an optimal alternative. Proper understanding of basic principles of OHS represents a precondition for a good quality of the design. These principles derive from safety requirements that concurrently help to identify the threats. The resulting risk, as a level of threat to workers, will depend on the conditions of the relevant project. Some of the safety requirements are further listed including related foreseeable risks:

– machinery can be operated only in accordance with the operating rules and operation manual, (risk of machinery selection which in terms of OHS may not suit conditions of construction, risk of unsuitable deployment and improper organization of works), and it is prohibited:
– to work with the machinery under reduced visibility and at night, in case the working area of the machinery and the work site are not sufficiently illuminated (risk of machinery selection which does not meet OHS during reduced visibility),
– to work with the machinery within dangerous vicinity of another machine or vehicle except for those that work in mutual interaction with the machinery (risk of selection of machines which may pose threat to each other in interaction),
– to move the machinery or its part above heads of persons and above an occupied car cabin of a driver of a vehicle (risk of a design of works organization of a machinery which will pose danger to other persons)),
– to work with the machinery at a site not providing good visibility from the operator's position and where persons carrying out building works including machinery operator may be endangered (risk of negligence of the principle of good visibility on the side of the machinery operators),
– to move the machinery or working tool of the machinery within the protection zone of a live power line (risk of a design of machinery whose reach of the working tool interferes with the protection zone of the power lines),

- an earth-moving machinery can move or work, depending on soil load capacity, within such a distance from the edge of a slope or trench as to avoid collapsing of the machinery (risk of a design of machinery which will not comply with the load capacity and composition of the soil),
- machinery can move or work below the wall or slope within such a distance as to avoid occurrence of danger of the wall's or slope's collapsing (risk of improper organization of works),
- in case a scraper is moving, within its close working area in the driving direction it is prohibited to remove rocks, roots as well as to carry out other works (risk of improper organization of works at the time of concurrently implemented other works),

and it is required that:

- during driving and operation of the machinery on the slope there is applied such a method of driving as to avoid any dangerous shift of the center of gravity and loss of stability of the machinery (risk of incorrect works procedure),
- under parallel operation of several machines at one site the distance kept between them is such as to avoid any risk of their operation (risk of machines posing mutual risk to each other under their parallel operation),
- during transportation of fresh concrete by a stable pump or a pump installed, must be secured safe arrival and movement of vehicles (risk of work threat due to spatial limitation),
- the pump for delivery of fresh concrete on a chassis (auto-pump) is positioned in such a way as to eliminate any obstacles in the area of jib and pipes handling, which could impede this handling, endanger occupational health and safety, and as to ensure observance of protection zones of air power lines (risk of incorrect works organization as well as improper positioning of the machinery),
- the person operating the pump has a good view of the work site (risk of improper spatial positioning of the machinery) etc.

5 ENVIRONMENTAL PROTECTION

The quality of the environment is becoming one of the major factors of success of companies, whereas the introduction of the system of environmental management is a process that increases ecological behavior of companies. The focus of the EU environmental efforts has been formulated in Art. 130r, sec. 1 of the Treaty of the European Community setting out the following objectives:

- preserving, protecting and improving the quality of the environment,
- protecting human health,
- prudent and rational utilization of natural resources,
- promoting measures at international level to deal with local, regional or worldwide environmental problems.

The environmental concept concurrently derives from the processes of planning in respect to which it is necessary to define the basic environmental characteristics (Szalayová 2012, Brno Tribun EU). These characteristics include the following:

- environmental aspects and impacts,
- legal and other requirements,
- long-term and short-term objectives, environmental programs.

5.1 *Analysis of environmental aspects at the construction site*

The construction sector plays an important role in the EU economy; nevertheless it concurrently significantly contributes to negative impacts on the environment. The Environmental Aspects (EA) are generated during the course of all the phases of construction activities. EA in the construction sector are divided to:

- direct environmental aspects,
- indirect environmental aspects.

The direct environmental aspect can be operatively managed and influenced by the organization. It concerns especially emissions to air and water, waste management, use and pollution of soil, use of natural resources as well as secondary raw materials including energy consumption, noise, vibrations, dust, odors, risk of extraordinary events and related environmental impacts, effects to biological diversity. The indirect environmental aspects are not fully controllable, it concerns mainly the behavior of contracting partners, contractors and subcontractors, designing, development, packaging, transport and use of products, production waste management, capital investment, loans, insurance, new markets.

The indirect EA include extraction of materials, production of building products, their transportation and energy supply. The indirect EA can be affected in several ways: one of the possibilities is to restrict the input in achieving the same quality of the output (oze). Another possibility is to purchase ecologically suitable building products and to consider criteria for such products already in the stage of selection and purchase. It is also possible to analyze the environmental aspects during the life cycle of a construction project, i. e.:

- during construction designing,
- during implementation of construction works itself,
- during the building use, possibly during its reconstruction and maintenance.

6 CONCLUSION

The analysis and application of currently used ways of designing the mechanized building processes and their integration in the system of construction technology designing, necessitates a more comprehensive and more objective definition of impacts affecting the assessment of mechanized building processes.

The given article contributes to the treatment of the given topic and in the conclusion it is possible to state that the optimal way of implementing mechanized building processes is not always the one achieving the maximum economical effect—minimal costs per production unit. Under certain conditions there may come to the forefront also other aspects that have to be considered and the overall perspective regarding the designing of mechanized processes acquires a different dimension. Such aspects can include:

- occupational health and safety,
- environmental protection,
- specialized works in the course of the process implementation,
- quality of the resulting product of the mechanized process,
- decreasing of energy intensity.

REFERENCES

Prokopčáková, K.: Mechanization and automation of construction work. International scientific conference. Neum, Bosnia and Herzegovina. Brno: *Tribun EU, 2008, s. 250–254.*
Prokopčák, L. 2015. Vplyv prostredia na mechanizované stavebné procesy. 25rd Annual PhD Student Conference on Architecture and Construction Engineering, Building Materials, Structural Engineering, Water and Environmental Engineering, Transportation Engineering, Surveying, Geodesy, and Applied Mathematics. Bratislava, SR, *CD-ROM, s. 356–362.*
Rockstroh, W. 1972. Technologické projekty I,II, *Alfa Bratislava.*
Szalayová, S. 2012. Environmentálne aspekty v stavebníctve I. *Brno: Tribun EU, 76 s.*
Waris, M. et al., 2014. Environmental implications of onsite mechanization in the Malaysian construction industry, WIT Transactions on Ecology and The Environment, Vol 181, *Press www.witpress.com, ISSN 1743-3541 (on-line) doi:10.2495/EID140131.*

Advances and Trends in Engineering Sciences and Technologies II – Al Ali & Platko (Eds)
© 2017 Taylor & Francis Group, ISBN 978-1-138-03224-8

Implementation of risk management theory for the development of lean thinking in a construction company

A.E. Radchenko & M.V. Petrochenko
Peter the Great St. Petersburg Polytechnic University, Saint Petersburg, Russian Federation

ABSRACT: This research article presents a new approach to the study of lean construction whose methods are widely used in modern countries. Lean construction today is one of the most advanced directions of the construction organization which regards the improvements of building site processes as fundamental for reducing the waste and decreasing the overall cost and time for construction. The main approach of lean is to identify construction waste in the value stream map and develop of methods to reduce them. However, which of them should be reduced firstly? What should the company which wants to become lean do during the initial stages? What main principles of lean should be accepted by this company? This article presents it.

1 THE REASONS TO USE RISK MANAGEMENT TOOLS IN LEAN CONSTRUCTION

Losses are the integral part of any construction process during the erection of the building. Different types of work can be associated with clarify loss, and today there is no construction project that is fully free from them. The work of identifying, monitoring and managing losses should be maintained constantly, during the entire life cycle of the project. The external and internal circumstances are constantly undergoing and change, and therefore, the reassessing should be continuously made for the identified losses and regular monitoring of the arrival and impact of new ones.

The presence of losses in the production of works does not necessarily mean that problems exist, because the problems occur only at the present time. However, we consider losses from the point of view of risk management, so attribute them to the future, and their appearance would be entirely probabilistic in nature (they may not appear). However, if not managed, they can become problems during the construction project. Thus, the identification of potentially occurring losses is a positive activity aimed at improving the activities of the construction company; improve efficiency and quality characteristics, increase productivity and reduce costs.

2 MODELLING LOSS MANAGEMENT ON THE BUILDING SITE

The successful management of losses during construction works is impossible without an adequate model, describing the whole process. The main purpose of its development is to obtain a clear sequence of decision-maker actions in determining the degree of influence of the losses on the production processes and to identify those whose impact is critical and must be reduced or eliminated altogether. The model should be applicable not only to a certain type (types) of work, but also to fully cover all construction operations at the site first, and then the construction zone, the big complex and up to across the company.

In order to achieve this goal it is necessary: to understand the essence of the arising losses at the construction site (the probability of their occurrence, the impact on the production processes); to develop the most effective ways of reducing the impact of each of them to an acceptable level or completely eliminate them; to implement interventions for the processing of losses according to the selected method. Therefore, the process of managing losses is a set of consistent targeted actions that focus on optimization of the construction process in the framework of maintaining the planned budget, reducing the duration of the works and along with the quality.

Considering a large number of works of predecessors, we can conclude that at present there is no universally accepted course of action for the management of losses during construction works. On the basis of the tools of risk management the following process was adopted of managing losses on a construction site that includes the main stages (Fig. 1).

2.1 The definition of the situation in the construction process of company, the development of the plan and analysis of production processes

At the first stage a clear structure of the process should be established to describe the process of losses identification, assessment and processing, as well as how to control these operations; the input data of management process should be obtained, i.e. the information set out in design and project documentation, calendar plan of construction, the scope of work, etc.

Based on this data, the project management staff should examine all construction processes which are planned from the normative point of view (the scope of work set forth in the norms and standards) and from the practical view, that means to determine the sequence of all possible operations on the construction of building "in nature". This stage is based on experts experience and observations over the construction of similar facilities. The obtained data should be organized and conducted with distribution by types of constructed structures.

2.2 Identification (detection) of losses

The next stage in the process of loss management is the identification of losses, i.e. the identification of the operations having a negative impact on the course of production processes. The purpose of loss identification on the construction site is to detect potential events on the site, which may occur with different probabilities and have different degrees of the consequences that affect the cost of the structures realization or construction time or the quality and performance of workers.

The results of the analysis form the list of losses, which increase costs and construction time and by further grouping 7 types in accordance with the lean management methodology (Liker, 2013; Ballard, 2003; Koskela, 2000; Walbridge Aldinger Co., 2000). For the purpose of identifying losses it should be formulated as random events that may not take place, therefore, a category such as the probability of occurrence should be proposed. The formulation of the single losses (source loss, event) should have some detailed information about it, and the formulation must be specific and unambiguous (Womack, 2000).

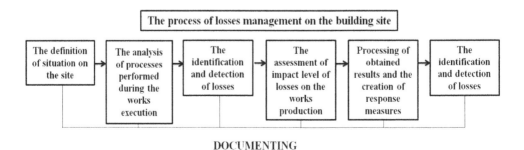

Figure 1.   The integrated model of losses management.

2.3 The assessment of the loss impact level on the processes of construction works

The purpose of this procedure is to identify and assess the priority losses, the reduction or elimination of the impact which is a priority for further action by the company. This operation includes the procedure of the probability of occurrence estimation for each of certain losses on the construction site and their impacts on manufacturing operations. The next step is to organize losses by priority in accordance with the assessment and identify which ones are the most influencing, taking into account such factors as the time limits and the tolerance levels of key stakeholders.

The most applicable is the method of expert assessment, as the simplest and easiest (Sheremet, 2011). One of the important characteristics in this method is the criteria of competence that can be defined considering the following qualities: the work experience; the level and profile of education; the profile of work performed; the level of solved problems; the number and level of projects in which the expert participated (Kravchenko, 2010).

The assessment is based on the pairwise comparison method (Kukushkin, 2007). For the further work on the assessment, the certain number of experts should be selected that will provide the necessary level of representativeness of the sampling. Each expert must have a certain set of qualities that are detailed on 4 levels: high, above-average, average and low (Morozova, 2011).

Then the direct estimate of losses using the method of expert assessment is initiated. The expert group conducts an estimation by assigning of each indicator's scores on a particular scale (Grebneva, 2012; Fedoseev, 2012; Romanovich, 2015). The processing of scores will be made taking into account the criteria of competence.

2.4 Expert's assessment

During the investigation 16 experts were assessing the losses. They estimated the losses probability P and the level their impact on the construction cost (I_c) and duration (I_t), the index R is determined for the each loss according to these indicators.

Index of the each loss is determined by the formula (Kravchenko, 2010; Voronova, 2003):

$$R_{ic,it} = P_i \cdot I_{ic,it} \qquad (1)$$

where P_i = probability of the each loss occurrence; I_{ic} and I_{it} = the impact level on the construction cost and duration.

The average statistic value is determined by the formula:

$$R_{iavc,iavt} = \frac{\sum_{j=1}^{n} R_{ij}}{m} \qquad (2)$$

where m = the sum of the expert competence levels.

The scores made by experts for each type of loss are obtained in the table in which the index of each loss is determined taking into account the level of expert's competence. Losses assessment is carried out by constructing so called matrix of "probability-losses". The main aspect when using this method is that the dimension of the matrix. The most common dimension is five by five, containing five numeric intervals of the losses occurrence probability and the degree of their impact on the construction cost and duration.

2.5 Cumulative integral coefficient of the losses and its impact on the technological processes on the construction site

The important aspect of executed operations is the question about the effectiveness of implemented changes management. For this purpose there is introduced a new characteristic measure: the indicator of losses impact on construction site. The principles of its definition are based on the methodology of the mathematical simulation and economic-mathematical methods. There are 3 main methods of its definition: additive, multiplicative and additive-multiplicative (Fedotov, 2006; Kaliszewski, 2016; Štreimikienėa, 2016). The significant drawback of additive and multiplicative transformations is the existence of unlimited opportunities of the compensation (Anfilatov, 2006; Klahna, 2016). While using the additive method if one of the individual criteria is equal or close to zero, then the final result does not suffer significantly. But the mutual compensation of the individual criteria can occur. It is necessary to impose restrictions for the minimum value of the individual criteria and their weights to mitigate this drawback (Klahna, 2016).

The cumulative coefficient of the losses impact can be calculated by the formula:

$$K_{cum} = \sum_{j=1}^{n} K_j \cdot \sigma_j \qquad (3)$$

where K_j = avalue of the j-th integral coefficient of the each identified losses, σ_i = a weigh factor of the i-th loss index (Beskorovayniy, 2005; Soboleva, 2012; Bolotin, 2010).

The integrated coefficient (indicator) of losses impact on the construction site will look like the following:

$$K_{loss} = \frac{K_{cum}}{z} \qquad (4)$$

where z = a multiplication of the maximum (minimum) value applied by the each expert for the probability of the loss and impact level.

The maximum (minimum) value of the indicator K_{cum} is equal the maximum (minimum) possible value of loss index $R_{c,t}$. The final appearance of the maximum (minimum) value of the integrated losses impact indicator has the following view:

$$K_{loss}^{max/min} = \frac{R_{c,t}^{max/min}}{z} \qquad (5)$$

When the value of integrated losses impact indicator is closer to 0, the level of losses impact on the construction site is lower and the efficiency of operations and productivity is higher. The approach to the value of 1 gives the decision-maker reason to concern measures of losses reduction and increase efficiency of production work.

2.6   *Matrix (map) of the losses*

Matrix (map) of the losses impact should be done in forms of the cost and duration value.

Similarly to the definition of the risk matrix from ISO Guide 73:2009, the matrix of losses influence is a tool for the classification and reporting the losses by ranking the impact degree for consequences and probability. The rules for its creation are the same. Interpretation of the losses influence matrix (map of losses) in graphic form will provide the visualization of the exposure index to losses relative to the tolerance level. It is possible to create it with the standard Microsoft Excel tools or by using the professional software.

It is necessary to use data which had been obtained before in the survey among the experts. The assessment scores which had been were given by experts during the survey are multiplied by the level of competence in a similar way:

$$P_{iL} = P_i \cdot m_j \qquad (6)$$

$$I_{iLc,iLt} = I_{ic,it} \cdot m_j \qquad (7)$$

where P_{iL} = the probability of the loss occurrence taking into account the level of expert competence; I_{iLc} and I_{iLt} = the level of the cost and duration impacts for the construction by the expert assessment taking into account the level of the expert competence; m_j = the level of the j-th expert competence.

Further calculations are repeating the statements from (4):

$$R_{icav,itav} = P_{iav} \cdot I_{icav,itav} \qquad (8)$$

where P_{iav} = is the average weight of the i-th loss occurrence probability; I_{icav} and I_{icav} = the average weight of the impact assessment value on the construction cost and duration taking into account the expert competence.

During the losses map creation it was necessary to set up the required tolerance levels indicating the maximum values of the losses impact on the construction cost and duration. It was needed to identify critical losses what required the priority response. Two tolerance levels were determined and the assessment results were defined the three groups of losses by their impact levels.

2.7 Measures for the impact of the identified losses reduction

In the final list of losses at the construction site all the losses are divided by the degree of their impact: the highest, the average and the lowest indexes. Losses with the low impact indexes can be reduced by the implementation of measures complex at the construction site by site engineers without the company management influence. Losses with the average and highest indexes require the more attention, because their reduction or elimination will lead to low costs of the works execution and production processes time, increase the productivity and quality, reduce injuries on the construction site. Measures to eliminate losses should be included in the enterprise standards for quality management (Adler, 2009; Mazur, 2010; Maslov, 2010).

At the initial stage decision makers should determine whether it is possible to prevent the occurrence of losses prior to the construction. The preventive organizational and financial measures should be taken within the existing construction technology. But if the execution of these measures is not possible, then there is necessity to change some of the construction technology to optimize or adopt alternative methods of works production.

It is possible that taken measures will not lead to the elimination of the certain losses. In this case, decision-maker should find the ways to reduce the losses impact. The main aim of this is to reduce the probability and/or consequences to the acceptable (low) values. The maintenance of the loss influence at the same level and development consequences elimination plans is one of the ways to cope with the losses. Decision-makers may follow two ways. The first is the passive acceptance, when nothing will be done in respect of the loss and the complex of individual measures is developed aimed to correction the consequences. The second way is the active acceptance, when the action plan is developed before the fact of the loss occurrence was recorded.

2.8 Monitoring, controlling and documenting the losses during the construction

The process of losses management at the construction site is an iterative. After making the changes, new losses can occur or the influence degree of the existing ones can increase. Therefore, it is based on the monitoring, documenting and controlling of ongoing changes. It should be noted that the monitoring provides more than the basic role as well as the documenting. They are the key processes in managing losses and without the monitoring and documenting the implementation of effective loss management system is impossible.

3 CONCLUSIONS

The presented research shows the methodology of losses management for the construction technology process, which includes all necessary steps. The main result of this investigation is the new approach of the losses estimation which proposes to use the integrated indicator of losses impact. It should be noted, that the loss management system of is individual for the each project but it is possible to use the proposed approach as the effective tool for the quick project estimation.

REFERENCES

Adler, Yu., Shechepetova, S. 2009. Process description the business foundations for the system of economy of quality. *Standards and quality No. 2*: 66–69.

Anfilatov, V.S. 2006. *The system analysis in management*. Moscow: Finances and statistics.

Ballard, G., Howell A.G. 2003. Lean project management. *Building Research and Information. Vol. 31. No. 2:* 119–133.

Belokorovin, E.A., Maslov, D.V. 2010. *Small business: ways of development*. Arkhangelsk: DMK Press.

Beskorovayniy, V.V., Trofimeko, I.V. 2005. Parametric identification of multiplicative models for multi-factor decision choice. *Proceedings of the Kharkov University of Air Forces named after Ivan Kozhedub conference. No. 5 (5):* 74–78.

Bolotin S.A., Dadar A.Kh. 2010. Оценка погрешности в аддитивном показателе качества календарного плана строительства The inaccuracy estimation of the additive quality indicator for the schedule of construction. *Bulletin of the Tyva state University. No.3:* 13–16.

Fedoseev, V.V. 2012. *Economic-mathematical methods and applied models*. Moscow: YUNITY.

Fedotov, Yu.V. 2006. Methods of constructing summary evaluations of complex production systems effectiveness. *Scientific reports No. 25(R)*. Saint-Petersburg: Grade school of management SPbSU.

Grebneva, O.A. 2012. *The theory of decision making*. Irkustk: IrSTU.

Kaliszewski, I., Podkopaev, D. 2016. Simple additive weighting - A metamodel for multiple criteria decision analysis methods. *Expert Systems with Applications. Vol. 54:* 155–161.

Klahna, Ch., Leuteneckerb, B., Meboldtb, M. 2015. *Design Strategies for the Process of Additive Manufacturing Procedia CIRP. Vol. 36:* 230–235.

Koskela, L. 2000. *An exploration towards a production theory and its application to construction, PhD Dissertation*. Finland: VTT Building Technology, Espoo.

Kravchenko, T.K. 2010. Expert evaluation in the decision-making process. *Actual problems of humanitarian and natural Sciences. No. 3:* 88–90. Moscow: Moscow state university.

Kukushkin, V.A., Morozova, T.F. 2007. *Scheduling in construction. Textbook and business game*. Saint-Petersburg: Saint-Petersburg energy institute of advanced training.

Liker, J. 2013. *Dao Toyota: 14 management principles of the world leading company*. Moscow: Litera.

Mazur, I.I., Shapiro, V.D. 2010. *Quality management*. Moscow: High school.

Medineckienea, M., Zavadskasb, E.K. Björka, F., Turskisb, Z. 2015. Multi-criteria decision-making system for sustainable building assessment/certification. *Archives of Civil and Mechanical Engineering. Vol. 15, Issue 1:* 11–18.

Morozova, T.A., Lapteva, N.A. 2011. The assessment of risks while implementation of investment construction project on the example of the business center. *Magazine of civil engineering, No.2(20):* 48–51. Saint Petersburg: Peter the Great St. Petersburg Polytechnic University.

Romanovich, M.A. 2015. *The increasing of organizational and technological reliability of monolithic housing construction on the basis of modeling the parameters of time schedule, PhD Dissertation*. Saint-Petersburg state university of architecture and civil engineering. Saint-Petersburg: SPbSUACE.

Sheremet, V.V. 1998. *Investment management: a guidebook for professionals and entrepreneurs*. Moscow: High school.

Soboleva E.V. 2012. Modification of the criteria for generalized utility in identification problems of multicriterion choice. *System Research & Information Technologies. No.3:* 58–65.

Štreimikienėa, D., Šliogerienėb, J., Turskisb, Z. 2016. Multi-criteria analysis of electricity generation technologies in Lithuania. *Renewable Energy. Vol.85:* 148–156.

Voronova, S.P. 2003. *The development of methods to assess investment and construction risks on the stages of the property life cycle. PhD Dissertation*. Saint-Petersburg: Saint-Petersburg state university of architecture and civil engineering.

Walbridge Aldinger Co. 2000. Lean Fundamentals. *Internal company document*. Detroit: Walbridge Aldinger Co.

Womack, P.J., Jones, T.D. 1996. *Lean thinking: banish waste and create wealth in your corporation*. New York: Simon and Schuster.

Advances and Trends in Engineering Sciences and Technologies II – Al Ali & Platko (Eds)
© 2017 Taylor & Francis Group, ISBN 978-1-138-03224-8

Definition of benchmarks for the assessment of the economic performance of buildings

D.A. Ribas
School of Technology and Management, Polytechnic Institute of Viana do Castelo, Portugal

M.M. Morais & P.B. Cachim
RISCO & Department of Civil Engineering, University of Aveiro, Portugal

ABSTRACT: Standard EN 16627:2015 was created to support the development of assessment methodologies for the economic performance of buildings, within the sustainability concept. It is based on a life cycle analysis where the impacts and aspects allow characterization of the economic performance, but do not exhibit valorisation methods nor define reference values (benchmarks). The purpose of this paper is to present the benchmarks used for the normalization of the parameters of the hierarchic structure of a Methodology of Assessment of Economic Performance—Residential Buildings (MAEP-RB). This methodology implements a systematic and simultaneous assessment of the performance and economic sustainability of buildings. The results of the economic performance are expressed in monetary units and economic sustainability is expressed by an Economic Sustainability Index (A$^+$, A, B, C, D, and E).

1 INTRODUCTION

The European Standards framework developed by CEN / TC 350 "Sustainability of construction works", proposes a system for assessing the sustainability of buildings based on Life Cycle Analysis (LCA). Based on LCA in the *before use* phase as defined in EN 16627:2015 (CEN, 2015), a new approach has been developed for systematically assess the economic performance of a building within the concept of sustainability *(Methodology of Assessment of Economic Performance—Residential Buildings: MAEP-RB)*.

The methodology follows the principle of modularity, where aspects and impacts that influence economic performance and building sustainability index during the *before use* phase, are assigned to the categories in which they occur. The hierarchical structure of the methodology directs the flow of information relating to aspects and impacts that influence the economic performance of the indicators, modules and stages of the life cycle, based on the quantification of the 65 parameters of the MAEP-RB methodology. In MAEP-RB, for assessing the economic performance in the *before use* phase, all parameters are expressed in monetary units, i.e. in euros. According to the hierarchical structure of the methodology and the flow of information therein defined, economic performance at the level of the indicators, of the modules and of the stages in the *before use* phase is obtained by direct aggregation, without any consideration of the results obtained in hierarchically lower levels. To obtain the sustainability index there is a need for standardization of all the values of the parameters and to develop a system of weights applied to the hierarchical structure.

This paper aims to present the methodology and calculation of the reference values (benchmarks) of the 65 parameters of the hierarchical structure of the MAEP-RB methodology.

2 METHODOLOGY MAEP-RB

The **MAEP-RB** methodology was developed in order to allow the assessment of economic per-formance and the level of economic sustainability of a residential building during the design phase, based on the expected behaviour for the entire building life cycle. It is a modular approach for the compilation of information throughout the building's life cycle including the four phases of the life cycle of a building: *before use* phase, *use* phase, *end of life cycle* phase and *beyond life cycle* phase. Each phase of the life cycle is divided into stages, modules, indicators and parameters. For the moment, **MAEP-RB** is developed only for the *before use* phase. The object of assessment is the building, including its foundations and landscaping within the building perimeter (Ribas, 2015). Table 1 shows the hierarchical structure of the method (stages, modules, indicators and parameters) that correspond to the *before use* phase.

At each level, information is obtained by aggregation of information at the lower level. For example, each of the twenty-one economic indicators (Level 3, indicators A0.1 to A0.2, A1.1, A2.1, A3.1, A4.1 to A4.2 and A5.1 to A5.14) is estimated by aggregating the results of one or more parameters (Level 4, parameters P1 to P65), following the hierarchical structure presented in Table 1.

The assessment of the economic performance of the *before use* phase is by obtained the aggregation of the results of each stage of the building's life cycle (Level 1). The parameters *(CPis)* are quantified by data obtained directly from the *"building record"* or *"building database"* evaluation generated by the methodology support software based on "general database" of **MAEP-RB** in accordance with the calculation process shown in Figure 1. The assessment of economic sustainability with assignment of a sustainability index for requires the standardization of parameters and the definition of a system of weights for the hierarchical structure of the **MAEP-RB** methodology.

The weighting system applied the hierarchical structure of the methodology defined by the application of Analytic Hierarchical Process (AHP) (Saaty, 1980) and presented in a previous article (Ribas, et al., 2015). The weighting system of the **MAEP-RB** methodology was assigned to all levels of the hierarchical structure, that is all parameters, indicators, modules and stage are relative weights assigned. Knowledge of the relative weights allowed determining the overall weight and to clarify the influence of a parameter, an indicator of a module or

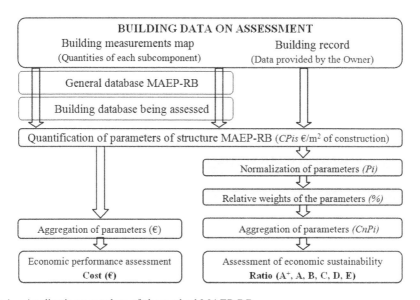

Figure 1.   Application procedure of the method MAEP-RB.

Table 1. Stages, modules, indicators and parameters of the MAEP-RB Methodology (Ribas, et al., 2015).

Level 1 Stages	Level 2 Modules	Level 3 Indicators	Level 4 Parameters
Pre-construction Stage	A0: Site and associated fees and counselling	A0.1: Cost of purchase and rental incurred for the site or any existing building. A0.2: Professional fees related to the acquisition of land.	P1: Costs with the site P2: IMT—Municipal tax on onerous transfer of property P3: IS—Stamp tax P4: Costs related to real estate P5: Costs of viability studies P6: Costs of legal support P7: Costs related to the notary fees P8: Costs related to the land registry fees
Product Stage	A1: Supply of raw materials A2: Transport of raw materials A3: Manufacturing	A1.1: Cost of raw materials A2.1: Cost of transportation of raw materials A3.1: Cost of transformation raw materials	P9: Percentage cost of each type of material used P10: Percentage cost of each type of material used P11: Percentage cost of each type of material used
Construction process Stage	A4: Transport	A4.1: Cost of transport of materials and products from the factory gate to the building site A4.2: Cost of transport of construction equipment such as site accommodation, access equipment and cranes to and from the site	P12: Percentage cost of each type of material used P13: Percentage of the cost of the building site
	A5: Construction—installation process	A5.1: Costs with exterior works and landscaping works A5.2: Cost of storing products including the prevision of heating, cooling, humidity etc. A5.3: Cost of transportation of materials, products, waste and equipment within the site A5.4: Cost of temporary works including temporary works off-site as necessary for the construction	P14: Cost for the earthmoving work P15: Cost of support structures and sealing P16: Cost concerning pavements P17: Cost relative to hydraulic networks P18: Cost related to outdoor lighting P19: Cost related to recreational equipment P20: Cost of sowing and planting P21: Percentage of cost for each type of material used P22: Cost of equipment related to the achievement of the subcomponents of the building P23: Cost construction site percentage of the total value of direct costs

(Continued)

Table 1. (*Continued*).

Level 1 Stages	Level 2 Modules	Level 3 Indicators	Level 4 Parameters
		A5.5: Cost on site production and transformation of a product	P24: Cost of hand labor P25: Cost of equipment P26: Cost of fuel P27: Cost of water
		A5.6: Cost of heating, cooling, ventilation, etc. during the construction process	P28: Cost of equipment P29: Cost of electricity
		A5.7: Cost of installation of the products into the building including ancillary materials	P30: Cost of hand labor P31: Cost of equipment P32: Cost of auxiliary materials
		A5.8: Cost of water used for cooling, of the construction machinery or on-site cleaning	P33: The cost of cooling water and cleaning
		A5.9: Cost of waste managing processes of other wasters generated on the construction site (RCD)	P34: Cost of the screening process of RCD P35: Cost of packaging of RCD P36: Tax amount
		A5.10: Transportation cost of waste RCD	P37: Cost of transporting the RCD
		A5.11: Costs of commissioning and handover related costs	P38: Cost of the extension of domestic wastewater sanitation. P39: Cost of the extension of sanitation storm water P40: Cost of extension of water supply P41: Cost of extension of electricity P42: Cost of extension of gas supply P43: Cost of extension of telecommunication P44: Cleaning cost
		A5.12: Cost for professional fees related to work on de project	P45: Fees of the project team P46: Fees of the inspection team P47: Fees the technical director P48: Fees of the health and safety at work team
		A5.13: Costs of the taxes and other costs related to the permission to build and inspection or approval of works	P49: Value of the license fee projects P50: Value of building permit fee P51: Exchange certifications gas project P52: Certification fee thermal design P53: Certification fee of electrical design P54: Rate design verification of fire safety P55: Certification fee of telecommunications project P56: National health service project certification fee P57: Certification fee of the gas network

(Continued)

Table 1. (*Continued*).

Level 1 Stages	Level 2 Modules	Level 3 Indicators	Level 4 Parameters
			P58: Rate of energy certification
			P59: Certification fee electricity grid
			P60: Certification fee telecommunications network
			P61: Rate survey of municipal services
			P62: Rate survey of the firefighters
			P63: Survey national health service fee
			P64: VAT rate
		A5.14: Incentives or subsidies related to the installation	P65: Value of the incentive

a step in the sustainability index obtained for the building at the *before use* phase. This information, when available in the design phase, is of the utmost importance, as it will serve as a guide for the design team, in order to improve sustainability, because there are parameters that have more influence on the final assessment than others.

The results of the assessment of economic performance and economic sustainability are broken down into multiple levels, namely the level of the *before use* phase of the building's life cycle, each stage of each module and each economic indicator. As the MAEP-RB assesses performance and economic sustainability of buildings, the result of economic performance is expressed in monetary unit and the sustainability of an economic sustainability index (A$^+$, A, B, C, D, and E).

3 NORMALIZATION

Normalization of parameters aims to set a dimensionless value that reflects the building's performance assessed in relation to the reference performance (benchmarks) *CPic* (Cost of conventional practice) and *CPim* (best practice cost) defined for each of the parameters hierarchical structure of the methodology.

The MAEP-RB methodology allows to perform a sensitivity analysis of the normalized values of the parameters, indicators, modules and stages, not limited to display their contents. More importance is given to the analysis of normalized values, because they reproduce the design options and determine the course of the outcome of the assessment of economic sustainability. Knowledge of these values already in the design phase allows designers to adopt design solutions that can achieve a certain goal in economic sustainability in the *before use* phase, eliminating or reducing all design solutions that contribute to normalized values of the worst parameters. In the standardization process Equation 1 was used to normalize the indicators (Diaz-Balteiro & Romero, 2004):

$$CnPi = \left(\frac{CPis - CPic}{CPim - CPic} \right) \qquad (1)$$

where: *CnPi* = *Pi* parameter normalization of results; *CPis* = Project solution cost (€/m² of building area); *CPim* = best practice cost (€/m² of building area); and *CPic* = Cost of conventional practice (€/m² of building area).

In Equation 1, *CPim* and *CPic* are the benchmarks *i* parameter, and respectively represent the levels of best practice and conventional practice. Their use converts the value of the parameters in a dimensionless scale, where the value 0 corresponds to the level of conventional

Table 2. Conversion of the standard value of a qualitative evaluation scale (SBTool, 2009).

Qualitative scale	Value normalized	Qualitative scale	Value normalized
A+	P > 1,00	C	0,10 < P ≤ 0,40
A (Best practice)	0,70 < P ≤ 1,00	D (Conventional practice)	0,00 ≤ P ≤ 0,10
B	0,40 < P ≤ 0,70	E	P ≤ 0,00

practice and the value 1 to the level of best practice. Should the performance parameter be greater than a best practice or lower than the conventional practice, the normalized value of the parameter will take, respectively, a value greater than 1 and less than 0. In any case, in order to avoid distortions in the aggregation of parameters/indicators, normalized values should not be smaller than −0,2 or bigger than 1,2.

To facilitate understanding of the results obtained, the normalized values are converted to a qualitative scale, graduated from "E" (lowest economic sustainability) to "A+" (bigger economic sustainability), using the equivalences shown in Table 2. In the qualitative scale presented, "D" index corresponds to conventional practice and the "A" to the best practice. The conversion range of normalized values used in the MAEP-RB methodology presented in Table 2 was adopted from SBTool[PT] system. The scale thus assumes a linear evolution in building performance between the index "D" and "A".

3.1 Definition of CPis—Application MAEP-RB

Bearing in mind the importance that the definition of 65 parameters (Pi) and their values (CPis) have in assessing the performance and economic building sustainability, there is a strong connection to the building under assessment in order to provide accurate information to each parameter (Pis). Therefore, each building under assessment is divided by the level of resources necessary for its materialization, and the data registered in the building database that contains all the information required for accurate determination of the parameters, i.e. the quantities and unit costs for each resource (such as water, cement, sand, bricks etc.). This database is constructed with use of the software developed for the methodology, after the introduction of data contained in the building measurements map evaluation.

The building database and the elements in the building record contain all the information needed to completely quantify the 65 parameters, and individual quantification of each parameter (CPis) conducted using the methodology.

Costs associated with A0 module (Site and associated fees and counselling) will be obtained directly from the "building record" that contains information provided by the Owner and are, therefore, treated as values that are always negotiated between the Owner and others, because they are extremely subjective: Costs with the site (CP1s); IMT—Municipal tax on onerous transfer of property (CP2s); IS—Stamp tax (CP3s); Costs related to real estate (CP4s); Costs of viability studies (CP5s); Costs of legal support (CP6s); Costs related to the notary fees (CP7s); Costs related to the land registry fees (CP8s). The costs associated with modules A1-A3 (A1: Supply of raw materials, A2: Transport of raw materials, A3: Manufacturing) and A4-A5 (A4: Transport, A5: Construction—installation process) are determined by the software developed for the methodology using the building database.

3.2 Definition of CPic and CPim benchmarks

Since there are no reference values (benchmarks) for the economic sustainability of buildings, it can be extremely difficult to find representative benchmarks for the 65 parameters of the MAEP-RB methodology. Definition of these values was carried out based on the actual costs observed in buildings already built and constructed by conventional methods. The values are expressed in Euros per square meter of construction (see Table 3). The results were defined

Table 3. Reference Values (Benchmarks).

Pi	CPic (€/m²)	CPim (€/m²)	Pi	CPic (€/m²)	CPim (€/m²)	Pi	CPic (€/m²)	CPim (€/m²)	Pi	CPic (€/m²)	CPim (€/m²)
P1	116,91	77,94	P17	11,36	2,84	P33	2,73	0,68	P49	5,80	1,45
P2	0,00	6,26	P18	8,36	2,09	P34	2,98	0,75	P50	33,68	8,42
P3	0,00	0,77	P19	12,54	3,14	P35	1,09	0,27	P51	4,20	1,05
P4	5,78	0,96	P20	10,85	2,71	P36	0,55	0,14	P52	6,35	1.59
P5	5,79	0,96	P21	4,05	1,01	P37	0,93	0,23	P53	6,46	1,62
P6	2,89	0,96	P22	5,82	1,46	P38	1,35	0,34	P54	4,25	1,06
P7	0,09	0,06	P23	70,34	17,59	P39	1,30	0,33	P55	6,25	1,56
P8	0,12	0,09	P24	130,26	32,57	P40	1,23	0,31	P56	6,38	1,60
P9	61,00	15,25	P25	18,35	4,59	P41	1,64	0,41	P57	10,25	2,56
P10	83,00	20,75	P26	13,35	3,34	P42	1,23	0,31	P58	25,69	6,42
P11	165,36	41,34	P27	6,30	1,58	P43	1,18	0,30	P59	18,87	4,72
P12	102,00	25,50	P28	6,10	1,53	P44	0,41	0,10	P60	13,45	3,36
P13	4,95	1,24	P29	1,87	0,47	P45	5,35	1,34	P61	10,15	2,54
P14	13,65	3,41	P30	21,30	5,33	P46	5,16	1,29	P62	10,36	2,59
P15	27,95	6,99	P31	3,85	0,96	P47	4,50	1,13	P63	9,89	2,47
P16	24,58	6,15	P32	1,05	0,26	P48	4,23	1,06	P64	35,68	8,92
–	–	–	–	–	–	–	–	–	P65	0,00	2,36

for the Portuguese *conventional practice*. The value calculated as *best practice* it was considered an improvement, that is, a reduction in the cost per square meter by 75%.

As can be observed in Table 3, the *CPic* and *Cpim* defined reference values is, for example, for the parameter; *P24: Cost of hand labour (CP24c=130,26€/m², CP24m=32,57€/m²).*

4 CONCLUSIONS

Quantification of the 65 parameters of the **MAEP-RB** methodology and the respective normalized values are decisive in assessing the performance and economic sustainability of buildings. As all other levels of the hierarchical structure of the methodology (indicators, modules and stage) are dependent on these, they are obtained by aggregation of the respective lower level.

The normalized values of the parameters (CnPi), are strongly influenced by CPic anf CPim reference values of each parameter and determine the evaluation result of the economic sustainability of the building. The development and improvement of CPic and CPim reference values for each parameter are of the highest importance for assessing the economic sustainability of buildings. This improvement is only possible through the implementation of **MAEP-RB** methodology to real cases, obtaining necessary information for the feedback loop of the reference values.

The relative weights system defined for **MAEP-RB** methodology allows for the determination of the overall weight, which is very useful in comparative studies of economic sustainability of buildings, since it clarifies how the sustainability index for the building at the before use phase is influenced by a parameter, an indicator, a module or a step. This information, when available in the design phase, is of the utmost importance, as it will serve as a guide for the design team, to improve sustainability, as there are parameters with bigger influence on the final assessment than others.

REFERENCES

CEN, 2015. EN 16627—Sustainability of construction works; Assessment of economic performance of buildings; Calculation methods. Brussels: CEN.

Diaz-Balteiro, L. & Romero, C., 2004. In search of a natural systems sustainability index. Ecological Economics, Volume 49, p. 401–405.

Ribas, D. A., 2015. Metodologia de Avaliação da Sustentabilidade Económica de Edifícios com Base no Ciclo de Vida. Dissertação de Doutoramento—Departamento de Engenharia Civil da Universidade de Aveiro, p. 206.

Ribas, D. A., Morais, M. M., Velosa, A. L. & Cachim, P. B., 2015. Application of an Analytical Hierarchical Process in the Assessment o Building Economic Sustainability. International Journal of Interdisciplinarity in Theory and Pratice, Volume ITPB—NR.:7, pp. 63–68.

Saaty, T. L., 1980. The Analytic Hierarchy Process. New York: McGraw Hill.

SBTool, 2009. http://www.iisbe.org/sbtool. [Online] [Acedido em 15 03 2015].

Advances and Trends in Engineering Sciences and Technologies II – Al Ali & Platko (Eds)
© 2017 Taylor & Francis Group, ISBN 978-1-138-03224-8

Baseline emission inventory – a methodology for municipalities

M. Rohlena & J. Frkova
Faculty of Civil Engineering, Czech Technical University in Prague, Czech Republic

ABSTRACT: In order for municipalities to work effectively on reducing emissions on their territories, it is necessary to identify the source of the emission, which is closely related to the energy consumption on their territory. An important presumption is the need to survey the energy demands of urban areas and to identify all key places of energy consumption within them. Let us call this document a Municipal Energy Consumption Inventory. The document presents a source base for the calculation of a Baseline Emission Inventory. How to elaborate these forms of documentation, what structure they should have and from where to derive information sources and data for the process are brought to the fore by this article.

1   INTRODUCTION

There is a rather broad set of problems involved in getting a fixed measure of energy consumption and subsequently CO_2 emissions for the whole territory of a municipality. Energy is consumed not only by buildings which are administered by the municipality but especially in private buildings of inhabitants and also in local industry and transport.

The article deals with a methodology for determining energy consumption on the territory of a town. This methodology provides the means and information sources through which this can be done at the municipal level. It was worked out according to the recommended structure proposed by The Joint Research Centre—Institute for Energy and Institute for Environment and Sustainability (JRC 2015). Such a document we will be labeled as the Municipal Energy Consumption Inventory (MECI). In this document it is necessary to divide the ascertained energy consumption by sectors and by mode of energy carrier (e.g. electricity, natural gas, coal etc.) the latter because each energy carrier has a different emission factor. Consequently, with the help of emission factors and the elaborated data in MECI there can be easily put together a balance of basic emissions on the municipal territory by way of a so called Baseline Emission Inventory. A Baseline Emission Inventory is a quantification of the amount of CO_2 emitted due to energy consumption on the municipal territory.

2   DEFINITION OF ESSENTIAL TERMS FOR SETTING THE INPUT DATA

Urban municipalities can be the most effective agents in the struggle against climate change, therefore it is very important that local level representatives are recognized as key actors and leaders in reducing greenhouse gas emissions and in taking measures for adapting to climate change. A municipality/town is a legally constituted public corporation, which means a legal person acting as a basis for public administration. In this sense it is a manifestation of self-government.

It is necessary to define municipality or town according to the methodological demands by the following: the number of inhabitants, the number of households, and the cadastral area of the municipality (Rohlena, Frkova 2013).

2.1   *Catchment area*

Municipalities at the secondary and higher level ensure services for adjacent municipalities such as offices, shops, schools and other functions which are not dealt with in the smaller municipalities. It is necessary to define the catchment area for the methodological require-ments by the following: the area of territories affiliated to municipalities of the highest level, the number of inhabitants, and the number of households.

2.2 *Energy carrier*

A substance or phenomenon that can be used to produce mechanical work or heat or to operate chemical or physical processes (ISO 13600:1997).

3 STRUCTURE OF THE MUNICIPAL ENERGY CONSUMPTION INVENTORY (MECI)

In order to establish an energy consumption inventory MECI it is necessary to divide energy con-sumption into the following sectors (this is according to the methodology of the Joint Research Centre (JRC), which after multiplying by emission factors for individual energy carriers will then give the resulting required Baseline Emission Inventory (BEI). (European Commission 2010) See Table 7. MECI divides energy consumption into a buildings sector and a transport sector.

3.1 *Determination of energy consumption in the buildings sector and the transport sector*

It is necessary to get reliable data to determine the energy consumption of individual sectors in a municipality. With the given set of problems this is easier for the property under actual municipal-ity ownership, because the municipality has concrete data on consumption in its own accounting records. For the private sector it is necessary to derive this from variously defined data sources.

3.2 *Municipal buildings*

This kind of property is obviously in the municipality/town ownership, and it directly serves administrative purposes. It especially relates to offices, schools and other institutions estab-lished by the town.

The term "equipment/facilities" includes objects consuming energy which are not deemed buildings (e.g. water treatment plants, waste recycling centres, and composting plants). It is necessary to include in the category "residential buildings" the residential buildings held in ownership by the local authority or an associated organization.

A municipality has the data on energy consumption for municipal property and it is able to carry out its own analysis. The received invoices for energy from suppliers contain the data not only on the price of these services but also on the consumption level. Alternatively, there can be used actual readings made at annual intervals from installed meters (Rohlena, Frkova 2013).

3.3 *Tertiary (non-municipal) buildings*

The tertiary sector includes commercial and public buildings which are not owned and administered by the municipality (e.g. private company offices, banks, small and medium sized enterprises, a commercial and retail area, hospitals etc.).

In order to gain data on the energy consumption of the tertiary sector it is necessary to determine the size of the catchment area and the number of inhabitants using services in a given municipality. The volume of provided services is directly dependent on the size of the municipality and the number of inhabitants in the catchment area. The size of the catchment area of a municipality depends on the size of the administrative district and the level of cen-tralization in provided services.

If we work from the monitoring of the Czech Statistical Office, which sets the final annual energy consumption for individual sectors, that will also include the service/tertiary sector. (CSO 2014) Dividing by the current number of inhabitants we determine the energy consumption of the tertiary sector per capita/inhabitant (see Table 1). By consequent multiplying the number of inhabitants from the catchment area there can be determined the annual energy consumption of the tertiary sector for the municipal territory of interest taking into consideration the amount of services provided.

Data from CSO and the Classification of Economic Activities CZ-NACE ("Nomenclature générale des Activités économiques dans les Communautés Européennes") are used for determining the energy consumption according to energy carriers (CSO 2009). The Czech Statistical Office monitors energy consumption in detail according to the division.

CZ-NACE and according to the type of energy carrier. From this data base we will calculate for the services which constitute the tertiary sector, and through which we will find out the energy consumption of the tertiary sector dividing into energy carriers according to Figure 1.

3.4 Residential buildings

These are buildings primarily determined for living. It is necessary to divide the energy consumption of the residential buildings sector into two statistically monitored sections—Electricity energy consumption of households and household energy consumption for heating.

3.4.1 Electricity energy consumption of households

The Czech Statistical Office monitors electric energy consumption for the household sector (CSO 2015). By dividing the number of inhabitants we get data of electric energy consumption per capita in the household sector—see Table 2 (CSO 2015, adjusted by authors). Electric energy consumption for residential buildings on the municipal territory is determined by multiplying the data and count the inhabitants.

3.4.2 Household energy consumption for heating segmented according to the heating source

A household energy consumption calculation on heating is derived from the statistics of the total household energy consumption (Table 3) lowered by electric energy consumption (Table 2). The resulting values of energy consumption for household heating converted to per capita measure are stated in Table 4.

For dividing up energy carriers for household heating there can be used data from the Census on people, houses and flats from 2011 – table: Inhabited flats—energy for heating (CSO 2011). These statistics will provide us with a basis for a percentage measure of energy outlay for household heating according to sources. We will divide the resulting values from Table 4 by these percentages.

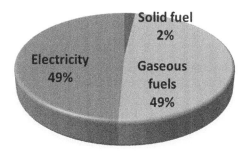

Figure 1.   Share of energy carriers in tertiary sector.

Table 1.   Tertiary sector energy consumption.

Year	2010	2011	2012	2013	2014
Tertiary sector energy consumption [PJ]	132.7	115.2	124.9	125.9	116.9
Per capita consumption [MWh]	3.430	2.974	3.236	3.252	3.028

Table 2. Electric energy consumption of households.

Year	2010	2011	2012	2013	2014
The electricity consumption [ths. TOE]	1292	1221	1253.7	1265.3	1214.5
Per capita consumption [MWh]	1.398	1.320	1.360	1.373	1.317

Table 3. Total energy consumption of households (source: CSO, adjusted by authors).

Year	2010	2011	2012	2013	2014
Final energy consumption [PJ]	265.9	259	256.81	267.28	236.74
Per capita consumption [MWh]	7.390	6.686	6.654	6.925	6.34

Table 4. Total energy consumption on heating after electric energy meter reading.

Year	2010	2011	2012	2013	2014
The electricity consumption [MWh]	1.398	1.320	1.360	1.373	1.317
Final energy consumption [MWh]	7.390	6.686	6.654	6.925	6.34
Per capita consumption [MWh]	5.99	5.37	5.29	5.55	4.82

3.5 Public lighting

Street public lighting owned and administered by the municipality. Energy consumption for lighting will be determined on the basis of invoice records or our own readings.

3.6 Industrial sector

According to the CZ-NACE the activities classified in categories such as forestry, mining, manufacturing industry etc. are considered as industrial. A register of emissions and air pollution sources from the industry is given by the Czech Hydrometeorological Institute to be found in the data base REZZO (Register of Air Polluters). A municipality or town can ask for a statement from the REZZO database in order to get information on air polluters on their territory.

4 TRANSPORT

This concerns energy consumption for three categories of transport on the municipal territory.

4.1 Public rolling stock

These cars are in municipal ownership. A municipality has in this category detailed information on energy consumption in its accounting records.

4.2 Public passenger transport

A municipality ensures transport services for the area by means of mass public transport and taxi services. In this subsector it is possible to get the necessary data from private carriers.

4.3 Private and commercial transport

If a busy road network transits the municipality, for the calculation of energy consumption in transport on the municipal territory it is possible to use data from the national Census of Transport, which is detailed by the Headquarters of Roads and Highways of the CR, the results of which are freely available on Directorate of Roads and Motorways. The results of the nationwide Census of Transport 2010 provide information on intensity of motor traffic on the highway and road network of the CR in 2010 and follow on from the results from the previous Census of Transport (2005 and earlier). (RSD 2010) In the table and map section there are the stated data: annual average of daily intensities—RPDI [vehicles/24h] in both directions. The table section contains vehicle categories: TV—heavy motor vehicles in total, O—passenger and delivery vans without trailers and with trailers. M—single track vehicles, SV—all motor vehicles in total (vehicles in sum total). If we take into account the length extent of the examined road network in the municipal territory of interest along with the knowledge of data from the transport Census, we can calculate fuel consumption for transit through the municipality and we will come out with average fuel consumption per 1 km according to the kind of a vehicle.

5 TABULAR ILLUSTRATION OF MUNICIPAL ENERGY CONSUMPTION INVENTORY MECI—THE RESULTS

In Table 5 there is given an illustrative example summarizing the results of Municipal Energy Consumption Inventory MECI for the town of Hlinsko (Rohlena, Frkova 2012). Total energy consumption of this municipality was 171,099 MWh in 2009.

5.1 Elaboration of CO_2 balance emissions

We will work from the document Municipal Energy Consumption Inventory MECI for elaborating the Baseline Emission Inventory BEI. The BEI has an identical structure with MECI, where there is determined energy consumption divided by sectors and also by energy source. We will then approach a conversion to an actually produced CO_2 emission. We will carry this calculation out on the basis of emission factors set out by a scientific panel IPCC (Intergovernmental Panel on Climate Change). (IPCC 2006) We will multiply the calculated energy consumption, structured according to individual resources and sectors, by the emission factors identified in Table 6.

Table 5. Total energy consumption of Hlinsko [MWh] (source: authors own calculations).

| Category | Electricity | Heat/cold | Fossil fuels | | | | Renewable energies | | Total |
			Natural gas	Gasoline	Lignite	Coal	Other biomass	Solar thermal	
BUILDINGS, EQUIPMENT/FACILITIES AND INDUSTRIES:									
Municipal buildings, equipment/facilities	1 540	957	4 777			210			7 483
Tertiary (non municipal) buildings, equipment/facilities		12 647							12 647
Residential buildings	30 035		107 423		11 520		955	478	150 410
Municipal public lighting	434								434
TRANSPORT:									
Municipal fleet				125					125
Total	32 008	13 603	112 200	125	11 520	210	955	478	171 099

Table 6. Emission factors for individual fuels according to IPCC.

Electricity	Natural gas	Diesel	Gasoline	Coal	Biofuel	Other biomass
0.95	0.2202	0.267	0.249	0.354	0.000	0.000

Table 7. Total CO2 emissions of Hlinsko municipality [t].

| Category | Electricity | Heat/cold | Fossil fuels | | | | Renewable energies | | Total |
			Natural gas	Gasoline	Lignite	Coal	Other biomass	Solar thermal	
BUILDINGS, EQUIPMENT/FACILITIES AND INDUSTRIES:									
Municipal buildings, equipment/facilities	1 463	106	965			74			2 608
Tertiary (non municipal) buildings, equipement/facilities		1 405							1 405
Residential buildings	28 533		21 700		4 193		10	5	54 440
Municipal public lighting	412								412
TRANSPORT:									
Municipal fleet				31					31
Total	30 408	1 511	22 664	31	4 193	74	10	5	58 897

In Table 7 as an example, there is given a resulting illustration of BEI in the way such as it was actually worked out for the town Hlinsko. It is evident from the Table that there is emitted approximately 59,000 tons of CO_2 per annum on the territory of this town.

6   CONCLUSION

The article informs readers on a methodology regarding the preparation and elaboration of documents which will enable municipalities to carry out an inventory of their energy consumption and subsequently to form the Baseline Emission Inventory. BEI allows us to identify the principal sources of CO_2 emissions and their respective reduction potentials. The proposed methodology has already been verified in practice. Using this methodology the authors worked out the BEI for the towns of Hlinsko, Lkáň and Úvaly u Prahy. BEI then served as the initiating document for creating the so called Sustainable Energy Action Plans (SEAP) for these municipalities. SEAP Hlinsko and Lkáň were accepted by the EU initiative—Covenant of Mayors, as a preliminary obligatory document committing the towns to emission savings of 20% by 2020 (fixed from the starting year 2009). According to this plan e.g. Hlinsko town will save on its territory 181,000 MWh of energy by 2020, which presents cost savings in the town circa CZK 46 million and an emissions reduction on the town territory of circa 75,000 tons CO_2. (Rohlena, Frkova 2012) The membership of Hlinsko and Lkáň in the pact Covenant of Mayors—the European initiative involving local and regional authorities in the fight against climate change and developing a more sustainable energy future—increases not only the prestige of both town and municipality, but it presents their own particular contribution for improving the living environment and quality of life on their territory.

ACKNOWLEDGEMENT

This work has been supported by the Grant Agency of the Czech Technical University in Prague, grant No. SGS16/023/OHK1/1T/11.

REFERENCES

CHI 2012. Czech Hydrometeorological Institute. 2012. *Emisní bilance České republiky 2012*. Available from <http://portal.chmi.cz/files/portal/docs/uoco/oez_CZ.html> [30 January 2016].
CSO 2009. Czech Statistical Office. 2009. *Klasifikace ekonomických činností (CZ-NACE)*. Available from < https://www.czso.cz/csu/czso/klasifikace> [20 January 2016].
CSO 2011. Czech Statistical Office, 2011. *Census on people 2011*. Available from <https://www.czso.cz/csu/sldb> [20 January 2016].
CSO, 2014. Czech Statistical Office. *Konečná spotřeba energie, v členění podle odvětví*. Available from <http://apl.czso.cz/pll/eutab/html.h?ptabkod = tsdpc320> [13 February 2016].
CSO 2015. Czech Statistical Office, 2015. *Statistical Yearbook of the Czech Republic—2015*.
European Commission 2010. *How to develop a Sustainable Energy Action Plan (SEAP)*. Luxembourg. Publications Office of the European Union.
IPCC 2006. Intergovernmental Panel on Climate Change. *Emission factors database 2006*. Available from <http://www.ipcc-nggip.iges.or.jp/EFDB/find_ef_s1.php> [28 January 2016].
IS 13600:1997. International Organization for Standardization. 1997. ISO 13600:1997 Technical energy systems. Available from < https://www.iso.org/obp/ui/#iso:std:iso:13600:ed-1:v1:en> [8 January 2016].
JRC, 2015. The Joint Research Centre, 2015. Institute for Energy and Institute for Environment and Sustainability. Available from <https://ec.europa.eu/jrc/en/institutes/iet> [28 January 2016].
Rohlena, M. & Frková, J. 2012. *SEAP of City Hlinsko*. CTU Prague.
Rohlena, M. & Frková, J. 2013. *Energy Consumption Methodology for Municipal Territory*. CTU Prague.
RSD 2010. Directorate of Roads and Motorways 2010. *Výsledky celostátního sčítání dopravy ČR v roce 2010*. Available from <http://scitani2010.rsd.cz/pages/results/default.aspx> [20 September 2015].

Turning grey into green – integration of living walls in grey water system

M. Rysulová, D. Kaposztasová & Z. Vranayová
Civil Engineering Faculty, Technical University of Kosice, Slovakia

J. Castellar & J. Morato
UNESCO Chair on Sustainability, Polytechnic University of Catalonia-BarcelonaTech, Spain

ABSTRACT: Living walls are nowadays becoming an essential part of building indoor or outdoor architecture. Through their ability of managing water, living walls are offering a considerable potential in waste water treatment technologies, also suitable for grey water. Integration of living walls into grey water system is relatively new concept. Design of functional grey water system using living wall, requires individual system parts specification, considering the treatment ability and options in addition to add this system only as a secondary treatment or focus on the characteristics improvement and apply only living walls treatment. According to the wide range of factors influencing the design of this technology, we focused on the system layout and variables affecting the system design.

1 INTRODUCTION

Water is considered as an essential source of living. No matter how scarce or abundant this source is, it is used for daily different activities and there is no doubt that growing demand for water is influencing and increasing the impacts of its quality. Therefore the current water management is mainly focusing on the necessity of more sustainable practices development and efficient use of water resources, as well as to focus on the protection of the ecosystems (WWDR, 2015; Vorosmarty et al., 2005; UNEP & WHO, 1996). Nonconventional water resources, such as water reuse and desalination, are being increasingly used and becoming more and more common as a tool of sustainable management (OECD, 2009). The main aim of these approaches is to substitute low-quality water for non potable purposes to preserve limited high-quality water resources, especially in arid regions or developing countries. The Vision 2050 of the treatment with water in sustainable world: "Water is duly valued in all its forms, with wastewater treated as a resource that avails energy, nutrients and freshwater for reuse. Human settlements develop in harmony with the natural water cycle and the ecosystems that support it, with measures in place that reduce vulnerability and improve resilience to Water-related Disasters." (WWDR, 2015). As an option of water management improvement in urban areas and the way how to reach this vision, could be ensured through water recycling, which nowadays usually integrates black, grey or rain water reuse systems (Pidou et al., 2008). According to, Ragheb, El-Shimy, & Ragheb (2016), "the protection and conservation of water throughout the life of a building may be accomplished by designing for dual plumbing that recycles water". However the grey water is considered as a light polluted wastewater, different types of contaminants can be presented in this type of water. The concentration of different contaminants, highlights the potential of health or environmental risks. These risks can be minimized by treating grey water to an appropriate quality, according to further utilization of this water (EHD, 2010). In regards to availability of water source suitable for reuse and light pollution rate, our main focus is on grey water systems, which can save particular part of potable water used for non potable purposes (EA, 2011). According to wide range of living walls utilization forms, the main aim of this paper is to introduce this system also in grey water treatment as an appropriate technology, which is able to ensure the required water quality, whether for different type of further reuse or for discharge this water back into environment.

Abbreviations used in text:

BOD5—Biochemical oxygen demand, CFU—Colony forming unit, COD—Chemical oxygen demand, FC—Faecal coliforms, ph—Potential hydrogen, TC—Total coliforms, TSS—Total suspended solids.

2 GREY WATER TREATMENT SYSTEMS

Grey water treatment technologies must be well designed to handle variations of pollution concentration in grey water and to consistently produce effluent of an appropriate and safe quality to manage the water and to meet required standards for reuse (Winward et al., 2007). In order to manage water quality, there are no international regulations for wastewater reuse, however, to preserve the potential health and environmental risks, many countries have individually produced their own guidelines (Pidou et al., 2007), some of the values are established only for reusing waste water in general and are not specified on grey water reuse. In Table 1 are shown limitations of contaminants concentration in different countries, according to reuse purpose.

In regards to establish the suitable and safe grey water reuse, have pollutants to be eliminated by appropriate treatment process. Treating wastewater properly before disposal or reuse is not a luxury but usually a necessity. Important is to define the quality of discharged grey water, along with required water quality according to further utilization.

There are numbers of different grey water treatment approaches, range from simple, low-cost devices or advanced treatment processes incorporating sedimentation tanks, bioreactors, filters, pumps and disinfections units (FBR, 2007). The treatment technologies, usually used for grey water system applications are based on physical, chemical and biological systems, or the combinations of them (Li et al., 2009), however as the most efficient systems are considered biological systems in combination with physical treatment technologies (FBR, 2007). As option of combined method in grey water treatment technologies can be considered relatively new concept in wastewater treatment—green walls or also called living wall systems, which are applied on empty spaces of building facades and besides the potential in the wastewater

Table 1. Limited values of contaminants concentration in different countries (modified from Li et al., 2009; Pidou et al., 2007; Slovak Government, 2005).

Country	ph (–)	BOD5 (mg/l)	COD (mg/l)	Turbidity (NTU)	TC (CFU/ 100 ml)	FC (CFU/ 100 ml)	Reuse
Slovakia*	5.5–9.0	5.0–7.0	8.0–35	–	100	–	– potable water, irrigation
Germany	–	5.0	–	1–2	<100	<30	– toilet flushing
Spain	–	10	–	2	2.2	–	– waste water reuse
USA	6.0–9.0	10	–	–	–	<200	– irrigation
Canada	–	10	–	2	2.2	–	– irrigation
Australia	–	20	–	–	100	–	– irrigation
Israel	–	10	–	–	–	<1	– waste water reuse
China	6.0–9.0	<10	–	<5	–	<3/100	– toilet flushing
Japan	5.8–8.6	<10	–	–	<1000	–	– toilet flushing

*Limited features established for freshwater sources used for potable purposes, irrigation and discharge into rivers.

treatment, can bring benefits in improving the face of urban areas (Sheweka & Magdy, 2011; Svete, 2012). Except the aesthetical benefits and ability to manage water, there are more advantages in living walls applications as reduction of buildings energy consumption and urban heat island, improvement of air quality, reclaiming urban wastelands etc (Hopkins and Goodwin, 2011) what makes this technology attractive.

2.1 *Conventional treatment systems*

As mentioned above, most common treatment technologies applied for grey water are based on the biological processes with addition of physical pretreatment or screening (Li et al., 2009). Focusing on these types of technologies, there are several most known treatment systems, including rotating biological contactor (Abdel-Kader, 2012) with more than 70% reduction of COD, BOD and turbidity (Friedler et al., 2005), sequencing batch reactor (Lamine et al., 2007) with 90% COD removal (Lamine et al., 2007), membrane bioreactor (Santasmasas et al., 2013), which in study of (Merz et al., 2007) approved more than 85% of COD, BOD and turbidity removal. Another treatment technology based on biological processes with relatively high pollutant removal are constructed wetlands (Gross et al., 2007), which are imitating the processes and ability in treating wastewater in natural wetlands (Vymazal, 2014) and according to study of (Gross et al., 2007) are able to remove more than 80% of COD, BOD and TSS from influent grey water.

2.2 *Living walls*

Living walls are representing the scale of indoor and outdoor walls covered by plants, also known as vertical gardens, green walls or biowalls (Hopkins and Goodwin 2011), which dispose with number of benefits following by environmental, economical and social, not just for the buildings, but for the whole urban areas (Sheweka & Magdy, 2011). There are available numbers of information about the living walls as a tool for reduce urban heat, island effect, improve carbon sequestration, thermal insulation and ensure energy savings by integration of more green spaces in urban areas (Coma et al., 2014; Cheng et al., 2010; Weinmaster, 2009; Marchi et al., 2015; Hoelscher et al., 2015).

However, considering that "Today, wastewater is a major factor for freshwater quality and human health" (Malik et al., 2015) and that there is a lack of studies dealing with utilization of the living walls as a wastewater treatment technology (Svete, 2012) underlined the importance of the concept of this study. According to Zhang (2013) "If cities are to become sustainable, they must reduce their use of all resources and decrease their waste outputs".

In regards to integration of living walls into grey water system, could be the grey water considered as a source of water for irrigation of living walls and walls can serve as a filter for this water before discharging or reusing (Weinmaster, 2009). In study of (Elmasry & Haggag, 2011) the grey water was used as a source for irrigation of vegetated walls, but not aiming achieve water quality for a further reuse but just as a water supply to reduce high quality potable water demand for this propose.

In order to design the grey water recycle process, there is an option to develop the wall as a linear wetland incorporated into wall system (Hopkins and Goodwin, 2011) and according to proved abilities and advantages of constructed wetlands in grey water treatment (Avery et al., 2007), there is a huge potential to integrate these features into the living walls system and create a well functioning treatment technology, without the large area requirement unlike for constructed wetlands (Wu et al., 2014). The wall construction treating a grey water could be designed whether in separated containers or in continual wall constructions (Figure 1.).

Besides the main construction supporting the wall, there are important features, which are influencing the functionality of the treatment system and efficiency of it, which could be considered as a crucial, especially filter media and vegetation selection (Svete, 2012). In these sense, which regards to build a living wall system a especial attention need to be given to species selection which regards to climate condition, water requirements, growing media as also to the their adaptation to the structure design (Victoria, 2013).

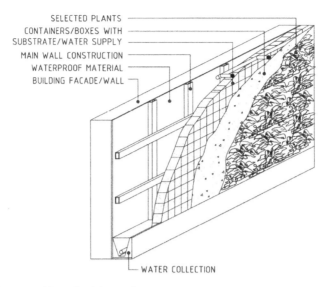

SELECTED PLANTS
CONTAINERS/BOXES WITH SUBSTRATE/WATER SUPPLY
MAIN WALL CONSTRUCTION
WATERPROOF MATERIAL
BUILDING FACADE/WALL

WATER COLLECTION

Figure 1. Basic composition of a living wall system construction.

3 FUTURE RESEARCH

In regards to the water treatment could be stated that "In the field of green wall character-istics, the main concerns are to find new strategies for a better performance and durability through the integration of water retention materials, drainage means and simpler assembly and maintenance processes." (Manso & Castro-Gomes, 2014), what confirms that there is a necessity to provide more researches dealing with integration of the living walls in water treatment management.

As mentioned above, the living walls have potential as an advanced technology in grey water treatment, however the system is still developing. Therefore the proposal of the future research is focused on the design of the living wall, based on the constructed wetland treatment prin-ciples, which ability of contaminants removal have been proved by different studies (Comino et al., 2013; Gross et al., 2007; Zhang et al., 2014). According to the concept of the walls as an integrated natural and sustainable element in urban areas, there is an intention to utilize the natural or waste materials as a wall components. However validation of the living walls proto-type will be according to ability to improve grey water quality for further reuse or discharge in comparison to currently most used biological treatment technology—membrane bioreactor. Real grey water discharged from the devices will be used as a source for the experimental tests.

4 CONCLUSION

Creating a new face of the urban areas incorporated with the nature could make living walls being considered as a tool of the future concept for green urbanization. Besides new architec-tural design living walls could provide, there are number of benefits, which could be applied as a sustainable management tool. Extensive conception of the technology is also point-ing on the potential risks which can occur during the system operation. This system brings together the advantages of treatment processes typical for constructed wetlands and applied them into the living wall system. However the combination of constructed wetland and living wall can play an important role which concerns the improvement of grey water treatment in sustainable way. Potential of these systems could be still developed and optimized through further investigations, what represent focus of our future research.

ACKNOWLEDGEMENTS

This work was supported by projects VEGA n. 1/0202/15: Sustainable and Safe Water Management in Buildings of the 3rd Millennium.

REFERENCES

Abdel-Kader, A. M. (2012). Studying the efficiency of grey water treatment by using rotating biological contactors system. *Journal of King Saud University—Engineering Sciences, 25*(2), 89–95. http://doi.org/10.1016/j.jksues.2012.05.003.

Avery, L. M., Frazer-Williams, R. a. D., Winward, G., Shirley-Smith, C., Liu, S., Memon, F. a. & Jefferson, B. (2007). Constructed wetlands for grey water treatment. *Ecohydrology & Hydrobiology, 7*(3–4), 191–200. http://doi.org/10.1016/S1642–3593(07)70101-5.

Coma, J., Pérez, G., Solé, C., Castell, A. & Cabeza, L. F. (2014). New green facades as passive systems for energy savings on Buildings. *Energy Procedia, 57*, 1851–1859. http://doi.org/10.1016/j.egypro.2014.10.049.

Comino, E., Riggio, V. & Rosso, M. (2013). Grey water treated by an hybrid constructed wetland pilot plant under several stress conditions. *Ecological Engineering, 53*, 120–125. http://doi.org/10.1016/j.ecoleng.2012.11.014.

Elmasry, S. K. & Haggag, M. A. (2011). Whole-building design for a green school building in Al-Ain, United Arab Emirates. *WIT Transactions on Ecology and the Environment, 150*, 165–176. http://doi.org/10.2495/SDP110151.

Environment Agency. (2011). *Greywater for domestic users: an information guide*.

Environmental Health Directorate. (2010). *Code of Practice for the Reuse of Greywater in Western Australia*. Retrieved from http://www.public.health.wa.gov.au/cproot/1340/2/COP Gretwater.pdf.

FBR. (2007). Greywater Low-load. *Greywater Recycling and Resue*. Retrieved from http://www.fbr.de/greywaterrecycling.html.

Friedler, E., Kovalio, R. & Galil, N. I. (2005). On-site greywater treatment and reuse in multi-storey buildings. *Water Science and Technology, 51*(10), 187–194.

Gross, A., Shmueli, O., Ronen, Z. & Raveh, E. (2007). Recycled Vertical Flow Constructed Wetland (RVFCW)-a novel method of recycling greywater for irrigation in small communities and households. *Chemosphere, 66*(5), 916–923. http://doi.org/10.1016/j.chemosphere.2006.06.006.

Hoelscher, M.-T., Nehls, T., Jänicke, B. & Wessolek, G. (2015). Quantifying cooling effects of facade greening: shading, transpiration and insulation. *Energy and Buildings*. http://doi.org/10.1016/j.enbuild.2015.06.047.

Hopkins, G. & Goodwin, C. (2011). Living architecture: Green roofs and walls. Collingwood, Australia: CSIRO Publishing.

Cheng, C. Y., Cheung, K. K. S. & Chu, L. M. (2010). Thermal performance of a vegetated cladding system on facade walls. *Building and Environment, 45*(8), 1779–1787. http://doi.org/10.1016/j.buildenv.2010.02.005.

Lamine, M., Bousselmi, L. & Ghrabi, A. (2007). Biological treatment of grey water using sequencing batch reactor. *Desalination, 215*(1–3), 127–132. http://doi.org/10.1016/j.desal.2006.11.017.

Li, F., Wichmann, K. & Otterpohl, R. (2009). Review of the technological approaches for grey water treatment and reuses. *Science of the Total Environment, The, 407*(11), 3439–3449. http://doi.org/10.1016/j.scitotenv.2009.02.004.

Malik, O. A., Hsu, A., Johnson, L. A. & de Sherbinin, A. (2015). A global indicator of wastewater treatment to inform the Sustainable Development Goals (SDGs). *Environmental Science and Policy, 48*, 172–185. http://doi.org/10.1016/j.envsci.2015.01.005.

Manso, M. & Castro-Gomes, J. (2014). Green wall systems: A review of their characteristics. *Renewable and Sustainable Energy Reviews, 41*, 863–871. http://doi.org/10.1016/j.rser.2014.07.203.

Marchi, M., Pulselli, R. M., Marchettini, N., Pulselli, F. M. & Bastianoni, S. (2015). Carbon dioxide sequestration model of a vertical greenery system. *Ecological Modelling, 306*, 46–56. http://doi.org/10.1016/j.ecolmodel.2014.08.013.

Merz, C., Scheumann, R., El Hamouri, B. & Kraume, M. (2007). Membrane bioreactor technology for the treatment of greywater from a sports and leisure club. *Desalination, 215*(1–3), 37–43. http://doi.org/10.1016/j.desal.2006.10.026.

OECD. (2009). *Alternative Ways of Providing Water: Emerging Options and Their Policy Implications. OECD Reports.*

Pidou, M., Avery, L., Stephenson, T., Jeffrey, P., Parsons, S. A., Liu, S. & Jefferson, B. (2008). Chemical solutions for greywater recycling. *Chemosphere*, *71*(1), 147–155. http://doi.org/10.1016/j.chemosphere.2007.10.046.

Pidou, M., Memon, F. A., Stephenson, T., Jefferson, B. & Jeffrey, P. (2007). Greywater recycling: A review of treatment options and applications. *Engineering Sustainability*, *160*, 119–131. http://doi.org/10.1680/ensu2007.160.3.11.

Ragheb, A., El-Shimy, H. & Ragheb, G. (2016). Green Architecture: A Concept of Sustainability. *Procedia—Social and Behavioral Sciences*, *216* (October 2015), 778–787. http://doi.org/10.1016/j.sbspro.2015.12.075.

Santasmasas, C., Rovira, M., Clarens, F. & Valderrama, C. (2013). Grey water reclamation by decentralized MBR prototype. *Resources, Conservation and Recycling*, *72*, 102–107. http://doi.org/10.1016/j.resconrec.2013.01.004.

Sheweka, S. & Magdy, N. (2011). The living walls as an approach for a healthy urban environment. *Energy Procedia*, *6*, 592–599. http://doi.org/10.1016/j.egypro.2011.05.068.

Slovak Government. (2005). *Slovak Government Regulation NV SR 296/2005*. Retrieved from www.zbierka.sk.

Svete, L. E. (2012). *Vegetated greywater treatment walls: Design modifications for intermittent media filters*.

UNEP, & WHO. (1996). Water Quality Monitoring—A Practical Guide to the Design and Implementation of Freshwater Quality Studies and Monitoring Programmes Edited Chapter 2—Water Quality (pp. 0–419).

Victoria, S. G. (2013). *Victoria's Guide To Green Roofs, Walls & Facades*.

Vorosmarty, C. J., Bos, R. & Balvanera, P. (2005). Fresh Water. In *Ecosystems and Human ....* Retrieved from http://books.google.com/books?hl=en&lr=&id=UFVmiSAr-okC&oi=fnd&pg=&dq=Fresh+Water+vorosmarty&ots=l7JzOW0ZSp&sig=PWAkL83Eq4Ltx2iWqy665MxXzXU\npapers3://publication/uuid/93F4835B-95CC-487B-A90D-18786F9394C0.

Vymazal, J. (2014). Constructed wetlands for treatment of industrial wastewaters: A review. *Ecological Engineering*, *73*, 724–751. http://doi.org/10.1016/j.ecoleng.2014.09.034.

Weinmaster, M. (2009). Are green walls as "green" as they look? an introduction to the various technologies and ecological benefits of green walls. *Journal of Green Building*, *4*(4), 1–18. http://doi.org/http://dx.doi.org/10.3992/jgb.4.4.3.

Winward, G. P., Avery, L. M., Frazer-williams, R., Pidou, M., Jeffrey, P., Stephenson, T. & Ab, A. (2007). A study of the microbial quality of grey water and an evaluation of treatment technologies for reuse, *2*, 187–197. http://doi.org/10.1016/j.ecoleng.2007.11.001.

Wu, S., Kuschk, P., Brix, H., Vymazal, J. & Dong, R. (2014). Development of constructed wetlands inperformance intensifications for wastewater treatment: A nitrogen and organic matter targeted review. *Water Research*, *57*, 40–45. http://doi.org/10.1016/j.watres.2014.03.020.

WWDR. (2015). *Water for a sustainable world*. http://doi.org/10.1016/S1366-7017(02)00004-1.

Zhang, D. Q., Jinadasa, K. B. S. N., Gersberg, R. M., Liu, Y., Ng, W. J. & Tan, S. K. (2014). Application of constructed wetlands for wastewater treatment in developing countries—A review of recent developments (2000–2013). *Journal of Environmental Management*, *141*, 116–131. http://doi.org/10.1016/j.jenvman.2014.03.015.

Zhang, X. (2013). Going green: Initiatives and technologies in Shanghai World Expo. *Renewable and Sustainable Energy Reviews*, *25*, 78–88. http://doi.org/10.1016/j.rser.2013.04.011.

Advances and Trends in Engineering Sciences and Technologies II – Al Ali & Platko (Eds)
© *2017 Taylor & Francis Group, ISBN 978-1-138-03224-8*

Evaluation of test methods for testing of sealants

J. Šlanhof, P. Liška & B. Nečasová
Faculty of Civil Engineering, Brno University of Technology, Brno, Czech Republic

ABSTRACT: The presented paper deals with methods of sealants testing, where the authors summarize their experience in a given field. It is a summary of four years of research focusing on the evaluation of sealant interaction with underlay materials. Focus is placed on official test methods including European technical standards; the benefits of individual methods, their advantages and disadvantages are discussed. Reviewing conclusions of practical application of test procedures applied for 10 commonly available industrially produced sealants on 8 various underlays, including e.g. façade tiling boards used for the systems of light building envelopes, are also a part of the presented paper. Moreover, the article includes general conclusions regarding individual test methods. Parts of the research include different methods of preparation and conditioning of test bodies that simulate the conditions of environmental effects and, in some cases, they are quite decisive for the final evaluation of the sealants.

1   INTRODUCTION

1.1   *The importance of sealants in modern building industry*

Joint sealants are materials applied in a condition not shaped in advance; after it is hardened or dried it can seal a joint by its adhesive or cohesive properties (EN ISO 6927, 2012). So the main task of sealants is the sealing function and this function is currently heavily used in building industry. In case of joint sealing in external constructions imposed to weather effects, the sealant used has a direct effect on protection of the building against damage due to leakage. Thus, it becomes an important factor of long-term functionality and reliability of the entire building, although it is basically only a solution of small extent of slits in building envelope. However, not only protection of humidity leakage but also penetration of insects or noise are concerned—proper sealing of all slits is also the basic presumption for improving resistance to undesired noise (Kantová & Řihák 2014), (Kantová & Motyčka 2014). Sealants are used in the construction of roofs (Selník et al. 2014). Usage of sealants within surface treatment of interiors when e. g. joining of various materials in corners is required, is one of the most typical and frequent areas. In many cases the sealants are used rather as adhesives for creating a construction joining that must also ensure necessary tightness.

1.2   *The issue of reliability of sealed joints*

If the amount of currently available sealants and the scope of their application are taken into account, it could easily be concluded that a wrong choice will relatively soon become evident by defects. In better cases, they will be aesthetic defects, when a hidden joint opens but does not cause any other economic damage. In worse case scenarios, penetrating humidity will cause secondary damages, elimination of which will require further costs in addition to the means required for repair of the damaged sealed joint; then the overall costs can be very high.

Sealants always have limited durability and are subject to ageing. This applies especially for applications in external environment when they are subjected to permanently changing climatic events. In many cases, defects due to incorrect use, or due to insufficient preparation of the underlay, occur. Often, even despite of proper preparation and correct application, a defect occurs; even if a sealant designated for a given underlay and environment was applied. The defect is usually the loss of adhesion or cohesion to the underlay. That is why it is appropriate to be concerned with the issue of reliability of sealed joints, which is directly connected with the method for testing sealants. The paper will deal with the assessment of the test methods together with practical applications.

2 TEST METHODS FOR TESTING SEALANTS, RESULTS AND ASSESSMENT

Identical test samples are the common base of all test methods. Current European standards are anchored to a test body consisting of two underlay boards interconnected by the tested sealant. The primarily considered material is mortar, aluminium or glass, the procedure is, however, also specifically applicable for any other underlay materials, which is a crucial assumption for verification whether the tested sealant is sufficiently reliable on a certain underlay. By application of the test method for various underlay materials it can be revealed, for which underlays the sealant is suitable and for which it is quite unsuitable. The geometry of the test samples is defined by relevant technical standards. However, they also allow to change the dimensions of underlay bodies when the dimensions of 12 x 12 x 50 mm of the tested sealant remain kept. Therefore, the standardized sample geometry was modified, see Figure 1. The test body with this modification should provide better properties as is e.g. higher resistant to breakage of underlay boards when strained during the tests. During the research, 1980 experiments have been performed and the test sample in accordance with Figure 1 has proved to be successful—the underlay boards broke not a single time.

Within the four-year research focused on reliability of sealed joints, the team of authors performed many experiments and proceeded in accordance with current European technical standards for testing sealants. The purpose of the research was not to classify the sealants themselves. The aim was to find the most suitable sealant for application on 8 particular underlays. Porous, e. g. high—pressure compact laminate or cold—rolled weather—resistant steel, as well as non—porous cladding materials, e. g. heat—treated Finnish pine, engineered stone or fibre—cement cladding boards, were tested in combination with selected sealants. Particular individual results of the tests of sealants have already been published (Nečasová et al. 2014, Liška et al. 2015, Wolf 1999, Chew & Zhou 2001, Chew 2003), nobody, however, has published generalised results focussed on assessment of the current test methods as is provided in the following chapters of the paper.

2.1 Determination of tensile properties at maintained extension

The test procedure is based on European standard EN ISO 8340: 2005. The principle of the test is extension of the tested sealant to pre-defined elongation, where the extension is maintained for 24 hours under certain conditions. The results are records of adhesion or cohesion defects for each test body; the results are stated in millimetres.

Within the assessment of the benefits of the method, the team of authors carried out 240 measurements at the temperature of $23 \pm 2°C$ and then 240 measurements at the temperature of $-20 \pm 2°C$ at consistent extension of the sealant by 100%. The 240 samples consisted of 80 different combinations comprising of 10 kinds of sealants and 8 types of underlay materials, where 3 test samples were tested for each of the combinations. The test results, in terms of the number of defects, are summarized in Table 1.

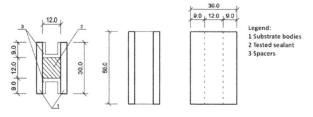

Figure 1.   Test sample.

Three measurements were carried out for each combination of the sealant and underlay material, in case any adhesion or cohesion defect occurred, even if only in one measurement, given combination was indicated as unsatisfactory. The test proved to be very effective, it excluded most samples. The results, in terms of the determination of unsuitable combinations of sealant/underlay, are summarized in Table 2.

2.2 Determination of tensile properties of sealants at maintained extension after immersion in water

The test procedure is based on European standard EN ISO 10590: 2005. Preparation of test bodies and principle of test is the same as in chapter 2.1, only before the test, the samples are put in a bath with water at the temperature of $23 \pm 2°C$ for 5 days. Within assessment of the benefits of this method, the team of authors performed 240 measurements consisting of 80 combinations of different sealants and underlays, always 3 samples at 100% extension of the sealant. The test results, in terms of the numbers of defects, are summarized in Table 3, in terms of determination of unsuitable combinations of sealant/underlay, in Table 4.

The test proved even more effective for evaluation of defects than the previous method, though the only difference was immersing the test bodies into water for 5 days. Thus, even more unsuitable combinations of sealants and underlay materials were excluded.

2.3 Determination of adhesion and cohesion properties of sealants at variable temperatures

The test procedure is based on European standard EN ISO 9047: 2003. The principle of the test is slowly changing cyclical strain by compression and extension in optional interval of the amplitude of 12.5%, 20% or 25%. During the test, maintained extension and compression alternate and subsequently the temperature changes in jumps in the values of $-20 \pm 2°C$ and $70 \pm 2°C$ for 2 weeks.

Within the search for benefits of the method, 240 measurements at 20% amplitude of extension and compression of sealant were carried out. The test results, in terms of the number of

Table 1. Number of defects during the test of sealants' tensile properties at maintained extension.

Test temperature	Number of samples	Number of defects	Number of unsatisfactory
°C	Piece	Piece	%
+23 ± 2	240	128	53.3
−20 ± 2	240	121	50.4

Table 2. Numbers of unsuitable combinations of sealants and underlays during the test of tensile properties of sealants at maintained extension.

Test temperature	Number of combinations	Number of defects	Number of unsatisfactory
°C	Piece	Piece	%
+23 ± 2	80	56	70.0
−20 ± 2	80	55	68.8

Table 3. Number of defects while testing tensile properties of sealants at maintained extension after immersing in water.

Test temperature	Number of samples	Number of defects	Number of unsatisfactory
°C	Piece	Piece	%
+23 ± 2	240	139	57.9

defects, are summarized in Table 5, in terms of determination of unsuitable combinations of sealant/underlay, in Table 6.

During the test not as many defects occurred as in previous types of tests. It is evident that it is due to significantly lower level of elongation. When defects occurred, it was particularly in the phase of extension, compression generally did not lead to defects.

2.4 Determination of adhesion/cohesion properties of sealants at constant temperature

The test procedure is based on European standard EN ISO 9046: 2002. The principle of the test is 100 cycles of gradually changing alternate extension and compression in optional interval of the amplitude of 7.5% or 12.5% at constant temperature of $23 \pm 2°C$. The results are records of adhesion or cohesion defects.

Within the method application, 240 measurements at the amplitude of extension and compression of sealant of 12.5% were carried out. Test results, in terms of the numbers of

Table 4. Number of unsuitable combinations of sealants and underlays while testing tensile properties of sealants at maintained extension after immersing in water.

Test temperature	Number of combinations	Number of defects	Number of unsatisfactory
°C	Piece	Piece	%
+23 ± 2	80	61	76.3

Table 5. Number of defects during the test of adhesion and cohesion of sealants at variable temperature.

Test temperature	Number of samples	Number of defects	Number of unsatisfactory
°C	Piece	Piece	%
−20 ± 2, +70 ± 2	240	109	45.4

Table 6. Number of unsuitable combinations during the test of adhesion and cohesion of sealants at variable temperature.

Test temperature	Number of combinations	Number of defects	Number of unsatisfactory
°C	Piece	Piece	%
−20 ± 2, +70 ± 2	80	51	63.8

Table 7. Numbers of defects while testing adhesion and cohesion of sealants at constant temperature.

Test temperature	Number of samples	Number of defects	Number of unsatisfactory
°C	Piece	Piece	%
+23 ± 2	240	28	11.7

Table 8. Numbers of unsuitable combinations of sealants and underlays at test of adhesion and cohesion of sealants at constant temperature.

Test temperature	Number of combinations	Number of defects	Number of unsatisfactory
°C	Piece	Piece	%
+23 ± 2	80	15	18.8

defects, are summarized in Table 7, in terms of determination of unsuitable combinations of sealant/underlay, in Table 8.

During the test, only some defects occurred, which is given by the laboratory temperature, preparation of samples without immersing in water environment and low amplitude of strain. For most assessed sealants, it was no problem to resist the applied strain without any damage.

2.5 Determination of tensile properties

In this chapter, two test methods can be mentioned. The initial test procedure is based on European standard EN ISO 8339: 2005. The principle of the test is elongation of sealant joint until it breaks. The tensile strength is recorded and the dependence of sealant joint elongation on applied force can be formulated into the graph. This test was carried out in two versions, namely at the temperature of $23 \pm 2°C$ and $-20 \pm 2°C$. The result for selected expansion or expansion at break can be presented by secant modulus that is calculated by the formula (1).

$$\sigma = \frac{F}{s} \tag{1}$$

where σ = secant tensile modulus [N/mm^2]; F = force at selected expansion [N]; and s = initial cross-sectional surface of the test body [mm^2].

Similar approach is included in EN ISO 10591 when the diagram of dependence of expansion on applied force is recorded, however, the test samples are immersed in water for 4 days before the tensile test is carried out. The result is the value of extension calculated by formula (2).

$$\text{Extension [\%]} = [(\text{final length—initial length})/\text{initial length}] \times 100 \tag{2}$$

Both tensile tests provide interesting knowledge of behaviour of a sealant on given underlay. Thanks to recorded course of deformations depending on the applied force, complete information on behaviour of the sealant in the interval of all its deformability is acquired.

The tests define properties of particular sealant in interaction with particular underlay, so they can be a very good guide when comparing sealants among themselves in case the most suitable type of sealant needs to be chosen for the underlay material. An adhesion or cohesion defect always occurs during the test; that is why this method cannot be assessed according to the number of defects as the previous one is. It is, however, evident that knowledge of deformability is an important characteristic of each sealant.

2.6 Determination of resistance to compression

The test is carried out according to EN ISO 11432 and it meaning lies in compression of tested sealants to 75% or 80% of their initial width, where the force required for given compression is recorded. Within the research, 60 measurements were carried out. No other measurements were carried out because this test has not provided any crucial information. All tested sealants survived the compression harmless, no defects that would provide any information on suitability of sealants for particular underlay occurred.

3 CONCLUSION

The presented text summarizes the results of individual test procedures that were applied during the four years of research of the authors' team focused on testing sealants in interaction with underlay. With regard to the number of completed experiments the authors can assess the existing test procedures for testing sealants. It became evident that not all sealants are optimal for selected underlays. If large extent of joint sealing with the risk of significant damage is concerned, it could be advised to have the sealant considered on particular underlay tested in the entire set of almost all tests described above. Only then the choice of the most suitable sealant can be guaranteed.

To illustrate possible results of such approach, results of the research can be presented: 8 different types of underlay material were available and the objective was to find the optimal type out of 10 assessed sealants. The only type of sealant (every time it was a different product) was satisfactory for 6 underlay materials; there was only 1 underlay, for which 2 types of sealants complied, and for 1 underlay no sealant out of the 10 sealants assessed complied. Complying sealant is characterized by the fact that no defects occurred in it during the tests according to chapters 2.1 to 2.4. The tests of maintained extension described in chapters 2.1 to 2.3 were of the greatest importance for final assessment. Other tests can be assessed as supplementary but not decisive.

Besides the method of strain, the way the test bodies were stored before the tests, played an important role too. Particularly when moisture—absorbing underlay materials are tested, the results are significantly influenced by previous exposure of the test bodies to water. In such materials, it is important to comply with standard's preparation requirements of the test samples since in some tested cases, the samples disintegrate themselves just during handling, even before the application of the test procedure.

ACKNOWLEDGEMENT

This research was carried out as a part of the FAST-S-16–3434 project, also as a part of the MPO FR-TI4/332 project and as part of a project FAST-J-15-2728.

REFERENCES

European Standard EN ISO 6927: 2012. Buildings and civil engineering works—Sealants—Vocabulary.
European Standard EN ISO 8339: 2005. Building construction—Jointing products—Sealants—Determination of tensile properties (Extension to break).
European Standard EN ISO 8340: 2005. Building construction—Sealants—Determination of tensile properties at maintained extension.
European Standard EN ISO 9046: 2002. Building construction—Jointing products—Determination of adhesion/cohesion properties of sealants at constant temperature
European Standard EN ISO 9047: 2003. Building construction—Jointing products—Determination of adhesion/cohesion properties of sealants at variable temperatures.
European Standard EN ISO 10590: 2005. Building construction—Sealants—Determination of tensile properties of sealants at maintained extension after immersion in water.
European Standard EN ISO 10591: 2005. Building construction—Sealants—Determination of adhesion/cohesion properties of sealants after immersion in water.
European Standard EN ISO 11432: 2005. Building construction—Sealants—Determination of resistance to compression.
Chew, M. Y. L. & Zhou, X. 2001. Enhanced resistance of polyurethane sealants against cohesive failure under prolonged combination of water and heat. *Polymer Testing* Vol. 21. Published by Elsevier Science Ltd. in 2001. PII: S0142–9418(01)00068-X. pp. 188–193.
Chew, M. Y. L. 2003. The effects of some chemical components of polyurethane sealants on their resistance against hot water. *Building and Environment* Vol. 38. Published by Elsevier Science Ltd. in 2003. doi: 10.1016/S0360–1323(03)00142–2. pp. 1382–1384.
Kantová, R. & Řihák, P. 2014. Noise Motor Working Tools and Their Measures for Noise Reduction During Construction of Buildings. In Electronic conference proceedings enviBuild 2014. *Advanced Materials Research*. Switzerland: Trans Tech Publications Ltd, 2014. pp. 424–427. ISBN: 978-80-214-5003-5. ISSN: 1022-6680.
Kantová, R. & Motyčka, V. 2014. Construction Site Noise and its Influence on Protected Area of the Existing Buildings. In Electronic conference proceedings enviBuild 2014. *Advanced Materials Research*. Switzerland: Trans Tech Publications Ltd, 2014. pp. 419–423. ISBN: 978-80-214-5003-5. ISSN: 1022-6680.
Liška, P., Nečasová, B., Šlanhof, J. & Šimáčková, M. 2015. Determination of Tensile Properties of Selected Building Sealants in Combination with High-pressure Compact Laminate (HPL). *Procedia Engineering*, 2015, vol. 2015, no. 108, pp. 199–205. ISSN: 1877-7058.
Nečasová, B., Liška, P., Šlanhof, J. & Šimáčková, M. 2014. Test of Adhesion and Cohesion of Silicone Sealants on Facade Cladding Materials within Extreme Weather Conditions. *Advanced Materials Research*, 2014, vol. 2014(1041), no. 23–26, pp. 23–26. ISSN: 1022-6680.
Selník, P., Nečadová, K. & Mohapl, M. 2014. Technological Analysis of the Orientation of the Pitched Icelandic Turf Roof to the Cardinal Points and their Effect on the Stability of the Vegetative Layers. *Advanced Materials Research*, 2014, vol. 2014(1041), no. 19–22, pp. 19–22. ISSN: 1022-6680.

Advances and Trends in Engineering Sciences and Technologies II – Al Ali & Platko (Eds)
© *2017 Taylor & Francis Group, ISBN 978-1-138-03224-8*

Thermal assessment of chosen foundation details of energy-efficient family houses constructed in Kosice, Slovakia

P. Turcsanyi
Institute of Architectural Engineering, Kosice, Slovakia

A. Sedlakova
Department of Building Physics, Kosice, Slovakia

ABSTRACT: Computer programs that allow determining accuracy of building design or its parts in advance from the thermal engineering point of view are on the rise. Finding correct solution for building (foundations from the heat-loses point of view) helps streamlining a process of energy utilization and significantly contributes to implementation of European Directive on Energy Performance of Buildings 2010/31/EU in Slovakia (also known as 20-20-20 Directive).

1 INTRODUCTION

The word "energy" gained a strong meaning over the last few years. Energy is among the principal factors of the social and economic development of our society, which is dealing with important issues, such as politics and the environment (V. Belpoliti, G. Bizzarri, 2015). An increasing interest in many aspects related to buildings energy efficiency has led to a growing amount of research and studies. Some of these aim at investigating the economic and financial feasibility of energy efficiency measures currently applied in the building sector, as well as deepening to what extent the energy performance of buildings could be able to affect the market price or the rent of real estate units (S. Copiello, 2015). One of the basic way of saving heat loses is sufficient building envelope insulation, however, efficiency of thermal insulation has its limits. Results indicate that adding insulation is not always beneficial, and thus in particular in the regions of Mediterranean climate as susceptible to anti-insulation behavior (W. A. Friess, K. Rakhshan, M. P. Davis, 2016).

2 EVALUATION OF A FAMILY HOUSE "A"

Family house, constructed in Košice, Slovakia, is placed by the Hornád River (approximately 30 meters), in the flood zone of this river. However, the risk of flood at this side of the river is minimal since the other side´s altitude is lower, thus in case of water level rise, naturally, Hornád River would flood the other side.

As it can be observed from the different views of the family house, A/V ratio, when designing energy-efficient building, plays significant role. As important as A/V ratio is proper house orientation in relation to "warm" and "cold" sides (south and north) and its openings. In today´s energy-saving era, to maximize solar gains and minimize heat loses means to design and use only high performance materials and parts. When talking about windows, triple-glazed windows with high Ug values are used. Great importance on openings distribution and position is being placed when designing energy-efficient buildings. Proper windows or fully-glazed curtain walls can provide enough solar gains through winter (or night) so no additional heating in required.

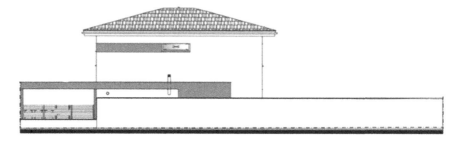

Figure 1.   Family house view A.

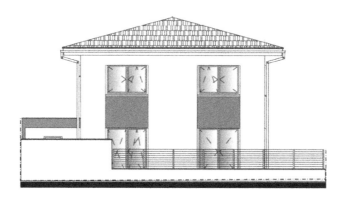

Figure 2.   Family house view B.

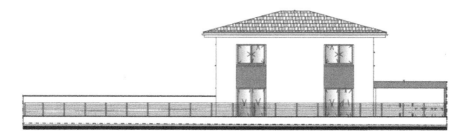

Figure 3.   Family house view C.

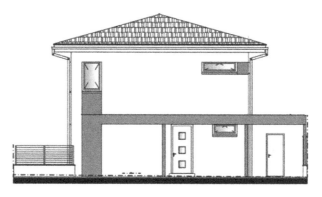

Figure 4.   Family house view D.

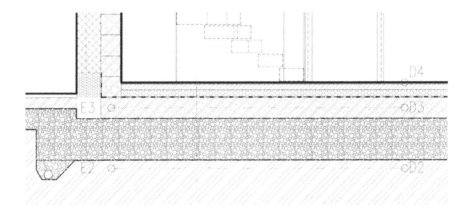

Figure 5. Foundation detail A.

Table 1. Floor on terrain layers.

	d [m]	λ [W/m.K]	c [J/kg.K]	ρ [kg/m³]	m [kg/m²]
Ceramic tiles	0.010	1.010	840.0	2000.0	200.0
Concrete	0.090	1.300	1020.0	2200.0	20.0
RFC slab	0.250	1.580	1020.0	2400.0	27.0
EPS NEO	0.200	0.031	1250.0	18.0	45.0
Stud membrane	0.0005	0.140	1100.0	1200.0	50000
Sand layer	0.050	0.950	960.0	1750.0	4
Gravel	0.400	0.650	800.0	1650.0	15

*RFC—reinforced concrete.

Table 2. Building envelope wall.

	d [m]	λ [W/m.K]	c [J/kg.K]	ρ [kg/m³]	m [kg/m²]
Gypsum plaster	0.005	0.570	1000.0	1300.0	10.0
RFC	0.070	1.580	1020.0	2400.0	29.0
Neopor insulation	0.330	0.031	1250.0	18.0	45.0
Adhesive mortar	0.005	0.800	920.0	1300.0	18.0
Silicon render	0.003	0.700	920.0	1700.0	37.0

*RFC—reinforced concrete.

Foundation detail was assessed in stationary, two-dimensional software for calculating thermo-physical parameters such as temperature distribution, surface temperatures, heat flux and isotherms. Results are shown graphically at Figure 5.

2.1 Partial conclusion

As it can be observed form the figure 5. thermo-physical analysis of the chosen foundation detail shows correctness in design from the energy efficiency point of view. The lowest surface temperature in critical detail is 19.75°C which is above the hygienic minimum according to STN 73 0540: 2012, thus there is no risk of mold occurring. From the top-left figure it is clear that −1°C isotherm do stays only in the insulation, which is the most correct option when insulating foundation, so there is no risk of RFC foundation slab degradation from the freezing.

651

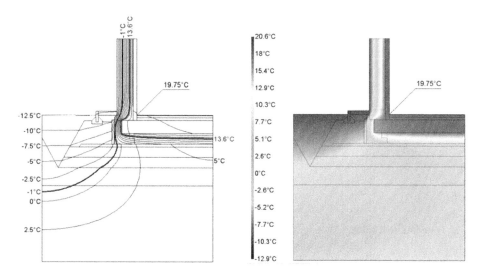

Figure 6. From the left: isotherms, temperature distribution.

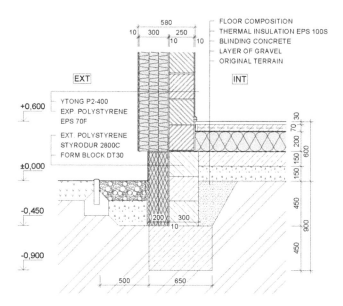

Figure 7. Foundation detail type B of a different energy-efficient family house (Geletka V. 2013).

3 EVALUATION OF A FAMILY HOUSE "B"

Type "B" energy efficient family is constructed in Lorincik, Kosice. It's A/V ratio is similar to type "A" family house for minimizing energy demand and heat loses and its opening also are designed to maximize solar gains through transparent constructions.

3.1 *Partial conclusion*

As it can be observed form the figure 8. thermo-physical analysis of the chosen foundation detail shows some incorrectness in design from the energy efficiency point of view. The lowest

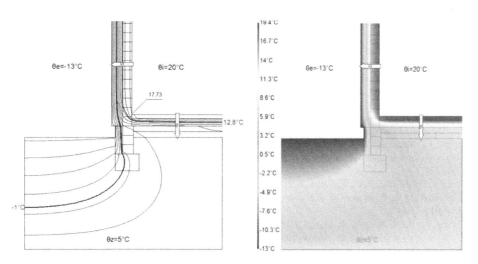

Figure 8. From the left: isotherms and temperature distribution.

surface temperature in critical detail is 17.73°C which is above the hygienic minimum according to STN 73 0540: 2012, thus there is no risk of mold occurring. However, from the top-left figure it is clear that −1°C isotherm intersects and transit through a strip foundation which indicates future problems with foundation over-freezing.

4 CONCLUSION

Outer local climate and indoor conditions were used in this analysis according to European and National directives. In this paper two construction details of foundation were assessed on surface temperatures in critical locations of each option. Today´s era of minimizing energy demands in construction sector starts with proper energy-efficient design. Thermal loses (thermal bridges) in wrongly designed thermal insulation detail of lower building parts (foundation, external wall, floor) result in a rapid rise of heating demand, as well as operating cost on energy. One of the basic steps when designing energy-efficient house is sufficient foundation insulation to provide comfortable surface temperatures in critical places (corners, etc.) and to prevent foundation against over-freezing. Type "A" family house is designed correctly in every way, surface temperature 19.75°C provides natural comfort for house users. At the same time, RFC foundation slab is insulated correctly with no foundation over-freezing. On contrary, type "B" family house's surface temperature is 17.73°C, which is also comfortable and acceptable surface temperature, however, insufficient foundation insulation can be observed, which indicates foundation over-freezing.

ACKNOWLEDGEMENTS

This study was financially supported by Grant Agency of Slovak Republic to support of projects No. 1/0307/16.

REFERENCES

Accessible on the internet: http://passivedesign/imager/stories/downloads/principles_of_ passive_design. pdf
Accessible on the internet: <http://izola.sk/pasivne-domy//referencie/nizkoenergeticke-a-pasivne-domy>
Accessible on the internet: <http://www.maphill.com/slovakia/kosice/okres-kosice-i/kosice/location-maps/blank-map/>

Geletka, V. 2013. *Výskumurčujúcichparametrovarchitektonickokonštrukčnéhonávrhubudov*, Dissertation, Košice, Slovak Republic.

S. Copiello, 2015. *"Achieving affordable housing through energy efficiency strategy,"* Energ. Policy, vol. 85, pp. 288–298.

STN 73 0540-2, 2012. *Thermal protection of buildings. Thermal performance of buildings and components. Part 2: Functional requirements.*

Sedláková, A., Vilčekován, S., Krídlová Burdová, E. 2015. *Analysis of material solutions for design of construction details of foundation, wall and floor for energy and environmental impacts*, Clean Technologies and Environmental Policy 17(5), 1323–1332, DOI: 10.1007/s10098-015-0956-3.

V. Belpoliti, G. Bizzarri, 2015. *"A parametric method to assess the energy performance of the social housing stock and simulate suitable retrofit scenarios: An Italian case study,"* Energ. Buildings, vol. 96, pp. 261–271.

Vilčeková, S., Sedláková, A., KrídlováBurdová, E., Vojtuš J. 2015. *Comparison of Environmental and Energy Performance of Exterior Walls, Energy Procedia,* 78, 231–236, DOI: 10.1016/j. egypro.2015.11.617.

W. A. Friess, K. Rakhshan, M. P. Davis, 2016. *"A global survey of adverse energetic effects of increased wall insulation in office buildings: degree day and climate zone indicators,"* Energ. Effic.

Advances and Trends in Engineering Sciences and Technologies II – Al Ali & Platko (Eds)
© 2017 Taylor & Francis Group, ISBN 978-1-138-03224-8

Heat exchange peculiarities in ventilated facades air cavities due to different wind speed

N.P. Umnyakova
Scientific Research Institute of Building Physics, Russian Academy of Architecture and Building Science (NIISF RAABS), Moscow, Russia

ABSTRACT: This article presents results of experimental investigations of dependence between the air velocity inside ventilated façades air cavities and external wind speed. New experimental results that were obtained by the author, and new objective laws and formulas, based on those results, allowed us to find the air velocity inside such an air cavity due to different external air speed. The article present results of investigations of convective and radiative heat transfer inside ventilated facades air cavities depending on different outside air temperature from −30°C to +30°C and on the different outer wind speed. These investigations and... derived from them Reynolds, Nusselt and Prandtl numbers were used to develop a methodology that is aimed to estimate heat resistance characteristics of ventilated facades air cavities subjected to different outer wind speeds and air temperatures in multistory buildings.

1 INTRODUCTION

Severe requirements for heat protection of building envelope necessitate the investigation of thermo technical properties of ventilated facade air cavities. The purpose of these investigations is to determine the thermal resistance of air cavities due to varying air velocity, which depend on wind speed.

2 EXPERIMENTAL SET-UP

For the solution of the formulated problems, the experimental investigations of ventilated facades were fulfilled in the climatic condition of Moscow (Russia). They were based on the simultaneous determination of air velocity in the air cavity of a ventilated façade and of wind speed near the outer surface of ventilated façade. For the implementation of this experiment, special-purpose holes were drilled in the ceramic or granite veneer of the ventilated façade. The diameter of the cutouts was 12–14 mm. A special device was put into these holes for measuring air velocity into the air cavities of the ventilated façade (Umnyakova, 2008). The wind speed measurements were done near the surface of the façade simultaneously with air velocity. As the result of the data manipulation the relationship between wind speed and air velocity in the air cavity of ventilated façade was obtained. It is illustrated on the Figure 1.

The conducted experiment provides an opportunity for the correlation of field data of certain experiments and get the estimated equation for determination of the air velocity in the air cavities of ventilated façade depending on the wind speed near the surface of the veneer. This relationship follows the Equation 1.

$$v_{a.c} = 0,327 \cdot v_{w.s.s}^{0.46}, \tag{1}$$

where $v_{a.c}$—air velocity in the air cavity of ventilated façade, m/s; $v_{w.s.s.}$—wind speed near the surface of ventilated façade, m/s.

The application area is limited by numerical data when the wind speed is less than 8 m/s.

It is known that wind speed increases with the height of the building. That is why on the level of each story the wind speed will be different. For its calculation we can use the investigations (Ariel, 1962), where this relationship is in the following form of Equation 2 or 3.

$$\frac{v_{w.s.s}}{v_{w.s}} = \frac{K_{w.s.s}}{K_{meteo}}$$ (2)

or

$$v_{w.s.s} = v_{w.s} \frac{K_{w.s.s}}{K_{meteo}}$$ (3)

where $v_{w.s}$—design wind speed, received on the base of meteorological search and statistical manipulation, m/s; $K_{w.s.s}$, K_{meteo}—coefficient of reduction to the wind speed on the level of the story and wind speed on the base of meteorological stations, measured on the level of 10 and 15 m, presented in Table 1.

Equation 2 be substituted into Equation 1 and then establish the new Equation 4 for the calculation the air velocity into air cavity of ventilated facades on different levels from the ground depending on different wind speed.

$$v_{a.c} = 0,327 \left(v_{w.s} \frac{K_{w.s.s}}{K_{meteo}} \right)^{0,46}$$ (4)

The calculations have shown that the air velocity in the air cavity of ventilated facades increases due to wind speed increase and decreases due to the reduction of wind speed. Also it is necessary to mention that with the number of stories increase in the building the air velocity rate grows and with the decrease of number of stories the air velocity reduces.

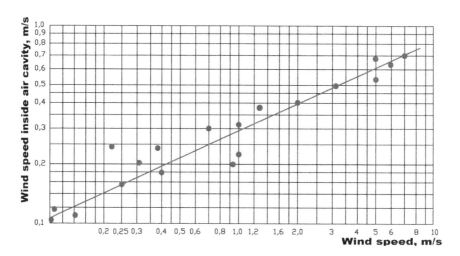

Figure 1.   The change in air velocity in the air cavity of ventilated façade with the wind speed.

Table 1.   Correction factors to the wind speed depending on the height.

The level from the ground, m	1	3	5	10	15	20	30	40	50	60	70	80	100
$K_{w.s.s}/K_{metio}$		0,36	0,47	0,53	0,61	0,66	0,69	0,74	0,77	0,8	0,83	0,86	0,88	0,91

2.1 Reynolds number

On the base of received equation, the state of the moving air is evaluated in the ventilated air cavity. For this purpose we implement the Reynolds number: $Re = v_{a.c} \cdot d/v$, where d—linear dimension of the diameter, m; $v_{a.c.}$—air velocity into the ventilated air cavity, m/s; v—coefficient of kinematic viscosity of the air, m²/s.

For the rectangular cross-section of the ventilated air cavity d changes into equivalent diameter d_{eq}, which can be calculated as ratio of the perimeter to area of the ventilated air cavity cut: $d_{eq} = 2ab/(a+b)$. The air velocity into real and phantom sections would be the same. The value of the Reynolds number for ventilated air cavity is stated in Equation 5.

$$Re = \frac{2ab}{v(a+b)} v_{a.c} \qquad (5)$$

where a and b—thickness and width of the ventilated air cavity, m.

The movement of the air flow when the value of the Reynolds number is less then $Re = 2 \cdot 10^3$ shows the laminar stream is in the cavity. The turbulent flow regime will take place when $Re > 1 \cdot 10^4$. The changing of the Reynolds number from $2 \cdot 10^3$ to $1 \cdot 10^4$ would conform the conditions of transient state.

The calculations of the Reynolds number showed, that inside the ventilated air cavity with the size of the cut $0,07 \times 1,0$ m and with the height 3 m the air flow can be in laminar, transient or turbulent state depending on the wind speed and air temperature (Table 2). The numerical values in the Table 2 shows that the type of the airflow into the cavity is determined by Reynolds number. The borderlines in the Table 2 divide the areas with laminar and transient state of airflow, as well as the area between transient and turbulent state of airflow.

The analysis of the Reynolds number shows that transient state of airflow inside the cavity is prevailing. The state of airflow in the cavity becomes turbulent only in the case of strong wind. The investigations of laminar and turbulent airflow can be done with help of the mathematic formula, received by (Miheev, 1973). As the transient state of airflow is prevailing on the base of this data, it is necessary to calculate the coefficient of convective heat exchange into the ventilated air cavity for this regime of airflow.

2.2 Transient airflow

The character of transient airflow is very complicated that creates many difficulties. According to investigations done by (Isachenko, 1981) that create much problems for quantitative description of the heat exchange process. The approximate estimation of the largest and the smallest value of the heat exchange coefficient can be done according with formulas for turbulent airflow.

Table 2. The value of the Reynolds number in a ventilated air layer thickness of 0.07 m and a width of 1.0 m.

Temperature of the air, t_{int}, °C	Kinematic viscosity of the air, m²/s	The Reynolds number in the ventilated air cavity at the (wind speed)/ (air velocity into ventilated air cavity)								
		$\frac{0,1}{0.1}$	$\frac{0,35}{0,2}$	$\frac{0,8}{0,3}$	$\frac{1,5}{0,4}$	$\frac{2,5}{0,5}$	$\frac{3,7}{0,6}$	$\frac{5,2}{0,7}$	$\frac{7,0}{8,0}$	$\frac{9,0}{0,9}$
30	16,0	811	1570	2355	3271	4056	4841	5626	6542	7327
20	15,06	863	1700	2485	3401	4317	5102	6018	6934	7719
10	14,16	915	1831	2747	3663	4579	5544	6411	7327	8242
0	13,28	985	1962	2878	3925	4841	5887	6803	7850	8766
−10	12,43	1046	2093	3140	4186	5233	6280	7588	8373	9420
−20	11,6	1123	2256	3271	4448	5626	6672	7850	8897	10074
−30	10,8	1203	2355	3532	4841	6018	7196	8373	9682	10856

The other investigator (Nachshekin, 2008) considers impossible the adoption of the equations, developed for turbulent regime, for the fields of transient airflow state. Figure 2 illustrates the curve trend in the transient airflow state of Reynolds number. The analysis of the curve development shows that in the area of transient regime under test points are not united into one consistent pattern. When the value of Reynolds number increases, the heat transfer becomes more intensive. Natural convection has a great influence on the heat exchange capacity. At the strong developed turbulent airflow, all the curves run into one line.

(Nachshekin, 2008) suggests a solution that convective heat exchange coefficient can approximately determine only with help of similarity Equation 6.

$$K_0 = Nu \cdot \mathrm{Pr}^{-0.43} \left(\frac{\mathrm{Pr}}{\mathrm{Pr}_{surf}} \right)^{-0.25} \tag{6}$$

The maximum value of K-complex is shown on the line graph in Figure 2. The approximation of this curve yield Equation 7.

$$K_0 = 0,008\,\mathrm{Re}^{0.9} \tag{7}$$

3 PROPOSED METHOD

If we set Equation 6 to Equation 7 with the provision of the Equation 4 we will receive Equation 8 for the calculation of the value of the mean coefficient of convective heat exchange.

$$\alpha_k = 0,003 \frac{\lambda_{air}}{d_{eq}} \left[\frac{2ab}{\nu(a+b)} \cdot \frac{d_{eq}}{l} \left(\upsilon_{wind} \frac{K_{a.s.s.}}{K_{meteo}} \right)^{0,46} \right]^{09} P_r^{0,43} \tag{8}$$

where λ_{air}—air thermal conductivity, Wt/m°C. The derived Equation 8 provides the opportunity to calculate the mean coefficient of convective heat exchange inside a ventilated air cavity of the ventilated façade for transient airflow state, depending on the wind speed (whether it

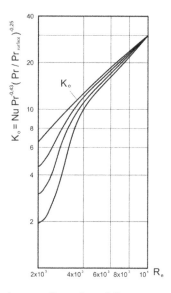

Figure 2. Mean heat transfer in the case of transient airflow.

is less than 8 m/s), air velocity inside the ventilated air cavity and versus the height of the calculation point, located on the facade of the building.

The radiated heat transfer coefficient for ventilated air cavity, when two surfaces are parallel, can be calculated with help the following Equation 9.

$$\alpha_{r.a.c.} = \frac{1}{\dfrac{1}{C_{fac.v}} + \dfrac{1}{C_{m.w.}} - \dfrac{1}{C_0}} \cdot \Omega \tag{9}$$

where $C_{fac.v}$—thermal emissivity coefficient of the polished face of façade veneer, equal 3,9 Wt/m ^{20} K^4; $C_{m.w.}$—thermal emissivity coefficient of the surface of the preformed mineral wool, equal 4,5 Wt/m^{20} K^4; C_0—black body emissivity coefficient, equal 5,76 Wt/m$^{2\,0}$K^4; Ω—temperature coefficient, determined with help of graph on Figure 3 on the base of temperature difference of two surfaces into the air cavity of ventilated façade.

According to Equation 9, the radiate heat exchange coefficients were calculated for air cavity of ventilated façade at the different outside temperature from 30°C till −30°C and are presented in Table 3.

It is clear from Table 3 that the value of the radiated heat exchange coefficients is directly proportional to an increase in the air temperature.

The thermal resistance $R_{air.c.}$ of the ventilated air cavity, taking into account the radiate and convective heat exchange, is possible to be write in form of Equation 10.

$$R_{air.c} = \frac{1}{2\alpha_{c.a.c} + \alpha_{c.a.c}} \tag{10}$$

Thermal resistance properties of the ventilated air cavity at different wind speed, at its different thickness and at different temperatures of outdoor air are calculated and listed in

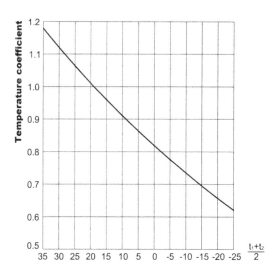

Figure 3. Temperature coefficient.

Table 3. The radiate heat exchange coefficients into the air cavity of the ventilated façade.

Air temperature, °C	30	20	10	0	−10	−20	−30
The radiate heat exchange coefficients, Wt/m^2°K^4	3,72	3,31	3,0	2,68	2,39	2,13	1,93

Table 4. Thermal resistance of ventilated air cavity with thickness 0,07 m and width 1,0 m.

Air temperature, t_H, °C	Thermal resistance of the ventilated air cavity at the (wind speed)/(air velocity into ventilated air cavity)							
	$\frac{0,1}{0,1}$	$\frac{0,35}{0,2}$	$\frac{0,8}{0,3}$	$\frac{1,5}{0,4}$	$\frac{2,5}{0,5}$	$\frac{3,7}{0,6}$	$\frac{5,2}{0,7}$	$\frac{7,0}{8,0}$
30	0,163	0,153	0,135	0,110	0,101	0,09	0,08	0,072
20	0,178	0,164	0,141	0,116	0,104	0,095	0,084	0,075
10	0,19	0,172	0,144	0,117	0,105	0,097	0,086	0,076
0	0,199	0,18	0,155	0,122	0,108	0,099	0,087	0,076
−10	0,215	0,19	0,155	0,126	0,11	0,102	0,085	0,075
−20	0,224	0,2	0,160	0,130	0,113	0,102	0,086	0,084
−30	0,23	0,208	0,163	0,135	0,114	0,101	0,088	0,083

the table 4. As followed from the table thermal resistance of the air cavity increases when the temperature becomes lower and decreases when the temperature becomes higher. For example, at the temperature $t_{out} = -30°C$, wind speed $\upsilon_w = 0,35$ м/s and air velocity into ventilated air cavity $\upsilon_{a.c.} = 0,2$ м/s the thermal resistance is equal $R_{air.\,c.} = 0,208$ м²°C/Wt. When the outdoor temperature changes from +30°C till −30°C at wind speed $\upsilon_w = 3,3$ м/s and air velocity into ventilated air cavity $\upsilon_{a.c.} = 0,6$ м/s the thermal resistance of air cavity will change from 0,09 till 0,101 м²°C/Wt.

4 CONCLUSION

On the base of analysis of calculated values, presented in Table 4, it is possible to conclude that thermal protection properties of ventilated air cavities should be applied in the calculations of the heat protection properties of ventilated facades. The value of thermal resistance of the ventilated air cavity should be taken into account for different climatic regions depending on design temperature and wind speed according to Equation 10, and corresponds with (Fokin, 2006) investigations.

In the base of data-proceeding operation of in-situ tests and thermo-technical investigations the method of the calculation of heat protection properties of ventilated air cavities depending on wind speed, air velocity into the cavity and on the level of the of the air cavity from the ground was developed. The calculations showed that thermal properties of air cavities of ventilated facades at an air velocity 0,5 m/s and more inside the cavity greatly decrease.

REFERENCES

Ariel, N.Z. & Kluchnikova, L.A. *Wind in the city*. Proceedings of GGO, issue 94. Leningrad, 1962.
Fokin, K.F. *Building termaltechnic of building envelope*. Moscow, 2006.
Isachenko, V.P., Osipova, V.A. & Suhomel, A.S. *Heat transmition*. Moscow, 1981.
Miheev, M.A. & Miheeva, I.M. *The base of thermal transmition*. Moscow, 1973
Nachshekin, V.V. 2008. *Technical thermodynamics and thermal transmition*. Moscow.
Umnyakova, N.P. *Condition of the thermal insulation in the ventilated facades in the Moscow*. Scientific conference "Modern ventilated façade systems: efficiency and durability. Book of reports. Moscow, 2008. pp. 261–268.

Advances and Trends in Engineering Sciences and Technologies II – Al Ali & Platko (Eds)
© 2017 Taylor & Francis Group, ISBN 978-1-138-03224-8

Impact of structural and technological optimisation on quality and durability of concrete car garages

V. Vacek & M. Sýkora
Klokner Institute, CTU in Prague, Czech Republic

V. Vančík
VIN Consult s. r. o, Prague, Czech Republic

ABSTRACT: The paper shows an example of a specific project built five years ago. It illustrates how it is possible to optimize a final quality while respecting a shape and layout of designs and construction methods. A combination of design and technology solutions together with better operating characteristics of the resulting object were achieved significantly simplified construction and reduced execution time. Using an advanced technology of hybrid (precast-monolith) concrete structures leads to considerable savings on the volume of the resulting structures when compared to the originally proposed solution of a monolithic reinforced concrete system. The case study is focused on such a building that after five years of an operation shows no visible defects and satisfactorily serves its purpose.

1 GARAGE CONCRETE STRUCTURES AND THEIR DURABILITY

Durability of building structures is to a certain degree conditioned by the quality of used materials and the quality of a workmanship. A good quality execution is therefore necessary in conjunction with the optimum project with respect to a structural and technological solution. The concept of the whole structure and its key details must be designed optimally with regards not only to structural but also to operational costs, in the context of the designed durability of the building.

The principal factor in design is the purpose of the building determined by requirements and conditions of the client. This implies loading of the structure during an execution and operation. It is not only stress by static or dynamic mechanical actions due to self-weight and imposed loading, but also actions caused by changes of temperature or humidity and the impacts of the surrounding environment, which usually leads to the deterioration of construction materials.

The Car Garages (Parking Houses) represent rather a specific type of civil engineering structures. Methods of their construction have been significantly improved over the last 20 years in the Czech Republic. They permanently show series of serious defects which are often leading to failures principally affecting their durability, then to expensive repairs and vast structural upgrades and even to their closure (Vacek, 2009).

This fact also represents a devaluation of initial costs spent on the construction and brings many subsequent operational complications when the operation cannot be provided in an originally designed extent. The most important issue which also finally initiates an interest of the owner or administrator of the building is damage of vehicles by leakage from cracks in ceiling structures.

The incidence and nature of cracks in relation to the parameters of the secondary protection floor (directly travelled waterproofing system) are essential for durability of the concrete structures of the garages (Vacek & Kostelecká 2013). No less important is a reliable and functional drainage system of other floors of the parking garages (Vacek, 2009).

It is appropriate to summarize in this relation issue of cracks in reinforced structures and special issues of open parking houses. Reinforced structure has a static impact on the cracks and these cracks are not a hindrance if their widths are within the approved limits. This relates to the approved cracks with the width of 0.30mm with respect to EN 1992-1-:2004 and Model Code (*fib* 2010) in relevant foreign standardsin the given case. The structure of the parking house was designed according to these criteria. The specific applied reinforcement is fulfilling the criteria of the calculation cracks width of 0.2mm caused by quasi-permanent combination of loading.

It is important to note that crack width is significantly dependent on a concrete cover and diameter of a reinforcing bar. When the assumptions accepted by (Marková & Sýkora 2016) are adopted, Figure 3 shows the variation of β (see EN 1990:2002), related to exceeding the limiting value of 0.3 mm, with the mean value of the concrete cover for different bar diameters. Obviously, reliability with respect to serviceability is sufficient when the target level of 1.5 is taken into account for a reference period of 50 years as is given in EN 1990:2002 for irreversible serviceability limit states.

Apart from static cracks, technological cracks occur in reinforced structures as a consequence of concrete shrinking (by impact of hydration during concrete hardening or even later), moreover open parking houses are exposed to thermal loading actions.

An occurrence of technological cracks in reinforced structures is a random quantity. Their width and classification can be suitably regulated by a minimum structural reinforcement, the actual course is however influenced by many factors (design of concrete, technology and weather conditions of concreting, subsequent concrete treatment), therefore the analytical design would not be able to secure that all cracks are within approved limits. Even cracks caused by secondary actions (volume changes and temperature) are not a hindrance if their width is within the approved limits or if they are additionally grouted.

Generally the structural design of division and width of cracks from forced cracks is always a compromise between a price of reinforcement and costs of an additional grouting of possible cracks (analogically this applies also to designs of waterproof white baths). In open parking houses for example in Germany, based on current status of technical knowledge, is recommended an additional application of crack reinforcement which is placed in the cover layer above static reinforcement and is metallized. This solution is however unacceptable for Czech investors due to its price.

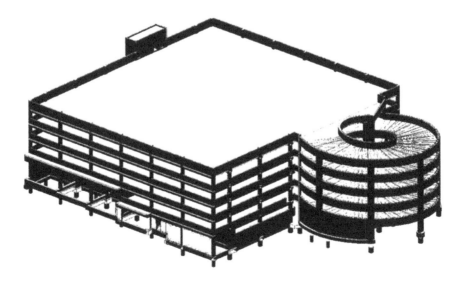

Figure 1.   View of the optimized garages.

2 EXAMPLE OF GOOD PRACTICE—PARKING HOUSE TELEFÓNICA O₂ IN PRAGUE

The parking house (car garage) was designed as a six-floor building (one basement and five floors, including a roof) with aground plan dimension of 52×69 m. The structural height of the floor level is 2850 mm, clearance under beams 2250 mm and clearance for walking by the column under the edge of the cantilever is 2100 mm.

Entrances, an exit and an arrival to the building are located next to the staircase core. Separated closed garages and a storage room were designed in the area of an underground atypical floor level. Internal vertical communication for vehicles between individual parking levels is secured by a two-way ramp in the shape of helix—rondel, with a new cylindrical fire reservoir designed in its centre. This structure of the parking house with a total capacity of 450 vehicles with a round ramp was designed in the tender documentation as a monolithic system (Vančík et al. 2011).

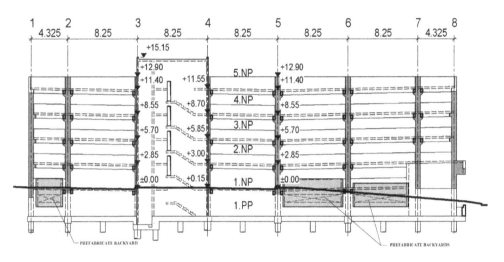

Figure 2.   The finished vertical section of the structure with apparently inclined ceilings.

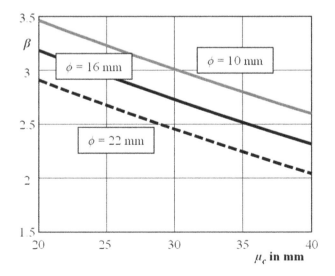

Figure 3.   Variation of reliability index β with the mean (nominal) value of concrete cover and different diameters of reinforcing bars.

Table 1. Comparison of volumes in the part of the structure under consideration [m³].

Part of structure	Original	Monolithic	Precast	Total
Vertical skeleton	530	227	212	439
Horizontal skeleton	3500	1682	1196	2878
Vertical rondel	280	180	8	188
Horizontal rondel	630	546	0	546
Total	4940	2635	1416	4051

2.1 Concept of technological optimization

Within the tender offer, the structural solution including the foundation system was changed, preserving the original layout solution and aiming to optimize the design in terms of costs, use of technologies of a contractor and reduction of structural volume. The height of the building was modified. The originally designed horizontal ceiling structures were changed to plates with a slope even at the price of increased initial costs of both the contractor and the designer (Figure 1).

The special attention was paid to the solution of the construction procedure and the speed of an assembly in the limited building site conditions. The dimensional classification of the building in monolithic and precast parts required careful structural solutions of their connection, but did allow for their independent construction process without collision points leading in its consequence to an optimum use of advantages of used technologies, the combination which significantly reduced the construction period of the whole building compared to its original design (Vančík et al. 2011).

2.2 Optimization of structural solution

The originally designed foundation had low load capacity for additional load 500 kN/m² and strengthening was not feasible due to an insufficient engineering-geological survey (IGS) during the previous phases of the project. An additional research of IGS proved layer of loose backfills of remarkably variable thickness up to 7m, so the possible alternatives of foundation had to be checked:

- Areal foundation on footing of increased dimensions—modified dimensions of footing were designed for characteristic additional load of the foundation joint 300 kN/m²
- Underground foundation on bored piles with foundation grids.
  After their critical comparison, bored piles with foundation grids were selected.

2.2.1 Structural system

The structural solution of the parking house was revised taking into consideration the technological optimisation for the selected contractor ensuring execution, to a hybrid concrete-concrete system. Main modifications were the following:

Crosswise gripped ribs of ceiling slabs were replaced by double continuous girders.

- Grid structures in the area of a skeleton connection to the rondel were simplified with omission of two columns at the same time and replaced with pre-stressed girders.
- Rondel (columns with girders and external parapets) was a cylindrically shaped structure with an opening.
- Horizontal ceiling structures were modified to plates with a slope.

The change of the horizontal to the sloped ceiling slabs was crucial for the final quality of the building (Figure 2).

2.2.2 Final modified design

The vertical bearing structure of the parking house is formed by the module system skeleton 8.25 x 7.60 m. The horizontal bearing structure was designed as a coupled hybrid

Figure 4.   Roof (left-hand side) and floor surface of the 3rd floor of the garage (right-hand side) after five years of service.

structure concrete-concrete with a maximum precast share. The major part of the structure is precast, the ramp is mostly monolithic. Further impermeable concretes for structures of the peripheral underground wall along the access communication, the fire tank and the underground collector for a connection of the fire tank to the main office building were used.

2.3   *Building construction*

The construction procedure was strongly influenced by the restricted conditions on the site with the only access communication, very limited possibilities of the material storage and the necessity to keep the fire tank in operation—first the existing one and then the new one. Possibility of a space division and therefore also an independent realisation of technologically and functionally separated parts of the structure was due to the condition to keep the time schedule considered even in the structural optimisation of the building. The final result of works was divided in the following phases:

– Construction of the collector of the buried services and the substructure of the water tank
– Planking and strutting of the excavation pit (first part), the underground foundation for the rondel and the first part of the skeleton
– Assembly of the rail crane in the field 4–5
– Construction of the staircase core and the substructure of the internal part of the rondel
– Assembly of the first part of the skeleton and the independent construction of the water tank and the monolithic rondel
– Reconnection of the original fire tank to a new water reservoir
– Demolition of the fire tank, planking and strutting of the excavation pit of the second part and the underground foundations for the second part of the skeleton
– Assembly of the second part of the skeleton
– Dismantling of the rail crane and completing of all floors in the field 4–5 of the skeleton.

An interesting technological part was the execution of additional concrete works in the field 4–5 after the dismantling of the rail crane at the end of the shell construction. Time for the execution of this section was limited to one week under an adverse weather conditions disabling use of common filigree slabs with assembly supports for relevant technological periods. Therefore filigree girder beams Montaquick with pressed concrete lines were additionally designed and used. The individual ceilings were concreted within a two-day working step. The shell construction under the condition of an uninterrupted operation of the fire tank was executed from 07/2010 till 01/2011, where the skeleton installation were taken only 4 months.

3 CONCLUSION

The parking house, finished in 2011, is serving successfully its purpose after 5 years of a daily operation. During an inspection, no defects or failures were found, even no tear and wear for example travelable layers, which might have an impact on a reduction of its durability or which might call for a need of the premature construction maintenance or repairs. The current condition of the structure is demonstrated by Figure 4.

Exceptionally successful outcomes of a workmanship resulted mainly from aclose cooperation of the designer and the contractor during the design phase and by the execution of the compact, aesthetic and highly economical structure. The design was based on:

- Respect to the operating requirements and conditions of the building
- Suitable modification of the shape—the slope of the surface of the parking areas
- Correct structural division of the building in the functionally and technologically different parts
- Consistent solution-making and execution of these parts—critical details of the building
- Optimisation of the structure with non-traditional details from the view of used technologies with impact on a minimisation of the risk of a defects occurrence during the execution
- Selection and a due execution of a technically optimal system of the protection of the horizontal structure and its water insulation with a suitable parking floor system
- Securing of reliability of the building function by a minimizing of demands on service and maintenance during the operating phase

ACKNOWLEDGMENT

This work has been supported by the Czech Science Foundation under Grant P105/12/G059.

REFERENCES

ČSN 73 6058. *Hromadné garáže (Parking garages, in Czech)*. Valid from 08/1988 to 09/2011.
EN 1990:2002. Eurocode—Basis of structural design.
EN 1992-1-1:2004. *Design of concrete structures—Part 1-1: General rules and rules for buildings*. Brussels: CEN. *fib* Model Code for Concrete Structures 2010. Lausanne: *fib*.
Marková, J., Sýkora, M. Uncertainties in Crack Width Verification of Reinforced Concrete Structures (accepted for publication). In *Proc. ESREL 2016*.
Vacek, V., 2009, Parking garages—defects and failures due to faulty structural design, Conference Proceedings XIXth International Symposium SANACE 2009, ISSN 1211-3700, pages 241–249.
Vacek, V., Kostelecká, M., 2013, The commentary to the issue of the thickness of coatings on concrete surfaces and its measurement, Conference Proceedings, 44thInternational Conference on Coatings Technology, ISBN 978-80-7395-627-1, pages 275–286.
Vančík, V., Svoboda, M., Ondrůšek, Z., 2011, ParkovacídůmTelefónoca O2 v Praze, časopis Stavebnictví 06-07 2011, ISSN 1802-2030, EAN 977180220300508326„ pages 24–31.

Advances and Trends in Engineering Sciences and Technologies II – Al Ali & Platko (Eds)
© 2017 Taylor & Francis Group, ISBN 978-1-138-03224-8

Software-based support to decision-making process regarding the selection of concrete suppliers

V. Venkrbec
Faculty of Civil Engineering, Brno University of Technology, Brno, Czech Republic

U. Klanšek
Transportation Engineering and Architecture, Faculty of Civil Engineering, University of Maribor, Maribor, Slovenia

ABSTRACT: An important issue that should be considered during the construction process is to organize the concrete supply for building site. However, there are several factors which have to be taken into account for the selection of a concrete supplier such as the transportation time, travel distance, traffic density, type of concrete, concrete additives and required amounts of concrete. In common practice, the effectiveness of supplier selection mainly relies on the intuition of a construction manager. The aim of this paper is to propose a software-based methodology to better support a decision-making process regarding the selection of concrete suppliers. At this point, Google Maps Directions API employing a hypertext transfer protocol requests for collecting sets of different routes was used. An application example is given at the end of the paper to demonstrate the advantages of proposed approach.

1   INTRODUCTION

The problem of a concrete supply can be defined as a distribution supply chain that is one of the three traditional stages of logistic supply chains (Thomas & Griffin 1996). The main aim of the logistic and its supply chains is to optimize material transport from point A to point B by using the most appropriate route for the specific conditions or limitations. The objective of the most appropriate route is usually to minimize the total cost of this supply.

The construction process comprises many decentralized operations which leads to the necessity for rigorous and critical decision-making points during the construction lifecycle (Naso, D. et al., 2007). These operations usually need to define the most appropriate routes from some production plants to specific places i.e. construction sites. In the research community a large number of papers bringing more different approaches to optimize this problem was generated. Genetic algorithms (Naso, D. et al., 2007); (Maghrebi et al. 2016), ant colony optimization (Bell & McMullen 2004); (Faria et al. 2006), using a combination of approaches (Silva et al. 2005) or solving the nonlinear discrete transportation problem (Klanšek 2015) were presented.

The decision to select a concrete supplier usually relies on the construction manager and his intuition. In current practice on site the shortest distance between the concrete plant and construction site is usually chosen as the most appropriate. This approach can cause increases in the total cost of the concrete since only one marginal condition is chosen (the shortest distance). The concrete transportation cost is derived from many aspects and characteristics of the concrete.

This paper presents a software-based approach for the construction manager to better understand the decision-making process for choosing the most appropriate concrete supplier. A given algorithm can also be taken into account at the decision point between buying and producing concrete by a builder's own concrete plant. It has been difficult for construction companies to solve this problem on site (by construction manager) until this time.

2 METHODS

The aim of the paper is to compare the total cost paid by the construction company (contractor) to the concrete plant for each possible concrete supply. The total cost consists of the price of fresh concrete and the price of transport.

The goal of the case study was to assess whether the distance is only one of the factors influencing the total cost, and to find out how the prices depend on the transport distance and on time of transport. There are 492 concrete plants on the area of 78,866 square kilometres in the Czech Republic. Only these concrete plants which are up to 90 minutes of transport time to the construction site can be taken into account due to the Czech national standard for concrete processing.

For Central European states (Czech Republic, Slovakia, Hungary) there are some existing possibilities to search for possible concrete suppliers by using online map-based applications such as Beton server (Lášek 2006), Cemex—Map of sales points (Cemex S.A.B. de C.V. 2016), Transbeton—Concrete calculator (Zepiko Group 2016), Transportbeton—Search plant (Českomoravský beton, a. s. 2016), Zapa—Find a plant (Zapa beton a.s. 2016). Only Beton server is independent and provides information about several concrete plants owned by different companies. The searching works based on the position of construction site and the results are displayed within driving distance of up to 50 km. Other applications are interconnected to some specific concrete supplier. For better understanding of the differences between the applications see Table 1.

The searching in applications is based on the position of the construction site. The transport time is not supported as an input parameter. Type of concrete as input parameter is supported only in two cases. The results are in most cases generated only as distances between the construction site and the concrete plant. In few cases the price of supply is generated. Time of transport is not generated in any case. Application Programming Interface (API) was used to collect the data. APIs are some sets of protocols and tools for creating software and applications. Web Mapping APIs (WMA) are online-provided applications, which use map servers (such as Google Maps, Bing Maps, MapQuest, HERE, Yahoo Maps) for better development of web applications with geospatial contents (GeoCMS). These applications are also known as "Map Mashups" (Batty et al. 2010) or "Web Mapping 2.0" (Haklay, Singleton, and Parker 2008). For decision-making process more than one possible route should be generated by API. This function is only available in Google Maps directions API. Google Maps direction API is a tool for finding multiple routes from point A to B, in mentioned case the transport of concrete from a concrete plant to the construction site. Output of request is distance and transport time in JavaScript Object Notation (JSON) format (recommended by Google). Each request to an API that is represented in the console must include a unique identifier. For HTTP requests Python language in version 2.7.10 was used for obtaining data from Google Maps Directions API using a script. Python is an object-oriented programming language, comparable to Perl, Ruby, Scheme, or Java (Lemburg 2015). The input data were collected from publicly available sources. The concrete plants positions are usually defined by the owners of the concrete plants as a coordination system—Global Positioning System and are converted to CSV (coma-separated values) format (Shafranovich 2005). A part of GPS coordinated and CSV converted input data are shown in the Table 2.

Table 1. Map-based applications—concrete plant searcher.

Application	Searching by			Results displayed		
	Cons. site position	Transport time	Type of concrete	Price	Distance	Time
Betonserver	YES	NO	YES	YES	YES	NO
Cemex	LIMITED**	GM*	NO	NO	GM*	GM*
Transbeton	YES	NO	YES	YES	YES	NO
Transportbeton	YES	NO	NO	NO	YES	NO
Zapa	YES	NO	NO	NO	YES	NO

*GM = provided by redirecting to Google Maps; **LIMITED = limited search only for region.

Returned data from the **API** thru Python contain combined information about a route from starting point to target point and the time of transport. Up to two or three different routes (e.g. r_{010101}; r_{010102}; r_{010103}) were received for each request. The number of obtained data is determined by options of Google Maps, respectively density of the road network in the area. The data were returned in combination considering the transport time (seconds) and transport distance (meters). The example of the combination of returned data is shown in Table 3.

It was necessary to define the price per m3 of concrete for comparing the profitability of each delivery. The prices were obtained from public web pages of various concrete mixtures suppliers. Prices of the three most common types of concrete were collected for the case study. Used types of concrete are shown in Table 4.

Transport prices were obtained as a price per unit of one m3 for a specific distance span. Distance spans are shown in Table 5.

3 RESULTS AND DISCUSSIONS

For analysing optional concrete supplier was chosen a real construction site in Brno city centre, position coordinates: 49.195934, 16.605238 (in CSV format). For the case study for this specific

Table 2. Part of input data.

Concrete Plant no. (CPxx)	GPS Coordinates		
Construction Site no. (CSyy)	Latitude	Longitude	CSV format
CP01*	49°08'31.434"N	16°35'56.183"E	49.142065, 16.598940
CS01**	49°11'45.362"N	16°36'18.856"E	49.195934, 16.605238

*CP01 = Concrete plant with index x = 01; **CS01 = Construction site with index y = 01.

Table 3. Example of returned data (selection of three routes, their distances and durations).

Direction	Route no. r	Distance (m)	Transport time (s)
CP01—CS01	010101	8024	938
CP01—CS01	010102	11273	1155
CP01—CS01	010103	15234	1235

Table 4. Used types of concrete.

Concrete class	Exposure classes		Consistence	d_{max}	CL
C16/20	X0	XC1	S3	16	0.2
C20/25	X0	XC1	S3	22	0.4
C25/30	X0	XC1	S3	22	0.4

Table 5. Example of transport prices.

Concrete plant	Distance span (km)								
	Unit prices/m3/span (Eur)								
CP01	0–5	5–10	11–20	21–30	31–40	41–50	51–60	61–70	71–80
	5.18	5.61	8.64	11.46	14.46	16.86	19.46	21.61	24.21
CP01	81–90	91–100	101–110	111–120	121–130	131–140	141–150	151–160	
	26.57	29.18	32.61	36.04	39.46	42.89	46.32	49.75	

construction site 74 concrete plants positions were analysed and 182 routes were generated by API. Only the first 25 concrete plants according to individual ranking criteria are displayed in the result charts. Results were sorted to rankings (Table 6) by distance from construction site.

The most favourable total price of 68.25Eur per m³ of fresh concrete was calculated by the 19th nearest Concrete Plant (CP19) from the construction site for C16/20. The nearest concrete plant is 33.59% more expensive than the Cheapest one (CP19). The first six nearest concrete plants are approx. 27–33% more expensive than the Cheapest one (CP19). The most favourable total price of 72.57Eur per m³ of fresh concrete was calculated by the 19th nearest Concrete Plant (CP19) to construction site for C20/25. The first four nearest suppliers are approx. 22–35% more expensive than the Cheapest one (CP19). The nearest Concrete Plant (CP01) is 35.19% more expensive than the cheapest one which is situated 29.48 kilometres away from the construction site. The most favourable total price of 82.50Eur per m³ of fresh concrete was offered by the 19th nearest Concrete Plant (CP19) for C25/30. The second cheapest concrete plant is CP24 with price of 86.36Eur per m³. CP24 is 35.899 kilometres away. First four nearest concrete plants are approx. 17–27% more expensive than the Cheapest one (CP19). The nearest Concrete Plant (CP01) is approx. 27% more expensive than the cheapest one which is situated 29.480 kilometres away from the construction site.

It may seem that the price of concrete is influenced by the distance of the concrete plant from bigger towns. The Tables 6–8 also show that the total cost can be lower from remotely placed concrete plants than from concrete plants in the town of the construction site in the case that the construction site is situated in bigger town e.g. Brno (350,000 inhabitants).

Table 6. Rating chart by distance from construction site for different concrete types.

Concrete plant	Distance meters	Transport time seconds	C16/20		C20/25		C25/30	
			TC	ITC	TC	ITC	TC	ITC
			EUR/m³	%	EUR/m³	%	EUR/m³	%
CP01	3099	627	91.18	33.59	98.11	35.19	105.00	27.27
CP02	3160	652	88.88	30.23	92.32	27.21	96.89	17.45
CP03	4464	660	87.21	27.79	88.50	21.95	94.54	14.59
CP04	5489	879	89.25	30.77	90.54	24.75	96.57	17.06
CP05	5565	857	78.46	14.97	82.57	13.78	88.82	7.66
CP06	7257	882	89.25	30.77	90.54	24.75	96.57	7.06
CP07	7798	1038	91.21	33.65	95.29	31.30	99.86	21.04
CP08	7877	1039	98.39	44.17	102.00	40.55	110.39	33.81
CP09	8024	938	92.93	36.16	95.96	32.23	100.50	21.82
CP10	15869	1204	90.29	32.29	93.11	28.30	96.79	17.32
CP11	15988	1192	97.93	43.49	104.68	44.24	114.43	38.70
CP12	19343	1310	81.29	19.10	85.61	17.96	89.50	8.48
CP13	20382	1569	78.82	15.49	83.57	15.16	90.07	9.18
CP14	23803	1490	84.04	23.13	88.36	21.75	92.68	12.34
CP15	27730	2224	100.29	46.94	102.00	40.55	108.04	31.00
CP16	24802	1707	92.89	36.11	96.79	33.37	100.25	21.52
CP17	26428	1741	94.75	38.83	99.32	36.86	108.04	31.00
CP18	27496	1620	94.86	38.98	99.61	37.25	106.96	29.65
CP19	29480	2043	68.25	0.00	72.57	0.00	82.50	0.00
CP20	30912	1963	93.24	36.61	99.00	36.42	108.68	31.73
CP21	33153	1815	94.41	38.34	97.01	33.67	100.89	22.29
CP22	34159	1737	94.11	37.89	95.43	31.50	100.61	21.95
CP23	34914	1832	99.46	45.74	101.39	39.71	108.32	31.30
CP24	35899	1911	77.71	13.87	82.04	13.04	86.36	4.68
CP25	36700	2236	104.79	53.53	109.54	50.94	114.07	38.27

TC = Total Cost; ITC = Increase from cheapest Total Cost.

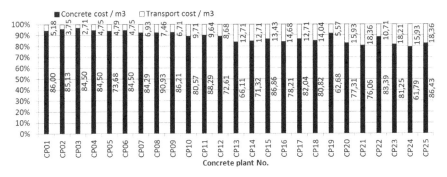

Figure 1.   Partial components (%) of the total price for concrete type C16/20.

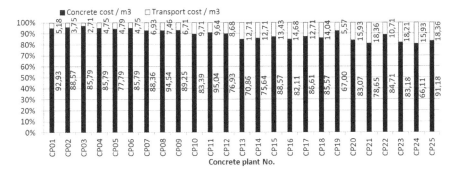

Figure 2.   Partial components (%) of the total price for concrete type C20/25.

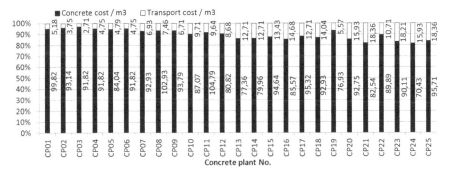

Figure 3.   Partial components (%) of the total price for concrete type C25/30.

Monitored dependence of time and distance compared to the total price can be seen in the Figures. In the bar graph (Figures 1–3) partial components (price of fresh concrete and price of concrete transport) of the total price are shown.

4   CONCLUSION AND FUTURE RESEARCH

The aim of this paper was to propose a software-based methodology to better support a decision-making process about the selection of concrete suppliers. In current practice on site the shortest distance between the concrete plant and construction site was usually chosen as the most appropriate criterion and in the decision about the concrete supplier the construction manager was usually using his intuition.

The study has its limitations. The analysis has not been counting with the use of additives to the concrete mixture that can have a significant impact on the extension of the period of concrete solidification. Also the concrete plants owned by the construction company have not been considered which can have an impact on the decision between buying and producing concrete. It is expected that the price of concrete produced by the company's own concrete plants will be lower, but the exact prices are not known. The authors will count with these facts in future research that can lead to creating a complete regularly updated database of concrete plants that will be useful for practical application in construction site operational management. The analysed data show that the mixer is not able to reach the speed limit in the city centres (50km/h). The data acquired from Google Map API count with the traffic density that is consistently updated by the users. Therefore, the acquired information about the transport time takes into account the most probable traffic situation.

The findings reveal that distance is not the only one appropriate criterion for the decision-making process if the lowest total cost is needed. As it can be seen in the results section that the price of concrete of different concrete suppliers can be crucial.

REFERENCES

Batty, M. et al., 2010. Map mashups, Web 2.0 and the GIS revolution. Annals of GIS, 16(1), pp. 1–13. Available at: http://www.tandfonline.com/doi/abs/10.1080/19475681003700831.

Bell, J.E. & McMullen, P.R., 2004. Ant colony optimization techniques for the vehicle routing problem. *Advanced Engineering Informatics*, 18(1), pp. 41–48.

Cemex S.A.B. de C.V., 2016. Betonárny, lomy a štěrkovny CEMEX. *Beton, lité směsi, kamenivo, cement | CEMEX Česká republika*. Available at: http://www.cemex.cz/mapa-prodejnich-mist.aspx [Accessed April 24, 2016].

Českomoravský beton, a. s., 2016. Betonárny—vyhledávání betonáren. *Českomoravský beton—výroba betonu, doprava betonu a čerpání betonových směsi*. Available at: http://www.transportbeton.cz/beton-arny.html [Accessed April 24, 2016].

Faria, J. et al., 2006. Distributed optimization using Ant Colony optimization in a concrete delivery supply Chain. In *2006 IEEE Congress on Evolutionary Computation, CEC 2006*. Vancouver: IEEE Xplore Dgital Library, pp. 73–80.

Haklay, M., Singleton, A. & Parker, C., 2008. Web Mapping 2.0: The NeogeoFigurey of the GeoWeb. *GeoFigurey Compass*, 2(6), pp. 2011–2039. Available at: http://doi.wiley.com/10.1111/j.1749–8198.2008.00167.x.

Klanšek, U., 2015. A comparison between MILP and MINLP approaches to optimal solution of Nonlinear Discrete Transportation Problem. *Transport*, 30(2), pp. 135–144.

Lášek, V., 2006. Beton server: Beton, vše z betonu a vše pro beton v ČR. Beton server: Beton—betonárny v ČR. Available at: http://www.betonserver.cz/betonarky [Accessed April 24, 2016].

Lemburg, M.-A. ed., 2015. The Python Wiki: BeginnersGuide Overview/Beginners Guide/Front Page. In *Wikipedia: The Python Wiki*. San Francisco (CA): Wikimedia Foundation. Available at: https://wiki.python.org [Accessed April 24, 2016].

Maghrebi, M., Travis Waller, S. & Sammut, C., 2016. Sequential Meta-Heuristic Approach for Solving Large-Scale Ready-Mixed Concrete–Dispatching Problems. *Journal of Computing in Civil Engineering*, 30(1), p.04014117-. Available at: http://ascelibrary.org/doi/10.1061/(ASCE)CP.1943–5487.0000453.

Naso, D. et al., 2007. Genetic algorithms for supply-chain scheduling: A case study in the distribution of ready-mixed concrete. *European Journal of Operational Research*, 177(3), pp. 2069–2099.

Shafranovich, Y., 2005. Common Format and MIME Type for CSV Files: Network Working Group. In *The Internet Engineering Task Force (IETF®): IETF-related tools, standalone or hosted on tools. ietf.org*. USA, California: SolidMatrix Technologies, Inc. Available at: https://tools.ietf.org/pdf/draft-shafranovich-mime-csv-05.pdf [Accessed April 24, 2016].

Silva, C.A. et al., 2005. Concrete Delivery using a combination of GA and ACO. *Proceedings of the 44th IEEE Conference on Decision and Control*, pp. 7633–7638. Available at: http://ieeexplore.ieee.org/lpdocs/epic03/wrapper.htm?arnumber=1583394.

Thomas, D.J. & Griffin, P.M., 1996. Coordinated supply chain management. *European Journal of Operational Research*, 94(1), pp. 1–15. Available at: http://linkinghub.elsevier.com/retrieve/pii/0377221796000987.

Zapa beton a.s., 2016. Find a plant. *ZAPA beton*. Available at: http://www.zapa.cz/find-plant/?l=eng [Accessed April 24, 2016].

Zepiko Group, 2016. Betonová kalkulačka. *Beton, doprava betonu, čerpání betonu | TRANSBETON*. Available at: http://www.transbeton.cz/betonova-kalkulacka [Accessed April 24, 2016].

Advances and Trends in Engineering Sciences and Technologies II – Al Ali & Platko (Eds)
© 2017 Taylor & Francis Group, ISBN 978-1-138-03224-8

Safety versus cost of linear structures in public procurement

J. Vlčková & S. Henková
Faculty of Civil Engineering, Brno University of Technology, Brno, Czech Republic

ABSTRACT: The article focuses on risk planning during the preparation stage of a public procurement procedure. Its aim is to show the necessity of the timely incorporation of costs of safety measures into bills of quantities as price is usually the most important evaluation criterion in public procurement. By incorporating safety elements, the same entry conditions are created for all applicants. An effort is made to provide a sufficient amount of arguments to aid in understanding the necessity of the timely production of an Occupation Health and Safety Plan for a construction site during the construction planning stage. The plan includes safety measures which will affect the prices quoted for the project during the public procurement procedure, and also the length of the construction process.

1 INTRODUCTION

Construction production is a field of industry characterized by specific preparation, organisation and execution. In contrast with other areas of manufacturing, the building industry has rather specific approaches to the preparation, organization and mainly the execution of the work it does. As in other branches, in the building industry the final result of any construction activity depends on many circumstances. The main aim is the creation of a structure which is of the highest quality possible for the most acceptable price. If we look at the realization of the construction plan through the eyes of the client, they are interested in obtaining the structure they have invested in as fast as possible, with the highest level of quality and user comfort. From the contractor's point of view, they need to build a structure for which they will be paid well. It needs to serve well, too, so that they don´t have many duties to fulfil during the warranty period.

In the research community lot of papers deals with different approaches to optimize OHS problems were published (Glendon, A.I. & Litherland, D.K., 2001) (Niu, M. et al. 2016). Also many information technology applications were presented (Skibniewski, M.J., 2014) in the OHS field. From the available sources, there was no contribution that solves OHS in relation with price and public procurement conditions.

Currently, a completely new public procurement law is being prepared which should take other priorities into consideration than just the bid price. Some of these aspects are, for example, protection of the environment (Kantová et al. 2014) or documentation aimed at ensuring occupational health and safety practices are of high quality during the execution of construction. We would therefore like to focus on one of these aspects of construction preparation in our article. It is something which influences all other areas: occupational safety.

Occupational safety needs to be dealt with as soon as the documentation for a project starts being prepared. Planners should incorporate the requirements for the safe execution of construction work into the project documentation after consultation with the occupational health and safety (hereinafter OHS) coordinator, and the needed financial sums should consequently be incorporated into the tender budget.

The attitude of a building company to how OHS is handled says a lot about the quality of the company and its approach to the values created. It is the duty of every employer to look out for risks, take measures to eliminate them, and learn from any deficiencies and accidents that may arise.

One large, high-risk group of structures is that of linear structures. During their construction, the probability that undesirable phenomena will occur is related mainly to insufficient safety precautions at the construction site, and to deficiencies in measures taken at excavations against workers falling into them (Čech 2013). There can also be inadequate information about the composition and quality of foundation soils, which is connected with the use of an insufficient amount of probes. This results in a lack of knowledge regarding the real geological conditions present at the construction site. During the realization of linear structures, new risks can also arise very often in relation to non-compliance with technical procedures, the proposal of unsuitable construction machines (Štěrba et al. 2013), (Štěrba 2014), a change in the financial situation of the construction firm or also to the occurrence of undesirable phenomena due to acts of God. It is therefore an essential part of risk management to continuously monitor the state of geological conditions and evaluate them in comparison with the state that is assumed to exist as found in the project documentation. The project documentation should also include various technical measures enabling soil behaviour to be kept within the required limits and thus provide a safe working environment. Financial reserves for the provision of these measures are also an inseparable part of the project.

2 PRECAUTION TO ELIMINATE RISKS

How should risk management be approached in the case of linear structures in order to ensure the safe realization of such structures? Unlike in the case of a construction site for buildings, where the site equipment is usually placed within one compact area, linear structures mainly feature zones with storage sites along the length of the constructed structure, while the buildings housing the site's equipment and facilities are concentrated in one suitable place. In the case of extensive linear structures, one independent construction yard is built where all the needed structures containing equipment and facilities are concentrated and from which workers are transported to individual parts of the construction site, i.e. individual work sites. Such a construction site must mainly fulfil technical requirements, which are:

- to prevent unauthorized persons from entering the site,
- to separate construction site operations from activities taking place around the site,
- to ensure the construction site is visibly marked even when visibility is low, or at night
- to provide safe entry and exit routes for machines using local roads,
- to select suitable safety measures that prevent employees from falling into excavations,
- to provide safe storage of construction material in a manner that does not endanger either workers or activities taking place in the surrounding area,
- to ensure workers and machines can move around safely during the execution of construction work,
- to provide healthy and sanitary conditions for workers at the construction site,
- to create basic conditions for the protection of the environment, which mainly involves the use of machines that are in good technical condition,
- to arrange waste sorting and collection.

Organizational measures include:

- to familiarize workers with selected work procedures,
- to familiarize workers with safety and fire risks,
- to ensure prescribed personal protective equipment is used, mainly high-visibility workwear, suitable work shoes and protective helmets,
- to notify workers concerning the prohibition against the use of alcoholic beverages and narcotics,

- to notify workers concerning the prohibition against movement under suspended loads or in the danger zones of machinery, and the prohibition against entering uncovered excavations and loading the edges of excavations within a minimum distance of 0.5 m, etc.,
- to appoint a person responsible for safe site operation and compliance with all set rules.

The technical and organizational rules stated above have been gradually introduced on the basis of experience obtained from repeated situations that have occurred at linear structures. Conformance to these rules is a prerequisite for ensuring that a place of work (construction site) is safe and harmless to health. The best way of ensuring that the systematic regulations detailed above are followed is via regular occupational safety training. Linear structures are in most cases public procurement which are funded by the state or multinational sector in the Czech Republic as well as abroad.

In recent years, when contractors have been chosen according to the lowest price offered, it often happened that contractors entered selection procedures with an intentionally underestimated price just to get the contract. Of course, lowering the price had the greatest impact on OHS. For these reasons, efforts are now being made to prepare procurement procedures in such a way that all participants compete on equal terms. One of the criteria should be the incorporation of specific OHS requirements into the project documentation, making it possible to ensure that the needed finances are taken into account in the tender budget, which the bids of participating would-be contractors are based on. It should be every investor's aim to have structures built which are both economically and safely planned.

It is therefore important that public procurements are organized so contractors are selected that have experience with the given kind of work, which is mainly visible from their references. Such a contractor is able to realize an order in a timely, high-quality and safe manner without unnecessary extra costs, if sufficient time is given for the preparation of the construction work. Good preparation for the selection of an experienced contractor should be part of every ordering party's investment plan, and this is especially true when the ordering party is holding a public procurement procedure. Well-prepared project documentation is needed, as is the production of an OHS plan for the preparations for construction work. When producing the plan, its creator must propose suitable safety measures aimed at risks which can occur during individual work procedures or during the individual stages of construction. These measures for the elimination of risks are subsequently incorporated into the tender budget. Selection procedure participants will decide on the pricing of the selected safety measures when producing their bids. In this way, the same terms and conditions will be provided for each contractor during the selection process. When determining the criteria for the selection of the contractor in public procurement proceedings, the ordering party should not only be interested in the lowest price but also in the quality as well as the safe and timely realization of the order.

It is stated in the proposal for the new public procurement law that building companies should include the costs of safety measures in their bids. It is also stated that it will be possible to exclude unreliable contractors which have failed to complete previous orders or have produced work of unsatisfactory quality, based on negative references.

Over the last year, with regard to EU grants related to the completion of utility networks in municipalities, and to grants for the development and improvement of the quality of roads, the number of linear structures built in the Czech Republic increased significantly. With the growing number of these structures in towns and densely populated areas, the risk of the occurrence of injuries resulting from the inadequate safeguarding and marking of these structures is also increasing. A technically well-prepared construction project includes a sufficient amount of safety elements which eliminate common safety risks.

After consultations with certain enlightened contractors who are not indifferent to the health and safety of their employees, we decided to create a set of safety measures for linear structures. With such structures it is important to classify the construction project correctly according to its location, i.e. whether it is in a built-up area with many people moving around, or if it is on the edge of a built up area or just in location where other human activity is not expected. Linear structures are very specific with regard to the realization of work and OHS measures. In addition, earthwork construction (and not only those related to linear structures) ranks among those tasks where possible injuries very often result in serious damage to the health of workers, or even death. When a worker or machine is buried in material, the pressure on the worker's body is so great that in most cases the result is instant death, or such serious injuries that death eventually follows. Safeguards in place at excavations are sometimes inadequate, mainly during the execution of short-term and less extensive earthworks which are performed by contractors with insufficient experience and whose managers are either unaware of all the risks or underestimate them. We want to create a new perspective on these issues and offer the proposal of safety measures aimed at fulfilling OHS requirements so that the risks involved in the execution and safeguarding of earthworks will not be underestimated and will be dealt with according to the specific conditions at the given construction site. Such risks are related mainly to activities taking place in the surroundings of the construction site, the need to propose suitable construction machines for the execution process, and also to the movement of persons at the locations where construction is expected.

In the case of linear structures, we are also interested in how excavations are safeguarded to prevent workers from falling into the depths (Figure 1), as well as whether the construction site is in a protected zone for utility networks, railway or trolleybus lines, roads, etc.

We also need to deal with material storage options from the aspect of the placement, stability and most of all size of storage areas, which are linked to requirements for the supply of material for the construction process. When cutting down green vegetation both in areas which have not been built on and those which have, it is necessary to guard against the presence of unauthorized persons in the danger area. Machines should be completely stopped when fuel is added. It is necessary to prevent oil products or oils from leaking into ground waters. When costs are high, they are mainly related to safety measures on roads. This is particularly the case when exceptional traffic control measures are needed, such as diversions, or when mobile light signalling devices are used (Figure 2), when a road is partially closed or narrowed (Figure 3), during the execution of a crosscut, while ensuring the needed amount of crossings are available, and during the provision of facilities allowing the safe movement of persons with reduced mobility and orientation, etc.

Figure 1. Incorrect fall-prevention measures, earth fall danger and short distance from machinery at an excavation.

Figure 2. The use of a mobile lighting signalling device and traffic signs.

Figure 3. Mobile traffic signs on a road.

3 CONCLUSION

OSH content of the plan is to establish working procedures addressing the security arrangements for the construction project with regard to the specific conditions of the planned buildings so that these measures ensure the safety of all persons involved in the construction and safe movement around public buildings. OSH problematics is very wide and is therefore where to direct further research. The authors also want to focus on objects for housing or industrial buildings. Our idea is to extend the methodology. We want to incorporate sorting according to classification of structures and set the documents for the pricing, especially financially challenging activities in terms of occupational health and safety. Our planned program risks will facilitate the correct valuation of linear structures in compliance with the requirement to secure implementation of the construction works of the highest quality and at the most reasonable price.

REFERENCES

Čech, D., 2013. Stavitelství do kapsy (in Czech). 1st edition, Prague: Published for the Czech Chamber of Certified Engineers and Technicians Active in Construction (ČKAIT) by the ČKAIT Information Centre.

Glendon, A.I. & Litherland, D.K., 2001. Safety climate factors, group differences and safety behaviour in road construction. *Safety Science*, 39(3), pp. 157–188. Available at: https://www.scopus.com/inward/record.uri?eid = 2-s2.0-0034841245&partnerID = 40&md5 = a05545b00e62533f0b686bcf2abba68b.

Kantova, R. & Motyčka, V., 2014, Construction Site Noise and its Influence on Protected Area of the Existing Buildings. ISSN 1022-6680, ISBN 978-80-214-5003-5. Switzerland: Trans Tech publications Ltd.

Niu, M., Leicht, R.M. & Rowlinson, S., 2016. Overview and Analysis of Safety Climate Studies in the Construction Industry. Construction Research Congress 2016: Old and New Construction Technologies Converge in Historic San Juan - Proceedings of the 2016 Construction Research Congress, CRC 2016, 2016, pp.2926–2935. Available at: https://www.scopus.com/inward/record.uri?eid = 2-s2.0-84976349168&partnerID = 40&md5 = 95a720a4bb37cf776bfb52b36c88963f.

Skibniewski, M.J., 2014. Information technology applications in construction safety assurance. *Journal of Civil Engineering and Management*, 20(6), pp.778–794. Available at: http://www.tandfonline.com/doi/abs/10.3846/13923730.2014.987693.

Štěrba, M., Čech, D. & Venkrbec, V., 2013. Návrh základních stavebních strojů pro zemní práce (in Czech). Silnice a železnice, 8(03/2013), pp. 88–89.

Štěrba, M., 2014. Effective Design Of Construction Machinery And Machine Assembly In Construction. In Integrated Approaches to the Design and Management of Building Reconstruction. Brussels: EuroScientia vzw, pp. 73–77.

Advances and Trends in Engineering Sciences and Technologies II – Al Ali & Platko (Eds)
© 2017 Taylor & Francis Group, ISBN 978-1-138-03224-8

The problem of sufficient/insufficient airflow in roofs of residential buildings

J. Vojtus & M. Kovac
Civil Engineering Faculty, Technical University of Kosice, Kosice, Slovakia

ABSTRACT: Flat roofs, namely double-shell roofs with low high of ventilated air gap typically have several faults in light of thermal and moisture regime. The low surface temperature of upper construction and presence of hot and moist air from interior in the roof air gap causes condensation of water vapour on the cold constructions. The analyzed flat roof construction does not contain any vapour barrier. The aim of this paper is to gain pictures of airflow in the air gap of the roof and to determine the surface temperatures in the roof air gap. Due to the obtained results there will be specified points in the double-shell roof where the condensation may form. On the basis of this knowledge we will be able to modify the geometric of the roof, what can improve the function of this exposed structure.

1   INTRODUCTION

It is a fact that flat and low-slope roofs demand careful detailing and good workmanship. While a flat or low-slope roof can offer a long service life (20 years or more) a small mistake can lead to a big leak. Flat roofs and low slope roofs also face potentially serious condensation problems that can in turn lead to costly rot or mold damage in buildings. The most common flat and low slope roof leaks occur at flashings and roof penetrations such as at plumbing vents, chimneys, and roof-mounted air conditioners or heat pumps. Roof flashing details that are not designed to absorb thermal or other building movement (thermal expansion of materials for a table of the coefficient of expansion of common building materials including brick, concrete, mortar, and stone) can lead to cracked broken metal flashings that leak badly into the building. While a well-installed flat or low slope roof can keep outside rain or snow-melt out of the building, water entering the roof cavity from inside the building in the form of water vapour can be more troublesome. For example, moisture collecting as condensation in fiberglass roof insulation may leave the insulation with serious mold contamination even though the insulation still looks "clean". Fear of condensation problems has led some roofers to add special breather vents to these compact roofs. Although breather vents are recommended by the National Roofing Contractors Association (NRCA)—one vent every 93 meter per square is specified—NRCA technical manager Wayne Tobiasson, who has studied flat roofs extensively for the U.S. Army Corps of Engineers Cold Regions Research and Engineering Laboratory (CRREL), goes further and says that these vents are "foolishness," particularly in flat roofs without vapour retarders. In these roofs, Tobiasson said, if the vents do anything, they will create problems by inducing airflow up through the ceiling from below. Roofs with non-permeable insulation tightly sandwiched between the deck and roofing are usually free of condensation problems except in the far north or in buildings with high moisture levels (InspectAPedia 2016).

However even a compact-roof with good indoor vapour barrier design can suffer from under-roof moisture condensation, that is, condensation under the roof inside the occupied space, if the building interior moisture levels are excessive and proper ventilation or dehumidification are not provided. Condensation within a flat roof mainly occurs during cold weather when moisture vapour in the air which has been generated within the heated building rises from the room below into the cold roof void above the ceiling. When the temperature of the vapour falls to or below its dew point the water vapour condenses on cold surfaces (BuildingRegs4Plans Premium 2016).

The text above describes flat and low slop roofs without ventilated air gap. In our climatic conditions were designed the flat roofs with ventilated air gap to protect roofs against condensation of air humidity. The construction of flat roof is split by the air gap in two parts. The upper construction protects the roof against rain and snow. The bottom construction of the roof contains the thermal insulation and that protects the building against thermal losses. In the external walls are situated circular openings for inlet and outlet of air. The ventilation efficiency of the air gap is influenced by many factors, for example roof area or number and position of openings for inlet and outlet of air. This article analyzes one type of flat roof with ventilated air gap. This construction of flat roof was commonly used for the apartment buildings in our country at the end of last century (Bludau, Ch. & Zirkelbach, D. 2008).

2 CFD ANALYSIS OF FLAT ROOF WITH VENTILATED AIR GAP

2.1 Geometrical and physical model

The geometric model of flat roof was created in the first step. The thickness of the air gap is 150 mm. Multilayer construction over the air gap of the flat roof protects the building against rain, snow etc. Multilayer construction under the air gap of the flat roof protects the building (heated space) against the thermal losses. The vapour barrier is not used in this flat roof construction. The next table (Table 1.) contains some parameters of the used physical model and some boundary conditions in CFD analysis (ANSYS CFX Introduction. 2012).

2.2 Analysis of results from CFD simulation

From the reference database for the location of Kosice, the incidence of wind speed is greater than 2.0 m/s and less than 5.0 m/s average of 43% during the year. The Figure 1 left shows the air velocity in the air gap of the roof by wind velocity at 2.0 m/s. In this case outdoor air temperature is 3°C, which is the average outside temperature during the winter (heating season) in Kosice. In the middle of the Figure 1 are showed surface temperatures (in degrees Celsius) at the bottom layer under air cavity of roof. We see that the surface temperatures in this case are in the range between 3.0 and 7.0°C. At the Figure 1 right are showed the surface temperatures (in degrees Celsius) of the upper layer over air gap of roof. There are the surface temperatures lower. This situation creates the risk of a possible condensation of warm and moist air from the heated space in the air gap of flat roof.

The Figure 2 illustrates places in the flat roof structure, which is a potential risk of condensation. For example a vertical shaft used for the distribution systems of building services (plumbing, ventilation ducts). Ventilation ducts are used to take the moisture load from sanitary facilities (toilet, bathroom) and kitchen out. The crossing of ventilation ducts through

Table 1.   Boundary conditions of the physical model parameter.

Thickness of air gap	150 mm
Diameter of air inlets and air outlets (circular openings) in external wall	60 mm
CFD model	3-dimensional analysis
Physical model	non-isothermal
Radiation model	surface to surface
Outside temperature	3°C
Air velocity of wind	2.0 m/s
Turbulent model	k-epsilon
Air temperature of heated space (room)	22°C
U-value (construction under air gap of roof)	0.52 W/(m^2.K)
U-value (construction over air gap of roof)	2.94 W/(m^2.K)

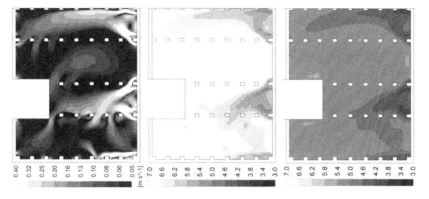

Figure 1.   Air velocity (left) surface temperatures (middle and right) in the air gap of flat roof.

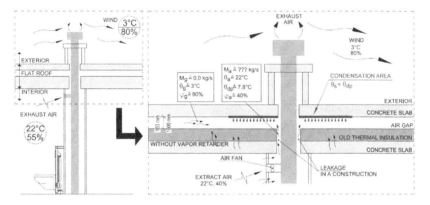

Figure 2.   Sketch of double skin roof with vertical shaft and leakages in building construction.

the ceiling constructions (and the flat roof) occur leakages in the building construction. These places represent the path of natural motion of warm and humid air from heated space into the air gap of flat roof. Air velocity in some parts of the air gap of roof is very low (0.05 m/s). Warm and moist air (air temperature $\theta_a = 22°C$, relative humidity $\varphi_a = 40\%$) from heated space of building could get into the air gap of roof and could condense in these parts of the air gap with cold surfaces.

In this situation we also analyzed the moisture regime of air gap of double skin roof. We know air velocity in the gap is low (0.05 m/s) in some places of the flat roof. If the warm and humid air enters from heated space through leakages in building construction into the air gap of flat roof, it can condensate on the cold surfaces.

There is a mixing process and in order to obtain new parameters of air after mixing process was used the psychrometric chart (Mollier h-x diagram). If we mix the air in the roof (velocity 0.05 m/s, temperature 3°C, relative humidity 80%) with the air from heated space (air temperature 22°C, relative humidity 40%) the final mixed air will have its dew point temperature at the value of 7.6°C. The result is that the water vapour contained in the air condensates on cold surfaces in the air gap of flat roof. The surface temperatures on the upper layer of the air gap are lower than 7.0°C. This value is lower than the dew point temperature 7.6°C. This is the situation where the warm and moist air could condense. It is very difficult to quantify the amount of warm and moist air that gets into the air gap of double skin roof. There are natural mechanisms, natural air motion from the temperature and pressure differences between the spaces in the building.

Double shell-roofs with ventilated air gap were designed in order to take warm and moist air away from the roof. These flat roofs did not contain any vapour barrier. Warm and moist

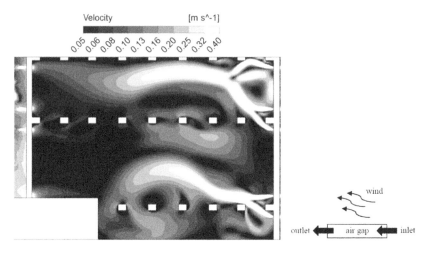

Figure 3.   Air velocity in the air gap of flat roof without additional ventilation tubes.

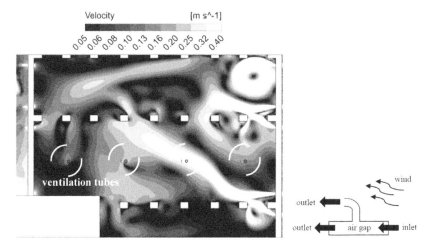

Figure 4.   Air velocity in the air gap of roof with ventilation tubes (outlets in direction of blowing wind).

air diffuses from heated space of building into the air gap of the roof. At this place the warm and moist air should mix with air from exterior environment. But the efficiency of ventilation in the air gap of roof is poor. There are many places with too much low velocity of motion air. Therefore we focused on options how to increase the motion air (its velocity) in the air gap of flat roof. We placed the ventilation tubes in the roof construction. These so called ventilation chimneys have a diameter of 100 mm. The Figure 3 shows motion air and its velocity in the air gap of roof without additional ventilation tubes (chimneys). We can see the large area with a low velocity of motion air.

The Figure 4 shows motion air and its velocity in the air gap of roof with additional ventilation tubes (chimneys). The outlets of ventilation tubes are oriented in direction of the blowing wind. If we compare these results with the previous figure 3 we can see the smaller area with a stagnation of air. The direction of outlets from ventilation tubes is oriented in direction of the blowing wind and helps to increase the ventilation of the air gap in the flat roof

Vice versa the outlets from ventilation tubes (chimneys) oriented opposite the blowing wind cause decreasing of ventilation in the air gap of the roof (Figure 5). In this case the blowing wind enters directly into the air gap through the openings of ventilation tubes. But

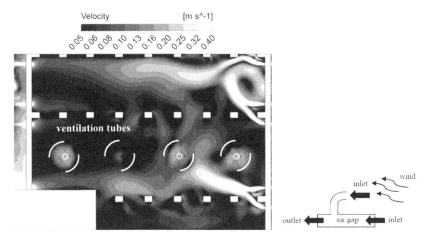

Figure 5. Air velocity in the air gap of flat roof with ventilation tubes (outlets opposite the blowing wind).

the thickness of the air gap is too low and this situation creates large turbulence and recirculation of air. The external air cannot get into the air gap effective. There is strong air recirculation in the air gap of roof under the ventilation tubes. We think that we could improve the ventilation efficiency in the air gap of the flat roof by the change of geometry shape of ventilation tubes inside the air gap.

3 CONCLUSION

The aim of this article was to analyze the air velocity, thermal and moisture regime of the flat roof with ventilated air gap. The velocity of wind around the building was modeled as a continuous stream with a constant value at 2.0 m/s. But the velocity of wind is not constant in real life. The change of wind velocity around the building causes the change of air velocity in the air gap of flat roof. The openings for inlet and outlet of air in the roof were simulated as the openings without net against birds. The results from CFD simulation showed that this flat roof with air gap has the main problem with efficiency of ventilation of the air gap. The warm and moist air in the air gap from the heated space of building can condense on cold surfaces in the air gap of roof. If we use additionally ventilation tubes we will increase the efficiency of ventilation in the air gap (outputs of ventilation tubes are in direction of the blowing wind). But it is needed to change the geometry of ventilation tubes inside the air gap of roof in order to improve the distribution of air in the air gap (outputs of ventilation tubes are opposite the blowing wind).

ACKNOWLEDGEMENT

This paper is the result of the Projects implementation: KEGA 052 TUKE-4/2013—The implementation of a virtual laboratory for designing energy-efficient buildings, and University Science Park TECHNICOM for Innovation Applications Supported by Knowledge Technology, ITMS: 26220220182, supported by the Research & Development Operational Programme funded by the ERDF.

REFERENCES

ANSYS CFX Introduction. 2012. *ANSYS Inc*. Canonsburg: USA.

Bludau, Ch. & Zirkelbach, D. 2008. Condensation problems in cool roofs. 11th International Conference on Durability of Building Materials and Components (DBMC), Instanbul, Turkey, 11–14 May 2008.

BuildingRegs4Plans Premium 2016. *Flat Roof—Condensation*. Available on http://www.buildingregs4plans.co.uk/guidance_flat_roof_condensation.php.

InspectAPedia 2016. Free *Encyclopedia of Building & Environmental Inspection, Testing, Diagnosis, Repair*. Available on http://inspectapedia.com/Energy/Flat_Roof_Moisture.htm.

Advances and Trends in Engineering Sciences and Technologies II – Al Ali & Platko (Eds)
© 2017 Taylor & Francis Group, ISBN 978-1-138-03224-8

Traffic calming as a means of improvement of the urban environment

M. Záhora, D. Vaněk & M. Peřinková
VŠB, Technical University of Ostrava, Ostrava, Czech Republic

ABSTRACT: The main subject of this paper is to survey traffic calming methodology in European cities. Traffic calming has a significant impact on improvement of the environment in city centres, particularly reduction of air pollution through emissions and restriction of noise from motor vehicle traffic. Emphasis is placed on the applications of the individual methods implemented in Rome, London, Stockholm and Munich. Through the long-term monitoring of the traffic situation, the pros and cons of the individual methods can be evaluated in these cities. The paper aims to find a broader context for implementation of the individual methods and their adaptation to the traffic calming conditions in Czech Republic. Pursuant to the facts obtained, a study of the current measures in Ostrava has been executed, together with a draft of suitable methods to reduce car traffic and improve the environment in the broader city centre

1 INTRODUCTION

The increase in motor vehicle traffic at the end of the 20th century brought new problems into the world's big cities. At the beginning of the 20th century, cities were particularly designed for pedestrian traffic; cars were exceptional in the streets. The rising popularity of cars, however, raised new questions which needed to be solved with regard to the concept of long-term urban planning. Thus municipal plans gradually included road and highway corridors, which uncompromisingly divide the landscape and cities into subparts. In the 1980s, there was a substantial increase in vehicle numbers in Western Europe. Motor affect cities negatively in the form of increased air pollution, noise as well as space problems in older estates. In the Czech Republic and other Eastern European countries, the development of car traffic was slightly delayed. Thus the time is approaching to look for suitable measures to calm traffic in city centres and tourist areas. A suitable method of seeking the correct solution is the analysis of solutions, which have already been applied and implemented in many Western European cities in the past.

2 TRAFFIC REGULATION AND CALMING POSSIBILITIES

2.1 *Direct methods*

Direct traffic regulation methods involve measures directly influencing car traffic by slowing down, segregating or changing its routes. This particularly includes systematic creation of city by-passes aimed at diverting transit traffic outside built-up areas of the city. In this way, negative effects are eliminated, including the occurrence of congestions in cities. Other methods involve pricing, or a complete ban on entry of vehicles into target locations, such as city centres, tourist areas, spa towns, etc. In traffic calming terms, this involves speed reduction by means of vertical road signs or other technical features (speed bumps, roadway narrowing through barriers, etc.). Less frequented streets gradually change into residential zones. In the course of time, busy arterials, originally intended for cars only, are transformed into city avenues with mutually balanced movement of vehicles and pedestrians.

2.2 *Indirect methods*

These are methods of support means aimed at motivating inhabitants and offering them available alternatives instead of car traffic. In the case of efforts towards regulation in city centres, such conditions are created as to make driving into the city more problematic than other modes of transport. The number of parking places is reduced; residents-only parking zones are introduced. Simultaneously, programmes are implemented in support of other transport modes. On the fringes of city centres, parking areas are created with direct links to public transport lines (car parks of the Park & Ride type). Public transport links running through the regulated area are reinforced. Cycle tracks are extended, or bicycle hire systems are introduced in cities.

The indirect methods often supplement those direct ones; practical experience, however, has revealed that they themselves can be a very functional method of efficient traffic regulation. At present, significant help is also provided by information technologies which can promptly inform inhabitants on the current state and possible alternative solutions, free parking places in the Park & Ride facilities. Moreover, they enable planning of the route if the vehicle is parked in a car park with a link to public transport, etc.

3 STATISTICAL DATA AND PREDICTIONS OF THE DEVELOPMENT OF CAR NUMBERS

In 2010 the number of cars in the world exceeded the limit of one billion (Table 1). According to the latest predictions, the two-billion limit should be achieved in 2035, which constitutes a twofold increase in a mere 25 years. Therefore, determination of the correct traffic planning and regulation concept is an essential tool for the future development and stability of big cities.

4 APPLIED METHODS OF TRAFFIC CALMING IN EUROPE

4.1 *London, charged entry into the city centre*

In 2003, a charge was introduced in London for entry into the closer city centre on working days in order to calm traffic and thus improve the air in the centre. Introduction of the charge was supplemented with detailed monitoring enabling more accurate analysis of the efficiency of the measures taken. With regard to the success of the measure, the paid zone was further extended westwards in 2007. Residents are not exempt from the charge; they are, nevertheless, granted a 90% discount. Entry is only free for the disabled (Blue Badge), motorcycles and cars conforming to the EURO 5 standard with emissions of no more than 75 g CO_2 per 1 km. The analyses reveal that the net profit from the collected charges and fines amounted

Table 1. Number of cars in the world (Source: Ward's automotive reports).

Year	Personal	Commercial	Total
2006	638,563,029	260,411,877	898,974,906
2007	658,111,828	271,448,901	929,560,729
2008	681,634,108	280,140,244	961,774,352
2009	684,569,516	295,114,868	979,684,384
2010	707,764,219	307,496,608	1,015,260,827
...			
2035			2,000,000,000*

*Estimation.

to GBP 90 million in 2005 (Transport for London 2006), and was further increasing with the rising daily charge. In 2007, the annual profit after the charge increase was GBP 137 million (Transport for London 2008). Moreover, profit from the charges collected was then used to improve technical infrastructure, streets and pavements in the centre and build new pedestrian ways and cycle tracks. The resultant effect was a reduction in car numbers of 15–20% between 2003 and 2013, as their owners chose other methods of entry into the city centre. Gradual restriction of the emission limits tolerated has resulted not only in increased profits, but also in reduced CO_2 emissions in the air by up to 16% (30,000 t annually).

4.2 *Rome, entry into the city centre for registered vehicles only*

Similar results were also achieved in Rome. Introduction of a controlled zone in the city centre in 1989 resulted in traffic reduction of up to 20%. The centre is only accessible to preregistered vehicles equipped with a dashboard unit which has also been electronically controlled by a system of cameras and sensors along the streets since 2001 (Musso & Corazza, 2006). In this case, exemption was granted to entrepreneurs with their registered office in the zone, residents, the handicapped, city transport and single-track motor vehicles. Introduction of the paid entry, however, also showed the necessity to check the exemptions granted. After the entry charge had been introduced, there was a significant increase in the number of entrepreneurs in the controlled area by reason of the exemption from the charges. Consequently, only 19% of the cars entering the centre pay the charge; the rest are exempt from payment (Řezáč & Fencl, 2009). In principle, the system forces tourists and occasional visitors to park their vehicles outside the city centre, using the Park & Ride car parks and public transport for further journey to the centre. Tourists in hotels inside the city centre are allowed to drive to the hotel upon previous notification to the dispatching centre.

4.3 *Stockholm, charged entry into the city centre with reinforced infrastructure*

In August 2005, a programme of charged entry into the city centre by means of toll gates was implemented in Stockholm. The programme also included a preparatory stage, within which bus services towards the charged city centre were reinforced in advance and sufficiently large parking areas for Park & Ride car parks were built on the boundaries of the paid zones. The positive effect of the programme was a decrease in traffic over the zone boundaries of 22% (Centre for Transport Studies, 2014). On the contrary, negative influences took effect on the circular routes along the zone perimeter, where the traffic intensity by contrast increased by 10%. Upon completion of the test programme, a referendum was held among the citizens, pursuant to which further operation of the programme was validated.

4.4 *Munich, city for cyclists*

Munich is a city with an elaborated public transport system involving buses, underground, trams and S-Bahn. The lines concur and the individual transport types are interconnected by means of transfer terminals. In 1996, the city set a target to increase the number of journeys ridden by bicycle from the then 6% to 15% in 2015. The strong media and financial support meant an increase to 14% in 2008, which enabled an increase in the original predictions and setting of a new target at the level of 17% for 2015. This effort resulted in a dense network of cycle tracks. Over 50% of the street network contains cycle lanes. There is a developing bicycle hire system operated by German Railways not only in Munich, but also in other German towns. Location of bicycle racks at railway and public buildings directly offers the possibility of the train-bicycle connection as an ideal mode of transport. In 2013, the total length of cycle tracks reached 1200 km, with a view of extension by another 200 km in the coming years. 60% of the inhabitants' journeys completed round the city are shorter than 5 km, therefore, ideal for a bicycle journey. According to statistics, 80% of Munich inhabitants own at least one bicycle (Zorn & Lonhard, 2013). Similarly, cyclists are also supported in Vienna, although they have not yet reached such significant successes as in Munich.

5 APPLIED METHODS OF TRAFFIC CALMING IN THE CZECH REPUBLIC

5.1 *Czech Republic*

Whereas in Western European countries the highest increase in the number of cars occurred in the 1980s, it was not until a decade later that similar development took place in the Czech Republic. For this reason, we are just in the period of primary interventions into the city structures, intended to resolve the capacity and ecological consequences of the increased car numbers in our cities.

According to statistical data (Figure 1), there was an apparent increase in car numbers of 63% in the Czech Republic between 1993 and 2012; in absolute numbers, this means over 2 million new cars put into operation. Therefore, implementation of suitable regulatory methods has become a topical issue for the biggest cities in the Czech Republic. In ecological terms, besides the total number, it is also the age of the cars that is essential. Owing to their technological design and physical age, older engines feature significant emissions in comparison with cars with newer, more environmentally-friendly engines.

The average age of cars in the Czech Republic is a relatively stable value. Statistical variations (Tab. 2) have been caused by interventions of the state in particular. These involve introduction of compulsory third party liability insurance, which led to mass decommissioning of unused cars, as well as the introduction of fees for the re-registration of vehicles.

5.2 *Employed systems of traffic regulation in the Czech Republic*

These rather involve partial modifications of city centres in an effort to resolve, through these interventions, the capacity problems with parking in city centres. The positive factor is the successive implementation of intelligent traffic control systems resulting in better traffic flow. In comparison with other European cities, the low utilization rate of systems for paid entry in city centres is an interesting phenomenon in the Czech Republic. They are used to some extent at tourist locations and spa towns, where they are in addition more used in the form of paid parking, which is the less expensive variant in technical terms. According to statistical

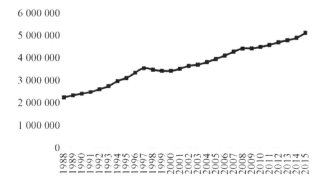

Figure 1.   Number of private cars in the Czech Republic between 1988 and 2015.

Table 2.   Average age of private cars in the Czech Republic.

Year	Av. age	Year	Av. age	Year	Av. age	Year	Av. age
1996	14.18	2001	13.61	2006	13.87	2011	13.83
1997	14.15	2002	13.72	2007	13.93	2012	13.90
1998	14.29	2003	13.49	2008	13.82	2013	14.20
1999	13.72	2004	13.53	2009	13.65	2014	14.49
2000	13.51	2005	13.82	2010	13.70	2015	14.53

Table 3. Selected statistics of Ostrava (Ostravské komunikace, a. s. 2011–2015).

Year	Cars	Length of streets	Bicycle network	Number of passengers in public transportation
2010	178,809	829,597 km	201 km	99,980,000
2011	181,845	819,840 km	203 km	108,710,000
2012	184,756	821,501 km	206 km	104,490,000
2013	176,100	823,382 km	209 km	93,480,000
2014	178,670	824,734 km	224 km	91,000,000

data of the Ministry of Finance, only CZK 22 million per year was collected on charges for entry in all towns and municipalities.

5.3 Traffic impact on air pollution in Ostrava

Ostrava is the 3rd largest city in the Czech Republic; with almost 300,000 inhabitants and is the largest city in the Moravian-Silesian Region. Since the 1990s, Ostrava has been undergoing gradual transformation from a purely industrial city to a cultural metropolis. As a result of its location and quantity of industrial sites, the Ostrava region is an area with the highest level of air pollution. In spite of ongoing efforts to impose stricter emission limits on the largest producers of pollution, the values of air pollutants are still above the determined limits in the long term. On a long-term basis, emission values are monitored and regulated, thank to which they have long been kept at a constant level. With the rising traffic intensity, the effect of emissions produced by cars starts to manifest itself more significantly.

According to an expert study from 2012 (Centrum dopravního výzkumu 2012), the share of traffic on the discharge of harmful nitrogen oxides NOx in the Ostrava-Město district was determined by calculation at 5% of the total emissions at the temperature of −7°C. With regard to the height at which the emissions are discharged in the case of cars, however, these are ground-level emissions, which have a more substantial impact on humans than the emissions released from the smokestacks of industrial sites. After considering this impact, the study indicates an expert estimate of the share of traffic in the total air pollution at the level of 16%. This is already such a significant value that it is appropriate to deal with the issue of regulation possibilities more intensely.

Traffic is a significant factor with regard to noise. According to the noise map elaborated, 11% of Ostrava inhabitants suffer from excessive noise. It is road traffic that is its largest source. In terms of noise and emissions, the main factor is the speed and flow of transport. The worst state is constituted by traffic congestions, at which emissions and noise produced by the ever-starting vehicles significantly rise. Therefore, in addition to traffic calming, it is also desirable to strive for traffic fluidity by appropriate planning in the area of transport engineering (Ostrava City Authority 2009).

5.4 Traffic regulation in Ostrava

None of the direct regulation methods (chapter 2.1) asserts itself to a larger extent within Ostrava. The north-western by-pass of the centre has been replaced by the completed D1 motorway, which is toll-free on this stretch, substituting for the bypass function in this way. Moreover, the Rudná street extension project is in progress; upon its completion, the I11 road line will be relocated outside the problematic stretch leading through the densely populated district of Ostrava-Poruba. Further modifications in this area are not planned in the near future; neither are they necessary with regard to further possible traffic regulation.

In the area of indirect traffic regulation, various supporting projects are implemented; in particular, new transfer terminals are built to facilitate interconnection of the city and suburban public transport. Successively, intelligent control systems are introduced, flexibly responding to the current situation and supporting the traffic flow.

Meanwhile, the traffic regulation in the city centre is only solved by restrictions of free car parking in the form of parking zones where only residents can park, having received and paid for an annual permit with the authorities. This provision, together with consistent monitoring and penalization of the ban violation, is a good method of discouraging some people from driving to the centre. On the other hand, there are still a certain number of places in which one can legally park free of charge, which is counter-productive—in the effort to find a free place, drivers drive as much as several kilometres, unnecessarily increasing the amount of emissions and noise in the densely populated parts of the city. If these places enabling free parking were abolished, people would gradually learn not to rely on being able to park directly in the city centre and use the Park & Ride car parks on the fringe of the centre more frequently.

Drafting the recommended new methods of regulation, one should take into account the financial capacity of the city in combination with the anticipated effect. The centre of Ostrava is not so congested with traffic that one should necessarily introduce a system of paid entry. From this point of view, the system of paid parking is more suitable and decidedly more economical in terms of operating costs; nevertheless, it has to be checked thoroughly, or supplemented with a media campaign accentuating the possibility of using the Park & Ride car parks.

Of the direct methods not yet used so much within Ostrava, traffic calming through the conversion of streets into pedestrian zones and city boulevards should be mentioned, which is, nevertheless, a costly solution. In the first stage, without high costs, it is only possible to set speed regulations on those streets with the highest housing density. In Munich, for example, 80% of the city streets outside the main trunk roads are limited to the maximum speed of 30 km/h, which is checked thoroughly. The resultant effects are only positive. The time needed to drive through the city is extended, which can be a cogent argument for the selection of another mode of transport. Moreover, the accident rate is lowered and the lower speed results in the reduction of emissions discharged.

6 CONCLUSION

In comparison with other European countries, the detailed analysis of the traffic regulation methods already implemented in the Czech Republic has revealed that partial small-scale interventions are preferably used in the Czech Republic, but without a more complex strategy which would join the individual elements into a comprehensive system. With regard to the state of economic advancement in the Czech Republic, it is primarily necessary to find financial resources which will enable not only drafting, but above all implementing a complex solution helping to restore the balance between cars and people in the centres of our towns.

REFERENCES

Centre for Transport Studies Stockholm. 2014. *The Stockholm congestion charges: an overview:* 9–10.
Centrum dopravního výzkumu. 2012. Stanovení podílu produkce emisí z automobilové dopravy vůči ostatním zdrojům znečišťování ovzduší na území Ostravské aglomerace: 4–10.
Musso, A. & Corazza, M.V. 2006. Improving Urban Mobility Management, Case Study of Rome: 54–59.
Ostrava City Authority. 2009. Strategic Plan for the Development of the City of Ostrava 2009–2015: 62–63.
Ostravské komunikace, a.s. 2011. *Informace o dopravě v Ostravě 2010:* 2–24.
Ostravské komunikace, a.s. 2012. *Informace o dopravě v Ostravě 2011:* 2–25.
Ostravské komunikace, a.s. 2013. *Informace o dopravě v Ostravě 2012:* 2–25.
Ostravské komunikace, a.s. 2014. *Informace o dopravě v Ostravě 2013:* 4–28.
Ostravské komunikace, a.s. 2015. *Informace o dopravě v Ostravě 2014:* 4–29.
Řezáč, M. & Fencl, I. 2009. *Vybrané otázky rozvoje dopravy ve městech.* Ostrava: Vysoká škola báňská—Technická univerzita Ostrava.
Transport for London. 2006. Impacts monitoring. Fourth Annual Report.
Transport for London. 2008. Impacts monitoring. Sixth Annual Report.
Zorn, E. & Lonhard, M. 2013. Masterplan "Bicycle Traffic in Munich": 2–7.

Advances and Trends in Engineering Sciences and Technologies II – Al Ali & Platko (Eds)
© 2017 Taylor & Francis Group, ISBN 978-1-138-03224-8

Possibilities of the utilization of natural fibers for insulation materials production

J. Zach, J. Hroudova, V. Novak & M. Reif
Faculty of Civil Engineering, Brno University of Technology, Brno, Czech Republic

ABSTRACT: Thermal protection of buildings as well as the usage of easily renewable resources of raw materials and secondary raw materials belong among the main topics currently being addressed by research organizations practically all over the world. The paper describes research dealing with the possibility of using different kinds of natural fibers for the production of thermal insulation materials. The main aim of the paper is to investigate the dependence of thermal insulation properties on the type and thickness of fibers and the dependence of thermal insulating properties on bulk density.

1 INTRODUCTION

Insulating materials based on natural, easily renewable materials currently represent one of the forms of implementing sustainable development in construction. These materials make it possible to improve the thermal insulation properties of the building envelope with minimal environmental harm in terms of CO_2 emissions and without the usage of non-renewable raw materials. Currently, the issue of increasing the thermal protection of buildings is currently very topical because of the need to implement the requirements of the Directive of the European Parliament 2010/31/EU legislation in European countries (including the Czech Republic and Slovakia).

Thermal insulation materials, which are based on natural and easily renewable raw materials, can also be instrumental in creating a healthy microclimate inside buildings and their usage can have a very positive impact on the health of users (provided the materials are appropriately considered in the structural design).

These materials are normally permeable to water vapor and may accumulate some amount of moisture from the air. The favorable properties of the vegetable matter and materials produced there from are the ability to absorb moisture in the internal pore system at elevated humidity and, conversely, at decreasing humidity gradually release moisture into the surrounding environment. This mechanism positively influences the properties of the internal microclimate in terms of humidity; especially in winter, when there may be a long-term reduction in humidity inside the building.

Another important characteristic of natural materials is their beneficial effect on the human senses. Many people today suffer from allergies and medical indispositions caused by exposure to substances, which were used during the construction. These problems can be avoided by using building materials from natural resources of raw materials. Natural building materials effectively regulate indoor humidity and their characteristic aroma is beneficial for the human psyche. The properties of insulating materials based on natural fibers are in most cases very good and are fully comparable with those of conventional insulators. The disadvantage of most crop-based materials is a certain degree of moisture sensitivity. Excessive moisture can cause long-term biological corrosion and degradation by bacteria or fungi. For this reason, these materials should be incorporated in the building structure and always kept separate from sources of moisture (or they must be allowed to dry rapidly). As long as a wall is properly designed and constructed, damp or another incidence of elevated moisture should

Table 1. Material composition and characterization of the fibers.

Type of fibers	Cellulose % wt	Lignin % wt	Hemicellulose % wt	Pectins % wt	Waxes % wt	Water % wt
Flax	71	2.2	18.6–20.6	2.3	1.7	10.0
Jute	61.1–71.5	12.0–13.0	13.6–20.4	0.2	0.5	12.6
Hemp	70.2–74.4	3.7–5.7	17.9–22.4	0.9	0.8	10.8
Kenaf	45.3–69.8	9.2–19.0	15.0–21.5	0.2	–	7.0–10.0

not pose any danger. Another possibility is to use impregnating preparations, which make it possible to reduce the moisture absorption of natural fibers. In most cases, however, natural insulation materials cannot be exposed to liquid water even after impregnation (Chybík 2009, Vejeliene, Gailius, Vejelis, et al. 2011).

2 THERMAL INSULATING MATERIALS BASED ON NATURAL FIBERS

Natural fibers are divided by their origin into animal (livestock) fibers which are obtained (mainly) from animal fur (e.g. sheep wool), and plant (crop) fibers, which are obtained from the stems, seeds, fruits and leaves of plants (e.g. hemp, flax, cotton fiber). The composition of sheep wool fibers depends on many factors; however on average it is composed from 75% of fibroin and 25% of sericin (Lancashire 2014).

Plant fibers usually have a high content of cellulose, often also high in lignin and hemicellulose. Based on their origin, plant fibers can be further subdivided into bast fibers (flax, hemp, jute, kenaf, ramie …), seed fibers (cotton, kapok …), fibers from plant leaves (sisal, abaca …) and fruits (coconut …) (Ibraheem, Aidy, Khalina 2011). The chemical composition of the selected types of plant fibers is in Table 1.

There are currently several technologies available worldwide, which focus on the production of insulating materials from natural fibers in the form of commercially available insulation mats and boards. These technologies differ in the methods of fiber bonding and fiber orientation. They are generally based on technologies established within the textile industry and are adapted for the manufacture of insulating materials (MacDougall 2008).

Depending on the choice of technology, the resulting properties and quality of the insulators differ. For some technologies, there are general restrictions such as the maximum thickness of the manufactured insulation, maximum or minimum bulk density of the insulation etc. Each production technology is also differently sensitive to the quality and purity of feedstock. In general, there are three basic types of technology: mechanical bonding (needle punching, strengthening water jets), chemical bonding and thermal bonding (bonding with bicomponent fiber, technology vertical laid nonwoven textiles /technology Strutto/).

The most widely used method for the production of insulation materials based on natural fibers is thermal bonding. The principle of the method is reinforcement by melting a deposited binder and subsequent cooling. Various synthetic and natural polymers with a variable melting point are used as binder. Selection of a suitable type of polymer depends on the properties of the processed raw fibers. In the Czech Republic, the so-called air-lay method is most commonly used. It is based on the aerodynamic shaping of the individual layers of the mat. Here, it is the effect of air on the individual fibers which causes their gradual composition into layers.

3 TEST SAMPLES

During the research, test samples of flax and hemp fibers were manufactured, each of two fiber types of different quality (with lower and higher fiber thickness). In addition, samples from sheep wool of 23 and 33 microns in thickness were made. A description of the raw fibers and their main parameters is in Table 2 below. The samples were thermally bonded using 15% polyester bicomponent fibers.

Table 2. Overview of the length and thickness of raw fibers.

Fiber	Description	Thickness mm	Deviation of thickness, mm	Length mm	Deviation of length, mm
H-F	Fine hemp (retted)	0.057	0.024	66.24	19.22
H-C	Coarse hemp (unretted)	0.079	0.018	64.03	24.46
F-C	Coarse flax	0.115	0.025	56.15	20.39
F-F	Fine flax	0.059	0.015	61.25	18.55
W-F	Fine sheep wool	0.024	0.005	28.11	12.04
W-C	Coarse sheep wool	0.041	0.008	92.01	20.22

Table 3. Composition of test samples.

| Sample | Type and content of fibers | | | | | |
	F-C	F-F	H-C	H-F	W-F	W-C
1	50%	0%	0%	50%	0%	0%
2	0%	50%	25%	25%	0%	0%
3	0%	50%	50%	0%	0%	0%
4	0%	50%	0%	50%	0%	0%
5	0%	0%	0%	100%	0%	0%
6	0%	0%	50%	50%	0%	0%
7	0%	0%	100%	0%	0%	0%
8	30%	0%	0%	70%	0%	0%
9	50%	0%	0%	50%	0%	0%
10	70%	0%	0%	30%	0%	0%
11	0%	0%	0%	0%	50%	50%
12	0%	0%	0%	0%	100%	0%
13	0%	0%	0%	0%	0%	100%

In this way, 13 types of test insulation materials were made in total. The composition of each test sample is shown in Table 3.

Specimens were made from each material sample (for details on composition, see Table 3) and their physical and thermal insulation properties were determined (minimum set of 3 samples for each test was used).

4 TEST RESULTS

The thickness of the materials was determined (using specimens measuring 200 × 200 mm) according to EN 823 at a nominal load of 50 Pa. Bulk density was determined and evaluated according to EN 1602 on specimens of 200 × 200 mm in size.

Next, thermal conductivity at a mean temperature of 10°C and a temperature gradient of 10 K was determined using the heat flow meter method in accordance with EN 12667 and ISO 8301 (using specimens measuring 300 × 300 mm). The specimens were dried at a temperature of 105°C to a constant mass and their dry-state thermal conductivity was measured. The results are shown in the Table 4.

The specimens were then additionally compressed so as to ascertain the influence of bulk density of each sample type on the final values of thermal conductivity.

As part of further result evaluation, samples 1–10 were examined with the aim of estimating the dependence of the thermal conductivity on bulk density. A very low correlation coefficient (about 20%) was found and it can therefore be stated that only the samples of sheep wool exhibited a very strong correlation between bulk density and thermal conductivity (see Figure 2). This is the reason why samples with dominant content of flax and samples with dominant content of hemp were evaluated separately:

Table 4. Overview of physical and thermal insulating properties.

Sample	Thickness mm	Density kg · m^{-3}	Thermal conductivity W · m^{-1} · K^{-1}
1	58	34.33	0.0399
2	58	34.55	0.0389
3	54	30.94	0.0380
4	57	33.90	0.0409
5	57	27.79	0.0417
6	55	32.17	0.0410
7	51	35.54	0.0418
8	58	31.85	0.0382
9	54	29.42	0.0393
10	58	28.82	0.0384
11	90	18.66	0.0376
12	90	23.48	0.0357
13	90	15.78	0.0429

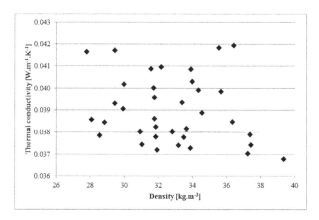

Figure 1. Dependence of thermal conductivity on the density of the samples based on hemp and flax.

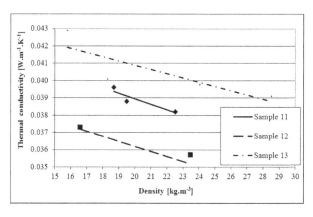

Figure 2. Dependence of thermal conductivity on the density of sheep wool-based samples with different thickness of fibers.

Flax insulations showed a high correlation of 88% between bulk density and thermal conductivity (see Figure 4). However, samples with a dominant content of hemp still exhibited a relatively low correlation of 32% and it is apparent that the purity and thickness of the fibers are also very important (see Figure 3).

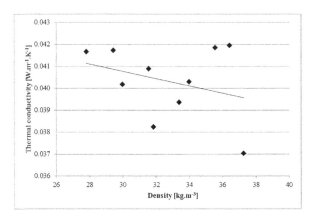

Figure 3. Dependence of thermal conductivity on density of samples based on hemp (samples 5, 7 and 8 with different compression).

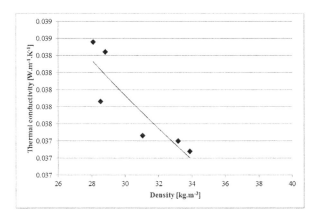

Figure 4. Dependence of thermal conductivity on density of samples based on flax (sample 10 with different compression.

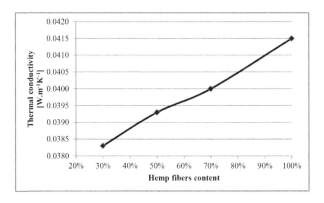

Figure 5. Dependence of thermal conductivity on hemp fiber content.

Based on the knowledge of basic models, dependencies between thermal conductivity and bulk density in fibrous materials (B. Meng 1994; Korjenic, A. Bednar, T. 2011) and the measured values, an equation for calculating thermal conductivity of an insulating material (for low bulk density of up to 50 kg.m^{-3}) was derived:

695

$$\lambda_{10,dry,wool} = d \cdot 3.51 \cdot 10^{-4} - \rho_v \cdot 3.30 \cdot 10^{-4} + 0.0355 \ [W \cdot m^{-1} \cdot K^{-1}] \qquad (1)$$

where: d – thickness [μm], ρ_v – density [kg · m^{-3}].

5 CONCLUSION

The research investigation focused on examining the thermal properties of samples of thermal insulations based on natural fibers made of flax, hemp and sheep wool. It was found that the dielectric properties are dependent on the type of fiber, its fineness (thickness) and density. Samples of sheep wool, with relatively high purity demonstrated a very clear dependence between bulk density and thermal conductivity of the fibers. A new model for the dependence of thermal conductivity on thickness of fibers and density was created.

In the case of bast fibers, additional parameters enter into this dependence such as the overall purity of the fibers (shives), followed by the chemical composition of the fibers (concerning various types of flax and hemp fibers). Overall, however, it was found that better thermal insulation properties were achieved by samples with a higher content of flax (see Figure 5).

Similarly to samples of sheep wool, the thermal insulation properties improve as the fiber thickness decreases. Considering a range of low bulk density values, the thermal insulation properties improve with increasing bulk density.

ACKNOWLEDGEMENT

This paper has been worked out under the project No. LO1408 "AdMaS UP - Advanced Materials, Structures and Technologies", supported by Ministry of Education, Youth and Sports under the „National Sustainability Program I" and under the projects GA 13–21791S "Study of heat and moisture transfer in the structure of insulating materials based on natural fibers"

REFERENCES

Chybík, J. 2009. Natural Building Materials. 1. Issue. Prague: Grada Publishing, a.s., 272 p. ISBN 978-80-247-2532-1.

http://www.vscht.cz/met/stranky/vyuka/labcv/labor/res_materialova_skladba_text_vlaken/teorie.htm

Ibraheem, S. A.; Ali, Aidy; Khalina, A. 2011. Development of Green Insulation Boards From Kenaf Fibres Part 1: Development and characterizations of Mechanical Properties. Fracture and strength of solids VII, Part 1 and 2, Book Series: Key Engineering Materials. In *8th International Conference on Fracture and Strength of Solids (FEOFS)*, Volume 462–463, p. 1343–1348,

Korjenic, A., Bednar, T. 2011. Developing a model for fibrous building materials, *Energy and Buildings*, vol. 43, 3189–3199.

Lancashire, R. J. 2014. Unit—Chemistry of Garments: Animal Fibres, The Univerrsity of the West Indies, The Departmemnt of Chemistry, study material course: Chemistry in our Daily Lives, http://wwwchem.uwimona.edu.jm/courses/CHEM2402/Textiles/Animal_Fibres.html.

MacDougall, Colin. 2008. Natural Building Materials in Mainstream Construction: Lessons from the U.K. *Journal of Green Building*. Volume: 3 Issue: 3, p. 3–14.

Meng, B. 1994. Calculation of moisture transport coefficients on the basis of relevant pore structure parameters, *Materials and Structures*, 27 125–134.

Vejeliene, Jolanta; Gailius, Albinas; Vejelis, Sigitas; et al. 2011. Evaluation of Structure Influence on Thermal Conductivity of Thermal Insulating Materials from Renewable Resources. Materials Science-Medziagotyra. In *National Conference on Materials Engineering*, Volume 17, Issue 2, p. 208–212.

Advances and Trends in Engineering Sciences and Technologies II – Al Ali & Platko (Eds)
© 2017 Taylor & Francis Group, ISBN 978-1-138-03224-8

Technical conditions of barrier-free housing and needs of people with physical limitations

R. Zdařilová
VŠB—TU Ostrava, Czech republic

ABSTRACT: The current state of requirements for specific forms of barrier-free housing leads us to consider how to provide quality and satisfactory housing that will be adaptable and versatile. What are the target users' real demands for space, equipment, technical solutions and other things? The answers can be found directly with the users. In cooperation with organizations of disabled persons, a questionnaire has been prepared, which contains all the relevant questions that can affect the quality of good and satisfactory barrier-free housing in a positive way. The task of the investigation is to find out the practical experience with the existing housing and the needs of potential users of barrier-free housing. The article aims to compare the current technical conditions of barrier-free flats as provided by legislation with the results of the research focused on the needs of the target groups of people with physical limitations, their practical knowledge and experience.

1 INTRODUCTION

Aging population, an increasing proportion of seniors is a trend all over Europe that has lasted for several decades. This trend is also observed for persons with disabilities. According to a survey conducted by the Czech Statistical Office in 2013 [CSO 2014], there are 1,078,673 persons with disablement in the Czech Republic, which is 10.2% of the population; over 29% of this group represent persons with physical disablement. It is evident from the above-mentioned facts that these are significant groups of citizens, and this fact creates the need to adequately respond to the creation of the necessary conditions of civil engineering construction, including housing.

The character of barrier-free housing should mobilize people and improve the general quality of their life. It should ensure the senior and disabled citizens' satisfaction with such housing and delaying the potential need for their retirement to a specialized facility.

2 LEGISLATIVE REQUIREMENTS FOR BARRIER-FREE HOUSING

2.1 Legal environment

Legal requirements for housing are governed in general in the Czech legal environment. The basic concept document for the housing policy of the Czech Republic is the Housing Concept [Ministry for Regional Development 2011], the objectives of which (among other things, increasing the supply and quality of housing) are specifically implemented by legislation and support tools in the area of housing. From this general perspective, there is currently no requirement with respect to the necessity of construction of barrier-free housing, if such housing is not supported by grant programs.

Specific technical requirements aimed at people with limited mobility and orientation are the subject of the Building Act implementing regulation. Regulation No. 398/2009 Coll., on general technical requirements ensures barrier-free use of buildings. Requirements for technical solutions to blocks of flats are only set down for blocks containing a special purpose flat.

Requirements for technical solutions of barrier-free flats are specified as requirements for the layout of modifiable flats and special-purpose flats.

Note: A special-purpose flat is a flat specially adapted for persons with disablement. A modified flat is a flat that can be used by persons with reduced mobility without further structural modifications.

Technical solution of barrier-free flats is the subject of the Annex to this regulation, which regulates the conditions for persons with reduced mobility in respect of the layout, requirements for sanitary facilities, loggias, balconies and terraces, the size of doors and windows, placement of controls. For people with visual impairments, conditions of technical solutions for energy distribution and equipment with electrical outlets are specified.

2.2 *Layout*

The layout of barrier-free flats shall be in accordance with the handling and maneuverability capabilities of persons in a wheelchair. Suggested smallest living area and the area of kitchen based on the size of the flat is provided by the Czech Technical Standard ČSN 73 4301 Residential buildings. For the layout, the minimal furnishings is also important, which is the subject of another technical standards ČSN 73 4305, and which, inter alia, in Article IV regulates furnishings for people with limited mobility. Following the request of recommended smallest kitchen area, it provides the recommended lengths of kitchen sets.

2.3 *Requirements for sanitary facilities*

The solution of barrier-free sanitary facilities set out in Regulation No. 398/2009 Coll. is based on partial general requirements for the solution of wheelchair accessible toilets, wheelchair accessible bathtubs, and wheelchair accessible showers and shower cubicles. On the other hand, the applicable ČSN 73 4305 Furnishing flats provides obsolete and invalid data of substandard small size hygiene facilities as of1 May 1989. Requirements for sanitary blocks with the barrier-free parameters—bathrooms with shower or bathtub are specified in another Czech Technical Standard ČSN 73 4108 Hygienic Facilities and Changing Rooms, which already responds to the requirements of applicable Regulation No. 398/2009 Coll. Although the standard is not applicable for designing sanitary facilities in residential buildings, it may provide some inspiration for solving these challenging areas.

2.4 *Doors and windows*

Entry doors, interior passages and door openings must have a width of at least 900 mm. It should be noted that the larger the width of the door leaf, the more difficult it becomes to handle with the leaf for a person on a wheelchair; at the same time, wider doors occupy a larger manipulation area within the flat layout, and thereby reduce the usable area of the individual rooms.

2.5 *Construction and technical requirements*

Construction and technical requirements of modifiable flats and special-purpose flats are largely demanding, spatial comfort is comparable with the parameters of special medical equipment and it is inadequate for many active seniors and persons with disablement. The disadvantage of these flats is a large surface area with a relatively unfavorable ratio of living/usable area. The buildings are then more costly in terms of both investment and operation, but also in the outcome less comfortable for many people with common types of disablement.

A sample survey and questionnaire surveys conducted among users listed below show that a significant proportion of users inhabit conventional housing stock, mainly because of the lack of barrier-free housing in the real estate market.

3 RESEARCH INTO THE NEEDS OF THE TARGET GROUPS

3.1 *The indicator of marital status*

As already mentioned in the introduction, in 2013, a sample survey of persons with disablement was conducted in the Czech Republic [Czech Statistical Office 2014]. One of the points of the survey was also the issue of housing for people with disablement. In this area, for further consideration of construction and technical conditions and the size of the flats, marital status of these persons is equally important indicator, as it gives some idea of the background of people with disablement. The survey shows that the largest group is the marital status of married; its share is almost 41%. The second, or the third place is occupied by very close groups amounting to 22–23% – widow/er and single.

3.2 *The indicator of housing*

The indicator of marital status is very closely followed by the indicator of housing with the necessary information on living in a household with another person and according to the type of housing. Living in a household with a husband/wife is represented by 36.4%, 18.7% live alone, 17.1% live with a son/daughter, and 13.8% live with parents (meaning adults with disablement). These figures give us an idea of what size categories of flats are needed in present-day real estate market.

Table 1 shows the most frequent type of housing reported by 80% of people with disablement is a conventional flat in a standard block of flats. The second position was occupied by barrier-free flats (6.7%), the third position by beds in social care facilities (6.4%). The remaining seven types of housing, covering a total of about 7% are quite marginal. If we consider the most widely represented form of housing in terms of age categories, without distinction of sex, then almost three quarters (exactly 76.3%) occur in three age categories (45 and over).

The data presented show us the true picture and the living conditions of persons with disablement. It is worth noting that, for example, less than 7% of these individuals use barrier-free flats. Moreover, if we take into account other types of housing, which can be considered as suitable for them to ensure a good quality of life, we get a share of only 2.3% (the high proportion of this type of housing is represented by flats with care service – 1.4%).

This sample survey of persons with disablement confirms that in the context of ensuring quality barrier-free housing, emphasis must be placed on standard housing with the possibility of adaptability to the specific conditions of the changing needs of target users.

3.3 *The indicator type of disablement*

Equally important data concern the frequency of type of disablement. This figure is a significant factor for the actual establishment of specific technical requirements, which are different for bodily, visual, and auditory, etc. disability, and which allow us to establish the necessary criteria for creating an adjustable housing—see Table 2.

Table 2 shows that the most frequent disablement is internal disability in a proportion of 41.9%. The second position is occupied by bodily disablement, in a proportion of 29.2%, which is quite a small difference from the previous value, and other types of disablement range in a significantly lower interval from 5.1 to 8.5%.

Following the data shown in Table 1, we can say that almost half of the persons with multiple disablement –49.6% live in the most strongly represented form of housing, a conventional flat. The second place is occupied by those with internal type of disability –22.3%, and in the third place there are people with bodily disablement, amounting to 16.5%. These data confirm the fact that technical requirements for barrier-free housing must be adapted especially to the users with bodily disablement, who also have the largest requirements for handling areas and clearways within the target groups of barrier-free housing. Therefore, these are the requirements that significantly affect the design of the flat layout itself.

Table 1. Housing by the type of flat and the type of disability.

	Flat in an conventional block of flats			Flat in a special block of flats		Protected housing	Supported housing	Bed in social care facilities	Other residential care services	Homeless	Unknown	In total
	Conventional flat	Barrier-free flat	Special-purpose flat	Special-purpose flat	Flat with care service							
Age group												
0–14	60729	1564	37	117	0	0	0	355	264	0	1241	64307
15–29	53922	4886	0	136	46	504	0	4040	172	192	1246	65143
30–44	89737	7247	23	473	433	676	56	6901	427	323	3650	109947
45–59	175920	12236	71	285	510	531	0	5837	1550	119	9831	206891
60–74	272296	21721	59	350	3645	2479	0	13992	442	87	18140	333211
75 +	209187	24898	373	938	10488	2452	0	37734	938	0	11165	298174
In total	861791	72551	562	2 299	15122	6643	56	68859	3794	721	45273	1077673
Type of disability												
Bodily	142277	16738	71	861	2412	158	0	3027	70	0	3748	169362
Visual	13466	1778	0	0	152	0	0	144	0	0	327	15867
Auditory	12013	426	0	177	46	0	0	195	86	0	994	13937
Mental	24664	1112	0	406	243	868	0	7441	341	0	1196	36271
Psychic	42721	784	37	0	46	526	16	3028	494	250	1344	49246
Internal	192483	4461	25	87	1558	0	0	2249	96	219	8687	209865
Other	6588	2762	0	0	0	0	0	19	0	0	0	9369
Multiple	427579	44491	429	767	10666	5090	40	52756	2708	252	28977	573756
In total	861791	72551	562	2 299	15122	6643	56	68859	3794	721	45273	1077673

Source: Sample survey of persons with disablement 2013 [Czech Statistical Office 2014].

Table 2. Type of disability*.

Age	Bodily	Visual	Auditory	Mental	Psychic	Internal	Other	In total
0–14	20077	6715	3811	15683	10538	31764	430	92890
15 29	29499	4501	2537	19394	12590	28521	1039	98081
30–44	38444	7892	5682	22216	21242	50285	2044	147805
45–59	89516	14851	8086	15660	30518	126966	11588	297185
60–74	157622	19796	17215	13919	31596	248118	14082	502348
75+	165009	48440	49144	17703	39032	231339	21272	571940
In total	500167	102195	86476	104574	145517	716993	54327	1710249

*Multiple answers possible—multiple disablement.
Source: Sample survey of persons with disablement 2013 [Czech Statistical Office 2014].

4 SURVEY ON BARRIER-FREE HOUSING AND SANITARY FACILITIES

The current state of requirements for specific forms of barrier-free housing leads us to consider how to provide quality and satisfactory housing that would be adaptable and versatile over time to satisfy the needs of all potential users. Sample survey of persons with disablement conducted in 2013 clarifies the situation of the current housing for target groups in respect of the type of housing. But what are the real needs with respect to the inner space, equipment and technical solutions of these users? Issues associated with this were the subject of the survey. Based on the cooperation with organizations of disabled persons, a questionnaire was prepared dealing with the issue of barrier-free housing and sanitary facilities, which contains all the relevant questions that can influence the quality of good and satisfactory barrier-free housing in a positive way. The survey aimed to identify practical experience with current housing and needs of potential users of barrier-free housing. Problem areas of the questionnaire related to the type of disability, the current type of housing, elevators and platforms in terms of suitability and requirements for minimum size, the size of door openings, handling areas to turn the wheelchair, size of rooms, the requirements for the kitchen, sanitary facilities and other issues to complement especially in the area of windows, balconies and floor surfaces, etc. The first round of the investigation took place in a paper form but given the interest of target respondents, the questionnaire was converted into the internet survey. In the mentioned first round of the research investigation only with regional impact the average age of the respondent was 43. All respondents were users of a wheelchair. The group consisted of 29% of paraplegics, 29% of quadriplegics and 42% of other form of health disability. Out of this sample 73% of people live in normal housing, which had to be adapted to a barrier-free one due to their handicaps. Only 27% of respondents live in adaptable (barrier-free) flats. The satisfactory width of 800 mm of the inner doors of the flat was stated by 49% of respondents. The width of 900 mm as it is defined by law is preferred by 51% of respondents. Regarding the width of corridors and possibility of rotation with the wheelchair 44% of people stated that 1200 mm is sufficient, 24% prefer at least 1300 mm, 13% require 1400 mm, 8% of respondents stated less than 1200 mm and further 8% on contrary the width 1500 mm, which is a requirement of construction regulations for barrier-free housing.

Valuable knowledge of this questionnaire is represented by specific requirements of users with disablement to solve sanitary facilities, which confirm the need for variable and versatile space that can be adapted to the specific user.

5 CONCLUSION

In the Czech Republic, demand for senior and barrier-free housing is constantly growing. This trend will continue to grow due to the expected demographic developments indicating permanent aging of the population. Simultaneously, senior and barrier-free housing in relation to persons with disablement has certain specifics. It is necessary to take into account the needs of the target groups and to realize the fact that these people largely use the flat all day and all year round, and for many of them, in certain periods of life, the radius of movement may radically diminish to the block of flat itself and its immediate surroundings, or even just the flat itself. Therefore, the living environment must be sensually stimulating, providing adequate space for different types of activities and levels of social contact. Physical living environment helps co-create a lively community where mutual support along with building modifications significantly prolongs stay stimulation in an environment that can be seen as their own home. It significantly delays and thus reduces the need for retirement to facilities providing institutional forms of care, which, in addition to reduced quality of life, also mean greater economic burden on our society.

The data obtained in the surveys conducted represent the fundamental basis for formulating the general principles of adaptable housing. Knowledge acquired from the questionnaire survey and its analysis is a prerequisite for achieving a state of equal conditions of access to housing. It can also enable to launch a system solution by defining model parameters with the basic principles of lifelong living in conventional housing construction. By reducing the requirements for e.g. the areal standard of rooms, we can achieve significant savings of up to 20% of the flat area. These outputs will be a stimulus for expert discussion with a subsequent revision of the current legislative environment. On the other hand, a detailed analysis of the requirements for sanitary facilities can provide persons employed in construction with clearer idea of the use of space around the individual furnishing objects.

ACKNOWLEDGEMENT

The works were supported from sources for conceptual development of research, development and innovations for 2016 at the VŠB-Technical University of Ostrava which were granted by the Ministry of Education, Youths and Sports of the Czech Republic and by the support of the Technology Agency of the Czech Republic in solving the project no. TD03000279 called Housing as a Factor for Social Inclusion.

REFERENCES

ČSN 73 4305:1989 *Flat furnishing*. Prague: Office for Standardization and Measurement.

ČSN 73 4301:2004 *Residential buildings*. Prague: Czech Standards Institute.

ČSN 73 4108:2013 *Sanitary facilities and changing rooms*. Prague: Czech Office for Standards, Technology and Testing.

Sample survey of persons with disablement in 2013, the Czech Statistical Office and the Institute of Health Information and Statistics of the Czech Republic, published on 30 April 2014, data source: https://www.czso.cz/csu/czso/vyberove-setreni-zdravotne-postizenych-osob-2013-qacmwuvwsb.

The concept of housing of the Czech Republic in 2020, the Ministry of Regional Development and the State Housing Development Fund, 2011.

Advances and Trends in Engineering Sciences and Technologies II – Al Ali & Platko (Eds)
© 2017 Taylor & Francis Group, ISBN 978-1-138-03224-8

Analysis of rainfall trends in Syria

M. Zeleňáková & I. Alkhalaf
Institute of Environmental Engineering, Technical University of Kosice, Kosice, Slovakia

P. Purcz
Institute of Construction Technology and Management, Technical University of Kosice, Kosice, Slovakia

P. Blišťan
Institute of Geodesy, Cartography and Geographical Information Systems,
Technical University of Kosice, Kosice, Slovakia

M.M. Portela
Department of Civil Engineering, National Technical University of Lisbon, Lisbon, Portugal

ABSTRACT: The Intergovernmental Panel on Climate Change (IPCC) provides a comprehensive review of the potential impacts on the hydrological variables of the man induced climate changes. It states that such changes will likely increase runoff in the higher latitude regions because of increased precipitation; also, the flood frequency is expected to change in some locations and the severity of drought events could also increase as a result of the changes both in precipitation and evaporation. Observations show that changes are occurring in the amount, intensity, frequency and type of precipitation. The frequency of extreme temperatures during the summer is likely to be higher. The Mann-Kendall test coupled with the Sen's slope was applied to identify the significant trends, positive or negative trend and the magnitude of those trends. The results achieved for rainfall revealed more frequent significant decreasing trends in Syria in the last 19 years.

1   INTRODUCTION

In arid and semi-arid areas, rainfall storms exhibit strong spatial variability, especially during heavy thunderstorms and localized torrential rainstorms (Ahmidat et al. 2014; Almazroui, 2011; Habib & Nasrollahi, 2009; Alyamani & Sen, 1993). However, many of these areas worldwide suffer from limited surface rainfall monitoring stations (Ragab & Prudhomme, 2002). The Inter-Governmental Panel on Climate Change (IPCC) in 20012 concluded that more intense precipitation events would very likely occur in the future over many areas, and that these would thus cause increased flash-floods, landslides, soil erosion and avalanches (IPCC, 2012). Syria, like the region as a whole, experiences high natural hydrologic variability. Over the past century (from 1900 to 2005), there were six significant droughts in Syria, where the average monthly level of winter precipitation the major rainfall season dropped to around one-third of normal. Five of these droughts lasted only one season; the sixth lasted for two (Gleick, 2014; Mohtadi, 2013). Starting in 2006, however, and lasting into 2011, Syria experienced a multiseason, multiyear period of extreme drought that contributed to agricultural failures, economic dislocations, and population displacement (Worth, 2010). Occasional heavy rainstorms occur on only a few days in a year and only in some parts of the country. This seldom received rain makes Syria one of the driest countries in the world. Hoerling et al. (2012) concluded that climate change is already beginning to influence droughts in the Mediterranean area by reducing winter rainfall and increasing evapotranspiration. Romanou et al. (2010) show statistically significant increases in evaporative water demand in the eastern Mediterranean region between 1988 and 2006, driven by apparent increases in sea surface temperatures. Mathbout & Skaf (2010) used the Standard Precipitation Index (SPI) and the Effective Drought Index (EDI) to identify an increasing tendency in annual and seasonal drought intensity in all regions corresponding with an increasing number of dry days in the rainy season.

This paper investigates rainfall trends over all Syria.

2   MATERIAL AND METHODS

We have investigated rainfall trends in Syria that is a country in the Middle East, bordering the Mediterranean Sea between Lebanon and Turkey.

Climate in Syria is mostly desert; it has hot, dry, sunny summers (from June to August) and mild, rainy winters (December to February) along coast. The climate in Syria is divided into two divisions: Mediterranean climate in the coastal region and adjacent areas; and droughts in other areas.

Data for evaluation were obtained from Aleppo University in Syria with collaboration of Ministry of Agriculture and Agrarian Reform and Meteorological Center in Syria. The evaluated period is from 1992 to 2010. Rainfall trend analysis was done for monthly average precipitation data for 75 climatic stations in Syria (Figure 2).

In this study non-parametric Mann-Kendall test is used for the detection of the trend in a time series. Mann-Kendall test (Mann, 1945; Kendall, 1975) is following statistics based on standard normal distribution (Z), by using Eq. (1).

where

$$Z = \begin{cases} \dfrac{S-1}{\sqrt{Var(S)}} & \text{if} \quad S > 0 \\ 0 & \text{if} \quad S = 0 \\ \dfrac{S+1}{\sqrt{Var(S)}} & \text{if} \quad S < 0 \end{cases} \tag{1}$$

where

$$S = \sum_{k=1}^{n-1} \sum_{j=k+1}^{n} \text{sgn}\left(x_j - x_k \right) \tag{2}$$

$$\text{sgn}\left(x_j - x_k \right) = \begin{cases} +1 & \text{if} \quad \left(x_j - x_k \right) > 0 \\ 0 & \text{if} \quad \left(x_j - x_k \right) = 0 \\ -1 & \text{if} \quad \left(x_j - x_k \right) < 0 \end{cases} \tag{3}$$

$$Var(S) = \left[n(n-1)(2n+5) - \sum_{i=1}^{m} t_i \left(t_i - 1 \right) \left(2t_i + 5 \right) \right] / 18 \tag{4}$$

where n = the number of data points, m = the number of tied groups (a set of sample data having the same value).

According to this test, the null hypothesis H0 states that the depersonalized data $(x_1, ..., x_n)$ is a sample of n independent and identically distributed random variables. The alternative hypothesis H1 of a two-sided test is that the distributions of x_k and x_j are not identical for all $k, j \leq n$ with $k \neq j$. The significance level is chosen as $\alpha = 0.05$ and $Z_{\alpha/2}$ is the value of normal distribution function, in this case $Z_{\alpha/2} = 1.95996$. Hypothesis H0 - no trend is if $(Z < Z_{\alpha/2})$ and H1 - there is a trend if $Z > Z_{\alpha/2}$. Positive values of Z indicate increasing trends, while negative values of Z show decreasing trends.

The magnitude of the trend was determined using Sen's estimator. Sen's method assumes a linear trend in the time series and has been widely used for determining the magnitude of trend in hydro-meteorological time series e.g. (Burn & Hag Elnur, 2002; Partal & Kahya, 2006; Yue & Hashino, 2003; Sen, 1968; Jain et al 2012). In this method, the slopes (β) of all data pairs are first calculated by

$$\beta = Median\left(\left(x_j - x_k\right)/(j-k)\right) \tag{5}$$

For $j = 1, 2,..., N$,. where x_j and x_k are data values at time j and k ($j > k$), respectively and N is a number of all pairs x_j and x_k. A positive value of β indicates an upward (increasing) trend and a negative value indicates a downward (decreasing) trend in the time series.

All mathematical relationships (1), (2), (3), (4) and (5) were programmed in Visual Basic in Microsoft Excel 2003.

3 RESULTS

Results of rainfall analysis—annual trends are presented in Table 1 for monthly average data. Monthly data series for the 19 years period, from 1992–2010, were considered for trend detection. The evaluation was done for the time period from October to May (X—V) as rainfall in other months was almost zero. Bold values indicate statistical significance at 95% confidence level as per the Mann-Kendall test (+ for increasing and – for decreasing). Magnitude of the trend is expressed by Sen's estimator of the slope of all the data points.

Table 1. Annual trends of rainfall in Syria, part 1.

Station	Trend (X-V) without autocorrelation	Trend (X-V) with autocorrelation
Idlib	−0.04338	−0.0282
Armanaz	−0.03931	−0.02129
Arihah	−0.08397	−0.06074
Addana	0	0.012419
Maarrat Misrin	−0.06	−0.04498
Ebla	−0.08397	−0.06074
Haram	−0.04116	−0.04116
Khan Shaykhun	−0.04958	−0.03869
Darkush	0.013492	0.023474
Saraqib	−0.04265	−0.03323
Kafar Takharim	−0.01029	0.005177
Kafr Nabl	−0.05313	−0.04483
Moh. Wadi Eldaif	−0.06618	−0.0413
Tell Beydar	−0.06618	−0.0413
Tell 3alo Althanya	−0.06618	−0.0413
Al Qunaytirah	−0.06923	−0.03031
Hader	−0.08333	−0.04296
Nada Alsakher	−0.04845	−0.03516
Azaz	−0.06597	−0.06597
Al Bab	−0.02727	−0.02727
Al Safira	−0.056	−0.056
Jarabulus	−0.03404	−0.03404
Aleppo Int. Airport	−0.00638	−0.00638
Kobani	−0.04583	−0.04583
Aleppo	0.057377	0.057377
Afrin	0.028261	0.028261
Manbij	**−0.06604**	**−0.06604**

Table 1. Annual trends of rainfall in Syria, part 2.

Station	Trend (X-V) without autocorrelation	Trend (X-V) with autocorrelation
Al Rastan	0.017895	0.010225
Al Quaryatayn	−0.00451	−0.00841
Al Qusayr	−0.01739	−0.01488
Al Mukharram	−0.0125	−0.01266
Al Nasrah	0.006667	0.016614
Palmyra	−0.00405	−0.00405
Taldou	−0.0125	−0.00712
Tal Kalakh	0.125316	0.085762
Homs	0.002532	0.018589
Shin	0.011702	−0.00101
Marmarita	−0.075	−0.03177
Izraa	0.001149	0.000328
Shaikh Maskin	0	0.009994
As Sanamayn	0	0
Al Musayrfah	0	−0.00076
Tal Shihab	0.002564	0.017253
Daraa	0	−0.00223
Nawa	−0.00417	0.006925
Al Tal	−0.02429	−0.01013
Al Qutayfah	−0.02128	−0.02128
Al Kiswah	−0.00769	−0.00583
Al Nabk	**−0.04615**	−0.03258
Duma	−0.01696	−0.01696
Al Zabadani	−0.068	−0.04096
Umuyyad Mosque	0	0.002736
Serghaya	−0.05862	−0.03799
Qatana	−0.04348	−0.02316
Maysaloon	−0.08	−0.0498
Yabrud	−0.02752	−0.02752
Abu Kamal	−0.00435	−0.00435
Al Busayrah	−0.01964	−0.01964
Al Tebni	−0.00562	−0.00816
Al Mayadin	−0.00625	−0.01042
Deir ez-Zur	−0.0063	−0.0063
Ash Shaykh Badr	−0.05741	−0.01148
As Sifsafeh	−0.12605	−0.06528
Al Qadmus	−0.11765	−0.08341
Baniyas	−0.01277	0.022233
Draykish	−0.13019	−0.0942
Safita	−0.03333	−0.03548
Tartus	−8.7E-17	−0.00321
Mashta Al Hilu	−0.13793	−0.0976
Ar Reqqah	**−0.04962**	−0.03841
Tell Abiad	**−0.07294**	**−0.07294**
Ain Issa	**−0.05556**	**−0.05556**
As Suwayda	−0.0413	−0.02409
Shahba	−0.04237	−0.04561
Salkhad	−0.03636	−0.02957

In Mann Kendall test the sign of Z must not be sufficient for trend definition (detection). In case a trend is detected, the existence of autocorrelation in the series will affect the result. The existence of serial autocorrelation will increase the probability of finding a significant trend at the end of the test even if there is no trend (Önöz & Bayazit, 2003; Bayazit & Önöz,

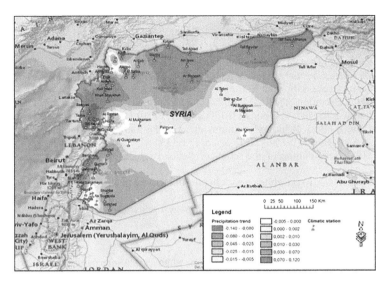

Figure 1. Resulted trends in rainfall in Syria (red color—decreasing trend. blue color—positive trend).

2007; Haktanir & Citakoglu, 2015). The serial autocorrelations of stations was performed in the presented study following Salas et al. (1980), who refer to Andersen (1941).

The surface interpolation technique in Arc-GIS was used to prepare a spatial precipitation data map over Syria from the point precipitation measuring in climatic stations (Figure 1).

Regarding trend analysis of monthly rainfall data (Table 1.) is proved clear decreasing trend in rainfall over all Syria. Statistically significant negative trends are identified in climatic station Manbij. Al Nabk. Ar Reqqah. Tell Abiad. Ain Issa (Table 1.) It means the decreasing of rainfall in almost all climatic stations in the country. especially north (mountainous) and western (coastline) parts of the country. Although some of the rainfall data series are autocorrelated so we can conclude significant decreasing trend at Manbij. Tell Abiad and Ain Issa climatic stations. The trend slope decrease of rainfall is up 0.07 mm/year.

4 CONCLUSION

Rainfall is a critical meteorological parameter and needs to be measured and evaluated accurately; many applications of rainfall data can be studied in depth through knowledge of the actual distribution of rainfall. Additionally. the amount of rainfall received over an area is an important factor in assessing the amount of water available to meet the various demands of agriculture. industry. and other human activities.

In the paper the trend analysis was applied to rainfall data series observed from 1992–2010 in 75 climatic stations in Syria. The Mann-Kendall test coupled with Sen's slope was applied to identify the significant trends. as well as the increasing or decreasing trends and the magnitude of those trends. Terrain in Syria is narrow coastal plain with a double mountain belt in the west; large. semiarid and desert plateau to the east. The results proved almost decreasing trend in rainfall in the whole country. These results should be of interest for agriculture and water management in Syrian Arab Republic.

ACKNOWLEDGEMENT

This work has been supported by the Slovak Research and Development Agency by supporting the project SK-PT-0007-15.

REFERENCES

Ahmidat. M.K.M., Káposztásová. D., Markovič. G. & Vranayová. Z. 2014. The effect of roof material on rain water quality parameters in conditions of Slovak Republic In: *Advances in Environmental Sciences. Development and Chemistry: Proceedings of the 2014 Conference on Water Resources. Hydraulics and Hydrology.* Greece. Santorini: Europment. p. 275–280.

Almazroui. M. 2011. Calibration of TRMM rainfall climatology over Saudi Arabia during 1998–2009. *Atmospheric Research.* 99. 400–414. doi:10.1016/ j.atmosres.2010.11.006.

Alyamani. M. & Sen. Z. 1993. Regional variations of monthly rainfall amounts in the Kingdom of Saudi Arabia. *J. King Abdulaziz Univ. Earth Sci.* 6. 113–133.

Anderson. R. L. 1941. Distribution of the serial correlation coefficients: *Annals of Math. Statistics.* 8(1): 1–13.

Bayazit. M. & Önöz B. 2007. To prewhiten or not to prewhiten in trend analysis?. *Hydrological Sciences Journal.* 52(4): 611–624.

Burn. D.H. & Hag Elnur M.A. 2002. Detection of hydrologic trends and variability. Journal of Hydrology. 255. 107–122.

Gleick. P.H. 2014. Water. Drought. Climate Change. and Conflict in Syria. *American Meteorological Society.* doi: 10.1175/WCAS-D-13–00059.1

Gleick. P.H. 1989. The implications of global climatic changes for international security. *Climatic Change.* 15. 309–325. doi:10.1007/BF00138857.

Habib. E.H. & Nasrollahi. N.. 2009. Evaluation of TRMM-TMPA satellite rainfall estimates over arid regions. In: *American Geophysical Union.* Fall Meeting 2009.

Haktanir. T. & Citakoglu. H. 2015. Closure to "Trend. Independence. Stationarity. and Homogeneity Tests on Maximum Rainfall Series of Standard Durations Recorded in Turkey" *Journal of Hydrologic Engineering.* 20(10).

Hoerling. M., J. Eischeid. J. Perlwitz. X. Quan. T. Zhang. and Pegion. P. 2012. On the increased frequency of Mediterranean drought. *J. Climate.* 25. 2146–2161. doi:10.1175/JCLI-D-11–00296.1.

IPCC. 2012. Climate Change 2012. Climate Change Impacts. Adaptation and Vulnerability. Cambridge University Press. Cambridge. 976 pp.

Jain. S.K., Kumar. V. & Saharia. M. 2012. Analysis of rainfall and temperature trends in northeast India. *International Journal of Climatology.* Published online in Wiley Online Library. doi: 10.1002/joc.3483.

Kendall. M.G. 1975. *Rank Correlation Measures.* Charles Griffin. London.

Mann. H.B. 1945. Non-parametric tests against trend. *Econometrica.* 13. 245–259.

Mathbout. S. & Skaf. M. 2010. Drought changes over last five decades in Syria. Economics of Drought and Drought Preparedness in a Climate Change Context. In: *A. Lopez-Francos. Ed. Mediterranean Seminars.* 95. CIHEAM. 107–112.

Mohtadi. S. 2013: Climate change and the Syrian uprising. Available online at: http://thebulletin.org/web-edition/features/climate-change-and-the-syrian-uprising.

Nations Online. 2016. One World. The Nations Online Project: Get In Touch with the World. http://www.nationsonline.org/oneworld/map/syria-map.htm.

Önöz. B. & Bayazit. M. 2003. The power of statistical tests for trend detection. *Turkish J. Eng. Env. Sci.* 27: 247–251.

Partal. T. & Kahya. E. 2006. Trend analysis in Turkish precipitation data. *Hydrological Processes.* 20. 2011–2026.

Ragab. R. & Prudhomme. C. 2002. Climate change and water resources management in arid and semi-arid regions: prospective and challenges for the 21st century. *Biosyst. Eng.* 81. 3–34.

Romanou. A.G., Tselioudis., C.S., Zerefos. C.A., Clayson. J.A., Curry. J. & Andersson. A. 2010. Evaporation-precipitation variability over the Mediterranean and the Black Seas from satellite and reanalysis estimates. *J. Climate.* 23. 5268–5287. doi:10.1175/2010 JCLI3525.1.

Salas. J.D., Delleur. J.W., Yevjevich. V.M. & Lane. W.L. 1980. Applied modeling of hydrologic time series: Littleton. Colorado. *Water Resources Publications.* 484 pp.

Sen. P.K. 1968. Estimates of the regression coefficient based on Kendall's tau. *Journal of the American Statistical Association.* 63. 1379–1389.

Worth. R.F. 2010: *Earth is parched where Syrian farms thrived.* New York Times. 13 October. New York ed.. A1.

Yue. S. & Hashino. M. 2003. Temperature trends in Japan: 1900–1990. *Theoretical and Applied Climatology.* 75. 15–27. 2003.

Zeleňáková. M., Purcz. P., Gargar. I.A.K. & Hlavatá H. 2013. Comparison of precipitation trends in Libya and Slovakia. In: *River Basin Management 7.* Southampton: WIT Press. p. 365–374.

Advances and Trends in Engineering Sciences and Technologies II – Al Ali & Platko (Eds)
© 2017 Taylor & Francis Group, ISBN 978-1-138-03224-8

Numerical analysis of cylindrical wedge-wire screen operation

M.Z. Zielina & A. Pawłowska
Cracow University of Technology, Cracow, Poland

A. Kowalska-Polok
Pol-Eko-Aparatura, Wodzisław Śląski, Poland

ABSTRACT: The article presents problems of fish entrainment and impingement at municipal water intakes. Cylindrical wedge-wire screen is one of the modern physical barriers helping to protect small fish against damaging at water intakes. This kind of physical protection visibly reduces entrainment and impingement of small fish mainly due to their small screen slot sizes and low slot velocities. The article presents cylindrical wedge-wire screen water intake designing methods and concerning them obligatory national regulations in some countries. At the end of article some visualization of water flow through the cylindrical wedge-wire screen based on computer fluid dynamics is shown together with the interpretation and possible application for designing process.

1 INTRODUCTION

Human intervention, including regulation of rivers and streams, construction of water intakes and dams poses a threat to ichthyofauna. Major threats to the migrating fish are hydroelectric power plant's water intakes. However, much smaller intake constructions supplying drinking water is also a problem for ichthyofauna. The water intakes may lead to injury or death of fish. One of the widely used methods to protect fish at water diversions is to provide a physical barrier that prevents fish from being entrained into the diversion (Department of Fisheries and Oceans 2006). Cylindrical wedge-wire screens are considered as a technology that has potential for effectively reducing entrainment and impingement of fish eggs and larvae at water intake structures. In order to protect fish from entrainment, special devices known as physical barriers are installed in water intakes. The water inlets are equipped with a cylindrical wedge-wire screen. Inside the wedge-wire screen there is a deflector. The purpose of the deflector is to equalize the velocity distribution on the surface (Jamieson et al. 2007). It is obtained thanks to properly designed differential size of slots (Taft 2000). The wedge-wire screens parameters should be properly designed what is crucial to ichthyofaunal protection. There are few important factors such as inlet velocity, mesh size and cylinder outer surface (Department of Fisheries and Oceans 1995). Since traditional designing of the cylindrical wedge-wire screens is quite imprecise, CFD modelling seems to be quite adequate. Analysis of flowrate velocity and streamline distributions are possible to conduct based on computer simulations.

2 MATERIAL AND METHODS

Since the full-scale and laboratory experiments are significantly time-consuming, CFD simulation seems to be suitable tool for designing and work analysis of wedge-wire screens at any conditions. However, results of computer simulations should be primary experimentally verified at laboratory conditions. That experiments verifying numerical simulations are planned to carry out at Cracow University of Technology in laboratory channel in hydraulic laboratory. The model of cylindrical wedge-wire screen shown in Figure 1 will be installed at the bottom of the channel. Water will be pumped with flowrate capacity guarantying maximum slot velocities according to US standards (Department of Fisheries and Oceans 2006) not exciding obligatory values of 0,15 m/s and approach velocities 0.06 m/s for fry smaller than 60 mm. The slot velocity should be measured inside the slots of screen, whereas the approach velocity should be measured 3 inches away from the surface of the screen. Inside the head there is located perforated deflector guaranteeing homogenous distribution of water inflow around the intake head. More homogenous distribution of water inflow velocity around the screen, higher flowrate capacity of the intake head. Figure 2 presents the perforation of the deflector.

CFD mathematic model was used to verify the velocity distribution around the laboratory model of water intake head. It was used the k-epsilon turbulence model in calculations. It is the most common model used in CFD to simulate mean flow characteristics for turbulent flow conditions (Toro 2009; Jiyuan et al. 2008; Middlestadt & Gerstmann 2014). It is a two equation model which gives a general description of turbulence by means of two transport equations. It is typically more accurate than the constant eddy viscosity model, but more computationally intensive and slightly less robust. It is not as resource intensive as the RNG model (based on Re-Normalisation Group methods), but still gives good results. The k-epsilon turbulence model include following equations:

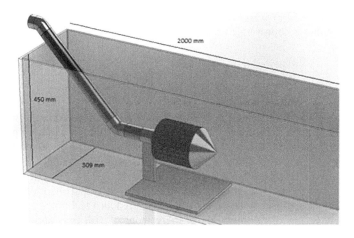

Figure 1. The laboratory model of cylindrical wedge-wire screen.

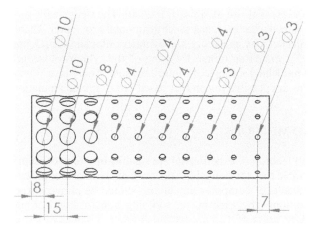

Figure 2. Deflector drawing.

For turbulent kinetic energy k:

$$\frac{\partial(\rho k)}{\partial t} + \frac{\partial(\rho k u_i)}{\partial x_i} = \frac{\partial}{\partial x_j}\left[\frac{\mu_t}{\sigma_k}\frac{\partial_k}{\partial x_j}\right] + 2\mu_t E_{ij} E_{ij} - \rho\epsilon \qquad (1)$$

For dissipation ϵ:

$$\frac{\partial(\rho\epsilon)}{\partial t} + \frac{\partial(\rho\epsilon u_i)}{\partial x_i} = \frac{\partial}{\partial x_j}\left[\frac{\mu_t}{\sigma_\epsilon}\frac{\partial\epsilon}{\partial x_j}\right] + C_{1\epsilon}\frac{\epsilon}{k}2\mu_t E_{ij} E_{ij} - C_{2\epsilon}\rho\frac{\epsilon^2}{k} \qquad (2)$$

where:
ρ – fluid density,
t – time,
x_i, x_j – coordinates,
u_i – velocity component in corresponding direction,
e_{ij} component of rate deformation,
μ_t – eddy viscosity,
σ_k, σ_ϵ, $C_{1\epsilon}$, $C_{2\epsilon}$ – adjustable constants.

Numerical calculations were conducted for the water intake head placed at the bottom of channel. Dimensions were the same as in the laboratory. The capacity of water intake head during the simulations equaled to 5 dm³/s. Calculations were conducted assuming lack of the water flow in the channel.

3 RESULTS AND CONCLUSIONS

The results of CFD simulation of water intake operation are presented in Figures from 3 to 6.

The flowrate velocity distribution at the distance of 3 inches away from the screen surface is presented in Figure 4. It can be seen that the highest value does not exceed maximum limited value (US Government 2006) and appears on the opposite side to the inflow pipe. The lowest value appears close to the pipe. Homogeneity of velocity distribution is quite high. Ratio of maximum to minimum values equals to 1.3. Figure 5 presents velocity distribution close to the outer surface of the screen. The velocity distribution was very rough. Inside the perforations velocity drastically increased, whereas between the holes velocities decreased and were around 20% lower. Almost 4 times higher values were noticed on the left side than on the right. As it was expected, higher velocities were observed in the deflector pores that were smaller in size and located closer to intake pipe. It is shown in Figure 5.

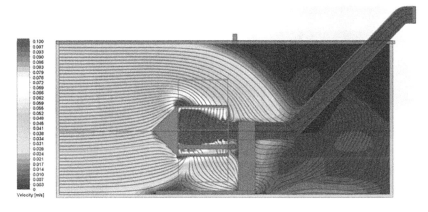

Figure 3.   Velocity distribution in vertical-section.

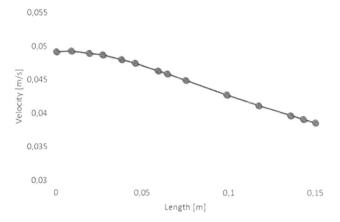

Figure 4.   The inflow velocity distribution in distance of 3 inches away from the top surface of the screen.

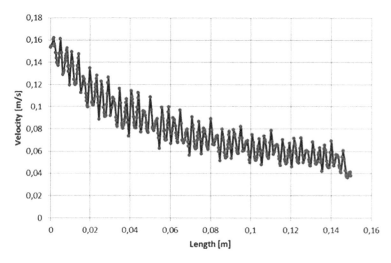

Figure 5.   The inflow velocity distribution in distance of 1 mm away from the top surface of the screen.

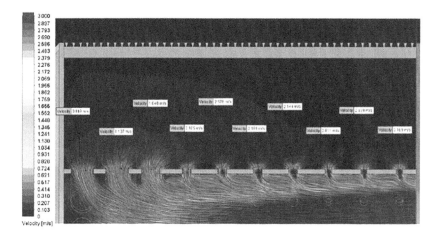

Figure 6.   The velocity distribution inside the deflator holes.

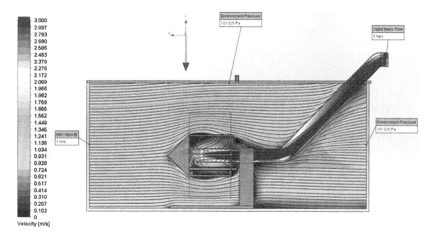

Figure 7. The velocity distribution in vertical-section for wedge-wire screen in the channel filled by water flowing with 1 m/s velocity.

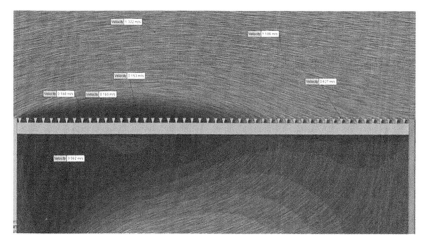

Figure 8. The velocity distribution in vertical-section for wedge-wire screen in the channel filled by water flowing with 1 m/s velocity.

Numerical calculations assuming flow rate of water in the hydraulic channel were also conducted. Water in the channel flowed with 1 m/s velocity. The results are presented in Figure 6. Few times higher velocities were observed on the left side than on right, what is shown in Figure 6.

CFD seems to be good tool for designing of cylindrical wedge-wire screens. Initial perforation of deflector gives quite homogenous distribution of flowrate velocities. However, perforation should be improved, since inside the screen slots velocities were few times higher on the left than on right side. Numerical results described in the paper should be verified experimentally. Assuming the flow of water in the channel, still quite high difference between velocities on opposite sides of the wedge-wire screen were observed.

REFERENCES

Department of Fisheries and Oceans 1995. *Freshwater Intake End-of-Pipe Fish Screen Guideline*. Ottawa: Ministry of Supply and Services Canada.

Jamieson D., Bonnett M., Jellyman D. & Unwin M. 2007. *Fish screening: good practice guidelines for Canterbury*. *NIWA Client Report: CHC2007-092*. Christchurch: National Institute of Water & Atmospheric Research Ltd.

Jiyuan, T., Guan, H. & Chaoqun L. 2008. *Computational fluid dynamics: a practical approach.* Oxford: Elsevier.

Middlestadt, F. & Gerstmann, J. 2014. *Numerical Investigations on Fluid Flow through Metal Screens*. 5th European Conference for Aeronautics and Space Science. Munich: EUCASS association.

Taft, E.P. 2000. Fish Protection Technologies: a Status Report. *Environmental Science & Policy* 3: 349–359.

Toro, E.F. 2009. Riemann Solvers and Numerical Methods for Fluid Dynamics: A Practical Introduction. Berlin: Springer.

US Government 2006. Fish Protection at Water Diversions. A Guide for Planning and Designing Fish Exclusion Facilities. Colorado: Department of the Interior Bureau of Reclamation Denver.

Part C: Geodesy, surveying and mapping, and Roads, bridges and geotechnics

Advances and Trends in Engineering Sciences and Technologies II – Al Ali & Platko (Eds)
© 2017 Taylor & Francis Group, ISBN 978-1-138-03224-8

The effect of light on the work of the electronic levels

A. Atroshchenkov, N. Belyaev, E. Kosyakov, E. Mikhalenko & M. Rodionova
Peter the Great St. Petersburg Polytechnic University, St. Petersburg, Russia

ABSTRACT: The article presents the comparison of performance of electronic (digital) and optical leveling. The advantages and disadvantages of working with electronic (digital) levels are described. The influence of external factors, such as the vibration of the instrument and leveling staff, non-vertical position of the staff, whether it is too strong or weak illumination etc., on the operation of the electronic (digital) levels is analyzed. Peter the Great St. Petersburg Polytechnic University performed field studies with different models of electronic levels under variable light conditions of leveling staff and different distances from the instrument to the rod. Illumination of barcode leveling staff was measured with a Lux Meter. The dependencies reflecting the influence of illumination degree within the devices safe operation are obtained. In addition, recommendations for their efficient use in conditions lacking daylight are given.

1 INTRODUCTION

Geodetic works are carried out at different stages of any building project as well as in solving problems related to land use (Mikhalenko et al. 2008a, Mikhalenko et al. 2008b). Levels of various designs are applied in geodetic works (Ermakov et al. 2001, Mikhalenko et al. 2012, Mikhalenko et al. 2014a, Mikhalenko et al. 2014b, Mikhalenko et al. 2014c).

Electronic (digital) levels with barcode staff were applied along with the usual optical devices in recent years. Exploitation of digital levels is simple and these levels allow for the making of high-speed and precise measurements of height and distance. Surveyor's functions are minimal during the work with the digital level. Digital levels can have their information read by the staff automatically without the participation of the land-surveyor; therefore, no errors occur during measuring connected with sighting, counting and lack of surveyor's experience in contrast to optical levels. Moreover, the surveyor can control distance between level and staff in every reading and improves the accuracy of distance measurements.

Digital levels fit perfectly for definition of slopes and profiling, measuring subsidence zones, observation for structure's deformations, leveling of a pavement, etc.

Digital levels can record obtained data in the internal memory of the device in addition to having the ability of automatic measuring. Therefore, they are the most effective in a measuring of large number altimetric points (Atroshchenkov et al. 2014, Belyaev et al. 2015, Trofimov et al. 2008).

2 REVIEW OF THE INFORMATION

Some disadvantages, along with obvious advantages, were found out in the field work with these devices (Abanmy et al. 2007, Adewale et al. 2014, Ali 1990, Atroshchenkov et al. 2014, Balashov et al. 2012, Belyaev et al. 2014, Elhassan & Ali 2011, Soboleva et al. 2007).

There are some requirements in the leveling instruction (Instructions 2004), which must be obeyed for levels of I and II classes. For example, a magnification of the telescope according to instruction must be at least 40-fold, but digital level has significantly less magnification. In addition, the leveling instruction requires making measurements from two sides of a staff but only one side of a staff is used with digital levels.

During field studies, it was found that digital levels are very sensitive to different kinds of hindrances, such as wind, ground vibration caused by building machines, inclined rod's position, sunbeam directed to the device objective and high or low level of illumination.

Normal operation of the level can be disturbed by hoarfrost on the staff at low temperatures. The level's battery can suddenly run down in the wintertime that makes leveling impossible.

Digital levels have one extra limitation for making measurements. If some obstacle is located in the way of the collimating ray, the optical level can take the reading from the staff through a slit width of 1 cm or even at half the staff. In addition, it is possible to tilt a staff and make measure with the slope correction. However, it is impossible to do when barcode staff is used. A staff must be held vertical and a big part of the rod must be seen in the level's eyepiece.

Along staff is waved back and forth to read the correct height in a measuring with the optical levels. The staff reading becomes the minimum when it is in the vertical position. In this way the staff's tilt error is excluded. Nevertheless, this operation is impossible with digital levels.

Optical levels are better than digital ones for location surveys, when the actual movements of the staff have to be seen.

In addition, some factors such as the unit weight, the price, the volatility, the temperature range of work, the complexity and cost of device repair, should be considered. The digital levels concede to the optical ones in all of these positions.

3 FORMULATION OF THE PROBLEM AND A LIST OF THE EQUIPMENT

The more detailed study of the influence of degree of illumination at the operation of the digital levels was the purpose of this work. Investigations were conducted in the Peter the Great St. Petersburg Polytechnic University. The work was carried out with three models of digital levels:

1. Leica Sprinter 50 (Figure 1) with 24-fold magnification of the telescope and barcode staff GSS 112.
2. Sokkia SDL 50 (Figure2) with 28-fold magnification of the telescope.
3. Leica Sprinter 250 M (Figure 3) with 24-fold magnification of the telescope and barcode staff GSS 112.

Figure 1. Leica Sprinter 50.

Figure 2. Sokkia SDL 50.

Figure 3. Leica Sprinter 250 M.

Table 1. Comparison of digital levels.

Technical data	Leica sprinter 50	Sokkia SDL 50	Leica sprinter 250 M
Height measurements. Standard deviation per km double run, mm: Electronic measurement with barcode staff:	2.0	1.5	1.0
Optical measurement with standard aluminum E-scale/Numeral staff:	2.5	2.5	2.0
Distance Accuracy (Standard deviation), mm	10 for D ≤ 10 m; Distance in m × 0.001 for D > 10 m	10 for D ≤ 10 m; ± (0.1% ×D) for 10 ≤ D ≤ 50 m ± (0.1% ×D) for D > 50 m	10 for D ≤ 10 m; Distance in m × 0.001 for D > 10 m
Range, m	2–100	5–50	2–100
Telescope Magnification (Optical):	24×	28×	24×
Field of view:	20'	1°20'	20'
Stadia ratio:	100	100	100
Display	128 × 104	128 × 32	128 × 104
Measuring time (in normal daylight condition), seconds	3 and les	2.5	3 and les
Operating temperature range, °C	−10 – +50	−20 – +50	−10 – +50
Dimensions, mm: Length	219	257	219
Width	196	158	196
Height	178	182	178
Weight (including batteries), kg	2.55	2.40	2.55

Comparison of digital levels is given in Table 1.

The research work took place in the park on Butlerov Street in Saint Petersburg. Digital Lux Meter GM 1010 (Figure 4) was used for the illuminance measurement.

Figure 4.   Measuring the illuminanceusing Lux Meter GM 1010.

The working range of device varies from 0 up to 200 000 lx. Accuracy: ± 3% rdg ± 0,5% F.S. (≤ 10,000 lx) and ± 4% rdg ± 10 digits (≥ 10,000 lx). The principle of its operation is based on the photoelectric effect. The light falling on the semiconductor photocell transfers its energy to electrons. As a result, the electron release in semiconductor occurs and photocell conducts current. The value of current strength is proportional to the illumination of the photocell (Operating principle, 2015).

4   EXPERIMENTAL STUDIES

The level was mounted fixedly on a tripod. Points for the staff installation were marked every 10 meters along the collimating ray over a stretch of 100 meters. The staff was established alternately on every leveling point from the nearest to the level. It was taken 20 staff readings at each point. While operating the illumination of staff was recorded by the Lux Meter. These operations were repeated at different illumination of territory and leveling staff, accordingly. This was achieved by carrying out the measuring at different times during dusk.

The number of failures among 20 measurements at every point, when the device «did not see» the staff, was fixed during the operation.

After completion of the measurement with one level, another one was established on a tripod and all actions were repeated. The amount of successful measurements was reduced with a decrease in illumination of the staff. It is shown on the Figure 5.

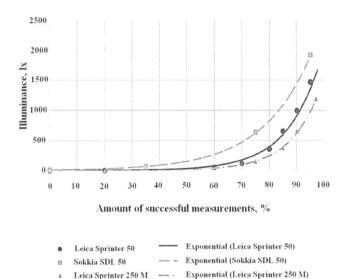

Figure 5. The dependence of the amount of successful measurements from the illuminance.

5 CONCLUSIONS

The following conclusions have been formulated in accordance with the results.

The distance between the level and staff declared by the manufacturer and equal to 100 meters corresponds to reality only for the model Leica Sprinter 250 M. This distance does not exceed 85 meters for the models Leica Sprinter 50 and Sokkia SDL 50.

The decrease of illumination reduces the amount of effective measurements for the all considered devices.

Sokkia SDL 50 is the most sensitive to a lack of illumination and Leica Sprinter 250 M – the least sensitive. If the amount of effective measurements for the first level with a degree of illumination 600 lx is 75% then for the second it reaches 90%.

For normal operation of the levels (with the amount of effective measurements80%), lighting should not be less than 250lx for Leica Sprinter 250 M, 400 lx for Leica Sprinter50, 800lxfor Sokkia SDL 50.

Usage of artificial illumination in the working area by means lighting devices is recommended to ensure trouble-free operation of the electronic (digital) levels in low light conditions of staff.

REFERENCES

Abanmy, F.A., Ali, A.E. and Alsalman, A.S. 2007. Optical, Laser and Digital Levelling: a Comparison of Accuracy in Height Measurement.*Sudan Engineering Society Journal*, Volume 53 No.48. 13 p.

Adewale, A., Emenari, U.S., Uwaezuoke, I.C. 2014. Comparison of Height Differences obtained from Automatic, Digital and Tilting Level Instruments. *Journal of Environmental Issues and Agriculture in Developing Countries*, Vol. 6, No. 1, pp. 85–92.

Ali, A.E. 1990. Comparative Evaluation of the Accuracy of Automatic Levels. *New Zealand Surveyor*, 33, No. 277, pp. 72–75.

Atroshchenkov, A.Y. et al. 2014. Nivelirovaniye trass s pomoshch'yu razlichnykh geodezicheskikh priborov [Levelling of tracks using various surveying instruments]. Week of the science of SPbPU: materials of the conference with international participation. SPb.: Izd-vo Politekhn. un-ta, pp. 118–120. (rus).

Balashov, B.V., Belyaev, N.D., Mikhalenko, E.B. and Tazeyev, T.A. 2012. Monitoring of fill dam constructed in Ust-Luga port on soft foundation soils. *Magazine of civil engineering*, 4, pp.10–16. (rus).

Belyaev, N.D. et al. 2014. Otsenka formy i obyema otvalov s pomoshchyu sovremennogo geodez-icheskogo oborudovaniy [Estimation of shape and volume of dumps with the help of modern surveying equipment]. Week of the science of SPbPU: materials of the conference with international participation. SPb.: Izd-vo Politekhn. un-ta, pp. 97–100. (rus).

Belyaev, N.D. et al. 2015. Satellite technologies in monitoring of ecosystems. Environment. Technology. Resources. Proceedings of the 10th International Scientific and Practical Conference. Rēzeknes Augstskola. Pp. 60–63.

Elhassan, I.M., Ali A.S. 2011. Comparative study of accuracy in distance measurement using: Optical and digital levels. *Journal of King Saud University – Engineering Sciences*, 23, pp. 15–19.

Ermakov, V.S. et al. 2001. Inzhenernaya geodeziya. Geodezicheskoye obespecheniye stroitelstva i ekspluatatsii morskikh i vodnotransportnykh sooruzheniy [The engineering geodesy. Geodesic support of construction and operation of ships and water transport facilities]. SPb.: Izd-vo Politekhn. un-ta, 71 p. (rus).

Fakhrutdinov, M.R., Yalyshev, A.I., Belyaev, N.D., Mikhalenko,E.B. 2011. Metodika provedeniya izmerenij na podkhodnoj dambe Kompleksa SUG v Ust-Luge [Method of measurement on Ust-Luga dam]. XL Week of Science. Part I. SPb.: Izd-vo Politekhn. un-ta, pp. 50–52. (rus).

Instructions for leveling I, II, III and IV classes. 2004. Moscow, CRIoGAS & P. 224 p. (rus).

Operating principleand application of luxmeter. Luxmeters and illuminosity measurement. 2015. Electronic resource: EcoUnit Ukraine—Electronic data.—Access mode: http://ecounit.com.ua/artikle_103.html – Title from the screen. (rus).

Mikhalenko, E.B. et al. 2008a. Inzhenernaya geodeziya. Geodezicheskiye razbivochnyye raboty, ispolnitclnyyc s'yemki i nablyudeniya za deformatsiyami sooruzheniy [Engineering geodesy. Geodesic work, executive survey and monitoring of deformations of structures]. SPb.: Izd-vo Politekhn. un-ta, 88 p. (rus).

Mikhalenko, E.B. et al. 2008b. Inzhenernaya geodeziya. Resheniye osnovnykh inzhenernykh zadach na planakh i kartakh. Polevyye geodezicheskiye raboty [Engineering geodesy. Solution of the basic engineering tasks on the plans and maps. Field geodetic]. SPb.: Izd-vo Politekhn. un-ta, 179 p. (rus).

Mikhalenko, E.B. et al. 2012. Inzhenernaya geodeziya: posobiye po uchebnoy geodezicheskoy praktike [Engineering geodesy: the manual for surveying practice]. SPb.: Izd-vo Politekhn. un-ta, 93 p. (rus).

Mikhalenko, E.B. et al. 2014a. Inzhenernaya geodeziya. Ispolzovaniye sovremennogo oborudovaniya dlya resheniya geodezicheskikh zadach [The engineering geodesy. Modern equipment for the solution of geodesic problems]. SPb.: Izd-vo Politekhn. un-ta, 98 p. (rus).

Mikhalenko, E.B. et al. 2014b. Kontrol kachestva stroitelstva. Geodezicheskoye obespecheniye stroitelstva i ekspluatatsii vodokhozyaystvennykh i gidrotekhnicheskikh sooruzheniy [Quality control of building. Geodesic support of construction and operation of water and hydraulic structures]. SPb.: Izd-vo Politekhn. un-ta, 143 p. (rus).

Mikhalenko, E.B., Belyaev, N.D. and Zagryadskaya, N.N. 2014c. Inzhenernaya geodeziya. Nablyudeniya za tekhnicheskim sostoyaniyem i deformatsiyami pri stroitelstve i ekspluatatsii sooruzhenii [Engineering geodesy. Monitoring of technical condition and deformations during building and exploitation of structures]. SPb.: Izd-vo Politekhn. un-ta, 80 p. (rus).

Soboleva, E.L. et al. 2007. Rezultaty proizvodstvennykh rabot s primeneniyem tsifrovogo nivelira [The results of the production of work using digital level]. *Journal INTEREXPO GEO-SIBERIA*. Issue 1, volume 1. 6 p. (rus).

Trofimov, S.V., Shpyrchenko, I.P. and Mikhalenko, E.B. 2008. Sravnitelnyj analiz geodezicheskikh priborov pri vypolnenii operacii nivelirovaniya [Comparison and analysis of different levelling instruments]. XXXVII Week of Science. Part I, SPb.: Izd-vo Politekhn. un-ta, pp. 18–20 (rus).

Advances and Trends in Engineering Sciences and Technologies II – Al Ali & Platko (Eds)
© 2017 Taylor & Francis Group, ISBN 978-1-138-03224-8

Cyclic load tests of driven pile base capacity

M. Baca, Z. Muszynski & J. Rybak
Wroclaw University of Science and Technology, Wroclaw, Poland

T. Zyrek
Silesian University of Technology, Gliwice, Poland

A.G. Tamrazyan
Moscow State University of Civil Engineering, Moscow, Russia

ABSTRACT: Since 2015 the Polish authors have developed an original method for assessing the capacity of closed-end pipe piles. A self-balanced appliance was built to test pile axial capacity. Currently, field tests are conducted to develop methods of evaluating the bearing capacity of the pile mobilized at its base and on its side. The studies considered different ways of application of the load on the pile base with regard to a different value and sign of stress transferred to the surroundings of the pile by its side surface. The cooperation with Moscow State University of Civil Engineering enabled a more sophisticated evaluation of the influence of cyclic loading to the pile capacity at various stages after pile completion. Simultaneously, various techniques of surveying were used for a fast and reliable measurement of displacement of many points in terms of pile testing.

1   INTRODUCTION—TESTING OF VIBRATORY DRIVEN PILE BASE

The advances in modern civil engineering make it possible to improve pile foundation techniques. This includes both the development of new pile technologies, and also more accurate ways of analyzing pile working conditions. Nevertheless, static load tests are still considered to be the most trustworthy method of pile capacity evaluation. This method makes it possible to measure the pile head displacement under physically applied load provided to the pile by a hydraulic jack. Despite its verified reliability, the static load test has some disadvantages. The most significant drawbacks are: a relatively long time of investigation and complications with construction of the reaction system.

Static load test, performed in natural scale, is considered the most reliable method of pile testing, despite some of its disadvantages. Different approach to a static load test is presented by the Osterberg test. This study requires an especially designed loading device called the Osterberg cell, which is placed inside tested pile. The load applied by the hydraulic jack creates an equal upward and downward force inside the pile. Therefore, the displacement of both sections under the applied axial load can be measured. In this test specially designed device, called the Osterberg Cell or the O-cell, is placed inside the pile during its construction. Load applied through the O-cell allows on simultaneous measurement of displacement of two sections of a pile—upper and below the O-cell. Therefore, the Osterberg test makes it possible to research separately the end bearing and the side shear of a pile. Despite some doubts about comparability of results between the Osterberg test and traditional static load test, there are many works confirming sufficient comparability between these two investigation methods (England 2009, Lee & Park 2008, Schmertmann & Hayes 1997). The Osterberg test is not only a different type of static load test. It is also a benchmark for many testing methods which enable pile bearing capacity estimation without the necessity to construct a complex reaction system. One of these approaches is presented in this paper. On the basis of the

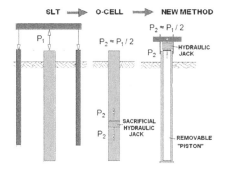

Figure 1. The idea of the new testing method.

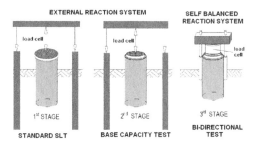

Figure 2. Subsequent steps of test-pile loading.

Osterberg test assumptions, a new pile testing method was proposed (Baca, Rybak & Zyrek 2015) for closed end steel pipe piles (Figure 1).

The force loading the pile base is transmitted from a hydraulic jack at an instrumented pile head by means of a removable piston. The hydraulic jack can be assembled on top of the piston and disassembled when the test is completed. The system consists of hydraulic jack with an electric pump, reaction beams, instrumentation for pile displacement control and the reference system for geodetic measurements (Baca et al. 2015, Baca et al. 2016).

2 PRINCIPLES OF CYCLIC LOAD TEST WITHIN TESTING PROGRAMME

The next phase of a huge investigation program is planned as the investigation of pile capacity, with the pile being fully embedded to the depth of 8 m. Three steps of loading are considered to investigate pile capacity. In detail, the research is divided into 3 steps shown in Figure 2. In the first stage, the pipe with a larger diameter and a steel toe will be installed in soil, and the investigation with traditional static test simulation will be performed. In the second stage, a smaller diameter pipe will be installed inside that pipe, and another simulation of static load test will be performed, with the load applied to the interior pipe. This will allow for measuring the pile base capacity, because the external pipe does not work. In the final stage of the investigation, the load will also be applied to the internal pipe but with the anchoring to the external pipe. During the pressurization of a jack, the internal pipe with the toe will be compressed inside the soil, while the external pipe, only working with its shaft, will be uplifted with the same load. This will model conditions similar to the Osterberg test, when the O-cell is placed at the pile base. When the load is applied, the displacement of two pipes may be measured using dial gauges. For each stage of the test, cyclic loading (multiple loading) is applied. After the investigation is completed, there will be few assumptions for verification. Primary hypothesis predicts that there is a strong correlation between pile bearing capacity in traditional static load test and in the case of the new proposed method. What is more, it is postulated that using model tests, site test and numerical simulations (Lee & Park 2008, Baca, Rybak & Zyrek 2015), it will be possible to create simple analytical correlation which will lead to an evaluation of ultimate capacity using only the method proposed.

3 INSTRUMENTED LOAD TEST—VERTICAL DISPLACEMENT CONTROL

The quality of pile load test is always strongly dependent on accuracy of displacement control. The basic measurement tools used for measuring vertical displacement are most often dial indicators with the readout precision of 0.01 mm. They are fastened to a stiff base, typically made of steel profiles. Such base should be restrained outside the zone of influence of the loaded pile settlement and outside the scope of subsoil deformation due to the raising of the anchoring piles. It was signaled in professional literature that when the piles are loaded with considerable forces, the settlement values thus measured are sometimes largely distorted due to the displacement of the whole base to which the dial indicators were fastened.

A way of overcoming the risk of distorted results is the simultaneous use of a leveling instrument with the appropriate readout precision on the measuring rod. In this way, the records of displacement by means of dial indicators might be supplemented, at the same time eliminating systematic errors brought about by the system of reference instability. The role of those techniques become crucial when the thermal factors (due to insolation) and the movements of the ground around the pile under test at the terminal phase of test affects the results of traditional dial gauges measurements. A short overview of most important methods of measurement applied in the testing of foundation pile capacity is presented below.

3.1 Basic method—precision geometric leveling

In the method of geometric leveling, the axis of sighting of the leveling instrument indicates in space a geometric horizon which is locally perpendicular to the direction of the gravity force. Leveling staffs are placed vertically on the examined points. By means of instrument, it is possible to read the distance (along the staffs) from examined points to the horizon of the level. For a measured pair of points, the difference of those distances depicts the difference in those points' altitude. Nowadays, two types of leveling instruments are used: optical (with manual readouts) and digital ones. Digital levels perform the readouts automatically via the analysis of barcode image, which is built into the staff. In terms of accuracy, levels can be classified into three groups: high-precision, semi-precision and technical levels. Newest high-precision digital levels (e.g. Leica LS15 - Figure 3a) allow to obtain the accuracy of 0.2 mm per 1 km of double run (with invar staffs). Semi-precision digital levels (e.g. Leica Sprinter 250 M used for this research—Figure 3b) allow us to obtain the accuracy of 0.7 mm or 1.0 mm per 1 km of double run. In the case of load tests of foundation piles, the control points are deployed on the head of the pile under load and on the heads of the anchoring piles. The points of reference are the benchmarks stabilized at a suitable distance, beyond the zone of deformation. Leveling measurements are taken cyclically for each particular degree of loading and/or time intervals, reading out the values from leveling staffs located at the control and the reference points.

3.2 Other methods of vertical displacement control

3.2.1 Precision tacheometry
Precision trigonometric leveling makes it possible to measure the control points located in inaccessible places. Those control points are signalled by means of a special target or prism and observed from special stations (e.g. concrete pillars enabling the forced centring of the instrument). The measurements of spatial distances and vertical angles can be performed synchronously from a couple of station, using motorized electronic tacheometers (e.g. Trimble S3 used for this research—Figure 4a) equipped with function of automatic target recognition. All readouts (for control and references points) are repeated for each particular degree of pile loading. The obtained accuracy of vertical displacement ranges between ±(0.5÷0.8) mm for a distance below 50 m.

3.2.2 Terrestrial laser scanning
The ground-based laser scanning is very popular in geodetic determination of displacements and deformations of engineering objects. Nowadays, this technique makes it possible to survey even a million points during one second. For each point the reflectorless measurement of the distance (up to 300 m) is taken, and the horizontal and vertical angles are recorded. It allows to calculate three dimensional coordinates for each measured point. These points can describe whole analyzed object or their selected parts indicated by HDS targets.

In the case of the pile load test the terrestrial laser scanning is used rather experimentally. Despite its relatively small accuracy of determining the position of a single point, the laser scanner (e.g. ScanStation C10 used for this research—Figure 4b) has a great advantage. In a short

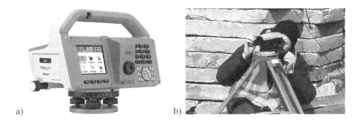

a)   b)

Figure 3. a) High-precision digital level Leica LS15 (source: http://surveyequipment.com), b) Semi-precision digital level Leica Sprinter 250M used for investigation.

a)   b)   c)

Figure 4. a) Motorized tacheometer Trimble S3, b) Terrestrial laser scanner ScanStation C10 manufactured by Leica Geosystems, c) Non-metric digital camera Nikon D-800.

time we are able to obtain the fully metric, three dimensional model of a pile, the testing structure and the ground in the vicinity of tested pile. The technology is remote and does not require contact with the structure under survey. Various examples of application were already published (Muszynski 2014, Muszynski 2015). Additionally, the use of HDS targets mounted on checked structure allow to control displacements of selected points with an accuracy of ±2 mm.

3.2.3 *Close-range photogrammetry*

Terrestrial photogrammetry is a well-known method, which enables fast and non-contact measurement of the analysed object. The application of one camera, which is stationary and oriented in the space, makes it possible to control the changes of object's geometry in one dedicated plane. Using two cameras mounted on the known base with known parameters of reciprocal orientation, it is possible to build a three-dimensional model of measured object. Nowadays, multiple-photogrammetry becomes increasingly more popular. In this technique, photos are taken from many places with unknown orientation of the camera. However, it is important to maintain the visibility of large common parts of object on the adjacent images. The advanced calculation algorithms allow us to use non-metric digital cameras with high-resolution matrices (e.g. 36 Mpix in Nikon D-800 used for this research—Figure 4c). The characteristic points of the controlled object can be indicated by special targets in order to automatic recognition of them on the pictures. As a result of calculation it is possible to get a three dimensional model of testing structure and to track the location of controlled points in each particular degree of pile loading.

4 PILE TEST LAYOUT AND PRELIMINARY CYCLIC TEST RESULTS

The reaction beam of testing appliance was anchored to 2 jet-grouting columns with the diameter of 800 mm. The range of the test was only limited by the capacity of the steel construction equaled to 1300 kN. The capacity of the beam was designed on the basis of estimated pile capacity computed from CPT testing in the close vicinity of tested pipe pile.

Figure 5. a) Location of instrument stations and reference points, b) Instrumented load test of closed-end pipe pile.

Figure 6. Cyclic loading of pile base—7 days after driving.

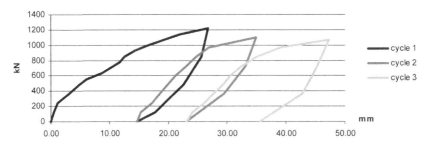

Figure 7. Cyclic loading of pile base—64 days after driving.

Table 1. Pile base capacities based on SLT and Chin-Kondner extrapolation method (1 week and 2 months after pile driving).

Pile base capacity	Subsequent cycles of pile base loading			
	Cycle 1	Cycle 2	Cycle 3	Cycle 4
R_{bk} (field results – 7 days)	732.7 kN	713.2 kN	778.5 kN	821.6 kN
R_{bk} (field results – 64 days)	1964.0 kN	1620.7 kN	1574.6 kN	–

All the measurements were done by means of dial gauges and controlled by four independent leveling techniques (semi-precision digital level, precision trigonometric levelling, ground-based laser scanning and terrestrial photogrammetry). The layout of these instruments may be seen on Figure 5a and pile testing appliance instrumentation on Figure 5b.

So far, only a few results were gathered by the authors, however they seem to be quite confusing. The test performed shortly after pile driving proves a consequent increase of pile stiffness and pile base capacity in subsequent cycles. After 2 months—the base capacity increases significantly but a reverse phenomenon may be noticed—the capacity decreases in every next load cycle. Displacement vs. Load graphs for fast and postponed load cycles are shown on Figure 6 and Figure 7, respectively.

727

5 PILE CAPACITIES DERIVED FROM CYCLIC TEST RESULTS

The first tests (1 week after pile driving) were in relatively good conformity with CPT-based capacity computations. After 2 months, only the shaft capacity confirms well with the results of calculations based on CPT. Pile base capacity and total capacity seem to be much bigger than the preliminary estimations. The results of pile base capacity were juxtaposed in Table 1.

6 CONCLUSIONS

We have presented the first results of cyclic load tests of tubular pipe piles with pile base capacity testing. The results are promising, however a remarkable increase of pile base capacity is far beyond the expectations. The last stage of research will include a bidirectional pile testing, which will let us measure separately the pile shaft and the base capacity. All the conducted tests will hopefully prove that the proposed method of pile testing can be a good alternative to the traditional pile static load test. Simultaneously, some remarks may be done about the increase of pile capacity with time.

ACKNOWLEDGEMENTS

The authors would like to express their gratitude to **PPI CHROBOK S.A.** for providing of the test field in Bojszowy and technical support in the assembling of the SLT appliance.

REFERENCES

Baca, M., Muszynski, Z., Rybak, J. & Zyrek, T. 2015. The application of geodetic methods for displacement control in the self-balanced pile capacity testing instrument. *Advances and trends in engineering sciences and technologies: proceedings of the International Conference on Engineering Sciences and Technologies, Tatranská Štrba, Slovakia, 27–29 May 2015*. CRC Press, cop. 2016, pp. 15–20.

Baca, M., Rybak, J. & Zyrek, T. 2015. Practical aspects of tubular pile axial capacity testing. *15th International Multidisciplinary Scientific GeoConference, SGEM 2015, Albena, Bulgaria, 18–24 June, 2015.* Book 1 Vol. 2, pp. 549–554, (DOI: 10.5593/SGEM2015/B12/S2.073).

Baca, M., Rybak, J., Tamrazyan A.G. & Zyrek, T. 2016. Pile foot capacity testing in various cases of pile shaft displacement. *16th Int. Multidisciplinary Scientific GeoConference, SGEM 2016, Albena, Bulgaria, 30 June - 6 July, 2016, 2016.* Book 1 Vol. 1, pp. 945–950, (DOI: 10.5593/SGEM2016/B11).

England, M. 2009. Review of methods of analysis test results from bi-directional static load tests, *Proc. of the 5th Int. Symp. on Deep Foundation on Bored and Auger Piles, Ghent, Belgium, 8–10 September 2008*, pp. 235–239.

Lee, J.-S. & Park, Y. H. 2008. Equivalent pile load-head settlement curve using a bi-directional pile load test. *Computers and Geotechnics*, 35, pp. 124–133.

Muszynski, Z. 2014. Assessment of suitability of terrestrial laser scanning for determining horizontal displacements of cofferdam during modernization works on the Redzin sluice. *14th International Multidisciplinary Scientific GeoConference SGEM 2014, Albena, Bulgaria, 17–26 June, 2014.* Book 2 Vol. 2, pp. 81–88, (DOI: 10.5593/SGEM2014/B22/S9.011).

Muszynski, Z. 2015. Displacement measurements of the Statue of Christ the King in Świebodzin—first results. *15th International Multidisciplinary Scientific GeoConference SGEM 2015, Albena, Bulgaria, 18–24 June, 2015.* Book 2 Vol. 2, pp. 227–233, (DOI: 10.5593/SGEM2015/B22/S9.028).

Osterberg, J.O. 1998. The Osterberg Load Test Method for Bored and Driven Piles—The First Ten Years, 7th International Conference 7 Exhibition on Piling and Deep Foundations. Deep Foundation Institute, Vienna, Austria, June 15–17, Deep Foundation Institute, Englewood Cliffs, N.J.

Rybak, J. & Zyrek, T. 2014. Time dependent factors in driven pile capacity control. *14th International Multidisciplinary Scientific GeoConference SGEM 2014, Albena, Bulgaria, 17–26 June, 2014.* Book 1 Vol. 2, pp. 995–1002, (DOI: 10.5593/SGEM2014/B12/S2.127).

Schmertmann, J. & Hayes, J. 1997. The Osterberg Cell and Bored Pile Testing—a Symbiosis. Proceedings at the Third Annual Geotechnical Engineering Conference. Cairo University, January 5–8, 1997, Cairo, Egypt.

Advances and Trends in Engineering Sciences and Technologies II – Al Ali & Platko (Eds)
© 2017 Taylor & Francis Group, ISBN 978-1-138-03224-8

The issue of barrier-free public places – the transport interchange Ostrava-Svinov

A. Bílková, B. Niemiec, P. Kocurová & D. Orsáková
VŠB-TUO, Ostrava, Czech Republic

ABSTRACT: This article deals with the issue of barrier-free public places in connection with the development of Czech legislation. The issue is presented in a case study of the significant regional transport junction in Ostrava-Svinov that connects the railway, bus and public transport. Due to the intensive use of these areas, it is necessary to assess these places not only from the viewpoint of transport, logistics, technology or economy, but also from the viewpoint of the movement of a person at all stages of his or her life and also from the viewpoint of a barrier-free access. The text focuses on appraisal and evaluation of completeness, continuity and correctness of barrier-free adaptations which were executed in the area of the transfer junction since 2000 in relation with Regulation number 398/2009 Coll. and the previous legislation (regulation no. 369/2001 Coll. and 492/2006 Coll.). The study pursues four basic criteria of barrier-free places, as well as the appropriateness and quality public space.

1 INTRODUCTION

The issue of barrier elimination in public spaces in relation to the development of legislative environment in the Czech Republic is one of many issues that must be addressed by municipal representatives and planners in the reconstruction and the development of public spaces of cities and municipalities, which includes public transport. The most important part of public transport are transport hubs, the locations with the highest occurrence and movement of people. Due to this intensive use, it is important to make an assessment in terms of technical execution and accessibility, to ensure the comfortable movement for people of all ages, as described by Bílková (2016). Fundamental changes in legislation are often very visible in the architecture and the design of modifications and additions to public spaces, as the new often does not follow the old and there is no comprehensive overview of a solution, which is very important for the correct execution of modifications to accessibility. Without maintaining the continuity of space, modifications to accessibility, although correctly executed and in compliance with the statutory legislation, become ineffective and therefore unsuitable for use by persons with limited mobility and orientation.

1.1 *Legal environment of the Czech Republic*

The legal environment of the Czech Republic has been addressing the issue of accessibility for over sixteen years. The first regulation was issued in 2001 in the form of Decree no. 369/2001 Coll., on General Technical Requirements Enabling the Use of Buildings by Persons with Limited Mobility and Orientation, which featured basic definitions of terms, determination of the scope, technical parameters and dimensions for ensuring accessible use of external environment and buildings. This decree was amended in 2006 to Decree no. 492/2006 Coll. Another, and most recent amendment of regulation addressing accessibility took place in 2009, when a significant change was made to the arrangement of content, which resulted in the new and comprehensive Decree no. 398/2009 Coll. on General Technical Requirements Ensuring Accessible Use of Buildings, which is in force so far. In addressing the issue of proper design and implementation of transport hubs, we must also not ignore other legal regulations and Czech technical standards, especially Act no. 13/1997 Coll., on Roads, Act no. 183/2006 Coll., on Territorial Planning and Building Regulations, CTS 73 4130 Stairs and Inclined Ramps, CTS 73 6110 Design of Roads and CTS 73 6425-2 Bus, Trolley and Tram Stops (part 2: Transfer hubs and stations), as described by Bílková (2016).

1.2   *Evaluation of accessibility*

To ensure accessible use it is also necessary to ensure not only compliance with valid legislation, but especially the basic criteria determining and defining the accessibility itself. These can be expressed using four words (terms)—security, availability, approachability and comprehensibility, as described by Jacura (2012).

In the environment of transport hubs this means that if we want the public space and its architecture to create a suitable and high quality environment that provides all users with all the comfort and possibilities of barrier-free movement, it is necessary to ensure those four criteria in the following form, as described by Jacura (2012) are met. This includes the creation of places for crossing roads, ensuring smooth movement of all participants over the shortest possible distances, provide colour, tactile contrasts and quality of all surfaces, design areas with the necessary elements for people with limited or impaired mobility or orientation, and last but not least, ensure production of comprehensible signs and information systems.

2   SOLUTION OF ACCESSIBILITY MODIFICATIONS OF THE OSTRAVA-SVINOV TRANSPORT HUB

The Ostrava-Svinov transport hub is located at the crossing of frequented transport routes (D1 motorway, Bohumín-Prague and Opava railway lines, roads 11 and 479). Thanks to this important location this area has the most important position among all the hubs in the Moravian-Silesian Region, as described by Bílková (2015). For the survey of this area, we chose a route along Svinov bridges upper stop to the Ostrava-Svinov railway station.

All areas of the Ostrava-Svinov transport hub have been reconstructed and supplemented by accessibility modifications based on valid legislation. The premises of the railway station with adjacent area in front of the station building underwent reconstruction in 2006. Reconstruction and modification of the area Svinov bridges upper and lower stop took place in 2013. Both areas have been modified independently and on the basis of different legislative environments. Due to this fact, this submission focuses primarily on the assessment and the evaluation of the completeness, consistency and correctness of all accessibility modifications.

3   MODIFICATIONS OF PUBLIC TRANSPORT STOPS—SVINOV BRIDGES UPPER AND LOWER STOP

The bus and tram stops Svinov bridges upper stop are located on an elevated road on Opavská street. The Svinov bridges lower stop suburban bus line station is located under the construction of the elevated road itself. The distance between the public transport stops and the railway station is about 400 metres and passengers must climb several levels, which can cause problems mostly to people with limited orientation ability, as described by Bílková (2015). An extensive reconstruction of the stop area was performed at the turn of 2012 and 2013. It consisted of repairs to the technical condition of the entire bridge structure, complete modification of platforms including the access to the stop area and the road leading to the station. The architectural concept utilizes modern surfaces and materials (glass, metal, concrete), and therefore corresponds to the glass building at the station.

All accessibility modifications were implemented according to Decree no. 398/2009 Coll., under which the elevated stops were made accessible using lifts, escalators and stairs. Contrast markings of the stair steps are executed incorrectly here; according to the above-mentioned decree the contrast markings should be only on the tread and not on the riser. Staircase railings are not equipped with signs in Braille anywhere. Lifts are equipped with all the necessary elements and they are easily accessible via guiding lines. Escalators are at all levels and have the necessary contrast markings. Locations of the digital voice beacons is also unsatisfactory, they are higher than 3000 mm above the staircase and at the highest level they are not present at all. The surfaces of the platforms are raised and provided with the necessary accessibility features. All indoor walkway spaces are made of concrete with red surface coating equipped with artificial guide lines in the form of white tactile strips. Given the uneven surface structure their tactile contrast is not sufficiently pronounced. Uniform colour schemes and materials used in a combination of red, white and grey and glass can cause significant problems in orientation for visually impaired persons. In the platform area obstacles are located in the form of roof structure pillars, seats, rubbish bins or insufficient contrast markings of glass surfaces. Pathways in the area are made of prefabricated concrete paving in black and white colour. Artificial guide lines are executed in two ways—raised curbs and black coloured paving with tactile grooves with a width of 400 mm (Figure 1).

The walkway between the public transport stops and the station is designed generously. Throughout its length there are located parking spaces for cars, but when parked they inappropriately interfere with the artificial guide line (Figure 2). Pedestrian crossings are executed

Figure 1.   Artificial guide line black coloured paving with tactile grooves.

Figure 2.   Cars inappropriately interfere the artificial guide line.

731

in accordance with the decree; all warning tactile elements are made of red paving with the required dimensions. The newly-established guide lines and accessibility elements are not connected to the premises in front of the station. This is due to the absence of guiding and warning tactile elements, which were not implemented during the modifications in 2006. This shortcoming is to some extent compensated by the possibility of using a wheelchair accessible bus service linking the area of the public transport stops and the railway station.

4   MODIFICATIONS OF THE OSTRAVA-SVINOV RAILWAY STATION

The railway station underwent extensive reconstruction in 2006. Completely new platforms and underpasses were developed. Further reconstruction was performed on the listed station building, supplemented with an interesting, attractive and award-winning glass area for passengers. A representative public space with water elements was created in front of the station.

All accessibility modifications were implemented pursuant to Decree no. 369/2001 Coll., according to which the platforms were made accessible via lifts, escalator and stairs. Railings are not equipped with signs in Braille anywhere. All entrance areas and platforms are equipped with digital voice beacons and information phrases. Platform areas are equipped with all accessibility elements. Doors into the station building open automatically. The interior of the new station hall is equipped with artificial guiding lines in the form of milled grooves in the floor of contrasting colour and they guide to the ticket office and waiting rooms in the old part of the building. Blind and visually impaired persons are left here to orientate themselves using a natural guiding line in the form of walls (Figure 3).

Fully-fledged use is hampered by numerous elements that create barriers and obstacles; they primarily consist of advertising and information stands and a piano, which complicate smooth movement for blind persons. The area in front of the station is architecturally designed. Orientation in it is not complicated. However, the main shortcoming is the choice of surface materials and their colours along with the absence of the necessary and logically arranged accessibility elements. The smooth surface of the paving made of grey-green sandstone is slippery under adverse climate conditions. The effect of time causes its erosion and flaking. The surface of the automobile road is made of grey granite blocks at the same height level as the walkways. These colour nuances are imperceptible for the visually impaired and may lead to an inadvertent entry into the road. The area in front of the entrances contains obstacles in the form of pillars (Figure 4) or the insufficiently marked glass surfaces of entrances, which cause problems especially to visually impaired and blind persons.

Figure 3.   Natural guiding line in the form of walls.

Figure 4.   Obstacles in the form of pillars.

5   CONCLUSION

When designing or reconstructing public spaces and especially transport hubs, we need to ensure that their use is pleasant and intuitive, free of complications and easy for both occasional and regular users with or without limitations. Safety and smooth operation must be ensured along with the necessary spatial requirements designated by the relevant statutory legislation. A measure of accessibility in these areas is necessary and paramount.

It is clear from the example mentioned above that it is not always easy to align technical, operational, architectural and aesthetic requirements. Both reconstructions are characterized by strong modern technical execution, which is welcome and necessary in such exposed public spaces, but often creates obstacles to fully-fledged barrier-free use. The survey also clearly shows that the value of a barrier-free space depends on the age of the construction, modification or reconstruction in connection with respecting and understanding the requirements arising from applicable laws. Although the wording of Decree no. 398/2009 Coll. does not significantly differ from the previous legislation (Decree no. 369/2001 Coll.) there are clearly visible differences in the execution of individual areas. The newer areas of public transit stops contain the necessary elements for barrier-free movement. On the other hand, the earlier reconstruction of the station contains (by current standards of accessibility) many errors and only the platform area can be considered to be without major defects. Many limitations in the premises of the old station building arise from an effort to preserve the original appearance of this listed building. A significant deficiency manifesting in both reconstructions, regardless of the year of implementation, is a number of small errors and shortcomings resulting mainly from misunderstanding of the regulations or omissions in the design itself. Here we encounter inappropriately designed materials for internal guiding lines, poorly placed acoustic elements, wrong choice of colours and contrasts of surfaces and materials used, improper placement of street furniture, and especially the incomplete execution and poor structuring of individual elements of barrier-free routes, which prevent continuous movement. It is therefore clear that if these spaces are to be considered comprehensively and coherently, they must be able to provide users with the required comfort, necessary for their proper use.

In conclusion, it must be emphasized that this space (considering its complexity) cannot be used by persons with limited mobility and orientation without prior familiarization or training in the company of another person. It is possible to utilize the option of easier transfer using the bus line connecting Svinov bridges and the railway station. This implies that the elements designed in this space, despite more than a few minor flaws and the lack of coherence between the individual reconstructions, will allow pedestrian movement for persons with reduced mobility and orientation.

ACKNOWLEDGMENT

The work was supported by the Student Grant Competition VŠB-TUO. Project registration number is SP2016/119.

REFERENCES

Bílková, A., Niemiec, B. & Orsáková, D. 2015. Case Study for Solution of Ostrava-Svinov Transfer junction Aimed at Movement and Orientation of Blind and Visually Impaired People. *International Journal of Engineering & Technology IJET-IJENS. Pakistan: Publisher—IJENS Publisher.* 15(06), (2077-1185), 58–63.
Bílková, A., Niemiec, B., Orsáková, D. & Zdařilová, R. 2016. Barrier-free use of transport interchanges. *Advances in Civil, Architectural, Structural and Constructional Engineering—Kim, Jung & Seo (Eds).* 2016, (978-1-138-02849-4), 65–68.
Jacura, M. 2012. *Optimal Layout of Public Mass Transport Transfers.* Prague: ČVUT in Prague, Faculty of Transportation Science, 2–5.
Regulation number 369/2001 Coll., *on General Technical Requirements Enabling the Use of Buildings by Persons with Limited Mobility and Orientation.* Prague: Collection of Laws of the Czech Republic.
Regulation number 398/2009 Coll., *on General Technical Requirements Ensuring Accessible Use of Buildings.* Prague: Collection of Laws of the Czech Republic.
Regulation number 492/2006 Coll., *amending Regulation of the Ministry for Regional Development no. 369/2001, on General technical requirements enabling the use of buildings by persons with limited mobility and orientation.* Prague: Collection of Laws of the Czech Republic.

Advances and Trends in Engineering Sciences and Technologies II – Al Ali & Platko (Eds)
© 2017 Taylor & Francis Group, ISBN 978-1-138-03224-8

Joint road safety operations

P. Danišovič & Ľ. Remek
Department of Construction Management, Faculty of Civil Engineering, University of Žilina, Žilina, Slovakia

ABSTRACT: The project "Effective and COordinated ROAD infrastructure Safety operation" (ECOROADS) deals with problems and differences of road safety audit/inspection and tunnel safety audit/inspection. The general objective of the project is to overcome the barrier established by a formal interpretation of the two Directives 2008/96/EC (on road infrastructure) and 2004/54/EC (on minimum safety requirements for tunnels in the trans-European road network longer than 500 m). The main problem is that the road is a unique linear infrastructure in open terrain and sometimes in tunnel, but Directives have a non-uniform approach to the infrastructure safety management outside and inside tunnels. Joint road safety operations are fundamental elements of this project.

1 INTRODUCTION

1.1 *Project ECOROADS*

To overcome the barrier (Figure 1) explained in the abstract, ECOROADS aims at the establishment of a common enhanced approach to road infrastructure and tunnel safety management by using the concepts and criteria of the Directive 2008/96/CE on road infrastructure safety management and the results of related European Commission (EC) funded projects. The project started in June 2015 and will run for 24 months. In the egg a kick-off meeting was organized by project coordinator (FEHRL—FORUM DES LABORATOIRES NATIONAUX EUROPEENS DE RECHERCHE ROUTIERE) and all partners and their third parties involved in the project were presented by themselves.

Objective should be achieved through the following specific activities:

– Workshops with European tunnel and road managers (stakeholders); the first one was held in September 2015
– Exchange of best practices and experiences between stakeholders, European tunnel safety experts and road safety professionals (seminar took place in November 2015)

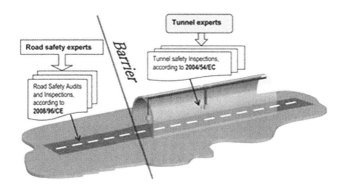

Figure 1. Illustration of ECOROADS objective (Ecoroads 2015).

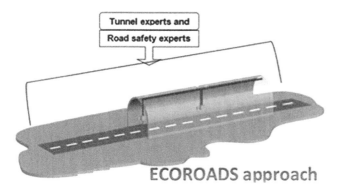

Figure 2. Illustration of ECOROADS outcome (Ecoroads 2015).

– Pilot joint safety operations in (at least) five European road sections with both open roads and tunnels (first two simulations of safety inspections have been performed yet)
– Recommendations and guidelines for the application of the RSA and RSI concepts within the tunnel safety procedures and operations (Ecoroads 2015)
– The expected outcome of the project is shown in following Figure 2.

1.2 Road safety audit and inspection definitions

The Directive 2008/96/EC defines a Road Safety Audit as "an independent detailed systematic and technical safety check relating to the design characteristics of a road infrastructure project and covering all stages from planning to early operation" (Article 2). RSAs therefore form integral part of the design process of an infrastructure project at the stages of draft design, detailed design, preopening and early operation, according to the criteria set out in Annex II of the Directive (Article 4).

It defines the Road Safety Inspection as "a dedicated periodical verification of the characteristics and defects that require maintenance work for reasons of safety" (Article 2). It continues, (Article 6) with the provisions that "the Member States shall:

a. ensure that safety inspections are undertaken in respect of the roads in operation in order to identify the road safety related features and prevent accidents and
b. ensure that periodic inspections are undertaken by the competent entity".

In the same article, it recommends that "such inspections shall be sufficiently frequent to safeguard adequate safety levels for the road infrastructure in question". While not providing specific criteria (as with a list of criteria annexed for Road Safety Audit), elements for evaluation of expert teams site visits are listed in Annex III of the Directive, referring to the safety ranking and management of the road network in operation (Article 5) (Miltiadou et al. 2016).

1.3 Tunnel safety definitions

The Directive on Tunnels (2004/54/EC) makes reference to periodic inspections carried out by the tunnel's Inspection Entity at maximum intervals of 6 years for any given tunnel. Parallelizing with the Road Safety process, the RSA during planning, design, construction and pre-opening and early operation phases, the Tunnel Directive requires a Safety Documentation (Annex II), which describes the processes for approval of the design, for opening of a tunnel, for modifications in the physical and operational characteristics of a tunnel and for performing periodic exercises for tunnel staff and emergency services, and includes the content and results of a Risk Analysis.

Risk analysis (Article 13) is an analysis of risks for a given tunnel, taking into account all design factors and traffic conditions that affect safety, notably traffic characteristics and type, tunnel length and tunnel geometry, as well as the forecast number of heavy goods vehicles per day.

The provisions of Annex II apply for all tunnels falling under the Directive: those whose design has not yet been approved and those whose design has been approved, but which are not yet open, and tunnels already in operation. For the latter, the compliance with the Directive's requirements is assessed, on the basis of the Safety Documentation, as well as on the basis of an inspection. As regards to the minimum safety requirements and countermeasures, those are described in Annex I of the Directive (Miltiadou et al. 2016).

2 ECOROADS IMPORTANT ACTIONS

ECOROADS procedure is divided into seven Work Packages (WP) and three Milestones (Figure 3):

WP1 Project Management—to carry out the effective management of the project and contact with the EC officers.

WP2 Overview of the application of the two directives (2008/96/EC and 2004/54/EC) - to have a clear picture of the current degree of practical application of the two Directives across the EU countries, as well as the other neighbouring countries and to select the main results of the studies related to road infrastructure safety management, tunnel safety and the road users' needs in order to pave the way to the workshops with the stakeholders.

WP3 Workshops with stakeholders

WP4 Exchange of best practises—to organize meetings, reciprocal training, exchange of best practices and experiences between European tunnel experts and road safety.

WP5 Joint road safety operations—to plan and perform joint safety operations in five European road.

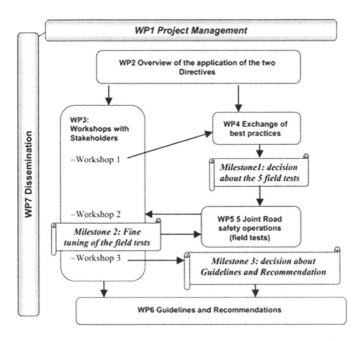

Figure 3.   Diagram of ECOROADS project activities and their relationships (Miltiadou et al. 2016).

WP6 Guidelines and recommendations—to provide recommendations and guidelines for the application of the RSA and RSI concepts within the tunnel safety operations.

WP7 Dissemination—the dissemination activities that ensure that the project's practical progress and outcomes.

2.1 Selection of test sites

The list of preliminary proposed field test sites was actualised, in the starting phase of the project, actualised and supplemented by other new test sites proposed by project partners and third parties. Totally 14 test sites was proposed from all over Europe:

1. Tunnel Rennsteig, Germany
2. Tunnel Demir Kapija, Macedonia (SEETO region)
3. Dobrovského tunnels, Czech Republic
4. Tunnel Katlanovo, Macedonia (SEETO region)
5. Kennedytunnel, Belgium
6. Klimkovice tunnel, Czech Republic
7. Krrabe tunnel, Albania (SEETO region)
8. Tunnel Murrize, Albania (SEETO region)
9. Tunnel Prima Porta, Italy
10. Tunnel Progon, Serbia (SEETO region)
11. Tunnel Rozaje, Montenegro (SEETO region)
12. Tunnel Straževica, Serbia (SEETO region)
13. Žilina tunnel cluster, Slovakia
14. Vegvesen tunnels, Norway.

The team of 10 road and tunnel experts from project consortium was created to evaluate and select 5 test sites (3 from SEETO region and 2 from other Europe countries). Based on census points from all experts five test sites were selected (Rennsteig, Kennedy, Krrabe, Straževica and Demir Kapija).

2.2 First workshop

First workshop was organized to gather views of key stakeholders on how to proceed with the joint road safety operations. It was devoted to a discussion on the results of the studies conducted in the previous Work Package and presentations of all locations of test sites which have already been preliminarily ranked by the project partners. The opinions of the participants were taken into account.

Breakout session about Roads & Tunnels Safety Management was one of the main parts of this workshop. All participants including the stakeholders were divided into three groups and on the basis of his/her professional experiences and information received from previous sessions of the workshop had an opportunity to suggest the best ways to conduct the joint visits.

2.3 Seminar for exchange of best practises

In November 2015 the seminar for tunnel safety experts and road safety experts was held to conjoin these groups of experts to discuss and explain their experiences with the safety operations in tunnels and open roads, major issues and best practices. After this workshop, specifically, on the basis of the outcomes of the first workshop and the seminar following very important issues that address fundamental principles for the accomplishment of the project's targets were specified in very important document Deliverable D5.1 Definition of Common Procedures for the performance of Joint Road Safety Operations:

1. Definitions of road and tunnel areas (open road, transition areas):
As mentioned above, the scope of ECOROADS project is to perform joint road safety operations (audit/ inspection) at specific sections of road infrastructure combining an open road

and a tunnel. Therefore, the length of the open road in each field test should be defined, taking into account the influence of the tunnel before the transition area of the tunnel. In this aspect, and despite the fact that a fixed length can be set indicatively based on experts' experience, this length cannot be rigidly defined horizontally for all field tests. The length of the open road section to be audited/ inspected shall be exactly defined based on the local conditions and particularities of each tested site, after receiving information from the infrastructure manager and local experts, and taking into account the distance of warning of road users about the existence/ approach to the tunnel (vertical signage and road marking) and the presence of adjacent interchanges, entry/ exit ramps, weaving manoeuvres, etc. As a general rule, the area of influence of each tunnel and the area of influence of other adjacent road infrastructures to each other should be considered and thus the relevant opinion of local authorities and experts is required.

The transition area between an open road and a tunnel is defined at least as the sum of the distance calculated as the distance covered in 10 seconds by a vehicle travelling at the speed limit before the tunnel portal and the stopping distance after the tunnel portal, for a vehicle travelling at speed limit, if not identical with design speed. This minimum rule obviously applies on the opposite direction and also—maybe slightly modified due to reduced speed within the tunnel—at the exit of the tunnel and on the same direction.

Consequently from the "transition area" definition, where part of the tunnel infrastructure is considered as part of the transition area, the "tunnel interior" section for the scope of ECOROADS project is defined as the remaining part of the infrastructure between the transition areas on both sides of a given tunnel.

2. Audit/Inspection process, roles and responsibilities for ECOROADS field tests:
On the basis of conclusions from first workshop and seminar the following categories of involvement in field tests process are foreseen:

– Infrastructure (Road/Tunnel) Manager(s)
– Host organization
– Audit/Inspection Group—the mixed international team of road/tunnel experts and other stakeholders; consists of Core Audit/Inspection Team, External observers, Facilitator, ECOROADS Internal observer, other External experts (stakeholders) and other ECOROADS involved parties.

The roles and responsibilities of each of the aforementioned categories of the people involved in the field tests are analytically described in Deliverable D5.1. It has to be stressed out that all of the involved parties have to accept the ECOROADS uniform approach, as a precondition to participate/be represented.

3. Composition of the joint team: the Core Audit/Inspection Team is consisted of at least three and preferably four experts, where two of them should be road safety experts and the other two should be tunnel experts. In case of three, two of them should be road experts and one tunnel expert. In both cases, one of the road safety experts should take over the role of the coordinator of this core team.

4. Execution of joint Audit/Inspection (preparation, equipment, data & documents, what to be audited/ inspected, reporting).

5. Feedback from the infrastructure manager: scheme of the written response the road authority should provide after having received and analysed the report.

6. Monitoring of the ECOROADS field tests (Miltiadou et al. 2016).

3 FIRST TWO JOINT SAFETY OPERATIONS

3.1 *Kennedytunnel*

In March 2016 the first one of ECOROADS joint safety operations was performed in Antwerp, Belgium. The Kennedytunnel is a 690 m two-tube tunnel with unidirectional traffic (three lanes in each tube) located under the Schelde river in Antwerp. The tunnel was built with an immersed tube concept and was opened in 1969 for road traffic. There is also a train tube and bicycle/pedestrian tube. Bicycle tube serves as the evacuation tube for car traffic and one of traffic tube serves as evacuation tube for train traffic. Traffic volumes in the Kennedy-tunnel are high and close to the theoretical capacity of the tunnel. Daily structural congestion is a common sight, in the morning in the direction of the Netherlands (towards the centre of Antwerp) and in the evening in the opposite direction. The total share of heavy good vehicle during a day is about 23%.

The field test started with a briefing meeting (involving all incorporated members) with presentation about tunnel Kennedy, discussion on accidents and other practical information. After the meeting the inspection team moved to the Traffic centre to check the most important and frequent incidents/accidents from live camera images. Then the inspection during daylight (infrastructure open for all road users) followed by cars to transit all possible entrance and exit ramps.

The inspection during night (one tunnel tube closed for traffic) was performing at the same time as tunnel tube maintenance. The day after inspection is devoted to discussion, preparation of RSI preliminary report and filling in evaluation forms of all participants.

3.2 *Tunnel Krrabe*

In April 2016 the second ECOROADS joint safety operation was performed in tunnel Krrabe near the capital of Albania—Tirana. Tunnel Krrabe is part of Albanian South Road Axe between Durres—Tirana and South of Albania, Elbasan. Tirana-Elbasan highway consists of two segments. First segment from Tirana to tunnel Krrabe is under construction phase and that second also with tunnel is fully opened for traffic. The length of tunnel tubes is approximately 2230 m and 2500 m. Both tunnel tubes are operated unidirectionally, but one of them with traffic only in one lane (safety reasons) because of traffic redirection to bidi-rectional directly behind tunnel tube. From these reasons the reduction of traffic to one lane is carried out in front of tunnel tube to reduce changing the lanes in the tunnel and avoid to possible incidents/accidents.

Field test also started with briefing meeting of many presentations about tunnel, road sections, accidents, etc. After the transition to site the inspection during daylight (infrastructure open for all road users) and visit of traffic control centre were performed. After sunset (minimum traffic volume) tunnel tubes were closed for more than two hours to carry out inspection during the night.

4 CONCLUSIONS

Until now two planned joint safety operations were performed. The results of inspections are not important from the ECOROADS point of view but can be useful for tunnel manager/administrator. The general object of the ECOROADS project is the process of joint (tunnel and road) safety inspection/audit. The second workshop will be held in June 2016 (after first two joint safety operations) in order to analyse the first results and fine tune the second set of joint operations. Recommendations and guidelines for the application of the RSA and RSI concepts within the tunnel safety operations will be issued after the final workshop and will be based on the results of all field operations.

REFERENCES

Ecoroads. 2015. Coordination & support action. Number—652821—ECOROADS.
Miltiadou, M., Cela, L. & Gjorgjievski, M. 2016. Effective and Coordinated Roads Infrastructure Safety Operations ECOROADS. *Deliverable D5.1 Definition of Common Procedures.*

Advances and Trends in Engineering Sciences and Technologies II – Al Ali & Platko (Eds)
© 2017 Taylor & Francis Group, ISBN 978-1-138-03224-8

Evaluation of pavement stochastic unevenness of Žilina airport RWY on the base of stationary ergodic process theory

M. Decký, E. Remišová & Ma. Kováč
Faculty of Civil Engineering, University of Žilina, Žilina, Slovakia

ABSTRACT: The article presents the results of an evaluation and objectification of the pavement surface of the Žilina Airport RWY according to surface irregularities on the base of stationary ergodic process theory. The longitudinal unevenness was objectified by International Roughness Index IRI and unevenness degree C. Parameter IRI is obtained using the Reference Quarter Car Simulation using 3D scanning of the RWY surface. Skid resistance of the RWY surface was evaluated using measurements by 3D scanner with high resolution.

1 INTRODUCTION

Airport pavements are required to fulfil their basic functions in order to provide adequate bearing strength, good riding qualities and good surface friction characteristics. Mechanical interactions between aircraft and runways are complex and depend on the properties of tire/surface contact. This small area is subjected to forces caused by the rolling and braking characteristics of the aircraft. To ensure tire/surface contact in wet conditions the slope and drainage of runway are important too.

According ICAO (2013) the surfaces of all movement areas including pavements (runways, taxiways) and adjacent areas shall be inspected and their conditions monitored regularly as part of an aerodrome preventive and corrective maintenance program with the objective of avoiding and eliminating any loose objects/debris that might cause damage to aircraft or impair the operation of aircraft systems. The surface of a runway shall be maintained in a condition such as to prevent a failure, degradation and formation of harmful irregularities.

The surface characteristic is the physical property of a pavement surface such as evenness, texture, skid resistance, and drainability (Čelko, J. et al. 2011). In general the current value of these parameters determines the pavement serviceability. There are a variety of methods for the identification and evaluation of a pavement condition (Iwański, M. et al. 2015). In the next chapters the longitudinal unevenness by International Roughness Index IRI and unevenness degree C based on the theory of stationary ergodic random process are evaluated and objectified; and the skid resistance of the surface by texture characteristics evaluated on the base of 3D scanning data.

2 SURFACE IRREGULARITIES IDENTIFICATION

Surface irregularities as a result of pavement deterioration can be investigated as waves with different wavelength and amplitudes. The scope of influence on interaction between tire and pavement surface is different according to values of wavelength and amplitudes. In term of these values are irregularities described as a micro, macro and mega texture and unevenness (Kováč, M. et al. 2012). The first two components of surface morphology are responsible for the pavement skid resistance and the second two are responsible for the pavement unevenness. As the limit value between unevenness and the texture is considered the wavelength of 0.5 m. There are many different methods for the investigation of the surface irregularities according to the purpose of evaluation in term of pavement serviceability.

2.1 Longitudinal unevenness

In term of pavement surface unevenness evaluation there are two kinds of parameters, which are usually observed. The first group is concerned with longitudinal unevenness and the second evaluates transversal unevenness (ruts). For airport Runway (RWY) surface the longitudinal unevenness and its evaluation is the most important parameter. Therefore, the longitudinal unevenness of Žilina airport RWY surface was objectified by International Roughness Index *IRI* and unevenness degree *C*, which is based on the theory of stationary ergodic random process (Slabej, M. et al. 2015).

An assessment methodology for pavement surface unevenness based on the pavement unevenness index *C* was created in former Czechoslovakia. This methodology enables us to quantify effects of longitudinal unevenness of pavements on moving vehicles from the point of view of safety and passengers' comfort, and also in terms of the burden on the structural components of vehicles. This diagnostics methodology is still used for diagnostics and the assessment of runways.

The pavement unevenness index C [rad.m.10^{-6}] of a pavement surface expressed from the basic relation and modified for Quarter Car Simulation device QCS (Figure 1) is calculated according to following formula:

$$C - \frac{D_y}{I \frac{1}{N} \sum_{i=1}^{n} v_i}$$ (1)

where D_y = dispersion of sprung mass acceleration, I = parameter of dynamic transfer, C = unevenness degree, v_i = digital values of a measured velocity.

Due to the fact that the *IRI* values were in line with the original methodology evaluated for a reference speed of 80 kmh^{-1}, a correlation formula between *IRI* and *C* will be used from which it is possible to express the pavement unevenness index *C* in accordance with the following formula (Čelko, J. et al. 2009):

$$C = {\left(\frac{1}{0.49}\right)} \sqrt{\frac{IRI}{1.99}}$$ (2)

Correlation parameters depending on pavement evenness were objectified on the basis of long-term research activities, which the authors carried out through direct simulation methods and identification of dynamic system of vehicle—road—environment.

Parameter *IRI* was obtained using the Reference Quarter Car Simulation via 3D scanning of the RWY surface. This model is defined mathematically by two second-order differential equations:

$$\ddot{z}_s m_s + C_s(\dot{z}_s - \dot{z}_u) + k_s(z_s - z_u) = 0$$ (3)

$$\ddot{z}_s m_s + m_u \ddot{z}_u + k_t z_u = k_t y$$ (4)

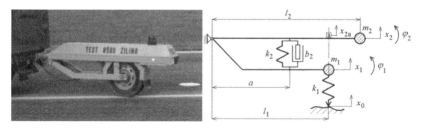

Figure 1. Real QCS and its mechanical scheme.

This system can be expressed as:

$$\ddot{z}_s + C(\dot{z}_s - \dot{z}_u) + k_2(z_s - z_u) = 0 \tag{5}$$

$$\ddot{z}_s + u\ddot{z}_u + k_1 z_u = k_1 y \tag{6}$$

where ms, mu = weight of the sprung mass and the unsprung mass [kg]; ks, kt = constant of the linear spring and the tire [kN.m^{-1}]; C_s = coefficient of linear damper [kN.s.m^{-1}]; z_s, z_u = displacement of the sprung and he unsprung mass [m]; $\dot{z}_s = dz_s/dt$, $\dot{z}_u = dz_u/dt$ vertical velocity of the sprung/unsprung mass [m.s^{-1}]; $\ddot{z}_s = d^2 z_s/dt^2$, $\ddot{z}_u = d^2 z_u/dt^2$ vertical acceleration of the sprung/unsprung mass [m.s^{-2}]; y(t) = profile elevation input [m]. The presented system can be expressed in the following compact matrix form:

$$Z(x)_{(i)} = \underline{S}Z(x)_{(i-1)} + R y'_{(i)} \tag{7}$$

where $Z^T(x)_{(i)} = (z_{1,i}; z_{2,i}; z_{3,i}; z_{4,i}) = (z'_{s,i}; z''_{s,i}; z'_{u,i}; z''_{u,i}) = (dz_{s,i}/dx; d^2 z_{s,i}/dx; dz_{u,i}/dx; d^2 z_{u,i}/dx)$ vector of spatial derivations; S = state transition matrix 4×4; R = partial response matrix 1 × 4; $y'_{(i)}$ = slope input; i = present step; i-1 = previous time step.

The profile slope input is computed for every measuring point:

$$y'_{(i)} = (y_{(i-1)} - y_{(i)})/dx, \; i = 2,3,\ldots, N \tag{8}$$

where $y'_{(i)}$ = smoothed profile slope input; $y_{(i)}$ = elevation of longitudinal profile [m]; dx = measurement interval dx = b = 0.25 m. The computation of vector $Z^T(x)_{(i)} = (z_{1,i}; z_{2,i}; z_{3,i}; z_{4,i}) = (z'_{s,i}; z''_{s,i}; z'_{u,i}; z''_{u,i})$ is realized by equation:

$$Z(x)_{(i)} = \underline{S}Z(x)_{(i-1)} + R y'_{(i)} \tag{9}$$

where $\underline{S}$ = state transition matrix 4 × 4; R = partial response matrix 1 × 4; i = present step; i-1 = previous time step.

Determination of the corrected profile slope is realized according to following formula:

$$Ti = (z_{3i} - z_{1i}), \; i=2,3,\ldots, N \tag{10}$$

The parameter *IRI* represents arithmetic average of the corrected slope. Values of parameter *IRI* can be appreciated for window of a discretionary length (conveniently 1, 10, 20, 100 m).

$$IRI = \frac{1}{N-1} \sum_{i=2}^{N} T_i \tag{11}$$

Results of measured and evaluated data depending on interval value are showed in Figure 2. On the basis of the realized measurements and evaluation of the pavement surface's longitudinal unevenness of the airport runway at Dolny Hricov (Žilina airport) the pavement is considered low quality but sufficient to remain in use as a runway (Kováč, M. et al. 2015).

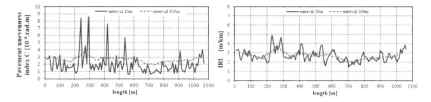

Figure 2.   Pavement unevenness index C and IRI evaluated with different interval.

2.2 Texture of pavement surface

The most important aspect of the pavement surface relative to its friction characteristics is the surface texture. The effect of surface material on the tire/surface coefficient of friction arises principally from the differences in surface texture. There are many methods of texture measurement in order to evaluation of pavement surface skid resistance.

Texture of the RWY surface was measured by Volumetric Patch Method (parameter *MTD*) and by 3D scanner with high resolution. The scanning process was performed with resolution 0.2 mm, and the scanned surface area was limited by the template with dimensions of 200×60 mm, which means that there were scanned 300 profiles with length 200 mm providing 600 baselines for texture parameters calculation. The family of all measured profiles derived from 3D scanned surface is shown in Figure 3.

The processing of all collected data measured by 3D scanner was performed in the MATLAB® software by algorithm created at University of Žilina. Using created algorithm it was possible to evaluate, besides *MPD* parameter (Mean Profile Depth), also the height (amplitude) parameters, such as Arithmetical mean deviation – *Ra*, Root mean square deviation —*Rq* etc., the height distribution shape parameters, such as Skewness—*Rsk*, Kurtosis—*Rku*, wavelength parameters, such as power spectral density—*PSD*, hybrid parameters, such as Texture ratio—*TR*, or spacing parameters, such as Mean width of the profile elements— *RSm*. Average values of all evaluated texture characteristics for evaluation interval 100 m are showed in the Table 1.

For better understanding of wearing course composition and its influence on tire-pavement interaction (Remišová E. et al. 2014) was also computed power spectral density of surface irregularities, which are homogenous from the point of view of construction and degradation conditions on the same pavement, and therefore can be evaluated through the medium theory of stationary random process. This type of random process can be best characterized by a correlation function or Power Spectral Density (PSD). The correlation function $K_h(\lambda)$ for this type of process is expressed in linear domain by equation

$$K_h(\lambda) = \int_{-\infty}^{\infty} \int_{-\infty}^{\infty} \left[h_1(l) - E_h \right]\left[h_2(l-\lambda) - E_h \right] f_2(h_1,h_2) dh_1 dh_2 \qquad (12)$$

where λ = linear lag, E_h = expected value of stochastic unevenness; $E_h = 0$, $h(l)$ stochastic unevenness, $f_2(h_1, h_2)$ = combination density of expectation.

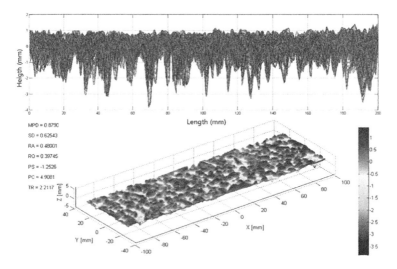

Figure 3.   The family of all measured profiles derived from 3D scanned surface.

Stochastic unevenness is computed as a difference between a real and theoretical profile. In our case we must identify elevations of longitudinal profile and profile irregularities are evaluated through the standardized correlation function $\rho_h(\lambda)$ as follows:

$$\rho_h(\lambda) = K_h(\lambda)/D_h \tag{13}$$

For the purpose of wave length spectrum investigation it is more appropriate to use Power Spectral Density (PSD) $S_h(\Omega)$, which can be expressed from the correlation function by means of Wiener Chinchine equation:

$$S_h(\Omega) = 2/\pi \int_0^\infty K_h(\lambda)\cos(\Omega\lambda)d\lambda \tag{14}$$

where D_h = dispersion of an stochastic unevenness; Ω = angular spatial frequency; L = unevenness wavelength.

The processing of all collected data measured by 3D scanner was performed in the MAT-LAB® software by algorithm created at UNIZA. Using created algorithm it was possible to evaluate, besides MPD parameter, also the height (amplitude) parameters, such as Arithmetical mean deviation—Ra, Root mean square deviation – Rq etc., the height distribution shape parameters, such as Skewness—Rsk, Kurtosis – Rku, wavelength parameters, such as power spectral density—PSD, hybrid parameters, such as Texture ratio—TR, or spacing parameters, such as Mean width of the profile elements—RSm. An example of results of 3D scanning as a selected profile and power spectral density of surface irregularities is showed in Figure 4.

Table 1. Average values of friction and texture characteristics for evaluation interval 100m.

Point	MTD (mm)	MPD (mm)	Ra (mm)	Rq (mm)	Rsk (−)	Rku (−)	RMS (mm)	TR (−)	RS (mm)
1	1.07	0.93	0.40	0.52	−0.03	0.62	0.52	1.79	10.37
2	0.88	0.73	0.27	0.37	−0.44	2.37	0.37	1.98	7.38
3	1.01	0.72	0.26	0.35	0.28	1.90	0.35	2.05	8.24
4	1.09	0.71	0.25	0.33	−0.34	2.47	0.33	2.15	7.79
5	1.15	0.85	0.29	0.39	−0.11	1.73	0.39	2.18	8.18
6	0.97	0.75	0.27	0.37	−0.40	2.43	0.37	2.04	8.81
7	1.19	0.90	0.35	0.45	−0.22	1.07	0.45	1.98	9.76
8	1.05	0.62	0.23	0.31	−0.72	4.37	0.31	1.97	8.20
9	0.86	0.64	0.22	0.29	0.13	1.08	0.29	2.21	8.45
10	0.96	0.71	0.26	0.34	0.06	0.95	0.34	2.07	9.48
φ	1.02	0.75	0.28	0.37	−0.18	1.90	0.37	2.05	8.67

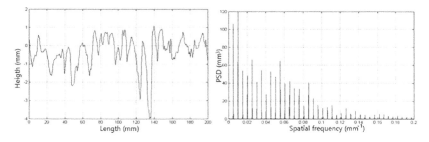

Figure 4.   Selected profile and power spectral density diagram.

On the basis of the realized measurements and evaluation of the pavement surface's skid resistance of the airport runway at Dolny Hricov (Žilina airport) according to surface texture characteristics the pavement is considered very good quality and sufficient to remain in use as a runway (Kováč, M. et al. 2015).

3 CONCLUSIONS

The article presented the results of evaluation and objectification of pavement condition of the Žilina Airport RWY according to surface irregularities on the base of stationary ergodic process theory. The longitudinal unevenness was objectified by International Roughness Index *IRI* and unevenness degree *C*. Parameter *IRI* is obtained using the Reference Quarter Car Simulation using 3D scanning of the RWY surface. Skid resistance of the RWY surface was evaluated using measurements by 3D scanner with high resolution.

On the basis of the realized measurements and evaluation of the pavement surface's longitudinal unevenness of the airport runway at Dolny Hricov (Žilina airport) the pavement is considered low quality but sufficient to remain in use as a runway.

On the basis of the realized measurements and evaluation of the pavement surface's skid resistance of the airport runway at Dolny Hricov (Žilina airport) according to surface texture characteristics the pavement is considered very good quality and sufficient to remain in use as a runway.

ACKNOWLEDGMENT

The research is supported by the project VEGA 1/0557/14 The influence selected surface characteristics on asphalt pavement serviceability.

The research is supported by European regional development fund and Slovak state budget by the project "*Research centre of University of Žilina*", ITMS 26220220183.

REFERENCES

Čelko, J., Decký, M. & Kováč, M. 2009. An analysis of vehicle—road surface interaction for classification of IRI in frame of Slovak PMS, *Maintenace and reliability*. Polish Maintenance society, Vol., No. 1, p. 15–21, ISSN 1507–2711.

Čelko, J., Kováč, M. & Decký, M. 2011: Analysis of selected pavement serviceability parameters, *Communications—Scientific Letters of UNIZA*. Vol. 13, No, p. 56–62, ISSN 1335-4205.

Ďurčanská, D, Decký, M., Licbinsky, R. & Huzlik, J. 2013: Project SPENS—sustainable pavement for European new member states, *Communications—Scientific letters of UNIZA*. Vol. 15, No. 2, p. 49–55, ISSN 1335-4205.

ICAO 2013. Aerodromes Volume I Aerodrome design and operations.

Iwański, M. & Chomicz-Kowalska, A. 2015. Evaluation of the pavement performance, *Bulletin of the Polish Academy of Sciences Technical Sciences*, Vol. 63, No. 1, DOI: 10.1515/bpasts-2015–0011.

Kováč, M., Remišová, E., Čelko, J., Decký, M. & Ďurčanská, D. 2012. *Diagnostika parametrov prevádzkovej spôsobilosti*, Žilinská univerzita v Žiline EDIS.

Kováč, M. et al. 2015. *Evaluation and reconstruction design of the Žilina Airport RWY*, Final report. Žilinská univerzita v Žiline.

Remišová E., Decký M. & Kováč M. 2014. The influence of the asphalt mixture composition on the pavement surface texture and noise emissions production, In *SGEM 2014* 14th international multidisciplinary scientific geoconference: geoConference on energy and clean technologies. Sofia STEF92 Technology, ISBN 978-619-7105-16-2.

Slabej, M., Grinč, M., Kováč, M., Decký, M. & Šedivý, Š 2015. Non-invasive diagnostic methods for investigating the quality of Žilina airport's runway, *Contributions to Geophysics and Geodesy*. Volume 45, No. 3, p. 237–254, ISSN: 1335-2806.

Road surface characteristics on experimental road section with crumb rubber additive and connection with road traffic noise

O. Frolova & B. Salaiova
Faculty of Civil Engineering, Technical University of Kosice, Kosice, Slovakia

ABSTRACT: The article presents the results of measurements of road surface properties on an experimental road section at Mala Ida village, Kosice region, Slovakia. Visual inspection was performed to detect the presence of potholes and cracks, the roughness of the road surface was measured too, and in the next step the road pavement was checked visually and from the friction coefficient point of view. The friction coefficient was calculated from the roughness value.

1 INTRODUCTION

Road users desire a road surface where they can drive safely and comfortably. This requires a pavement structure with enough stiffness, fast run-off for rainwater, even road surface, good reflection of (artificial) light from the road surface, and limited production of noise in the contact area between the vehicle tyre and the road surface. These properties should be present over a long period of time, preferably throughout the life of the pavement structure.

The main parameters and characteristics that determine the traffic capacity and operational condition of the road are:

– Geometrical parameters (the width of the roadway, the width of the wayside (reinforced and unreinforced), longitudinal slopes, radius of curves, visibility distance);
– Skidding resistance;
– Texture of the road surface;
– Aggregate at the road surface;
– Condition of the road surface.

Nowadays in road construction porous materials absorbing noise (for example, PA—Porous asphalt pavement) are used (Bernhard et al. 2005), but this road cover type is difficult from the maintenance and cleaning points of view, especially in winter when the pores are clogged with dirt and snow and the surface does not serve as acoustic cover anymore (Gschwendt 1999). For this reason asphalt mixtures with the addition of crumb rubber from waste tires with better porosity are now attracting much attention. These mixtures have good slip resistance and acoustic properties (Kudrna & Dasek 2009). The latters' experiments show good fatigue resistance and track-copying resistance on road sections with crumb rubber additive. Vasiljev et al. (2013) in turn found interaction between road traffic noise and the friction coefficient, and increase in noise levels with increasing road friction coefficient.

In the next section the desired road surface characteristics at Mala Ida village, Kosice region, Slovakia (with addition of crumb rubber to the mixture) are discussed in more detail.

2 DESCRIPTION OF EXPERIMENTAL ROAD SECTION

The experimental road section at Mala Ida consists of asphalt concrete with added crumb rub-ber. The studied asphalt mixture was developed within the project "NFP 26220220051

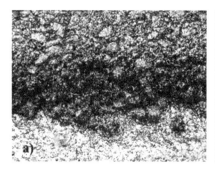

Figure 1.   Road section with addition of crumb rubber a) view of the mixture; b) view of the experimental road section after realization of the road.

Development of progressive technologies for utilization of selected waste materials in road construction engineering, supported by the European Union structural funds" at the Department of Geotechnics and Traffic Engineering, Technical University of Kosice. After the asphalt mixture went through laboratory testing, an experimental road section was constructed. The original surface layer of this road was replaced with asphalt mixture with addition of crumb rubber. Crumb rubber is elastic material and its behavior could improve classic asphalt mixtures in loaded roadways. Asphalt mixtures can be modified with crushed rubber in different ways. Properties of asphalt mixture with rubber additive have been monitored using wet as well as dry methods. The wet method is based on mixing crushed rubber and asphalt binder at elevated temperature, and the final product is asphalt rubber binder. The dry method includes replacement of the aggregate part with crumb rubber. The wet method for asphalt rubber pavement composition was used in preparation for this study. The rubber content should be chosen with regard to the aggregate and binder content (Olexa et al. 2014). Crumb rubber from waste tires was supplied by the company V.O.D.S. a. s., and this product fulfils all current environmental and health standards in the EU. The experimental section of road is 200 m long and the surface layer is 40 mm thick. The weather during the laying process was acceptable with air temperatures from 15°C to 25°C. The temperature of the mixture after output from the plant was 160°C. The laying and compaction were done one hour after mixing and the temperature of the pavement was over 135°C. These processes were the same as for ordinary mixtures. Views of the road section with addition of crumb rubber are shown in Figure 1.

3   ROAD PAVEMENT CHARACTERISTIC MEASUREMENTS

On this experimental road section the following road surface properties were measured:

– Visual inspection of road section (presence of cracks, potholes, degradation of cover);
– Roughness of the cover.

3.1   *Visual inspection of road section*

Correct identification and classification of defects are the basic conditions for the correct design of repair or service of road cover. Another condition is to determine the area, fullness and quantity of defects. Visual inspection of the road section was carried out in accordance with the Technical Specifications 2014, with graphical recording to the paper blank with stationing every 10 m. During visual inspection defects of four defect groups were detected. These included potholes, cross cracks, longitudinal cracks, network cracks. On the basis of the data collected assessment of the road cover condition was performed. Examples of defects on the experimental road section are presented in Figures 2, 3.

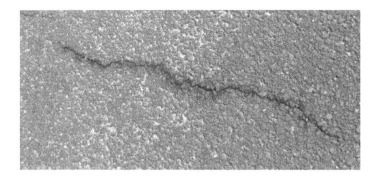

Figure 2. Longitudinal crack on experimental road section.

Figure 3. Network cracks on experimental road section.

3.2 *Evaluation of defects according to weighting coefficient*

According to the Technical Specifications 2014, the integrated assessment of defects is based on the percentage impact on the overall quality of the road depending on the road category. Assessment of the state of the road surface is based on the IPSV parameter, and for a 3rd-class road this is calculated in the following way:

$$IPSV_p = 5.03 - 0.07(P+O) \tag{1}$$

where $IPSV_p$ = index of road conditions breaches; P = area of surface defects; O = area of repairs.

Based on calculation form area of surface defects P = 182.91 m², O = 0%, area of the road segment is 1306.5 m².

$$IPSV_p = 5.03 - 0.07(14+0) = 4.05$$

Based on this $IPSV_p$ value and according to Table 3 in the Technical Specifications 2014, the road cover is assessed as "very good".

3.3 *Measurements of roughness of road cover*

According to the Analytical Review 2011, roughness is the road surface characteristic which provides interaction between the tire and road cover. If we consider roughness from the geometrical point of view, in fact we are talking about the texture of the surface, i.e. the morphology that is the location of single aggregate particles on the surface (macrotexture) and the presence of projections on the surface of the aggregate particles (microtexture). If we consider roughness as the road surface characteristic which provides skid resistance of the vehicle wheel on the surface, we are talking about tangential reaction of the road cover serving to transfer braking power or driving power to the perimeter of the tire.

On the experimental road section the roughness was measured using the equipment in Figure 4. This is a standard device for measuring the values of skid resistance based on the normalized energy loss in the rubber base after test surface sliding.

3.4 Measurement procedure

According to the Analytical Review 2011, skid resistance is measured over a small area (at about 0,01 m²). The test surface must be cleared of loose particles with a brush and then rinsed clean with water. The surface must not be icy. The pendulum test device is placed on a hard surface so that the pendulum swings in the direction of the traffic.

On the experimental road section the roughness measurement was carried out in accordance with the Analytical Review 2011. During the procedure the **PTV** (Pendulum Test Value) value was measured at seven points along the road (every 40 m). It was measured at least five times at each road point. If the first few readings differed by more than three units, the measurement was repeated until three consecutive scores recorded had the same value, which was then recorded.

3.5 Measurement results and discussion

Calculates the value of PTV as the average of five oscillations according to equation:

$$PTV = \frac{\sum (v_1 + v_2 + v_3 + v_4 + v_5)}{5} \tag{2}$$

where v_1 to v_5 = the individual values for each oscillation.

Results of the PTV value measurement are in Table 1.

Based on these results a bar graph summarizing the measurement data was created. The bar graph is shown in Figure 5.

The friction coefficient is determined on the basis of loss of kinetic energy measured with a friction rubber component pendulum on a wet road surface. The representative value of the coefficient of friction is determined as follows Celko et al. (2000).

Figure 4.   Equipment for the roughness measure.

Table 1. Results of the PTV value measurement.

Direction	0.00 m	40 m	80 m	120 m	160 m	200 m	Σ
Mala Ida—Kosice Šaca	62	73	79	80	82	85	76.8
Kosice Šaca—Mala Ida	–	81	79	80	79	–	79.8

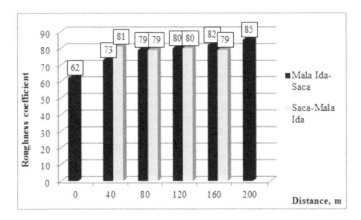

Figure 5. Summarizing data measurement.

$$f_k = f_k' - 0.2 \times (30 - t_{vm}) \tag{3}$$

where f_k' = average measured value; t_{vm} = temperature of wet road surface, °C.

$$f_k(M.Ida - Saca) = 76.8 - 0.2 \times (30 - 10) = 72.8 > 55$$
$$f_k(Saca - M.Ida) = 79.8 - 0.2 \times (30 - 10) = 72.8 > 55$$

According to STN 73 6195, evaluation of skid characteristics is done as follows: if the value of friction coefficient is greater than 55, the road pavement it is satisfactory from the skid resistance point of view.

4 CONCLUSION

The experimental road section was laid in the summer of 2013 as a project of the Department of Geotechnics and Traffic Engineering at the Technical University of Kosice. In the spring of 2016 we carried out detailed visual inspection and roughness measurements in order to identify any changes in the state of road cover during that period of use. Evaluation of the test results showed that the road cover was in a satisfactory condition. Based on the visual inspection, the cover quality was termed "good", while the assessment based on the friction coefficient rated the cover as "satisfactory".

Long-term monitoring of road pavement properties and connections between road pavement properties and road traffic noise is the subject of research in the Department of Geotechnics and Traffic Engineering.

ACKNOWLEDGEMENTS

The contribution was incurred within Centre of the cooperation, supported by the Agency for research and development under contract no. SUSPP-0013-09 by businesses subjects Inžinierske stavby and EUROVIA SK.

The research has been carried out within the project NFP 26220220051 Development of progressive technologies for utilization of selected waste materials in road construction engineering, supported by the European Union Structural funds.

REFERENCES

Analytical review. University of Žilina. Aktualizácia hodnotiacich kritérií pre protišmykové vlastnosti vozoviek v zmysle európskych noriem (EN) [Actualization of assessment criteria for antiskid properties according to European standards (EN)]. In *University of Žilina, Civil Engineering faculty, Department of Highway Engineering*. October 2011.

Bernhard, R., Wayson R. L., et al. An introduction to tire/pavement noise of asphalt pavement. Final Research Report Number: SQDH 2005-1. In *Purdue University, USA*, 2005.

Celko, J. et al. 2000. Povrchové vlastnosti vozoviek. Prevádzková spôosobilosť vozoviek. [Road surface characteristics. Operational capacity of roads], DIS: 2000. ISBN 80-7100-774-9.

Gschwendt, I. 1999. Vozovky. Konštrukcie a ich dimenzovanie [Road pavements. Constructions and dimensioning]. In: *Jaga Group.*

Kudrna, J., Dasek, O. Zpráva o měřeni protismykových a protihlukových vlastností obrusných vrstev s asfaltem modifikovaným pryžovým granulátem [Report about measurements of skid resistance and acoustic properties of the road pavements from asphalt with crumb rubber modified bitumen]. In *Vysoké učení technické v Brně*.

STN 73 6195: 1974. Evaluation of skid resistance of pavement surface. UNM, Prague.

Technical specifications 06/2014. Implementation and evaluation of detailed visual inspections of asphalt road pavements. Ministry of transport, construction and regional development of Slovak Republic, section of road transport and over land communication.

Vasiljev, J. E., Beljakov, A. B., Subbotin, I. V., Malofeev A. S. 2013: Shum kak pokazatel scepnych svojstv dorozhnogo pokritija [Noise as road pavement friction properties indicator]. In: *Moscow Automobile and Road Construction State Technical University*. http://naukovedenie.ru/PDF/74TVN613.pdf.

Assessment of slope stability in interaction with the subsoil

S. Harabinová, E. Panulinová, E. Kormaníková & K. Kotrasová
Faculty of Civil Engineering, Institute of Structural Engineering, Technical University of Košice, Košice, Slovakia

ABSTRACT: For the proposal of a comprehensive assessment and possible remedial action it is necessary to know the actual barrier, engineering and geological conditions of the body and subsoil of a levee. This paper considers the failure of a levee along the right bank of the Laborec River caused by floods in 2010, addressing the assessment of the stability of the levee in interaction with the subsoil and its possible remediation. The calculation of the factor of safety was made using GEO 5 software. The critical factors of safety have been determined by Petterson and Bishop Methods. The paper presents various solutions to ensure slope stability of protection levee. The best solution was proposal additional load of the slope heel with well graded gravel (GW)—Variant IV.

1 INTRODUCTION

Currently, frequent flood events are taking place in our area. In recent years the intensity and frequency of rainfall, which are capable of causing flooding have significantly increased in Slovakia. Floods are destructive and cause damage to human society. They are associated with high water levels and extreme water flow in the river beds (Ondrejka Harbuľáková, V. & Zeleňáková, M. 2013). In 2010, severe floods hit most of the territory of the Slovak Republic. It was the most severe flooding in the last 50 years (Zelenakova, M. et al. 2009).

The recurring threat of flooding necessitated the construction of protective structures. Levees are water constructions, which serves to protect the area from flooding. They are built transversely to flow of the river or along it.In Slovakia there are more than 2270 km of levees (Ondrejka Harbuľáková, V. & Hudáková, A. 2014). They are artificially created walls, embankments or mounds, formed from local materials like soils or stone and other building materials.

The Laborec River is 129 km long, flowing through almost the entire Zemplín territory from north to south. The total catchment area of Laborec is 4 522.5 km², while the left catchment has area of 4 076.7 km² and the right catchment area is 445.8 km². This difference is conditioned by basin morphology and by the tributary UhRiver, which significantly increases areas of the river basin on the left. On the right side of the Laborec creates a riverbed almost parallel to the flow of Ondava River and it also influences the size of the catchment area (Ondrejka Harbuľáková, V. & Hudáková, A. 2014).The right levee of the Laborec River is located south of the village Oborín and was built in 1964.

2 GEOLOGICAL CONDITIONS OF THE LABOREC LEVEE

According to the geomorphological division of Slovakia there is the territory located in the Eastern Slovak Lowland. Into the geological structure are involved quaternary river sediments of the Laborec River.

The levee is homogeneous and is built from materials from the surrounding area. The height of the levee body is up to 5 m. In terms of use it is understood as a major levee.

According to information from the catchment area, a landslide of downstream face of a levee was originally seen on 21st May 2010, at 3.4 kilometer, along the right bank of the Laborec River in length about 20 m, with a decrease of soil about 30 cm. Occurrence of landslides were observed during the following days.

Between 2nd of June and 5th of June 2010 further decline in soil was occurred. At this stage, landslide was associated with expanding on both slopes along the levee. Within hours of 20:00, the 3rd degree of flood activity at critical water level was subsequently declared. The water level was about 1 m from the levee crest at the site of the landslide. At the time of inspection failure (27th of July 2010) the edge of detached area intersects the crest of the levee that was breached in a length of about 26 m. Decrease of the levee at the site of detached edge was 30 to 40 cm (Harabinová, S. & Ondrejka Harbuľáková, V. & Zeleňáková, M. 2014).

For the calculation and assessment of slope stability it is necessary to know the geological profile of levees which was detected by exploration activities.

In the place of disturbance have been made two boreholes (JO-3 and JO-4), the dynamic penetration test (DP-1 DP-2, DP-3, DP-4 and DP-5) and the static penetration test SPS-7- (Ondrejka, J. & Sycevova, M. 2010).Situation of the area and geological profiles of the levee body and its subsoil are shown in Figure 1.

Thickness of the levee body at the site of the landslide is about 5 m and consists of anthropogenic sediments. Level of groundwater was found at a depth of 7.2 m below the levee crest. This water is percolating from sands verified at a depth of 10.0 m below the levee crest. Groundwater level is increasing mainly during the flooding.

Geotechnical parameters of soils located in the levee and in the subsoil of the levee were detected by laboratory testing (Ondrejka, J. & Sycevova, M. 2010).

The levee body consists mainly of very high plasticity clay (CV) and less of high plasticity clay (CH) or medium plasticity clay (CI). The consistency of the soil at a depth of 1.0 to 1.9 m below the levee crest is soft and stiff, less hard. There is stiff consistency at a depth of 2.0 meters. Limit and average values of the soil properties are listed in Table 1.

According to STN 72 1001, the body of levee is indicated as loose earth material by a symbol CIY and CHY. The body of the levee is identified as insufficiently compacted earth construction, with low values of dry density ρ_d, wet density ρ_n and high values of porosity n. The mechanical properties of the body of levee in the place of disturbance are shown in Table 2.

Under the levee crest are found fine soils created by high plasticity clay (CH), very high plasticity clay(CV),medium plasticity clay (CI), sandy clay (CS) and clayey sand (SC). Their consistency is stiff, very stiff or soft. Limit and average values of the soil properties of subsoil are listed in Table 1. The mechanical properties are shown in Table 2.

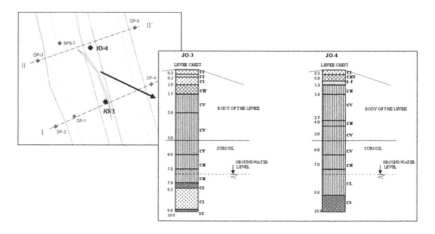

Figure 1.   The place of disturbance and the geological profile of the levee body and subsoil.

Table 1. Limit and average values of the soil properties.

Properties	Symbols	Body of levee			Subsoil		
		$X_{min.}$	X	$X_{max.}$	$X_{min.}$	X	$X_{max.}$
Water content (%)	W	32.6	35.3	38.0	31.0	35.9	40.3
Wet density (kg.m^{-3})	ρ_n	1745	1791	1859			
Dry density (kg.m^{-3})	ρ_d	1264	1324	1402			
Density (kg.m^{-3})	ρ_s	2682	2700	2721			
Porosity (%)	n	48.3	50.9	53.5			
Degree of saturation (%)	S	89.7	91.6	94.6			
Liquid limit (%)	W_L	72	76	80	34	71	87
Plastic limit (%)	W_P	23	27	31	20	27	31
Plasticity index (%)	I_P	49	49	50	13	44	57

Table 2. The mechanical properties of the levee body and of subsoil.

The mechanical properties of the levee body

Properties	Symbols	For softconsistencyclay		For stiff consistencyclay	
		Min.	Max.	Min.	Max.
Deformation modulus (MPa)	E_{def}	2.1	2.7	3.6	6.0
Undrained shear strength –cohesion (kPa)	c_u	23	30	40	60

The mechanical properties of subsoil

Properties	Symbols	For soft consistency clay		For stiff consistency clay		For very stiff consistency clay	
		Min.	Max.	Min.	Max.	Min.	Max.
Deformation modulus (MPa)	E_{def}	1.5	3.0	3.3	5.1	5.7	6.6
Undrained shear strength –cohesion (kPa)	c_u	16	33	36	56	63	63

According to STN 72 1001 high plasticity clay (CH) and very high plasticity clay (CV) are classified in Group F8. Medium plasticity clay (CI) is classified in Group F6 and sandy clay (CS) is classified in Group F4.

3 ASSESSMENT OF SLOPE STABILITY OF THE LABOREC LEVEE

Currently several methods of slope stability calculation based on the balance of forces, moments or energy balances are used. Most often methods are based on the assumption that failure occurs towards slip surface. The shape of slip surface depends mainly on the mechanical properties of soils or their arrangement in the profile.

Calculation and assessment of slope stability of the Laborec levee was carried out by using program "Slope stability", which is a sub-program of GEO 5 by company FINE Ltd. The GEO 5 program is designed for the slope stability analysis of a generally sliced earth embankment, which suits the purpose of heterogeneous earthen levees analyses. The slopes stability calculations are done by using Bishop and Petterson Methods. These methods are so-called limit equilibrium techniques, i.e. based on the equilibrium principle of moments above a selected slip surface. They are derived from the existence of stress condition in the surrounding environment, while the surface where slip may occur is sought (so-called the critical slip surface). The result is a factor of safety (*Fs*) determining the ratio between the active and passive forces (Panulinová, E. 2011). In general, if the factor of safety of a slope is within the interval between 0 and 1.0, the slope is actively unstable. The value over 1.0 indicates that the slope is considered stable. The body of the levee is made up of fine-grained soils, it assumed that the slip surface is formed a circular shape. That is a reason why Petterson and Bishop Methods were chosen for calculation and assessment of slope stability of the levee.

Using the methods of calculation was found the minimal degree of stability for the critical shear surface for all proposed variants. The calculation was carried out in accordance with the Slovak standards. The factor of slope stability can be defined as the ratio of the forces contributing to the stability, the stability of the shrinking forces, i.e. the ratio of active and passive forces. Calculation of slope stability has been made according to STN EN 1997. The calculated the factor of safety has been compared to limit value of stability degree.

The levee slope stability assessment was made on the downstream slope of the levee, i.e. at the area of the levee failures, in four cross-sections 1, 2, 3 and 4 (in river km 3.4 to 4.0).

Calculation of slope stability of downstream face was realized in several variants:

– Variant I - The original condition of the levee (i.e. before the flood).
– Variant II -Levee with water (i.e. after the flood).
– Variant III - Remediation - slope stabilization using local material (very high plasticity clay CV).
– Variant IV - Remediation - slope stabilization using gravel material (well graded gravel GW).
– Variant V - Remediation - regulation of the levee shape.

3.1 *Analysis of stability*

The first calculation was made for the original condition of the levee, i.e. before the flood in 2010 (Variant I), further calculation was carried out for flood conditions that arose in 2010 (Variant II). Due to the location of probes (Figure 1) was considered for the slope stability assessment in cross sections 1 and 2 the geological profile JO-3, and for the cross-section 3 and 4 the geological profile JO-4. The results of the calculation of mentioned variants for the selected cross sections are shown in Table 3.

Based on the calculations and the results listed in Table 3can be seen that the stability of slopes did not satisfy the assessment of slope stability before the flood situation (Variant I), except the cross-section 2. These results point to the remediation of levee requirement probably not only in the part where the failures are visible after the flood.

Table 3. Calculated factors of safety for a levee cross-profile (Variant I and Variant II).

Variant	The cross-section	Bishop method	Petterson method	Evaluation (*Fs*> 1.5) (Bishop method / Petterson method)
Variant I	1	1.54	1.43	OK / NO*
	2	1.66	1.54	OK / OK
	3	1.32	1.24	NO / NO
	4	1.43	1.33	NO / NO
Variant II	1	0.99	0.96	NO / NO
	2	1.06	1.02	NO / NO
	3	0.85	0.85	NO / NO
	4	0.91	0.90	NO / NO

*OK = satisfactory, NO = unsatisfactory

Calculations results:

- The greatest factors of safety were obtained using Bishop Method.
- The original condition of the levee can be regarded as safe and stable ($Fs>1.0$).
- Because all calculated factors of safety are greater than 1 (Variant I), we can assume that the break of stability would have happened if it were not extremely high water levels in the Laborec River.
- Variant II, which reflects flood situation showed that the levee during flooding with water up to the crest is unstable and prone to failure (assessed in all cross sections $Fs<1.0$). It confirms the flood situation in 2010.

3.2 *The design of remedial measures*

As a first method of remediation measures to slope stability of a levee was proposed additional load of the slope heel. Stabilization slope with local material is simple to implement (availability of the material). This is the financial undemanding solution. The very high plasticity clay (CV) is the most common local material. This material is not very suitable, but given the fact that the entire levee was built so far from local (readily available) materials has been as one of the alternatives for the design of remediation measures used for this, although not well suitable material (Variant III).

For comparison there was chosen other option - well graded gravel (GW) - Variant IV, which is the best for making the riprap slope stabilization (Figure 2).

After the modelling of various shapes and heights of riprap for slope stabilization of the downstream face, the most suitable solution, the double riprap of the same material has been proposed. First riprap is 1.0 m below the crest and the second riprap is 2.0 m under the levee crest. The water level was considered at two levels in the design. Like as Variant I, i.e. water level before the flood situation and further calculations were made in consideration of the water level as in Variant II, i.e. during flood events. In this way, we can verify the stability

Figure 2. The place of disturbance and the geological profile of the levee body and subsoil.

Table 4. Calculated factors of safety for a levee cross-profile (Variant III, Variant IV and Variant V).

Variant	The cross-section	Bishop method	Petterson method	Evaluation ($Fs> 1.5$) (Bishop method / Petterson method)
Variant III	1	1.57	1.51	OK / OK
	2	1.57	1.50	OK / OK
	3	1.81	1.83	OK / OK
	4	1.75	1.79	OK / OK
Variant IV	1	1.58	1.51	OK / OK
	2	1.57	1.50	OK / OK
	3	1.80	1.83	OK / OK
	4	1.76	1.76	OK / OK
Variant V	1	1.68	1.55	OK / OK
	2	1.68	1.55	OK / OK
	3	1.67	1.62	OK / OK
	4	1.68	1.61	OK / OK

of the proposed remediation action in the event of another flood situation. For the selected cross sections, the results of the calculation of mentioned variants are shown in Table 4.

As another method of remediation of a levee stability was proposed the regulation of the levee shape, i.e. increase levee crest and the downstream face of a levee (Variant V). This remedial measure is designed only for the downstream face of a levee (Figure 2). Extension on the upstream side would reduce the potential flow area profile of the river. This type of remediation measures could prevent any future overflow levee. To increase the levee crest was used medium plasticity clay and for drainage of the heel on the downstream face of the levee was used gravel. New levee crest is from the original on a meter above. Width of the levee crest stayed 3.0 meters (Figure 2). The water level in the design was considered at two levels too, i.e. water level before the flood situation and during flood events. For the selected cross sections, the results of the calculation of mentioned variants are shown in Table 4.

Remediation proposal was made because of the availability of the materials used for the rehabilitation and financial undemanding of this solution. The aim of the riprap slope stabilization of the heel on the downstream face of the levee was to achieve a higher degree of stability than 1.5, thereby improving the stability of the slope itself. Calculations results:

- The greatest factors of safety were obtained using Bishop Method.
- Variants III and IV, in which material were varied, improved slope stability.
- Variant IV is more appropriate than Variant III.
- Variant III will not be sufficient to solve for potential further flood situations. In modelling such a new flood situation similar to that of 2010, when the water level reached the crest of the levee, this measure has been shown insufficient.
- Increase levee crest and the downstream face of a levee (Variant V), improved slope stability.
- Variant V will not be sufficient to solve for potential further flood situations. In modelling such a new flood situation similar to that of 2010, when the water level reached the crest of the levee, this measure has been shown insufficient.
- This measure reduces the degree of stability on the upstream slope of the levee. This is because the increase of the levee has been the extension of the slope of the upstream face of the levee, reducing its stability.

4 CONCLUSION

The levees are highly vulnerable structural elements of a water body. The main purpose of artificial levees is to prevent flooding of the landscape nearby territory and to slow the natural course changes in a waterway which also confines the flow of the river, resulting in higher and faster water flow.

As a result of an imbalance there are often excessive distortions and landslides of slopes of the levee. Violations can be induced by stature of active factors (e.g., load of slope by construction, own weight raising caused by increase of soil moisture, seepage of water pressure formation, etc.) or by a decrease of passive factors (e.g., reducing the shear strength of soil, deterioration of consistency of cohesive soil by moisture increase, etc.).

Calculation of slope stability of downstream face of levee was realized in three variants, in which we analyzed several possibilities to solve and improve the stability downstream slope of the levee after the flood. The best variant was Variant IV, in which was proposed additional load of the slope heel with well graded gravel (GW).

Since 2010, when the fault occurred, the levee became completely grassed, and therefore can be identified only by the irregularity of the slope. Given that the flood and landslide occurred 6 years ago, a new tensile stress state of the levee was created and the slope can now be considered "stabilized". In the event of another flood of similar or larger scale, this stability would again be disturbed. In view of the fact that the body of the levee is made of fine soil group F8, i.e. unsuitable soil, it will be necessary to deal with remediation measures, which could possibly improve the stability across the levee. However, it is already a cost and time-consuming solution, which we want to address in the future.

Levees are considered as significant structures and therefore it is needed to pay high attention not only to the choice of materials intended for their construction but also for the design and assessment of these constructions (Bulatov, G. & Vatin, N. & Nemova, D. & Ibraeva, Y. & Tarasevskii, P. 2015). Because of their importance, safety and reliability throughout their lifetime remain the top priority.

ACKNOWLEDGMENT

This work was supported by the Scientific Grant Agency of the Ministry of Education of Slovak Republic and the Slovak Academy of Sciences under Project VEGA 1/0477/15.

REFERENCES

Bulatov,G. & Vatin, N. & Nemova, D. & Ibraeva, Y. & Tarasevskii, P. 2015. Ensuring The Reliability Of Earth Dams In Complex Hydrogeological Conditions. *Applied Mechanics and Materials.* Vols. 725–726 (2015): 342–348.

Harabinová, S. & Ondrejka Harbuľáková, V. & Zeleňáková, M. 2014. Assessment of the levee slope stability at Laborec river, Slovakia. In. Brno Univeristy of Technology, Faculty of Civil Engineering (ed.), *People, Buildings and Environment 2014 (PBE2014). Vol. 3.*: 568–576. *Kroměříž, 15–17 October, 2014.* Czech Republic: Brno.

Ondrejka Harbuľáková, V. & Hudáková, A. 2014. Hydraulic model for determination of flood hazard area. *Journal of Landscape Management.* Vol. 5 (2): 65–70.

Ondrejka Harbuľáková, V. & Zeleňáková, M. 2013. Technical measures of riverbank stabilization in engineering practice. Visnik: *Teoria i praktika budovnictva.* No. 756 (2013): 200–206.

Ondrejka, J. & Sycevova, M. 2010. *Oborín - Right bank landslide of Laborec levee in rkm 3.4*: Final evaluation. Slovak Republic: Kosice.

Panulinová, E. 2011. *Shear strenght.* In. Technical Univeristy of Košice, Faculty of Civil Engineering (ed.),Slovak Republic: Kosice.

STN-EN 1997–1. Eurocode 7. Geotechnical design, Part 1: General rules. 2005.

STN 72 1001 Classification of soils and rocks. 2010

Zeleňáková, M. 2009. Preliminary flood risk assessment in the Hornád watershed. In. Wessex Institute of Technology (ed.), *River Basin Management 5: intern. conference Ramla Bay Resort, Malta,7–9 September 2009.* Southampton: Wessex Institute of Technology.

Advances and Trends in Engineering Sciences and Technologies II – Al Ali & Platko (Eds)
© *2017 Taylor & Francis Group, ISBN 978-1-138-03224-8*

Guidelines and recommendations for vibration control in the case of rapid impulse compaction

A. Herbut & J. Rybak
Wroclaw University of Science and Technology, Wroclaw, Poland

ABSTRACT: As the soil improvement technologies are the area of a rapid development, they require designing and implementing novel methods of control and calibration in order to ensure the safety and reliability of geotechnical works. The results and observations make it possible to delineate specific modifications to the parameters of technology applied (e.g. hammer drop height or vibrator frequency). On the basis of numerous case studies of practical applications, already summarized and published, researchers are able to formulate the guidelines for work on the aforementioned sites. This work presents a specific aspect of an active design and technology calibration, using the investigation of the impact of vibrations that occur during the Rapid Impulse Compaction on this adjacent structures. A case study entails the impact of construction works on the historic brick chimney in a ruined old factory.

1 INTRODUCTION-FACTORS CONSIDERED BY LEGISLATION REQUIREMENTS

New techniques of site monitoring are continually developed at Wroclaw University of Science and Technology (Poland) with an aim to provide the appropriate tools for the preliminary design of work process, as well as for the further ongoing on-site control of geotechnical works (steel sheet piling, pile driving, soil improvement). They include measurements and continuous histogram recording of shocks and vibrations and its dynamic impact on engineering structures (steel, concrete and masonry structures) in the close proximity of the construction site.

Athanasopoulos and Pelekis gave recently a brief review of the existing standards and limits for the man-induced vibrations (Athanasopoulos & Pelekis 2000). In other works those authors presented the effects of soil stiffness in the attenuation of Rayleigh-wave motions (Athanasopoulos & Pelekis, 2000). The control of vibration is usually related to pile driving works. A brief overview of legal basis for ground vibration control was also presented in work of Brzakala (et al. 2014). In general, one may notice that sharing of information and increasing globalization has led to the gradual replacement of many national codes by international ones. The national codes or annexes to international codes usually contain specific data, such as the levels of earthquake excitation for the region under analysis, allowable noise and vibration limits and references to other national legislations. The criteria of vibration in buildings are described by Bachmann and Ammann (1997) and Srbulov {2011) with regard to the following factors: humans (physiological effects on people), equipment (overstressing of machinery and impediment of technological process) and finally the whole structure (the destruction of structural parts). For the purpose of this work only the last criterion (structural) will be analyzed.

2 STRUCTURES CRITERIA USED IN POLISH AND INTERNATIONAL STANDARDS

The intensity of vibrations affecting structures is usually described by peak particle velocity measured at ground level close to the structure. Permissible values of velocities are usually given in form of curves, depending on the values of dominant frequencies, the type of structure and duration of the vibrations. Generally, the allowable values of the peak particle velocities components range from 3 to 70 mm/s. The corresponding frequency range is usually from 1 to 100 Hz. This recommendations are usually limited to an ordinary type of structures and typical ground conditions. In more complicated cases engineers need to seek advice and help from specialists in soil dynamics—more advanced numerical simulations are necessary to predict the possible structure damage. Different standards and codes give various vibration criteria to prevent buildings damage for various construction materials and technologies. The comparison studies and summary for wide range of approaches was recently made by Srbulov (2011), Skipp (1998), Hwang (2006), Pieczynska and Rybak (2014), Brzakala (et al. 2014) and Athanasopoulos (2000).

2.1 *Guidelines used in Polish standards*

In accordance with Polish Standard PN-85/B-02170, the effects of vibration on structures can be determined by complex analysis based on the numerical model for each type of structure. Simplified analysis based on special charts describing scales of the dynamic effects on structures (so called: SWD-I—generally for compact buildings with height of less than two floors and horizontal dimension of less than 15m; SWD-II charts—for the building height of less than five floors and vertical dimension of less than the building's doubled horizontal dimension).

The displacement/acceleration amplitudes of structure vibrations measured at the ground level in horizontal directions are compared with the threshold values. The simplified analysis can be used in the case of typical brick-made buildings with the height of less than five floors. On Figures 1 and 2, the vibration criteria for dangerous cases are presented. The allowable values of displacements/accelerations are given in frequency domain for different types of situations depending on:

- the structure condition (there are lower limits for old and damaged buildings than for the recently built ones) and the materials used in buildings and their quality
- type of soil and foundation
- type of vibrations (lower limits for long-term vibrations (> 30 minutes) than for the temporary ones (< 30 minutes)

The vibration criteria depend also on the assumed acceptable effects:

- A—limit is the lowest limit for vibration affecting structures, bellow these values vibrations do not influence the buildings
- B—limit is the lowest threshold for cracks in structural elements (structure damage)
- C—limit is a strength limit for the material used in constructional building elements, the lowest limit for a serious damage
- D—limit is the lowest limit for the structure stability; exceeding this value causes global damage of the structure

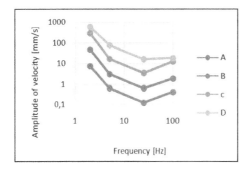

Figure 1. Acceptable vibration level SWD-I based on PN-85/B-02170 in logarithmic scale.

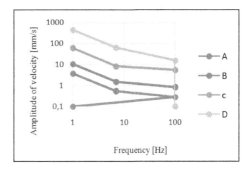

Figure 2. Acceptable vibration level SWD-II based on PN-85/B-02170 in logarithmic scale.

Basic information about the limits established by the above-mentioned standards were juxtaposed in work (Brzakala et al. 2014). In general, in all the cases, the maximum component of the vibration velocity on the foundation level caused by dynamic loads is compared with the given criteria. The time-history of the vertical and horizontal vibration is recorded in one of the directions parallel to the side wall of the building. The limits of the vibration velocities are given in the form of curves in a frequency domain for three different types of structures (industrial, residential and sensitive to vibrations). According to DIN, for short-term vibrations, the threshold values related to industrial/commercial purpose buildings vary from 20 mm/s for low frequencies to 50mm/s for high frequency range. Lower bound values refer to dwellings (5 mm/s-20 mm/s) and the strict criteria are related to structures sensitive to dynamics effects (3 mm/s-10 mm/s). The intensity of vibrations may be amplified significantly along the height of the buildings, that is why the measurements made on particular storey should be also verified with relevant codes and with the allowable values for the human exposure to vibrations (Brzakala at al. 2014). For long-term vibrations the requirements are much more rigorous.

The Association of Swiss Highway Engineers distinguishes in their Standard SN 640312 four different categories of buildings, according to the type of structure from the least (I-industrial and commercial buildings) to the most sensitive to dynamic effects (IV-sensitive structures and monuments). The threshold values of peak vibration velocities are given for each structure category, depending on the frequency range and the source of vibration. Generally, for the vibrations induced by machinery, traffic or construction works (source M of vibrations), the values vary from 3mm/s (structure category IV, low frequencies) to 18mm/s (structure category I, high frequencies). For the vibrations induced by blasting operations (source S of vibrations), the acceptance criteria are less restrictive, because this factor occur less frequently than M source of vibration. Threshold values vary from 12 mm/s (structure category IV, low frequencies) to 40mm/s (structure category I, high frequencies), so greater values are accepted for source S than for source M of vibrations for the same frequency range. Compared with German DIN 4150-3 (1999), the requirements given by SN 640312 are similar for the excitation frequencies close to 10 Hz and residential and sensitive buildings. For industrial buildings and excitations frequencies greater than 10Hz SN 640312 there seems to be a more restrictive approach (Athanasopoulos & Pelekis 2000).

British standard BS 5228-2 (2009) recommends the allowable peak particle velocities for minor or architectural (i.e. non-structural) damage for different types of structures. The category 1 relates to reinforced or framed structures, industrial and heavy commercial buildings, category 2 - to non-reinforced or light frame structures, residential or light commercial buildings and category 3 - to underground services.

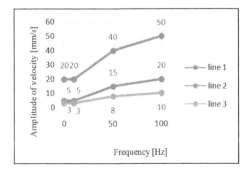

Figure 3.   Acceptable vibration level according to DIN 4150-3:1999-02.

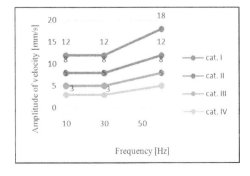

Figure 4.   Acceptable vibration level according to SN 640312.

The limits presented by AASHITO and Eurocode 3 are similar. The acceptable vibration level in form of peak particle velocities does not depend on the frequency range. However, compared with the Eurocode 3 threshold, the values given by AASHITO are greater. In Eurocode 3, threshold values of peak particle velocities do not depend on the frequency range as in SN 640312 or DIN 4150-3. Acceptable vibration level to prevent structure damage depends on the structure type as in other international standards. The bound values of velocity amplitudes vary from 2mm/s for the most sensitive structures, like architectural merits, to 25 mm/s for buried structures.

3 DYNAMIC COMPACTION—MOST HARMFUL MAN-MADE GROUND VIBRATIONS

There are two sources of ground vibrations—natural phenomena or human activities. Ground vibrations induced by human activities are characterized by a much lower intensity compared with the natural phenomena like earthquakes, although due to the high infrastructure and urban development these effects belong nowadays to the rapidly developing area of soil dynamics described by various authors (Athanasopoulos & Pelekis 2000, Hwang & Tu 2006). Potentially the most invasive (harmful) technologies of soil improvement are related to single dynamic impacts like in the case of micro-blasting, dynamic consolidation and continually developing technology of Rapid Impulse Compaction.

Human activities may generate significant ground vibrations and have the influence on the structures especially in urban environment, due to the close distance of buildings and structures from the source of vibration. Human activities that can generate ground vibrations were classified by (Athanasopoulos & Pelekis 2000, Bachmann & Ammann 1987) into the following categories:

– operation of machines,
– road and railway traffic,
– and construction activities.

The last from the above-mentioned factors is the main topic of this paper. The main construction or demolition activities causing dynamic problems in close proximity of structures are, according to works (Hiller & Hope 1998, Srbulov 2011), as follows:

– pile or sheet pile driving and dynamic soil compaction,
– demolition of structures,
– rock excavation and soil deep compaction by explosives.

Table 1. Acceptable vibration level according to AASHTO and Eurocode 3.

Type of situation (AASHTO)	Amplitude of the velocity [mm/s]	Building type (Eurocode 3)	Amplitude of the velocity [mm/s]	
			permanent	transient
–	–	–	permanent	transient
Historic sites or other critical locations	2.54	Architectural merit	2	4
Residential buildings, plastered walls	5.08–7.62	Residential areas	5	10
Residential buildings in good repair with gypsum board walls	10.16–12.7	Residential areas	5	10
-	–	Light commercial	10	20
Engineering structures, without plaster	25.4–38.1	Heavy industrial	15	30

The second factor is the main objective of case studies presented in the paper. The dynamic compaction is the ground improvement method, which allows to densify the underlying soils by dropping heavy weight. It may excite significant dynamic effects in the form of the surface propagating ground waves. In that case, vertical and radial displacement/velocity components are much more significant than tangential ones, so the R-waves are dominant. The attenuation rate in radial direction is the highest, in vertical direction—a little smaller, and in tangential direction—the lowest. That is why for a large distance from the vibration source, the peaks of the velocity/acceleration values are very similar in each direction. The aspects of the vibration attenuation in the case of dynamic compaction are discussed in detail by Hwang (Hwang & Tu, 2006) The authors observed also that soil compaction generates greater vibrations but with a faster attenuation rate the greater the distance, compared to pile driving. Similar remarks were previously published in the work of the authors (Pieczynska &Rybak 2014).

4 CASE STUDY—SITUATION AND TEST RESULTS

The main recording device was Minimate® Pro4 vibration control tool made by Instantel.® The Sample Rate was 2048 Hz. Interval duration was set to 15 s. The minimum recorded value of vibration velocity was 0.1 mm/s. Minimate® Pro4 sensor was consequently placed on construction elements of monitored objects. The objects of major importance were newly constructed residential buildings (S20, S19) and a ruined factory with a tall brick chimney (of historical value).

4.1 First stage of works—maintained impact compaction at M9 site

Test points were chosen in order to evaluate the intensity of dynamic impact (vibrations) on neighboring structures. Test measurements and continuous vibration control began on December 2nd, 2015. The basic layout of test points is shown on Fig. 5. The construction company TERRA-MIX was informed on-line about the results of the vibration survey. Even though the machine operator was able to change hammer drop height immediately; because of relatively low levels of dynamic impact there was no necessity to act like that. After trial tests, the whole work period could be carried out without any threats concerning vibrations.

4.2 Second and third stage of test—maintained impact compaction at M8a and M7 sites

Test points were chosen again in order to evaluate the intensity of dynamic impact (vibrations) on neighboring structures. Test measurements and continuous vibration control were carried out on December 8th, 2015. Basic layout of test points is shown on Fig. 6. The construction company was again informed on-line about the results of the vibration survey and the machine operator altered the drop height from 100% to 80% of the possible energy. When the construction works moved to M7 site in the close proximity of the chimney, a continuous monitoring was applied and for the duration of 3 working days (December 19th,

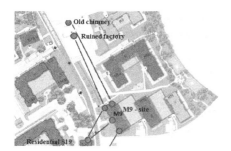

Figure 5. Work site (M9) and monitored objects S20, S19, ruined factory and chimney.

Figure 6. Job sites (M8a, M7) monitored objects S16, S15, ruined factory and chimney.

21st, 23rd), the whole process was controlled and calibrated. The drop height varied from 100% to 40% of the possible energy. All the results were later on gathered in "Vibration testing reports U-series no. 145/2015, 153/2015, 2/2016" published at Wroclaw University of Technology.

5 BASIC CONCLUSIONS

Modern impact hammers let us change hammer drop height for some part or the whole duration of the works. Such policy was applied during the compaction works for soil improvement at the construction site in Wroclaw. Such procedure made it possible to reduce considerably the amplitudes of vibrations observed on the adjacent structures. However, it must be remembered that the "switching" of the hammer may limit the efficiency of soil compaction (depth) and prolong the time of works. The above presented case study showed the situations in which only after the technology was selected and the works started, the investigation was carried out in order to determine whether the technology may be further applied or whether it should be switched to some other one. In other cases, when the necessary precautions impose the works to come to a standstill, the investment costs increase (both of the standstill and the potential technology changes). That often results from the passive attitude on the part of investors and contractors, while—with a small cost of the survey, both might be saved: time and money.

REFERENCES

Athanasopoulos G.A., Pelekis P.C., 2000. Ground vibrations from sheetpile driving in urban environment: measurements, analysis and effects on buildings and occupants, *Soil Dynamics and Earthquake Engineering 19*, pp. 371–387

Bachmann H., Ammann W., 1987. Vibration in Structures Induced by Man and Machines. *International Association for Bridge and Structural Engineering*

Brzakala W., Herbut A., Rybak J., 2014. Recommendations for ground vibrations survey in course of geotechnical works. 14th International Multidisciplinary Scientific GeoConference SGEM 2014: GeoConference on science and technologies in geology, exploration and mining, Albena, Bulgaria, 17–26 June, Vol. 2, Hydrogeology, engineering geology and geotechnics. pp. 747–754

Hiller, D.M. & Hope, V.S., 1998. Groundborne vibration generated by mechanized construction activity. *Proc., Inst. Civil Engineers (131)*, pp. 223–232

Hwang J.H., Tu T.Y. 2006. Ground vibration due to dynamic compaction. *Soil Dynamics and Earthquake Engineering 26*, pp. 337–346

Pieczynska-Kozlowska J., Rybak J.,2014.Vibration monitoring as a tool for a calibration of geotechnical technologies. 14th International Multidisciplinary Scientific GeoConference SGEM 2014: GeoConference on science and technologies in geology, exploration and mining, Albena, Bulgaria, 17–26 June, Vol. 2, Hydrogeology, engineering geology and geotechnics. pp. 1043–1050

Skipp B.O., 1998. Ground vibration—codes and standards. In: Skipp (ed) Ground dynamics and man-made processes. The Institution of Civil Engineers.

Srbulov M., 2010. Ground Vibration Engineering. Simplified Analyses with Case Studies and Examples., Springer

Srbulov M., 2011. Practical Soil Dynamics. Case Studies in Earthquake and Geotechnical Engineering., Springer

Advances and Trends in Engineering Sciences and Technologies II – Al Ali & Platko (Eds)
© 2017 Taylor & Francis Group, ISBN 978-1-138-03224-8

Determination of tunnel face stabilizing pressure using numerical and analytical methods

E. Hrubesova, L. Duris, A. Melichar & M. Jaafar
VŠB—Technical University of Ostrava, Ostrava, Czech Republic

ABSTRACT: One of the important tasks in mechanized tunneling using tunnel boring machines is to determine the required face pressure ensuring its stabilization. For the solution of this task various methods can be used. Analytical calculations to determine the required stabilizing face pressure are mainly based on limit equilibrium methods and plasticity theory. Although numerical methods (finite element method, etc.) to determine the tunnel face pressure facilitate the modelling of this problem with more general calculation assumptions (inhomogeneous environment, etc.) as compared with the above analytical methods, the creation of the model and the actual calculation are more time consuming. The paper presents the comparison of face pressure values obtained from analytical models of different authors, and from the numerical model. The paper concludes by presenting the resulting comparative charts for various computational methods and an analysis of obtained results.

1 INTRODUCTION TO THE TUNNEL FACE STABILIZATION

One of the fundamental tasks in tunnel excavation by tunnel boring machines is the determination of the optimum face support pressure, which ensures the safety of the excavation itself, minimizes surface settlement and ultimately prevents the collapse of the entire tunnel. The value of the face support pressure generally depends on many factors, primarily on the size of the workings, the depth below the surface, cohesion and internal friction angle of the rock environment, unit weight of the soil environment, permeability of the rock environment, machine type, excavation speed, depth of groundwater level, etc.

Research concerning the determination of the optimum tunnel face pressure is globally conducted by experimental methods, including centrifugal tests (e.g. Kamata 2003, Kimura 1981), as well as by various mathematical (analytical and numerical) methods Broere (2001). Presented results confirm, that in case of larger centrifugal accelerations, the failure area ahead of the tunnel face can be propagated to the surface above the tunnel.

2 BASIC PRINCIPLES OF ANALYTICAL CALCULATION METHODS TO DETERMINE FACE PRESSURE

The methods to calculate pressure on the tunnel face differ in their mathematical essence, calculation assumptions, the degree of conservatism of results, required computing time and calculation efficiency, which is a very important factor, particularly when flexibly solving practical problems.

Analytical calculation methods to determine the minimum tunnel face pressure are either based on the Limit Equilibrium Method (LEM) or the Limiting Analysis (LA). Analytical calculation methods to determination of minimum face support pressure based on limiting analysis can be divided into two basic groups. The first group of methods is based on the so-called "lower bound theory". These methods allow for the determination of a statically acceptable solution for minimum face pressure that satisfies both balance and rock failure conditions. The basic underlying assumptions for these types of methods imply that they

Table 1. Basic analytical methods to determine minimum tunnel face pressure.

	Type of analytical method	Basic describing equation
Caquot (1934)	LA—lower bound	$p = \gamma^* z + q_s - 2\,c\,\ln(1+h/R)$
Broms and Bennermark (1967)	LEM	$p = \gamma(h+R)+q_s-Nc_u$, where $N = 6$
Atkinson and Potts (1977)	LA—lower bound	$p = 2\,K_p \gamma R/(K_p^2-1)$, $K_p = (1+\sin(\varphi))/(1-\sin(\varphi))$
Atkinson and Potts (1977)	LA-upper bound	$p = (\gamma D)/(4^*\cos\varphi)^*(1/\tan\varphi+\varphi-\pi/2)$
Davis (1980)	LA-upper bound	$p = \gamma(h+R) - 4c_u\ln(h/R+1)+q_s$
Krause (1987)	LEM	$p = (D\gamma/9-\pi c/2)/\tan\varphi$ half sphere mechanism $p = (D\gamma/6-\pi c/2)/(0.5+\tan\varphi)$ half circle mechanism $p = (D\gamma/3-\pi c/2)/\tan\varphi$ quarter circle mechanism
Leca and Dormieux (1990)	LA	See Leca and Dormieux (1990)
Anagnostou and Kovári (1994)	LA-upper bound	See Anagnostou and Kovári (1994)
Jancsesz and Steiner (1994)	LEM, including infiltration	See Jancsesz and Steiner (1994)
Vermeer and Ruse (2001)	LEM	$p = -(c/\tan\varphi)+2^*\gamma^*R(1/(9^*\tan\varphi)-0.05)$
Broere (2001)	Equation determined based on a numerical model using the finite element method (Plaxis)	See Broere (2001)
Mollon (2009)	LA—improved multiblock mechanism	See Mollon (2009)

Legend to table: γ—unit weight of rock material, h—overburden height, R—tunnel radius, D—tunnel diameter, c—cohesion, c_u—undrained shear strength, φ—internal friction angle, q_s—surface surcharge, z—depth below the surface.

allow obtaining conservative value of the minimum tunnel face support pressure. The second group of computational methods is based on the so-called "upper bound theory", which assumes a kinematically acceptable solution subjected to a certain accepted rock failure mechanism. Methods of this type assume that, given a certain rock failure mechanism, the compatibility of displacements between individual blocks of material in a rock failure process is not breached. Unlike solutions based on lower bound theory, these computational approaches to obtain a kinematically acceptable solution allow obtaining a non-conservative value of face pressure.

Computational schemes which foresee the emergence of a particular rock failure mechanism ahead the face are based on the Horn model (Horn 1961), which replaces the circular shape of the tunnel face with a square shape, and approximates the corresponding rock failure area before the tunnel face using rigid blocks—wedge and the follow-up prism (chimney).

Table 1 presents an overview of the most commonly used analytical methods to the determination of the minimum support pressure for tunnel face stabilization.

3 CHARACTERISTICS OF THE NUMERICAL MODEL FOR THE DETERMINATION OF FACE SUPPORT PRESSURE

For the analysis of the required minimum face support pressure the numerical method (FEM etc.) can be used also. The authors of this paper worked out the model of an excavated tunnel of a circular cross-section with a diameter D = 5 m at different depths h = 10, 20 resp.

Table 2. Material properties of the rock mass.

	E [kPa]	ν	γ [kN/m³]	c [kPa]	φ [deg]
Rock mass	45 000	0,35	20	2	Alternatives 20, 30, 40

30 meters below the surface. For the modelling the commercial software **MIDAS GTS** was used. To the more objective comparison with analytical methods, the assumption of hydrostatic state of primary stress was accepted, i.e. the coefficient of lateral pressure is equal to one. The influence of groundwater is not accounted in the calculation. The considered material properties of the rock environment (Young's modulus E, Poisson's ratio ν, unit weight γ, cohesion c, internal friction angle φ in which the tunnel is excavated are given in Table 2. The model also assumes stabilization of the tunnel excavation by concrete lining with a thickness $d = 35$ cm, modulus of elasticity $E_c = 20\ 000$ MPa, Poisson's number $v_c = 0.2$ a unit weight $\gamma_c = 24$ kN/m³. To the numerical modelling the tetrahedral finite elements were used, the behavior of rock mass was described by Mohr-Coulomb elastic—perfectly plastic constitutive model.

4 ASSESSMENT OF MINIMUM TUNNEL FACE SUPPORT PRESSURE OBTAINED BY ANALYTICAL AND NUMERICAL METHODS

For the assessment of minimal tunnel face pressure both analytical and numerical methods were applied. Figure 1 shows the comparison of the results of the selected analytical approaches to determining the minimum stabilization face pressure. The comparison includes methods that allow for the influence of the internal friction angle, which is the critical parameter of shear strength for the planned geological environment in the vicinity of the tunnel, and are described in Table 1 by a defining equation. Due to the practically non-cohesive nature of the investigated rock environment (cohesion is equal to 2 kPa), methods working with undrained shear strength cu are not considered in the presented study. The analyzed methods don't reflect the impact of tunnel depth.

A detailed dependency of minimal stabilizing tunnel face pressure and friction angle of a rock mass according Krause's half sphere method is presented in Figure 2.

Based on the results of the calculations, it is possible to formulate the following conclusions:

a. The range of the minimum face support pressure for all the compared analytical methods is 20 to 80 kPa for internal friction angle $\varphi = 20°$, values of 15 to 50 kPa for $\varphi = 30°$ and 10 to 38 kPa for $\varphi = 40°$.
b. The most conservative values of the minimum face support pressure result from the Krause method, considering the quadrant rock failure mechanism.
c. In terms of the face support pressure, the least conservative analytical method is the Krause method considering hemispherical rock failure mechanism.
d. The model indicates more significant dependence of the face support pressure on the internal friction angle in the case of lower values of the internal friction angle.

The 3D numerical model was developed by using **MIDAS GTS** software, which is based on the finite element method. The basic dimension of the model 70×50 m (length x width) was assumed, the height of the model was considered 75 m. A circular underground working with a diameter of 5 m was considered at the three different depths (10 m, 20 m and 30 m respectively). The ratio of the overburden height (h) and the diameter of the working (D) thus reached the different values of 2, 4 and 6 respectively. In accordance with the technology of mechanized excavation, the concrete lining with a thickness of 35 cm is installed immediately after the excavation (a non-stabilized part of the excavation behind the face is not considered).

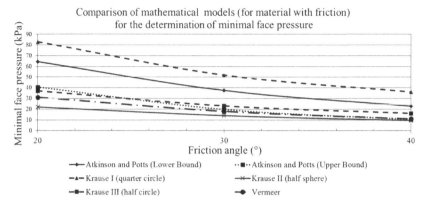

Figure 1. Comparison of the values of minimum face support pressures, that were determined by analytical computational methods.

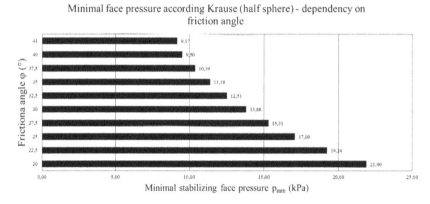

Figure 2. A detailed relationship between minimal stabilizing face pressure and friction angle (Krause half sphere method).

The final tetrahedron-shaped elements and elastic-perfectly plastic Mohr-Coulomb constitutive model were used in the model. Material input characteristics of the rock environment are given in Table 2 of Chapter 3. The elastic constitutive model was used for the behavior of concrete lining only. The standard deformation boundary conditions ("rigid container") were assumed.

The different uniform loads have been applied to the face area, the load varied from 10 to 800 kPa. The entire excavated area of the length of 25 m behind the face is stabilized by the concrete lining with the previously specified characteristics.

Based on this above mentioned numerical model the following values were assessed:

1. Horizontal face displacements corresponding to the face extrusion into the space behind the face
2. Plastic points on the face and on the area ahead the face
3. Plastic strain in the area ahead the face

Figure 3 illustrates the dependence of the maximum horizontal face displacements depending on the face pressure for three different values of internal friction angle and for the tunnel depth below the surface of 30 m. This dependence allows us to conclude, that the horizontal face displacements are not very significant up to a certain face pressure (there is no significant extrusion into the free space behind the face). A reduction of the pressure to a certain level leads to qualitatively different development of horizontal displacements. The face pressure,

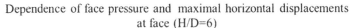

Dependence of face pressure and maximal horizontal displacements at face (H/D=6)

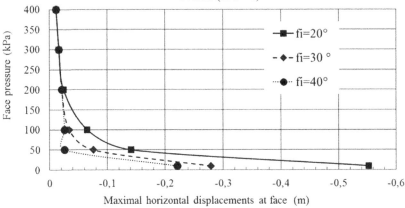

Figure 3. Dependence of the maximum horizontal face displacements on the face pressures (for different internal friction angles).

Table 3. Results of numerical modeling to determine the minimum face support pressures and support pressures eliminating face extrusion.

Tunnel depth below the surface (m)	h/D	Internal friction angle (°)	Minimum face support pressure (kPa)	Support pressure eliminating extrusion (kPa)
10	2	30	40	200
20	4	30	50	500
30	6	20	100	600
		30	75	600
		40	50	600

which causes both a qualitative change in the developments of horizontal displacements and more significant face extrusion, is considered as the required minimum face pressure. Based on this approach, the minimum face support pressure determined by the model, reaches of 50 kPa for the highest considered internal friction angle of 40°, approximately of 75 kPa for $\varphi = 30°$ and of 100 kPa for $\varphi = 20°$. Face pressure of approximately 600 kPa leads to a unsignificant face extrusion and a zero horizontal displacements are occured. At pressures exceeding 600 kPa, the calculation model shows a development of horizontal face displacements in the opposite direction (the face is pushed back by the pressure).

Based on the results of the numerical model, it is possible to assess the minimum support face pressures, as well as the face support pressures, which can eliminate a face extrusion. The results of numerical modeling in terms of face pressures are summarized in Table 3.

5 CONCLUSION

The results of analytical computational models and numerical 3D model using MIDAS GTS (finite element method) indicate minimum face pressure values during the excavation of the specified full-face excavated tunnel with circular cross-section. Using the finite element method for a depth of 30 m, the model results show, that the minimum face support pressure is 20% higher, than the most conservative analytical Krause method. For the tunnel depth of 20 m, the values of the maximum face support pressures are comparable for the most conservative analytical method as well as for the finite element method. For the tunnel depth of

10 m, the value of the minimum face support pressure is smaller in comparison with the most conservative analytical method, but the value is greater than the value determined by the least conservative analytical method. When formulating these conclusions, we must take into account the fact, that the finite element method is a deformation method. The results depend not only on strength parameters of the soil environment, but also on deformation parameters (Young's modulus and Poisson's ratio). However, although the quantitative values of face deformations (equivalent to the horizontal displacements on the face) can be expected different for different values of the Young's modulus of the soil environment, they would not be very different from the qualitative point of view. For the verification of mentioned particular conclusions, more extensive numerical analysis and comparison with a real face pressure during an actual tunnel excavation is necessary.

ACKNOWLEDGEMENT

The paper was prepared with the support of the Competence Centers of the Technology Agency of the Czech Republic (TAČR) within the project Center for Effective and Sustainable Transport Infrastructure (CESTI), project number TE01020168.

REFERENCES

Anagnostou, G., Kovari, K. 1996. The face stability of slurry—shield driven tunnels. *Tunneling and Underground Space Technology*, 9(2). pp. 165–174.

Atkinson, J. H., Potts, D. M. 1977. Stability of a shallow circular tunnel in cohesionless soil. *Geotechnique* 27(2). pp. 203–215.

Broere, W. 2001. Tunnel Face Stability & New CPT Applications. *Dissertation*, TU Delft 2001.

Broms, B. B., Bennermark, H. 1967. Stability of clay at vertical openings. *ASCE Journal of Soil Mechanics and Foundation Engineering Division 93*. pp. 71–94.

Carranza-Torres, C. 2004. Technical Report to Geodata S.p.A.: *Computation of factor of safety for shallow tunnels using Caquot's lower bound solution*. Itasca Consulting Group. Minneapolis.

Davis, E. H. 1980. The stability of shallow tunnels and underground openings in cohesive material. *Geotechnique* 30(4). pp. 397–416.

Horn, M. 1961. Horizontaler Erddruck auf senkrechte Abschlussflächen von Tunneln (Horizontal earth pressure on vertical tunnel fronts). *Landeskonferenz der ungarischen Tiefbauindustrie* (*National Conference of Hungarian Civil Engineering Industry*). Translation into German by STUVA, Düsseldorf.

Jancsecz, S., Steiner, W. 1994. Face support for large mixshield in heterogeneous ground conditions. *Proc. Tunnelling '94*. London: Chapman & Hall, 1994. pp. 531–550.

Kamata, H., Mashimo, H. 2003. Centrifuge model test of tunnel reinforcement by bolting. *Tunnelling and Underground Space Technology 18*. Elsevier. pp. 205–212.

Kimura, T., Mair, R. J. 1981. Centrifugal testing of model tunnles in soft clay. *Proceedings of the 10th ICSMFE*, Vol. 1. pp. 319–322.

Krause, T. 1987. Schildvortrieb mit flüssigkeits—und erdgestützter Ortsbrust. *in Mitteilung des Instituts für Grundbau und Bodenmechanik*, Technische Universität Braunschweig, No. 24.

Leca, E., Dormieux, L. 1990. Upper and lower bound solutions for the face stability of shallow circular tunnels in frictional material. *Geotechnique* 40(4). pp 581–606.

Mollon, G., Dias, D., Soubra, A. 2009. Two new limit analysis mechanisms for the computation of the collapse pressures of circular tunnels driven by a pressurized shield. *In: EURO:TUN 2009 (2nd International Conference on Computational Methods in Tunnelling)*, Ruhr University Bochum, 2009.

Vermeer, P. A., Ruse, N., Marcher, T. 2002. Tunnel heading stability in drained ground. *Felsbau* 20(6), pp. 8–18).

Evaluation of scenic route choice factors

Ç. Kara, Ş. Bilgiç & K.B. Akalın
Eskişehir Osmangazi Üniversity, Eskişehir, Turkey

ABSTRACT: The importance of the roads is high in terms of touristic, cultural and socio economic structure for the developing countries. Travelling by vehicles has fast grown consistently for several centuries. Road constructions are expected to not only transmit people from one place to another, but also present fascinating historical sites, spectacular views, or distractive wildlife for travelers' en route. Scenic byways are roadways which provide people an opportunity to travel through lands of scenic beauty, historical, natural, recreational utility or cultural interest. In this paper, scenic byway regulations, laws and scenic systems existed in western countries are compared with applications and laws in Turkey. Six factors (historic, natural, archeological, cultural, scenic and recreational) which is used to consider project planning of Scenic Roads in U.S. National Scenic Byway Program are investigated. Hence, it is to be encouraged to enhance sustainable applications and be instructive for future studies in Turkey.

1 INTRODUCTION

Tourist travelling by vehicle has fast grown abidingly for several centuries Generally, roads transfer people from one place to another, however drivers do take to the roadway for the pleasure of driving (Richard L Kent, 1993).Furthermore, drivers expect a road to offer fascinating historical sites, spectacular views, or distractive wildlife in enroute. Scenic byways and highways are roadways which provide an opportunity for recreational and travel through lands of significant scenic or cultural interests. It might also offer access to an excitement of outdoor activities or discover cultures, art or structures. Hence, developing, explicating, designating and planning management for scenic routes has become an important issue of interpretive application offered by most land management agencies (Brunswick, 1995). The purpose of this study was to using a benefits-based approach, investigate community perceptions of the advantage or disadvantage of cultural tourism opportunities along a scenic road (Besculides, Lee, & McCormick, 2002).In a survey study, which was conducted for determining different public expectations and opinions about the use of scenic road, so as to promote the aim, an alternative road, Erzurum—Çaykara—Of route, which connects Eastern Black Sea Region to Eastern Anatolia Region, was evaluated. In the conclusion of the paper it was determined that study route choice should be evaluated as a scenic byway or road (Sezen & Yilmaz, 2010). The research which is presented by Clay and Smidt evaluates descriptor variables used by agencies to assess scenic quality along roads. The aim was to determine whether these descriptors, when applied in an expert evaluation, could produce results indicative of public choice for highway conditions (Clay & Smidt, 2004). Investigation done by Kent and Elliott focused on local scenic roads to determine whether the designation process factors were appropriate or not. Preference ratings obtained from statistical analysis indicated that cultural features had a high correlation with landscape choice. Eight criteria, natural and cultural, potentially affecting reactions of the observer were selected for comparison. Thirty-six slides sampling these factors were shown to groups of town residents who rated them for choice (R. L. Kent & Elliot, 1995). Especially for recreational travelers and tourists, generating or construct a scenic route from one region to another is a common trip planning work. In a study which was used to model for route selection, Path Size Logit model, their relative importance and the relevant attributes are identified. In order to determine the effects of different attributes on scenic route choice, three hypotheses are formulated and tested with three Path Size Logit models. The results define several variables of the surrounding environment as significant contributors to planning scenic routes after checking for road type.

Landscape experience is significantly associated with drivers' emphasis route choice viewpoints such as a beautiful scenery from the road, visiting attractions/places, seeing a mountain plateau, travelling through a different landscape, an interest in a less-travelled road, geology and history and appreciation of nice rest areas (Kristian Steen Jacobsen, 1996). Eby and Molnar evaluated data obtained from a survey in the US of motorists in order to better understand the characteristics of people who use scenic byways. The study showed that when planning a route to a destination on an overnight automobile trip, the driving tourist is most concerned with features of the route related to the driving, such as safety, directness, amount of congestion and distance (Eby & Molnar, 2002). Brown describes a study which is conducted in Alaska to evaluate highway qualities for use in scenic byways planning. With this study, while planning and designing scenic route, it's emphasized that road selection and design is done not only oriented toward safety and cost reducing, but also it is very important to encourage people to tourism, scenic and cultural resources, preserve and understand our countries heritage.

2 SCENIC ROADS OF WESTERN COUNTRIES AND TURKEY

The adoption of legislative measures for scenic byways, systematic conservation work about nature and historical areas, doing scientific studies about these kind of projects coincide with the 20th century (Yücel & Babuş, 2005). A National Scenic Byway Program is firstly administrated by the U.S. Department of Transportation. In July 1988, the Federal Highway Administration (FHWA) cosponsored a conference which is called Scenic Byways '88: A National Conference to Map the Future of America's Scenic Roads and Highways. In content of the conference, the FHWA published Scenic Byways as a guide and reference for participants. The program was established by Congress in 1991 to protect and preserve the scenic but often less-traveled roads and elevate tourism and economy (https://en.wikipedia.org; https://www.fhwa.dot.gov).

Too much studies done especially in U.S. and European countries demonstrate that they have paid special attention to Scenic Byway application as planning and designation. They have also a lot of model practices which have visual landscape for scenic byway in its varied regions as seen in Figure 1.

There is no law about planning and construction of scenic roads in Turkey. However there is only the technical procedure communication about scenic roads in the forest road planning, construction and maintenance notification published by Ministry of Environment and Forestry: "While determining the route for national parks and picnic areas, forest area lose have

Figure 1. A lot of model practices which have visual landscape for scenic byway in various countries (https://scenicbyways.info/).

to be minimum, usage and development plans have to considered by taking into account aim of establishment and properties, protecting environment and nature (Anonymous, 2008).

In other words, although there are so many scenic byway applications, in Turkey, there is no specific regulation or standard about routing and planning these roads which is especially in protected areas and national parks. As in U.S. and other European countries which dwell on scenic roads planning. In accordance with this subject, legal and administrative arrangements are required for scenic roads applications in Turkey.

The priority in scenic byway construction should always be protection and enhancement value of the nature. Being a natural bridge between Asia, Europa and Africa, diversity of geological structure and having various climate zones and ecological wealth, Turkey is very rich in terms of biological, visual and historical diversity (Figure 2).Turkey has some regulations on urban planning in the protected area but there is no standard about construction in these kind of areas.

Protected areas can be defined as locations which have to receive protection because of their recognized natural, ecological, cultural, social or architectural values. Protected areas can be examined in the following subjects (Gazette, 1983);

Urban site: is a place which has cultural and natural properties such as architectural, historical, esthetic or artistic.

Archaeological site: is a place in which evidence of past activity is preserved, and which represents a part of the archaeological record such as cultural, social, economic etc. Sites may range from those with few or no remains visible above ground, underground or underwater, to buildings and other structures still in use.

Historic site (heritage site): is an official location where pieces of military, political or social history should be preserved due to their cultural heritage value.

Natural site: is location where a piece of either prehistoric or historic or contemporary nature should be protected due to its precious value (https://en.wikipedia.org). Numerical information about protected areas is given in Table 1.

Figure 2. Some of applications which have visual landscape for scenic roads in U.S. (http://www.milliparklar.gov.tr).

Table 1. Registered site throughout Turkey (http://www.kulturvarliklari.gov.tr, 2015).

Type of site (protected area)	Number of site
Archaeological site	13947
Urban site	267
Historic site	159
Urban and archaeological site	32
Mixed site	77
Convergent site with natural site	358
Total	14840

In Turkey, designs should be created in accordance with the principles and procedures issued by the Ministry of Development.

These following provisions come to the forefront in the principles and procedures which issued in 2014 (DOKAP, 2014).

"In the projects:

1. Protection of natural and historical values,
2. Ensuring compliance with local architecture and materials,
3. Observance of the balance between conservation and usage,
4. The development of sustainable tourism practices,
5. To minimize adverse effects on the environment issues are sought."

Although there are so many scenic road applications in Turkey, there is no regulation or standard about these roads as in U.S. and other European countries. In accordance with the following subjects, legal and administrative arrangements are required for applications in Turkey.

3 VISUAL EFFECTS ON ROAD PROJECT

Roads connect people, life zones, cultures each other. Roads can also be described as not only places that we pass through but also places where we live in it. Travelling by the vehicle has fast grown abidingly for several centuries. Roads constructed is expected to not only transmit people from one place to another but also present fascinating historical sites, spectacular views, or distractive wildlife in travelers' route. Scenic byways are roadways which provide an opportunity for travel through lands and roadside of significant scenic, historical, natural or cultural interests. Rapid increase of the population and fast vehicles which are produced by the help of the new technology make large scale road constructions a current issue on the other hand reveal that the roads are not only technical structures. Roads which cause negativity for the nature should be compatible with the nature and become integrated with the people (Erdem, 2004). Thus, it is necessary in road projects to make assessment from the point of landscape and environment besides architecture, engineering and economic approach (Altınçekiç & Altınçekiç, 1999). There are two important steps in the road planning (Öztürk & Şehir):

1. To ensure adequate, safe and the rapid transport system,
2. To perform these attributes taking into consideration the landscape motifs.

Drives and passengers get visual information about environment while they are travelling. This is important to recognize the environment and enjoy the travel. Also it is necessary for roads to take the travelers to the destination without monotony and emergency brake needs (Özgüç, 1999).

Main factor in road design is speed. Speed affects road width, super elevation and minimum curb radius. There are three main road type which have different landscape character. These are:

1. Low traffic roads which blend into the landscape and serve local needs,
2. One lane roads which need safe stopping sight distance,
3. Both long sight distances and encounter places needed multi line traffic routes.

It is very difficult to adapt the latter road type to the landscape and scenic road standards. Also this kind roads are costly and their potential hazards are higher than others (Seçkin, 1986).

4 PLANNING OF SCENIC ROUTE

The importance of the roads is high in terms of touristic, cultural and socio economic structure for the developing countries. Therefore, there are some standards that must be considered in designing the roads. Roads should be in harmony with the region where they pass through the natural and historical places while they are responding the needs of modern traffic and tourism. Therefore, according to Jellicoe various professional disciplines should work together as shown below in order to ensure reliable, fast and modern traffic patterns (Ertekin & Çorbacı, 2010):

– City planner determines route of the road as needed.
– Engineer draws the route.
– Landscape architect corrects mistakes on the route.
– Engineer draws the vertical curvatures.
– Landscape architect corrects mistakes on the levelling.
– Engineer draw parallel boundaries, bridges and other elements.
– Landscape architects make an organization for the plants which they want to build both side of the road and examine the road for the arrangement for the view. They analyze existing specifications in the boundaries of the road. Thus, instead of splitting landscape into two parts they build a road which is integrated with landscape.

In planning route in scenic roads with regard to landscape; important variables such as vividness, naturalness, unity, and variety come to the forefront. These factors were chosen from a large research (Table 1), which was compiled from many of state and federal highway documents(Clay & Smidt, 2004).

5 CONCLUSIONS AND SUGGESTIONS

Road design is most often oriented toward safety and cost reduction. However, while planning it is very important to encourage people to tourism, understand and preserve our countries heritage, scenic and cultural resources through visual landscaping. Scenic byways are roadways which provide an opportunity for recreational and travel through lands of significant scenic or cultural interest.

In a road project, the planner and decision-maker should pay high attention tothe scenic byway/highway development of Turkey in terms of principles, targets, image positioning, recreational service facilities planning and arranging visual landscape. Highway engineers, urban and regional planners, traffic engineers and landscape architects should work together during scenic roads route determining, designing project and application. In this process, the target should be minimizing the destruction of the nature as much as possible and presenting a safe and enjoyable journey to passengers and drivers (Öztürk & Şehir).

Although there are so many scenic byway applications in Turkey, there is no regulation or standard about these roads as in the U.S. and other European countries. In accordance with the following subjects, legal and administrative arrangements are required for scenic roads applications in Turkey.

– It should be taken into account that the road is to be close to historical places, improving bad buildings near the road, designing the route passing near the mountain slopes instead of mountain foot or summit,
– Six factors (historic, natural, archeological, cultural, scenic and recreational) should be considered in the construction of scenic byway,
– Nature should be both protected and well used.
– Buildings in the highlands should be integrated with nature,
– Rural renewal should be provided as urban renewal,
– Controlled, planned and environmental friendly routes should be created. Also visual landscape should be considered while creating SCENİC ROADS routes.

There are just a few studies on the improvement and application of Scenic Byway in our country. Too many studies done in the U.S. and European countries demonstrate that they have paid special attention to Scenic Byway application. With this study, putting emphasize on relation between scenic roads routing, visual landscaping, it is tried to be encourage to enhance sustainable applications and it will be instructive for future studies in Turkey.

REFERENCES

Altinçekiç, S. Ç. & Altinçekiç, H. (1999). Karayolları peyzaj düzenleme çalışmalarında bitkilendirme esasları. *Journal of the Faculty of Forestry Istanbul University (JFFIU), 49*(1–2-3–4), 99–104.

Anonymous. (2008). *Forest Road Planning, Construction And Maintenance* (B.18.l.OGM.0. 11.04–010.05/).

Besculides, A., Lee, M. E. & McCormick, P. J. (2002). Residents' perceptions of the cultural benefits of tourism. *Annals of Tourism Research, 29*(2), 303–319.

Brunswick, N. A. (1995). *Visitor preferences and scenic byway interpretive design and planning in Logan Canyon.*

Clay, G. R. & Smidt, R. K. (2004). Assessing the validity and reliability of descriptor variables used in scenic highway analysis. *Landscape and Urban Planning, 66*(4), 239–255. doi: http://dx.doi. org/10.1016/S0169–2046(03)00114–2.

Dokap. (2014). Doğu karadeniz turizm master plani ve "Yeşil Yol" bilgileri.

Eby, D. W. & Molnar, L. J. (2002). Importance of scenic byways in route choice: a survey of driving tourists in the United States. *Transportation Research Part A: Policy and Practice, 36*(2), 95–106. doi: http://dx.doi.org/10.1016/S0965–8564(00)00039–2.

Erdem, N. (2004). E-5 otoyolunda ekolojik ve peyzaj yönünden bozulmalar. *Journal of the Faculty of Forestry Istanbul University (JFFIU), 54*(1), 89–104.

Ertekin, M. & Çorbacı, Ö. L. (2010). Landscape Planning and Plantation Studies of Highway. *e-Journal of New World Sciences Academy, 5.*

Gazette, O. (1983). Cultural And Natural Heritage Protection Law. *Prime Publications, 18113,* 5879–5900.

http://www.kulturvarliklari.gov.tr. (2015). *Türkiye Geneli Sit Alanları İstatistikleri.* The Ministry Of Culture And Tourism.

http://www.milliparklar.gov.tr. *National Parks in Turkey.* Ministry Of Forestry And Water Affairs.

https://en.wikipedia.org. National Scenic Byway.

https://scenicbyways.info/. AMERICA'S SCENIC BYWAYS.

https://www.fhwa.dot.gov. History of Scenic Road Programs.

Kent, R. L. (1993). Determining scenic quality along highways: a cognitive approach. *Landscape and Urban Planning, 27*(1), 29–45.

Kent, R. L. & Elliot, C. L. (1995). Scenic Routes Linking and Protecting Natural and Cultural Landscape Features—a Greenway Skeleton. *Landscape and Urban Planning, 33*(1–3), 341–355. doi: Doi 10.1016/0169–2046(94)02027-D.

Kristian Steen Jacobsen, J. (1996). Segmenting the use of a scenic highway. *The Tourist Review, 51*(3), 32–38.

Özgüç, İ. M. (1999). TEM Hadımköy-Kınalı arası peyzaj planlaması üzerinde görsel araştırmalar. *Journal of the Faculty of Forestry Istanbul University (JFFIU), 49*(2), 115–132.

Öztürk, B. & Şehir, E. Ü. M. F. Kent İçi Ve Kent Dışı Karayolu Ulaşım Sisteminde Bitkilendirmenin Trafik Tekniği Yönünden İşlevleri.

Seçkin, Ö. B. (1986). Karayolu ve peyzajı. *Journal of the Faculty of Forestry Istanbul University (JFFIU),* 45–53.

Sezen, I. & Yilmaz, S. (2010). Public opinions about the use of highways as scenic roads: The sample of Erzurum aykara of route. *African Journal of Agricultural Research, 5*(8), 700–706.

Yücel, M. & Babuş, D. (2005). Doğa korumanin tarihçesi ve türkiye'deki gelişmeler.

Advances and Trends in Engineering Sciences and Technologies II – Al Ali & Platko (Eds)
© 2017 Taylor & Francis Group, ISBN 978-1-138-03224-8

The large roundabout adjustment's assessment of effectiveness

V. Krivda, K. Zitnikova, J. Petru & I. Mahdalova
VSB—Technical University of Ostrava, Ostrava, Czech Republic

ABSTRACT: The article describes adjustments of the original large two-lane roundabout and deals with the assessment of effectiveness of these adjustments from several perspectives. This roundabout is made with a spiral arrangement of lanes on the circular line, instead of the original two-lane orbital lane and some of entrances and exits. Moreover, this roundabout was equipped with traffic lights at some entrances. Pedestrian and bicycle traffic were also changed. To assess the effectiveness of these adjustments road safety was monitored by using video analysis of conflict situations and the capacity of the intersection by using of valid Czech technical conditions and simulation software. Some of the capacity assessment calculations of this spiral roundabout with traffic lights had to be modified because calculations in technical conditions are used separately for roundabouts and intersections with traffic lights.

1   INTRODUCTION

The rising intensity of traffic on the road is a constant problem in many developed countries. Traffic engineers along with designers of transport structures must seek for solutions that shall comply with the requirements of safety and traffic fluidity and also the needs of all road users (e.g. Vehicle drivers, cyclists and pedestrians). Intersections in general are the most problematic places of the road. For lower traffic intensity are suitable uncontrolled intersections. In the case of larger traffic intensity it is necessary to proceed e.g. to the construction of roundabouts, implementation of traffic lights, eventually to expensive construction of interchange. It is obvious that every solution has its advantages and disadvantages in safety, capacity, financial or space requirement.

In the case of roundabouts, the construction of which has seen in the Czech Republic (and not only here) in recent years a great boom, you can choose from a number of different types of this particular intersection (single-lane, two-lane, spiral, etc.). The presented article deals with the reconstruction of a two-lane roundabout with six arms into the spiral roundabout, which is also controlled by the traffic light.

It is a roundabout at the entrance to Havirov city in the direction from Ostrava (see Figure 1 left). The traffic intensity of this intersection was track before the reconstruction from the two-lane roundabout (see Figure 2 left) and after the reconstruction into the traffic light controlled spiral roundabout (see Figure 2 right, detail see Figure 1 right).

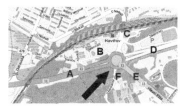

Figure 1.   Surroundings of monitored roundabout and detail of the traffic light entrance to the spiral roundabout (right) (https://www. openstreetmap.org).

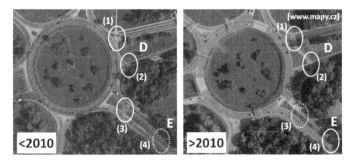

Figure 2. Monitored roundabout before and after reconstruction (for detail see Figure 1 right).

The outer diameter of the roundabout is 140 meters. Before the reconstruction there was two-lane circulatory roadway and some exits with two lanes. After the reconstruction circulatory roadway has spirally arranged three lanes and entrances to the roundabout (A, C and E) are controlled by traffic light (see Figure 1, left).

2 USED METHODS

2.1 Videoanalysis of conflict situations

A conflict situation is a situation on the road, when one of the parties may be at greater risk than usual (Folprecht 1995). In other words, this is a situation that could cause a traffic accident, but we can prevent it, thanks to evasive action or braking. Conflict situations can be divided in many ways. Division by the severity of the conflict is a basic one. A simple violation of traffic rules, usually by solitary participant in traffic is called Potential conflict situation (we called it a severity no. 1). Severity no. 2 indicates that conflict situations, which will take place between two or more participants potentially causing a traffic accident is prevented by small maneuvers or undistinguished braking. If these maneuvers are large or sudden braking is used, in terms of the degree of severity, we call it as conflict situations no. 3. If the accident happened we call these situations as no. 4.

There are many methods for the analysis of conflict situations. The results given in this article is based on research, for which has been used the so called Folprecht Video analysis of conflict situations. Basically this is a technique in which traffic on the monitored intersection (or any other place on the road) is recorded on a video camera and the video is then analyzed. Each situation is described by appropriate symbol, which describes the parties to the conflict, type, origin of the conflict and its severity. The analysis is done as tabular and graphical marks on the ground plan of intersection, where individual marks symbolize frequency of conflict situations and the progress leading to such situations. Subsequently, the statistical evaluation is made, while you also are aware of so called coefficient of relative conflict k_p which specifies the number of observed conflict situations in view of the number of passing vehicles.

An analysis of the conflict situation should be followed by proposals of possible traffic-engineering or construction work on monitored intersection. Besides the conflict situation it is possible to get from the video some other traffic-engineering information (intensity, composition and routing of traffic flow, speed, etc.). Videoanalysis can also be used e.g. during safety inspections on roads. The details on the aforementioned videoanalysis of the conflict situation can be found in (Krivda 2013).

2.2 Capacity assessment according to technical conditions

The capacity of entrances and exits on roundabouts in the Czech Republic is assessed according to technical conditions TP 234 (TP 234, 2011) all these capacities must comply. The capacity of entry is determined using the following equation (TP 234, 2011).

$$C_i = 3600 \cdot \left(1 - \frac{\Delta \cdot I_k}{n_k \cdot 3600}\right)^{n_k} \cdot \frac{n_{i,koef}}{t_f} \cdot e^{-\frac{I_k}{3600}\left(t_g - \frac{t_f}{2} - \Delta\right)} \; [\text{veh/h}] \tag{1}$$

where: I_k = intensity on circulatory roadway [veh/h], n_k = number of lanes on the circulatory roadway [–], $n_{i,koef}$ = coefficient taking into account the number of lanes at the entrance [–] (1.0 for single-lane and 1.5 for two-lanes entrance), t_g = critical time interval [s], t_f = subsequent time interval [s] and Δ = minimum time gap between vehicles traveling on the circulatory roadway behind each other [s].

The capacity on the considered entry must always be greater than the intensity of vehicles (the resulting reserve is positive). The capacity of exiting the roundabout is determined using the following equation (TP 234, 2011).

$$C_e = \frac{3600 \cdot n_{e,koef}}{t_f} \; [\text{veh/h}] \tag{2}$$

where: $n_{e,koef}$ = coefficient taking into account the number of lanes at the exit [–] (1.0 for single-lane and 1.5 for two lanes entrances) and t_f = subsequent time interval [s].

If the exit is followed by a crosswalk and the number of pedestrian exceeds a set limit (according to TP 234 (TP 234, 2011)), then the above mentioned equation will be complement by the influence of the length of the crosswalk, the speed of vehicles and pedestrians, etc. (See TP 234, 2011). The capacity of exit will then be assessed by the degree of load, which is calculated as the ratio of the intensity at the exit and its capacity. If the load factor is less than 0.9, then the capacity of the exit from the roundabout meets its requirements.

2.3 Assessment of traffic using simulation software

Many of the problems in the direct traffic is difficult to analyze, and therefore it is advisable to use sophisticated tools such as after simulation software. PTV VISSIM software was used to create traffic models of various traffic-engineering and construction arrangement of monitored roundabout was. It is a simulation program for the creation of microsimulation models, where we observe certain parameters of the traffic flow. Traffic flow can be considered as system consisting of individual elements, for example vehicles (Celko & Drliciak & Gavulova 2007). Each vehicle is defined by its properties, of which the most important is acceleration (or deceleration) in the immediate surroundings (in the simplest case, the relative speed and distance to the preceding vehicle). Some of the used algorithms use known properties of real flow, others are trying to emerge from properties of real vehicles (their dynamics and the physiological and psychological capabilities of the driver). Thereafter it is possible to use this software to obtain various traffic-engineering data, such as travel time, delay time, queue length, queue stops etc. The data can be exported e.g. to MS Access or MS Excel and processed using appropriate statistical methods (see e.g. Janas Krejsa & M. & V. Krejsa 2015).

3 RESULTS OF ANALYZE OF CONFLICT SITUATIONS

It is generally known that roundabouts with two-lines on circulatory roadway are usually the most dangerous of all types of roundabouts. The following conflict situations were traced during monitoring of the traffic (only a selection of the most frequent) According to general professional experience it corresponds to the respective traffic accidents that occur in this type of intersections:

- Limitation of the space of vehicle traveling in the right lane of circulatory roadway by the vehicle leaving the circulatory roadway from the left lane (see Figure 3-a); both vehicles were going to leave the circulatory roadway in the same exit, but there is only a single-lane for exit.
- An alternative of the previous conflict situation, but with the vehicle in the right lane wanting to continue further along the circulatory roadway (see Figure 3-b); note: there

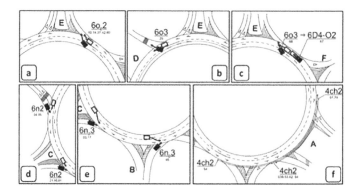

Figure 3.   Selected conflict situation on the roundabout, with two-line circulatory roadway.

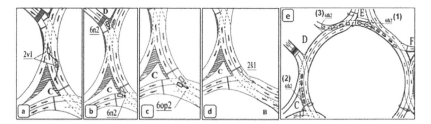

Figure 4.   Selected conflict situation on spiral roundabout.

was a traffic accident during a traffic survey, the driver in the right lane was forced to brake hard and there was a collision of two other vehicles traveling behind it (see Figure 3-c)
– Not giving priority to the vehicle moving on circulatory roadway (see Figure 3-d)
– An alternative of the previous conflict situation, the vehicle on circulatory roadway changed lanes (Figure 3-e)
– Pedestrians walking in close proximity to the circulatory roadway due to the absence of pedestrian crossings (see Figure 3-f).

On spiral roundabout (at a time without active traffic lights) were selected following conflicts (Valecek 2012):

– Driving vehicles on circulatory roadway across the solid line (Figure 4-a)
– Not giving way on circulatory roadway (see Figure 4-b)
– Conflict situations caused by changing lanes on circulatory roadway (see Figure 4-c)
– Passing another vehicle in a lane that is not designated for passing (Figure 4-d).

On spiral roundabout (at a time when traffic lights were activated) there were similar conflict situations as when there were no active traffic lights, while moreover, there are situations associated just with traffic lights (Valecek 2012):

– Vehicle on circulatory roadway was forced to stop due to the lighting of a red signal on the traffic light (see Figure 4-e—No. 1 and 2); Resulting into the queue of vehicles that prevented continuous traffic in another part of the circulatory roadway
– Queue no. 3 (see Figure 4-e) is not related with monitored roundabout, because it is a queue of vehicles, which belongs to the close second roundabout (the so-called not proper conflict situation—see Krivda 2013).

4   CAPACITY CALCULATION

Also, capacity was assessed on all the above mentioned types of roundabouts (on the roundabout with two-lane circulatory roadway, spiral roundabouts without an active traffic light and

Table 1. Results of capacity calculations.

Arm	Roundabout with two-line circulatory roadway		Spiral roundabout		Spiral roundabout with traffic light
	Entrance (reserve)	Exit (degree load)	Entrance (reserve)	Exit (degree of load)	Entrance/circulatory roadway (reserve)
A	Negative	0.80	Negative	0.57	16.2%/51.6%
B	47.9%	0.90	52.7%	0.08	Without traffic lights
C	Negative	0.45	22.6%	0.40	50.5%/49.9%
D	26.7%	0.46	39.6%	0.40	Without traffic lights
E	Negative	1.20	Negative	0.87	10.1%/35.9%
F	22.1%	0.06	33.0%	0.04	Without traffic lights

Table 2. The results from the microsimulation model PTV VISSIM—delay time in seconds.

Arm/Type of roundabout		Exit A	Exit B	Exit C	Exit D	Exit E	Exit F
Entrance A	I	–	27.4	22.3	17.1	11.9	7.8
	II	–	25.4	19.7	15.4	10.0	7.2
	III	–	41.5	37.8	29.9	25.4	21.0
Entrance B	I	6.9	–	26.3	21.2	18.0	10.9
	II	6.1	–	24.2	18.9	16.4	9.4
	III	12.5	–	32.1	23.4	24.9	13.2
Entrance C	I	12.2	7.0	–	25.1	20.7	16.2
	II	10.1	6.7	–	23.4	17.1	14.0
	III	24.3	22.2	–	40.7	38.7	28.1
Entrance D	I	18.1	10.8	8.2	–	28.4	20.3
	II	16.4	8.9	7.8	–	26.6	19.0
	III	24.9	23.4	19.1	–	31.4	27.7
Entrance E	I	22.3	17.7	10.8	7.8	–	29.0
	II	19.7	15.9	9.0	7.0	–	27.5
	III	36.3	30.1	24.5	23.2	–	39.4
Entrance F	I	26.4	23.3	16.1	11.7	6.8	–
	II	24.4	18.6	14.1	10.2	6.2	–
	III	31.6	29.7	22.4	19.7	14.7	–

with an active traffic light. Table 1 provides selected results of calculations by the relevant technical conditions, ie. TP 234 for the roundabouts (TP 234, 2011) and (Methodological Guideline, 2004) and TP 81 and TP 235 for the traffic lights. (TP 81, 1996) (TP 235, 2011).

5 RESULTS OF MICROSIMULATION MODEL

All these types of roundabouts have been modeled in **PTV VISSIM** software, and the traffic in the simulation was analyzed in terms of average delay time for the different traffic flows. Delay time is defined as the difference between actual travel time and the theoretical travel time. Results obtained in microsimulation models are presented in Table 2, where I = Roundabout with two-line circulatory roadway, II = spiral roundabout without active traffic lights, III = spiral roundabout with active traffic lights.

6 EVALUATION AND CONCLUSIONS

In terms of safety assessments it is apparent that roundabouts with two-line circulatory roadway have more serious potential conflict situation than the other two types. A spiral roundabout with a traffic light has the same conflicts as a spiral roundabout without traffic light, but also problems with a possible worsening of traffic flow by stopping vehicles on the circulatory roadway.

Changes were also made for on the pavement of pedestrian roads. Before adjusting the roundabout, there was only one crosswalk on arm D (see Figure 2 left—number 1 and 2) at the shoulder E there was only place for the crossing (see Figure 2 left—figure 3). After adjusting the intersections crosswalk no.2 was also adjusted (see Figure 2 right), Crosswalk no. 1 and 2 are supplemented by crossing for cyclists. Place for crossing (no. 3) was completely cleared and a new crosswalk has been set up no. 4 (see Figure 2 right), which is significantly offset from the roundabout. All transitions already fulfill requirements arising from the intensities of pedestrians and their needs.

Capacity calculations were made according to valid Czech technical conditions and also the values of delay times obtained from the microsimulation models in PTV VISSIM program shows similar results. Roundabout with two-line circulatory roadway has comprehensively worst results. Spiral roundabout without traffic lights (except for two entrances) meet the requirements. Spiral roundabouts with traffic lights are already suitable for all entrances and exits. Putting all the results of capacity calculations (incl. Results from the PTV VISSIM) would be outside the scope of this article, and therefore the Tab. 1 and Tab. 2 contains only the most important results.

Finally, it must be pointed out that the observed intersection, in terms of capacity, traffic fluency and safety is suitable in a different arrangement for example as an interchange (after all of the pictures mentioned at the outset, it is clear, that the width dimensions of arms A and D are allowed with this option). In the 50 s of the 20th century it was prospectively calculated with interchange alternative, but this project never happened. A spiral roundabout with traffic lights is a compromise between safety and capacity on the one hand and financial capabilities on the other. However, due to the increasing traffic intensities, the construction of the interchange will soon be a necessity.

ACKNOWLEDGEMENTS

The work was supported by the Student Grant Competition VSB-Technical University of Ostrava, Czech Republic, in 2016. Project registration number is SP2016/96.

REFERENCES

Celko, Jan & Drliciak, Marek & Gavulova, Andrea 2007. Transportation Plannig Model, In *Komunikacie*, Vol. 9(3): 28–32. ISSN 13354205.

Folprecht, Jan 1995. Method of Monitoring and Assessment of Conflict Situations in Road Transport and Its Importance for Improving Traffic Safety. In *International scientific conference Transport*, Ostrava *2015,* 227–231.

Janas, Petr & Krejsa, Martin & Krejsa, Vlastimil & Bris, Radim 2015. Structural reliability assessment using direct optimized probabilistic calculation with respect to the statistical dependence of input variables. In *25th European Safety and Reliability Conference, ESREL 2015*, 4125–4132. ISBN: 978-113802879-1.

Krivda, Vladislav 2013. Analysis of Conflict Situations in Road Traffic on Roundabouts. In *Promet—Traffic & Transportation*. Vol. 25(3): 295–303. DOI: 10.7307/ptt.v25i3.296.

Methodological Guideline "Large Roundabout", Transport Research Centre, 2004. Available from: http://www.ibesip.cz/data/web/kampane/legislativa/besip-01-VOKMPPROSINEC2004POVTISK.pdf.

TP 81 Design of traffic lights for control of road traffic. Czech technical standards. 1996. ISBN 80-902141-2-6 (in Czech).

TP 234 Assessment of Roundabouts Capacity. Czech technical standards. 2011. ISBN 978-80-87394-02-01 (in Czech).

TP 235 Assessment of Signalized Intersections Capacity. Czech technical standards. 2011. ISBN 978-80-87394-03-8 (in Czech).

Valecek, Lukas 2012. Analysis of Road Traffic Safety on Roundabout Ostravska—Zeleznicaru—U Nadrazi—Orlovska—Hlavni trida—U Koupaliste in Havirov-City. VSB-Technical University of Ostrava (in Czech).

Advances and Trends in Engineering Sciences and Technologies II – Al Ali & Platko (Eds)
© *2017 Taylor & Francis Group, ISBN 978-1-138-03224-8*

Laboratory models of a railway track with the subgrade layer based on rubber granulate

M. Lidmila, V. Lojda & O. Bret
Faculty of Civil Engineering, Czech Technical University in Prague, Prague, Czech Republic

ABSTRACT: This article deals with laboratory verification of a rubber granulate obtained from recycling the tires in the track substructure. The main idea of testing the rubber granulate in a subgrade layer was to verify its influence to parameters of the track substructure. The observed parameters included load-bearing capacity, vertical deformation, frost protection of subgrade surface and its permeability. This paper describes the specification of the rubber granulate origin as well as the geotechnical characteristics. There is the process of construction of the geotechnical models M1 and M2 in an experimental box described and the laboratory tests performed on layers. The conclusion of this article is based on comparison of two different types of geotechnical models with the rubber granulate - one with separate layer and one with mixture of the rubber granulate and crushed stone. The results of load-bearing capacity and the deformation are presented as the main contribution of this paper.

1 INTRODUCTION

The present use of the rubber granulate in railway structures is common in the form of vibration damping mats, rail dampers or noise absorbing boards (Lidmila & Krejčiříková 2014). All of the mentioned applications are based on mixing the rubber granulate with any binder component. Polyurethane (PU) is the most common binder. This production procedure allows prefabricating elements with different shape; on the other hand these solutions are considered as expensive and significantly increase the final costs. Crumbed rubber is usually used in the production of vibration-damping mats for the trackbed. However, the disadvantage worth mentioning is the necessity to interlock mats and avoid gaps between them. The authors were focused on a solution which minimizes connections or joints, ignores the requirement of PU binder and allows the application of crumbed rubber (production of a sub-ballast layer) on site. The hypothesis of the contribution was verified by experimental measurement in laboratory as well as in Horníček & Lidmila (2010). The similar research was also performed in Moghaddas Tafreshi et al. (2014).

2 LOAD-BEARING CAPACITY DETERMINATION OF LAYERS IN THE CZECH REPUBLIC

In order to achieve the required level of load-bearing capacity of the subgrade, other layers and the sub-ballast layer, the substructure is designed in the form of structural layers. The basic functions of structural layers are as follows: to transmit the effects of service load and permanent way load onto the sub-ballast layer surface; to increase the load-bearing capacity of the substructure; to provide rainwater drainage from the sleeper subsoil structure; to improve temperature regime (Lidmila & Krejčiříková 2001).

Mechanical and physical parameters, as well as basic types of structural layers are described in the SŽDC S4 Instruction (2008). The design of structural layers of the substructure is based on the methodology of designing a multi-layer system. The multi-layer system methodology applies the following preconditions: the resulting equivalent modulus of deformation of the entire designed structure on the subgrade level must exceed or equal the required minimum modulus of deformation E given in SŽDC S4 Instruction; the modulus of deformation of the material belonging to the upper structural layer exceeds that of the lower layer.

The following parameters are of essential significance for designing the structural layers of the substructure body: thickness of layer; material type; presumed line speed; value of the modulus of deformation on sub-ballast surface.

The basic criterion used for the identification of the load-bearing capacity of the subgrade and sub-ballast layer surface is the relaxed modulus of deformation. The relaxed modulus of deformation is identified by means of a load test carried out in accordance with the SŽDC S4 Instruction or with the standard ČSN 72 1006 (2015).

The relaxed modulus of deformation E_0 is identified as:

$$E_0 = \frac{1,5pr}{y} \tag{1}$$

where E_o = the relaxed modulus of deformation; p = specific pressure applied on the loading plate; r = loading plate radius (0.15 m); y = load plate push.

The pressure applied on the plate is exerted by a hydraulic press supported on a counterbalance by means of extensions of varying length. In order to load the plate, the selected specific pressure (0.2 MPa) is added incrementally in 0.05 MPa. The loading cycle is repeated two times. The values of the modulus of deformation are defined for the maximum mass per axle—22.5 t (SŽDC S4 Instruction 2008). The minimum required values of the modulus of deformation are given in SŽDC S4Instruction.

3 CHARACTERISTICS OF THE RUBBER GRANULATE AND GEOCELLS

The rubber granulate have not been described in any technical standard or other regulation in countries of the European Union, neither in the field of material properties nor nomenclature. In the current research and in industry worldwide this material is considered as the bulk material generated from crushing tires at ambient temperature. The grain size averaged 5 with a 10 mm maximum. The rubber granulate consisted of 2% textile fibres and contained no metal wires and thus the risk of ground currents spreading should be eliminated (Montstav CZ 2016). The grain size distribution of the used rubber granulate was determined in accordance to the standard EN ISO 14688 (2002) intended for soils. On the base of the test the characteristic grain size curve (Fig. 1) was plotted.

The grain size curve showed that 95% weight of the rubber granulate grains was the size from 2 to 8 mm as well as that 50% weight of the rubber granulate were grains with size from

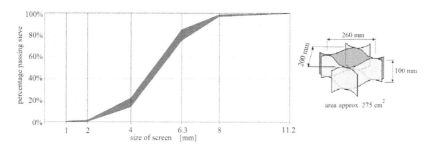

Figure 1. The grain size curve of the rubber granulate and geocells used in the sub-ballast layer.

4 to 6 mm. On the base of the performed grain size test in further text this material is identified as the rubber granulate 2/8 (RG2/8).

The used geocells filled with theRG2/8 (Fig. 1) are called Geokrata tabbos and are made of high-density polyethylene (HDPE) with the height of 100 mm. The tensile strength declared by its producer is 15 kN·m⁻¹.

4 TESTING THE RUBBER GRANULATE IN SUB-BALLAST LAYER

The laboratory test was performed in a geotechnics laboratory box (Fig. 4) at Faculty of Civil Engineering, Czech Technical University in Prague which enables the laboratory model of construction in the scale of 1:1. The box's inner dimensions are 1.8×1.3 m.

There were total of two laboratory models of the railway track with the RG2/8 in this research (Figs. 2–3). The difference between these two models was based on the mixing of the RG2/8 with crushed stone. The first one was based on the RG2/8 and the crushed stone (grain size 0–32) both placed separately (model M1). The model M2 was based on the mixture of the RG2/8 and crushed stone (grain size 0–32). This model followed the research mentioned in Moghaddas Tafreshi et al. (2014).

The geotechnical model M1consisted of 200 mm of cohesive soil simulating the active zone of the trackbed, geocells with a height of 100 mm filled with 50 mm of the rubber granulate the RG2/8 and 50 mm of crushed stone (grain size 0–32), 100 mm of crushed stone (grain size 0–32) simulating the sub-ballast layer.

In relation to the geotechnical model M1 the geotechnical model M2 varied in the way of the geocells filling. There was a mixture of the RG2/8 and crushed stone (grain size 0–32) with a total height of 100 mm. The rest of the model M2 was the same as in the model M1.

Soil simulating the active zone of the trackbed was placed on the bottom of the experimental box. In accordance to EN ISO 14688 (2002) standards this soil is classified as a sandy

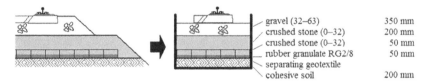

Figure 2. The geotechnical model M1 based on placing the RG2/8 and crushed stone in two separate layers.

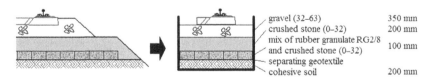

Figure 3. The geotechnical model M2 based on mixing the RG2/8 and crushed stone in one layer.

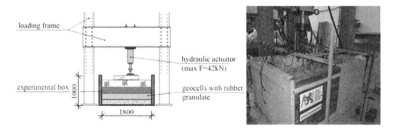

Figure 4. The test set-up of the geotechnical models M1 and M2 and the experimental box.

silty clay, which is not permeable and is highly susceptible to frost. On the subgrade were performed three static load tests (as described in chapter 2). Then the subgrade was covered with separating geotextile to avoid blending with upper layer.

The construction of the next layer consisting the RG2/8, crushed stone (grain size 2–8) and geocells differed in the geotechnical model M1 and M2. The geocells in the geotechnical model M1 were filled with the RG2/8up to the half of geocells' height (50 mm before compacting). The remaining volume of the geocells was filled with crushed stone (grain size 2–8) moistened to optimal moisture in accordance to standard EN 13286-2 (2011).The ratio of thickness of crushed stone and RG2/8 in the geocells was 1:1 before compaction and 1:2 after compaction. Since the filling of the geocells was compacted, there was an extra layer of crushed stone with a height of 20 mm on top of the surface of the geocells.

The geocells filling in the geotechnical model M2 was based on the same components as in the geotechnical model M1 but theRG2/8 and crushed stone were turned into the mixture (ratio 1:2). The whole volume of the geocells was filled with this mixture. On the base of the deformation modulus resulting from the static load test performed on the geocells layer it was estimated that for the sufficient deformation modulus (SŽDC S4 Instruction 2008)it is necessary to add a layer of crushed stone (grain size 0–32) with 100 mm high. The final layer of both the geotechnical models M1 and M2consisted of railway ballast containing gravel with a depth of 350 mm (grain size 31.5–63) prepared with accordance to standard EN 13450 (2004). There was a half of concrete sleeper type B 91 S with elastic fastening W14 and rail 60 E1.

The loading of the geotechnical models M1 and M2 was applied with actuator Inova AH 1000 (Fig. 4). The actuator simulated the dynamic loading caused by rolling stock axles. The force varied in the range of 2 and 42 kN with a frequency of 3 Hz. The maximum force 42 kN was determined from the calculation of axle load distribution in railway track (Krejčiříková, unpubl.). A total of 500 000 loading cycles were performed. Due to measuring the consequence of dynamic loading, the loading cycles were paused at the point of 100, 1 000, 10 000, 200 000, 250 000 and 500 000 cycles and the vertical displacements of subgrade (with the RG2/8), sub-ballast layer and the top of sleeper were measured. When the dynamic loading was stopped the vertical displacements were measured at steps of static force 0, 10, 20, 30, 42 kN.

5   DISCUSSION

The primary criterion was the assessment of load-bearing capacity in accordance to SŽDC S4 Instruction (Fig. 5) and the comparison of the models M1 and M2 as well as the vertical displacement development of layers (Figs. 6–7)measured before and during dynamic loading.

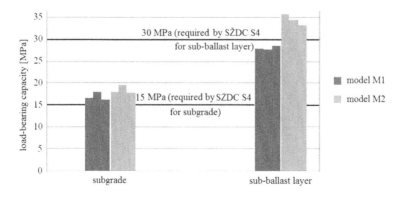

Figure 5.   Load-bearing capacity of the geotechnical models M1 and M2 complemented with requirement according to SŽDC S4 Instruction.

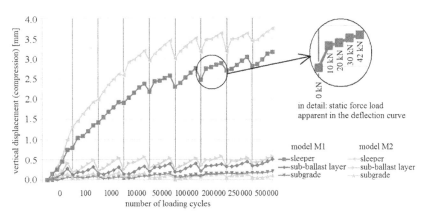

Figure 6. Increase of the vertical displacements in certain moments of the cyclic load.

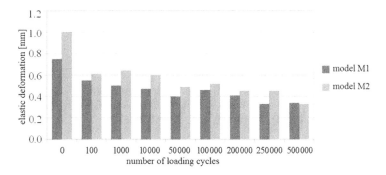

Figure 7. The decreasing trend of elastic deformation related to the amount of performed loading cycles.

The subgrade of both geotechnical models M1 and M2 achieved comparable deformation modulus (17.0 MPa and 18.5 MPa in average) and thus the load-bearing capacity was suitable in accordance to SŽDC S4 Instruction (15 MPa required). The deformation modulus of sub-ballast layer required by SŽDC S4 Instruction is 30 MPa. The results of the geotechnical models M1 and M2 (28.1 MPa and 34.5 MPa in average) were considered as satisfactory.

The observation of the vertical displacement on the subgrade, sub-ballast layer and on the top of the sleeper measured in certain moments of the cyclic load is shown in Figure 6.

There is shown that the geotechnical model M1 resulted lower values of the deformation after dynamic loading than the M2. The most significant deformation was in the initial cycles, where the value of the deformation for 100 and 1 000 cycles was even greater, than the permanent deformation added with the elastic deformation under load of 42 kN from the previous cycle.

The development of the elastic deformation measured on the top of sleeper is described in Figure 7. Elastic deformation was always determined as a difference of deformation between maximal load, i.e. 42 kN and unloaded model, i.e. 0 kN after certain loading cycles.

The more loading cycles were performed, the more the linear deformation decreased. Based on this comparison the more appropriate model was again the geotechnical model M1.

6 RESULTS

The evaluation of this research proved the estimation that the layer of the rubber granulate (the RG2/8) with a thickness 40 mm filled in plastic geocells does not deteriorate the load bearing capacity required with SŽDC S4 Instruction. The use of the RG2/8 in the sub-ballast layer may improve the properties of railway track such as reduction of vibration from the railway transport to the environment, the frost protection of the sub-ballast surface and the water drainage.

The two geotechnical models M1 and M2 dissimilar in the way of the RG2/8 application also proved that there is different behaviour in relation to placing the rubber granulate. Based on requirements for the structure of sub-ballast layers the application of the RG2/8 as separate layer (M1) was more favourable. The application of the rubber granulate in the form of mixture resulted as less appropriate.

When comparing the results achieved in this research and in the experiment referred in Moghaddas Tafreshi et al. (2014) the authors observed similar behaviour and results.

Authors estimate that the different thickness of the rubber granulate may influence the overall elastic deformation of the railway sub-structure. This option has potential for the application in a railway track with a sudden change of rigidity of the sub-ballast layer, e.g. in transition area of railway bridges, railroad crossings, turnouts or ballastless track system. At present such spots of the railway tracks are treated with anti-vibration mats, under sleeper pads or railway ballast glued with polymers. TheRG2/8 filled in geocells would be a technical alternative to mentioned measures.

ACKNOWLEDGEMENT

The paper was prepared with the support of the program Competence Centres of Technology Agency of the Czech Republic (TAČR) within the Centre for Effective and Sustainable Transport Infrastructure (CESTI), project number TE01020168.

REFERENCES

Horníček, L. & Lidmila, M. 2010. Simulation of long-term behaviour of antivibration mats from rubber recyclate by means of cyclic. In Proceedings of International Conference on Modelling and Simulation 2010 in Prague; *Proc. International Conference on Modelling and Simulation 2010, Prague, 22–25 June 2010*. Prague: Czech Technical University in Prague.

Krejčiříková, H. unpublished.*Calculation of load distribution in relation to the type of super-structure, sub-structure.*, Prague: Faculty of Civil Engineering, Czech Technical University in Prague.

Lidmila, M. & Krejčiříková, H. 2001.Classification of the Load-Bearing Capacity of Substructure in Laboratory Conditions. In CTU Reports 2-3/2001: Contributions to Experimental Investigation of Engineering Materials and Structures; Proc. Workshop 2001—Contributions to Experimental Investigation of Engineering Materials and Structures, Prague, 5–7 February 2001. Prague: Czech Technical University in Prague.

Lidmila, M. & Krejčiříková, H. 2014.Noise barrier of recycled rubber. In Týden výzkumu a inovací pro praxi 2014; Proc. Týden výzkumu a inovací pro praxi 2014-, Hustopeče, 23–25 April 2014. Hustopeče: Vladimír Študent.

Moghaddas Tafreshi S. N., Khalaj, O., Dawson A.R.2014.Repeated loading of soil containing granulated rubber and multiple geocell layers.*Geotextiles and Geomembranes* 42(1): 25–38.

Montstav CZ, 2016. *Recyklace—Montstav CZ.* http://www.montstav.cz/recyklace (accessed Apr. 20, 2016).

SŽDC. 2008. SŽDC S4 Instruction: 2008: Railway substructure.

ÚNMZ. 2002. EN ISO 14688-1: 2002: Geotechnical investigation and testing—Identification and classification of soil—Part 1: Identification and description.

ÚNMZ. 2004. EN 13450: 2004: Aggregates for railway ballast.

ÚNMZ. 2011. EN 13286-2: 2011: Unbound and hydraulically bound mixtures—Part 2: Test methods for laboratory reference density and water content—Proctor compaction.

ÚNMZ. 2015. ČSN 721006: 2015: Compaction control of engineering fills.

Advances and Trends in Engineering Sciences and Technologies II – Al Ali & Platko (Eds)
© 2017 Taylor & Francis Group, ISBN 978-1-138-03224-8

Natural asphalt additives for road surface layers

J. Mandula & T. Olexa
Civil Engineering Faculty, Institute of Structural Engineering, Technical University of Košice, Košice, Slovakia

ABSTRACT: New progressive construction materials use environmentally-friendly, recycled additives to improve material properties. The aim of these additives is to decrease the price and to increase the bearing capacity of the construction structure. Asphalt mixtures are also modified with different kinds of additives. This study focuses on natural bitumen additives. Natural bitumen was used as the first kind of binder in asphalt mixtures. It is a high-class material and its deformation properties are better than those of refined bitumen. Sources of this material are limited, which makes the price of natural bitumen much higher than refined bitumen. In recent times different kinds of additives based on natural bitumen have become available, which should increase the quality of classic refined road bitumen with economic efficiency. The study focuses on three different additives based on Trinidad natural bitumen, tested for chosen deformation parameters.

1 INTRODUCTION

The endurance of pavement depends on its component resistance to different negative influences of traffic or weather (Demjan I. & Tomko M. 2013). Bitumen binder is the most frequently used binder for road constructions because of its viscous elastic behavior. Viscosity behavior during high temperature periods causes rehabilitation of bitumen mixtures, but on the other hand it also causes permanent deformation of pavements. Bitumen behavior at low temperatures changes to elastic, and in this case bitumen mixture inclines to frost cleft of the surface road layer because of its high stiffness. Negative influences of bitumen in the mixture can be reduced by using different kinds of additives. During the last decade there have been different types of additive like crumb rubber, polymers, synthetic additives and also natural bitumen. This paper focuses on a combination of natural asphalt with synthetic adhesive additive. This combination should sustain the full potential of natural bitumen additives.

1.1 *Natural bitumen*

Natural bitumen has been involved in many research projects, but these studies concentrated mainly on the chemical and structural composition of natural bitumen. The definition of natural asphalt according to a study of Richard F. Meyer (Meyer & Witt 1990) describes natural asphalt as bitumen created in the process of crude oil processing. Natural bitumens are native substances of variable color, hardness and volatility, composed principally of the elements carbon and hydrogen, and sometimes associated with mineral matter and essentially free of gas. The non-mineral constituents are largely soluble in carbon disulfide.

The definition of natural bitumen according to T.F. Yen (Yen 1984) describes this material as semisolid or solid mixtures of hydrocarbons and as much as 50 percent heterocyclic compounds constituted largely of carbon and hydrogen but having substituents of sulfur, oxygen, nitrogen, and trace metals, especially iron, nickel, and vanadium, in the carbon network. In sedimentary rocks, natural bitumen is the soluble (in organic solvents) portion of disseminated organic matter. Part or all of the bitumen may be expressed from sedimentary rock (the source bed or rock) through primary migration to form natural bitumen deposits. The natural bitumen deposits may be subsequently altered to form different varieties of natural bitumen, crude oil, or natural gas. Some crude oil or gas also can evolve in the source rocks and subsequently migrate to form oil or gas reservoirs (pools). The oil or gas thus accumulated may be altered to create reservoir bitumen or, ultimately, natural gas or graphite (Meyer & Witt 1990).

The best known source of natural asphalt is Pitch Lake in Trinidad where natural asphalt has been extracted since 1792. Trinidad natural asphalt was first used as material for water-proofing of boats and bowls. Later when the first asphalt pavements were constructed, natu-ral asphalt was used as the binder for asphalt mixtures. The properties of natural asphalt make it suitable for using in asphalt mixtures but because of the limited sources its price has risen up and refined bitumen albeit with worse properties has replaced it. Natural asphalt is stiffer than refined bitumen and its softening point is also higher. Nowadays natural asphalt is offered as an additive to refined bitumen which is a more ecological way of improving asphalt properties than using other additives (Mandula 2015).

1.2 *Natural bitumen additives*

Trinidad natural bitumen is processed into three different types of additive. Basic natural bitumen is black sticky or stony material which has to be customized for the purpose of reducing clumps.

The first used type of additive in the study was Trinidad Epuré—Z (TEZ). TEZ is natural bitumen pulverized into particles of size 0–8 mm and coated with stone powder. This type has a wide field of application, for example it could be used in mastic asphalt mixtures, asphalt con-crete mixtures or stone mastic asphalt mixtures. A close-up of a sample can be seen in Figure 1.

Another kind of natural bitumen additive is Trinidad Epuré with cellulose fibers (NAF). NAF is a combination of natural Trinidad bitumen (45%), stone powder (38.5%) and also cellulose fibers (16.5%). The cellulose fibers in this type of additive serve as elastic filler in the mixture and also prevent clump creation. The content of stone powder is lower than in TEZ additive and because of this manipulation with this material is simpler. This type of natural bitumen could be used in stone mastic asphalt and also asphalt concrete mixtures. A close-up showing the structure of NAF can be seen in Figure 2.

The last used type of additive is Trinidad Epuré with paraffin particles (TENV). In gen-eral this type is similar to TEZ, but it also contains paraffin balls and stone powder, and the

Figure 1.   Trinidad Epuré—Z 0/8.

Figure 2.   Trinidad Epuré with cellulose fibers.

Figure 3. Trinidad Epuré with paraffin particles.

volume of natural bitumen is 54%. The purpose of the paraffin balls in this additive is to improve viscosity parameters of the asphalt mixture during low temperature mixing. Low temperature mixtures reduce production heating by 25%. TENV is mainly used in mastic asphalts and it could be combined with polymer modified bitumen or road bitumen binder. Production temperature in this study was set to 160°C for each additive and this temperature should not influence the quality of mixture with TENV addition. A close-up of TENV additive with paraffin balls can be seen in Figure 3.

2 EXPERIMENTAL PARAMETERS

2.1 Asphalt mixture

The studied additives were chosen to improve the deformation resistance of asphalt mixtures in the road surface layer. The surface layer is the top layer of the road and it is influenced not only by traffic loading but also by climatic extremes. The reference mixture was asphalt concrete with an 11 millimeter maximum aggregate sizeand continuous gradation curve. The binder used in the mixture was road bitumen with penetration 50/70 and softening point 49°C. Based on previous research results, the mixture also contained WetFix BE (WFB) synthetic additive which improves the adhesive properties of binders and also improves the coating of aggregate. The aggregate gradation curve was set beyond the minimum limit curve and below the Fuller curve for the purpose of the proper distribution of aggregate grains.. The composition of the asphalt mixture is illustrated in Figure 4.

The amount of additive was the same for each type for the purpose of similar comparison conditions. Two percent of additive is higher than the recommended amount, but previous research (Mandula 2015) clearly showed that 1.5% of natural bitumen is not enough. All mixture components were heated to 160°C for four hours and during the last hour the additive was added to the aggregate of fraction 8–11 mm. After heating all the components were mixed for 5 minutes and treated. All studied mixtures are described in Table 1.

The second reference mixture had the same gradation curve and parameters, but the polymer modified bitumen binder was used. This mixture was considered as a high-class variant and this study should show if natural bitumen can improve an ordinary mixture to the same level.

2.2 Studied parameters

Standard testing of bitumen-bonded mixtures includes only a few empirical tests. The aim of this study was to test not only the empirical parameters of the mixture, but also permanent deformation resistance and complex modulus as bearing parameters of the mixture.

Empirical tests were done according to European standards and Slovak technical manuals. Testing of maximum density, volume of air void content, volume of gaps filled with bitumen was done using standard scales and vacuum pump to get maximum density of the material. Two types of samples were made, the first being Marshall samples with 100 mm diameter and

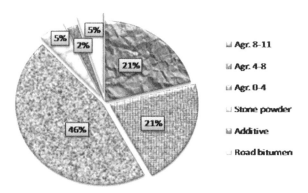

Figure 4.   Structure of the studied asphalt mixture.

Table 1.   Studied mixture variations.

Mark	Binder content	Binder type		Additives
R1	5.6%	Road bitumen 50/70		WFB 0.2%
R2	5.6%	Polymer modified bitumen	25/55	
A1	4.5%	Road bitumen 50/70		WFB 0.2% + NAF 2%
A2	4.5%	Road bitumen 50/70		WFB 0.2% + TEZ 2%
A3	4.5%	Road bitumen 50/70		WFB 0.2% + TENV 2%

the second was bulk asphalt mixture material. Marshall samples were also used for testing the Indirect Tensile Strength (ITS) of compacted mixture. This kind of test was done for dry and wet samples, wet samples were tempered for 72 hours in water at a temperature of 40°C. The ratio of wet to dry sample strength is called the indirect tensile strength ratio (ITSR) which describes the water resistance of the mixture.

Measurement of permanent deformation resistance required plate samples. These plates were loaded by a passing wheel over 10,000 cycles in a chamber with temperature of 50°C. The testing temperature was same as the softening point of classic road bitumen binder. This fact caused deep permanent deformations in the mixture without any additives.

Complex modulus measurement is the only test where limit values are not fixed. Testing conditions were set according to the standards for asphalt mixture testing, and for this study the indirect tensile stiffness method was chosen. This method required Marshall samples with a 100 mm diameter and average height of 50 mm. The sample was loaded in five loading cycles with half-sine shape with frequency around 8 Hz and temperature 20°C.

3   STUDY RESULTS

3.1   *Empirical parameters*

The presented empirical parameters of reference and modified mixtures fulfilled all standard requirements, which meant that each of them could be used for low-traffic road constructions. In Table 2 it can be seen that the addition of natural bitumen improved mainly indirect tensile strength, but also the air void content was good enough. The ITSR values indicate that mixtures with natural bitumen addition should be more resistant to degradation caused by water. In this case a frost resistance test could also be done, but it is not considered as a standard test.

3.2   *Wheel tracking deformations*

The results of the wheel tracking test clearly show that the reference mixture without natural bitumen inclined to permanent deformation. Test results for the reference mixture with road bitumen

Table 2. Results of empirical testing of studied mixtures.

Mix	Air void content (%)	ITS dry (MPa)	ITS wet (MPa)	ITSR (%)
R1	3.6	1.3	1.1	86.9
R2	3.8	1.4	1.4	100
A1	3.4	1.4	1.4	100
A2	3.8	1.8	1.8	100
A3	3.3	1.9	1.7	89.3

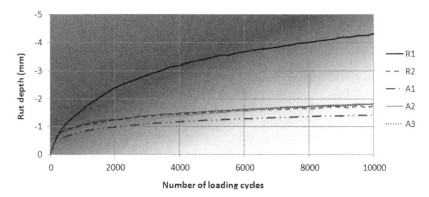

Figure 5. Development of rut depth in the studied mixtures.

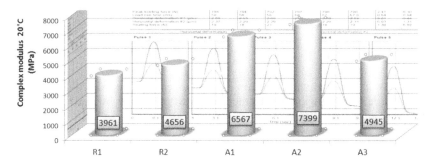

Figure 6. Complex modulus of the studied mixtures.

were beyond the limited standard values, but other mixtures were below the maximum limit. The wheel tracking test simulates the real situation of road construction loaded by wheels of heavy vehicles. Figure 5. Shows the development of tracking depth. Depending on rut depth values, other parameters like wheel tracking slope and proportional rut depth could also be calculated. Based on the presented results, the combination of WFB and NAF additives was the most effective.

3.3 Complex modulus

The second functional test was a measurement of complex modulus (stiffness modulus) using the method of indirect tensile stiffness. Complex modulus is calculated from a real element of complex modulus and a virtual element, and it could be considered as the elastic modulus of the material. According to the graph in Figure 6., the most resistant mixtures were A1 and A2. In comparison with mixture R2, which should have the best results, the complex modulus of natural bitumen modified mixtures was much higher, almost 40%. For example high modulus mixtures used in base layers of pavement have complex modulus values of around 9000 MPa and the studied mixture A2 had a complex modulus of 7399 MPa. Mixture A3 had lower values, but even so it was better than R2 mixture stiffness.

The functional tests used here should indicate which mixture variant is most suitable for using in the surface layer of pavement. Based on the good results of these tests, measurement of the fatigue behavior of these modified mixtures could also be done. High complex modulus values indicate that the fatigue lifespan of these mixtures could be more than average values.

4 CONCLUSION

The positive influence of natural Trinidad bitumen on asphalt mixtures has already been discussed in a study by R. Paul (Paul et. al. 1985), but in this study the content of natural bitumen was much higher and natural bitumen was also used in pure form. Other studies (Min-Chin Liao et. al. 2014; Rui Li et. al. 2015) mostly focused on the influence of natural Trinidad bitumen on the properties of road bitumen binders. In those studies it was clearly proved that this kind of additive improves mainly viscous elastic parameters, and it can also effectively ameliorate the resistance performance of thermal oxidative and UV aging.

The combination of deformation test results in this study highlighted mainly additive NAF (Natural asphalt Trinidad Epuré with cellulose fibers), which clearly improved permanent deformation resistance, but its complex modulus was also considered as suitable, and its indirect tensile strength was 1.42 MPa. Its empirical parameters were absolutely suitable according to European standards, for example the air void content in the mixture was only 3.4% and water sensitivity indicated by the indirect tensile strength ratio was around 100%. Our study conclusions indicate that this asphalt mix can well resist a combination of negative climatic and traffic influences.

Almost all studied mixtures were suitable for using in pavement surfaces, except the mixture with only WetFix addition. On the other hand, using WetFix additive helped natural asphalt additives to react with the road bitumen binder, and it is known from other studies that a combination of road bitumen binder with natural asphalt additive is not sufficient.

ACKNOWLEDGEMENTS

The contribution was incurred within Centre of the cooperation, supported by the Agency for research and development under contract no. SUSPP-0013-09 by businesses subjects Inžinierske stavby and EUROVIA SK.

The research was performed within project NFP 26220220051, Development of Progressive Technologies for Utilization of Selected Waste Materials in Road Construction Engineering, supported by the European Union Structural Funds.

REFERENCES

Demjan, I. & Tomko, M. 2013. Frequency of the analysis response of reinforced concrete structure subjected to road and rail transport effects. *In: Advanced Materials Research: SPACE 013: 2nd International Conference on Structural and Physical Aspects of Civil Engineering:* High Tatras, Slovakia, 27–29 November 2013. Vol. 969 (2014), p. 182–187.—ISBN 978-303835147-4 - ISSN 1022-6680.

Mandula, J. & Olexa, T. 2015. Permanent deformation resistance of asphalt mixtures with addition of natural asphalt. *International Multidisciplinary Scientific GeoConference Surveying Geology and Mining Ecology Management.* Volume 1. Issue 5 (2015). pp. 623–630.

Meyer, R. F. & Witt, W. Jr.1990. Definition and World Resources of Natural Bitumens. *Geological Survey Bulletin. U.S., No. 1944.* pp. 13.

Min—Chin, L. & col. 2014. Rheological behavior of bitumen mixed with Trinidad lake asphalt. *In: Construction and Building Materials 66 (2014).* pp. 361–367.

Paul, R. H. & Kemp, S.F. 1985. Laboratory evaluation of Trinidad lake asphalt—Final report. *Louisiana department of transportation and development. U.S.* pp. 44.

Rui, L. & col. 2015. Experimental study on aging properties and modification mechanism of Trinidad lake asphalt modified bitumen. *In: Construction and Building Materials 101 (2015).* pp. 878–883.

Yen, T.F. 1984. Characterization of heavy oil. The future of heavy crude and tar sands—International conference 2D. Caracas. (1982). pp. 412–423.

Advances and Trends in Engineering Sciences and Technologies II – Al Ali & Platko (Eds)
© 2017 Taylor & Francis Group, ISBN 978-1-138-03224-8

Asphalt belt sliding joint – possible extension of use due to electric target heating

P. Mateckova, M. Janulikova & D. Litvan
VSB—TU, Ostrava, Czech Republic

ABSTRACT: Rheological bitumen sliding joints are utilized for the elimination of friction in footing bottoms of areas affected by underground mining where horizontal terrain deformation is expected. At the Faculty of Civil Engineering, VSB—Technical University of Ostrava different types of asphalt belt have been tested since 2010. The influences of vertical and horizontal load and the effect of temperature in a temperature controlled room have been examined. Currently targeted heating of an asphalt belt used in a specimen of sliding joint was tested. Asphalt belt heating is provided by a power grid with electric resistance wire. A sliding joint specimen is exposed to a vertical and a horizontal load and heated repeatedly in stages. The test results are being used to prove wider utilization of a sliding joint e.g. to eliminate the horizontal friction in the case of foundation prestressing and also for mathematical modelling of soil-structure interaction. Sliding joints should contribute to the design of more durable and sustainable building structures.

1 INTRODUCTION

Building structures in areas affected by underground mining demand specific treatment due to expected terrain deformation. Terrain deformation comprises subsidence, declination, curvature, and horizontal deformation. The most demanding, and also most expensive, are requirements for terrain horizontal deformation. One of the reasons for this is that, through the friction between subsoil and foundations, the foundation structure has to resist significant normal forces.

A simple method for appointing the shear stress between a foundation and subsoil without a sliding joint is offered in the valid Czech code (CSN 730039, 2015). Shear stress between a foundation and subsoil and consequently the normal force in a foundation structure are settled as a function of terrain horizontal deformation ε, the dimensions of the foundation structure and oedometric modulus of subsoil.

The idea of sliding joints between subsoil and foundation structure, which eliminates friction in the footing bottom, comes from the 1970's; see Figure 1 (Balcarek & Bradac 1982).

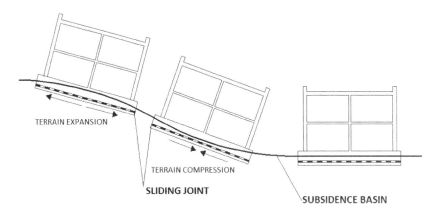

TERRAIN EXPANSION

TERRAIN COMPRESSION

SLIDING JOINT

SUBSIDENCE BASIN

Figure 1. Sliding joint function.

In the beginning there were several materials considered, (e.g. the use of cardboard with ash, isinglass, graphite). Finally, the bitumen asphalt belt, a widely available and reasonably priced material given its rheological properties, has been proven as an effective material for sliding joints. Next, when the term sliding joint is mentioned, a bitumen asphalt sliding joint is considered. Bitumen sliding joints were applied in a few buildings situated in a region affected by underground mining (Cajka et al. 2014).

Application of a sliding joint is effective also where there is horizontal deformation of foundations due to shrinkage, creep, pre-stressing and temperature variation.

2 ASPHALT BELT TESTING

2.1 Description

When using an asphalt belt as a sliding joint the shear resistance is the main material characteristic. In the primary testing in the 1980's it was discovered that the shear resistance of sliding joints was primarily dependent on deformation velocity (Balcarek & Bradac, 1982). When the deformation velocity is slow the shear resistance of the bitumen sliding joint is low.

New testing began at the Faculty of Civil Engineering in 2008. Deformation velocity was measured for different shear stresses. Shear resistance for expected deformation rate is appointed using linear regression.

Asphalt belt specimens are placed in between concrete blocks with a dimension of $300 \times 300 \times 100$ mm. Specimens are exposed to a vertical load. A horizontal load is applied after a one-day delay. Displacement of the middle concrete block is measured; see Figure 2. The specimens were exposed to 100 kPa and 500 kPa vertical loads, i.e. values that correspond to the expected stress in the footing bottom. The horizontal load is between 1.0 and 2.0 kN, such that the deformation velocity corresponds to expected terrain deformation velocity.

2.2 Different types of asphalt belt

Primitive asphalt is refined with oxidization or modified with an admixture of polymers. Depending on the type of admixture polymer, there are asphalt belts modified with rubber, usually Styrene-Butadiene-Styrene (SBS asphalt), and thermoplastics, mostly amorphous polypropylene (APP asphalt). Oxidized and modified asphalts possess different temperature sensitivity, elasticity and plasticity or adhesiveness also in correlation with the amount of admixture.

Figure 2. Shear resistance testing.

Consequently, the asphalt belts show different rheological shear characteristics for the group of oxidized bitumen asphalt belts, SBS asphalt belts and APP asphalt belts.

Until now, fourteen types of different trademark bitumen asphalt belts have been tested since the year 2008, four of them oxidized, nine types SBS modified and 1 type APP modified. The thickness of asphalt belts is between 3–5 mm, and they have been predominantly covered with mineral gritting. Generally, tested SBS asphalt belts show higher deformation than oxidized asphalt belts. Particular experiment results were published in several papers, e.g. (Cajka & Manasek 2006, 2007, Cajka et al 2011, 2012, Janulikova 2014, 2015, Janulikova et al. 2013, 2014).

2.3 Temperature-controlled room

One of the important factors that affect the rheological shear resistance of a sliding joint is temperature. Selected materials have been tested in a temperature- controlled room, and exposed to the temperatures from −10°C to +20°C. It was proven that higher temperatures lead to lower shear resistance, both for the group of oxidized and SBS modified asphalt belts (Cajka & Mateckova 2011).

The research team's aim was to prove the slide joint shear resistance for temperatures expected in a footing bottom (Cajka & Fojtik 2013, Cajka 2014, Cajka et al. 2014). Both the temperature during the first days after concreting, and also the long-term temperature, are important. The temperature during the first few days after concreting is influenced not only by environmental temperature but also by heat caused by hydration. Sliding joint shear resistance during the time after concreting is important for the elimination of cracks in the footing bottom due to shrinkage and possible pre-stressing. Long-term temperatures are important for calculating the shear resistance of the sliding joint to eliminate long-term horizontal deformation due to undermining or creep.

Favourable asphalt belt shear characteristics at higher temperatures lead to the idea of the temporary target heating of a slide joint, e.g. during the foundation pre-stressing.

3 TARGET HEATING

3.1 Description

An asphalt belt sliding joint is part of the foundation structure and its target heating during the period of increasing horizontal deformation, (in the case of undermining after months or years after the building's commissioning), is a complex problem. The heating device's arrangement has to resist significant vertical and horizontal pressure and also endure load during the foundation's execution.

It was decided to examine an SBS modified specimen with heating provided by a power grid with electric resistance wire, in between two layers of asphalt belt; see Figure 3. Safe low electrical voltage was used.

3.2 Particular test results

The testing principle was described in Chapter 2.1. Specimens with a power grid were exposed to a vertical load of 500 kPa. After a one day delay, a horizontal load of 1.0 kN was also applied, which causes a shear stress of 5.6 kPa. Measuring equipment was placed in a temperature-controlled room with a steady constant temperature of 4°C. Horizontal deformation of the middle concrete block was measured for over 100 days and, during this period, specimens were heated in four stages to a temperature of 20°C. The deformation rate responded to the specimen heating and increased as expected. Visible in Figure 4 is the increasing deformation rate during the first and second heatings at 2 days and 7 days, while in Figure 5 it was the third and fourth heating at 83 days and between 94 to 97 days.

Figure 3.   Specimen target heating using a power grid.

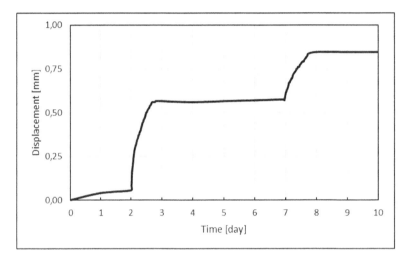

Figure 4.   Particular test results during first 10 days, first and second target heating.

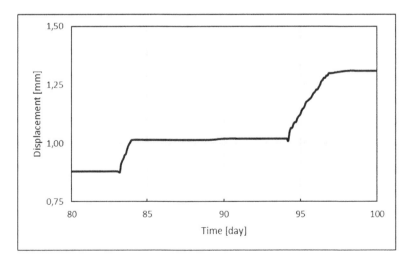

Figure 5.   Particular test results between 80–100 days, third and fourth target heating.

Table 1. Assessment of test results.

	Deformation rate mm.day^{-1}
Without warming	0.005
1st warming	0.758
2nd warming	0.324
3rd warming	0.180
4th warming	0.099

3.3 *Discussion*

With respect to the heating of the specimen there was the danger of the middle block slipping and devaluing the specimen and, consequently, the entire test. To reduce the risk of slipping the test was carried out with a higher value of vertical load and a lower value of horizontal load. This fact, in combination with an environmental temperature of 4°C, resulted in a small deformation velocity during the stage without heating, e.g. between 10 to 80 days the deformation was nearly zero; see Figures 4, 5.

The deformation rate increased while the specimens were heated, i.e. shear resistance decreased. However, with repeated heating the deformation rate decreased. Deformation rates are sequenced in Table 1.

This measurement will be complemented by testing of the same specimen exposed to different horizontal loads, i.e. for different shear stresses, so that the shear stress per expected deformation rate may be calculated; see Chapter 2.1.

The results of asphalt belt testing can also be used for a soil-structure interaction model (Cajka 2013, Labudkova & Cajka 2014).

4 CONCLUSION

In this paper, the idea of using an asphalt belt as a sliding joint for the elimination of friction between a foundation structure and subsoil is introduced. It has been proven that higher temperatures decrease shear resistance and this finding leads to the idea of temporary target heating of a slide joint, e.g. during foundation pre-stressing or at a time when subsoil deformation due to underground mining is expected.

It was decided to examine an SBS modified specimen with heating provided by a power grid with electric resistance wire, in between two layers of an asphalt belt. Specimens were exposed to a vertical load and also to shear stress via a horizontal load for over 100 days and they were heated repeatedly over 2 days, 7 days, 83 days, and 94 to 97 days. The deformation rate increased while the specimens were heated; however, with repeated heating the deformation rate decreased. This measurement will be complemented with testing of the same specimen exposed to different shear stresses, so that the shear stress for an expected deformation rate may be calculated.

Multi annual testing and analysis of shear characteristics of bigger amounts and different types of asphalt belt exposed to mechanical loads and temperatures shows that it is necessary to take into account also the age of the material at the time of load exposure.

The aim of this research is to design more durable and sustainable building structures.

ACKNOWLEDGMENT

The paper has been supported by the project of "Conceptual development of science and research activities 2016" on the Faculty of Civil Engineering, VSB—Technical University of Ostrava.

REFERENCES

Balcarek, V., Bradac, J. 1982. Utilization of bitumen insulating stripes as sliding joints for buildings on undermined area. *Civil Engineering* 1982/2 (In Czech).

CSN 730039, 2015. Design of constructions on the mining subsidence areas. Czech code, UNMZ Prague (In Czech).

Cajka, R. 2013. Analytical derivation of friction parameters for FEM calculation of the state of stress in foundation structures on undermined territories. *Acta Montanistica Slovaca* 18.4: 254–261. WOS:000343184100006.

Cajka, R. 2014. Numerical solution of temperature field for stress analysis of plate structures. *Applied Mechanics and Materials* 470: 177–187. DOI:10.4028/www.scientific.net/AMM.470.177.

Cajka, R., Mateckova, P., Janulikova, M. 2012. Bitumen sliding joints for friction elimination in footing bottom. *Applied Mechanics and Materials* 188: 247–252. DOI: 10.4028/ www.scientific.net/ AMM.188.247.

Cajka, R., Fojtik, R. 2013. Development of temperature and stress during foundation slab concreting of National Supercomputer Centre IT4. *Procedia Engineering* 65: 230–235. DOI: 10.1016/j.proeng.2013.09.035.

Cajka, R., Janulikova, M., Mateckova, P., Stara, M. 2011. Modelling of foundation structures with slide joints of temperature dependent characteristics. *Proceedings of the Thirteenth International Conference on Civil, Structural and Environmental Engineering Computing*, Crete, Greece. DOI:10.4203/ ccp.96.208.

Cajka, R., Labudek, P., Burkovic, K., Cajka, M. 2014. Construction of a green golf club buildings on undermined area. *4th International Conference on Green Building, Materials and Civil Engineering.*

Cajka, R., Manasek, P. 2006. Physical and finite element shear load response modelling of viscoelasticity materials. *Proceedings of the Eighth International Conference on Computational Structures Technology.* Gran Canaria, Spain. DOI:10.4203/ccp.83.240.

Cajka, R., Manasek, P. 2007. Finite element analysis of a structure with a sliding joint affected by deformation loading. *Proceedings of the Eleventh International Conference on Civil, Structural and Environmental Engineering Computing*, St. Julians, Malta. DOI:10.4203/ccp.86.18.

Cajka, R., Mateckova, P. 2011. Temperature distribution of slide joint in reinforced concrete foundation structures. *17th International Conference on Engineering Mechanics.* Svratka, Czech Republic. WOS: 000313492700017.

Cajka, R., Mateckova, P., Fojtik R. 2014. Calculation of temperature in sliding joint designed as a part of foundation structure. *Transactions of the VSB-Technical University of Ostrava, Civil Engineering Series*, 14(1): 1–7. ISSN (Online) 1804–4824, ISSN (Print) 1213–1962. DOI: 10.2478/tvsb-2014–0004.

Janulikova, M., Stara, M. 2013. Multi-layer rheological sliding joint in the foundation structures. *Transactions of the VSB-Technical University of Ostrava, Civil Engineering Series*, 13(2): 41–46. ISSN (Online) 1804–4824, ISSN (Print) 1213–1962, DOI: 10.2478/tvsb-2013–0008.

Janulikova, M. 2014. Comparison of the shear resistance in the sliding joint between asphalt belts and modern PVC foils. *Applied Mechanics and Materials* 501. DOI: 10.4028/www.scientific.net/ AMM.501–504.945.

Janulikova, M., Stara, M., Mynarcik, P. 2014. Sliding Joints from Traditional Asphalt Belts. *Advanced Materials Research* 1020. DOI: 10.4028/www.scientific.net/AMR.1020.335.

Janulikova, M. 2015. The New Options to Reduce Shear Stress into Foundation Structure. *Procedia Engineering* 114: 514–521. DOI: 10.1016/j.proeng.2015.08.100.

Labudkova, J., Cajka, R. 2014. Comparison of Measured Displacement of the Plate in Interaction with the Subsoil and the Results of 3D Numerical Model. *Advanced Materials Research*. Vol. 1020. DOI: 10.4028/www.scientific.net/AMR.1020.204.

Advances and Trends in Engineering Sciences and Technologies II – Al Ali & Platko (Eds)
© 2017 Taylor & Francis Group, ISBN 978-1-138-03224-8

Interaction of the asphalt layers reinforced by glass-fiber mesh

J. Olsova, J. Gašparík & Z. Stefunkova
Faculty of Civil Engineering, Slovak University of Technology, Bratislava, Slovakia

P. Briatka
COLAS Slovakia, Košice, Slovakia

ABSTRACT: The goal of this paper was to better understand the stresses which are induced in glass-fiber meshes while the following loads are applied: 1) traffic of site trucks; 2) compaction of pavement layers, by study of the mechanical behavior of interface between the reinforced layer and their substrate in order to optimize the characteristics of the interfaces (choice of the type of emulsion, dosage of emulsion, application technology, etc. Manufactured and tested samples enabled the assessment of the quality of the bonding layers by tensile test. The objective was to evaluate the effect of the interposition of the mesh in between two layers of asphalt on the bonding and to determine if this insertion could affect this characteristic which is essential to the service life of the structure.

1 PRESENT STATE OF THE ART

Nowadays the reinforcing meshes (in following just meshes) made out of glass fibers still more and more utilized for reinforcement of asphalt pavements, extension of their service lives, decreasing pf thickness and last but not least for the improvement of tensile strength and resistance to crack propagation.

In comparison with other conventional techniques, usage of meshes prioritize the economical aspect. This technique strengthens the pavement layers, slows or even mitigates crack propagation in upper layers and enhances performance of the pavement in flexure. (Aldea & Darling 2004)

Nowadays, there are plenty of products in the form of meshes developed for this purpose on the market. In this study, we used composite meshes of glass fibers and fabric. The French group COLAS has vast experiences in utilization of reinforcement of the pavements with glass-fiber meshes. Since 1990 they have been successfully using the technique of reinforcing the pavements developed in cooperation with the French producer of meshes. In June 2014 they applied meshes on a taxiway Whiskey, secondary runway and aircraft parking areas of the airport Beauvais (FR). (R. Vesin, 2014) As an example of usage of the meshes in Slovakia, we can identify at least the transition between concrete and asphalt pavements adjacent to the portals of Tunnel Sitina in 2006.

Despite relatively frequent application of the meshes, the knowledge regarding the design are mostly empirical. With respect to a lack of information and laboratory analysis on the effect of the meshes in pavements, it is essential to improve the concept of mesh-reinforcement and laboratory procedures of determining mechanical resistance and to better understand the stresses which they are exposed to during construction of the pavement.

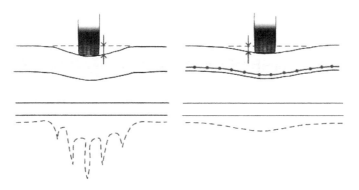

Figure 1. The deformation of the pavement under the load (without (on left) and with (on right)) inserted mesh showing also peaks of stresses.

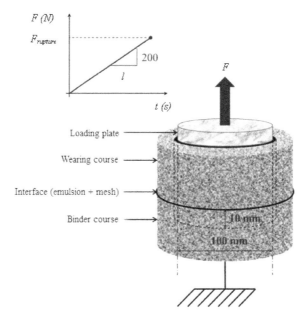

Figure 2. Scheme of determination of adhesion—tensile strength of bond according to Pr EN 12697–48.

2 THE APPROACH

The complex combination of loads, for a better understanding, was simplified to fundamental principles. In this case, one of them was plain/direct tension.

The determination of tensile strength lies in the application of force increment (200 N/s) on samples with partial core drills of diameter 100 mm till failure of the sample (Figure 2). The maximum force recorded indicates the adhesion.

3 MATERIALS USED

The asphalt mixture used for manufacturing of all samples (binder and wearing course of thickness 5 cm) for tensile tests was designed as AC 11 O. For good bond between the courses (in this case also for application of the mesh) was used the tack coat of Type R 69 which was made out of bitumen 160/220. The composite meshes of both types, either A 50 kN and B 100 kN consisted of epoxy bound glass fibers and non-woven synthetic fabric. Both glass-fiber meshes

Table 1. Manufacturing of the samples.

Designation of the samples' variants	V0 ref.	V1	V2	V3
Inserted mesh	50 kN	100 kN	without	50 kN
Number of layers of inserted meshes	1	1	1	1
Temperature of asphalt mixture (°C)	160	160	160	160
Temperature at testing (°C)	20	20	20	20
Dosage of task coat (g/m²)	300	300	300	300
Type of tack coat	R60 160/22	R60 160/22	R60 160/22	R60 160/22
Shape of samples	Prisms	Prisms	Prisms	Cylinder
Number of specimens	8	8	8	4

were of the same openings 40 mm × 40 mm. The reference configuration was made out of mesh A 50 kN in one layer, sticked by tack coat R 69 160/220 in dosage 300 g/m² of residual binder (breaking time 2 hours). As for discussion and analysis, we refer to type of samples V0.

Standard Pr EN 12697-48 determines the reference cylindrical samples with diameter 150 mm and internal diameter of the groove 100 mm. For the purposes of saving the material and time, we chose prismatic shape (simplified manufacturing) with centrically drilled grooves. Subsequently, the samples were cured (24 ± 4) hours at temperature (23 ± 5) °C. Later on, after dry sand blasting of the upper surface of the area inside the grooves, the steel tension plates were glued on using epoxy. Then, before the tests, the samples were cured at least 12 hours at temperatures (0 ± 1) °C and (10 ± 0.5) °C.

Based on the recorded results of the tests, using formula 1, as introduces Pr EN 12697-48, the tensile strength βt (MPa) of the bond, can be derived as simple ratio of maximum recorded load F_{max} (kN) and area inside the groove A (mm²) on which F_{max} was applied.

$$\beta_t = \frac{F_{max}}{A} = \frac{4 F_{max}}{\pi D^2} \tag{1}$$

where D is diameter of the area inside the grove, in mm.

The manner of rupture of the sample and appearance of the after failure was recorded for each specimen according to scale listed in Pr EN 12697-48.

4 DISCUSSION OF RESULTS

From each of the tests, the working diagram (load against time) was recorded. All of them are of similar shape to those presented in Figure 4. Growth of the load till maximum is in line with mechanics of braking of the bond between layers. After the failure, the load dramatically falls. Sometimes we can observe repeated differential raises of bonds between layers (Figure 3 V1 – 6), which is most likely caused by the interposition of the glass fibers since this phenomenon was not observed in samples without meshes.

At early stage of loading, we can sometimes observe a slight deformation of working diagram (sudden drop and immediate raise). These are caused by imperfections of the samples´ surface and creep o fit during raising load. It is recommended to justify the edges of the prismatic samples (see Figure 4) before testing.

It was proved that the cylindrical samples are of the same tensile strength as the prismatic ones. Based on this, we assume that the shape of the samples have no impact on tensile strength, as was expected at the beginning.

4.1 Effect of presence of the mesh on tensile strength

As presented in Figure 5, the samples without the mesh exhibit the highest tensile strength. The explanation is that the surface of the mesh reduces contact area between two layers and

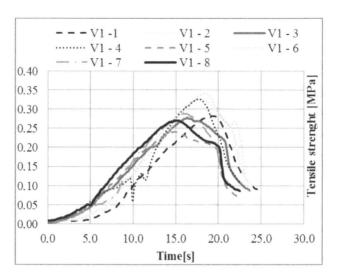

Figure 3. Evolution of the tensile stresses of variant V1.

Figure 4. Imperfection of the surface, loading steel plate bond to the sample.

therefore also quality of bond—reduces adhesion. In these particular areas, the tack coat binds the glass fibers instead of two adjacent asphalt layers. The fabric, as a part of composite mesh, should enable the emulsion to penetrate through and bind the asphalt layers together.

Presence of the fibers, however, catches part of emulsion for binding (Figure 6). Mesh A 50 kN, of which glass fibers occupy a smaller area, reduces the tensile strength (bond) less than Mesh B 100 kN.

This statement is supported by difference of the mean value of stresses at moment of failure of the samples between set of mesh A 50 kN and B 100 kN. The fabric as such causes a reduction of strength by approximately 0.2 N/mm² (the average value without any mesh 0.78 N/mm² minus average strength with mesh A 50 kN (on area excluding this fibers) 0.58 N/mm²).

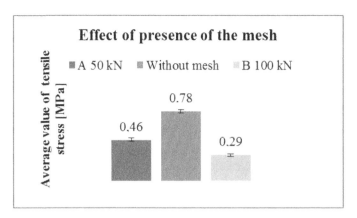

Figure 5.   Comparison of results of variants V0, V2 and V1.

Table 2.   Stresses at failure.

Set of samples	Average stress at failure [N/mm²]	Total length of fibres in core drill [cm]	Area of the fibres in core drill [cm²]	Area of the core drill [cm²]	Area of the core drill except of area of the fibres [cm²]	Average stress at failure over are without fibers [N/mm²]
V0 ref. A 50 kN	0.46	36.2	14.5	78.5	64.0	0.57
V1 B 100 kN	0.29	36.6	32.9	78.5	45.6	0.50
V2 WM*	0.78	0.0	0.0	78.5	78.5	0.78

*WM = without inserted mesh.

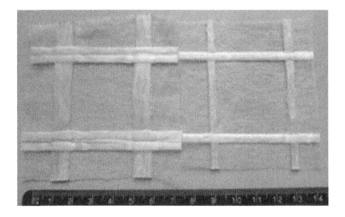

Figure 6.   Used meshes (100 kN on left, 50 kN on right).

5   CONCLUSIONS

By inserting the mesh between two asphalt layers, their adhesion can be reduced. The highest adhesion is reached without any mesh, i.e. approximately 0.8 N/mm². In case of inserted mesh A 50 kN, the average adhesion falls down to approximately 0.5 N/mm². In case of inserted mesh B 100 kN, the reduction can be even more significant, down to 0.3 N/mm². It is obvious, that specific surface area of the mesh fibers inserted into the structure reduce is inversely proportional to resulting adhesion.

Since plain tension, as load of the pavement, is rarely a case, these results cannot be interpreted as a negative feature of usage of the meshes in asphalt pavements. The tensile tests were performed just for better understanding and potentially quantification of adhesion of two asphalt layers.

ACKNOWLEDGEMENT

This paper was worked out based on results of research project of Campus for Science & Technique, unit of COLAS group.

REFERENCES

Aldea, C. and Darling, J. 2004. Effect of coating on fiberglass geogrids performance. *In Proceedings of 5th International RILEM Conference on Reflective Cracking in Pavements: Mitigation, Risk Assessment and Prevention*, Limoges.

Chémia—Servis, 2016. Chémia servis. In *Chémia—Servis, a.s.* In: *[Online]*. Available: http://www.chemiaservis.sk/referencie/vystuzovanie-vozoviek.

Maccaferri Central Europe, 2015. Maccaferri Slovakia. In *[Online]*. Available: http://www.maccaferri.com/sk/products/geogrid-macgrid/.

Pr EN 12 697–48: Bituminous mixtures—Test methods for hot mix asphalt—part 48: Interlayer bonding. 2015.

Vesin, R. 2014. Rénovation de l'aéroport de Beauvais avec Colgrill. *In Film*. Ostinato Production.

Advances and Trends in Engineering Sciences and Technologies II – Al Ali & Platko (Eds)
© 2017 Taylor & Francis Group, ISBN 978-1-138-03224-8

Extinct settlement identification using small format aerial photography – methods and accuracy

J. Pacina, J. Popelka & M. Tobisch
J.E. Purkyně University, Usti nad Labem, The Czech Republic

ABSTRACT: Small Format Aerial Photography (SFAP) is an excellent data source for the creation of high resolution orthophotos and digital terrain models. SFAP collection by UAVs is restricted by several regulations, battery durability and weather conditions. We have developed a method that uses the same principle for SFAP collection as UAVs (camera stabilization, time lapse) but instead, the camera is mounted on a small aircraft and thus we may avoid most of the limitations. This paper is focused on the identification of extinct settlement residuals in the Czech—German borderland, where several municipalities vanished after World War II due to the political situation and large water dam projects. The selected sites were scanned from different flight altitudes and the resulting data were tested both for position accuracy and extinct settlement identification possibilities. Data accuracy was examined using advanced methods of geostatistics to show the thorough quality of the data acquired from different altitudes.

1 INTRODUCTION

Settlement extinction was a common phenomenon in North-West Bohemia (the Czech Republic). The Ore Mountains (in the Czech—German borderland) were depopulated after the Second World War, when German speaking inhabitants were deported to Germany. The rest of this region has been affected by heavy industry (large open-pit mines) and many other types of anthropogenic activity causing large landscape changes (see Brůna et al., 2014). Large water dam projects are the second type of anthropogenic activity that accelerated settlement extinction and landscape transfigurations. The large water dams have been used as a water support for heavy and chemical industry in the region or in some cases, as drinking water supplies.

Several types of analyses and 3D reconstructions using old maps and archival aerial photographs were performed on these sites (Brůna et al., 2014, Pacina, Novák, 2014, Duchnová, 2015 or Pacina, Sládek, 2015). Identification of settlement residuals in flooded areas complicated locating sites in forests or meadows (Pacina, Holá, 2014, Pacina, Havlíček, 2016). Several methods allow studying submerged areas with LIDAR (Doneus et al., 2013), (Wang, Philpot, 2007) or bathymetric mapping (Pacina, Sládek, 2015). These remote sensing methods have limits (clear and shallow water for LIDAR), required special equipment and permission for operating a boat within the drinking water reservoir.

In autumn 2015, large construction works within the water reservoir Fláje took place and so the water level was lowered by about 10 meters. This was a unique chance to survey a sunken village that had been inundated in 1960 when the dam was finished.

1.1 *Area of interest*

The Fláje dam was built between the years 1951–1964 using a unique piling method (hollow damming). Thanks to its special structure, the dam is on the list of the historical landmarks of the Czech Republic. The main purpose of the reservoirs is to support the towns at the foothills with drinking water and to protect the valley from flooding.

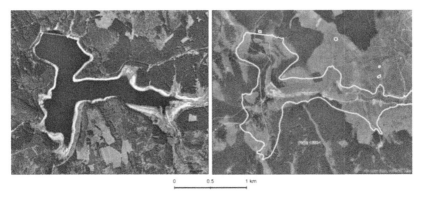

Figure 1. Area of interest with the low water-stand in autumn 2015 and the original landscape in 1953.

The village Fláje was established in the 14th century. In 1921, there were 263 houses and 1099 inhabitants. After the Second World War (in 1950), there were only 100 houses and 192 inhabitants. The last inhabitants left the village in 1961, when the village was inundated. Only the village wooden church was moved and re-built in the nearby village of Český Jiřetín.

2 DATA AND METHODS

The standard method of bathymetric mapping was performed on the nearby dam Nechranice in 2014 (Pacina, Sládek, 2015). The results were satisfactory for studying the bottom sedimentation within the scanned transects but the density of the data made complete bottom mapping impossible (with respect to the dam size). Underwater LIDAR could have been performed, but only in clear and shallow water (Mandlburger et al., 2011). The running water in the Ore Mountains contains a high amount of humid compounds and the water is brown and opaque.

2.1 Archival data sources

The entire area is covered by processed old maps and archival aerial photographs offering a retrospective view on the vanished village. The processed old maps are described in detail in (Brůna et al., 2014). The archival aerial photographs that describe the area of interest in the best way originates from the year 1953—just several years before the village was extinct. The aerial photographs were processed into the form of an orthophoto using the standard ways of photogrammetry (Kraus, 2007).

2.2 Small format aerial photography

The short-period absence of water made it possible to use Small Format Aerial Photography (Aber et al., 2010) based on the structure from motion 3D modelling. A small aircraft carrying a Nikon D810 with a Sigma 35mm f/1.4 DG HSM Nikon lens mounted on a stabilization device (gimbal) was used for the photo survey instead of an UAV. UAV usage would have been problematic as the size of the surveyed area was over 500 hectares. Aircraft usage was a safer solution in this case with respect to the drinking water protection area. The aircraft is as well capable of carrying a heavier "camera-lens" set assuring high spatial resolution of the produced data.

The aircraft survey was performed from three different altitudes - 300 m, 600 m and 1000 m. The "300 m altitude" survey covers only the area of the former village Fláje and the purpose was to create a high resolution orthophoto and 3D model that will be further used for vanished settlement identification and landscape change analysis. The 300 m elevation is the lowest altitude the small aircraft may operate at. The "600 m altitude" survey was used

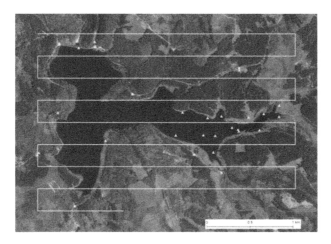

Figure 2. Flight plan for 600 m altitude and measured ground control points.

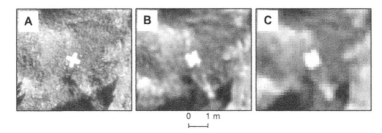

Figure 3. Uncertainty in ground control points identification: A—300 m, B—600 m, C—1000 m altitude.

for the creation of the whole reservoir orthophoto (and 3D model) and the "1000 m altitude" survey was performed just to test the limits of this method as the 1000 m altitude is used for standard photogrammetric surveying.

Within the surveyed area 31 Ground Control Points (GCP) in a form of white crosses were placed and surveyed using the GPS-RTK method. Altogether 675 digital photographs were taken from the 300 m altitude, 1167 from 600 m and 851 from 1000 m. The surveying flight plan along with GCPs locations is presented in Figure 2. All of the SFAP images were processed using Agisoft PhotoScan (v. 1.5.0) software. The images were aligned using the *high* setting, pair selection *generic, keyp oint limit* 40 000 points and *tie point limit* 4000 points. The automatic camera calibration was used during this step. The *sparse point cloud points* with reprojection error larger than 1 were removed from the point cloud to assure the high accuracy of the resulting data. The camera locations were optimized during this step as well. The *dense point cloud* was generated using the *medium* setting for the "300 m altitude" and *high* for the "600 m and 1000 m" altitude. This setting produces results suitable for the process of identification (see Table 2). The *mesh* was created using the *height-field* mode with the face count setting set to 0 (this corresponds to the optimal number of created faces). The orthophoto was generated using the default PhotoScan settings.

3   IDENTIFICATION AND ACCURACY

The resulting data were further tested for accuracy and then used for the identification of vanished settlements and rescue archaeological research. The identification uses the orthophoto and Digital Surface Model (DSM) derived from the SFAP images. The accuracy is tested by a comparison of the resulting DSM with GPS-RTK surveyed points.

3.1 *Accuracy of the surveyed data*

An important part of image based 3D modelling is quality and accuracy evaluation. The exposed shores along the reservoir were measured using the GPS-RTK method with more than 90 points used as "etalon" data. The differences (errors) computed from GPS measurements and from the Digital Surface Model derived from SFAP were subject to geostatistical analysis. The global overview from the resulting data quality offers the Root Mean Square Error (RSME)— see Table 1. Here we may note that RMSE grows with flight altitude. This is caused by the lower spatial resolution of the obtained SFAP dependently at a higher altitude. A related higher error in the point cloud produced by the Structure from Motion algorithm is directly related with this fact and eventually uncertainty of the GPCs identification as well (see Figure 3).

The three main attributes of model errors were evaluated and tested.

A. The errors should not be spatially auto correlated (should be randomly distributed in the study area). The Global Moran's Index (Moran's I) was used to measure the spatial auto correlation of errors. Z-score and p-value evaluate the significance of Moran's I in addition. Local Moran's I was calculated for each point and clusters of high/low errors were mapped for a better interpretation of eventual randomness violation. All of the above mentioned tools were calculated using ArcGIS 10.3.1 software.

B. Errors should be normally (or at least symmetrically distributed). The Jarque-Bera test along with p-value calculated by the Monte Carlo method and the graphical analysis of errors (including a histogram of error distribution and kernel density) were used to evaluate the normality of errors. As an additional distribution statistic, the skewness of errors was also calculated.

C. There should be no dependency (neither linear nor monotone) between the elevation and model errors—the model should be "elevation stable". Spearman's coefficient of correlation and its p-value are appropriate to evaluate this attribute. All normality, correlation characteristics and tests were calculated using Past 3.11 software.

The 3D model based on data from the highest flight altitude (1000 m) violates both randomness and normality of errors (see Table 1 and Figure 4). This is mainly due to the

Table 1. Root Mean Square Error (RMSE) calculated for different flight altitudes.

Flight altitude (m)	300	600	1000
RMSE (m)	0.039	0.086	0.134
Spatial autocorrelation of errors—Moran's I (z-score)	0.290 (3.388)[†]	0.098 (1.128)[***]	0.216 (2.613)[†]
Normality of errors (Jarque-Bera test)	9.928[*]	0.315[***]	111.8[†]
Skewness	0.111	−0.095	1.082
Elevation stability (Sperman's r)	0.025[†]	−0.212[†]	−0.126[†]

Statistical significance: [†]not significant, [*]p-value < 0.1, [**]p-value < 0.05, [***]p-value < 0.01.

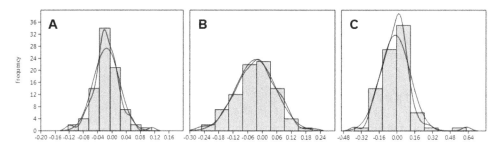

Figure 4. Distribution of model errors (histogram), kernel density and normal density: A—300 m, B—600 m, C—1000 m altitude.

positive skewness of error. This means that the 3D model overestimates the elevation. The model from 300 m data also seems to have troubles. When analyzing the clusters using Local Moran's I statistics we discovered one cluster of low values, a place where the 3D model overestimates the elevation. This cluster consists of five points, all of them belong to ten points with the highest value of error (in absolute value). This model deficiency is probably caused by the GPS measurement error. Both models from low flight altitude (300 m and 600 m) show normally distributed errors. A small violation in 300 m model (slightly positive skewness of value 0.111) is caused by the same reasons as mentioned above.

There is no dependency between elevation and model errors (see Table 1). From this point of view, all the models are elevation stable. This means, they don't overestimate or underestimate the model elevation in relation to the real elevation.

3.2 *Identification of vanished settlement*

Identification of the vanished settlement is performed on the orthophoto and DSM derived from the SFAP images. Thanks to the high quality camera and lens, we are able to obtain an orthophoto and DSM of high spatial resolution—see Figure 5 and Table 2. Note that DSM spatial resolution reflects the quality setting within the Agisoft Photo-Scan software.

The main vanished structure identification may be performed using the processed archival aerial photographs or georeferenced old maps (Pacina, Holá, 2014). Within the area of interest, there are still identifiable roads, bridges, house foundations and other remains of the former settlement—see Figure 6. The resulting data, important for archaeologists and interesting for the wider public, are available (as the web-mapping application created in ArcGIS API for JavaScript) online within the http://fzp.maps.arcgis.com/home/ map portal.

Figure 5. Orthophoto derived from Small Format Aerial Photography: A—300 m, B—600m, C—1000 m altitude.

Table 2. Spatial resolution of the resulting orthophoto and Digital Surface Models (DSM).

Flight altitude (m)	300	600	1000
Orthophoto spatial resolution (cm/pixel)	2.7	8.4	14.1
DSM spatial resolution (cm/pixel)	5	33	56

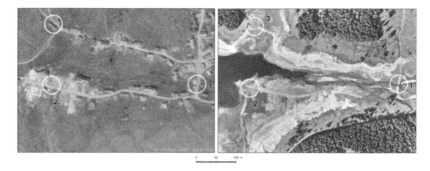

Figure 6. Settlement residuals identified within the area of interest—orthophoto year 1953 (left), orthophoto derived from small format aerial photography—year 2015 (right).

4 CONCLUSION

The aircraft based SFAP method for the identification of vanished settlements in the Czech-German borderlands has been used in this paper. The advantage of this method is the large area that is possible to survey with the aircraft. Flights at three different altitudes were performed during the survey to evaluate the accuracy of the resulting data. Based on our expectations, the best results were achieved within the 300 m altitude, where the RMSE and other geostatistical indicators showed high accuracy of the resulting data. The possibility to survey such area appears only couple of times within a life-cycle of the water reservoir (in this case the first time since 1960) and thus the produced data will be further used by archaeologists, hydrological experts and the wider public. With respect to the size of the resulting data, all of the processed data are published as map services (addable as map layers into GIS) and as an interactive web-mapping application where the user may browse and compare the data within the web-browser independently on their own operating system (Android and iOS compatible). The web-mapping application is available within the university web-map portal at http://fzp.maps.arcgis.com/home/.

ACKNOWLEDGEMENT

The research was supported by J. E. Purkyne University internal grant "Observing landscape changes using the methods of geoinformatics".

REFERENCES

Aber, JS. & Marzolff, I. & Ries, JB. 2010. *Small-Format Aerial Photography: Principles, Techniques and Geoscience Applications*, Amsterdam—London: Elsevier Science. ISBN: 978-0-444-53260-2.
Brůna, V. & Pacina, J. & Pacina, J. & Vajsová, E. 2014. *Modeling the extinct landscape and settlement for preservation of cultural heritage*. Città e Storia. 2014, IX, no. 1, pp. 131–153. ISSN: 1828-6364.
Doneus, M. & Doneus, N. & Briese, C. & Pregesbauer, M. & Mandlburger, G. & Verhoeven, G. 2013. *Airborne Laser Bathymetry—detecting and recording submerged archaeological sites from the air*. Journal of Archaeological Science 40, pp. 2136–2151. ISSN: 0305-4403.
Duchnová R. 2015. The reconstruction of the vanished town of Pressnitz [online]. URL: http://prisecnice.eu/
Kraus, K. 2007. Photogrammetry: Geometry from Images and Laser Scans. Berlin: Walter de Gruyter. ISBN: 978-3-11-019007-6.
Mandlburger, G. & Pfennigbauer, M. & Steinbacher, F. & Pfeifer, N. 2011. *Airborne Hydrographic LiDAR Mapping—Potential of a new technique for capturing shallow water bodies*. In: Chan, F., Marinova, D., Anderssen, R. (Eds.), MODSIM2011, 19th International Congress on Modelling

and Simulation. Modelling and Simulation Society of Australia and New Zealand, December 2011, Perth, pp. 2416–2422. ISBN: 978-0-9872143-1-7.

Pacina, J. & Holá. M. 2014. *Settlement identification in abandoned borderland.* 14th International Multi-disciplinary Scientific GeoConference, SGEM 2014, Conference Proceedings. 2014, Book 2, Volume I, pp. 769–776. DOI:10.5593/sgem2014B21. ISBN: 978-619-7105-10-0/ISSN: 1314-2704.

Pacina, J. & Novák, K. 2014. *Lake Most—How A Royal Town Can Be Transformed Into A Lake*: Georelief analysis 1938–2012. Land use/cover changes in selected regions in the world. 1st pub. Asahikawa: IGU-LUCC, 2014, 105–112. IGU-LUCC research reports. ISBN: 9784907651107.

Pacina, J. & Cajthaml. J. 2014. *Historical Data Processing, Modelling, Reconstruction, Analysis and Visualization of Historical Landscape in the Region of North-West Bohemia.* DOI: 10.1007/978-3-319-07926-4_36 In book: Modern Trends in Cartography, Edition: Selected Papers of CARTOCON 2014, Chapter: 477, Publisher: Springer International Publishing, Editors: Jan Brus, Alena Vondráková, Vít Voženílek, pp. 477-488. ISBN 978-3-319-07925-7.

Pacina, J & Sládek, J. 2015. *Observing Landscape Changes Using Distant Methods.* Civil Engineering Journal. 2015, (1): 1-11. DOI: 10.14311/CEJ.2015.01.0004. ISSN 1805-2576.

Pacina, J. & Havlíček, J. 2016. *A vanished settlement in the Ore Mountains—the creation of 3D models.* In Advances and Trends in Engineering Sciences and Technologies: Proceedings of the International Conference on Engineering Sciences and Technologies, 27–29 May 2015, Tatranská Štrba, High Tatras Mountains—Slovak Republic, May 27–29, 2015, pp. 341–346. ISBN 978-1-138-02907-1.

Wang, CK. &, Philpot, W.D. 2007. *Using airborne bathymetric lidar to detect bottom type variation in shallow waters.* Remote Sensing of Environment 106, pp. 123–135. ISSN: 0034-4257.

Advances and Trends in Engineering Sciences and Technologies II – Al Ali & Platko (Eds)
© 2017 Taylor & Francis Group, ISBN 978-1-138-03224-8

Deformation measurement and using of GPR on the Charles Bridge in Prague

K. Pavelka, M. Faltýnová, E. Matoušková & J. Šedina
Department of Geomatics, Faculty of Civil Engineering, Czech Technical University in Prague, Prague, Czech Republic

ABSTRACT: A constructional and a historical research including assessment of protected buildings state are essential parts of cultural heritage preservation. Currently, there is a number of non-invasive techniques that open up new possibilities of expanding the constructional and historical survey of architectural monuments, including stone structures. Non-invasive methods can be used to identify priorities and to prepare documents for the restoration and reconstruction of the object. Combination of several methods seems to be very useful in this case. In proposed article a deformation of two arches of the Charles Bridge in Prague using laser scanning and photogrammetry is described. As a non-destructive method GPR (Ground Penetrating Radar) for internal structures monitoring is shown. Results from the on-site measurements and measurement analysis are the major themes of this paper.

1   INTRODUCTION

A constructional and a historical research including assessment of protected buildings state are essential parts of cultural heritage preservation. An assessment of internal structure of stone blocks in the historical structures, which can be accomplished using probes, belongs amongst the most complicated issues of the constructional and the historical research. This method is destructive by its nature and there is an intention to minimize the number of probes that must be performed. In last decades more attention is being laid on non-destructive methods which provide sufficient data without a need for physical intervention into the surveyed object (Reynolds, 2011). The identification and correct interpretation of the structure state is crucial for an extension of a structures life cycle and selection of an appropriate way of renovation.

Figure 1.   GPR SIR 3000.

The goal of Constructional and Historical Research (CHR) is, beside the documentation of object and its evolution, also an objective evaluation of its state, prioritization and preparation of fundamentals for reconstruction (Pešta, 2012). During last years, a significant progress in technology has been achieved, which leads to an expansion of number of non-invasive techniques used for the non-destructive CHR of the masonry structures (Hanzalova, 2014, Pavelka, 2013). Most of these techniques (ultrasonic techniques, Ground Penetrating Radar (GPR) and impact-echo) are based on an interaction of specific electromagnetic radiation with the surveyed object. The ultrasonic is a technique well known and commonly used in medicine. The ground penetrating radar is a geophysical method that uses radar pulses to image the subsurface. The GPR can detect voids, cracks and changes in material properties. The impact-echo is an acoustic method that uses impact-generated stress (sound) waves which propagate through concrete and masonry and are reflected by the internal flaws and external surfaces (Pérez-Gracia, 2013).

1.1 *Ground penetrating radar*

New methods of 3D objects documentation and research are tested at the CTU (Czech Technical University) in Prague, Laboratory of Photogrammetry. A ground penetrating radar GSSI SIR—3000 with two different antennas (400 MHz and 1.6 GHz) was used for documentation of the Charles Bridge. The GPR uses electromagnetic radiation in the microwave band of the radio spectrum and analyses signals reflected from the subsurface structures. The antenna transmitter emits impulses of sine wave that travels through the material. These impulses are reflected back to the antenna receiver by boundaries between the layers or the objects with different dielectric characteristics. The velocity of the radiation depends on the electromagnetic characteristics surveyed material.

2 SURVEY OF MASONRY—CHARLES BRIDGE

In the fall of 2012 and 2013 a basic comparative research of selected part of the Charles Bridge masonry was accomplished. First measurements were acquired using GPR SIR—3000 with the 400 MHz antenna characterized by a maximal range up to 2 m. This antenna is usually used for applications in archaeology and for power line detecting. Basically it is not suitable for a masonry survey due to the low resolution, antenna size and measuring wheel size. Sensitivity, resolution and depth range (limited by length of the pulses) of the instrument depend on the surveyed object, local conditions and humidity. The basic parameter of the instrument setting is dielectric constant which influences the velocity of the signal traveling through the material and consequently determines a depth of discovered objects. The dielectric constant could not be determined in laboratory and the value was set experimentally to 7 (implicit setting). Other parameters of measurement were set accordingly (Weber et al, 1996). The measurements were taken in two modes:

1. time mode—detection method—the shift of the antenna over the masonry was realized by hand;
2. using measuring wheel, which enables to register the data in profiles with defined distance scale—impossible to use on vertical walls or vaults without an assistance of more people, possible to use only on roadway or pillars surroundings.

The structure or objects hidden under the surface are in GPR data usually display as parabolas, which is caused by the radar motion, the width of the signal and the echo. Regular parabolas represent small cylindrical objects (e.g. power lines, pipelines, ironing), in some cases also voids or material changes. Vertical resolution is not crucial in this case, main issue is the detection of the searched features (voids, cracks). The GPR can be very useful nondestructive technique, but its use is limited by site situation as well its results should not be always faithful because of some unknown conditions on site. The outputs were processed using a software Radan 3D. The software enables data visualization, post-processing and calibration using reference measurement. The basic radargramm is adjusted—correction of zero point, application of nonlinear gain coefficient. The software disposes with tools *Migration* and *Hilbert transformation*, both tools serve for transformation of basic parabolas into real shapes of features hidden in the object structure.

2.1 *Ground penetrating radar—configuration with 400 MHz antenna*

The antenna was used to determine the limitation of its use for object survey, especially in masonry survey. In number of projects, the GPR with similar antenna was successfully used in masonry survey. The system was useful in cases searching for large objects (crypts, different masonry etc.). In 2012 it was proved that the antenna should be success-fully used on stone blocks from the Charles Bridge in limited way, the measurements were taken in static mode at depositions site Šutka. The repairs of blocks by covering desks/patches (inferior joining) were discovered (Přikryl, 2004). Other measurements were taken on the Charles Bridge in Prague. Part of pillar n.10 and vault (10–11) on part of the bridge near Malá Strana were chosen (first pillar and first vault over the Vltava River). The Figure 3 confirms that the

Figure 2.   Detail of stone; antenna, scale (depositions site Šutka).

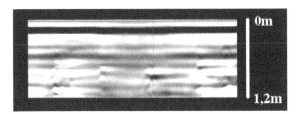

Figure 3.   Antenna 400 MHz. Part A1—profile 1, measurement A, depth range 20ns (approx. 1 m), in depth of 75 cm—visible features in regular distance (the third can be void). Right side display change of material—may be northern string wall.

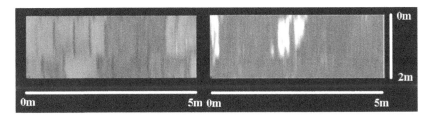

Figure 4.   Measurement C; left—profile in depth 10 cm (gaps among stone blocks); right—profile in depth 25 cm (discontinuities in blocks).

400 MHz antenna can be used for masonry survey despite the limited resolution (tens of cm) and missing detail. It is possible to detect basic structures.

2.2 Ground penetrating radar—configuration with 1.6 GHz antenna

The GPR with antenna 1.6GHz was tested on selected parts of the Charles Bridge—pillar n.11 (second pillar over the Vltava river, side Malá Strana). This part was tested by different methods to determine the kind of used material. The GPR with antenna 1.6GHz is easy to use for masonry survey (Neal, 2004). The measure trolley enables to define the profile length, mark the positions of findings by cursor and get back to the findings. The disadvantage of the system is in quite tough connecting cables. They make the operation with the system inconvenient. The results have high resolution (cm). The depth range is up to 60 cm. There are discontinuities (probably end of stone blocks) clearly visible in depth 25–30 cm (Fig. 5). Regularly spaced parabolas visible from surface into depth display the gaps among the stone blocks. The measurements were accomplished repeatedly (parts A, B, C, D—to layers of stone blocks, Figs. 4–8). The Figure 6 (profile C in depth 29 cm) offers the best visualization of the state of the surveyed object (back sides of the stone block and material changes are clearly visible).

2.3 Laser scanning

The Surphaser laser scanner was used for documentation and analysis of a part of the Charles Bridge. This instrument has accuracy 0.6 mm on 10 m and it is an important tool for the deformation analysis. Measurement with this laser scanner is user friendly, a common

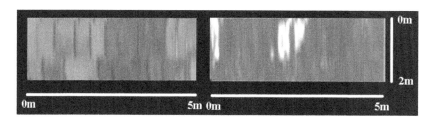

Figure 5. Measurement C; left—profile in depth 27 cm (more discontinuities—small difference in depth, high difference in findings—there is a significant change in stone structure between 25–30 cm depth); right—profile in depth 25 cm (discontinuities in blocks); right—profile in depth 28 cm.

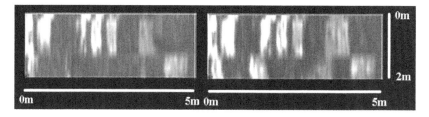

Figure 6. Measurement C; left—profile in depth 29 cm; right—profile in depth 34 cm.

Figure 7. Row of stone blocks B, measured by GPR with antenna 1.6 GHz; determination of stone layer thickness and stone damage.

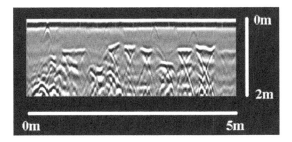

Figure 8. Measurement B—row of stones B1-B8—GPR with antenna 1.6 GHz. Data without software compensation. White and black line in upper part illustrate significant change of environment—air and stone.

Figure 9. Part B.

laptop for data storing is necessary. In this case project, measurements on five scan stations were made (vault and bridge walls between pillar 10 and 11).

A plane was fit to point cloud and the deviation between the point cloud and the plane was displayed hypsometrically. This procedure enables to display imperfections or inaccuracies of the structure. By comparing the point clouds from different epochs it is possible to provide a long-term monitoring of structure or deformation which are caused by temperature variations or load (Pavelka et al, 2013).

3 CONCLUSION

Presented techniques—ground penetrating radar and laser scanning—belong to the very important non-invasive techniques for a constructional research. In case of masonry survey, it can serve for assessment of surface layers of stone, identification and localization of damaged parts. GPR can be used for the construction detail detection (e.g. differentiation between stone blocks or thin stone layers/patches). Based on performed testing, both of these techniques can be recommended for survey of stone masonry as a part of constructional historical research or preparation of fundamentals for reconstruction. Currently, more attention is being laid on non-destructive methods which provide sufficient data without a need of physical intervention into the surveyed object.

The testing of GPR technique was focused on two main tasks—determination of stone blocks' thickness and identification parts of stones that are damaged. It is necessary to keep in mind that the GPR outputs are not explicit and measured data and its interpretation don not lead to clear results in each case.

ACKNOWLEDGEMENTS

This research has been supported by the Ministry of Culture under project NAKI—DF13P01OVV002.

REFERENCES

Ground-Penetrating Radar, 2011. U.S. Environmental Protection Agency. http://www.epa.gov/esd/cmb/GeophysicsWebsite/pages/reference/methods/Surface_Geophysical_Methods/Electromagnetic_Methods/Ground-Penetrating_Radar.htm (accessed Aug 25, 2014).

Hanzalová, K., Pavelka, K. Using radar data in archaeological sites. In: *14th International Multidisciplinary Scientific Geoconference SGEM 2014*, Conference Proceedings vol. III. Sofia: STEF92 Technology Ltd., 2014, vol. 3, pp. 315–322. ISSN 1314-2704. ISBN 978-619-7105-12-4.

Neal, A. 2004. Ground-penetrating radar and its use in sedimentology: principles, problems and progress. *Earth-Sci. Rev.*, (66), 261–330.

Pavelka, K., Pavlík, M., Řezníček, J. The Chateau Chapel in Smirice: The Uses and Limits of Modern Technology in the Analysis of Architectural Composition and the Question of Authorship. *Umění.* 2013, roč. 61, č. 1, čl. č. 2, s. 23-+. ISSN 0049-5123. (in Czech).

Pavelka, K., Řezníček, J., Bílá, Z., Prunarová, L. Non Expensive 3D Documentation and Modelling of Historical Object and Archaeological Artefacts by Using Close Range Photogrammetry. Geoinformatics. 2013, vol. 2013, no. 10, p. 53–66. ISSN 1802-2669.

Pešta, J. 2012. *Zkoumání historických staveb*; Národní památkový ústav: Praha.

Pérez-Gracia, V., Caselles, J. O., Clapés, J., Martinez, G., Osorio, R. 2013. Non-destructive analysis in cultural heritage buildings: Evaluating the Mallorca cathedral supporting structures. NDT & E International, (59), 40–47.

Přikryl, R. 2004. Jaký přírodní kámen vybrat pro opravu Karlova mostu? *Oprava Karlova mostu*, 1, 50–54.

Reynolds, J.M. 2011. An Introduction to Applied and Environmental Geophysics, 2nd ed.; John Wiley & Sons Ltd.: Chichester.

Řezníček, J., Pavelka, K. 2010. Photogrammetrical measuring of the dynamical deformation of the joint and the column web panel at elevated temperature. In: Proceedings of 31st ACRS Conference. Hanoi: ACRS, 2010, vol. 1, pp. 1859–1863. ISBN 978-1-61782-397-8.

Weber, J., Köhler, W., Bayer, K. 1996. Stone material and construction history of the main portal of Saint Weber, J., Köhler, W., Bayer, K. 1996. Stone material and construction history of the main portal of Saint Stephan's Cathedral: non-destructive measurements. In: *Proceedings of 5th International Conference on non-destructive testing, microanalytical methods and environmental evaluation for study and conservation of works of art*; Ed.; Budapest.

Advances and Trends in Engineering Sciences and Technologies II – Al Ali & Platko (Eds)
© 2017 Taylor & Francis Group, ISBN 978-1-138-03224-8

Modeling of the impact of a pedestrian crossing on resulting capacity of a roundabout

J. Petru, V. Krivda, K. Zitnikova & M. Kludka
VSB—Technical University of Ostrava, Ostrava, Czech Republic

ABSTRACT: This article deals with problematics of the impact of a pedestrian crossing on the resulting capacity of a roundabout. The issue of capacity calculation is described according to valid Czech technical conditions. The influence of pedestrian crossing is only considered when calculating the capacity of the exit. Moreover, only cases which are taken into account are the ones with certain minimum number of pedestrians' passes over the pedestrian crossing. Results obtained by calculations according to these technical conditions can be properly compared with results of realized simulations. The article compares the outputs that have been identified through a traffic microsimulation model created in PTV VISSIM software.

1 INTRODUCTION

Assessing the effectiveness of a specific type of intersection safety and continuity of traffic depends on many aspects. These aspects are for example traffic accidents, the number of conflict situations (Krivda 2013), traffic flow intensity, vehicle speed, impact on public transport (Horak, Ivan & Fojtik 2015) etc. Another important aspect is the resulting capacity of the intersection, which is especially influenced by the type of intersection (uncontrolled, roundabouts, traffic-light controlled, etc.), by traffic volume of vehicles and pedestrians, their speeds etc. (Perdomo, Bergman) The resulting capacity can be determined by estimation according to the type of intersection, by calculation or by using the specialized simulation software.

In the Czech Republic—according to CSN 73 6102 (CSN 73 6102, 2012)—is generally estimated the maximum hour capacity:

– of uncontrolled intersections and mini-roundabouts on 1500–2000 vehicles/hour,
– of single-laned roundabouts on 2000–2700 veh/h,
– of two-laned and spiral roundabouts on 2500–3500 veh/h,
– of traffic-light controlled intersections on 2000–6400 veh/h.

For more accurate calculations of the capacity appropriate technical conditions that corresponds to the specific type of intersection are used i.e. for uncontrolled intersections— technical conditions TP 188 (TP 188, 2007), for roundabouts—technical conditions TP 234 (TP 234, 2011) and for traffic-light controlled intersections—technical conditions TP 235 (TP 235, 2011).

However, those technical conditions only deal with simpler intersections—which means in some cases other aspects such as the impact of pedestrians or cyclists, the impact of tram traffic, the human factor and so on are not taken into account. The effect of these impacts can be taken into account, for example, by using the microsimulation modeling software PTV VISSIM (Gavulova 2012). Hence, this is the main topic of this article—comparing the capacity calculations according to technical conditions and the outputs that have been identified through the microsimulation model created in the PTV VISSIM software, exactly the impact of pedestrians on the resulting capacity of the roundabout.

2 USED METHODS

2.1 *The theoretical calculation according to technical conditions*

In the Czech Republic there are used technical conditions TP 234 (TP 234, 2011) for roundabouts. These technical conditions assess the capacity of intersection not only on the basis of the entry-lane capacity, but also exit-lane capacity. Entry lanes and exit lanes must fit the requirements to satisfy the resulting capacity of roundabout. Crossing pedestrians have a main impact on vehicles leaving the roundabout, so the following text focuses on exit-lane capacity assessment.

Exit-lane capacity is determined according to the following formula (TP 234, 2011).

$$C_e = \frac{3600 \cdot n_{e,koef}}{t_f} \, [\text{veh/h}] \tag{1}$$

Where: $n_{e,koef}$ = coefficient taking into account the number of exit lanes [–] (1.0 for single-laned and 1.5 for two-laned exit lanes) and t_f = subsequent time interval [s], which is determined by the exit-lane radius (TP 234, 2011):

- Re < 15 m $t_f = 3.0$ s
- 15 m ≤ Re ≤ 30 m $t_f = 3.6 - 0.04 * $ Re
- Re > 30 m $t_f = 2.4$ s.

If pedestrians cross the exit, then we must take into account the impact of these pedestrians. This impact is taken into account only if the crossing pedestrian volume I_{ch} is more than 250 ped/h or the sum of pedestrians which are crossing the exit lane, and vehicles which are leaving the roundabout ($I_{ch} + I_e$) is more than 800 (ped + veh)/h. Exit-lane capacity C_e is determined as (TP 234, 2011):

$$C_e = \frac{3600 \cdot n_{e,koef}}{t_f} \cdot e^{-\frac{I_{ch}}{3600}\left(t_g - \frac{t_f}{2}\right)} \, [\text{veh/h}] \tag{2}$$

where $n_{e,koef}$ and t_f (see above), I_{ch} = pedestrian crossing volume [ped/h], t_g = critical time interval [s], according to formula:

$$t_g = \frac{d_p}{v_p} + \frac{d_v}{v_v} + t_{bezp} \, [\text{s}] \tag{3}$$

where: d_p = length of the pedestrian crossing [m], v_p = pedestrian speed [m/s]; $v_p = 1,6$ m/s, d_v = vehicle length [m], ($d_v = 6,0$ m), v_v = vehicle speed [m/s]; for $R_e \leq 15$ m is $v_v = 5,56$ m/s (i.e.. 20 km/h) and for $R_e > 15$ m is $v_v = 8,83$ m/s (i.e. 30 km/h), t_{bezp} = safety distance between vehicle and pedestrian [m]; $t_{bezp} = 1,7$ s.

The assessment of the exit-lane capacity depends on the degree of capacity utilization, which is calculated as a ratio of the entry-lane volume and entry-lane capacity. If this degree of capacity utilization is less than 0.9, the entry lane fits the requirements. If all of entry lanes fit the requirements, than the resulting capacity is satisfying. According to the technical conditions TP 234 (TP 234, 2011) the impact of the backwater of vehicles entering the ring road belt is recommended to check by using microsimulation model. This is the subject of this article.

2.2 *Automated data collecting—volumes, speed, vehicle types*

To determine volumes, speed and vehicle types there were used a video record and results of automatized data collecting. There were used the counting cards **NUMETRICS NC-200** (see left Figure 1), which our workplace—Traffic engineering laboratory (www.kds.vsb.cz/ldi/ 2013)—owns. The counting card is installed directly on road, to the middle of the watched lane. The card must be covered by the special rubber cover, which is screwed by several screws (see right

Figure 1.　The counting card NU-METRICS NC-200 with accessories including card installation on the road (www.kds.vsb.cz/ldi 2013).

Figure 1). The counting card NC-200 uses the **VMI** technology (Vehicle Magnetic Imaging) to detect number, speed and type of vehicles. The **HDM** software (Highway Data Management) is used for setting the parameters of the measuring device, to export the measured data and for data evaluation. The counting card allows the classification into 13 groups according to length and 15 groups according to speed. The counting card NC-200 is able to detect vehicle speed of 13 to 193 km/h. The measuring device is placed in a durable aluminium case. The case dimensions are 181 x 118×12.1 mm and its weight is 0.59 kg. The device is able to collect data in the temperature range of $-20°C$ to $+60°C$. The memory capacity is 3MB, which allows to save data of about 300 000 vehicles. Lithium rechargeable battery should stand about 21 days of measurement.

2.3　*Microsimulation modeling*

Model is generally simplified form of reality (Brozovsky & Jasek & Maluchova 2014). Modelling of the traffic can be used in various ways—e.g. modelling the traffic to monitor the traffic flow all over (Bezak & Neumannova 2011), the traffic flow on intersections (Gavulova 2012), modelling of the demand and supply of mass transport (Olivkova 2015), of interchanges and so on. To create a model which would be used to simulate traffic of monitored roundabouts, there was used a specialized simulation software **PTV VISSIM**. This software is designed for microsimulations. The principle of microsimulation is modelling of the driving of each vehicle over the road network. There are taken to account all the parameters of the infrastructure and traffic means, including driving behavior. To enter details about vehicles are used their size, achieved speed, acceleration and deceleration, weight or power. Another part of the inputs are characteristics of all drivers or other road users (pedestrians, cyclists, trams etc.). Further, it is necessary to enter details about total intensity of road network, the number of freight vehicles and other data. The output is standard data that can be collected within other traffic surveys. It may be a capacity of roads or intersections, delays, queue length, section speed, average speed and so on.

3　THE ASSESSMENT OF EXIT-LANE CAPACITY OF ROUNDABOUT UNDER DIFFERENT CONDITIONS

As a part of the scientific research there were monitored many roundabouts. For the purpose of the research of the impact of pedestrian crossing on resulting capacity of roundabout using microsimulations, there were chosen two-laned roundabout, exactly single-lane exit, which occurs a problematic pedestrian crossing (see Figure 2). There were designed new adjustments at this intersection (the length of the pedestrian crossing, exit-lane radii etc.—see below.

First, this exit-lane was assessed without the impact of pedestrians according to the calculations in the technical conditions ("calculation 1"—see Table 1), then it was taken to account the impact of pedestrians again according to the calculation in the technical conditions ("calculation 2"—see Table 2) and finally, using the microsimulation software PTV VISSIM ("simulation"— see Figure 3), where were simulated vehicle flow and also pedestrians, which were crossing. In calculation 1 and 2, there was monitored the degree of capacity utilization. In simulation, there were monitored travel time and delay time. Travel time is real time of passing the vehicle between

Figure 2. The monitored exit-lane (www.mapy.cz 2016).

Table 1. Results of the capacity calculation no. 1.

	Variant A	Variant B	Variant C
I_e [veh/h]	600	600	600
$n_{e,koef}$ [-]	1	1	1
R_e [m]	60.4	82.0	35.0
t_f [s]	2.4	2.4	2.4
C_e [voz/h]	1500	1500	1500
a_v [–]	0.4	0.4	0.4

Table 2. Results of the capacity calculation no. 2 (I_e, $n_{e,koef}$, R_e and t_f see Table 1).

	Variant A	Variant B	Variant C
I_{ch} [ped/h]	368	368	368
d_p [m]	25.5	7.5	4.0
v_p [m/s]	1.6	1.6	1.6
d_v [m]	6.0	6.0	6.0
v_v [m/s]	8.83	8.83	8.83
t_{bezp} [s]	1.7	7	1.7
C_e [veh/h]	215	823	1029
a_v [–]	2.8	0.7	0.6

selected road profiles (in our model between the point on the orbital lane in sufficient distance from the pedestrian crossing and the point 5 meters far from the pedestrian crossing). Delay time is the difference between optimum (i.e. theoretical) travel time and actual travel time.

Each of these cases was assessed for different conditions, i.e. for different actual length of the pedestrian crossing, which is de facto credited to the calculation (for more details see Table 1 and Table 2):

– variant A—current setup (see Figure 2); in the Czech conditions the total length of the pedestrian crossing is measured between both corners (because the pedestrian crossing is not split). That means it includes that part of the pedestrian crossing which pass through the entrance to the roundabout,

– variant B—the pedestrian crossing is split into two parts; the pedestrian crossing is split in the place where is located the traffic island. Calculations include only the part which pass through the roundabout's entry lane; moreover, there was made an increase of the exit-lane radius and the number of orbital lanes was reduced from two to one.

– variant C—the pedestrian crossing is split into two parts again; there were made significant reductions of the exit-lane radius and the exit lane width which is related to the length of the pedestrian crossing; this option has also a single-laned orbital lane.

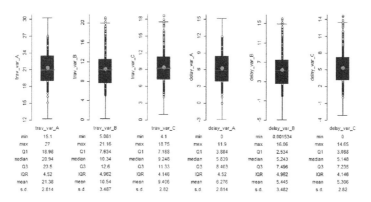

	trav_var_A		trav_var_B		trav_var_C		delay_var_A		delay_var_B		delay_var_C
min	15.1	min	5.081	min	4.1	min	0	min	0.001534	min	0
max	27	max	21.16	max	18.75	max	11.9	max	16.06	max	14.65
Q1	18.98	Q1	7.634	Q1	7.188	Q1	3.884	Q1	2.534	Q1	3.088
median	20.94	median	10.34	median	9.248	median	5.839	median	5.243	median	5.148
Q3	23.5	Q3	12.6	Q3	11.33	Q3	8.403	Q3	7.496	Q3	7.235
IQR	4.52	IQR	4.962	IQR	4.146	IQR	4.52	IQR	4.982	IQR	4.146
mean	21.38	mean	10.54	mean	9.406	mean	6.276	mean	5.445	mean	5.306
s.d.	2.814	s.d.	3.487	s.d.	2.82	s.d.	2.814	s.d.	3.482	s.d.	2.82

Figure 3.   Results obtained by simulation in the PTV VISSIM software, processed in the GeoDa software (units = seconds).

Results obtained from the **PTV VISSIM** microsimulations were statistically processed in GeoDa software and they were represented in the so-called Box Plots (see Figure 3). The first three graphs show results of the measurement of travel times for each variant (A to C). Next three graphs show results of the measurement of delay times (A to C again). In each graph are empty circles which shown every single value of measured units. Some of statistical values are eliminated (i.e. those above, respectively below the horizontal lines in the graph). The horizontal lines show a maximum, respectively a minimum of not eliminated values. The rectangle shows interquartile range (IQR) - i.e. the difference between the 3rd quartile (Q3) and the 1st quartile (Q1). The 2nd quartile is denoted as a median (see the light horizontal line in the graph). Mean is denoted as a full light circle. Standard deviation is denoted as s.d.

4   EVALUATION AND CONCLUSION

Results from the capacity calculation no. 1 (see Table 1), i.e. condition without the impact of pedestrians, shows the same data inputs. The difference between all variants are only in the exit radius Re, which is the largest in the variant B (82 m) and the smallest in the variant C (35 m). Based on the calculation according to the formula 1 we obtain the same results of exit-lane capacity Ce and of the degree of capacity utilization av. It is due to the fact that the subsequent time interval t_f for exit radii larger than 30 meters is the same, so proportions of this radius don't affecct the equation.

In capacity calculation no. 2 (see Table 2), i.e. conditions with the impact of pedestrians, there are other parameters, which significantly influence the resulting exit-lane capacity Ce and the degree of capacity utilization av. The most important parameter is the number of pedestrians Ich and the length of the pedestrian crossing dp. The number of pedestrians is the same for each variant, the length of pedestrian crossing has changed due to an appropriate variant (details are written above). The resulting values shows how the length of the pedestrian crossing significantly influences the resulting exit-lane capacity Ce. The variant A (the length of pedestrian crossing is 25.5 m) disposes the lowest capacity (215 veh/h) and the reserve of capacity is even negative (that means: the degree of capacity utilization av is higher than 1, exactly 2.8). Shortening of pedestrian crossing to 7.5 m (in the variant B), respectively to 4.0 m (in the variant C) will raise the resulting capacity to 823 veh/h (in the variant B), respectively 1029 veh/h (in the variant C). It will also reduce the degree of capacity utilization av to 0.7 (in the variant B), respectively to 0.6 (in the variant C). Both of these values satisfy the condition of sufficient capacity for roundabout's exits according to TP 234. The assessment of the influence of pedestrian crossing of the roundabout exit with the use of simulation model is based on different results. These results are the average travel time and the delay time. Even in these simulation models the pedestrian impact was taken into account. It is obvious that the travel time and the delay time in variant A are higher than in the other variants. Travel time reaches twice better average value in the variant A (21.38 s) than variants B and C (10.54 s and 9.406 s). Delay time is in average the longest in the variant A (6.276 s), in variants B and C values are similar (5.445 s and 5.306 s). In case of the variant A, the higher values are caused by the obligation to allow pedestrians to cross pedestrian crossing, which, in the variant A, have to surpass longer distance than variants B and C.

This article shows the option of using not only sophisticated capacity calculations, but also using modern microsimulation software. Results of both assessments are similar, although the ways of assessment were different, and can be used not only for the described effect of pedestrians crossing the roundabout. Sometimes modeling can be faster than the capacity calculation. A significant advantage is that it is possible to create one model and then simply and quickly change the input data.

ACKNOWLEDGEMENTS

The work was supported by the Student Grant Competition VSB-Technical University of Ostrava, Czech Republic, in 2016. Project registration number is SP2016/96.

REFERENCES

Bergman, Astrid, Johan Olstam a Andreas Allstrom. Analytical Traffic Models for Roundabouts with Pedestrian Crossings. *Procedia—Social and Behavioral Sciences* [online]. 2011, **16**, 697-708 [cit. 2016-08-25]. DOI: 10.1016/j.sbspro.2011.04.489. ISSN 18770428. Available from: http://linkinghub.elsevier.com/retrieve/pii/S1877042811010354.

Bezak, Bystrik & Neumannova, Miroslava 2011. Modeling of Cross-Border Interaction in central Danube Region. In *8th International Conference on Environmental Engineering, ICEE 2011*, p: 854–859.

Brozovsky, Jiri & Jasek, Marek & Maluchova, Marketa 2014. Modeling of Strengthening of Masonry Pillars with Carbon Fabric. In *Advanced Materials Research*, Vol. 969: 253–258. DOI: 10.4028/www.scientific.net/AMR.969.253.

CSN 73 6102 Design of Intersections on Highways—edition 2. Czech technical standards, 2012.

Gavulova, Andrea & Drliciak, Marek. Capacity Evaluation of Roundabouts in Slovakia 2012. In *Transport and Telecommunication*. Vol. 13(1): 1–10. DOI: 10.2478/v10244-012-0001-0.

GeoDa [online]. [cit. 2016-08-25]. Available from: https://geodacenter.github.io/

Horak, Jiri & Ivan, Igor & Fojtik, David 2015. Time of day dependency of public transport accessibility in the Czech Republic. In *11th Symposium on Geoinformatics for Intelligent Transportation, GIS Ostrava 2014*, Lecture Notes in Geoinformation and Cartography, Vol. 214: 93–108. DOI: 10.1007/978-3-319-11463-7_7.

http://kds.vsb.cz/ldi—Laboratory of Traffic Engineering, Department of Transport Constructions, Faculty of Civil Engineering VSB—Technical University of Ostrava, Czech Republic http://www.mapy.cz - maps server.

Ivan, Igor 2016. Interchange Nodes between Suburban and Urban Public Transport: A case study of the Czech Republic. In *Acta Geographica Slovenica*. Vol. 56(2): 221–233. DOI: 10.3986/AGS.754.

Krivda, Vladislav 2013. Analysis of Conflict Situations in Road Traffic on Roundabouts. In *Promet—Traffic&Transportation*. Vol. 25(3): 295–303. DOI: 10.7307/ptt.v25i3.296.

Olivkova, Ivana. Model for Measuring Passenger Satisfaction and Assessing Mass Transit Quality. In *Journal of Public Transportation*. Vol. 18(9): 52–70. DOI: 10.5038/2375-0901.18.3.4

Perdomo, Mario, Ali Rezaei, Zachary Patterson, Nicolas Saunier a Luis F. Miranda-Moreno. Pedestrian preferences with respect to roundabouts—A video-based stated preference survey. *Accident Analysis & Prevention* [online]. 2014, 70, 84–91 [cit. 2016-08-25]. DOI: 10.1016/j.aap.2014.03.010. ISSN 00014575. Available from: http://linkinghub.elsevier.com/retrieve/pii/S0001457514000761.

TP 188 Assessment of Unsignalized Intersections Capacity. Czech technical standards. 2007. ISBN 978-80-902527-6-9 (in Czech).

TP 234 Assessment of Roundabouts Capacity. Czech technical standards. 2011. ISBN 978-80-87394-02-01 (in Czech).

TP 235 Assessment of Signalized Intersections Capacity. Czech technical standards. 2011. ISBN 978-80-87394-03-8 (in Czech).

Advances and Trends in Engineering Sciences and Technologies II – Al Ali & Platko (Eds)
© *2017 Taylor & Francis Group, ISBN 978-1-138-03224-8*

Research of selected factors of safety in road tunnels for practice

M. Rázga, E. Jančaříková & P. Danišovič
Faculty of Civil Engineering, University of Žilina, Žilina, Slovakia

ABSTRACT: Safe operation of tunnels is very important, because tunnels are specific engineering structures constructed in order to shorten transport routes and improve road safety. Risk analysis for Slovak road tunnels is carried out according technical specification TP 02/2011. Model of risk analysis examines the personal risk of tunnel users and evaluates statistically the expected number of victims per year. For a more detailed examination of the risks of transporting dangerous goods through the tunnel, it is necessary to use a special risk model. A suitable example is a specific risk model DG-QRAM (Dangerous Goods—Quantitative Risk Assessment Model). Tunnel Traffic & Operation Simulator at the University of Žilina in combination with unique software is conducive to research of the possible operating conditions during normal service and model emergency situations. The Simulator manages technological equipment of a virtual two-tube highway tunnel interconnected with simulation of vehicle tunnel traffic.

1 RISK ANALYSIS

Risk analysis model (TP 02/2011) examines the personal risk of tunnel users, which means that all the parameters applied thereto are solely for accidents to persons. To assist the rating valuation limits are implemented and according to them, tunnels are classified during project preparation and some adjustments can be realized in existing tunnels if it's allowed by technical and operating conditions. Risks can be classified according to mechanical damage, fire and transport of dangerous goods and the effects of dangerous goods are examined in an extremely simplified manner, and therefore the model is not suitable for deeper exploration of the risks of accidents (in the transport of dangerous goods).

Given the results of the parametric study the influence of individual input parameters it can be concluded that the simplified model of the effects of dangerous goods should be on that model completely excluded or should be determined by objective parameters for the use of this parameter in the risk analysis. It should not be the case in view of the fact that the TP from authorized transport of dangerous goods recommended for deeper exploration risk to use a special risk model DG-QRAM (Dangerous Goods—Quantitative Risk Assessment Model), developed in partnership with the OECD-PIARC (Danišovič 2011). Nowadays, this document is used for a safety assessment of the tunnels operation.

1.1 *The structure of the risk model*

The structure of risk model requires a systematic approach and it consists of two main parts (Figure 1):

- quantitative analysis of frequency,
- quantitative analysis of the consequences of accident.

1.2 *Risk assessment*

The calculation was realized by researchers—engineer students (Ing. Martin Rázga) who worked on topics related to the scientific tasks solutions and PhD students (Ing. Peter Danišovič, PhD, Ing. Ján Filipovský, PhD). Remarks of the solvers of the research task oriented to the security of the operation in tunnels (Transport Research Centre) were used within a processing of a doctoral thesis and the results were presented in the doctoral thesis.

Evaluation of the results of quantitative analysis is made from the relative risk evaluation comparison with a reference tunnel and absolute risk evaluation assigned to hazard classes. Relative risk assessment is that the tunnel complying with all relevant technical safety characteristics cannot show a higher risk than a similar reference tunnel which satisfies relevant minimum requirements. In this way, it should be possible to prove that using prescriptive measures can keep a minimum level of security. However, experience shows various weaknesses in the model for calculating risk reference tunnel (Schlosser et al. 2013).

Manual TP 02/2011 and TP 04/2014 is used for solving scientific tasks and the realized evaluation serves as a basis for the preparation of tunnels operation prediction models processing. For each solution it is created an event tree according to the incident´s nature (an accident or a fire in the tunnel) for specific tunnel conditions (Figure 2) (Danišovič & Schlosser 2014).

Absolute risk evaluation complements the relative evaluation and its goal is to provide information about the absolute risk margin. Based on the expected value of risk (specified risk analysis) the studied tunnel is assigned to risk (hazard) class according to (TP 02/2011) (Table 1):

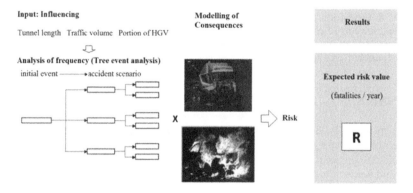

Figure 1.   Structure of the risk model, (TP 02/2011).

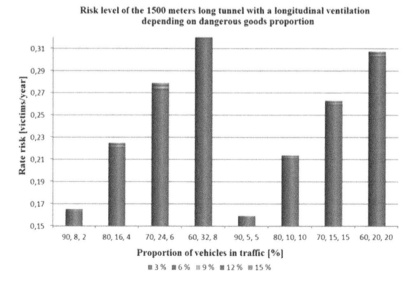

Figure 2.   Risk level of the 1500 meters long tunnel with longitudinal ventilation depending on dangerous goods proportion (Source: own).

Table 1. Distribution of hazard classes, (TP 02/2011).

Expected risk value		
Lower thresholds	Upper threshold	Hazard classifications
–	= 0.02	I
>0.02	= 0.10	II
>0.10	= 0.50	III
>0.50	–	IV

2 TRANSPORT OF DANGEROUS GOODS

A prescription TP 02/2011 in combination with QRAM software is used for the parameter´s evaluation. In the Figure 2 we can see an example of the parameter of quantity of vehicles with dangerous goods of the 1500 m long tunnel according to the methodology of the Centre transport research.

According to the prescription TP 02/2011 it is necessary to use a QRAM methodics because of small part of the influence of vehicles with dangerous goods. In some tunnels—by the degree of danger—the passage of vehicles with dangerous goods is prohibited. Nowadays, it is reviewing the transit possibility of these vehicles, so we are accomplishing some calculations.

For a more detailed examination of the risks of transporting dangerous goods through the tunnel, it is necessary to use a special risk model. Suitable is for example QRA software, which was developed at the OECD-PIARC. This program is part of the equipment of the department of technology and management structure.

2.1 Exploration of ADR

Transport Research Centre of University of Žilina made the exploration of proportion of dangerous goods in transport through the tunnel and it was found out how many of these vehicles use a detour of tunnel, and how many these vehicles drive through the tunnel despite the entry ban. The research was carried out three days in the period from 5.3.2013 to 7.3.2013 from the place before the tunnel, where were two observers 24 hours a day. Result of the research showed that proportion of dangerous goods in this section is 2.19% from the transport of HGVs. It was also found that approximately 26.32% of these HGVs with dangerous goods (marked as ADR) go through the tunnel despite the entry ban the transport of dangerous goods in road tunnels in Slovakia and 73.68% (Figure 3) use the detour (Schlosser & Rázga & Danišovič 2014).

2.2 Risk analysis of one tube tunnel with transport of dangerous goods

Based on the experimentally determined ratio of the transport of dangerous goods was done re-calculation of risk analysis One tube tunnel according to TP 02/2011, but with modified input data, with the exact proportion of the transport of dangerous goods (2.19%). The result shows Figure 4.

The Figure 4 shows that as soon as the transport of dangerous goods passed through the tunnel, the number of fatalities per year will increase from 0.07389 to 0.07415 which is an increase by about 2.8%. It seems that this difference is negligible, because proportion of transport of dangerous goods is not very large, but the danger is mainly that when vehicle with dangerous goods in an accident risk of fatalities is very big. With the increase in the number of highway tunnels in Slovakia in the near future, this fact can be important in the safety of road users.

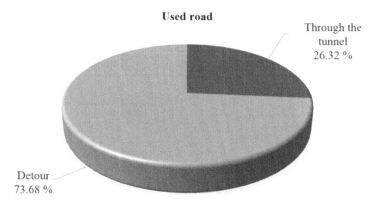

Used road

Through the
tunnel
26.32 %

Detour
73.68 %

Figure 3.   The use of the tunnel with the transport of dangerous goods (Source: own).

Number of death per year

With transport
of dangerous
goods
0.07415

Without
transport of
dangerous
goods
0.07389

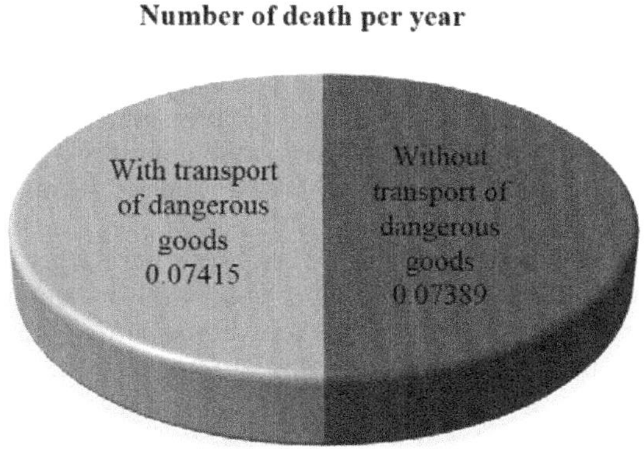

Figure 4.   Number of death per year in the tunnel with and without the transport of dangerous goods
(Source: own).

3   TUNNEL TRAFFIC & OPERATION SIMULATOR

Basic composition of the Simulator consists from central control system (CCS—part of automatic tunnel equipment control), manual control module (MCM—part of manual control
as separated module, it serves to a manual tunnel operation in the case of the central control
system's accident), software for prediction of phenomena (EMUT)—evidence of tunnel incidents. A workplace of one of two operators is in the figure 5.

It is possible to verify operating conditions at the Tunnel Traffic & Operation Simulator by
real operation at the visualization on each operating condition. An operating condition with
a formation of fire and the possible extension to other vehicles in the tunnel is in preparation too. Verification of operators response in operating conditions in the tunnel tube, also
throughout a formation of fire, is possible only on the Tunnel Traffic & Operation Simulator
in the Transport Research Centre. In the case of operating condition with dangerous goods
transportations vehicles it is very necessary and required by the TP 04/2014 a preventive verifying of operators professional competence.

3.1   *Software for evidence of tunnel's incidents/accidents (EMUT)*

EMUT is a special program for evidence and evaluation of incidents in tunnels in Slovakia.
It is registered every change compared to the tunnel's operating conditions under normal
operating conditions, when an intervention of technology's operator and traffic's operator
is not required. Important output of this software is a graphic representation of incidents by

Figure 5.   Tunnel Traffic and Operation Simulator (Source: own).

time (day, month, year), by type of incident or by tunnel, and this can be useful for planning, statistics and assessment of the level of security.

4   CONCLUSIONS

It is possible to assess qualitatively each highway tunnel´s parameter in the project preparation and in the operation using a risk analysis and special software for the evaluation of tunnel's emergencies. The outputs from the research of scientists PhD students and practice specialists allow for a direct application of outputs in practice.

Risk analysis, special operations and the simulation of operating conditions on the Tunnel Traffic & Operation Simulator increases professional competence of operators, who manage the highway tunnels operation. Lessons learned from the solutions are useful in practice and contribute significantly to improving the environment by optimizing the highway tunnel tubes design.

ACKNOWLEDGEMENT

The research is financed by: This contribution is the result of the project implementation: "Centre of Transport Research" (ITMS 26220220135) supported by the Research & development Operational Programme funded by the ERDF. "We support research activities in Slovakia/project co-founded from the resources of the EU".

The research is supported by the European Regional Development Fund and the Slovak state budget by the projects "Research Centre of the University of Žilina"—ITMS 26220220183.

REFERENCES

Danišovič, P. 2011. Analýza rizík a návrh opatrení na prevádzku tunelov. Dizertačná práca. Žilinská univerzita v Žiline, Stavebná fakulta. Žilina, 2011.; in english: Risk Analysis and proposal of measures for the tunnels operation. Dissertation. University of Žilina, Faculty of civil engineering, Žilina.
Danišovič P., Schlosser F. 2014. TUNELY. Riadenie a prevádzka tunelov. Návody na cvičenia. EDIS ŽU v Žiline, 2014, ISBN978-80-554-0962-7.; in english: Tunnels. Management and operation of tunnels. Instructions for exercises. EDIS University of Žilina.
Nariadenie vlády SR 344 zo dňa 24. mája 2006.; in english: Slovak Republic Government Regulation 344 from may 24, 2006.

Schlosser F. a kol. 2008. Stavba a prevádzka tunelov/Tunnelbau und Tunnelbetrieb. EDIS ŽU, 2008.; in english: Construction and operation of tunnels. EDIS University of Žilina.

Schlosser F., Danišovič P., Rázga M. 2013. Zvýšenie bezpečnosti počas prevádzky v cestných tuneloch, Zborník z konferencie Q-2013, Žilina, 2013, ISBN 978-80-554-0758-6.; in english: Improve safety during operation in road tunnels, In: Q-2013 Conference, Žilina.

Schlosser F., Danišovič P. 2014. Tunnel Traffic & Operation Simulator. In: International Scientific Conference "Construction Technology and Management". Bratislava, September, 09–11.

Schlosser F., Rázga M., Danišovič P. 2014. Risk Analysis in Road Tunnels. In: XXIII R-S-P Seminar, Theoretical Foundation of Civil Engineering (23RSP).

TP 02/2011 Analýza rizík pre slovenské cestné tunely. MDVRR SR, október 2010, platnosť od 1.6.2011.; in english: Risk Analysis for slovak road tunnels. Ministry of transport, construction and regional development SR, october 2010, validity from 1.6.2011.

TP 9C-1/2005 Prehliadky, údržba a opravy cestných komunikácií. Tunely—stavebné konštrukcie, [10] MDPT:2005.; in english: Inspection, maintenance and repair of roads. Tunnels—Building construction.

TP 02/2002 Katalóg porúch asfaltových vozoviek.; in english: List of failures of asphalt pavement.

TP 10/2005 Katalóg porúch tunelov na pozemných komunikáciách, MDPaT SR, 2005.; in english: List of failures of tunnels on roads.

TP 4/2014 Prehliadky, údržba a opravy cestných komunikácií. Tunely—technologické vybavenie. MDVRR SR, 2014.; in english: Inspection, maintenance and repair of roads. Tunnels—technological equipment.

Urdová A., Rázga M. 2014. Simulátor riadenia tunelov ako nástroj zvyšovania bezpečnosti v cestných tuneloch. In: Časopis "Pozemné Komunikácie A Dráhy", Roč. 10, č. 2/2014, ISSN 1336-7501, Košice.; in english: Tunnel Traffic & Operation Simulator as a tool for improving safety in road tunnels. In: Journal "Roads & tracks", grade 10, n. 2/2014.

Advances and Trends in Engineering Sciences and Technologies II – Al Ali & Platko (Eds)
© 2017 Taylor & Francis Group, ISBN 978-1-138-03224-8

Effects of biaxial and thermal loads on rail-slab track interaction behavior

V. Stančík & P. Ryjáček
Department of Steel and Timber Structures, Faculty of Civil Engineering,
Czech Technical University in Prague, Prague, Czech Republic

ABSTRACT: The number of slab track applications is nowadays increasing mainly due to its assumed lower life cycle costs and its suitability for High Speed Railway (HSR) tracks. Moreover the bridge deck self-weight and structural height may be significantly reduced in the case of performing bridge decks with slab tracks. On the other hand the rail-slab track interaction phenomenon has to be taken into account especially when designing bridge structures as it may significantly affect the longitudinal stress distribution in the continuously welded rail or the bridge substructure dimensions. The main goal of this contribution is to investigate the longitudinal and vertical stiffness of the direct fastening system which is the key element for describing the interaction phenomenon. In order to obtain these data a small scale experiment on a laboratory sample was conducted and the results were subsequently numerically verified. The whole analysis provides useful remarks for bridge design in practice.

1 INTRODUCTION

1.1 *Continuously welded rails and slab track*

Railway infrastructure went through a significant development in recent decades. Most importantly jointed rails were replaced with the Continuously Welded Rails (CWR), which naturally ensure a lowering of noise and dynamical impact on both the railway track and the rail vehicle. Using CWR consequently allows for higher travelling speeds and passenger comfort. However the removal of expansion joints results in significant normal stress increases, because the rail expansion movement is restricted. Reaching the ultimate fatigue limit state by cyclic loading in tension may cause rail breakage. Reaching the critical value of stress in compression will result in the railway track buckling. The value of the stress increment is generally affected by the parameters of railway sub- and superstructure and acting horizontal and vertical loads.

Simultaneously with the development of CWR new possibilities have been searched to create competitive and sustainable railway tracks. A good alternative for classical ballasted tracks were found to be the slab tracks. Its main advantage is the reduction of the railway track life cycle costs. In slab tracks the durability of the small steel parts of the fastenings is longer. Also, thanks to the rigid substructure the track gauge doesn't change in time, which significantly saves maintenance costs. This is especially suitable for construction of HSR. Slab tracks also allow reduction of substructure construction height, which is convenient especially when considering railway bridges. Higher lateral stiffness prevents the slab track from buckling. Unfortunately the initial costs related to newly built slab tracks are much higher in comparison with traditional ballasted tracks. Ballastless tracks generally consist of CWR directly fastened to in-situ poured or precast concrete slab, or a steel bridge deck plate. The most frequently used slab track type in Europe is the system Rheda 2000 using the Vossloh DFF 300 direct fastening system.

1.2 *Bridge-slab track interaction phenomenon*

Unlike the ballasted track, the horizontal and vertical stiffness of the slab track doesn't depend on whether the track is located on bridge or in an open track. However on the bridge we can surely predict higher stress increments due to diverse expansion movement possibilities of the bridge structure and the CWR located in the sections above sliding or fixed bearings. Bridge-track interaction evaluation principles are shown on Figure 1. It is obvious that the longitudinal forces coming from bridge-rail interaction may affect not only rail stress distribution, but also the whole bridge design, especially the dimensions of substructure and bearings. The longer the bridge span, the higher the influence of the interaction.

Fastening system is the crucial component, which determines the transmission of horizontal and vertical forces between the substructure and the CWR. The coupling interface is represented with the horizontal spring, introducing the nonlinear dependence of longitudinal resistance r_x against the relative bridge-rail displacement u_x. Generally this nonlinear function has the shape of the dotted line showed in Figure 2, but for the purposes of numerical analysis, the function is usually approximated with bilinear idealization, where the elastic resistance grows linear till it reaches the limit value of elastic displacement u_0. Above the value u_0 the resistance doesn't grow anymore and remains constant with the value of plastic resistance r_0. Longitudinal stiffness k_x of the fastening is defined with the slope of the elastic part of the graph. Under vertical load the longitudinal resistance grows. Hence different interaction functions for loaded and unloaded track need to be considered. Last but not least the vertical interaction is defined with the linear vertical stiffness k_z and is variable for compressive and tensile loadings.

Values of interaction parameters for balastless tracks (Figure 2) introduced by European national codes do not cover modern fastening systems, including low resistance systems. Moreover the effects of thermal loads on the direct fastening elastomer pads have not been checked yet. Existing regulations requires only testing of the resistance under normal condition according to EN 13146-1. We may presume that the horizontal and vertical coupling functions will change with temperature. Therefore the main goal of this contribution is to investigate the interaction parameters of one of the most used slab-track fastening systems under various thermal and biaxial loading conditions and to provide data for rail-slab track interaction assessment in practice.

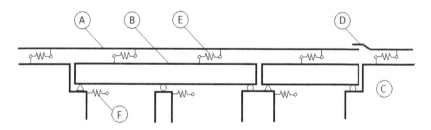

Figure 1. Interaction principles (A = railway track; B = bridge structure; C = embankment; D = expansion device; E = spring representing nonlinear horizontal interaction functions and vertical stiffness of the track; F = horizontal stiffness of the substructure and bearings).

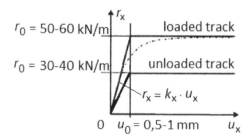

Figure 2. Longitudinal resistance (solid line = bilinear idealization; dotted line = experimental data).

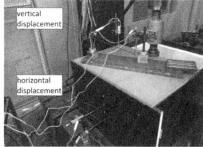

Figure 3. Experiment layout.

2 EXPERIMENT

In order to investigate the interaction behavior of the ballastless track there was conducted an experiment on a small scale laboratory sample which involved a $500 \times 500 \times 20$ mm steel plate, one node of the Vossloh DFF 300 fastening system equipped with standard tension clamp skl 15, in which the 60 E1 rail was clamped. The basic principle of the experiment corresponds to the EN 13146-1. The tested sample was placed into the insulated box first and clamped to the grid support with the bolts. The dimensions of the sample were designed with the intention to reduce the maximum needed space so that the insulated box could be smaller and the cooling and heating procedures more effective. Between the vertical cylinder and the rail head there was placed a small sliding bearing to reduce the influence of friction. The rail was then loaded with the horizontal and vertical hydraulic cylinders according to particular load cases. Observed quantities were the loading forces, the longitudinal displacement between the rail face and the steel plate, vertical displacement between the rail foot and the steel plate and the temperature in various spots of the tested sample.

A series of load cases were designed in order to achieve the goals of the experiment. The experiment was performed step by step considering four temperature values: $T = 20, 55, 0$ and $-15°C$. Each temperature was tested under two levels of vertical loading. In unloaded track load cases the value of vertical force was set to be $F_z = 0$ kN, while the load cases for loaded track considered the value of vertical force $F_z = 62.5$ kN, determined according to the load model 71. Each load case was terminated when the relative displacement reached the value of $u_{x,max} = 7$ mm. The process of tempering and maintaining certain level of temperature was quite interesting part of the experiment. Higher temperatures were obtained and maintained quite easily using the heating cable that was wrapped around the rail. The heating procedure took about 7 hours and resulted in quite uniform heat distribution in the insulated box. By the cooling procedure the temperature was reduced using the liquid nitrogen, which was delivered into the insulated box through a small tube from the vessel. To ensure more uniform spreading of the nitrogen vapors, there were also placed two small ventilators. Procedures of cooling and subsequent maintaining of required temperature level demanded very large amount of liquid nitrogen because of the high thermal losses caused by the test sample anchoring and the big dimensions of the box. However we managed to hold the uniform temperature around the level of $-15°C$ for about 3 hours, which was satisfactory for reaching the required temperature of the fastening components.

3 FINITE ELEMENT ANALYSIS

Obtained experimental data was verified using the numerical analysis. For this purpose a complex finite element model was created in the software Dlubal RFEM 5. The model geometry and anchoring was modeled with respect to the real sample. Numerical model consists of several components. The steel HEB profiles, which are used for anchoring the sample and for elevating the rail face against the horizontal cylinder, are modeled using the shell elements. Similarly the steel plate and the DFF 300 steel base plate are modeled using the shell elements too. All contacts

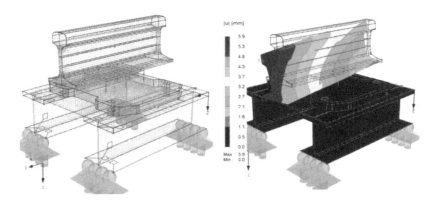

Figure 4.   Finite element model of the tested sample (left = mesh of finite elements; right = typical global deformation under horizontal and vertical loading).

between the plates or solid materials are modeled using solid contact elements. The rail itself has been modeled with the material solid elements, not because the rail assessment was needed, but because the accurate geometry of the rail allowed proper modeling of the contact between the rail and the fastening. Bolts, tension clamps and elastomer pad were modelled as beam members.

The essential part of the modelling procedure was the contact setting between the rail and the fastening system. The rail rests on steel distribution plate, which is supported by the elastomer pad. The pad is represented by four fictive rigid beams which are connected to the distribution plate through nonlinear joints, introducing the experimentally evaluated functions of R_x and vertical stiffness in compression K_z. This contact couldn't be modeled using the solid contact elements because they don't allow modelling of nonlinear shear between two plates, but the simpler solution using the fictive beams appeared to be accurate enough and clear to understand. The tensile vertical stiffness of one DFF 300 fastening node $K_z{}^+$ was introduced using the nonlinear joints of the four fictive rigid beams representing the tension clamps, which connected the rail foot to the steel distribution plate. In order to set the global rotation stiffness properly all the solid contact elements transmitted only compression and the tensile forces were transmitted through the bolts.

The nonlinear analysis using the various load steps was performed on the numerical model. Because of the relative big number of load cases and quite high complexity of the numerical model the analysis was quite time consuming. Therefore only two or three loading steps were checked for each load case.

4   RESULTS

4.1   *Verified experimental data*

The attached graphs below show the evaluated experimental data and their comparison with numerical results, which are in the figures represented by the signs related to certain load case. As can be seen the FEA results are in very good agreement with the experiment, thus the evaluated interaction functions may be considered verified. Figure 5 shows nonlinear functions of one DFF 300 node longitudinal resistance R_x categorized into loaded and unloaded cases, so that the effects of temperature would be more obvious. Figure 6 shows the vertical stiffness K_z of one node in compression.

4.2   *Evaluation of coupling parameters*

First it has to be mentioned that the observed functions are related to one node of fastening only (marked with capitals), but the interaction parameters in codes are generally offered related to one meter of the railway track. This is why the obtained data was transferred according to EN 1991-2 first. For the load cases with vertical loading it has been also assumed a small lowering of the plastic resistance which was caused by the friction in the sliding bearing. By the numerical parametric analysis it has been determined that the friction causes about 5–10 kN increase in the node longitudinal resistance. Table 1 shows an evaluated data after exclusion of friction effect.

The values of longitudinal stiffness are determined by finding the most appropriate coefficients of the Equation 1, which is used for describing nonlinear interaction functions in a more general way.

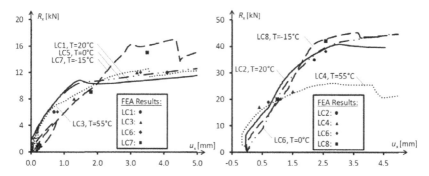

Figure 5.   Verified nonlinear resistance functions of one DFF 300 node (left = unloaded track; right = loaded track).

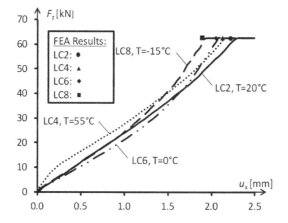

Figure 6.   Verified vertical stiffness of one DFF 300 node in compression.

Table 1.   The evaluated interaction parameters.

	T	F_Z	R_0	r_x	u_0	K_x	k_x	K_z
Load case	°C	kN	kN	kN/m	mm	kN/m	kN/m²	kN/m
LC1	20	0	11	36.7	0.59	19,000	62,150	–
LC2	20	62.5	37	80.0	1.08	38,000	84,000	26,850
LC3	55	0	12	40.0	0.89	13,500	45,000	–
LC4	55	62.5	21	53.3	1.20	33,000	48,000	27,850
LC5	0	0	12	40.0	0.63	19,000	63,500	–
LC6	0	62.5	42	91.7	1.70	27,500	60,200	31,000
LC7	−15	0	17	56.7	1,78	10,000	31,800	–
LC8	−15	62.5	42	97.1	1,46	32,000	73,800	35,850

$$r_x = r_0 \cdot \left(1 - e^{-\frac{k_x}{r_0} \cdot u_x}\right) \tag{1}$$

where r_x = longitudinal resistance; r_0 = plastic longitudinal resistance; k_x = longitudinal stiffness; u_x = relative displacement between the rail and the substructure.

5 CONCLUSION

As can be seen in the graphs and table, the values of longitudinal resistance and vertical and horizontal stiffness are affected by temperature. The vertical stiffness is increasing with decreasing temperature. Nevertheless it is also surprisingly slightly increasing with higher temperatures. Regarding the functions of longitudinal resistance, it may be stated that the plastic resistance r_0 is quite significantly varying with the temperature. Observing the data it can be stated that the value of plastic resistance decreases with temperature growth. Longitudinal stiffness k_x is also changing with temperature, however it seems to change irregularly because no pattern has been found in the temperature-longitudinal stiffness relation. This means that the temperature dependent changes in longitudinal interaction functions are more likely connected with some secondary effects of temperature change. Interestingly the detected values of plastic resistance are higher than usually recommended. This remark should be considered for potential further monitoring on existing bridges.

ACKNOWLEDGEMENT

Research reported in this paper was supported by Competence Centres Program of Technology Agency of the Czech Republic (TA CR), project Centre for Effective and Sustainable Transport Infrastrucutre (no. TE01020168).

REFERENCES

European Committee for Standardization (CEN). 2003. *Actions on Structures; Traffic loads on bridges* Eurocode 1, Part 2 (EN 1991-2). Brussels: CEN.
Esveld, C. 2001. *Modern Railway Track, 2nd ed.* Zaltbommel: MRT-Productions.
Freystein, H. & Geissler, K. 2013. (Rail/Bridge interaction on Steel Bridges with Examples. In *Steel Structures* 82(2)): 78–86. Berlin: Ernst & Sohn. [in German].
Ryjáček, P. & Vokáč, M. 2014. Long-term monitoring of steel railway bridge interaction with continuous welded rail. In *Journal of Constructional Steel Research* 99: 176–186.
Ryjáček, P. et al., 2015. The Behaviour of the Embedded Rail in Interaction with Bridges. In *IOP Conference Series Materials Science and Engineering* 96(1). IOP Publishing.
Shah, B.J. & Surti S.K. 2015. The Need for Rail Structure Interaction Analysis for Bridges: A State-of-the-Art Review. In *Journal of Structural Engineering* 8(3): 47–61. Hyderabad: IUP Publications.
Stančík, V. et al., 2014. In-situ testing of railway bridges with CWR. In *Proceedings of the 12th International Conference on steel, Space & Composite Structures*: 327–335. Singapore: CI-Premier Pte Ltd.
Stančík, V. & Ryjáček P. 2014. Comparison of the accuracy of several bridge/rail coupling models with experimental and numerical results. In *Proceedings of the Second International Conference on Traffic and Transport Engineering* (ICTTE): 257–264. Belgrade: City Net Scientific Center Ltd.
Union Internationale des Chemins de fer (UIC). 2001. *Track/bridge interaction, Recommendations for calculations* 774–3R. Paris: UIC.

Advances and Trends in Engineering Sciences and Technologies II – Al Ali & Platko (Eds)
© 2017 Taylor & Francis Group, ISBN 978-1-138-03224-8

Using of photogrammetric methods for deformation measurements and shape analysis

J. Šedina, K. Pavelka & E. Housarová
Department of Geomatics, Faculty of Civil Engineering, Czech Technical University in Prague, Prague, Czech Republic

ABSTRACT: This paper focuses on an accuracy testing of photogrammetrical methods for deformation measuring of wooden and concrete beams as well as the construction parts at sequential loading or shape analysis of mechanical parts and historical objects. Classical photogrammetrical methods like stereophotogrammetry and intersection photogrammetry were compared the new precise semi- or full automated technology by Photomodeler and Linearis 3D company. Sub-mm range photogrammetric systemsbased on coded and non-coded self-adhesive targets and calibrated camera with fixed lens and special software have been used. At the Laboratory of photogrammetry at Czech Technical University in Prague, we tested the accuracy and the resolution of the photogrammetrical systems on laser interferometric optical bank, which can measure very small changes in one direction. After the testing, we conducted a real measurement as a case project, which can demonstrate an accuracy, time requirements and usability of this method.

1 INTRODUCTION

There are several photogrammetric solutions for a deformation measurement. Their usage depends on a type of deformation—there are usually dynamical and piecewise deformations. For the dynamical deformation it is necessary to take photos in very short intervals. Single photogrammetry can be used in special cases (deformation in one direction only, in 2D). If 3D measurement is necessary, measuring system must consist at least of two synchronized cameras or stereo photogrammetry which can be used as an universal method. If the deformation arises in steps, single camera system can be used (with photo-stations changing). Photogrammetrical systems were tested in many forms (Hampel, Maas, 2003). Nowadays an image-based modeling is very popular. It is simple and easy to use, it needs only a good digital camera and asoftware. A dense point cloud can be computed (e.g. Agisoft Photoscan, Insight3D, Bundler, 3DF Zephyr Pro and other) from the set of overlapped images (tens of photos), based on image correlation technology. Output is unselected set of points, often in a grid (the same as by laser scanners). This software allows you to automatically reconstruct 3D models from photos. The process is completely automatic and no coded targets, manual editing or special equipment is needed.

For the deformation measurement, images must be taken for example in one load case (dynamical deformation monitoring should be possible with an adequate number of synchronized cameras only). The results depend on the texture of the measured and documented objects. Natural wooden and concrete parts have ordinarily a good texture; with other materials (e.g. steel, glass, and painted parts) it is necessary to use synthetic textures; it can be done by using a special texture projecting on a measured object with a projector. Other system for dynamical deformation monitoring consists of at least two (calibrated) digital high resolution cameras. All cameras, mounted on tripods, must be controlled remotely by the computer and synchronized taken photographs should be fast stored on a disk space based on deformation progress. For sequential deformation, one camera system can be used (common case). For precise measurements (and in case that on the objects there are not suitable visible parts which

Figure 1. Moving part of interferometer with signalized object points and coded targets.

can be used as an object points) it is necessary to mark a special targets (e.g. small adhesive rings or coded targets)on the object. The measurement is performed on selected points only. It is a typical task for intersection photogrammetry, which can be processed manually by operator (in case of tens images only) or fully automatic. Photomodeler or Linearis standard can be mentioned as examples of low cost solutions. The two above mentioned recent versions of these close range photogrammetric systems were tested during this project.

2 REFERENCE MEASUREMENT

For photogrammetric precision comparing and testing, other measurement should be carried out, independent on photogrammetry, with comparable or higher accuracy. For example, thedirect measuring sensors can be placedon tested concrete or timber beam fastened on a load case stool. They measure vertical deformations with a very high frequency and high precision (Litoš et al. 2013). In our case study, we used an interferometric system. The deformations were simulated by small movement of movable measuring table tracking through a prism by an interferometer. This technology and instrument in available kit have accuracy approx. 10 μm which is good enough for photogrammetrical systems testing. All measurements were carried out in Laboratory of geodesy, department of geomatics, Czech Technical University in Prague, Faculty of Civil Engineering. Five positions were performed with different steps (three times approx. 3 mm and once 10 mm). For photogrammetrical technology (Hamouz et al. 2014), signalized object points on small adhesive labels were used for precision of a moving measurement (Pavelka et al. 2011).

3 LINEARIS 3D

Linearis (Linearis 3D GmbH, Germany). First tested system was Linearis 3D in a standard version, which can be used for example for laser scanners calibration and testing. The Linearis 3D Professional photogrammetry system is designed for an application in industry and research. The system computes three dimensional points from the photographs using a bundle adjustment algorithm. A professional version is more expensive, it has direct deformation analysis implemented and other modules, and reaches accuracy approx., 30 μm. Precision is defined using the German guideline VDI 2634. This guideline defines a cube as shown in the first picture with seven scales. Precision is influenced among others by object size, object visibility, camera, camera positions, number of photos, software, user as well as length and number of scales. The configuration of the cube according to VDI 2634 is optimal for photo-grammetric precision (Luhmann, 2003).Whole system consistsof digital DSLR camera (with fix lens 24 mm), carbon or invar scales, coded targets for camera position orientation, adhesive small targets for marking object points and software. It is easy to use and fully automated.

Working with Linearis 3D is easy, it is necessary to look at photogrammetrical basics and rules (overlapping, intersection angles, distance, camera settings etc.). First, the coded targets must be installed on the object and in the near neighborhood (at least four targets must be visible on each photo) such as labeled small object points (adhesive targets). In the measured space a special cross and scale should be placed. The best results are reached with more than 20 photographs of one object (depend of course on the object and its position and size). After taking photographs they are stored in a computer and fully automatically processed. In standard version only limited functions are at disposal. It is necessary to set scale dimension (depend on scale size) One additional distance for scale verification is recommended.

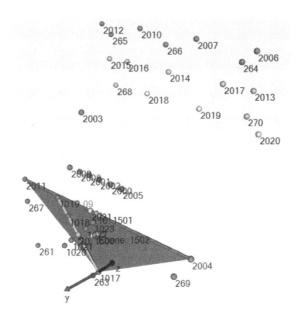

Figure 2.   Moving table of interferometer with signalized object points.

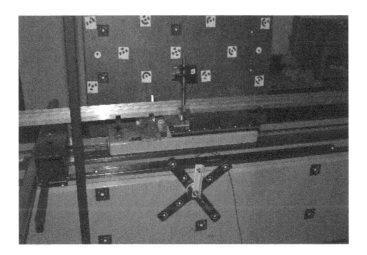

Figure 3.   Moving table of interferometer with signalized object points.

In our case project a problem with coordinate system occurred. Based on an orientation and a scale cross, the coordinate system should be the same in all steps of measurements (only small moving table was moved). The object points moving (represented deformation) have increasing significant error in mm instead of tens of μm. The coordinates were ok in x direction(already on the cross), e small differences were in z direction, but in y direction (spatial coordinate) there were differences in mm. This may be due to the object, points and coded targets configuration (too planar) or by other unknown cause. Based on marked visible points, control cross and stable scale with defined visible outer points, additional spatial affine under-transformation were applied on state 5 and 9 (only three from five states are computed because of the time); they were transformed to the state 1, which was original position of the moving table on the interferometer. After this, very good accordance's were reached and differences between measured points by interferometer and photogrammetric system reached unbelievable values from 0,5 μm to 1,5 μm (mean of six measured moved object points) with spatial standard deviation 0,012 mm. The occurred errors were in rotations of coordinate system in each step of measurement.

As a case project concrete beam was measured directly on construction site—before and after load (30 kg). Maximal observed deformation reaches 3,2 mm in the middle of beam. Entire measurement procedure took about an hour (Strejček et al., 2011).

Table1. Final calculation.

Position – points	x [mm]	y [mm]	z [mm]	Empirical pixel STDV	Mean pixel deviations	Distance [mm]	Interfer-ometer [mm]	[mm]
1								
2005	375,1012	−55,3162	71,8314	0,0018	0,0263	0,00000	0,00000	0,00000
2007	338,4103	−55,5333	121,8874					
2002	305,5333	−55,5409	171,5114					
2001	270,0762	−55,8387	220,0899					
2000	231,2597	−56,0368	273,2277					
2003	195,6972	−56,1419	323,7637					
5								
2002	371,6239	−55,3292	76,71127	0,0013	0,0241	5,99202	5,9915	−0,00046
2005	334,9364	−55,5511	126,7676			5,99045		
2003	302,0543	−55,5610	176,3924			5,99400		
2000	266,6029	−55,8601	224,9681			5,98841		
2001	227,7809	−56,0571	278,1080			5,99333		
2007	192,2211	−56,1621	328,6431			5,99103		
9								
2009	364,0643	−55,3816	87,3228	0,0080	0,0312	19,0211	19,0214	0,00147
2008	327,3725	−55,6010	137,379			19,0218		
2001	294,4939	−55,6139	187,0037			19,0233		
2002	259,0373	−55,9060	235,5797			19,0200		
2000	220,2216	−56,1063	288,7173			19,0203		
2005	184,6598	−56,2116	339,2549			19,0212		

Table2. Standard deviations [mm].

Position	Mean STDV x[mm]	Mean STDV y[mm]	Mean STDV z[mm]
1	0,018032	0,012848	0,011284
5	0,039364	0,020925	0,010668
9	0,020943	0,012896	0,007350
		STDV 0,017	Spatial STDV 0,012

4 PHOTOMODELER

PhotoModeler (EOS Systems, Canada). Photomodeler photogrammetric system is typical close-range photogrammetrical software, based on intersection photogrammetry. It was a leader in close range photogrammetry after 2000. Since 2010 the coded targets have been used for semi or fully automated photographs orientations; other modules such as SIFT and image correlation technology have been implemented directly into software to derivate the dense point clouds.

In our testing, coded targets (special by Photomodeler) were used such as small adhesive targets for object points marking (same as for Linearis 3D). First, camera calibration should be carried out; it consists of taking a set of convergent photographs of calibration field and automatic calibration procedure. After this, automatic orientation of all photographs has been performed and object points has been measured with high accuracy.

The outputs were four distances (mean from object points marked on moving table of inter-ferometer); the precision is characterized by spatial standard deviation 0,04 mm, differences between distances measured by interferometer and computed by Photomodeler varied from 0,021 to 0,074 mm. The results are very good, it is a little bit more time consuming as by using Linearis3D due to the camera calibration and the steps in processing (Reznicek, Pavelka, 2010).

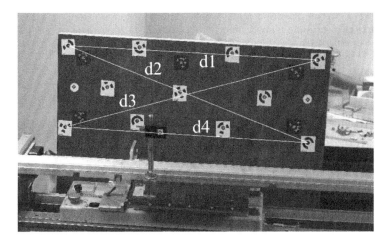

Figure 4.   Photomodeler configuration—it uses coded targets too.

Table 3.   Results from photomodeler [mm].

1	2	3	4	5	6	7	Mean	SD	Interfero meter	[mm]
3,01	2,98	2,96	2,99	2,96	2,99	2,94	2,974	0,024	2,995	0,021
6,05	6,03	6,02	6,05	6,03	6,04	6,03	6,036	0,011	5,992	−0,044
9,07	9,08	9,06	9,07	9,10	9,16	9,03	9,081	0,041	9,007	−0,074
19,07	19,04	19,08	19,09	19,07	19,09	19,11	19,079	0,021	19,020	−0,059

Table 4.   Standard deviations [mm].

1	2	3	4	5	6	7	STDV	Spatial STDV [mm]
0,015	−0,013	−0,039	−0,007	−0,036	−0,007	−0,057		
0,055	0,035	0,032	0,062	0,039	0,045	0,041		
0,063	0,068	0,053	0,066	0,095	0,153	0,019		
0,048	0,023	0,062	0,071	0,046	0,073	0,087	0,06	0,04

5 CONCLUSIONS

Results of the close range intersection photogrammetrical methods are good enough and have a better accuracy level than the aforementioned laser scanning. The advantage is, that only selected points are measured. Nowadays, fully automated methods are required. Both above mentioned and tested systems fulfill the requirements. Our case project reaches accuracy better than 0,05 mm by using coded targets and adhesive object labelled points by using special and universal software for close range photogrammetry. In respect to the fully or nearly fully automated process, the results seem to be very good.

ACKNOWLEDGEMENTS

This research has been supported by the CTU in Prague SGS grant Nr. SGS16/063/OHK1/1T/11 and RPMT 2016 grant.

REFERENCES

Hamouz, J., Braun, J., Urban, R., Štroner, M., Vráblík, L. 2014. Monitoring and evaluation of pre-stressed concrete element using photogrammetric methods. *In 14th International Multidisciplinary Scientific Geoconference SGEM 2014*, Conference Proceedings vol. III. Sofia: STEF92 Technology Ltd., p. 231–238. ISSN 1314-2704. ISBN 978-619-7105-12-4.

Hampel, U., Maas, H.G. 2003. *Application of Digital Photogrammetry for Measuring Deformation and Cracks During Load Tests in Civil Engineering Material Testing*. Optical–3D.

Linearis 3D web page (cited 30-4-2016) http://linearis3d.com/professional%20photogrammetry.html.

Litoš, J., Vejmelková, E., Konvalinka, P. 2013. Monitoring of Deformation of Steel Structure Roof of Football Stadium Slavia Prague. In: *Measurement Technology and its Application.* Uetikon-Zurich: Trans Tech Publications, p. 622–630. ISSN 1660–9336. ISBN 978-3-03785-545-4. http://www.scientific.net/AMM.239-240.622

Luhmann, T. 2003. *Nachbereichsphotogrammetrie*, Wichmann, Heidelberg.

Luhmann, T. 2005. On the determination of objects rotation and translation in 3-D space (6 DOF) by a single camera. *Optical 3-D Measurement Techniques*, 2005.

Luhmann, T., Robson, S., Kyle, S., Boehm, J. 2014. *Close-Range Photogrammetry and 3D Imaging*. 2nd edition, ISBN 978-3-11-030269-1, de Gruyter, pp. 684.

Pavelka, K., Řezníček, J. 2007. 3D documentation of sculptures by using image correlation techniques. In *Proceedings of ACRS 2007*. Kuala Lumpur: ACRS, 2007, p. 222–229. ISBN 978-983-43550-0-5.

Pavelka, K., Šedina, J., Bílá, Z. 2015. Non-contact deformation measurement using photogrammetry. *Interdisciplinarity in Theory and Practice.* vol. 3, no. 7, art. No. 41, p. 227–231. ISSN 2344–2409.

Pavelka, K., Šedina, J., Bílá, Z. 2016. Use of close range photogrammetry for documentation of structure parts deformation. *In Advances and Trends in Engineering Sciences and Technologies: Proceedings of the International Conference on Engineering Sciences and Technologies*, 27–29 May 2015, TatranskáŠtrba, High Tatras Mountains—Slovak Republic. London: Taylor and Francis, ISBN 978-1-138-02907-1.

Řezníček, J., Pavelka, K. 2010. Photogrammetrical measuring of the dynamical deformation of the joint and the column web panel at elevated temperature. In *Proceedings of 31st ACRS Conference.* Hanoi: ACRS, 2010, vol. 1, p. 1859–1863. ISBN 978-1-61782-397-8.

Strejček, M., Wald, F., Řezníček, J., Tan, KH. 2011. Behaviour of column web component of steel beam-to-column joints at elevated temperatures. *Journal of Constructional Steel Research.* 2011, vol. 67, no. 67, p. 1890–1899. ISSN 0143-974X.

Advances and Trends in Engineering Sciences and Technologies II – Al Ali & Platko (Eds)
© 2017 Taylor & Francis Group, ISBN 978-1-138-03224-8

Creation of ground plans in terms of data structure and map bases

L. Teslikova Hurdalkova & D. Kuta
Department of Urban Engineering, VŠB-TUO Faculty of Civil Engineering, Ostrava

ABSTRACT: Planning documentation is one of the fundamental coordination instruments for sustainable development of the territory. The territory of the Czech Republic is topographically mapped in two ways—as the basic state map series and thematic state map series. The raster base mostly consists of the cadastral map on a scale of 1:2880, which is provided in the form of scanned maps of the stable cadastre. The actual records of buildings is an essential element for the urban development of the area. Extending the database facilitates the process of acquisition of the plan and flexible response to change in the area under the multi-year acquisition process. This article concentrates on the creation of ground plans in terms of data structure and map bases.

1 INTRODUCTION—USE OF THE COORDINATE SYSTEM S-JTSK

From the early 1920's, a new and more accurate trigonometric network was created in the new country—the Czechoslovak Republic (CSR). In this context, the requirement was raised to create a cartographic view that would not distort the angles and distances when the image of the surface of the CSR territory was transferred into the plane. The proposal by Ing. Josef Křovák (head of the triangulation office at the Czechoslovak Ministry of Finance) was chosen as a default, and the least distortionary solution., In 1922 it was accepted as a provisional proposal and since 1933 has been used as a definitive view which is the basis for the system of plane coordinate system S-JTSK (Baranová, 2004).

It is the case of so-called double projection, which shows the Bessel ellipsoid on a sphere using the Gaussian orthomorphic projection in the first step, and this sphere to the cone surface in the second step (orthomorphic conic projection in the general position with cartographic pole with ordinates $\phi = 59°45'27''$, $\Lambda = 24°50'$ east of the Greenwich.

As a result of the meridian convergence, the upper edge of the map in Křovák's projection is not oriented to the north. C convergence towards the west continues to grow according to the equation $C = (24°50'-\lambda)/1.34$, therefore, the difference between the northern direction towards the vertical direction is $4°28'$ in the east of the Czech Republic, and up to $9°30'$ in the west of the Czech Republic (Bláha, 2014).

In Czechoslovakia, this projection was introduced for the first time in 1922 for cadastral maps, and from 1933 also for the maps of so-called military mapping. At the same time, a new coordinate system of integrated cadastral trigonometric network—S-JTSK was defined for Czechoslovakia. Although the territory of the Czech Republic has been reduced several times, Křovák's projection has still been used for maps of the state department; it was legalized by effective Government Decree No. 116/1993 Coll.

2 THE STATE MAP SERIES IN DIGITAL FORM

The territory of the Czech Republic is continuously processed into individual map sheets. Processing state maps is subject to mandatory Government Regulation No. 430/2006 Coll. (*dated 16 August 2006 on establishing geodetic reference systems and state map series obligatory for the country and the principles of their use, as amended by Amendment no. 81/2011*). There are basic state map series and thematic state map series. The publisher of the basic

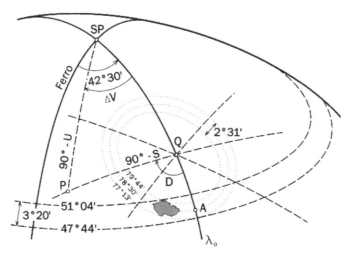

Figure 1.   Location of the cartographic pole in Křovák's projection.
(Source: Bláha, 2014).

state map series is the Czech Geodetic and Cadastral Office (ČÚZK), which ensures that
they are updated. The distribution of these map series is currently underway in digital form,
usually in exchange dgn and vfk formats. Raster maps are scanned analogue documents that
are distributed mostly in TIFF files, including the placement files for S-JTSK.

The following maps are available in digital form:

– State map 1:5,000
– Raster Base Map of the Czech Republic 1:10,000
– Raster Base Map of the Czech Republic 1:25,000
– Raster Base Map of the Czech Republic 1:50,000
– Raster Base Map of the Czech Republic 1:200,000
– Raster map of the Czech Republic 1:200,000
– Raster map of the Czech Republic 1:1,000,000

For the needs of spatial planning, the State Maps 1:5,000 (SM5) are used. The State Map
1:5,000 is a national large scale map series. It shows the Czech Republic in a continuous series
of map sheets on 16,301 map sheets. The dimensions of the map face with a projected area
of 5 km² are 50 x 40 cm. The dimensions and designation of map sheets are derived from the
state map sheets in the scale of 1:50,000, divided into 10 columns and 10 layers, the sheet title
is supplemented with the column and layer number. Maps include topography, elevation and
description.

The subject of topography are points of the geodetic point field, the boundaries of the
plots, types of plots and methods of their use, structures, administrative and cadastral
boundaries, protected areas and other elements of topography. The subject of altimetry is
the terrain relief displayed by means of contours describing the altitude and terrain features,
and the description consists of geographical names, names of public spaces and map type
identification. Map sheets are supplemented with the plane rectangular coordinate network
(S-JTSK). Today, there are four versions of the State Map 1:5,000 (CUZK).

2.1   State Map 1:5,000 – derived (SMO 5)

The fundamental topographic basis were cadastral maps, the contour basis usually the Base
Map of the Czech Republic 1:10,000. The topographic map content was supplemented with
other elements of the available graphic materials, mainly from aerial photographs. Since
2001, the SMO 5 has gradually been replaced by the State Map 1:5,000, and since 2009 with

the new form of the State Map 1:5,000 in response to the digitized set of geodetic cadastral information, see Fig. 2. The state map was widely used in raster format in processing ground plans of cities, towns and villages until 2000.

2.2 State Map 1:5,000

The State Map 1:5,000 (SM 5) was processed from 2001 and it gradually replaced the existing analogue State Map 1:5,000 – Derived. It was only created in areas where there was a digital cadastral map (DKM) or digitized cadastral map (KM-D) (CUZK). SM 5 comprises three components:

– The cadastral component, based on DKM or KM-D,
– The contour component, which is based on a set of contours of the Primary Geographic Data Base of the Czech Republic—ZABAGED®,
– The topographical component is an orthophoto.

2.3 State Map 1:5,000—raster

State map 1:5,000—raster (SM 5 R) was processed from 2003 to 2007. It was created by scanning the planimetry print base from the last edition of SMO 5. It is made up of three components:

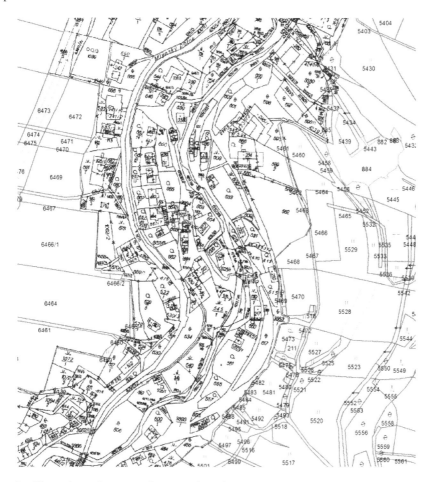

Figure 2.   The register of vector and raster cadastral maps
(Source: author's archive).

849

– The cadastral component is a raster planimetric projection of the last edition of the State Map 1:5,000 – Derived (SMO 5),
– The contour component is a raster contour projection of the last edition of SMO 5,
– The topographic component is an orthophoto.

2.4 *State Map 1:5,000 – new form*

This product represents a new form of the State Map 1:5,000 in vector form, whose advantage is up-to-dateness and colour processing. The new form of **SM 5** is designed as an automatic visualization of selected objects derived or taken over from cadastral maps, **ZABAGED,**® Geonames and the geodetic point field database. The new form of the State Map 1:5,000 is repeatedly generated twice a year only in the part of the territory of the Czech Republic where the vector cadastral map form is already available (CUZK).

The map in vector form, whether **DKM** or **KM-D**, however, is not available for all cadastral areas in the Czech Republic. For a large area, maps are still available only on paper or in the raster format. As at 31 December 2015, approximately 81% of the country was digitalized, in other words, as at 31 December 2015, it was possible to obtain the cadastral map of 81% of the territory in the exchange format. Availability of the cadastral map in digital form for a specific cadastral area can be found on the Czech Geodetic and Cadastral Office website.

3 STRUCTURE OF THE LAND REGISTRY INFORMATION SYSTEM (ISKN) EXCHANGE FORMAT

The Exchange Format (VF) is designed to enhance the transfer of data between the ISKN system and other data processing systems. A set of descriptive information and the structure of the set of descriptive information have been imported into the DBF database files used by the land registry. The current object of exchange is an exchange format **ISKN** in **VFK** text format, which is used by real estate records; it includes descriptive and graphical information including management data according to the set combination of blocks.

The format can be created in these time modes:

– Raw data (from any point in time after the introduction of **ISKN**),
– Changes in the specified time period (you can enter the date from—to including the time).

The scope of the data provided:

– Territorial unit (cadastral area, municipality, district, the Czech Republic),
– Authorized body,
– Selection of plots,
– Selection of plots by a polygon on the map.

The VFK set contents can be optional pluralities of data blocks, which means that the set can optionally contain either descriptive information, a cadastral map, or both. The following table shows which pluralities of data blocks can be ordered at the Land Registry, and which ones can be used either within the module *KN Data Import in the Exchange Format* or within the module *Working with Descriptive KN Information*. The column *Import Mode* indicates which pluralities of data blocks correspond to the import regime in the opening dialog module *KN Data Import in the Exchange Format*. For the creation of ground plans, it is necessary to request data for the cadastral map and descriptive information, or export of all of the data blocks (without geometrical plans).

4 USING VFK FOR TERRITORIAL PLANNING

The digitization of maps enables more efficient processing of ground plans. The information contained in the database enables the effective use of information from the Land Registry,

Table 1. Example of data blocks of the VFK format, (Source: http://www.gisoft.cz/ Moduly/ImportVFK).

Plurality of data blocks	Import mode	Description
Elements of the cadastral map	Map	plurality of data blocks needed to create cadastral map drawings
Bonited soil-ecological units	Map	plurality of data blocks needed to create map drawings for bonited soil-ecological units
Definition points of plots and buildings	Map	plurality of data blocks needed for loading definition points of plots as parcel numbers (land registry with raster maps). Without a data block of the Piece of real estate, this block has no meaning
Real estate	Descriptive information	plurality of data blocks needed for displaying descriptive information
Affluent parts of plots	Descriptive information	plurality of data blocks needed for displaying infor-mation on bonited soil-ecological units in descriptive information
Ownership	Descriptive information	group of data blocks needed for displaying descriptive information

mainly data on the type of structure or the use of a building or the type of land. Information on easements influences the proposal of the functional use of the particular territory, which significantly affects the urban concept. The database of plot numbers, the data of the title number accelerates the identification of the requirements of natural or legal persons to change the ground plan, which are often an impulse for processing new zoning documentation. Currently, new zoning documentation is processed not due to changes in the base map, but due to changes in the functional use. Based on these requirements, changes to ground plans are processed, and the legal situation after the release of the territorial plan change is processed subsequently. These two documents are often processed on a different map base. Registering the approved documentation of changes to the original planning documentation can lead to erroneous decisions of the building authority, which works with the plots of the current map work. Therefore, the derived state maps 1:5,000 often contained information different from map bases in VFK formats. These new formats are updated for each cadastral area no later than the sixth day of each month. The distribution of these current formats is provided by the Czech Geodetic and Cadastral Office portal. The combination of the raster map base in built-up areas and of the vector map base in rural areas leads to inaccuracies in the demarcation of plots on the edge of built-up areas/rural areas—see Figure 2. These topographic duplications or gaps can lead to misguided decisions of the building authority.

5 CONCLUSIONS

Based on the described map bases, it is obvious that the planimetry processing is an essential step in making good decisions not only on the part of the building authority, but also with regard to the creation of urban planning documentation. Combining several planimetry formats (raster/vector) is topographically very inaccurate and misleading. We cannot expect raster maps to have the same explicitness as vector maps. Future processing of urban planning documentation should follow the sufficiently processed database of the description of the particular territory to enable feedback which could lead to the identification of plots, including regulations and functional applications specified in the urban planning documentation. Such information would represent a good basis for the decision making of building authorities.

ACKNOWLEDGMENT

The article was supported by the Student Grant Competition at VŠB—Technical University of Ostrava. The project registration number is SP2016/46.

REFERENCES

Baranová, M. 2004. Projections used for Czechoslovakia and the Czech Republic Prague: Math cartography.

Bláha, J. 2014. Effects of the use of Křovák's projection in the geographic information system on Czech users, ArcRevue, page 10, 4, Prague.

Czech Geodetic and Cadastral Office, [online],2016, Available at: http://www.cuzk.cz/Zememerictvi/Geograficke-podklady/Tistene-mapy/Mapy-velkych-meritek.aspx.

Advances and Trends in Engineering Sciences and Technologies II – Al Ali & Platko (Eds)
© 2017 Taylor & Francis Group, ISBN 978-1-138-03224-8

Verification of pavement construction characteristics measured with non-destructive methods

K. Zgútová, M. Trojanová, J. Šrámek & M. Blaško
Faculty of Civil Engineering, University of Zilina, Slovakia

ABSTRACT: The condition and specification of pavement subgrades are essential for road design. Incorrectly determined subgrade deformation modulus can cause a collapse of the construction or permanent deformation in each course of the pavement. On the territory of Slovakia the modulus of deformation is mainly identified by static load test and progressively extended to other non-destructive tests which are not so time demanding and more efficient. Non-destructive methods for assessing the quality of earth structures are nowadays commonly used and the implementation progressive methods will ensure accurate diagnostic of existing pavements for the extension of life cycle. The paper is focused on theoretical modelling for specifying the suitability of non-destructive methods for a particular construction element and for the possible future determining of the correlation between each ordinary non-destructive test used in Slovakia.

1 INTRODUCTION

Generally, the static deformation modulus is obtained from indirect methods of control e.g. Static Load Plate Test. Nowadays innovative test technologies are successfully extended, the evaluation of which is based on the reaction of the tested layers to dynamic loads. Prior to the application of indirect methods of control, it is useful to determine their suitability, which is verified by parallel measurements during the control process or by a compacting test. The extent of the compaction test will be determined by the designer. Observing the parameters $E_{def,2}$ or ratios $E_{def,2}/E_{def,1}$ not only enable the determination of the degree of compaction but also the condition necessary for the required compaction of further pavement layers, for site transport and for the long-term behavior of the pavement. Indirect testing methods that are typically used are listed in Table 1.

This investigation sets out to perform the numerical model of Dynamical Load Test in order to determine the theoretical compaction parameter which is described by value of a dynamic module of deformation E_{vd}.

Table 1. Indirect testing methods to detect degree of compaction in accordance with STN 73 6133.

Method	Type of material
Radiometric methods	Si, Sa, BoCo
Static Load Test	Si, Sa, Gr, BoCo
Dynamic Load Test	Si, Sa, Gr, BoCo
Geodetic compaction control	Gr, BoCo
Dynamic method of compaction control (compaction meter)	Sa, Gr, BoCo
Penetration test (static, dynamic)	Si, Sa, Gr, BoCo

*legend: Si fine-grain soil, Sa sandy soil, Gr gravelly soil, BoCO rocky and boulder backfill.

2 DYNAMIC LOAD TEST

A dynamic load test is performed according to the principles of STN 73 6192. The Annex G of standard STN 73 6133 contains the following stipulations. The essence of the test is to load the tested surface of subsoil (compacted layer) with a contact strain (power pulse) for a defined loading time, depending on the layer being tested and the type of equipment used. When using a large FWD (Falling Weight Deflectometer), contact stress p must be between 0.2 MPa to 0.3 MPa. When using a small dynamic FWD, contact stress p must be equal to, or greater than, 0.07 MPa. The Dynamic load test is used to determine the degree of compaction which includes the determination of the dynamic modulus of elasticity E_{vd}, or the dynamic modulus of deformation E_{def}, according to STN 73 6192; or the ratio of moduli from the second and first loading impulses in the same test. Values determined in this manner are a binding value applicable only to the specific soil type and compaction control project. Dynamic load test should not be repeated in the same location. When checking compaction using dynamic load tests, it is necessary to establish a correlation between the values of the moduli from the static and dynamic tests from measurements made during the terrain compactability test. With regards to the use of single-purpose light-shock devices, praxis has shown the LDD 100 device (Light falling weight deflectometer) (Figure 1) to be the most commonly applicable device used for the rapid control of the dynamic parameters of compacted backfill.

When measuring with LDD an impact load is distributed to the soil contact area via a circular measuring plate of diameter d = 300 mm with built-in sensor to measure deflection. An impactor is mounted on the measuring plate, consisting of a guiding rod weighing 5 kg ± 0.5 kg, weights of 10 kg ± 0.1 kg, ring to facilitate handling, shock absorber, locking pins, safety locks to secure the weights in the upper part of the guiding rod and hand-holder to ensure

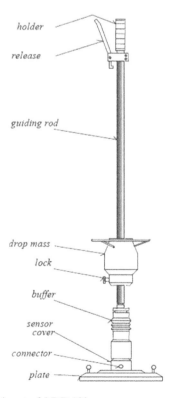

Figure 1.   Scheme of mechanical part of LDD 100.

vertical position of the device. After releasing the safety locks, the falling mass slides down and impacts the buffer. One impact of a weight is not measured, then three strokes with automatic measurement of surface deflection are performed progressively, after each stroke the weight is captured and secured in the upper position. Each completed measurement is followed by the automatic evaluation of the dynamic modulus of deformation E_{vd}, calculated from the deflection value, which is the arithmetic average of the three deflections obtained from three strokes.

3 DYNAMIC METHOD USING IMPACT LOAD PRINCIPLES

The test is carried out on a simple mechanical device, which generates a controlled strength impulse I_p with particular intensity and duration. A subdued impact is created by the motion of the impacted device m_0, which is falling from definite height h to base plate characterized with stiffness k. Force effects and dynamic displacement is appropriate to divide into static and dynamic elements:

$$P(t) = P_{st} + P_{dyn}(t)$$
$$w(x,t) = w_{st} - w_{dyn}(x,t) \tag{1}$$

Reaction $R_{(t)}$ of mechanical device, which in the sense of action and reaction loaded tested construction has also impulse character and is defined by known stiffness of buffer $k_{z,pad}$:

$$R(t) = k_{z,pad} * \left[w_o(t) - w(t) \right]$$
$$w(t) << w_o(t) \tag{2}$$

where $w_{o(t)}$ = buffer deformation; $w_{(t)}$ = vertical displacement of base plate. The equation of mass m_0 after impact on plate has shape of simple wave (Figure 2):
The basic equation for calculating the dynamic modulus of deformation E_{vd} is as follows, according to STN 73 6192:

$$E_{vd} = \frac{\pi \cdot d \cdot \sigma}{4 y_{ml}} \cdot (1 - \mu^2) \tag{3}$$

In the LDD 100 manual the following formula is mentioned for calculating modulus of elasticity:

$$M_{vd} = \frac{F}{d \cdot y_{el}} \cdot (1 - \mu^2) \tag{4}$$

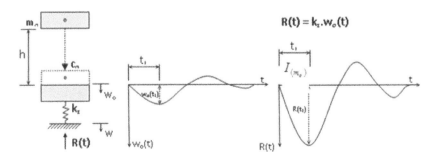

Figure 2.   Presumption of impact force process.

855

In the amended standard ČSN 73 6192, the dynamic modulus of deformation M_{rz} for embankment subsoil and backfill is calculated:

$$M_{rz} = 1{,}57 \cdot \frac{a \cdot \sigma}{y_c} \cdot \left(1 - \mu^2\right)$$

(5)

Formulae (3) to (5) use the following variables: a = radius of loading plate; d = diameter of loading plate; F = Impact power; σ = contact strain; $y_{ml,}$ $y_{el,}$ y_c = deflection amplitude under the centre of loading plate, μ = Poisson's ratio

4 NUMERICAL STUDY

The Numerical model was created for specifying suitability and verifying the relevance of this test method on various construction courses. The impact load on the surface of the soil is analyzed via simulation that contains two substructures: the model of the soil and the model of the impact device. Model of the compacted soil is modeled as a half-sphere which geometry is defined by employing of 3D Solid elements.

The isoparametric elastic material includes the material characteristics of the soil so there are requested only three constants. Young modulus, Poisson's ratio and density. At the edge of the half-sphere are defined boundary conditions in the form of zero displacements in each direction. The mechanic characteristics of the impact device is defined via the mass-point with the weight of 10 kg. The elastic characteristics are defined via the spring. The contact area is defined by the circle plate which is at the bottom of the spring.

The contact between the impact device and the surface of the soil is defined by using the contact algorithm based on the constraint function. Bathe algorithm is used to solve this dynamic simulation. The results of the simulation are presented in form of vertical displacement changed in time. The observation point, in which the vertical displacement is analysed, is located in the place where the plate of impact device hits the surface of the soil. The results are in the Figure 5.

The following Figure 6 shows the vertical displacement of the 10 kg weight. The weight falls from the height of 0.8 m, then bounce back to the height of 0.6 m. The theoretical height of the bounce received from numerical model was compared with the real measurement and results are very similar. It means that, the model of the soil has similar elastics characteristics as the real soil, because it's elastic characteristics influence the intensity of bounce of the impact device.

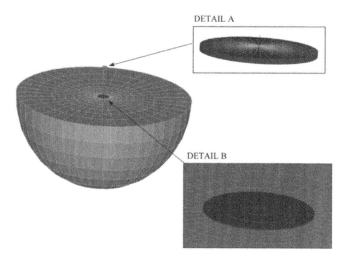

Figure 3. Model of the soil as a half-sphere and the model of the impact device.

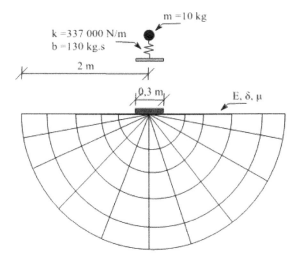

Figure 4. Geometry and material characteristics of the numerical model.

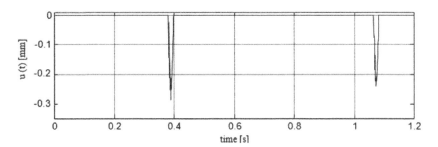

Figure 5. The vertical displacement on the surface of the soil in the place where the impact device hits it.

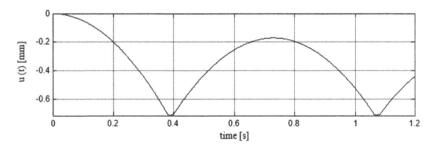

Figure 6. The movement of the 10 kg weight impact device.

Two main goals of numerical model can be listed as follows:

1. The gradually creating of the theoretical results on various construction layers and soils, will be verified and compared with ours in-situ database of non-destructive tests results. Then will be possible improve the quality of the numerical model and result would be more valuable.
2. Define more precise correlation between the values of the moduli from the static and dynamic tests and possible relations between different test devices which works in principles of Dynamic Load Test.

5 CONCLUSIONS

The basic aim of this research paper is related to the description of dynamic process on FWDs in the sense of the simplified standard evaluation procedure. The final evaluations of the numerical model results will be possible only with the precise adjusting on various material parameters. In following steps, the numerical model will be prepared also for other non-destructive test, which is frequently used in conditions of Slovakia—Humboldt. Then the potential addition will be in establishing the correlation dependence for different non-destructive test devices and for various earth structures materials. All observed correlation dependence will be gradually recorded and precise compared with in-situ measurements in order to create the Matrix of dependence. According to correlation, which will refer to particular soil and test device subsequently can be determined suitability of using this test method and described recommendations for using. By implementation indirect non-destructive methods and deduced correlation will ensure more precise and not so demanding diagnostic of construction layers in order to extend the life cycle.

ACKNOWLEDGEMENTS

This article is created with support of Slovak grant agency VEGA 1/0254/15 Implementation of new diagnostic measurements for the project of the optimize life cycle of roads.

REFERENCES

Adam, D., Adam, C., Kopf, F. The dynamic Load Plate Test with the Light Falling Weight Device: Experimental and Numerical Investigation. Proceedings of 11th International Conference on Soil Dynamics and Earthquake Engineering 2004.

Decký, M., Drusa, M., Pepucha, Ľ., Zgútová, K. Earth Structures of Transport Constructions. Pearson.

Education Limited 2013, Edinburg Gate, Harlow, Essex CM20 2JE. Edited by Martin Decký, p. 180, ISBN 978-1-78399-925-5.

LDD 100 Light Dynamic Plate. User manual, ZBA GeoTech, Nové Město nad Metují.

Paulmichl, I., Kopf, F., Adam, C. Numerical simulation of the dynamic load plate test with the light falling weight device by means of the boundary element method. 2005 Millpress, Rotterdam, ISBN 905966 033 1.

STN EN 73 6133. 2010: Road Building. Roads embankments and subgrades.

STN EN 73 6192. 2011: Plate bearing test of pavements and subgrades by FWDČSN 73 6192. 1996: Impact load tests for road surfaces and subsurfaces.

Advances and Trends in Engineering Sciences and Technologies II – Al Ali & Platko (Eds)
© 2017 Taylor & Francis Group, ISBN 978-1-138-03224-8

Author index

Talian, J. 561
Tamrazyan, A.G. 723
Taskin, K. 3
Tazky, L. 279
Tekieli, M. 519
Teslikova Hurdalkova, L. 847
Timčaková, K. 291
Tobisch, M. 809
Tomko, M. 285
Topolář, L. 267, 291
Tóth, S. 329, 335, 459
Trojanová, M. 853
Turcsanyi, P. 649

Umnyakova, N.P. 655
Urban, V. 63
Usanova, K. 297
Úterský, M. 413

Vacek, V. 661
Václavík, V. 389, 413, 419
Valašík, A. 45
Vančík, V. 661
Vaněk, D. 685
Vaník, Zs. 165
Vatin, N. 81, 297
Venkrbec, V. 667
Veselý, V. 309
Vlčková, J. 673
Vodička, R. 171
Vojtus, J. 679
Vokáč, M. 187
Vovesný, M. 303
Vranay, F. 605
Vranayová, Z. 365, 605, 637

Vůjtěch, J. 303
Vyhlídal, M. 309

Zaborova, D.D. 531
Zach, J. 691
Záhora, M. 685
Zdařilová, R. 697
Zeleňáková, M. 703
Zgútová, K. 853
Zielina, M. 371
Zielina, M.Z. 709
Zitnikova, K. 779, 823
Živner, T.J. 27, 33
Zubkova, M. 513
Zupan, D. 433
Zyrek, T. 723

Printed and bound by CPI Group (UK) Ltd, Croydon, CR0 4YY

24/10/2024

01778295-0011